Television Technology Today

OTHER IEEE PRESS BOOKS

Insights Into Personal Computers, *Edited by A. Gupta and H. D. Toong*
The Space Station: An Idea Whose Time Has Come, *By T. R. Simpson*
Marketing Technical Ideas and Products Successfully! *Edited by L. K. Moore and D. L. Plung*
The Making of a Profession: A Century of Electrical Engineering in America, *By A. M. McMahon*
Power Transistors: Device Design and Applications, *Edited by B. J. Baliga and D. Y. Chen*
VLSI: Technology Design, *Edited by O. G. Folberth and W. D. Grobman*
General and Industrial Management, *By H. Fayol; revised by I. Gray*
A Century of Honors, *an IEEE Centennial Directory*
MOS Switched-Capacitor Filters: Analysis and Design, *Edited by G. S. Moschytz*
Distributed Computing: Concepts and Implementations, *Edited by P. L. McEntire, J. G. O'Reilly, and R. E. Larson*
Engineers and Electrons, *By J. D. Ryder and D. G. Fink*
Land-Mobile Communications Engineering, *Edited by D. Bodson, G. F. McClure, and S. R. McConoughey*
Frequency Stability: Fundamentals and Measurement, *Edited by V. F. Kroupa*
Electronic Displays, *Edited by H. I. Refioglu*
Spread-Spectrum Communications, *Edited by C. E. Cook, F. W. Ellersick, L. B. Milstein, and D. L. Schilling*
Color Television, *Edited by T. Rzeszewski*
Advanced Microprocessors, *Edited by A. Gupta and H. D. Toong*
Biological Effects of Electromagnetic Radiation, *Edited by J. M. Osepchuk*
Engineering Contributions to Biophysical Electrocardiography, *Edited by T. C. Pilkington and R. Plonsey*
The World of Large Scale Systems, *Edited by J. D. Palmer and R. Saeks*
Electronic Switching: Digital Central Systems of the World, *Edited by A. E. Joel, Jr.*
A Guide for Writing Better Technical Papers, *Edited by C. Harkins and D. L. Plung*
Low-Noise Microwave Transistors and Amplifiers, *Edited by H. Fukui*
Digital MOS Integrated Circuits, *Edited by M. I. Elmasray*
Geometric Theroy of Diffraction, *Edited by R. C. Hansen*
Modern Active Filter Design, *Edited by R. Schaumann, M. A. Soderstrand, and K. B. Laker*
Adjustable Speed AC Drive Systems, *Edited by B. K. Bose*
Optical Fiber Technology, II, *Edited by C. K. Kao*
Protective Relaying for Power Systems, *Edited by S. H. Horowitz*
Analog MOS Integrated Circuits, *Edited by P. R. Gray, D. A. Hodges, and R. W. Brodersen*
Interference Analysis of Communication Systems, *Edited by P. Stavroulakis*
Integrated Injection Logic, *Edited by J. E. Smith*
Sensory Aids for the Hearing Impaired, *Edited by H. Levitt, J. M. Pickett, and R. A. Houde*
Data Conversion Integrated Circuits, *Edited by D. J. Dooley*
Semiconductor Injection Lasers, *Edited by J. K. Butler*
Satellite Communications, *Edited by H. L. Van Trees*
Frequency-Response Methods in Control Systems, *Edited by A.G.J. MacFarlane*
Programs for Digital Signal Processing, *Edited by the Digital Signal Processing Committee, IEEE*
Automatic Speech & Speaker Recognition, *Edited by N. R. Dixon and T. B. Martin*
Speech Analysis, *Edited by R. W. Schafer and J. D. Markel*
The Engineer in Transition to Management, *By I. Gray*
Multidimensional Systems: Theory & Applications, *Edited by N. K. Bose*
Analog Integrated Circuits, *Edited by A. B. Grebene*
Integrated-Circuit Operational Amplifiers, *Edited by R. G. Meyer*
Modern Spectrum Analysis, *Edited by D. G. Childers*
Digital Image Processing for Remote Sensing, *Edited by R. Bernstein*
Reflector Antennas, *Edited by W. Love*
Phase-Locked Loops & Their Application, *Edited by W. C. Lindsey and M. K. Simon*
Digital Signal Computers and Processors, *Edited by A. C. Salazar*
Systems Engineering: Methodology and Applications, *Edited by A. P. Sage*
Modern Crystal and Mechanical Filters, *Edited by D. F. Sheahan and R. A. Johnson*
Electrical Noise: Fundamentals and Sources, *Edited by M. S. Gupta*
Computer Methods in Image Analysis, *Edited by J. K. Aggarwal, R. O. Duda, and A. Rosenfeld*
Microprocessors: Fundamentals and Applications, *Edited by W. C. Lin*
Machine Recognition of Patterns, *Edited by A. K. Agrawala*
Turning Points in American Electrical History, *Edited by J. E. Brittain*
Charge-Coupled Devices: Technology and Applications, *Edited by R. Melen and D. Buss*
Spread Spectrum Techniques, *Edited by R. C. Dixon*

Television Technology Today

Edited by
Theodore S. Rzeszewski
Member of Technical Staff
AT&T Bell Laboratories

Associate Editors
Harold J. Benzuly, Matsushita Industrial Company
Barry G. Haskell, AT&T Bell Laboratories
Ronald L. Hess, General Electric, Video Products Division
Akinao Horiuchi, Sony Corporation, Advanced Engineering Division
Joseph L. LoCicero, Illinois Institute of Technology
Dimitrios P. Prezas, AT&T Bell Laboratories
Anthony Troiano, RCA David Sarnoff Research Center

A volume in the IEEE PRESS Selected Reprint Series, prepared under the sponsorship of the IEEE Consumer Electronics Society.

IEEE
PRESS

The Institute of Electrical and Electronics Engineers, Inc., New York

IEEE PRESS

1984 Editorial Board

M. E. Van Valkenburg, *Editor in Chief*
M. G. Morgan, *Editor, Selected Reprint Series*
Glen Wade, *Editor, Special Issue Series*

J. M. Aein
J. K. Aggarwal
James Aylor
J. E. Brittain
R. W. Brodersen
B. D. Carroll
R. F. Cotellessa
M. S. Dresselhaus
Thelma Estrin
L. H. Fink
S. K. Gandhi
Irwin Gray
H. A. Haus
E. W. Herold
R. C. Jaeger
J. L. Limb
E. A. Marcatili
J. S. Meditch
W. R. Perkins
A. C. Schell
Herbert Sherman

W. R. Crone, *Managing Editor*
Hans P. Leander, *Technical Editor*
Teresa Abiuso, *Administrative Assistant*
Emily Gross, *Associate Editor*

Copyright © 1985 by
THE INSTITUTE OF ELECTRICAL AND ELECTRONICS ENGINEERS, INC.
345 East 47th Street, New York, NY 10017
All rights reserved.

PRINTED IN THE UNITED STATES OF AMERICA

IEEE Order Number: PC01818

Library of Congress Cataloging in Publication Data
Main entry under title:

Television technology today.

(IEEE Press selected reprint series)
Includes index.
1. Television—Addresses, essays, lectures.
I. Rzeszewski, Ted. II. Benzuly, Harold J.
TK6630.T368 1985 621.388 85-103
ISBN 0-87942-187-8

Contents

Preface

THE papers contained in this book form eight parts designed to cover a number of important topics of current interest in television. It was tempting to add parts on the video disk, monitors, cameras, etc., but space restrictions prevented their inclusion. However, the eight parts contained here should provide a good description of "what's happening" in television today. The parts contained in the book are:

1) CATV and Broad-Band Communications
2) Direct Broadcast Television from Satellites
3) Advanced Television Systems
4) Digital Television
5) Teletext
6) Multichannel Television Sound
7) Projection Television
8) Videotape

The first part, on CATV and broad-band communications, covers topics ranging from standard CATV to new facets that are of interest in this area, such as fiber distribution with switched video and other services. One of the papers considers some of the problems in using multichannel television sound on CATV. Finally, many of the problems and areas of interest in CATV are covered in the remaining articles.

Direct broadcast satellites (DBS) are the topic of the second part. This is a service that is ready to happen, and we should be hearing a great deal more about it, including its impact and potential in the future. The future of this new broadcast service may be intimately linked with the topic of the third part of this book because it can more readily handle a new transmission format than established services that have a large commitment to the National Television System Committee (NTSC) system.

The third part is on advanced television systems. An industry committee was formed in 1983 called the Advanced Television Systems Committee (ATSC). This committee should form standards in this area in a manner similar to that in which the NTSC formed the present broadcast standard. The committee is considering the following three areas: high definition television (HDTV), enhanced definition television (EDTV), and improved NTSC. Many new system ideas are surfacing in this exciting new area. The end result should be higher quality images that provide the viewer with a much better feeling than is presently available.

Digital television, the topic of the fourth part, is a large area that is introduced by first covering the fundamentals of coding video signals digitally. If fiber-optic cable systems become common, it is likely that they will eventually use digital transmission. Next, digital television sets are covered by looking at some IC's that may be used. This exciting new area can significantly change the television set as we know it, and the quality of the reproduced image.

Teletext is the topic of the fifth part. This area has been under standards consideration for some time. It was decided that there would be a marketplace decision and now there are several systems that are vying for position.

The FCC has recently adopted the BTSC Multichannel Television Sound System recommended by the Electronic Industries Association (EIA) and our sixth part contains a description of that system together with related papers of interest to this area. This system is not only capable of providing stereo sound for television, it can also provide a second audio program (bilingual sound) and a professional channel.

Most of the topics of the previous six parts, except for CATV, are new and have yet to significantly impact the consumer market. The areas covered by the next two parts have been around for a while and their presence has been felt. However, they are vital areas that are still considered current. Projection television is the topic of the seventh part. It can also benefit from progress in several other areas covered in this book such as ATS, DBS, and digital television.

The final part is devoted to videotape. The video cassette recorder (VCR) as we know it today is covered by papers on the VHS, Betamax, and the Philips systems. In addition, newer developments in this area such as extremely small recorders and the recording video camera are also covered. This final part ends with a coverage of various magnetic tape duplication techniques.

THEODORE S. RZESZEWSKI
Editor

Part I
CATV and Broad-Band Communications

EQUIPMENT technology in CATV has improved dramatically since inception of the industry in 1948. Manufacturers have continued to develop new products in an effort to keep pace with industry needs. The first part of this book strives to review some of the recent technology improvements, such as addressability, signal scrambling, multichannel sound, feed-forward amplification, fiber-optic cabling, and data transmission techniques. It is composed of eight articles, five dealing with CATV and three with data communications.

The lead paper in this part deals with feed-forward techniques in distribution amplifiers and was written by J. C. Pavlic. It is entitled "Some Considerations for Applying Several Feedforward Gain Block Models to CATV Distribution Amplifiers" and discusses the advantages of feed-forward techniques to improve the dynamic range and reduce distortion of CATV distribution amplifiers. Pavlic analyzes four feed-forward circuit models and concludes with an optimum trunk station configuration.

The second paper investigates intermodulation distortion and provides, what the authors believe are, improved prediction and measurement techniques. The paper is entitled "An Intermodulation Prediction and Measurement Technique for Multiple Carriers Through Weak Nonlinearities" and was authored jointly by T. Sasaki and H. Hataoka. Intermodulation products are generated in CATV or communication systems when two or more carriers are passed through a common nonlinear amplifier. The authors conclude that their prediction technique does not require the single carrier transfer function of the amplifier, and therefore is more useful than other intermodulation prediction techniques.

The third paper was authored by M. T. Hayashi and is entitled "Effectiveness of Static Scrambling vs Dynamic Scrambling Systems—A Classification Method." The paper discusses the aspects of CATV scrambling systems, and strives to classify available methods and evaluate the effectiveness of each.

The fourth paper is entitled "The Impact of Multichannel Sound on CATV Systems" and was written by J. Van Loan. It considers the performance requirements of major CATV functions and discusses potential problems which could be encountered when multichannel sound is transmitted.

The fifth paper, authored by J. Staiger, is entitled "Return System Set-Up and Maintenance." Staiger discusses the ramifications of implementing a two-way cable system and how to minimize some of the problems encountered.

The sixth paper was authored by T. E. O'Brien, Jr. and is entitled "A Unified Approach to Data Transmission Over CATV Networks." This is the first article on data communications and reviews some important characteristics of data transmission such as bandwidth, noise performance, modulation and encoding techniques, and error detection methods. O'Brien concludes with a proposal for a standard set of system parameters for CATV data transmission.

The next paper investigates one of the newer transmission mediums, fiber-optic cable. The paper was coauthored by four individuals, G. Guekos, H. P. Berger, B. Illi, and H. Melchior and is entitled "Potentials of Fiberoptic Multichannel Television Transmission by Analog Modulation." The authors present a theoretical analysis of AM and FM modulation techniques used for transmission over optical fibers and provide measurements of a test transmission through a laboratory link.

The last paper of this part is entitled "An Experimental Digital Video Switching Architecture" and was coauthored by D. Vlack and H. R. Lehman. This paper discusses several architectural features of a new wide-band video switching system, including a system architecture, the new switching element power-switching emitter-coupled logic (PSECL), network philosophy, and a modulation method with the capability of transporting digital voice/data and analog baseband video over direct fiber links.

RONALD L. HESS
Associate Editor

BIBLIOGRAPHY

[1] O. Shimbo, "Effects of intermodulation, AM–PM conversion and additive noise in multicarrier TWT systems," *Proc. IEEE*, vol. 59, Feb. 1971.

[2] E. Imboldi and G. R. Stetle, "AM-to-PM conversion and intermodulation in nonlinear devices," *Proc. IEEE*, vol. 61, June 1973.

[3] T. Sasaki and H. Hutuoka, "Generation of inter or cross-modulation distortion in nonlinear devices," *IEEE Trans. CATV*, vol. CATV-2, July 1977.

[4] M. Horstein and D. T. LaFlame, "Intermodulation spectra for two SCPC systems," *IEEE Trans. Commun.*, vol. COM-25, Sept. 1977.

[5] J. L. Pearce, "Intermodulation performance of solid-state UHF class-C power amplifiers," *IEEE Trans. Commun.*, vol. COM-25, Mar. 1977.

[6] G. N. Watson, *Theory of Bessel Functions*. London, England: Cambridge Univ. Press, 1958.

[7] H. Hodora, "Statistics of thermal and laser radiation," *Proc. IEEE*, vol. 53, July 1965.

[8] C. E. Shannon, "Communication in the presence of noise," *Proc. IRE*, vol. 37, 1949.

[9] A. B. Carlson, *Communication Systems*. New York: McGraw-Hill, 1975.

[10] T. Kanada, K. Hakoda, and E. Yoneda, "SNR fluctuation and nonlinear distortion in PFM optical NTSC video transmission systems," *IEEE Trans. Commun.*, vol. COM-30, Aug. 1982.

[11] P. F. Pander, *Modulation, Noise and Spectral Analysis*. New York: McGraw-Hill, 1965.

[12] Bell Telephone Laboratories Staff, *Physical Design of Electronic Systems*, vol. 1. Englewood Cliffs, NJ: Prentice-Hall, 1970.

[13] G. M. Masson, "Upper bounds on fanout in connection networks," *IEEE Trans. Circuit Theory*, 1973.

SOME CONSIDERATIONS FOR APPLYING SEVERAL FEEDFORWARD GAIN BLOCK MODELS TO CATV DISTRIBUTION AMPLIFIERS

John C. Pavlic
Engineering Manager, Distribution Products

C-COR ELECTRONICS, INC.
State College, Pennsylvania

Abstract

Many systems are being designed with amplifiers containing Feedforward technology because of its improved dynamic range over conventional push-pull hybrids. All Feedforward amplifying stages achieve this improved distortion performance by cancelling the distortion created in the stage's main amplifier. During the process of designing a Feedforward amplifying stage for use in trunk amplifiers, four circuits were modeled that would fulfill the basic requirements of a Feedforward amplifier stage. These four circuit models are presented with the operational advantages and disadvantages of each. In addition, the performance characteristics of several trunk stations using the most advantageous Feedforward circuit models are compared to each other and to conventional push-pull type trunk stations. The performance characteristics of several line extenders using the most advantageous Feedforward circuit models are also presented.

1. Introduction

Since the advent of Feedforward technology, its operational benefits and usefulness have not been well defined. We believe some insight into Feedforward theory would be helpful. Our purpose is to answer three basic questions. These are:

1. What is the optimum Feedforward gain block configuration?
2. What is the optimum Trunk Station configuration using Feedforward gain blocks or standard push-pull hybrids in conjunction with a Feedforward gain block?
3. Is the optimum trunk Feedforward gain block also the optimum for line extender amplifiers?

2. Designing an Optimum Feedforward Gain Block

The first consideration was to evaluate several Feedforward gain block configurations. We weighed the following characteristics: Gain, Noise Figure, Distortion Performance, Maximum Reach, and Power Consumption. Figure 1 illusrates the signal path of a Feedforward stage from input to output. Each characteristic of the Feedforward amplifying stage will be discussed separately to demonstrate how performance is determined.

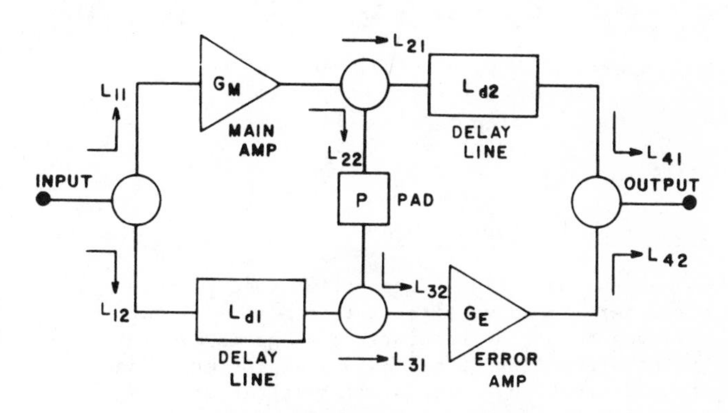

FIGURE 1
FUNCTIONAL BLOCK DIAGRAM OF A FEEDFORWARD GAIN BLOCK

2.1 Gain

The first design rule is that gain of a Feedforward Amplifying Stage equals the gain in the signal path minus incurred losses.

$$G_{FF} = G_M - L_{11} - L_{21} - L_{d2} - L_{41} \qquad (1)$$

Where: G_{FF} = gain of the Feedforward Stage,
G_M = gain of the main amplifier,
L_{11}, L_{21}, L_{41} = coupler losses, and
L_{d2} = second delay line loss

The second rule is that gain of the signal path input to output in the stage equals gain of the error path. This assumes $L_{21} = L_{31}$. That is:

$$G_M - L_{11} - L_{21} - L_{d2} - L_{41} = G_E - L_{12} - L_{d1} - L_{31} - L_{42} \qquad (2)$$

Where: G_E = error amplifier gain,
L_{12}, L_{31}, L_{42}, = coupler

Reprinted with permission from *Technical Papers, 32nd Annual Convention and Exposition, National Cable Television Association*, 1983, pp. 297-302.

Copyright © 1983, The National Cable Television Association.

losses, and

L_{d1} = First delay line loss

Equipped with Equations 1 and 2, the circuit designer can model several Feedforward gain stages using standard hybrids for G_M and G_E. The four circuit models in Figure 2 represent the only possible configurations left to the designer, since Equation 2 limits the circuit losses.

2.2 Noise Figure

Noise Figure of Feedforward gain stage is determined by the noise performance of the error amplifier leg. Since the noise produced by the main amplifier is canceled by the first loop, the Noise Figure of a Feedforward stage can be calculated in the following manner.

$$NF_{FF} = NF_{GE} + L_{31} + L_{D1} + L_{12} \qquad (3)$$

Where: NF_{FF} = Noise Figure of the Feedforward stage, and

NF_{GE} = Noise Figure of the error amplfier

2.3 Distortion Performance

Assuming that the limiting performance parameter is composite triple beat, distortion performance of the Feedforward stage is determined by the distortion produced by amplifier G_M and the distortion improvement factor K_D. Distortion improvement factor is a measure of the increase in output capability of the Feedforward stage as compared to the output capability of G_M. The amount of distortion cancelation achieved by the error amplifier loop is directly proportional to the amplitude and phase balance within loop. It has been determined that 24 to 25 dB of cancelation can be realized if the amplitude balance is within .25 dB peak-to-valley and the phase error is held within 2 degrees.[1] K_D is the distortion cancelation accomplished in the loop minus the circuit losses incurred between G_A and the Feedforward stage output. The distortion improvement factor, therefore, is:

$$K_D = \frac{24 - 2(L_{21} + L_{d2} + L_{41})}{2} \qquad (4)$$

Where: K_D = the distortion improvement factor, dB

2.4 Maximum Reach

Maximum reach is the longest cascade in dB that the gain blocks can be cascaded given a specific noise and distortion performance.

The hybrids used in the Feedforward circuits have a noise figure of 6 dB and a 56 dB carrier-to-composite beat performance at +46 dBmV flat output, loaded to 450 MHz with 60 channels. Maximum reach is a system with a desired 43 dB carrier-to-noise ratio and 59 dB carrier-to-composite triple beat ratio. Maximum reach can be calculated from the following equations.

$$R_{max} = N \times G_{FF} \qquad (5)$$

$$N = 10^{X} \qquad (6)$$

$$X = \frac{Vspec - Vopt}{10} - \frac{59 - CCTBspec}{20} \qquad (7)$$

$$Vopt = Vspec + \frac{43 - CNRspec}{2} - \frac{59 - CTBspec}{4} \qquad (8)$$

Where: R_{max} = maximum reach in dB,

N = number of gain blocks in cascade

Vspec = specified gain block output level,

CCTBspec = Specified gain block carrier-to-composite triple beat ratio, and

CNRspec = specified gain block carrier-to-noise ratio

2.5 Feedforward Circuit Models

To evaluate gain block performance (Figure 2), we assume the same noise and distortion assigned to G_M and G_E gain blocks. The blocks use standard values of G_M and G_E. Then gain, K_D, N_F, power consumption, and reach are calculated using Equations 1, 3, 4, 5, 6, 7, and 8. Table 1 gives comparisons.

	FF1	FF2	FF3	FF4
Gain, dB	23	18	18	24
NF, dB	9	16	9	12
K_D, dB	9	9	2	6
R_{max}, dB	1725	846	1134	792
Power, W	16.3	13.4	13.4	16.3

TABLE 1
Comparison of Performance of Several Feedforward Gain Blocks

FF1 is, therefore, the optimum gain block; it simultaneously produces minimum noise and maximum distortion cancelation. FF1 also provides maximum cascade when analyzed for trunkline use. FF2 is also attractive, even though its noise is high. The performance of each gain block is given in Table 1.

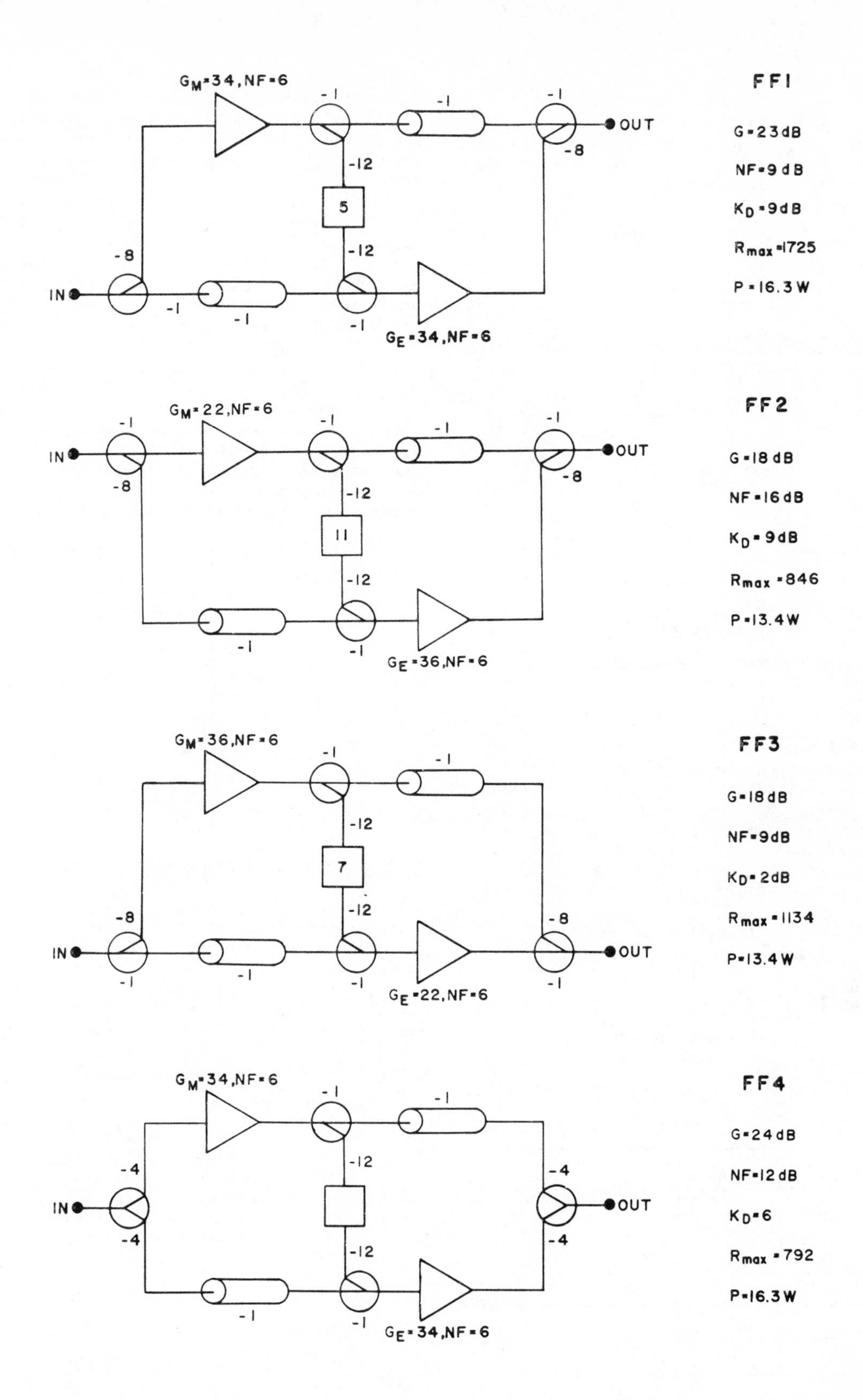

FIGURE 2

SEVERAL FEEDFORWARD GAIN BLOCKS

3. Configuring a Trunk Station.

Figure 3 illustrates a generic trunk station with two amplifying stages, G1 and G2, and losses from housing, slope, gain, PIN Diode attenuator, and an automatic level control/bridger amplifier sampling circuit.

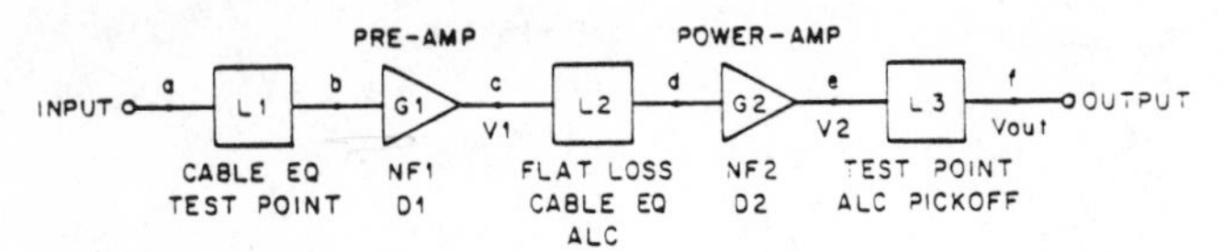

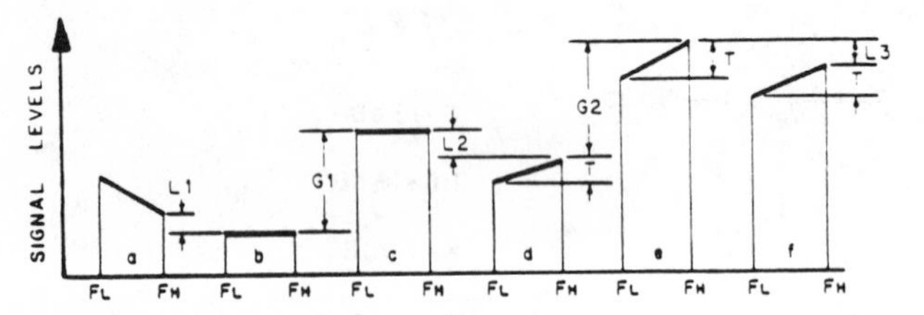

FIGURE 3

TRUNK STATION CONFIGURATION

Since FF1 and FF2 have advantages previously noted, we will model those gain stages into the trunk station in Figure 3. The distortion specifications in Table 2 were calculated by applying the distortion improvement factor, K_D, to the following equation.

$$D_{FF} = D_{GM} - 2K_D \quad (9)$$

Where: D_{FF} = distortion of the Feedforward gain block at 46 dBmV out, 60 channels flat, and
D_{GM} = distortion of G_M at 46 dBmV out, 60 channels flat

Gain Block	Gain, dB	CCTB* dB	NF, dB	Power, Watts
FF1	23	74	9.0	16.3
FF2	18	74	16.0	13.4
GB12	12	58	8.0	4.8
GB18	18	58	6.0	5.8
GB22	22	56	6.0	5.3

*46 dBmV out, flat, 60 channels, 450 MHz

TABLE 2
Comparison of Performance of Several Gain Blocks to be Used in Trunk Design

Table 2 lists distortion performances of FF1 and FF2 as well as several standard push-pull hybrids specified at +46 dBmV out, 60 channels flat. With information from Figure 3 and Table 2, we modeled several trunk amplifiers and evaluated their performance. Trunk model evaluation was based on specifications of trunk spacing, optimum output signal level, carrier-to-composite triple beat ratio, carrier-to-noise ratio, noise figure, maximum cascade in dB, maximum number of amplifiers in cascade, and power consumption. These specifications were drawn as outlined below.

3.1 Trunk Spacing

Trunk spacing is the maximum cable distance in dB at the highest operating frequency at which the station can be placed. Measured at 70°F, spacing includes all circuit losses and the reserve gain required for automatic level control.

$$GT = G1 + G2 - L1 - L2 - L3 \quad (10)$$

Where: GT = trunk spacing,
G1 = gain of G1,
G2 = gain of G2,
L1 = 2.5 dB,
L2 = 10.0 dB, and
L3 = 1.5 dB

or,

$$GT = G1 + G2 - 14 \quad (11)$$

3.2 Optimum Output Signal Level

Optimum output signal level is the station output level that permits maximum cascading of amplifiers while still meeting system performance requirements for both carrier-to-noise and carrier-to-composite triple beat. The trunk stations (Table 3) were optimized for a system with a carrier-to-composite triple beat ratio of 59 dB and carrier-to-noise ratio of 43 dB. To calculate the optimum output voltage for a trunk station, use Equation 8.

Where: Vspec = specified trunk station output level,
CNRspec = specified trunk carrier-to-noise ratio, and
CCTBspec = specified trunk carrier-to-composite triple beat ratio

3.3 Distortion Calculations

The station carrier-to-composite triple beat ratio, carrier-to-noise ratio, and noise figure are all determined by inserting Table 2 gain blocks into Figure 3 and calculating, on either a voltage or power basis, their distortion effect on station performance.

3.4 Maximum Trunk Reach

Maximum reach is defined as the maximum length in dB that trunk stations can be cascaded and still meet the trunk system requirements of 59 dB carrier-to-composite triple beat ratio and 43 dB carrier-to-noise ratio. To calculate maximum cascade, substitute trunk station performance for the Feedforward gain block performance in Equations 5, 6, 7, and 8.

Where: N = number of trunk stations in cascade

From Table 3, we conclude:

1. For the 31-32 dB Spaced Units. With a push-pull hybrid pre-amplifier and (FF1) output amplifier, trunk number 2 performs better than the 32 dB spaced trunk with two Feedforward stages. Power consumption in trunk station 2 is 11 watts less than trunk station 1.

2. For the 27 dB Spaced Units. The reach and power consumption of Model 4 is superior to that of 3. Model 4 requires 7.6 watts less.

No	Trunk Spacing dB	G1	G2	Maximum * Reach dB	Maximum* Cascade	Hybrid Power Watts	Vopt dBmV	Carrier-to-CTB dB	Carrier-to-Noise dB	Noise Figure dB
1	32	FF1	FF1	608	19	32.6	41.0	84.7	55.5	12.5
2	31	GB22	FF1	682	22	21.6	37.5	86.0	56.5	9.0
3	27	FF1	FF2	783	29	29.7	38.1	88.3	57.5	12.5
4	27	GB18	FF1	918	34	22.1	36.1	89.8	58.4	9.7
5	21	GB12	FF1	1050	50	21.1	34.5	93.0	59.9	12.6
6	22	GB18	GB18	704	21	12.5	29.4	89.3	58.1	8.3

*Using 43 dB CNR, 59 dB Composite Triple Beat Ratio

TABLE 3
Trunk Station Model Specifications

3.5 Power Consumption

The power consumption listed in Table 3 represents only the DC power consumed by G1 and G2.

3.6 Trunk Comparison

Table 3 summarizes performance of the trunk station models generated by installing Table 2 gain blocks into Table 3. All models are loaded to 450 MHz with 60 channels operating with a 7 dB output tilt.

Table 3 lists 21-22, 27, and 31-32 dB as three trunk spacing categories. Performance calculations assume that G1 operated at a distortion level 5 dB higher than normally encountered. We could, therefore, buffer the final trunk station performance calculation, since the input hybrid contributes to station distortion performance.

3. For the 21-22 dB Spaced Units. The dynamic range improvement of the FF1 gain block in Model 5 is reflected in the significantly improved reach.

From the Feedforward gain block modeling and performance data of the six trunk station models, C-COR proceeded to develop trunk station Models 2, 4, and 5. These stations are configured with an FF1 output gain block and a push-pull hybrid pre-amplifier of either 12, 18, or 22 dB to achieve spacings of 21, 27 and 31 dB.

4. Line Extenders

The line extender presents a different problem than the trunk because a gain of 34 dB is required. That gain spans the gap between +16 dBmV--the typical input level of the line extender--and the output capability limit of +50 dBmV of a Feedforward line extender. The 16 dBmV input level is mandated by the minimum signal level required on the feeder line; the 50 dBmV signal level is dictated by the non-linearity of Feedforward gain blocks operated above +46 dBmV out. Figure 4 illustrates a Feedforward line extender, and Table 4 lists performances of two models. One uses an FF1 output gain block; the other uses an FF2. For evaluation, both extenders were loaded to 450 MHz with 60 channels operating with a 7 dB output tilt. Extender distortion characteristics were calculated using the same methods for calculating trunk amplifier performance.

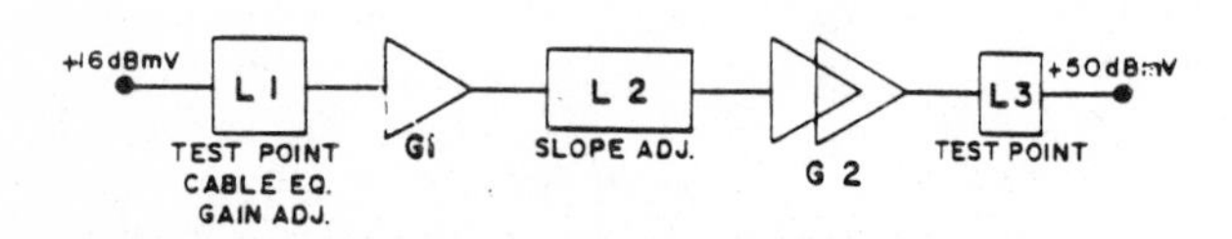

FIGURE 4

FEEDFORWARD LINE EXTENDER CONFIGURATION

From Table 4, we can conclude that Model 1 performed best, even though extender Models 1 and 2 both failed to produce the desired 34 dB extender spacing. G1 limited the distortion performance of Model 2 because of the low-gain characteristics of FF2. Although its power consumption is high, Model 1 or some similar design is the most desirable Feedforward line extender.

5. Conclusions

Answers to questions about Feedforward and its application to CATV distribution equipment follow.

1. The optimum Feedforward gain block configuration is FF1 (Figure 3).

2. The optimum Feedforward trunk station contains a standard push-pull hybrid as a pre-amplifier and an FF1 gain block as an output amplifier.

3. The optimum Feedforward gain block in line extenders is the same FF1 required for trunk application.

Model Number	Spacing dB	G1	G2	Output Level dBmV	Input Level dBmV	Carrier-to-CTB	Carrier-to-Noise	Noise Figure	Power
1	33	GB18	FF1	50	17	69.0	66.8	9.2	22.1
2	32	GB22	FF2	50	18	66.8	67.3	9.7	18.7

TABLE 4
Line Extender Model Specifications

References

[1]Meyer, Eschenbach and Edgerley, "A Wideband Feedforward Amplifier" IEEE Journal of Solid State Circuits, December 1974.

[2]Preschutti, Joseph P., "Feedforward Gain Block Performance Criteria" Engineering Report #695, C-COR Electronics, Inc., 1981

[3]Haney, Alan P., "Feedforward Trunk Gain Block Configuration" Engineering Report #704, C-COR Electronics, Inc., 1982.

An Intermodulation Prediction and Measurement Technique for Multiple Carriers through Weak Nonlinearities

Tai Sasaki and Hiroshi Hataoka
National Defense Academy, Yokosuka, Japan 239

Abstract— The theory of intermodulation distortion is discussed and a direct intermodulation measurement technique is proposed for the lower order components when many carriers are amplified by a weakly nonlinear amplifier. The test signal consists of a number of sinusoids fewer than that originally required and an amplitude modulated signal whose envelope distribution is given by a Rayleigh function. This technique simplifies the test equipment required and permits weaker nonlinearities to be studied than with the previous single carrier transfer technique.

In addition, the theory which permits the prediction of intermodulation level when n carriers are present from a measurement made with m carriers (n>m) is presented.

I. Introduction

In CATV or FDM satellite communication systems, a large number of carriers pass through a common amplifier. When the sum of these carriers encounters the nonlinearity of an amplifier, intermodulation (IM) products are generated. The generation of these products is a significant feature in determining the overall system performance. For this reason, numerous papers concerning IM problems have been published [1–5].

Most of these papers have carried out calculations with the aid of a single carrier transfer function (SCTF). Usually, the SCTF has been determined by measuring the output amplitudes and phases of the amplifier, which is driven by a single sinusoid. However, in some devices such as transistors, a temperature dependency of the SCTF is often observed; this phenomenon is especially noticeable in the phase characteristics of the SCTF. In such devices, determination of the SCTF by using the ordinary "point by point measurement method" becomes difficult if not impossible because the device temperature changes with varying input power.

In order to avoid this difficulty, a dynamic characteristics measurement technique was deviced by Pearce [6]. The basic idea of this method is to insert a pulsed amplitude modulated signal of low duty cycle into the test amplifier and then to determine the SCTF with the aid of a network analyser. This technique requires elaborate instrumentation. However, if the number of carriers is small (e.g. two or three carriers), it is not difficult to directly measure the IM performance by applying multi-carriers to the test amplifier. Of course, the direct measurement runs into difficulty with increasing the number of carriers. Therefore, it is desirable to predict the IM products for many carriers by using a test signal consisting of a number of sinusoids less than the number of carriers of interest. This approach allows prediction of the IM products even in temperature dependent amplifiers because the measurement using multi-carriers is considered as a sort of dynamic measurement technique.

It is known that the generation of IM products is not affected uniformly by the nonlinearities in each input level but is affected by nonlinearities in certain input regions which are determined by the number of carriers and their amplitudes[3]. For example, the IM produced by two equal sinusoids is significantly affected by the nonlinearities near the peak envelope value of the input. As we can see in the section III, the IM for many carriers is affected by the nonlinearities near the mean power associated with the original input carriers, and the effect of each input level varies corresponding to a Rayleigh function. Accordingly, a measurement made with unsuitable test signals results in emphasis of the nonlinearities which are not relevant to the actual case of interest.

This article includes the theory to analyse the IM for n carriers in memoryless nonlinear amplifiers when the IM for carriers less than n is known. First, we introduce conversion coefficients which represent the relationship between outputs for different carrier numbers, and it is shown that the coefficients are not affected by the amplifier characteristics and can be calculated beforehand. Once the coefficients are tabulated, it is possible to predict the IM for many carriers from a measurement made with fewer carriers.

The theory is applicable to an intermodulation measurement technique in the case of many carriers, and a direct IM measurement technique using a lesser number of carriers is proposed. Finally, the theory is experimentally confirmed by using a UHF transistor amplifier.

Manuscript Received (revised version) January 4, 1980

Reprinted from *IEEE Trans. Cable Telev.*, vol. CATV-4, pp. 146–154, Oct. 1979.

II. General Theory

When n carriers pass through a nonlinear amplifier, output components at frequencies $(k_1\omega_1+k_2\omega_2+ \ldots +k_n\omega_n)$ are produced, where $\omega_1, \omega_2, \ldots, \omega_n$ are the input angular frequencies. The k's are integers and the sum $\Sigma k_i = 1$ in the fundamental zone. Although the output contains an infinite number of IM components, for our practical interests it will be sufficient to consider only the cases in which the order of the IM product $\Sigma|k_i|$ is low, because the lower order components usually dominate in amplitudes. For instance, the third order IM products such as $(2\omega_1-\omega_2)$ and $(\omega_1 + \omega_2 - \omega_3)$ components which correspond to $\Sigma|k_i| = 3$ are known as significant disturbing noises. In such cases, if the number of carriers is large, there are many k's with zeros and only a few k's remain with nonzero values.

It is not necessary to arrange each carrier frequency in the order of magnitude, because the amplifier is assumed to be without frequency dependency within the pass band. We can divide the carriers into two parts, one containing the carriers corresponding to nonzero k's and the other containing zero k's, and then

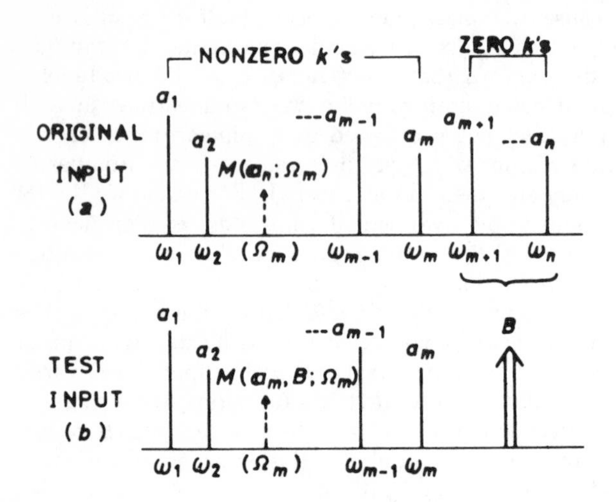

Fig. 1. (a) A set of rearranged original input carriers.
(b) Test input with a substitutional carrier.

renumber again. Fig. 1a illustrates a set of the rearranged input carriers; the carriers corresponding to nonzero k's are located from the first to mth. The angular frequency of the IM component of interest is assumed to be $\Omega_m=k_1\omega_1+k_2\omega_2+\ldots+k_m\omega_m$, and where m denotes the number of the carrier frequencies involved in the IM component.

Let the amplitudes of the input carriers be $a_1, a_2, \ldots, a_n$, then the magnitude of the output Ω_m component $M(\boldsymbol{a}_n;\Omega_m)$ can be represented by the following [3].

$$M(\boldsymbol{a}_n;\Omega_m) = \int_0^a F(\boldsymbol{a}_n, v; \boldsymbol{k}_m)\cdot G(v)dv \qquad (1)^*$$

$G(v)$ represents the SCTF for a single input sinusoid with an amplitude v. Since amplifiers operating in high frequency regions exhibit both amplitude and phase nonlinearities, $G(v)$ is generally given by a complex function. $F(\boldsymbol{a}_n,v;\boldsymbol{k}_m)$ may be considered as a weighting function for $G(v)$ and is given by the following, under the condition $k_s=0$ (s=m+1, m+2, ..., n):

$$F(\boldsymbol{a}_n, v; \boldsymbol{k}_m) = \int_0^\infty yv^2 J_1(vy) \prod_{\ell=1}^{m} J_{k_\ell}(a_\ell y) \prod_{S=m+1}^{n} J_0(a_s y)\, dy, \quad (2)$$

where J_k is the k-th order Bessel function. The upper limit a in the integral of (1) denotes the maximum envelope voltage of the original input multi-carriers and it is given by

$$a = a_1 + a_2 + \cdots\cdots + a_n. \qquad (3)$$

It is found that $F(\boldsymbol{a}_n,v;\boldsymbol{k}_m)$ can be calculated irrespective of the amplifier characteristics and if $G(v)$ is given with sufficient accuracy, the IM products can be known by making use of (1). However, as explained earlier, an accurate measurement of $G(v)$ is often difficult, especially in weak nonlinearities. Consequently, the estimation of the IM performances by using (1) becomes ambiguous.

Now, let us consider the substitution of a single carrier with the amplitude B for the cluster of the carriers from the (m+1)th to the nth, as shown in Fig. 1b. First, we can show that the product of the Bessel functions in the integrand of (2) is expressed as follows (cf. Appendix A):

$$\prod_{S=m+1}^{n} J_0(a_s r) = \int_0^\infty J_0(Br)\, P(\boldsymbol{a}_{m+1,n}, B)\, dB \qquad (4)^*$$

$P(\boldsymbol{a}_{m+1,n}, B)$ is a probability density when the envelope voltage of the sum of the carriers from the (m+1)th to the nth is B, and is given by the following [8]:

$$P(\boldsymbol{a}_{m+1,n}, B) = B\int_0^\infty r J_0(Br) \prod_{S=m+1}^{n} J_0(a_s r) dr. \qquad (5)$$

Equation (4) means that the product of (n-m) Bessel functions can be obtained by averaging $J_0(Br)$ with respect to the envelope voltage of the sum of (n-m)

$^*\boldsymbol{a}_n = (a_i), (i = 1, 2, \cdots, n)$ and $\boldsymbol{k}_m = (k_i), (i = 1, 2, \cdots, m)$.
$\boldsymbol{a}_{m+1,n} = (a_i), (i = m+1, m+2, \cdots\cdots, n)$.

sinusoids.

Let $a_{m+1,n}$ be the maximum value of (n-m) sinusoids, that is;

$$a_{m+1,n} = a_{m+1} + a_{m+2} + \cdots + a_n . \qquad (6)$$

Since $P(a_{m+1,n}, B)$ for $B > a_{m+1,n}$ is always zero, the infinite upper limit of the integration in (4) can be reduced to $a_{m+1,n}$. Substituting (4) into (2) and changing the order of the integration, then the following equation can be deduced:

$$F(a_n, v\,; k_m) = \int_0^{a_{m+1,n}} \left[\int_0^{\infty} yv^2 J_1(vy) \prod_{\ell=1}^{m} J_{k_\ell}(a_\ell y)\, J_0(By)\,dy\right] \times P(a_{m+1,n}, B)\,dB. \qquad (7)$$

It is obvious that the inside of the parenthesis in (7) is equal to the weighting function for (m+1) carriers with the amplitudes a_m and B. Rearranging (7) and multiplying both sides of its expression by G(v) and then integrating, we can finally get the following equation:

$$M(a_n\,; \Omega_m) = \int_0^{a_{m+1,n}} M(a_m, B\,; \Omega_m) \cdot P(a_{m+1,n}, B)\,dB. \qquad (8)$$

$M(a_m, B; \Omega_m)$ in the integrand of (8) is the output for (m+1) carriers which consist of the original m carriers with the amplitudes a_m $(=a_1, a_2 \ldots, a_m)$ and the substitutional carrier with the varying amplitude B. $P(a_{m+1,n}, B)$ may be considered as a conversion coefficient from the output for (m+1) carriers to that for n-carriers. From (5), it is found that $P(a_{m+1,n}, B)$ is independent of the amplifier characteristics and can be calculated beforehand if the input signals are known. Thus we can compute the IM for n carriers with the aid of the IM for (m+1) carriers, one of which has the varying amplitude up to $a_{m+1,n}$.

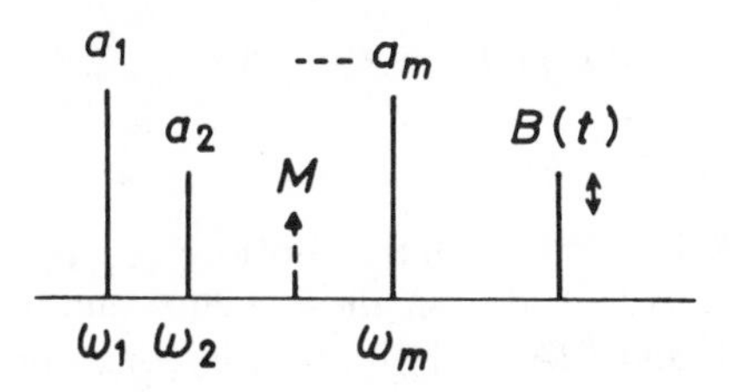

Fig. 2. Test input with an amplitude modulated carrier.

Next, we consider an arbitrary modulated signal with an envelope B(t), as shown in Fig.2. The probability density of this envelope is assumed to be $P(a_{m+1,n}, B)$ at B(t)=B; that is, the same probability as that of the envelope of the original (n-m) carriers. In general, the envelope B(t) may be a multiple-valued function and may take the same value at many different times. Let the period of this envelope be T_0 and the sum of each time interval within the period where the envelope voltage falls into the infinitesimal range $B \sim B+dB$ is assumed to be dt', then the probability of finding the envelope in its range is dt'/T_0. On the other hand, this probability can also be represented by $P(a_{m+1,n}, B)dB$. Accordingly, we have

$$P(a_{m+1,n}, B)\,dB = \frac{1}{T_0}\,dt' \qquad (9)$$

Substituting (9) into (8), the output for n carriers can be obtained as follows:

$$M(a_n, \Omega_m) = \frac{1}{T_0}\int_0^{T_0} M(a_m, B(t); \Omega_m)\,dt \qquad (10)$$

Equation (10) means that the simple time average of the output component for (m+1) carriers is equal to the same component for n carriers. The cluster of carriers from the (m+1)th to the nth can be represented by only one signal, provided that the probability density of this signal is the same as that of the cluster. This result is applicable to an IM measurement technique for the case of many carriers which will be discussed later.

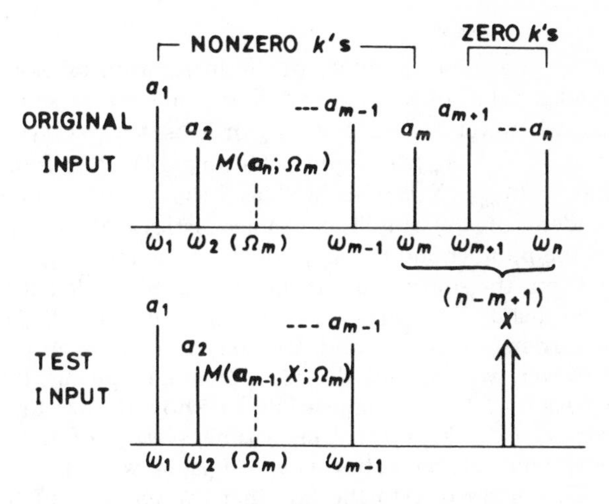

Fig. 3 Test input carriers with a substitutional carrier.

Now, further reducing the number of test carriers by one, as in Fig.3, let us express the output for n carriers in terms of the output for m carriers. In order to do this, the following formula with respect to the product of the Bessel functions is useful [7]:

$$J_{k_m}(a_m y)\, J_0(By) = \frac{1}{2\pi}\int_{-\pi}^{\pi} J_{k_m}(Xy)\, e^{jk_m\varphi}\, d\theta \qquad (11)$$

where

$$\left.\begin{aligned} X &= \sqrt{a_m^2 + B^2 + 2a_m B\cos\theta}, \\ \varphi &= \sin^{-1} \frac{1}{2a_m X}\sqrt{(X^2-(B-a_m)^2)((B+a_m)^2-X^2)} \end{aligned}\right\} \quad (12)$$

Making further calculation of (2) by using (4) and (11), and using (1), the following relation can be deduced:

$$M(a_n, \Omega_m) = \int_0^\infty M(a_{m-1}, X; \Omega_m)\cdot P_1(a_{m,n}, X; k_m)\,dX \quad (13)$$

$M(a_{m-1}, X; \Omega_m)$ is the output Ω_m component for m carriers with the amplitudes a_{m-1} and X. The coefficient $P_1(a_{m,n}, X; k_m)$ represents the relationship between two outputs for the carrier numbers of n and m, and may be called the conversion coefficient and is given as follows:

$$P_1(a_{m,n}, X; k_m) = \frac{2}{\pi}\int_0^{B_1} P(a_{m+1,n}, B)\frac{X\cos k_m\varphi}{\sqrt{(X^2-(B-a_m)^2)((B+a_m)^2-X^2)}}dB, \quad (14)$$

where the upper limit B_1 of the integration takes the smaller value of $a_{m+1,n}$ and $X+a_m$. It can be shown that whenever $X \geqq a_m + a_{m+1,n}$ or $X \leqq |a_m - a_{m+1,n}|$, $P_1(a_{m+1,n}, X; k_m) = 0$. From equation (14), it is found that $P_1(a_{m,n}, X; k_m)$ as well as $P(a_{n+1,n}, B)$ is independent of the amplifier characteristics and depends on the number and the amplitudes of the input carriers and on the output frequency assigned. Calculating beforehand this coefficient, we can predict the IM for n carriers with the aid of the m carrier output. However, we can not deduce a formula equivalent to equation (10). It is impossible to know the IM for n carriers by taking the simple time average of the m carrier output, no matter what test signal is used.

This arises due to the fact that the phase φ_i of the ith sinusoid causes the phase shift by $k_i\varphi_i$ to the output with the angular frequency $(k_1\omega_1 + k_2\omega_2 + \cdots + k_i\omega_i + \cdots)$ [5]. In the above discussion, we carried cut the calculation of the IM by substituting the cluster of many carriers by only one signal. In general, the cluster of sinusoids can be written by the form of a single sinusoid (that is, of the form of $B\sin(\omega_0 t + \varphi)$). The amplitudes B and phase φ are both slowly time-varying functions. In the first example, the substitutional signal which represents the cluster of the original (n-m) carriers is assigned at the (m+1)th and as $k_{m+1} = 0$, the phase of the substitutional carrier makes no contribution to the output. Thus, we can obtain the IM by taking the average over the amplitude B alone, as seen in equation (10). While in the second example, the substitutional signal is assigned at the mth and as $k_m \neq 0$, the output is also affected by the phase of this signal. Accordingly, the output must be calculated by taking account of both amplitude B and phase φ. Thus we can not obtain a formula equivalent to equation (10).

III. Calculation of Conversion Coefficients

a) Calculation of $P(a_{m+1,n}, B)$

By making use of (5), we calculate the conversion coefficients for the cases of n-m=2 and 3. The calculated results are shown in the normalized forms by using the following relations:

$$\left.\begin{aligned} A_i &= a_i / a_{m+1,n} \quad (i = m+1, m+2, \cdots, n) \\ Y &= B / a_{m+1,n} \\ W(Y) &= P(a_{m+1,n}, B)\cdot a_{m+1,n}. \end{aligned}\right\} \quad (15)$$

$n - m = 2$

$$\left.\begin{aligned} W(Y) &= \frac{2Y}{\pi\sqrt{(1-Y^2)(Y^2-(A_n-A_{n-1})^2)}} \\ & \qquad |A_n - A_{n-1}| < Y < 1 \\ &= 0 \quad \text{elsewhere.} \end{aligned}\right\} \quad (16)$$

For instance, if $(2\omega_1 - \omega_2)$ component for three carriers is known, the same component for four carriers can be calculated by making use of (8) and (16).

$n - m = 3$

$$\left.\begin{aligned} W(Y) &= k\,C\,K(k), \quad 0 < k < 1 \\ &= C\,K(1/k), \quad 1 < k \\ & \qquad Y_0 < Y < 1 \end{aligned}\right\} \quad (17)$$

where K is the complete elliptic integral of the first kind, and Y_0 is the smallest value among $(2A_n - 1)$, $(2A_{n-1} - 1)$ and $(2A_{n-2} - 1)$ under the constraint that Y_0 is no less than zero. The modulus k and the constant C are given by

$$k = \frac{4\sqrt{A_n A_{n-1} A_{n-2} Y}}{\sqrt{(1-Y)(1+Y-2A_n)(1+Y-2A_{n-1})(1+Y-2A_{n-2})}} \quad (18)$$

$$C = \frac{1}{\pi^2} \sqrt{\frac{Y}{A_n A_{n-1} A_{n-2}}} \tag{19}$$

where $A_n + A_{n-1} + A_{n-2} = 1$.

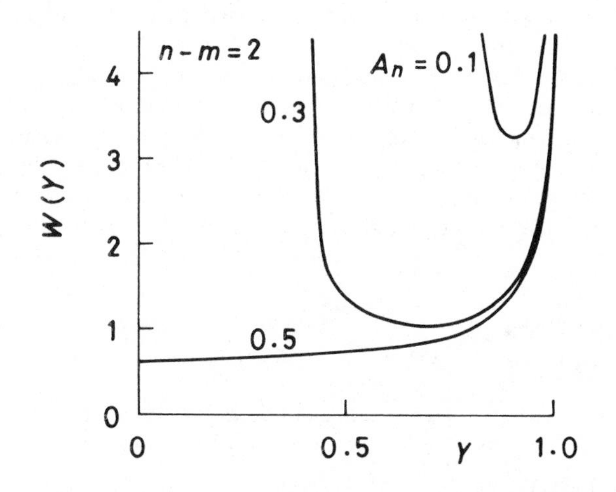

Fig. 4. Calculated results of $P(a_{m+1,n}, B)$ for n-m = 2.

Fig.4 illustrates the calculated results of W(Y) for n-m=2. From this figure, it is apparent that the IM product for n carriers does not depend uniformly on the IM for (m+1) carriers, but a strong dependency occurs in certain regions. Especially, W(Y) takes infinity at two points of Y=1 and $|A_n - A_{n-1}|$. Accordingly, the n carrier output can be roughly estimated from the (m+1) carrier outputs near these two input levels. As is clear from the symmetricity of (16), W(Y) remains unchanged when interchanging A_n and A_{n-1} with each other.

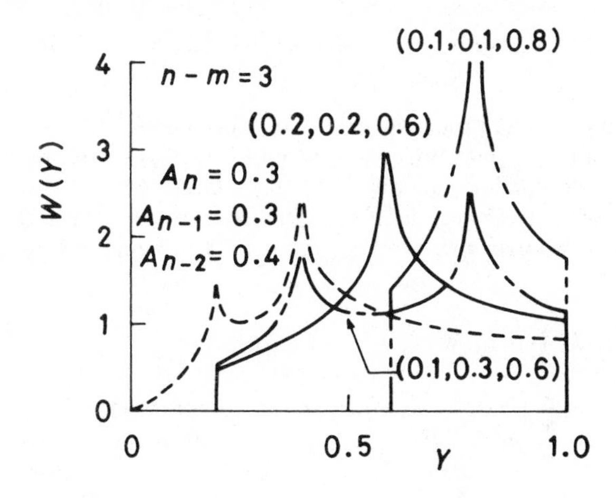

Fig. 5. Calculated results of $P(a_{m+1,n}, B)$ for n-m = 3.

Fig.5 illustrates the calculated results of W(Y) for n-m=3. W(Y) takes zero in the region where Y is smaller than any of $(2A_n - 1)$, $(2A_{n-1} - 1)$ and $(2A_{n-2} - 1)$. As Y>0, whenever all of A_n, A_{n-1}, and A_{n-2} are smaller than 0.5, there are no regions for W(Y)=0. Furthermore, W(Y) takes large values near the points of $Y = (1-2A_n)$, $(1-2A_{n-1})$ and $(1-2A_{n-2})$, unless Y<0.

Case of many carriers

Let us consider the case in which the number of carriers is large and in addition, we assume that each carrier is not dominant in its amplitude from the others. These conditions will be satisfied in most CATV systems. In this case, the following approximation can be made (eq. (39) in [1] or p.421 in [7]):

$$\prod_{S=m+1}^{n} J_0(a_s r) \cong \exp\left\{ - \sum_{S=m+1}^{n} \frac{(a_s r)^2}{4} . \right\} \tag{20}$$

Substituting (20) into (5), we can obtain $P(a_{m+1,n}, B)$ for a large (n-m).

$$P(a_{m+1,n}, B) \cong \frac{B}{S_{mean}} \exp\left\{ - \frac{B^2}{2 S_{mean}} \right\}, \tag{21}$$

where S_{mean} is the mean power associated with the input carriers from the (m+1)th to the nth, namely:

$$S_{mean} = \frac{1}{2}(a_{m+1}^2 + a_{m+2}^2 + \cdots + a_n^2). \tag{22}$$

Equation (21) is well known as a Rayleigh distribution function. P takes the maximum value at $B=\sqrt{S_{mean}}$. It means that the output for m original carriers plus the substitutional carrier with the power S_{mean} contributes significantly to the output for n carriers. For instance, the $(2\omega_1 - \omega_2)$ component for many carriers has strong dependency with the $(2\omega_1 - \omega_2)$ component produced by the two original carriers at ω_1 and ω_2 plus the third carrier with the power near S_{mean}; that is, intermodulation for many carriers is strongly affected by the nonlinearities near the mean power associated with the input signals, not near the peak power.

b) Calculation of $P_1(a_{m,n}, X; k_m)$

As an example, let us calculate the conversion coefficient $P_1(a_{2,3}, X; k_2)$ at $k_1 = 2$, $k_2 = -1$ which denotes the relationship between $(2\omega_1 - \omega_2)$ components for two carriers and for three carriers. In this case, we substitute n=3, m=2, $B = a_3$ and $k_m = -1$ into (14) and by making use of (5), (12), the following equation can be obtained:

$$W_1(Y) = \frac{1}{\pi A_2} \frac{Y^2 + A_2^2 - A_3^2}{\sqrt{(Y^2 - (A_3 - A_2)^2)(1 - Y^2)}}, \tag{23}$$

where $W_1(Y)$ is shown in the normalized form by using the following relations:

$$\left.\begin{aligned} W_1(Y) &= P_1(a_{2,3,}\ X\ ;-1)\cdot(a_2+a_3) \\ Y &= X/(a_2+a_3) \\ A_2 &= a_2/(a_2+a_3) \\ A_3 &= a_3/(a_2+a_3) \end{aligned}\right\} \quad (24)$$

Even though equation (23) is deduced under the condition of m=2, (23) can bé also held whenever $k_1=\pm 1$, n=m+1 are satisfied. For instance, if $k_1=k_2=1$, $k_3=-1$, and m=3, (23) represents the relation between $(\omega_1+\omega_2-\omega_3)$ components for three carriers and for four carriers. And if $k_1=1$ and m=1, $W_1(Y)$ represents the relation between ω_1 components for a signal carrier and for two carriers. In this case, the analysis arrives at a conventional method based on the SCTF.

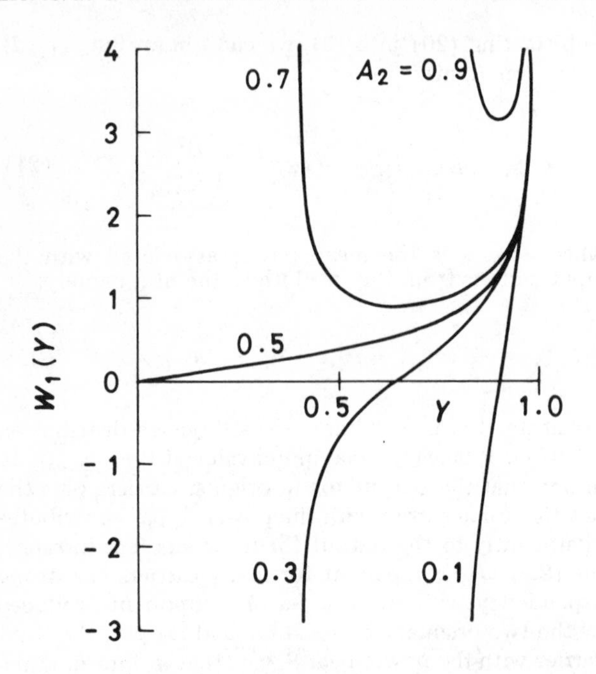

Fig. 6. Calculated results of $P_1(a_{m,n}, X; k_2)$ for $k_1=2$, $k_2=-1$, m = 2 and n = 3.

The calculated results of $W_1(Y)$ are shown in Fig.6. $W_1(Y)$ takes large values when the normalized input Y takes the values near 1 and $|A_3-A_2|$. $W_1(Y)$ differs from each of the W(Y) discussed before, because the values can be assumed negative as well as positive, when A_2 is less than 0.5. For instance, when $A_2=0.3$, $W_1(Y)$ takes large negative values near Y=0.4 ($=A_3-A_2$). Accordingly, when we intend to predict $(2\omega_1-\omega_2)$ component for three carriers from the measured IM for two carriers, we need to multiply the large coefficient values to the measured IM near Y=0.4 and 1.0, and then to subtract. This means that much care must be taken with the measurement near these points because the measurement errors around these points contribute significant errors for the calculated IM. On the other hand, it is also found that if we trim the two carrier performances to cause the cancellation between the parts of the intermodulations which are produced from the positive region of $W_1(Y)$ and the negative region of $W_1(Y)$, the three carrier IM can be improved.

c) Conversion coefficient between different output components

In the above discussion, we carried out the calculations of the IM with the aid of the same output component for a different carrier number. Making further generalization, we can deduce the conversion coefficient which represents the relationship between two different components for different carrier numbers.

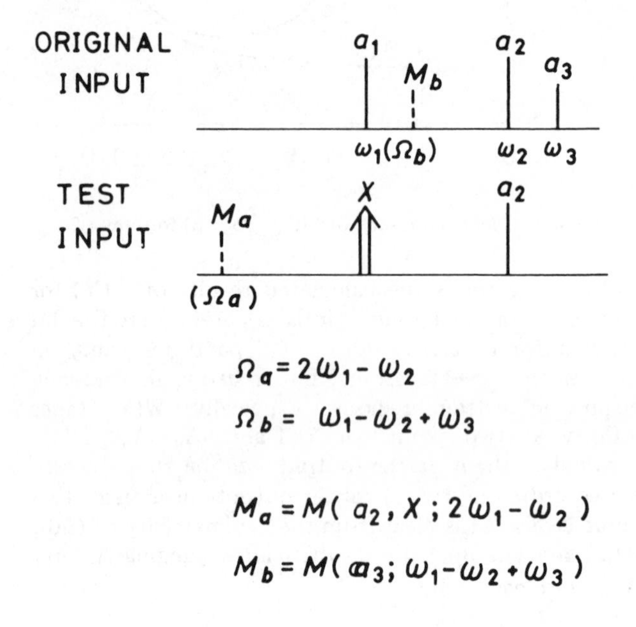

Fig. 7. A set of test input carriers with a substitutional carrier.

As an example, we calculate the coefficient representing the relationship between $(2\omega_1-\omega_2)$ component for two carriers and $(\omega_1-\omega_2+\omega_3)$ component for three carriers, as shown in Fig.7. In order to calculate the above coefficient, we may use the following relation [7]:

$$J_1(a_1y)\,J_1(a_3y) = \frac{1}{\pi}\int_0^{\pi} J_2\left(\sqrt{a_1^2+a_3^2+2a_1a_3\cos\phi}\ y\right) \times \frac{(a_1^2+a_3^2)\cos\phi+2a_1a_3}{a_1^2+a_3^2+a_1a_3\cos\phi}\,d\phi \quad (25)$$

From (1), the magnitude of $(\omega_1-\omega_2+\omega_3)$ component for three carriers is written by

$$M(a_3\,;\,\omega_1-\omega_2+\omega_3)$$

$$=\int_0^a G(v)dv\int_0^\infty yv^2J_1(vy)J_1(a_1y)J_{-1}(a_2y)J_1(a_3y)dy. \tag{26}$$

Applying (25) into the second integrand of (26) and rearranging the result, then the right side of (26) becomes

$$\int_0^\infty M(a_2,X\,;\,2\omega_1-\omega_2)\cdot P_2(X)dX \tag{27}$$

where

$$P_2(X) = \begin{cases} \dfrac{(a_1^2+a_3^2)X^2-(a_1^2-a_3^2)^2}{\pi a_1 a_3\sqrt{((a_1+a_3)^2-X^2)(X^2-(a_1-a_3)^2)}} \\[2ex] \qquad (a_1-a_3)^2 \leqq X^2 \leqq (a_1+a_3)^2 \\[1ex] = 0 \qquad \text{elsewhere} \end{cases} \tag{28}$$

$M(a_2,X;2\omega_1-\omega_2)$ is the output $(2\omega_1-\omega_2)$ component for two carriers with the amplitudes X and a_2. $P_2(X)$ is also independent of the amplifier characteristics and we can estimate the $(\omega_1-\omega_2+\omega_3)$ component for three carriers from $(2\omega_1-\omega_2)$ component for two carriers. Even if equation (27) is deduced for n=3 and m=2, (27) can be also valid for the conversion coefficient between $(\omega_1-\omega_2+\omega_3)$ component for n carriers and $(2\omega_1-\omega_2)$ component for (n-1) carriers.

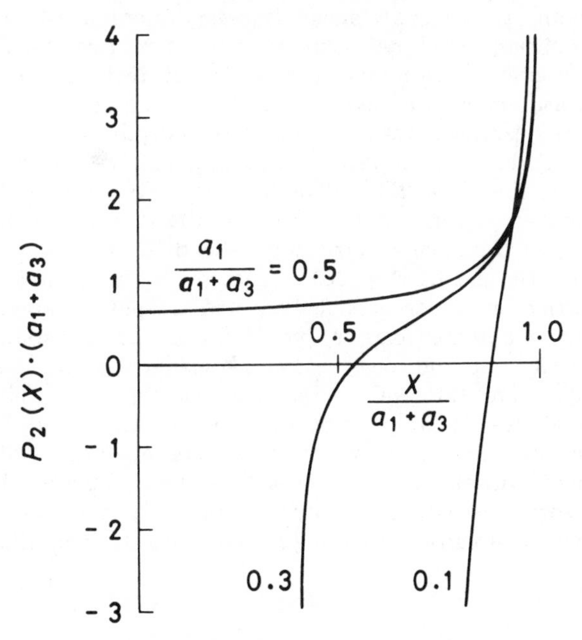

Fig. 8. Calculated results of $P_2(X)$ for $k_1 = 2$, $k_2 = -1$, m = 2 and n = 3.

Fig.8 illustrates the calculated results of $P_2(X)$. From this figure, it is found that $P_2(X)$ takes large values near $X=(a_1+a_3)$ and $|a_1-a_3|$ and can take both positive and negative values.

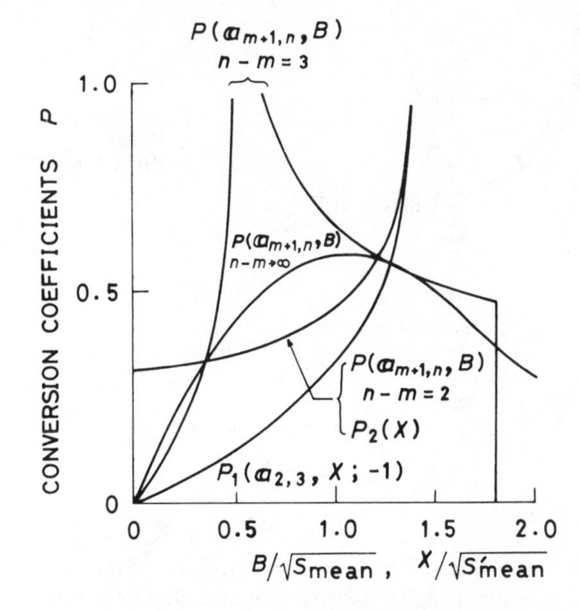

Fig. 9. Calculated results of conversion coefficients in the case of equal input amplitudes. Calculated values are shown in the normalized forms by using the following relations:

$$P=\begin{cases} P(a_{m+1,n},B)\cdot\sqrt{S_{mean}} \\ P_1(a_{2,3},X\,;-1)\cdot\sqrt{S'_{mean}} \\ P_2(X)\cdot\sqrt{S'_{mean}} \\ S'_{mean}=S_{mean}+a_m^2/2 \end{cases}$$

Fig.9 shows the calculated results for the several conversion coefficients when all carriers have equal amplitudes.

IV. Measurement of IM for Many Carriers

The theory is applicable to an IM measurement technique for a large number of carriers. Note that from (10), the average value of Ω_m component for (m+1) carriers is equal to the same output component for n carriers. Although the average time is taken to be T_o in (10) and even if T_o may be unknown in most applications, we can obtain the IM by averaging over sufficient long time instead of T_o.

One example for producing the test signal is shown in Fig.10. At first, (n-m) low frequency oscillators with the amplitudes a_{m+1}, a_{m+2}, ..., a_n are added. The envelope of these signals should have the same probability density as the envelope of the original (n-m) RF carriers, even if slowly varying with time. Then the (m+1)th RF carrier is modulated so as to have the

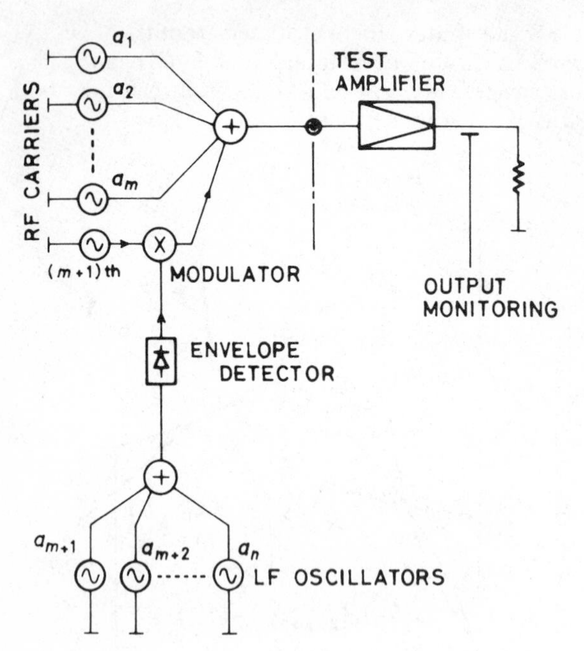

Fig. 10. IM measurement method for a large number of carriers.

same amplitude of this low frequency envelope. Finally, the original m RF carriers are added. These signals can be used as the test input.

In this manner, when (n-m) is large, many low frequency oscillators are needed. However, it is well known that the envelope of many sinusoids has a Rayleigh distribution function. It is also well known that this approximation is valid for the number of sinusoids with equal amplitude larger than three [8]. Therefore, even if (n-m) is large, it is sufficient to use no more than three LF oscillators, provided that the mean power associated with this test input is equal to that of the original input.

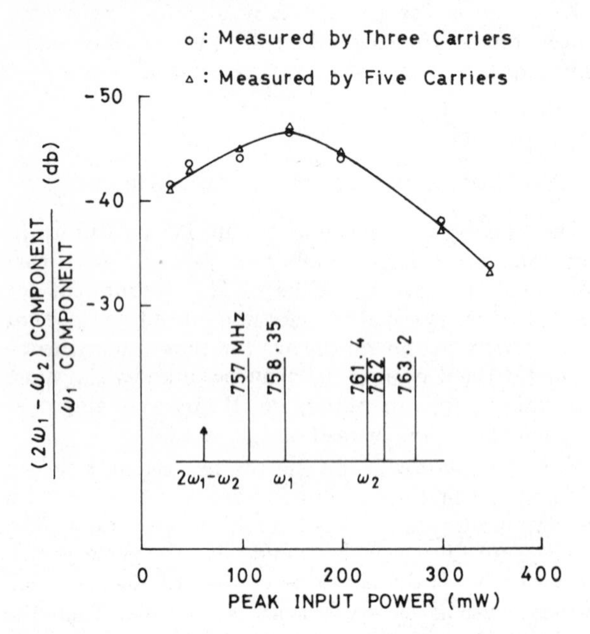

Fig. 11. Measured IM in a UHF transistor amplifier (2SC1042).

Experimental results in a UHF transistor amplifier are shown in Fig.11. The triangles represent the directly measured intermodulation at the angular frequency $(2\omega_1 - \omega_2)$ by using five equal carriers, and the circles represent the measured results by using two original carriers at ω_1 and ω_2 plus the RF carrier which is modulated by the envelope of three LF oscillators. The horizontal axis is a peak input power. Both intermodulations are measured by using a spectrum analyser. In this figure, it is apparent that the measured results by the different methods nearly agree.

V. Conclusion

We have discussed the prediction of intermodulation distortion with the aid of the output for a number of carriers less than the total number in which we are interested. In order to represent the relationship between two outputs, we introduced coefficients which were determined irrespective of the amplifier characteristics. Calculating these coefficients between several outputs for different carrier number beforehand, the theory predicts the intermodulation when the carrier number is increased.

Calculated examples of the coefficients were presented. Especially, the coefficients representing the relationship between the same components for different carrier numbers n and (m+1) can be expressed by a simple form, that is, the probability density function of the envelope of the input from the (m+1)th to the nth.

This approach does not require any information about the single carrier transfer function of the amplifier, therefore it is more useful than others for IM prediction in nonlinear devices in which the accurate measurement of SCTF is difficult; e.g. in weakly nonlinear devices with temperature dependency.

In addition, we have pointed out that the theory is applicable to an IM measurement technique for a large number of carriers. For instance, the $(2\omega_1 - \omega_2)$ component for many carriers can be directly measured with the aid of a spectrum analyser by inserting two carriers at the frequencies ω_1 and ω_2 plus an amplitude modulated carrier, provided that the probability density and the mean power associated with this test signal are the same as in the original input. This technique is useful for the estimation of the IM performance in nonlinear amplifiers operating with many carriers such as in CATV systems. Finally, the theory was experimentally confirmed for the case of five equal carriers by using a UHF transistor amplifier.

Appendix A

Let us express the sum of (n-m) sinusoids by

$$\sum_{S=m+1}^{n} a_s \sin(\omega_s t + \lambda_s) \qquad \text{(A-1)}$$

where $\lambda_{m+1}, \lambda_{m+2}, \ldots, \lambda_n$ are the phase of each sinusoid and are uniformly distributed over the continuous range $(-\pi, \pi)$. (A-1) can be rewritten by the form of the single sinusoid, namely;

$$B_{m+1,n} \cdot \sin(\omega_0 t + \varphi_{m+1,n}) \qquad \text{(A-2)}$$

where $B_{m+1,n}$, $\varphi_{m+1,n}$ are the envelope and phase of the cluster of (n-m)-sinusoids, and $\omega_s = \omega_0 + \Delta_s$.

Now, $B_{m+1,n}$ can be also written in terms of $B_{m+1,n-1}$ and a_n as

$$B_{m+1,n} = \sqrt{B^2_{m+1,n-1} + a_n^2 + 2B_{m+1,n-1} \cdot a_n \cos(\lambda_n + \Delta_n - \varphi_{m+1,n-1})} \qquad \text{(A-3)}$$

where $B_{m+1,n-1}$ is the envelope of the cluster except the nth sinusoid. Since $B_{m+1,n-1}$, φ_{n-1} are independent of λ_n, we get

$$\int_{-\pi}^{\pi} J_0(B_{m+1,n} \cdot r) d\lambda_n = (2\pi) J_0(B_{m+1,n-1} \cdot r) J_0(a_n \cdot r) \qquad \text{(A-4)}$$

Expressing $B_{m+1,n-1}$ in terms of $B_{m+1,n-2}$ and a_{n-1}, we iteratively get

$$(2\pi) \int_{-\pi}^{\pi} J_0(B_{m+1,n-1} \cdot r) J_0(a_n \cdot r)\, d\lambda_{n-1} = (2\pi)^2 J_0(B_{m+1,n-2} \cdot r) J_0(a_{n-1} \cdot r) J_0(a_n \cdot r) \qquad \text{(A-5)}$$

By repetition of this process, we deduce that

$$\prod_{S=m+1}^{n} J_0(a_s r) = \left(\frac{1}{2\pi}\right)^n \int_{-\pi}^{\pi} \int_{-\pi}^{\pi} \cdots\cdots \int_{-\pi}^{\pi} J_0(B_{m+1,n} \cdot r)\, d\lambda_{m+1} \cdot d\lambda_{m+2} \cdots\cdots d\lambda_n \qquad \text{(A-6)}$$

Remembering that the probability density of each λ is $1/2\pi$, it is apparent that the right side of (A-6) denotes the average value of $J_0(B_{m+1,n} \cdot r)$ over $\lambda_{m+1}, \lambda_{m+2}, \ldots, \lambda_n$. On the other hand, the average of $J_0(B_{m+1,n} \cdot r)$ can be taken over the envelope of the sinusoids instead of λ's. Thus we get

$$\prod_{S=m+1}^{n} J_0(a_s r) = \int_0^{\infty} J_0(Br)\, P(a_{m+1,n}, B)\, dB \qquad \text{(A-7)}$$

where $P(a_{m+1,n}, B)$ denotes the probability density of the envelope of (n-m) sinusoids at $B_{m+1,n} = B$.

Acknowledgement

The authors wish to thank Dr. Y. Shibata for his useful discussions and suggestions.

References

1. O. Shimbo, "Effects of intermodulation, AM-PM conversion, and additive noise in multicarrier TWT systems", Proc.IEEE, Vol.59, No.2, pp.230—238, February 1971.
2. E. Imboldi, G.R.Stette, "AM-to-PM conversion and intermodulation in nonlinear devices", Proc.IEEE, Vol.61, No.6, pp.796—797, June 1973.
3. T. Sasaki, H.Hataoka, "Generation of inter-or cross- modulation distortions in nonlinear devices", IEEE Trans. on CATV, Vol.CATV-2, No.3, pp.131—137, July 1977.
4. M. Horstein, D.T.Laflame, "Intermodulation spectra for two SCPC systems", IEEE Trans. on Communications, Vol.COM-25, No.9, pp.990—994, September 1977.
5. T.Sasaki, "Generation mechanisms of distortions in nonlinear devices", JIECE (JAPAN), Trans. IECE, Vol.61-A, No.9, pp.872—879, September 1978.
6. J.L. Pearce, "Intermodulation performance of solid-state UHF class-C power amplifiers", IEEE Trans. on Communications, Vol.COM-25, No.3, pp.304—310, March 1977.
7. G.N. Watson, "Theory of Bessel Function", London, England: Cambridge Univ. Press, 1958.
8. H.Hodara, "Statistics of thermal and laser radiation", Proc. IEEE, Vol.53, No.6, pp.696—704, July 1965.

EFFECTIVENESS OF STATIC SCRAMBLING VS DYNAMIC SCRAMBLING SYSTEMS A CLASSIFICATION METHOD

MICHAEL T. HAYASHI

PIONEER COMMUNICATIONS OF AMERICA, INC.
COLUMBUS, OHIO

ABSTRACT

Various scrambling systems have been introduced to the market place as a possible solution to the industry wide problem of theft of service. The effectiveness of scrambling is often a very confusing and difficult factor to determine. Classification of the various scrambling systems available today yeilds two basic forms -- static and dynamic. The effectiveness of the two varies depending upon the key signal and reference signal used in the descrambling process. Majority of scrambling systems are designed by slightly deviating from NTSC TV standards. Thus, from its initial design concept, these scrambling systems are vulnerable to pirate designs. Complete video encryption may certainly be the solution. However, the associated price is not affordable today. What compromises can be made to design an affordable ultimate scrambling system?

INTRODUCTION

Theft of service is a major concern for the entire cable industry. This concern has increased in proportion to the increase in value and quantity of products to be "stolen." Various methods of service protection have evolved in recent years, with more expected to surface. TV signal scrambling (as we all know) is one of the techniques used to deter potential theft of TV signals. It should be made clear that scrambling is merely a protective mechanism for premium TV pictures. TV signal scrambling is not a means to avoid other forms of service theft to which our industry is exposed. The operator will still need to contend with theft of hardware, hardware tampering and security of operation (installers), etc. However, with the considerable increase in consumer value of cable programming and the number of channels offered, cable signals must be protected.

This paper discusses specific aspects of all forms of TV scrambling/signal encoding. The flood of various scrambling methods being introduced -- RF vs baseband, sine wave sync vs gated sync, jamming, dynamic switching, random rate, encrypted scrambling, just to name a few -- can be very confusing, especially when one is attempting to judge the relative security provided by each method.

DEFINITION OF TV SIGNAL SCRAMBLING

From a technical perspective, a scrambling system has two purposes -- to prevent reception by "normal" television and to be capable of restoring the scrambled signal for reception by "normal" television. In the United States, television sets are designed to National Television System Committee (NTSC) standards. Therefore, deviation from the NTSC standard processing of the signal in most cases will accomplish scrambling. However, the term, "normal TV," provides its own share of confusion due to the technical advances in television receivers to accommodate such relatively unstable signal sources as home VTR's and video game machines. In addition there is cable ready, component TV, and digital TV. Scrambling methods, therefore, must be carefully chosen to be sure they introduce a scrambling factor beyond what the TV reciever may consider a tolerable variance from its standard (NTSC).

The subjective effectiveness of scrambling is another factor to consider when selecting a scrambling method. The tolerance level for a scrambled picture of someone recieving that picture "free" may be very high. Therefore, the scrambling level should be sufficient to render any TV picture received without a decoder subjectively unacceptable to an audience group, even if they are recieving it "free." Depending on program content, what one wants to see or

Reprinted with permission from *Technical Papers, 32nd Annual Convention and Exposition, National Cable Television Association*, 1983, pp. 309-313.
Copyright © 1983, The National Cable Television Association.

hear will vary, making it very difficult to determine a guideline as what is acceptable scrambling in all cases.

The final factor in determining secure scrambling is the level of difficulty required to defeat the scrambling method. The level of difficulty is tied directly to the cost of preventing defeat. Obviously, the ideal is authorized descrambler would have the highest difficulty level possible at the lowest possible cost.

STATIC VS DYNAMIC SCRAMBLING SYSTEMS

Based upon method of application, TV signal scrambling or encoding can largely be classified into two major categories: static scrambling and dynamic scrambling.

Static scrambling processes the signal in a constant and predictable manner with respect to time. Dynamic scrambling, on the other hand, takes away the element of predictability within the scrambled signal itself. Dynamic scrambling is thus more secure in most cases than static scrambling forms because it introduces an added element to be decoded -- time.

In addition to the actual scrambling itself, all active scrambling systems may incorporate one or two types of information for proper descrambling. A reference signal may be required to re-establish proper descrambling levels and/or timing. A key signal may be used to determine when and what type of encoding method may be taking place. In certain instances, the same signal may carry both two types of information creating a situation where the signal function can be easily misinterpreted or misunderstood.

Obviously, knowing the built-in reference/key signals is vital to decoding dynamic scrambling systems. This knowledge is not absolutely essential in static systems since a reference can be recreated once the scrambling method is determined. Thus, in a static system, a potential pirate designer has a choice of generating his own reference signal or utilizing the reference signal available within the scrambling system; whereas, in a dynamic scrambling system, the pirate designer is forced to retrieve the reference signal to descramble, restricting his choice of approach.

One variation of the static system is the combining of static form of scrambling with a varying reference signal. This method would seem to provide added security due to its protected reference signal, but it is essentially still weak due to the fact that the actual scrambling mechanism is static. Dynamic scrambling with an unprotected reference signal is likewise not absolutely secure once the relationship between the reference signal and the scrambled picture is established.

This brings us to the combination of dynamic scrambling and protected reference/key signal. To steal service, a pirate designer will now have to determine how the reference/key signal is protected. This type of scrambling system depends greatly on two factors -- the "dynamic"-ness of the scrambling method itself and the level of reference/key signal protection. Dynamic scrambling, as we determined earlier, depends on the level of upredictability with respect to timing of the scrambling itself. Perhaps it would be more understandable to say that the larger the number of possible scrambling patterns or modes, the more unpredictable the system will be. No matter how dynamic the signal, the scrambling system itself loses its effectiveness against pirate designers if the reference signal is easily decoded. Therefore, analogue protection of reference signal with its limited number of variations is not as desirable as digital encoding which potentially has a significantly greater number of combinations.

In order to better understand the differences in static vs dynamic scrambling and the relationship to reference/key signal, let us look at a generic example.

A TV picture consists of synchronization pulses required to center the picture onto its CRT. Elimination of these pulses theoretically causes scrambling by preventing the TV set from stabiling the picture. Sine-wave sync suppression systems and gated sync suppression systems are all designed to achieve this effect. For the sake of illustration, let us use sync suppression for our exercise design of a secure scrambling system.

The first form we might use is constant video sync suppression with a fixed reference signal AM modulated on the aural carrier. Refering to our definition, this method is static scrambling in its most basic form. The second step we may take is to vary the reference signal timing so it does not corelate to the actual sync suppression timing. Still, the scrambling is a constant video sync suppression which is static. It is therefore, vulnerable to pirate design by bypassing the reference signal all together. Understanding that even a varying reference signal does not adequately protect a scramble picture because it can be bypassed, we can probably safely conclude that all forms of static scrambling offer approximately the same amount of protection from pirate designs.

Dynamic scrambling when applied to sync suppression offers a wide variety of scrambling combinations. Alteration of the depth of sync suppression, variation of the suppression frequency in a manner that makes it a harmonic of the sync frequency, random sync suppression by frame and random sync suppression by line, all have the potential to qualify as dynamic scrambling if they meet the criteria of unpredictability with respect to time. These methods are in many cases an improvement over static methods. However, even here an unprotected reference signal makes dynamic scrambling just as vulnerable to theft as static systems. For example, if the timing information for random sync suppression were directly AM modulated on the aural carrier, all the pirate would have to do is reapply that signal to the scrambled video. The timing reference for random sync suppression can be digitized. A digital data word corresponding to suppressed or not suppressed is an added layer of protection requiring data detection and decoding. Although considerably more secure than our starting point of basic static sync suppression, there is still a vulnerability factor in the "dynamic"-ness of the scrambling method and the decoding of the reference signal.

ENCRYPTION DEFINITION AND POTENTIAL

A constant "game" is currently being played in the cable industry with regard to theft of signal. One day a very powerful scrambling method is announced. The next day it is defeated. The cable operator wants a secure signal, but cannot relay on claims made because pirate designers are keeping up with the pace of vendor technology. In an environment like this, encryption of signal is an ideal form of signal protection. Encryption technology assumes, given time, all codes, will eventually be broken. This is philosophy recognizes the present scrambling games played between the pirate designers and the cable industry. The difference is that most encryption systems allow an astronomical number of variations for possible key codes to the encrypted signal. An anology can be made to a door lock and its key. The mechanism of a door lock is common knowledge; however, if you do not have the key that fits, you will not be able to open that door lock. Suppose you finally duplicated the key by carefully studying the door lock, but that lock can be easily changed to let a different key work. . . . The door lock is like a scrambling method which can be made public knowledge because there are a billion variations of possible keys and the internal components of the lock are continually changing.

The advantages of scrambling systems using encryption are numerous. Descrambling devices could be sold directly to the subscriber without fear that they would be used as a potential theft tool. The majority of today's pirate devices are add-on descrambling bases made up of actual manufacturers' products which have been either stolen or sold indirectly to the pirate houses. If the descrambler can be properly activated only by entering a unique key code which will vary from time and which is given only to paying subscribers, problems associated with the distribution of descramblers can potentially be solved.

The benefits of descrambler standardization as a result of encryption, coupled with TV standardization, may eventually allow the cable operator to eliminate a significant amount of hardware investment in the home. Of course, the operational aspect of this possibility will have to be carefully studied. With cable penetration over the 35% mark, making the descrambler a direct consumer product is not an unreasonable proposal. Encryption algorithm must be chosen so that it allows viewing only by a valid paying subscriber. The problem of paying subscribers disclosing encryption keys must be resolved both in operational system design and hardware design. A customized unique decrypting number for specific subscriber hardware may exist on a monthly billing basis, service basis, or even per program basis.

Now that we have seen a some idea of what encryption can possibly do for us, we can explore what is to be encrypted. Let us continue the evolution of the product design we started in our earlier appraisal of static scrambling. Our next step will be to encrypt the key signal associated with a dynamic scrambling method. If random sync suppression within a TV picture frame were the dynamic scrambling method chosen, the suppressed or not-suppressed timing is encrypted. Detection of digital key signal cannot be used to directly decipher the random occurrences of sync suppression unless the algorithm and decryption code are determined.

This form scrambling is particularily powerful since a decryption code may have a million possible combinations in addition to the dynamically changing patterns of sync suppression. In addition to the signal security of dynamic scrambling, encryption of key signal now provides opportunity to design systems which could safely allow standalone descramblers. The descramblers in this type of system could be made unique relative to each other. Changes in algorithm factors from systems to system will automatically resolve the cross system theft problem.

The last step in this design exercise is to encrypt video content. So long as the scrambled information does not alter basic video information, all non-encrypted video scrambling methods carry the possibility of being defeated. A variety of methods exists for encrypting video. These methods range from a simple a simple line randomization to time randomization of picture content, just short of digital video transmission.

STATE OF TECHNOLOGY

A true encrypted scrambling system is currently only available to the satellite industry due to the cost associated with encrypted scrambling. Satellite descramblers can afford to carry a price tag of several thousand dollars. Scramble/ descramble systems for the CATV industry certainly will have to maintain current price levels, eliminating direct application of satellite descramblers in the home. However, with the advent of charge coupled device (CCD) technology, digital television technology and advances in other semiconductor technology, the cost associated with complicated video processing can significantly drop, and true encrypted scrambling may some day be a viable technique for CATV signal protection.

Probably the most advanced form of scrambling systems available today within a competitive price range are the hybrid systems which use dynamic scrambling and encrypted digital key codes. Certainly not expected to last forever undefeated as long as these systems are designed within the realm of NTSC standards. Dynamic scrambling methods all maintain the basic rules set forth in the NTSC standards. For example, the deviation from NTSC standards of sync suppression and video inversion are relatively very minor. Significant deviation is not possible from the reasons associated to cost of product and ease of design. And for the very same reason, the vulnerability to pirate designers remains.

CONCLUSION

Scrambling system as they exist today are certainly not the ultimate solution to theft of service. The degree of difficulty in descrambling may vary from method to method; however, no method available to the industry can guarantee it will never be defeated. Some new TV sets are designed to be capable of tuning to semi-scrambled signals. Certain TV sets, for example, can automatically descramble static sync suppression. Less simple but a likely possibility for defeating all regular scrambling systems including dynamic scrambling, are other modification method using the TV set as a descrambling tool. Such modifications are possible since descramblers, to be price competitive, are designed with components commonly found insider the TV set itself.

Furthermore, many scrambling systems do not take into consideration other factors which impact theft of service. A system may develop a very powerful secure scrambling method which is ultimately defeated because it is housed in a descrambler device which lacks proper hardware security.

Theft of service can be greatly reduced by eliminating incentives that induce theft. Hardware construction of descrambler units should be secure to protect the internal components. Mechanical locks, access traps, custom chips, etc., should be used. Even the all outdoor delivery methods base signal security on lack of incentive for a potential thief to climb a pole or break a pad lock to steal service. While secure scrambling is certainly desireable, it is often overemphasized in the total theft of service scene. Strong scrambling methods are needed but equal attention must be placed on the operational aspect of the design so that the incentive to steal is eliminated.

Similarly, scrambling methods should minimize theft incentives. However, all video scrambling methods available to the cable industry today are only minor deviations from the NTSC standards, and thus, remain vulnerable to pirate designs. The issue is then the relative strength of the system against pirate designs. The question remains to be answered as to how much value does an ultimate scrambling system, designed within the realm of minor deviation from NTSC standard, have. Today, short of complete video encryption, the dynamic scrambling with encrypted key signal is most secure alternative one can offer.

THE IMPACT OF MULTICHANNEL SOUND ON CATV SYSTEMS

Joseph Van Loan

Viacom Cable

Pleasanton, California

A proposed Multichannel Sound (MCS) System for television is described, and differences between three proponent systems are discussed. Each of the proponent systems includes provision for transmission of:

1. L+R information to maintain compatibility with existing receivers
2. L-R information for stereo
3. A Second Audio Program (SAP) channel for a second language, quadraphonic sound or as a tutorial channel; and
4. A non-public channel for voice and/or data telemetry.

Although immediate wide-scale implementation is not expected, the planned system may have a substantial financial impact on the CATV industry. Headend equipment (including processors, modulators, demodulators and microwave transmitters and receivers) may have to be modified or replaced before passing an MCS signal. Present day descramblers will likely suffer deteriorated performance when used on MCS signals, and the MCS signals themselves may suffer degradation when used with descrambling equipment. The introduction of MCS carriers with their increased bandwidth and possible higher levels on a CATV system raise questions about the ability of present day receivers to trap out sound carriers on lower adjacent channels.

The MCS Subcommittee of the NCTA Engineering Committee is conducting tests in cooperation with the Electronics Industries Association and the National Association of Broadcasters to determine the impact MCS signals will have on the CATV industry and to recommend which of the three proponent systems, if any, should be selected.

INTRODUCTION

Multi-Channel Sound for Television and its Implications for CATV

In 1979 the Consumer Electronics Group of the Electronics Industries Association, the National Association of Broadcasters and the Joint Council of Inter-Society Coordination formed the Multichannel Sound Subcommittee to study proposals to introduce multichannel sound (stereo) into the U.S. television system. In August, 1982 the Committee published a 1,000 page report which contained results of tests conducted on three proponent systems.[1] EIA-J, Telesonics and Zenith have each proposed systems which are similar in many respects and different in others.

Each system will maintain compatibility with the existing monophonic system by transmitting an L+R channel with the same characteristics as the present monophonic channel. Each system will transmit stereophonic information using an L-R channel and a corresponding pilot in the spectrum above the L+R channel. A channel called a Second Audio Program (SAP) channel will permit the simultaneous transmission of additional aural information such as a second language or a narrative. The performance of this channel is limited both in fidelity and noise performance when compared with the stereo channel. Finally, a channel called the non-public channel is proposed to transmit data and voice information.

Figure 1 depicts the baseband spectra of each of the proposed systems. Each of the systems has undergone changes to improve performance as test results are analyzed. Most noteworthy among the differences are: the EIA-J systems uses an FM subcarrier for L-R transmission while the Telesonics and Zenith systems use double sideband suppressed carrier amplitude modulation (DSB-SC AM) for L-R transmission. Note, current practice for FM broadcast stereo in the U.S.A. uses DSB-SC AM for stereo transmissions. In the EIA-J and Zenith systems the L-R subcarrier is centered at twice the horizontal scanning frequency ($2f_H$), but in the Telesonics system, the L-R subcarrier is positioned at 2.5 the horizontal frequency. The SAP channel has been subjected to numerous changes. Figure 1 indicates the proposed configurations at the time of this writing.

In August, 1982 the NCTA Engineering Committee formed the MCS Subcommittee to study the impact that MCS will have on CATV system operation and

[1]"Multichannel Television Sound: The Basis for Selection of a Single Standard", by Electronics Industries Association's BTS Committee, Published by the National Association of Broadcasters, Vol. I, July 16, 1982; Vol. II, August 6, 1982.

Reprinted with permission from *Technical Papers, 32nd Annual Convention and Exposition, National Cable Television Association*, 1983, pp. 254-260.
Copyright © 1983, The National Cable Television Association.

to recommend one of the three systems if deemed appropriate. The Subcommittee set a precedent in early 1983 when professional help was retained to supplement the volunteer activity in studying the impact. The EIA/NAB Committee has made its laboratory facilities, located near Chicago, available to aid in the study. Tests are now being conducted in that facility with joint cooperation between the two groups.

TECHNICAL IMPLICATIONS FOR THE CATV INDUSTRY

A. Audio Signal/Noise Ratio[2]

In Volume I of the Multichannel Television Sound report, it was indicated[3] that for high quality sound reception, the signal to noise ratio should be at least 60dB, preferably 70dB or better, for the principal community of viewers. For comparison, EIA RS-250B specifies for monaural sound that the minimum unweighted audio SNR for end-to-end television relay facilities be 56dB, including buzz. We should deliver at least a 60dB SNR to the TV receivers connected at the extremities of our systems. To determine the expected SNR (thermal noise only) in cable systems, calculations were performed by several members of the NCTA Subcommittee.

For the case of 36 dB NCTA Video RF-SNR[4], with the sound carrier 15dB below the video peak envelope power, we get the following unweighted audio SNR's.

	Monophonic	Stereo L or R	SAP
Separate Mixing	63.8dB	49.2	39.6
Intercarrier, Video at Blanking Level	63.6	49.0	39.4
Intercarrier, Video at White Level	59.0	44.4	34.8

Obviously we are in trouble if there are subscribers whose ears demand stereo but whose eyes will tolerate 36dB. In addition, the Separate Audio Program SNR will be approximately 10dB less than stereo signal. If we assume that a more typical situation is for the cable system to deliver a 43dB Video SNR, then we find ourselves dealing with a stereo SNR or approximately 55dB and a SAP SNR of approximately 45dB. These are certainly more tolerable conditions and could be improved even more if companding is used. (A separate working group is developing a companding system).

From the calculations it appears that threshold margins are adequate, both for new sets with sound IF bandwidths sufficient to support the broadband signal and for old sets.

B. Visual/Sound Ratio

It is recognized that one way to improve the multichannel sound SNR is to increase the sound carrier level. Above threshold the audio SNR is increased one dB for each dB we raise the sound carrier level. In Volume I[5] of the EIA report, use of the highest feasible sound power is encouraged.

It must be recognized that increasing the sound carrier level in a cable system is undesirable and would create a multitude of problems, as explained in the following:

1. TV Receiver Adjacent Channel Rejection

All cable systems operate with TV channels adjacent to one another. In the early days of cable it was determined that most TV receivers lacked sufficient selectivity to provide for beat-free pictures, unless the lower adjacent sound carrier was reduced in level below the visual/sound ratio being transmitted by broadcast stations. Although the NCTA committee recognized that receiver selectivity has been considerably improved over the years, one must recognize that the proposed multichannel sound standards with their increased deviations (approximately 70kHz) creates sound carriers whose energy occupies a much wider bandwidth than before (approximately 320 kHz for 70 kHz deviation as compared to 80kHz for 25kHz deviation). In a paper written in 1972 by Will Hand of Sylvania entitled "Television Receiver Requirements for CATV Systems", tests were performed on a number of color receivers which represented collectively over 50% of receiver designs in the industry at that time. The data shown below is the weighted results of these measurements. The data was weighted to reflect the proportion of the market held by the receiver manufacturers.

[2]In order to simplify analysis and due to similarities between systems the Zenith was used for discussion and analysis except where clear differences are noted.

[3]Multichannel Television Sound: "The Basis for Selection of a Single Standard", Volume I, The National Association of Broadcasters, July 16, 1982, p. 51.

[4]Minimum allowed CNR for Cable Systems. Part 76 of FCC Regulations.

[5]Multichannel Television Sound Report, p. 164.

	Frequency (kHz)
Adjacent Sound null to 41dB point on low frequency side of IF trap	94
Adjacent Sound null to 41dB point on high frequency side of IF trap	115
The 41dB rejection bandwidth of this trap is 94 + 115 = 209kHz	

One can only conjecture what the rejection of this trap would be to a sound carrier adhering to the proposed multichannel sound standards; however, it goes without question that raising the sound level would only aggravate what already may be an unacceptable operating condition.

Set-top converters with adjacent channel traps would help, although only a few presently in the field have this feature. Separate traps connected to the output of converters and tuned to the lower adjacent sound would help; however, they suffer from the following problems:

a. AFC and fine tuning errors would diminish effectiveness.

b. They would introduce additional group delay errors.

c. The overall cost would be substantial.

d. Subscribers might confuse them with pay TV traps and remove them.

2. Headend Equipment (Signal Processors, Strip Amps and Modulators)

The majority of these devices operate at an output level between +50dBmV to +60dBmV (+60dBmV = 1 volt rms @ 75 ohms) for the video carrier with the sound carrier typically 15dB below this level. Most manufacturers also quote a specification which states that when the unit is operated at maximum output (this is not uncommon), all spurious outputs will be at least 60dB below the desired video carrier. One component of this distortion falls 1.5MHz above the video carrier of the lower adjacent channel. Any attempt to raise the sound carrier level causes a one dB rise in this undesired signal for each dB the sound level is increased. This is a particularly sensitive area of the video spectrum and any interfering carrier must be 55dB to 60dB below the desired carrier not to be perceptible. This problem is only aggravated when scrambling systems are employed which amplitude-modulate the sound carrier with descrambling timing information. The NCTA Subcommittee recognizes that this potential problem can be solved with highly selective bandpass filters on the output of these devices but not without an increase in the envelope delay distortion inherent in these filters. Equipment already in place may have to be replaced or undergo major modification to meet new performance standards.

3. AML Equipment

AML equipment is used to transmit channels of television from one area of a community to another. Increases in aural carrier levels would cause corresponding increases in the lower 4.5MHz products, as described in 2. above. For systems operating at full rated power, it would likely be necessary to reduce output power thereby reducing fade margin. The performance of this equipment is impaired by scrambling systems which cause a 6dB or more increase in the aural level during transmission of descrambling timing pulses. This practice creates a marginal condition to exist with today's practices. Increasing levels for Multichannel Television Sound would be unacceptable.

4. Distribution Equipment

Measurements were conducted to determine the perceptibility of sound carrier beats when the aural carrier levels were increased above 15dB below the visual carriers. From these measurements, perceptibility of sound-carrier beats occurs at a system level 3 to 3-1/2dB above that level which produces barely perceptible video carrier beats (CTB-composite triple beat distortion) when not phase locked. In the same test the system level could be elevated approximately 5dB when phase locked before background images (modulation cross-over) could be seen. These tests were performed with 54 channel Harmonically Related Carriers (HRC) loading. The conclusion to be drawn from these tests indicate if aural levels must be increased to accommodate Multichannel Sound, then the advantage gained from phase locking is lost.

Another problem may surface if aural levels are increased. Many cable operators find it possible to use certain channels in the aeronautical radio service bands by lowering aural carrier levels to +28.75dBmV maximum. In fact, several channels can sometimes only be used this way due to conflicts with both the visual and aural carriers. If aural levels are raised, this practice would be eliminated resulting in the loss of these channels for stereo service.

C. Increased Deviations

In order to achieve the highest possible stereo SNR, plus provide for auxiliary services, all proponents of the multichannel sound systems intend to increase the peak deviation to approximately 70kHz. By doing so, technical problems may be created in both cable headends and set-top converters, not to mention the selectivity problems of TV receivers. The following section discusses the nature of these problems in detail.

1. TV Receiver Selectivity (See Section B.1.)

2. Headend Equipment

a. Signal Processors

Two types are currently in use. One type uses a split sound system where the aural carrier is trapped and processed separately. The second type processes the video and sound combined and uses adjustable traps to reduce the sound carrier level. Both systems will suffer when deviations are increased from 25kHz to 70kHz.

For split sound units, the sound notch must not introduce amplitude or group delay errors in chroma information while attenuating the sound carrier and its sidebands at least 40dB. This has been achieved for 25kHz deviation with a Carson bandwidth of 80kHz. When the new Carson bandwidth for stereo (approximately 350kHz) is measured, attenuation of the upper sideband is on the order of 10dB. With incomplete trapping, a portion of the aural carrier passes through the visual processing circuitry. When this signal recombines with the processed aural signal, impairment of the received aural signal will result. This impairment could result in a substantial reduction in the amount of separation between the left and right channels. More testing needs to be conducted to confirm or dispel this concern.

One processor measured had 2% AM on the FM sound carrier with 25kHz deviation and 16% AM with 70kHz deviation. This would clearly cause a problem when using analog descrambling information on the aural carrier.

In addition to the notch problem, the 3dB bandwidth of the sound path was measured on one type of processor to be approximately 350kHz. No attempt is made to control delay characteristics at the band edge. One member of the Committee described the use of unequalized filters in home stereo receivers. This practice results in stereo separation of 40dB, a number considered adequate. General practice has been to control amplitude characteristics of filters, but no special care is used in controlling delay characteristics. This practice has not been reported to have caused any problems of which the Committee is aware.

In general, the NCTA Subcommittee believes IF circuitry in all existing processors would have to be redesigned for successful MCS operation.

b. Modulators

If external MCS signals are generated and pre-emphasis circuits are removed from modern modulators, it may be possible to use them for MCS transmission without difficulty. Older designs may have problems caused by transformers used for coupling and uncontrolled amplitude and phase characteristics in filters.

c. Demodulators

Volume I of the referenced MCS report discusses the impact MCS will have on TV receivers; since most demodulators use similar circuitry, they will experience many of the same problems. This includes insufficient sound IF bandwidth, problems with buzz caused by intercarrier sound processing and inadequate baseband response to pass MCS subcarrier components.

3. Set-Top Converters(Scrambling/Descrambling)

Of all the possible problems, this one has the potential for the greatest impact on the CATV industry. Devices with the greatest proliferation use either pulse or sinewave sync suppression. Problems with both systems can arise when FM to AM conversion occurs in headend processors, modulators and set-top converter bandpass filters. These products manifest themselves in different ways. In the case of pulse systems, the AM pulse on the sound carrier, and any spurious AM products, fire a trigger circuit, usually a monostable multivibrator. Any stray AM products or noise can cause descrambling pulse jitter due to slicer uncertainty.

An attempt was also made to simulate stereo susceptibility to stray tagging and descrambling pulses. Measurement data seems to indicate compatibility between MCS and pulse scrambling systems; however, reports from the field indicate buzzing is being experienced on some sets imported from Japan which include stereo demodulator circuitry using the Japanese standard. These reports must be investigated prior to reaching any final conclusions.

The sinewave sync suppression system may be most susceptible to FM to AM conversion products. These systems will undoubtedly suffer from the almost certain increase in AM products on the aural carrier. Additional testing is to be performed in this area by the Subcommittee.

Several manufacturers have baseband systems in the field. Since these units have demodulators, the performance and problems will be similar to those experienced with TV sets. Special MCS units will likely be similar to the tuner, IF amp and detector circuitry developed for TV receivers intended for MCS operation. This suggests units already in the field can continue to be used for subscribers with monophonic TV receivers, while units of a new, compatible design would be required for subscribers wishing to avail themselves of the opportunity to have multichannel sound. The one unknown, to be investigated, was discussed earlier in section B.1. of this report. The adjacent channel sound traps in these products could be expected to behave like those in TV sets. One manufacturer uses crystal stabilized local oscillators with AFC and SAW IF filters. When compared with older tube-type television sets with unstable IF circuitry, these units may offer satisfactory performance. This must be verified before concerns can be relaxed.

D. Phase Noise Considerations

In Volume I of the Multichannel Sound Report, mention was made that a return to split-sound TV receiver design could eliminate sound buzz due to incidental phase modulation of the sound carrier generated by mixing with the video carrier. This would certainly be the case; however, all concerned must also recognize there are additional problems created by this technique which will likely be made worse by equipment presently used in cable systems. Indicated below are several areas which need additional investigation before split-sound receiver design should be considered.

1. AML Equipment

On page 161 of Volume I of the MCS report, the EIA Committee discusses a problem with translators in Japan. "During the rebroadcast tests of MCS signals, a large increase in buzz interference within the subchannel was observed. The interference was especially noticeable in split-carrier reception. This was due to amplitude and phase nonlinearities of traveling-wave tube (TWT) power amplifiers used in old translators and also due to cascaded use of such translators." There is reason to believe a similar problem may exist with AML equipment. Tests are underway to determine what impact this equipment will have on future TV receiver design.

Although most operators use phase locked receivers for AML transmission, some operators occasionally use receivers with free running local oscillators. In such cases, the levels of IPM increase substantially. One cable operator measured substantial increases in noise levels using these receivers with synchronous demodulators. MCS TV receivers using split sound systems would experience impaired noise performance as a result of this phenemonon.

2. Headend Equipment

Most signal processors and modulators use oscillator designs which are crystal controlled; however, there are presently on the market several multichannel units (typically used for standby purposes) which use oscillators that are either frequency synthesized or under AFC control. These devices will almost certainly increase the amount of incidental phase modulation present on the sound carrier.

3. Set-Top Converters

Converters which use synthesized tuning systems introduce substantial levels of incidental phase modulation (ICPM) on the TV signal. Older varactor tuned products also introduce ICPM, but to a lesser degree. TV receivers using intercarrier sound detection are not affected by high levels of ICPM when both the visual and aural carriers are subjected to the introduction of ICPM: however, in TV receivers using split sound IF and detection systems ICPM products introduced by CATV converters can contribute to deteriorated noise performance. It is not uncommon for varactor converters to introduce frequency modulation on the TV signal in excess of 10kHz. This suggests a serious problem for the cable industry if split sound TV receivers are introduced in the marketplace.

SOME ADDITIONAL THOUGHTS ON MULTICHANNEL SOUND

One must ask if the course of action the television industry is about to embark on is appropriate now. We are introducing 40-year old technology at a time when a quantum leap in digital technology is about to take place. The TV sound system in the United States is already being strained to capacity with performance which can best be described as marginal. The course of action we are about to embark on appears to take a marginal situation and make it worse. This action is taking place at a time when the telephone industry is moving in the direction of digitized voice communications, at a time when the consumer stereo industry is on the verge of moving away from the traditional analog methods into digital technology using laser disks and tape, and when satellite delivery mechanisms such as Public Broadcasting Service (PBS), Home Box Office (HBO) and Direct Broadcast Satellite (DBS) services are moving to digital audio.

The CATV industry is facing some serious problems with the present methods of premium program security. Our premium TV security systems face serious threats not only by tampering but also by units manufacturered and sold on the black market. The industry is also facing pressure in the regulatory and legislative area to make the sound on certain channels unavailable to the owners of cable-ready sets and users of conventional converters. The task of keeping our premium signals secure eventually may come to depend on scrambled digital audio. If the TV industries could successfully develop standards for digital stereo audio which are compatible with the existing monophonic system, then a quantum leap will have been realized which would be a direct benefit to the American viewing public. It will be a challenge to develop cost-effective digital technology compatible with the TV sound system already in place, but engineers using the latest technology have the capacity to overcome difficult challenges. It is true there are difficult compatibility issues to be overcome, but this is a real opportunity to enhance the quality of television in the United States.

CONCLUSION

This paper has described several major elements in CATV systems where the carriage of multi-channel sound might create problems. At the time of this writing the NCTA Engineering Committee has outlined a test plan to investigate these topics. The tests, which began in February and will last about six months, will be performed by representatives of the cable industry working in conjunction with EIA Committee engineers. The results will be described in a comprehensive report including both subjective and objective evaluations of each area of concern. If testing results are negative, it may be appropriate for the industry to lead the drive to develop standards for a digital audio system.

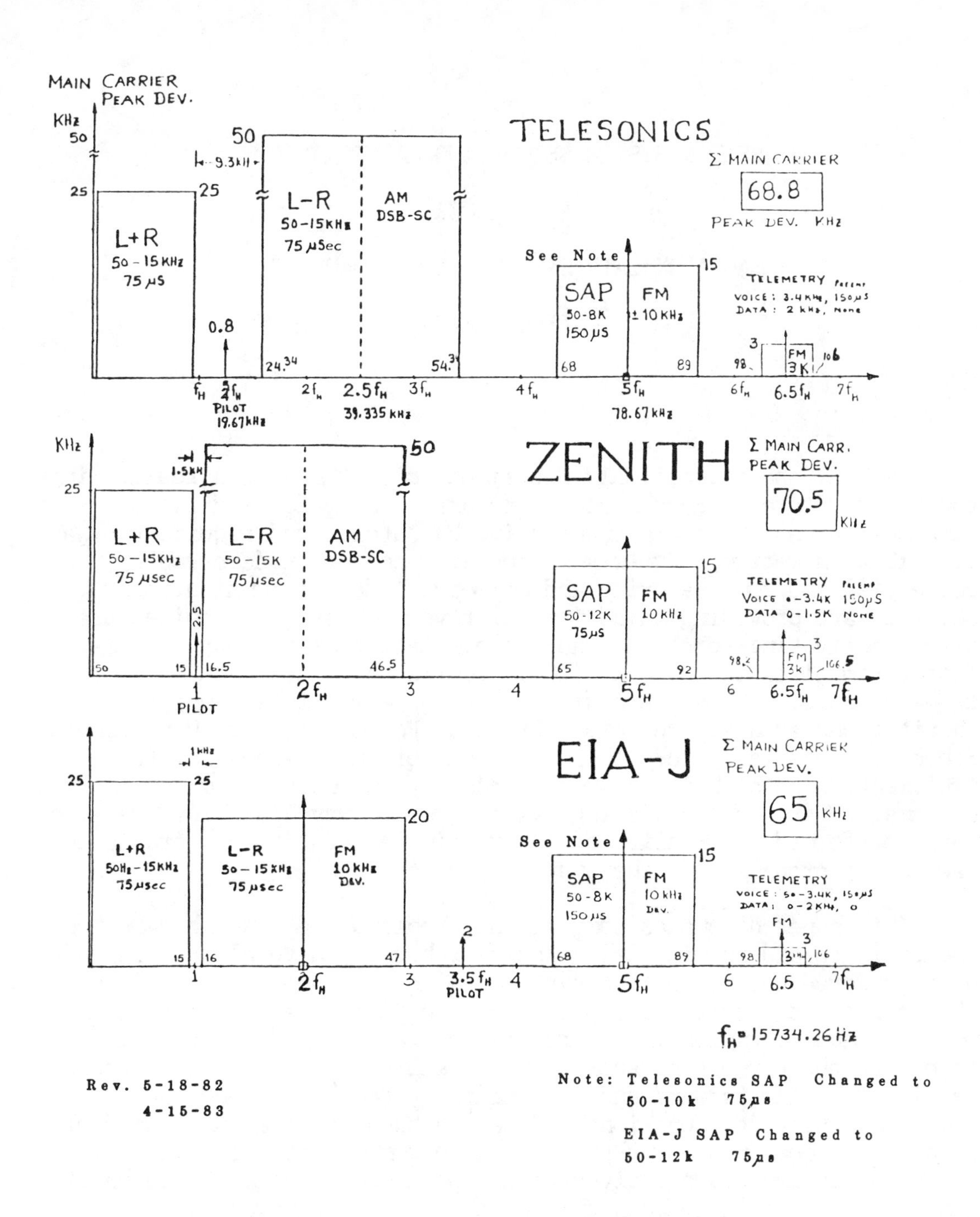

Figure 1

Baseband Spectra (Proposed Systems)

Source: Multichannel Television Sound, Volume I, p. 57

RETURN SYSTEM SET-UP AND MAINTENANCE

JAY STAIGER

MAGNAVOX CATV SYSTEMS, INC., MANLIUS

UNITED STATES

INTRODUCTION

Two-way operational CATV distribution systems are a reality in the United States. System operators are looking for additional revenue sources which are interactive by nature, and require two-way distribution medium. Two-way addressable converter descrambler, commercial and home security and commercial data communication services are providing an additional revenue base. Franchises are also a motivating force for the implementation of two-way systems. Institutional cable networks (ICN) require the activation of a two-way cable. ICN's are dedicated cables which interconnect institutions such as, Municipal Offices, Police and Fire Departments, Schools, Libraries, etc. These are some of the reasons for the implementation which is presently underway in most new, major market systems. Upgrades of existing systems to two-way operation are also being performed. A reality of two-way implementation has presented a new set of problems for the systems engineers.

This paper discusses some of the problems and describes methods and tools to aid in minimizing these problems. Several factors must be considered during the pre-system engineering and design. After the design and construction phases are complete the return system must be made operational. There are several methods which have been tried for the purpose of setting up a return system. Some of these require that the forward and the return system be balanced as completely separate functions. Other methods require that two field people, minimum, plus over-the-air radio communications be used. Set-up procedures have evolved to the point where only one technician can set-up and balance both the forward and return systems simultaneously without the need for two-way radio communications. This method, along with the test equipment required, is presented.

The maintenance of a return system has been found to require significant effort, more so than the forward system. This problem is due mainly to RF interference, or ingress. The tools for minimizing the maintenance effort in return system maintenance and trouble shooting are also presented.

Reprinted with permission from *13th International Television Symposium*, Montreux, Switzerland, 1983, CATV Sessions, vol. 3, pp. 366-390.

DESIGN OBJECTIVE

One approach is to design for unity gain in the return system. Return system unity gain is defined as 0 dB gain from input of any amplifier in the distribution system to output of the return system at the headend. (See Figure #1). This means that if at amplifier #7 of Figure #1 there is an input level of 20 dBmV, an output level measured at the return system output would be 20 dBmV. Again, Figure #1 illustrates a 20 dBmV input at amplifier #4, and, again, it results in a 20 dBmV level measured at the return system output.

Normally, the CATV distribution sytstem is optimized based on lowest initial cost. Lowest initial cost is realized when the number of active devices is minimized. This requires, in some CATV architectures, that a less than unity gain system must be planned. (See Figure #2 as an example.) You will note that the overall system as measured from amplifier #7 to the return system output has a -5dB gain; or stated differently, a net 5 dB loss.

The typical system architecture applied in the United States is the forward tree structure, and the return system can be thought of as a funnel architecture. The system is segmented into two systems:

1. The trunk system
2. The feeder system

Some system designs set-up the feeder system as having unity gain, and the trunk system as having unity gain, but the combination of the feeder and trunk system might have a net negative gain, or loss. This is as illustrated in Figure #2.

When planning the CATV system, the available gains from the feeder amplifiers and the trunk amplifiers in the return path, must be considered. In some cases, the available gains are not sufficient to overcome CATV losses. The reason for this net loss is the optimization of the forward system design to use the minimum number of active devices. Amplifiers typically require periodic maintenance and require a source of energy which are both ongoing operational expenses. The minimization of these devices serves the purpose of reducing the initial expense of building the system, and also the ongoing expense of maintaining the system. Therefore, the return system is secondary during the design process. Where possible, without adding unreasonable expense, unity gain should be achieved.

RETURN SYSTEM CONSIDERATIONS

There are a few considerations that must be made during the planning of the CATV return system. These are listed below and will

Figure #1

Unity Gain

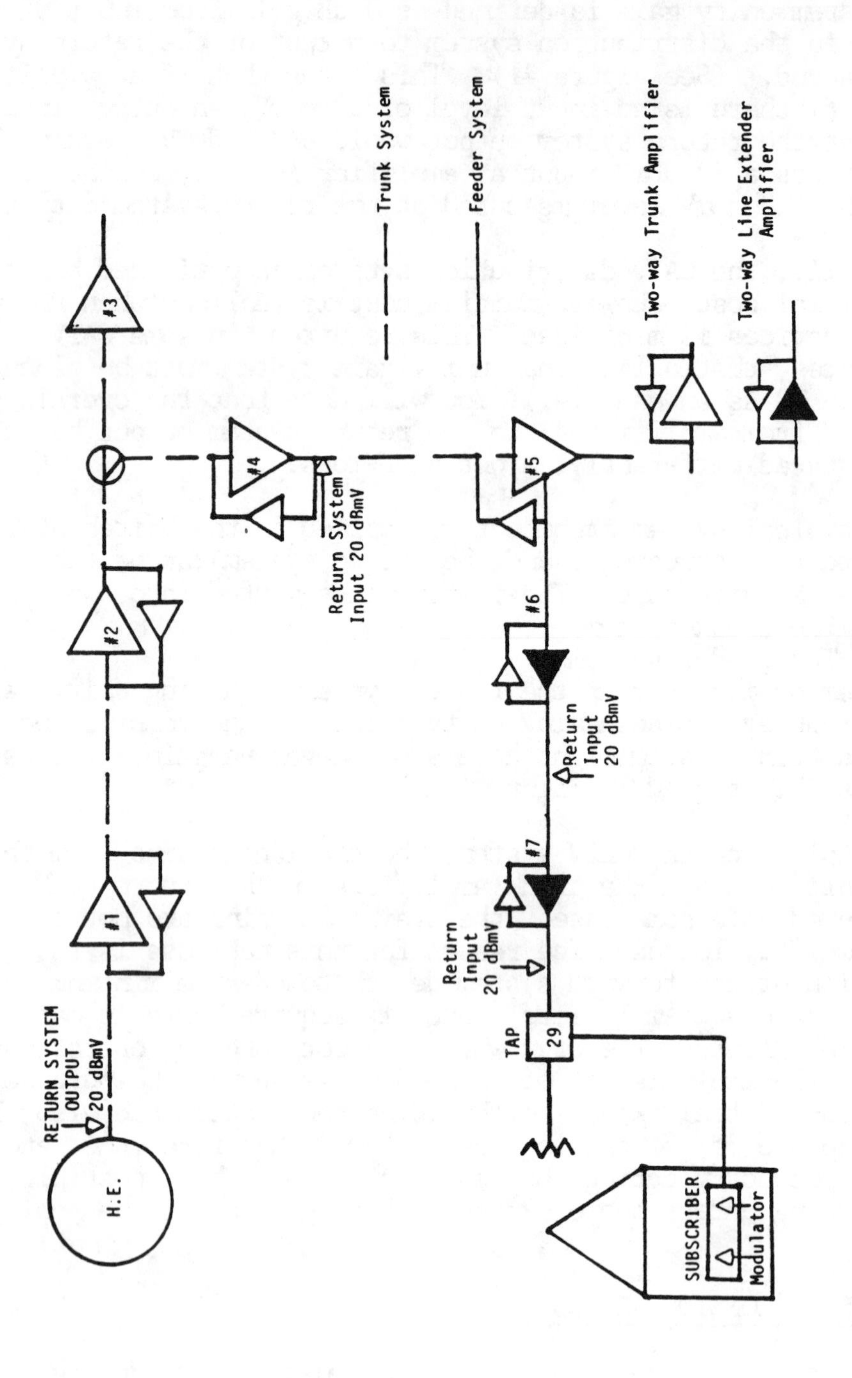

Figure #2

Not Unity Gain

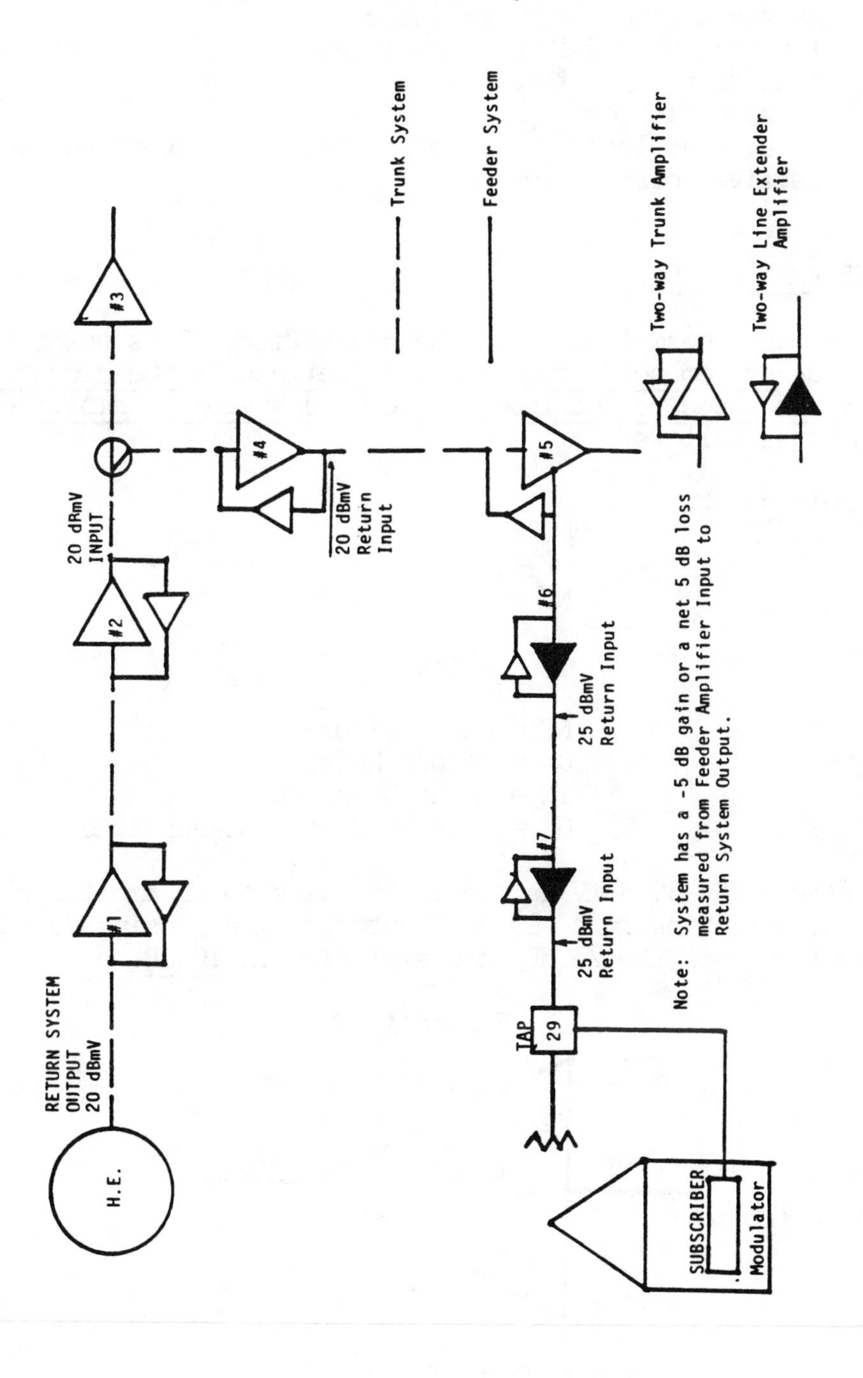

be addressed individually in the succeeding text:

1. System thermal noise (system noise figure)
2. System carrier-to-noise ratio
3. Terminal or modulator output level
4. Amplifier input level
    a. Cable loss
    b. Flat loss (taps, splitters, directional couplers)
5. Receiver noise tolerance

SYSTEM THERMAL NOISE

The system thermal noise can be quantified. This quantity is known as the system noise figure and is defined as the amount of thermal noise added to the input noise by a system or amplifier. The equation is as follows:

Formula #1

$$NF = ON - IN - G$$

Where:

NF = Noise Figure
ON = Output Noise
IN = Input Noise
G = Gain of System or Amplifier

In Figure #3 -59 dBmV is the input noise to an amplifier having 20 dB gain, resulting in a -20 dBmV output noise. Using Formula #1, the noise figure of the 20 dB gain amplifier is 10 dB.

Figure #3

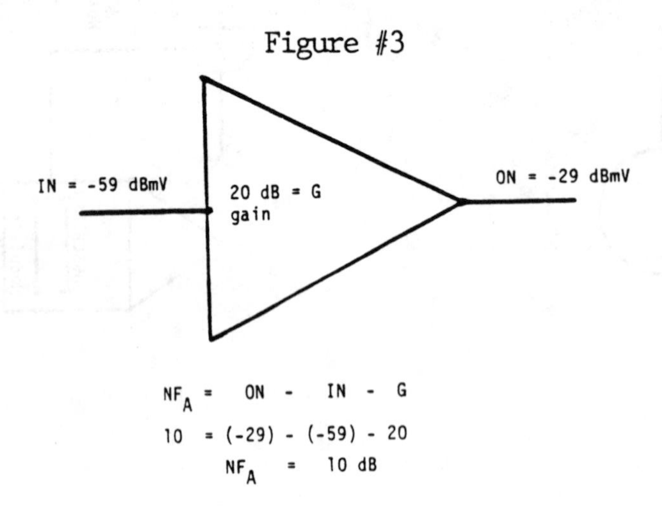

$$NF_A = ON - IN - G$$

$$10 = (-29) - (-59) - 20$$

$$NF_A = 10 \text{ dB}$$

The amplifier noise figure is a specification that defines the thermal noise characteristics of the CATV amplifier, and is specified by the manufacturers. This number can be used to define the entire system noise figure. Referring to Figure #4, we will now apply the formula to determine the noise of the entire system. Figure #4 shows a hypothetical system consisting of 16 trunk amplifiers which are numbered from 1 to 16. The thermal noise output from the return system is measured at -36.96 dBmV. The noise input at amplifier #13 is equal to -59 dBmV. Additionally, the return system is capable of being set-up with a unity gain (0 dB gain). Applying the input and output noise measurements and the system gain to the formula results in a 22.04 dB system noise figure.

The system noise figure can be calculated, if the individual amplifier noise figure is known, by the following formula:

Formula #2

$$NF_S = NF_A + 10 \text{ Log } (N)$$

Where:

NF_S = The system noise figure
NF_A = The noise figure of one (1) amplifier which is identical to every other amplifier in the system.
N = Total number of identical amplifiers.

Use Formula #2 to calculate the noise figure of the hypothetical system in Figure #4 as follows: Referring back to Figure #3, the hypothetical amplifier noise figure was calculated to be 10 dB. Applying this noise figure in the the formula yields:

Calculation #3

$$NF_S = NF_A + 10 \text{ Log } (N)$$

$$= 10 + 10 \text{ Log } (16)$$

$$= 10 + 12.04$$

$$NF_S = 22.04 \text{ dB}$$

You will note that calculation #'s 2 and 3 above result in the same system noise figure of 22.04 dB.

Figure #4

SYSTEM NOISE FIGURE

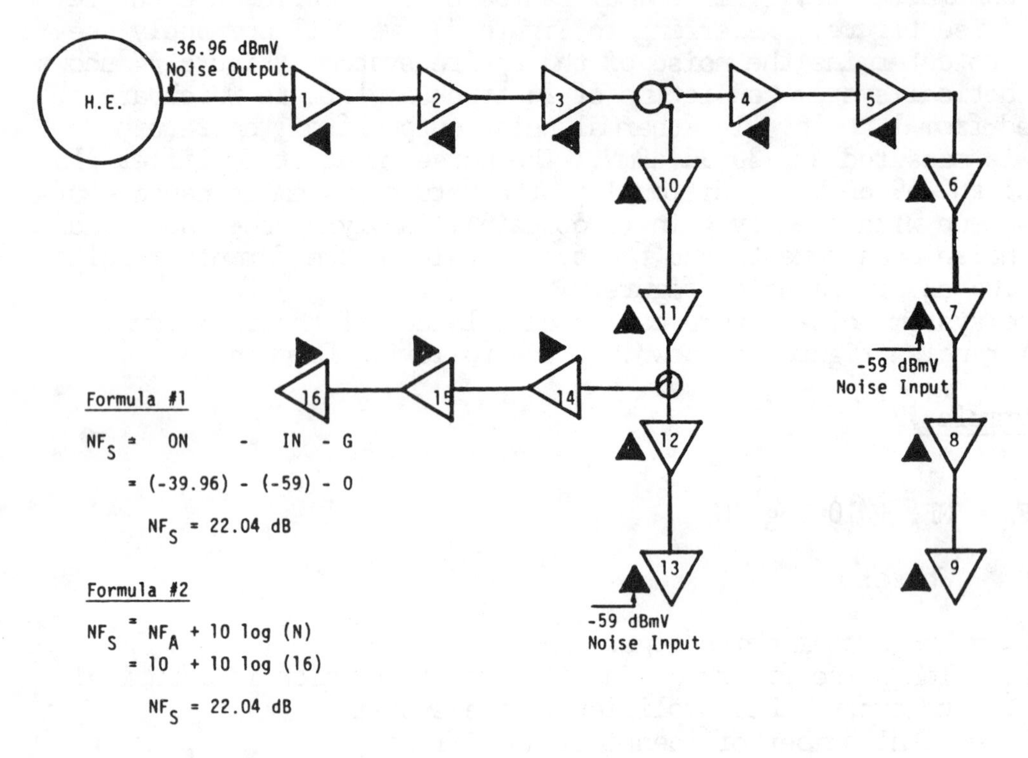

NOISE FUNNELLING

CATV distribution systems having the typical tree architectures with implemented return systems, must deal with the factor of noise funnelling. This is the summation of the noise which is generated by each active amplifier in the return system, and accumulates at the common trunk output port at the headend. This is unlike the forward system noise which accumulates as a result of the number of amplifiers in cascade. In the forward system there was only one path for noise to accumulate. This is from the first amplifier through each successive amplifier to the termination. (See Figure #5.) The noise of amplifier #1 adds to #'s 2 through 9. The noise from amplifier #'s 10 to 13 is added only to the noise generated by amplifier #'s 1 to 3. The noise generated in amplifier #'s 4 to 9 do not, in the forward system, contribute to the noise accumulated in amplifier #'s 10 to 13. Again, it is stated that noise in the forward system having a tree architecture, accumulates only as a result of sequential amplifiers in cascade. Therefore, the noise figure of the forward system is calculated as follows:

Formula #3

$NF_S = NF_A + 10 \text{ Log } (CSD)$

Where:

CSD = The number of amplifiers in cascade

Figure #5

Forward Noise Figure

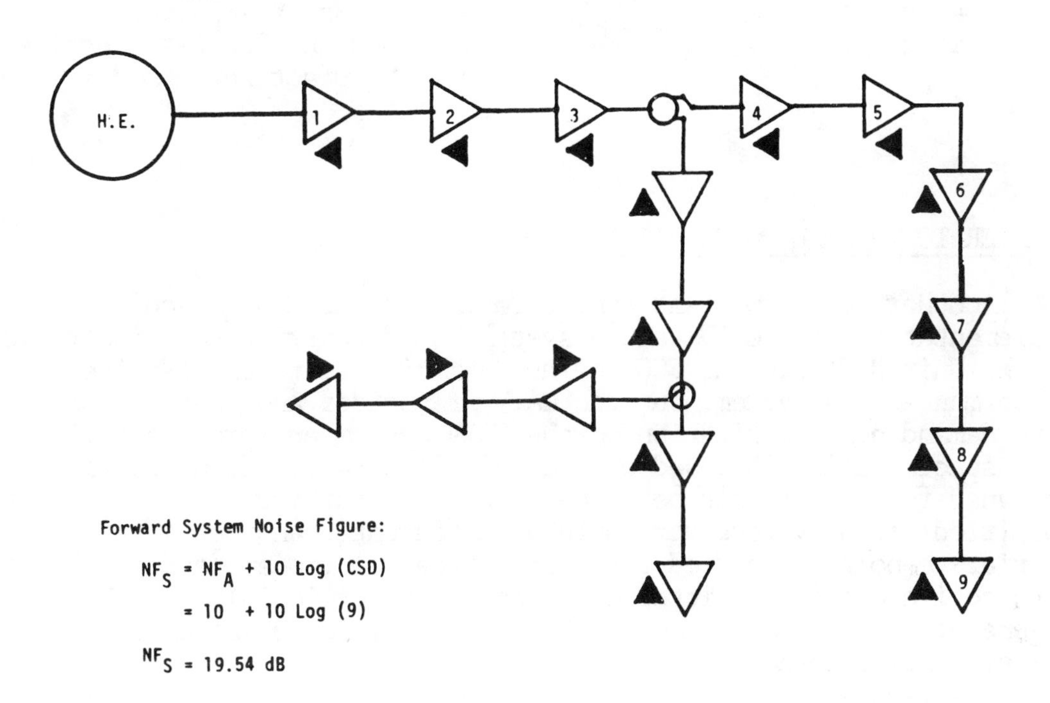

Forward System Noise Figure:

$NF_S = NF_A + 10 \text{ Log } (CSD)$

$= 10 + 10 \text{ Log } (9)$

$NF_S = 19.54 \text{ dB}$

Some earlier CATV designs did not consider the noise funnelling problems. When the return path was implemented, it was found that the total noise accumulated back at the headend was excessively high, and resulted in poor TV picture quality. There are some solutions to this problem and tools to apply the solutions. These will be discussed later in this paper.

Typical Carrier Ratio
Operational Specifications

Acceptable Carrier/Noise Ratio	Type Receiver
43 dB	4 MHz TV Receiver
15 dB	300 KHz FSK Data Receiver Manufacturer - Model Type

CARRIER TO THERMAL NOISE RATIO

Another very important consideration in the design and implementation of a CATV return system is the carrier-to-noise ratio. In the United States the FCC defines a minimum carrier-to-noise ratio performance of a system, but the CATV franchiser and operator set a more demanding specification on the CATV system equipment supplier. This specification is based on the noise tolerance of the receiver designed for the signals being carried. The carrier-to-noise specified for a TV receiver would be different than the carrier-to-noise specified for a data receiver (modem, addressable interactive converter, interactive security terminals). If the noise figure of the system is known, it is very simple to calculate the system carrier-to-noise:

Formula #4

$$C/N = TN - IN + NF$$

Where:

C/N = The carrier-to-noise ratio.
TN = The thermal noise level generated in a 75 ohm resistance at 68°F.
IN = The the input level to the system at an amplifier.
NF = The noise figure of the amplifier or system.

Assuming that the noise figure of an individual amplifier in Figure #5 is 10 dB, the system noise figure will be calculated as follows:

Calculation #4

$NF_S = NF_A + 10 \text{ Log (CSD)}$

$= 10 + 10 \text{ Log } (9)$

$= 10 + 9.54$

$NF_S = 19.54 \text{ dB}$

Unlike the forward system, the return system noise is accumulated at the common output port located at the headend from all amplifiers in the system (Refer to Figure #6). Noise from branch #1 and noise from branch #2 accumulate at combining point #1. The accumulated noise at combining point #1 adds to the noise of amplifier #11 plus amplifier #10, which make up branch #3. At combining point #2, total accumulated noise from branch #'s 1, 2, & 3 sum with noise from the amplifiers in branch #4. This total accumulated noise adds to branch #5, and results in all amplifiers in the system contributing to the noise measured at the return output of the system. The calculation for noise figure in the return system is shown by formula #2.

Figure #6

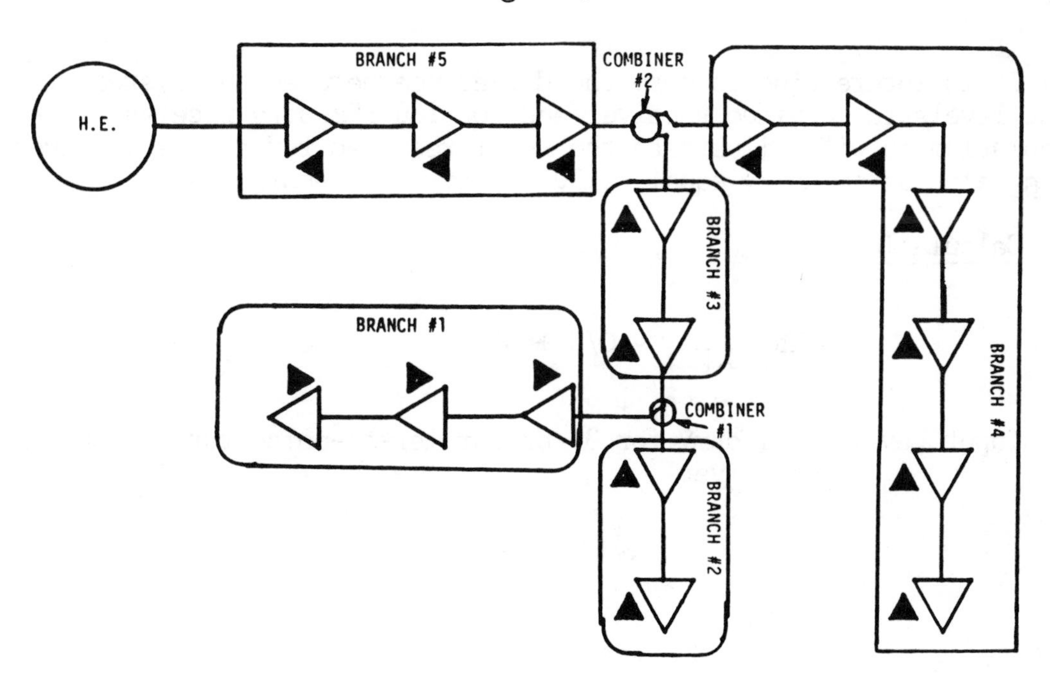

It is important to make the statement that the above calculation is a simplification and is more complicated in a system containing components with various gains and losses. The scope of this paper is to make the reader aware of the factors involved in return systems. The actual specific methodology in determining the proper operating levels and actual noise figures of the system are beyond the scope of this paper. Formula #4 is only accurate where the system has a unity gain.

The necessary input level to the system can be determined if the minimum carrier-to-noise of a receiver and noise figure of the system is known. The operational carrier-to-noise ratio specification of the receivers are available from their manufacturer for CN in calculation #5. Using this specification, with some additional headroom added, the proper inputs to the system can be determined.

Rearranging Formula #4 the input level to the system having video receivers requiring a 43 dB carrier-to-noise ratio, is:

Calculation #5

$$IN = TN_{4\ MHz} - CN + NF$$

$$18\ dBmV = 59 - (-43) + 34$$

Input Level = 18 dBmV for 43 dB carrier-to-noise (video receiver)

It is interesting to note the difference between the system input levels for a video receiver and for 100 KHz data receiver. Calculation #6 will illustrate the level required for a data receiver having a noise immunity equal to 30 dB carrier-to-noise ratio.

Calculation #6

$$IN = TN_{100\ KHz} - C/N + NF$$

$$-11 = -75\ (-30) + 34$$

Input Level = -11 dBmV for 30 dB carrier-to-noise ratio (100 KHz receiver)

It is significant to note that system input level requirements are lower for data applications. There is one other consideration which should be mentioned at this time. The -11 dBmV input level was based on thermal noise immunity and did not consider dynamic range on receiver input levels. CATV systems must deal with other types of noise, such as other ambient radio frequency signals, or industrial noise. This problem can be summarized as general electro-magnetic interference (EMI) or more commonly referred to as ingress. The CATV system operator would want to operate the inputs into his return system at the highest possible level in order in minimize the degree of interference caused by ingress.

There are two reasons that can be noted in Calculation #6 for the reduction in the input levels necessary for the data signal. One reason is the better noise tolerance of the data signals. In the example the data signals had 13 dB better noise tolerance. The other reason is a narrower power bandwidth. The data receiver bandwidth in this example is 100 KHz (-75 dBmV thermal noise level compared to -59 for dBmV 4 MHz bandwidth). The RMS noise voltage for the input of a receiver having a characteristic impedence of 75 ohms is dependent on the bandwidth. The narrower the bandwidth, the lower the RMS power. The RMS voltage can be calculated by Formula #5.

Formula #5

$$e_N = \sqrt{4 \times R \times B \times K}$$

Where:

EN = RMS Thermal Noise Voltage
R = Resistance in ohms
B = Bandwidth in MHz
K = Constant @ 40×10^{-16} @ 68°F

Refer to Calculation #'s 7 & 8 which show the difference between a 4 MHz video bandwidth and 100 KHz data bandwidth.

Figure #7

Thermal Noise

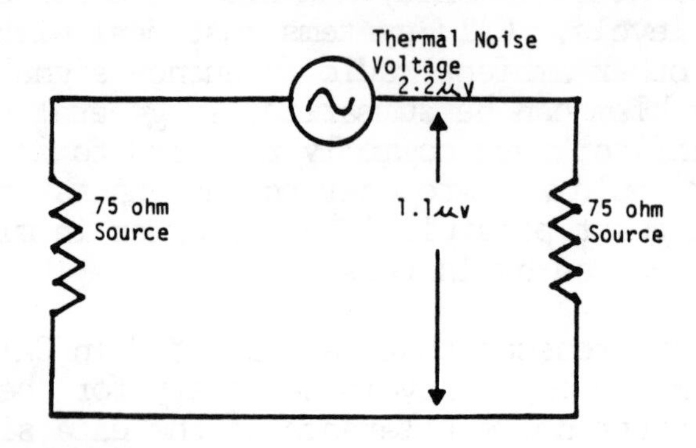

Calculation #7 (Refer to Figure #7)

$$E_N = \sqrt{4 \times R \times B \times K}$$

$$= \sqrt{4 \times 75 \times 4 \text{ MHz} \times 40 \text{ X}10^{-16}}$$

$$E_N = 2.2 \ \mu V$$

Expressed in dBmV:

$$20 \log \left(\frac{1.1 \ \mu V}{1 \ MV}\right)$$

$$= - 59.17 \text{ dBmV}$$

Calculation #8

$$E_N = \sqrt{4 \times R \times B \times K}$$

$$= \sqrt{4 \times 75 \times .1 \text{ MHz} \times K}$$

$$= .346 \ \mu V$$

Expressed in dBmV:

$$20 \log \left(\frac{.346 \ \mu V}{1 \ MV}\right)$$

$$= - 75.23 \text{ dBmV}$$

DATA TERMINAL AND VIDEO MODULATOR OUTPUT LEVELS

As demonstrated in the previous text, the system input level from a data modem can be relatively low to meet carrier to thermal noise requirements for data receivers when compared to the carrier-to-thermal-noise of a video receiver. The system designer, however, must be able to achieve the required system input levels in order to meet the necessary carrier to noise ratios. In some instances it may be difficult to achieve the video signal input levels. Refer to Figure #8 which will illustrate the factors involved in determining the achievable system input level. For the purpose of discussion, the system input level is defined as the level input to any return amplifier port in the system.

Figure #8

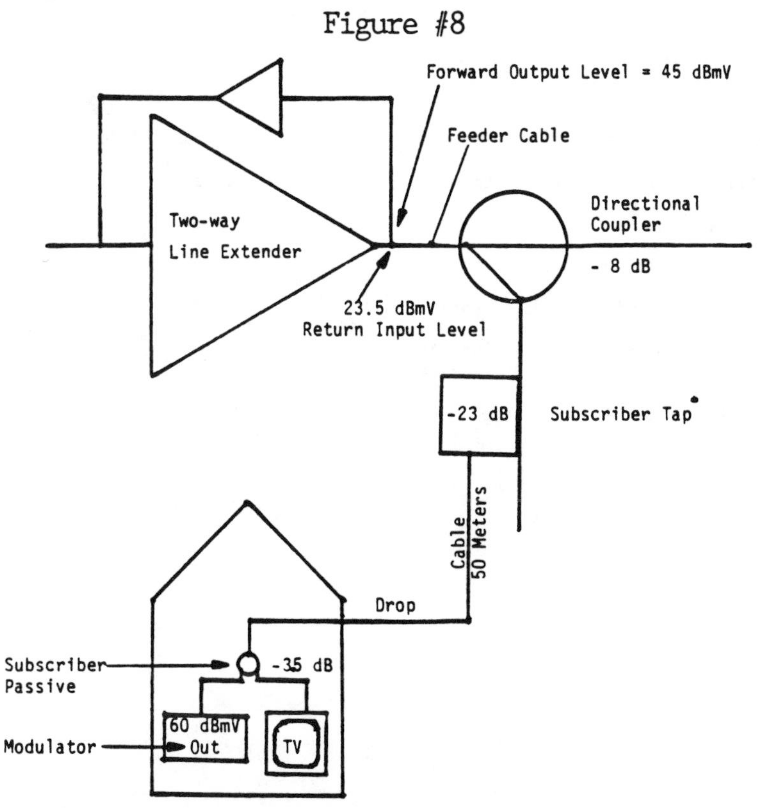

Figure #8 shows a diagram of a typical case which could exist in a feeder design. It shows a line extender amplifier having an output level of 45 dBmV in the forward direction. Directly at the output of the amplifier is a directional coupler which routes the signal in two different directions. Directly at the output of the tap leg of the directional coupler, a subscriber tap is installed. The subscriber tap connects to the home terminals via a 50 meter drop cable. The drop is split two ways: One of the outputs connects to a TV set; the other output connects to a return modulator. In consideration of all the components just described, the maximum return system input level shown by the example in Figure #9 is 23.5 dBmV.

In a unity gain system this input level would be the maximum input level achievable for a return system. This level could be used to apply to Formula #4 for the carrier-to-noise calculation for a video receiver. The same procedure for calculating carrier-to-noise of a data signal can be applied by substituting the data modulator output level in line #1 of Figure #9.

Figure #9

FACTORS	
1. Operational Modulator Output Level	+ 60 dBmV (Video)
2. Operational Subscriber Passive Loss	- 3.5 dB
3. Operational Drop Cable Loss	- 2 dB
4. Operational Tap Loss	- 23 dB
5. Operational Directional Coupler Loss	- 8 dB
6. Operational Feeder Cable Loss (Including In-Line Passives)	0 dB
7. Maximum Return System Input Level	+ 23.5 dBmV

Table #1 shows the maximum input level to the return system for various architectures which might be typical of a CATV distribution system. Part A of Table #1 shows return system input levels in the last column for the architecture shown in Figure #8. The table varies the directional coupler loss and the tap loss, showing the resulting return system input level change. In Table #1, Part B, the output level of the amplifier increased by 3 dB to 48 dBmV. Then the same combinations of directional couplers are applied to determine the system input level. Take note that in Table #1, Part B the lowest maximum input level to the system is 3 dB less than in Table #1, Part A. The worst case input levels of Table #1, Parts A & B, are enclosed in a box. Again, take note that the modulator output levels shown by the left most column are indicated as video or data modulators. Another indication given by this table is the fact that the higher the forward levels, the lower the input levels to the return system. If the carrier-to-noise requirement of the headend receivers cannot be achieved with these input levels, it is obvious that certain modifications in either the system architecture or the modulator output level will be required. The system architecture could be modified by a limitation of the forward system design. The net effect of the design modification would be a short spacing of system amplifiers resulting in increasing the initial cost and operating expense. The system designer in these cases would limit the loss of all the factors shown in Figure #8 so that the necessary input level to the system can be achieved.

The system engineer charged with responsibility of setting up and maintaining the return system, must specify separate output levels for both data and video modulators, when contracting for system design services. A separate video and data receiver carrier-to-noise expected from the system also must be specified. If this information is not provided, the designer will not be able to plan for the most critical factors.

NOISE FUNNELLING SOLUTIONS

As stated earlier under the heading of Noise Funnelling, some CATV systems in operation today would have an accumulated thermal noise which would interfere with good picture quality. Newer CATV system designs consider this problem by limiting the total number of amplifiers funnelling back on a single trunk.

TABLE 1

PART A

Note: Forward tap selected to meet:

a) Minimum tap output level @ 440 MHz = 12 dBmV
b) Minimum tap output level @ 50 MHz = 6 dBmV
c) Forward amplifier output level @ 440 MHz = 45 dBmV
d) Forward amplifier output level @ 50 MHz = 37 dBmV

Refer to Figure #____

Modulator Output (dBmV)	Tap (dB)	Directional Coupler (dB)	Return Input
+ 60 (Video)	-23	- 8	+ 23.5
+ 35 (Data)	-23	- 8	+ 5.5
+ 60 (Video)	-29	- 0	+ 25.5
+ 42 (Data)	-29	- 0	+ 7.5
+ 60 (Video)	-17	-12	+ 25.5
+ 42 (Data)	-17	-12	7.5
+ 60 (Video)	-14	-16	+ 24.5
+ 42 (Data)	-14	-16	+ 24.5

Part B

Note: Forward tap loss selected to meet:

a) Minimum tap output level @ 440 MHz 12 dBmV
b) Minimum tap output level @ 50 MHz 6 dBmV
c) Forward amplifier output level @ 440 MHz 48 dBmV
d) Forward amplifier output level @ 50 MHz 40 dBmV
e) Sum of subscriber passive, drop cable, feeder span loss = 5.5 dB

Modulator Output (dBmV)	Tap (dB)	Directional Coupler (dB)	Return Input
+ 60	- 26	- 8	+ 20.5
+ 42	- 26	- 8	+ 2.5
+ 60	- 32	0	+ 22.5
+ 42	- 32	0	+ 4.5
+ 60	- 20	- 12	+ 22.5
+ 42	- 20	- 12	+ 4.5
+ 60	- 17	- 16	+ 21.5
+ 42	- 17	- 16	+ 3.5

One way to deal with noise funnelling is to apply return trunk and return bridger switching. This is a product which has been applied for several years. Magnavox CATV Systems incorporates an on/off switch, as well as a 6 dB switchable pad in both the trunk and feeder return signal paths. These switches and/or pads can be remotely controlled from the headend via a computer system. The trunk and bridger switches can also be controlled manually at each of the amplifier locations. Bridger switching is effective where there is a requirement for video type return. With the bridger switch, all the noise contributing amplifiers in the return feeder system can be turned off, except for those carrying active TV channels. A sub-split system having 5-30 MHz return will necessitate having only four (4) bridger switches on at one time. A hypothetical application has a total of 500 amplifiers in the feeder systems, and leaving only four legs of that feeder system open, you can reduce the number of noise contributing amplifiers to, maybe, four noise contributing amplifiers. This would result in a reduction in the feeder system noise figure of approximately 21 dB. The 500 feeder amplifiers and the 4 feeder amplifiers were selected just to illustrate the drastic improvement in feeder noise figure and in some systems, may be representative of a system design. (Architectures vary drastically from system to system.)

Figure #10

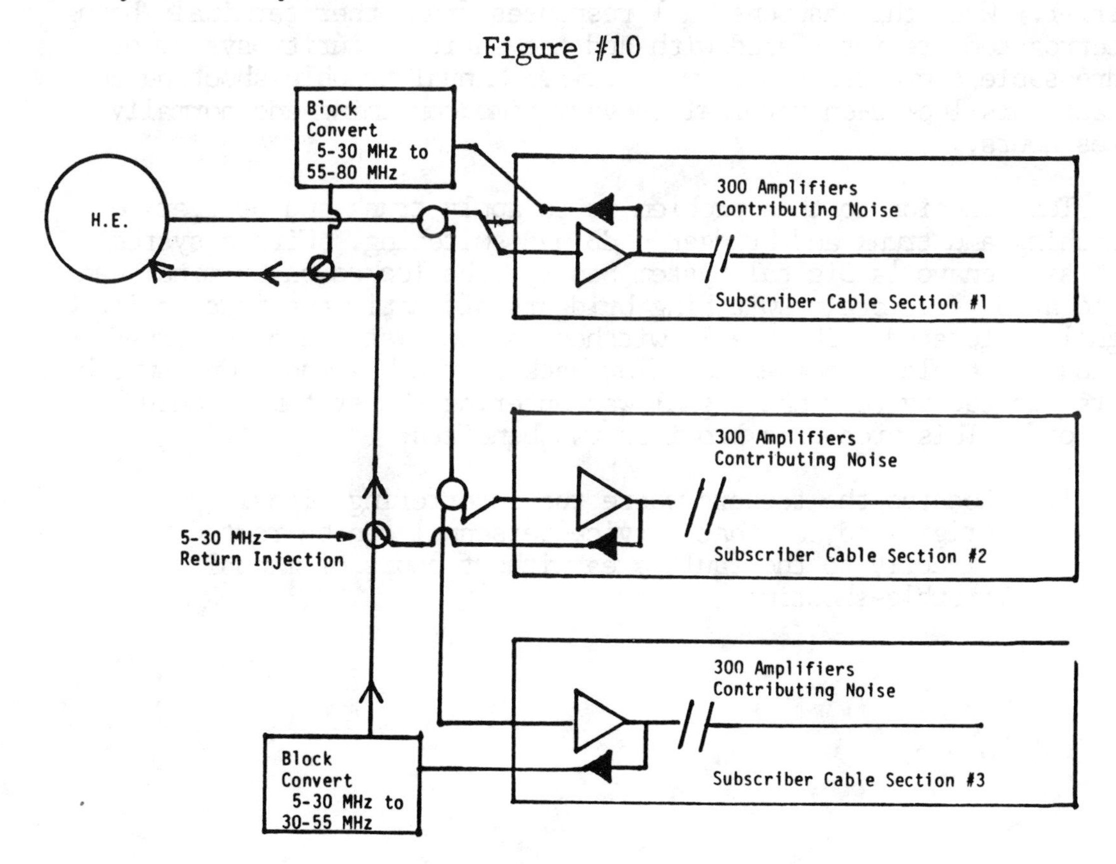

Another solution to the noise funnelling problem is a block segmented trunk. (Refer to Figure #10 for a block diagram.) This is simply a transportation trunk that originates at noise combining points along the return trunk. At the combining points in a two-way distribution system the return path is extracted and block converted up to an unused frequency band in a separate transportation trunk cable. Referring to the block diagram, it shows that funnelling of a system containing 900 return amplifiers can be limited to a maximum noise funnelling of 300 amplifiers. This would result in a tolerable video carrier-to-noise ratio.

LOCKED-ON HOME TERMINALS

Another problem which needs to be dealt with during return system operation is the potential failure of data terminals that are normally polled for a response. These are the terminals that are applied for two-way addressable converters and home security systems. These terminals have an individual address and are interrogated by a central computer in the headend. When these terminals are addressed, they respond by turning on a transmitter modulated with a digital response. One failure mode of these terminals is a locked-on carrier. When this happens, all responses from other terminals being interrogated are interfered with and the entire security system or addressable converter system will fail. Manual trouble-shooting to locate this locked-on terminal is very time consuming and normally takes hours.

The solution to this problem is to apply trunk and bridger switching and trunk and bridger 6 dB pad switching. With a system such as Magnavox's Digital System Sentry, the locked-on terminal can be located by remotely switching bridgers off until the interfering signal is located. Then, all switches in the system can be turned on so that the pole responses can flow back to the headend. One switch where the faulty terminal signal was entering the system, should be left off. This procedure provides two benefits:

1. Locates the feeders where the interfering signal originated, so that service personnel can be routed directly to the fault area without hours of manual trouble-shooting.

2. Isolates the interfering signal from the rest of the system so that the other terminals can respond to the headend.

All of this can take place in a matter of minutes as opposed to a matter of hours when manual methods are used.

ELECTRO-MAGNETIC INTERFERENCE (EMI)

Electro-magnetic interference is caused by ambient RF signals for broadband noise created by industrial machinery leaking into the return path of a CATV system. If the CATV system is constructed and maintained in the proper manner, leakage into the cable system (ingress) is of minimal consequence. From time to time, over changing environmental conditions, the cable system tends to become less EMI immune. Tools are necessary to detect, locate and repair EMI leaks in an expeditious manner.

The subscriber output port from the cable TV system is another possible entry point for EMI. A subscriber may disconnect a cable from his TV set, thus leaving an unterminated cable and a source of EMI entry. Other sources are unterminated subscriber tap ports. It is mandatory in a two-way operational system to terminate all unused tap ports. The problems with unterminated subscriber drop cables can be minimized by installing highpass filters where two-way services are not required to the home. Magnavox CATV Systems provides a filter for this purpose. The cable TV drop must be of high quality, usually quad shield is used for maximum immunity. Feeder systems that applied low quality drop cables will have to invest in a higher quality cable.

Figure #11

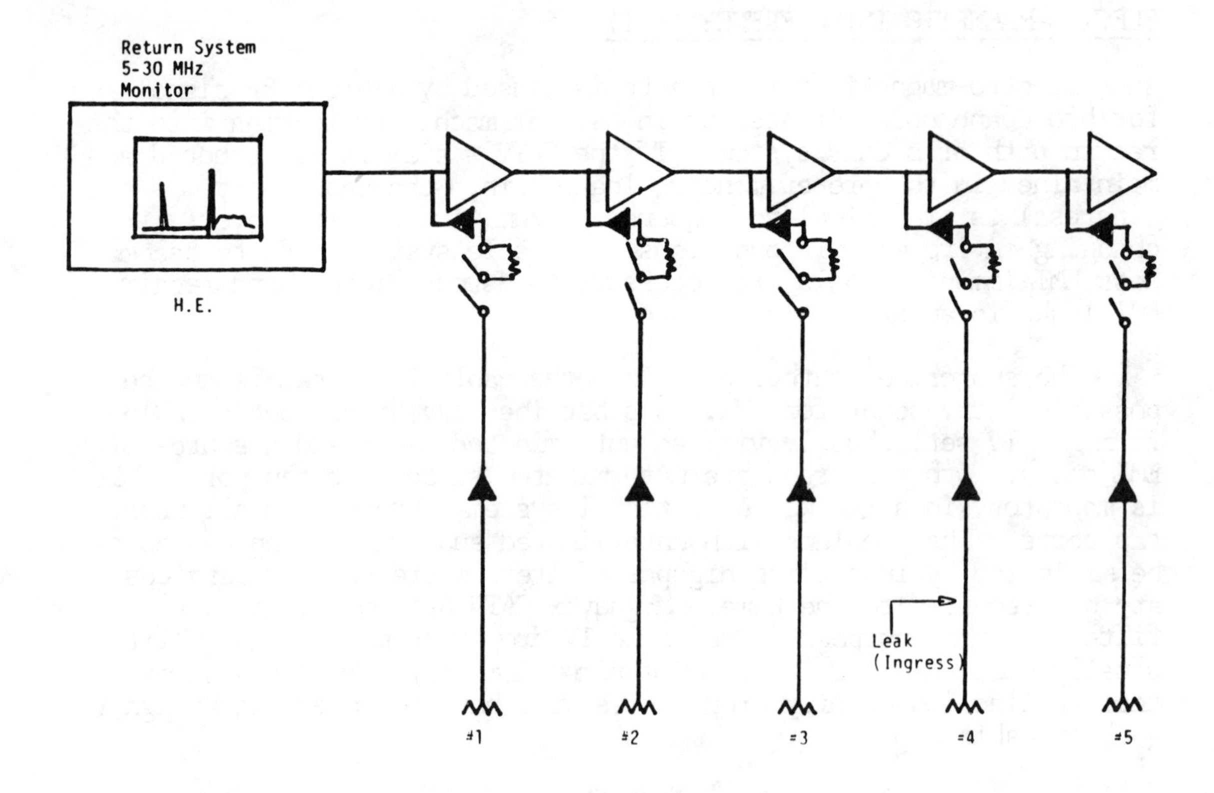

Since EMI requires a continuous ability to efficiently locate ingress and repair the leak, a tool such as status monitoring and trunk and bridger switching is necessary. Using a switchable 6 dB pad in the return system located at the trunk and bridger input, facilitates a tool used to locate the ingress. The 6 dB pads can be used to locate a source of ingress without disrupting normal signal flow in the return system. Figure #11 is used to illustrate the technique for applying the 6 dB switch. Consider that each return feeder input to the bridger has a switchable pad. Also, consider that there is a source of ingress or leak in feeder #4. This leak is monitored back at the headend by a spectrum analyzer display. The status monitoring computer is instructed to insert a 6 dB pad in all the return feeder paths. When this is done, the ingress on the monitor will drop by the amount of the pad (6 dB in Magnavox's DSS system). Now the operator of the computer removes the 6 dB pad from feeder leg #1, and notes the monitor display. If the monitor display has increased in level by 6 dB, the operator would know that the source of the ingress is coming from the feeders emanating from bridger #1 switch. If an increase in signal level is not noted, a 6 dB pad's switched back into bridger #1 path and then select the next bridger switch for feeder #2 and repeat the sequence of removing the pad, and then reinserting the pad. In the example shown by Figure #11, when the operator selected the pad in feeder #4 a 6 dB increase in the level on the monitor is noted. This indicates the leak is coming from feeder leg #4. The Service Department will then be directed to amplifier #4 to futher pinpoint the leak. The computer operator at the headend would then cycle all the 6 dB pads so that they would be switched out of the system.

If the level of the ingress from feeder line #4 is such that it interferes with signals coming from all other feeders in the system, the computer operator could use the on/off switch to isolate the source of ingress from the rest of the system. When the repair technician solves the problem, the switch can be closed, thus connecting feeder line #4 again to the system.

BALANCING THE RETURN SYSTEM

Early in the implementation of 2-way systems, the balancing of a forward and return system was completed as two separate steps. The forward system would be balanced first and become operational; then the return system would be implemented as a separate procedure. The return system set-up requires two technicians, two sets of test equipment, two-way radios for communication and two sets of transportation. As operators gained experience, procedures and equipment were developed so that both the forward and return systems can be implemented and balanced at the same time using only one field technician. Figure #12 shows the test equipment necessary for the single man set-up of a return system.

A system should be set-up in an outgoing manner starting with the amplifier closest to the headend and proceeding outward until terminations. The objective is to set-up a unity gain return system. Before proceeding with return system set-up, a reference level would be established on the spectrum analyzer at the headend. The test signal generator should be input to the combiner/diplex filter through a 30 dB attenuator. Once the spectrum analyzer reference level is achieved, the controls on the signal generator must not be changed. The technician should then proceed to the first amplifier, remove the 30 dB padding from the signal generator and insert the signal output into the 30 dB test point at the input of the return amplifier. The gain and slope of the return amplifier should be adjusted so that the signal levels received at the headend match the reference established on a spectrum analyzer. The spectrum analyzer is viewed on the portable TV receiver which is to be connected to an unused port on the chassis or connected to the input test point of the amplifier. The technician should then proceed to the first line extender off of the first trunk amplifier balanced, and in the same manner continue until all amplifiers in the system are sequentially balanced.

The above set-up procedure is a very cursory explanation of the method used. A more detailed explanation can be obtained by writing Magnavox CATV Systems, Inc.

Figure #12

Test Set-up for Return System Balance

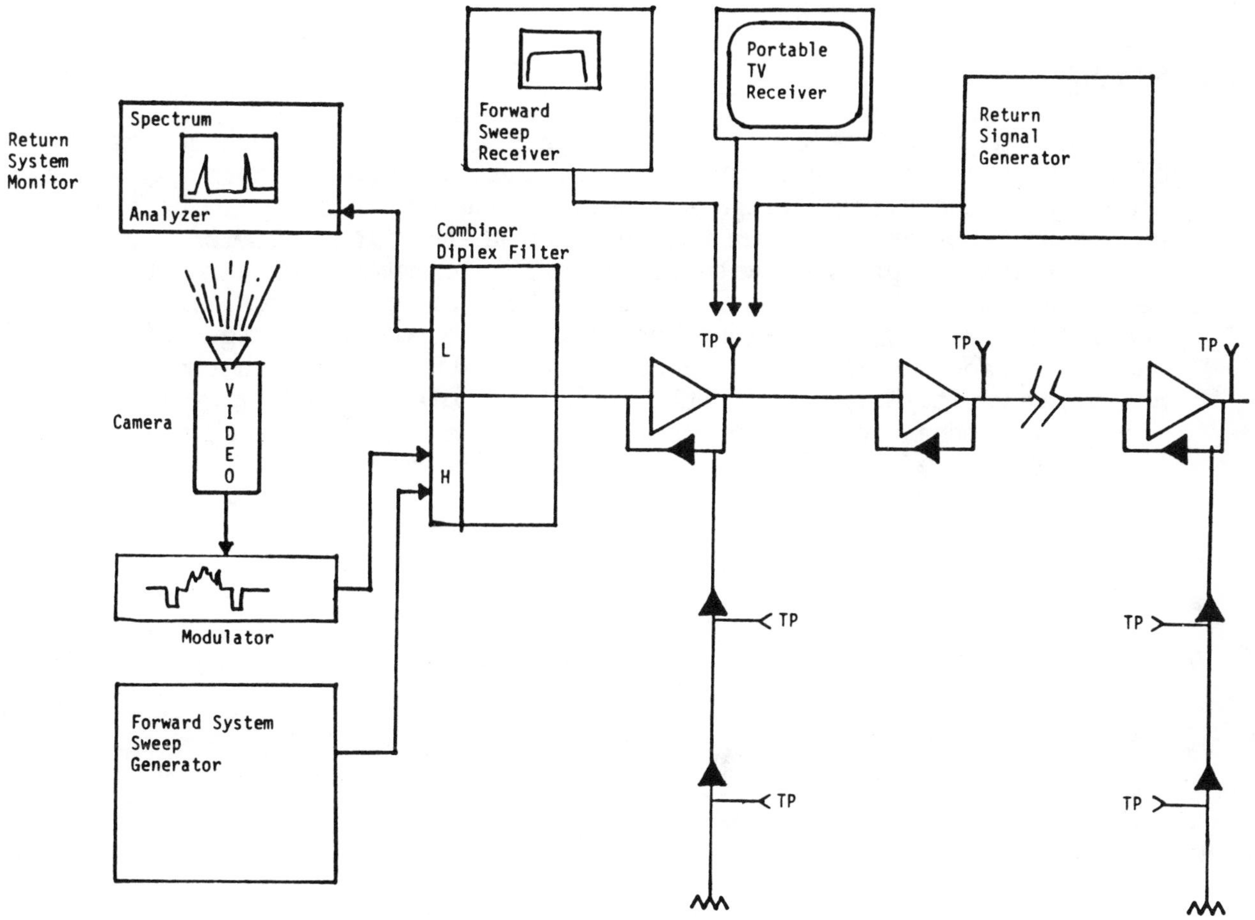

Return System Monitor

Spectrum Analyzer

Forward Sweep Receiver

Portable TV Receiver

Return Signal Generator

Combiner Diplex Filter

L

H

Camera

VIDEO

Modulator

Forward System Sweep Generator

TP

TP

TP

TP

TP

TP

TP

Test equipment manufacturers are presently developing specialized sweep systems which will enable the reduction of the components shown in Figure #12. This specialized test equipment will enable the sweeping of both the forward and return systems simultaneously, and display the response at the technicians remote location.

SUMMARY

Return system implementation requires a complete understanding of all factors before attempts to apply them are made. Thorough planning of the system design and performance is necessary to have a usable transportation medium for return signals. The systems designer must be provided with additional information; data input level requirements are much lower than those of video signals, and special tools must be provided to the field technician charged with the set-up and repair of the return system. Unity gain, thermal noise, carrer-to-noise ratio, terminal or modulator output levels, and electro-magnatic interference (EMI) are all factors which must be considered.

A UNIFIED APPROACH TO DATA TRANSMISSION
OVER CATV NEWORKS

THOMAS E. O'BRIEN JR.

JERROLD DIVISION, GENERAL INSTRUMENT CORPORATION

ABSTRACT

Many feel that data services will compel both rural and urban cable systems to form a national communication network. Concurrance of video and data services can be achieved through establishment of CATV data transmission standards. The characteristics of data transmission, including channel capacity, bandwidth, signal to noise ratio, modulation, error rate, error detection and encoding methods are investigated, a distributed processing complex supporting multiple service offerings is detailed and finally, a full set of quantitative parameters for CATV data transmission is recommended.

INTRODUCTION

Through the 1980's and beyond, the quality of life will be improved in many different ways. A large part of the anticipated enrichment will accrue through progress in communication technology, as it applies to the common man. CATV networks, initially constructed for the purpose of bringing entertainment to rural areas, are now being built in major cities. By the end of this decade, interconnection between individual systems will enable the carriage of information over a national communication network.

The CATV manufacturer stands today, on the threshold of a challenging opportunity. As the emphasis within the network shifts from entertainment toward information, the equipment and techniques offered to support the needs of CATV operators must remain viable. Engineering philosophies now being developed, must be designed to fulfill not only the current requirements, but also the needs of the future. It is, therefore, prudent to establish a set of guidelines for data transmission over CATV networks. The individual characteristics embodied in such a philosophy, are too numerous to explore in detail; however, it is instructive to identify the major technical implications and the resulting benefits of the approach.

INFORMATION TRANSMISSION

All information passed through a communication system is degraded to some extent by distortion and the addition of interfering signals and noise. The degradation results in decoding uncertainty (error rate) whose tolerance is somewhat application dependent and, in general, may be improved by reduction of information rate and/or higher system cost. A suitable philosophy should, therefore, allow for variations in modulation method, bandwidth, error rate and carrier frequency assignments, while providing established guidelines for channel usage, signal levels, interference with other signals and compatability with other equipment and services on the CATV network. Before proposing parameters for CATV networks, it will be helpful to review the general problem of information transmission.

Figure 1 illustrates the data transmission system as it applies to CATV networks. Although a one-way system is shown for simplicity, the concept is the same when extended to two-way.

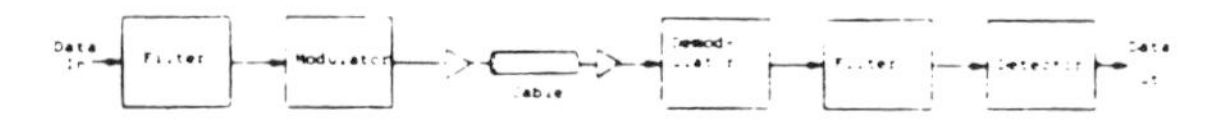

Figure 1
Data Transmission System

The system consists of a modulator, transmission path (CATV cable network) and demodulator. The filters, whether part of the transmission process, or used intentionally, are necessary to assure adequate signal to noise ratio. However, they also reduce the ability to separate the individual transmitted bits, which is called intersymbol interference. The information handling capability of the transmission system, or the maximum rate of transmission of data over the channel, is referred to as the channel capacity. The maximum possible rate of transmission of binary digits over a channel limited to bandwidth W, with mean signal power S and mean noise power N, was found by Shannon[1] to be given by

$$C = w \log_2 (1 + S/N)$$

Reprinted with permission from *Technical Papers, 32nd Annual Convention and Exposition, National Cable Television Association*, 1983, pp. 119-123.
Copyright © 1983, The National Cable Television Association.

It should be noted that in order to achieve this rate, the information must be coded in the most efficient manner which will generally involve highly complex circuitry and incur large time delays in transmission. It is evident that for a specified channel capacity, bandwidth and signal power can be exchanged for each other. The modulation method is essentially a means for effecting this exchange, however, the process is highly inefficient since one must increase the power exponentially to effect a corresponding linear decrease in required bandwidth. It should be obvious that in CATV systems, direct transmission of baseband data is not practical due to the amount of bandwidth required. Instead the whole spectrum is shifted to a higher frequency by modulating an RF carrier. This process gives rise to upper and lower sidebands and hence, the required bandwidth is doubled.

As indicated above, virtually error free digital transmission could be achieved (provided channel capacity is not exceeded), by appropriately coding the binary message sequence. Specifically, at a binary transmission rate of R bits/sec., if R<C it may be shown that the probability of error is bounded by

$$Pe \leq 2^{-E(R,C)T} \quad R<C$$

as shown in Figure 2. As the transmission rate R approaches the channel capacity the probability of error approaches 1. The parameter T indicates the time required to transmit the encoded signal. With the transmission rate and channel capacity fixed, the probability of error may be reduced by increasing T.

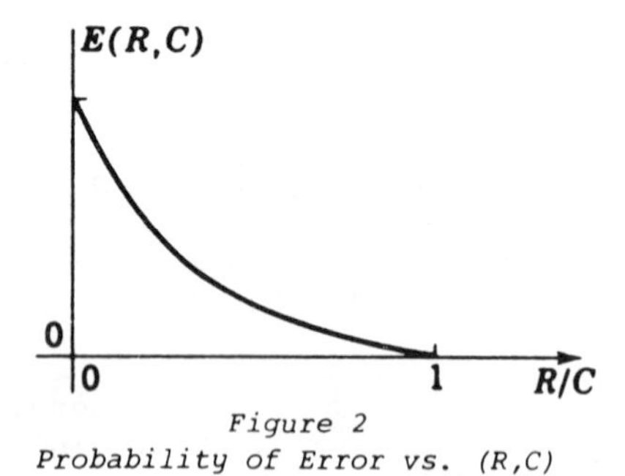

Figure 2
Probability of Error vs. (R,C)

A great deal of research activity is being devoted to the investigation of various modulation and encoding methods. This work froms the basis of communication theory.

MODULATION METHODS

There are essentially three ways of modulating a sine wave carrier: variation of its amplitude, frequency or phase in accordance with the transmitted information. These are commonly known as ASK, FSK and PSK respectively. FSK systems perform better than ASK, while PSK systems perform still better. The major factors affecting the selection of modulation method lie in the demodulation or detection process. The two commonly used detection methods are envelope detection and synchronous detection. ASK may use envelope detection, FSK may use differentiation (to convert frequency variation to amplitude variation) followed by envelope detection, while PSK requires synchronous detection. Synchronous detection requires a locally generated receiver clock of the same frequency and phase synchronized or slaved to the transmitter clock to within much less than a fraction of a cycle. This is difficult and costly to achieve in practice, for example, at a data rate of 3.5 MHz the required accuracy is much less than 60 nanoseconds. The signal to noise ratio of AM versus FM is also important. As indicated previously, widening the transmission bandwidth (as is required for wideband FM) improves the signal to noise ratio. With AM, the signal to noise ratio is linearly dependent on carrier to noise ratio and cannot be improved. In fact, any bandwidth increase beyond what is actually required serves only to increase noise, thereby, lowering the signal to noise ratio. With FM, as illustrated in Figure 3, it becomes obvious that significant improvement in signal to noise ratio is possible by increasing the modulation index at the expense, of course, of increased bandwidth. Notice that the signal to noise ratio beyond the 12 dB carrier to noise point, results in a constant linear improvement of $3\beta^2$, where β is the modulation index - the ratio of FM deviation to modulating frequency.

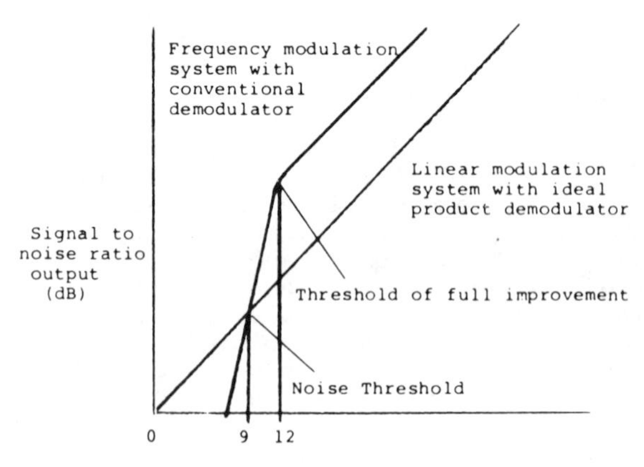

Figure 3

The resulting increase in bandwidth is illustrated in Figure 4. Notice that the AM spectrum consists of only one pair of sidebands per sinusoidal component of the modulation signal and has an effective bandwidth of 2Fm. The FM spectrum has multiple pairs of sidebands, and an effective bandwidth of 2Fm + 2ΔF (where ΔF is the frequency deviation). The noise improvement factor of FM is proportional to the ratio of ΔF to Fm. This improvement corresponds to wide transmission bandwidth (β>>1) and better than 10 dB carrier to noise ratio. With narrowband FM (β<1) the deviation is constrained to produce a bandwidth of 2Fm (as in AM), therefore, no signal to noise ratio improvement over AM can be obtained.

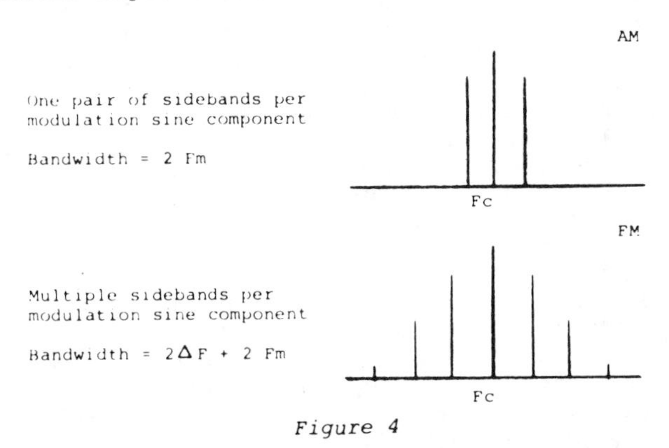

Figure 4

With the previous emphasis on signal to noise ratio, it is important to note that various methods have been devised for coding, modulation, and demodulation of digital signals for the purpose of matching the data integrity with information quality. The selection of technical parameters should, therefore, be tied to the application of the information. For example, in a Teletex application, a decoding error in received data may show up on the screen as a missing letter. The viewer will seldom object to this, because the value of the missing character can usually be implied from the context of remaining text. On the other hand, an error byte in a banking transaction could have drastic implications.

FSM vs. TDM

The two methods of simultaneous transmission of several band limited signals on a channel are frequency and time division multiplexing. In frequency division systems, all of the signals are modulated on different carriers and transmitted continuously. In time division systems, all of the signals are mixed in time and modulated on a single carrier, each signal occupying a distinct time interval. TDM seems to offer a cost advantage, in that, only one carrier need be generated and relatively simple circuits can separate the data intended for each distination. With TDM, the data intended for any individual receiver will occur in bursts as shown in Figure 6, whereas with FDM once the channel is selected, all the data is intended for the individual receiver. Channel capacity, as previously indicated, is a function of channel bandwidth and signal to noise ratio. Consider the 6 MHz bandwidth allowed for a TV signal. With 10 dB signal to noise ratio, the highest usable data rate using ASK is about 3.5M bits/sec. Compared with FSK at β = .6, the available 6 MHz accomodates 62 channels of 56K bits/sec. which is 3.47M bits/sec., or roughly the same channel capacity. There are numerous "holes" in the cable spectrum as shown in Figure 7, which are too narrow for conventional video services, for example, the FM band. Virtually, every cable system has FM channels which are unassigned, and are essentially wasted bandwidth. Data services in such holes can provide a new revenue source, while not reducing the capacity for carriage of traditional services. Since no two cable systems have the same spectral holes, however, a mechanism is needed to allow the terminals to tune themselves to the desired channel. Recent advances in integrated circuit technology make accurate and inexpensive frequency control systems not only possible, but cost effective as well.

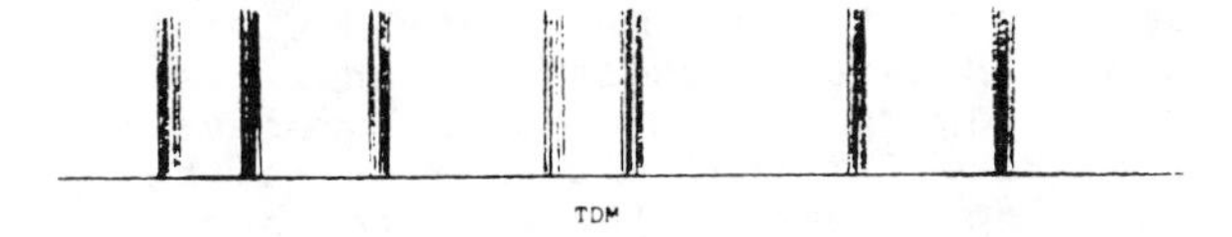

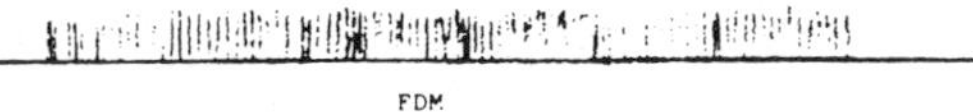

Figure 6
Data Rate Receiver vs. Multiplex Method

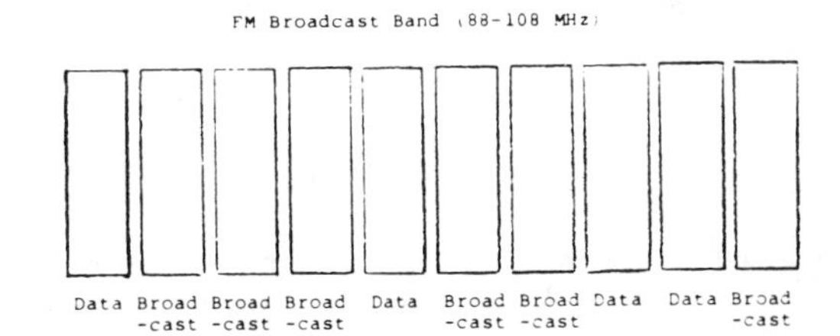

Figure 7

FDM FM is particularly interesting from the viewpoint of interfering carriers. Figure 8 illustrates the allowable interference carrier level versus frequency for an FM transmission with a minimum signal to noise ratio of 10 dB and modulation index of 5. Notice that an interfering carrier within a ±200 KHz range of the desired carrier need be only 6 dB down to be essentially rejected. In fact, an interfering carrier ±400 KHz away can be as high as 25 dB above the desired carrier without degrading data reception.

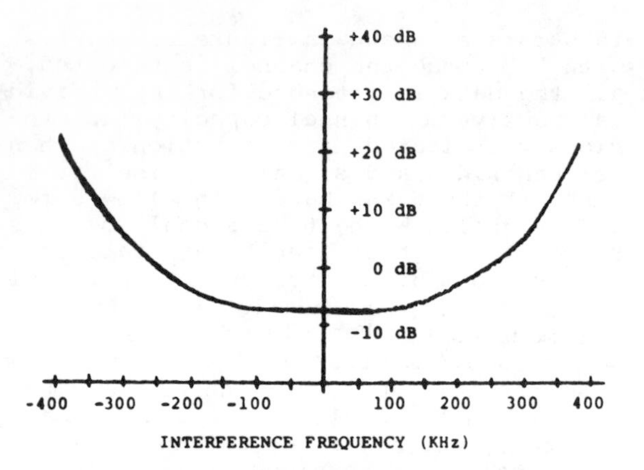

Figure 8
FDM FM Interfering Carrier Rejection

The multiplex method, therefore, should be selected on the basis of cost, available space, and desired error rate.

DATA ENCODING

The maximum baud rate inferred from Shannon's capacity is 2W elements/sec. or 2 bauds/sec. for every hertz of available bandwidth. The signalling rate (or baud rate as it is commonly known) which is defined as the minimum elapsed time interval between successive signal elements, places an upper bound on the achievable data rate. This limit relates to the data encoding scheme in terms of the number of transitions per data cell or baud/bit. Figure 9 shows some commonly used data

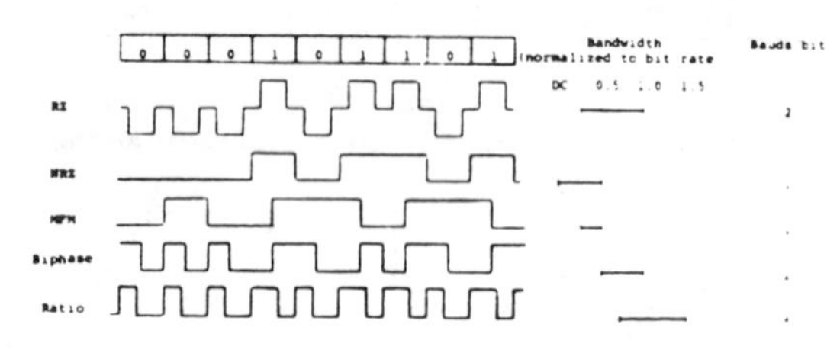

Figure 9

encoding schemes. Three factors are involved in the selection of an encoding scheme:

1) the ratio of maximum to minimum frequency which defines the detector passband filter characteristic

2) the inverse of the minimum time between transitions or baud rate which defines the upper bandwidth limit

3) the presence of a DC component which defines the bandwidth lower limit, precludes the use of AC coupling in the detector, and requires the use of a separate transmission method for the bit clock.

It should be noted that the MFM scheme makes use of previous bit history to reduce the baud rate, with no apparent increase in bandwidth. The advantage of this technique is somewhat negated by the requirement for a preamble to acquire clock synchronization. A preamble is a known bit pattern appended to the front of each message.

ERROR DETECTION

As we have seen above, error probability in digital transmission is a direct function of signal to noise ratio. If, for a given application, the signal to noise ratio is maximized and the error rate is still unacceptably high, then error control coding can provide the solution. Error control coding is simply the calculated use of redundancy, where extra bits or words (or both) are added to the message. They convey no new information themselves, but make it possible for the receiver to detect or even correct errors in the information bits. A multitude of error detecting and correcting codes have been devised to suit various applications. They may be used alone or in combination, for example, a single parity bit on each transmitted word, and a longitudinal checksum on the entire message. For many applications, errors can be rendered harmless if they are simply detected with no attempt at correction. In a two-way communication link, the fact that an error has been detected can be sent back to the transmitter for appropriate action, namely, retransmission.

HEADEND

There are many application similarities regarding data communication when viewed from the data processing end of the CATV network. Figure 10 shows the headend

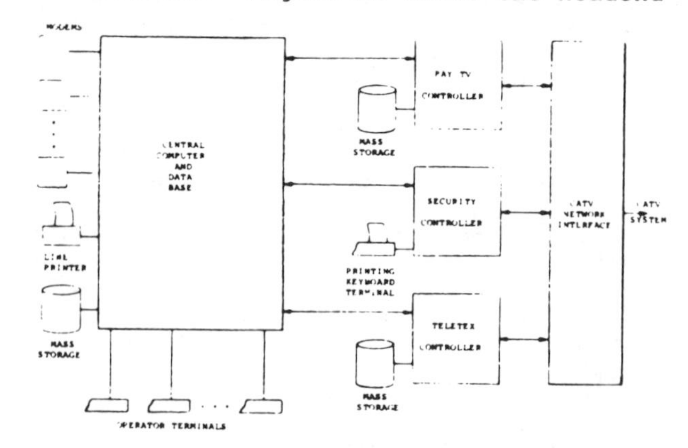

Figure 10

equipment complex for controlling a variety of two-way services. Note that separate controllers are used for each service type, along with the peripheral compliment required by the application. This functional distribution of computing power leads to the following advantages:

1) the controllers can operate standalone because of minimum application requirements (ie: cost), of the central computer is shut down for repair or maintenance.

2) if a controller needs repair, only one service is affected.

3) services may be added through standard interface modules

4) the central computer need not occupy the same physical premises as the controllers

5) one central computer may serve multiple cable systems through the use of more than one controller of the same type

The data path between the central computer and each controller uses a standard communication protocol such as X.25. This allows each controller to use the same type interface hardware and, in a non co-located application, permits the use of standard telco communication links.

Figure 11 shows the recommendations for the central computer and the interface between the central computer and the rest of the system. Figure 12 lists the controller to central computer and controller to digital communication hardware interface recommendations. Note that the signalling speed is either 3.58 MHz or a submultiple of it. This technique permits the use of low cost crystals in subscriber terminals and simplified clock regeneration in coherent detection schemes.

Central Computer

CPU Type:	mini or midi
Word Length:	16 or 32 bit
Mass Storage:	Winchester disk
Capacity:	1K bytes/sub/service
Terminals:	Intelligent
Capacity:	64
Speed:	9600 BPS
Electrical:	RS-232C
Protocol:	X.25

Figure 11

Controller

Central computer	interface
Data link:	Serial line
Capacity:	16
Speed:	Selectable (up to 38.4 KBPS)
Electrical:	RS-424C
Protocol:	X.25
CATV Network Interface	
Data link:	Serial line
Impedance:	75 Ohm co-ax
Voltage:	TTL (0=0.8V, 1=2.0V)
Speed:	Selectable (up to 3.58 MBPS)
Coding:	Manchester biphase

Figure 12

A list of recommended bit rates is shown in Figure 13.

Divide Ratio	Bit Rate
1	3.58M
2	1.79M
4	895K
8	447K
16	224K
32	112K
64	56K
128	28K
256	14K

Figure 13

The CATV network interface consists of the modulators necessary to convert the preformatted digital data to RF sine wave carriers and demodulate the upstream carriers in a two-way system. Figure 14 describes the recommended RF parameters applied to the CATV network, showing the application dependence described above.

APPLICATION	TV GAMES	PAY TV	SECURITY	TELETEX	HOME SHOPPING
DOWNSTREAM					
Multiplex	FDM	-	-	TDM	FDM
Modulation	FSK	FSK	FSK	ASK	FSK
Bandwidth	±200 KHz	±200 KHz	±200 KHz	6 MHz	±100 KHz
Coding	Biphase	Biphase	Biphase	NRZ	Biphase
Error Method	parity + checksum	parity + checksum	parity + checksum	CRC	CRC
Protocol	unique	unique	unique	Prestel	HDLC
Bit Rate	14 KHz	14 KHz	14 KHz	3.58 MHz	28 KHz
UPSTREAM					
Multiplex	N/A	-	FDM	-	FDM
Modulation	N/A	PSK	PSK	PSK	FSK
Bandwidth	N/A	±200 KHz	±200 KHz	±100 KHz	±100 KHz
Coding	N/A	Biphase	Biphase	Biphase	Biphase
Error Method	N/A	parity + checksum	parity + checksum	parity + checksum	CRC
Protocol	N/A	unique	unique	unique	HDLC
Bit Rate	N/A	14 KHz	14 KHz	28 KHz	28 KHz

Figure 14

SUMMARY

With CATV equipment manufacturers rushing to respond to the demand for data services, now is the right time to establish standards for data transmission. The development of guidelines must begin with an understanding of transmission schemes, CATV networks, and the equipment used to provide data services. The parameters of data transmission have been discussed along with some comments and recommendations on channel usage, data rates, encoding techniques and the various tradeoffs involved. The philosophy embodied herein is intended to provide incentive toward the establishment of universally acceptable data transmission standards, designed to achieve harmony between data and video services. Once established, standards will enable CATV manufacturers to produce fully compatible equipment for both current and future systems, while avoiding product rejection or early obsolescence.

1 C.E. Shannon, Communication in the Presence of Noise, Proc. IRE, Vol. 37, pp. 10-21, January 1949

POTENTIALS OF FIBEROPTIC MULTICHANNEL TELEVISION TRANSMISSION BY ANALOG MODULATION

G. Guekos, H.P. Berger, B. Illi and H. Melchior
Swiss Federal Institute of Technology, Institute of Applied Physics
CH-8093 Zurich, Switzerland

Summary

The capabilities of analog modulation for multichannel television transmission over graded-index multimode optical fibers have been investigated theoretically and experimentally. TV channels were transmitted in laboratory links by amplitude (AM) or frequency modulation (FM) of the carriers. Measurements of the signal quality at the receiver end as a function of fiber length are shown and compared with theoretical predictions. Two TV channels were transmitted over the same fiber without repeater with high picture quality over 4.4 resp. 7 km using AM resp. FM and a laser diode at 0.83 μm as light source. The respective figures with a 1.3 μm LED as source are 3 and 8.3 km. Finally, four channels were transmitted over a single fiber with the laser as source over 1.1 km.

Introduction

The transmission of high quality television signals over optical fibers for distances of several kilometers is an interesting application for wideband distribution systems. A number of possibilities for this transmission can be envisaged. These possibilities pertain mainly to the optoelectronic components that can be employed and to the various modulation schemes. Light sources and detectors have reached a high degree of technical maturity and commercial availability at wavelengths around 0.8 μm where most of the systems installed for telephone and data transmission work. However, worldwide interest for TV transmission in the infrared is rapidly growing because the fibers exhibit very low attenuation and dispersion around 1.3 μm and 1.55 μm. Digital or analog methods can be applied for signal modulation. Pulse code modulation (PCM) yields to a high picture quality at the receiver,

Reprinted with permission from *13th International Television Symposium*, Montreux, Switzerland, 1983, CATV Sessions, vol. 3, pp. 192–202.

provided that enough quantisation steps are used in the coder/decoder. Thus, the influence of system noise and distortion is minimized and high quality TV programs can be transmitted, by using repeaters if necessary, over tens of kilometers fiber length. This technique requires, however, a large system bandwidth and costly encoding and decoding circuits. On the other hand, the cost of electronics is expected to diminish with the advent of modern large-scale integrated circuits. Until recently, the application of analog modulation techniques was often plagued by noise and distortion problems [1,2,3]. These problems were generally caused by the non-linearities of the laser ligth-current transfer characteristic, the optical noise produced by the laser and by the laser-fiber interaction and the optical detector circuit noise. The significant improvement in laser linearity of the last few years and the development of effective methods for reducing the influence of laser-fiber interaction make analog systems become increasingly interesting for TV transmission. The main advantage of analog modulation is that it is simple to implement and therefore less costly than digital methods. In the following we show the signal quality measured as a function of fiber length by applying amplitude (AM) or frequency modulation (FM) of the carriers in two experimental transmission systems. The first uses laser diodes at 0.83 μm and the second light-emitting diodes (LED's) at 1.3 μm [1,4].

Theoretical considerations

The straightforward approach is to bias the light source at a dc current and to modulate the light intensity by modulating the current amplitude through the TV signal (AM-IM). At the dc bias, the laser or LED emits the optical power P_0 which is modulated with the modulation index m per TV channel. The modulation index is defined as the ratio of peak-peak optical power divided by two times the power P_0. The mean power coupled into the fiber P_F has statistical fluctuations which are caused by the light generation process in the semiconductor light source. These fluctuations can be described for communication purposes by the relative intensity noise factor [5] RIN:

$$\mathrm{RIN} = \overline{\Delta P_F^2} \,/\, P_F^2 \tag{1}$$

$\overline{\Delta P_F^2}$ being the density of the mean square optical power fluctuations. The RIN values of present day laser diodes having passive dielectric waveguidings like buried-heterostructure and channeled substrate types, and exhibiting single transverse mode emission are as low as 10^{-16}s for currents well above the lasing threshold [5,6]. Lasers with active gain guiding, like proton-bombarded and V-groove types, have usually higher RIN values, typically an order of magnitude higher well above threshold [5]. The RIN is the main factor limiting the signal quality at short fiber lengths, typically up to about 4 km, as we shall see later. LED's have much less optical noise than lasers and their RIN values are usually less than 10^{-18}s. Deviations of the values given above may be expected with commercial lasers due to possible laser instabilities. Another problem encountered in analog fiberoptic systems is the laser source instability caused by the light reflected from the fiber. This problem can be circumvented by using an optical isolator between laser and fiber or by using lasers with a large number of axial modes, f.ex. 10 to 20, as can be the case with V-groove lasers. An additional advantage with multi-axial mode lasers is the low interference noise (modal noise) produced in the fiber. These advantages are of particular importance in AM-IM systems, which are most sensitive to noise and non-linearities, for they largely compensate the penalty of the higher RIN of that type of laser.

The fiber is characterized by its attenuation coeffisient α, length ℓ and transfer function H(f) given by [7]:

$$H(f) = 10^{-\frac{\alpha}{10}\cdot\ell} \cdot \delta(f,\ell) \qquad \alpha[\mathrm{dB/km}] \tag{2}$$

where $\delta(f,\ell) = \exp - \left| \frac{f}{f_0} \left(\frac{\ell}{\ell_0}\right)^{\gamma} \right|^{z}$.

Here, f denotes the TV carrier frequency, the index o the reference values determined by the fiber and γ and z are exponents. We can now calculate the carrier-to-noise ratio (CNR) at the receiver output.

The receiver may have as detector a photodiode or an avalanche photodiode (APD) with gain factor G and the photocurrent can be further amplified with an electronic circuit. The CNR determined by the optical link alone, i.e., without consideration of the noise introduced by the modulation/demodulation equipment, is given by:

$$\mathrm{CNR} = \frac{P_S}{\text{noise power}} = \frac{\left(I_O \cdot \frac{m}{\sqrt{2}} \cdot \delta \cdot G\right)^2}{\overline{i_L^2} + \overline{i_D^2} + \overline{i_C^2}} \tag{3}$$

where P_S = carrier signal power at the photodiode output

$\overline{i_L^2}$ = mean square noise current at the photodiode output due to the laser power fluctuations = $\mathrm{RIN} \cdot (I_O \cdot \delta \cdot G)^2$

$\overline{i_D^2}$ = mean square noise current at the photodiode output due to the photodiode itself = $2e \cdot (I_O + I_D) \cdot G^2 \cdot F$

$\overline{i_C^2}$ = mean square noise current at the photodiode output due to the receiver circuit

I_O = mean photocurrent of the photodiode at unity gain

F = noise factor of the APD

I_D = dark current of the APD

For an amplitude modulated TV channel we obtain an unweighted video signal-to-noise ratio SNR_{AM}:

$$\mathrm{SNR}_{AM} = \frac{K_{AM}}{b} \cdot \mathrm{CNR} \tag{4}$$

where b is the video bandwidth and K_{AM} a factor resulting from the video SNR definition for TV channels [8].

Instead of using AM-IM, one can apply frequency modulated TV-carriers to the laser (FM-IM). The linearity requirements on the light-current characteristic of the source are now less severe than with AM. The modulation depth of the laser can thus be increased and consequently longer transmission distances can be achieved. The only drawback of the FM as compared to AM is that it necessitates the use of more complex and costly signal processing equipment. For a FM TV channel with frequency deviation ΔF the SNR becomes:

$$SNR_{FM} = SNR_{AM} \cdot 3K_{FM} \cdot (\Delta F/b)^2 \quad (5)$$

where K_{FM} is a multiplication factor given by the modulation signal and the pre- and de-emphasis [9].

Taking into account the signal-to-noise ratio SNR_{md} due to the modulation/demodulation equipment, we obtain a total SNR at the output of the system:

$$SNR_{tot} = \frac{SNR_{AM/FM}}{1 + \frac{SNR_{AM/FM}}{SNR_{md}}} \quad (6)$$

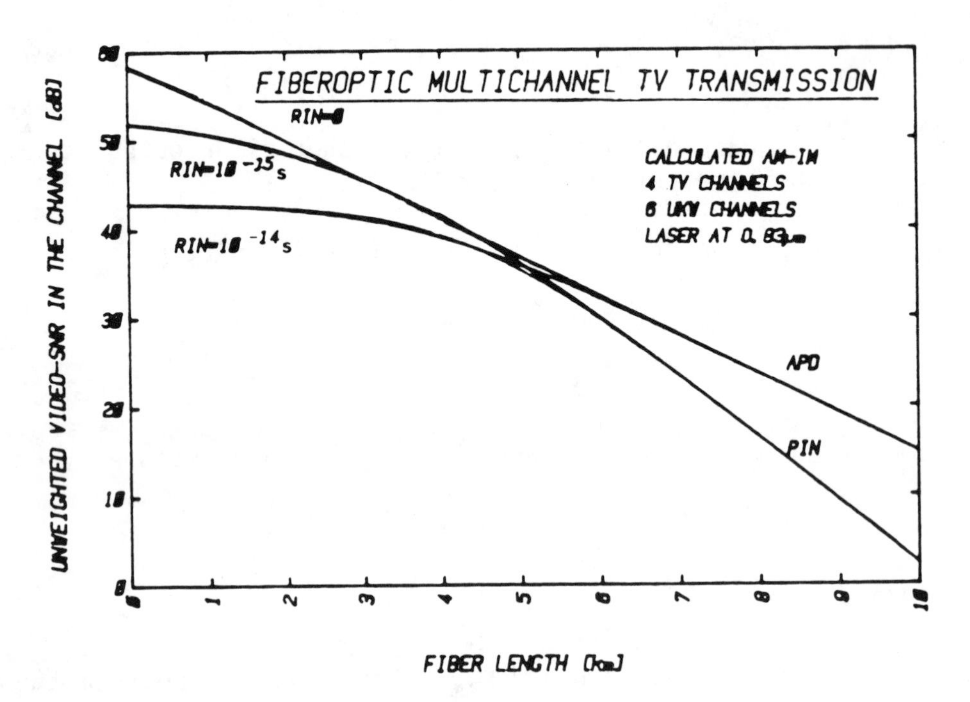

Fig.1 Calculated video SNR in the channel (unweighted) versus fiber length for AM transmission of 4 TV and 6 audio channels over a fiber. Note the importance of the optical noise of the source (given by RIN, see text) on the SNR.

Fig.1 shows the calculated dependence of the video SNR for the optical link in the channel (unweighted) on fiber length for AM-IM assuming a point-to-point transmission without repeater, a diode laser at 0.83 µm and a Silicon pin-photodiode or an APD detector. Further assumptions are: 4 TV and 6 FM audio channels, m = 0.1 per TV channel,

P_F = 3 mW, α = 3.5 dB/km including splice losses, fiber bandwidth 800 MHz•km. As is evident from Fig.1, the SNR can reach high values when the laser noise is low (low RIN) and the transmission distance does not exceed a few kilometers. The difference of the SNR due to the detector type becomes significant at low SNR values only, which is of no practical interest for high quality transmission. In the experimental situation, the transmission length achieved for multichannel TV is less than in Fig.1 because of intermodulation distortion. Results are given in more detail in the experimental section.

System arrangement

The experimental system consists of customary TV pattern generators (for CCIR 625 lines test patterns), the electronics for carrier modulation, demodulation (vestigial side band VSB), channel addition and separation and the equipment for the measurement of the video signal-to-noise ratio, intermodulation distortion (IM), differential gain and phase (DG,DP) in the channels. The optical link consists of a linearly driven V-groove multilongitudinal mode laser at 0.83 μm or alternatively of a LED at 1.3 μm (edge emitter, P_F = 80 μW), a graded-index fiber cable (attenuation 3 dB/km at 0.83 μm and 1.3 dB/km at 1.3 μm, 50/125 μm) of variable length and a receiver using PIN- or avalanche-photodiodes. The system bandwidth (3 dB points) without fiber extends from 0 Hz to 250 MHz for the laser diode system (limited by the receiver circuit) and to 90 MHz for the LED system (limited by the LED). We used commercial AM-TV-modulators and demodulators in the intermediate-frequency (IF)- and in the VHF-band having an unweighted video SNR of 54 dB connected directly, i.e. without fiber link. The FM-modems at 16, 90 and 132 MHz IF were low cost commercial equipment with a peak-to-peak frequency deviation of 8 MHz, whereas the channel at 140 MHz was a high quality modem with 25 MHz peak-to-peak deviation. The unweighted video SNR of the modems connected directly, i.e. without fiber link, was 55, 59, 61 and 70 dB for the 16, 90, 132 and the 140 MHz IF modem, respectively.

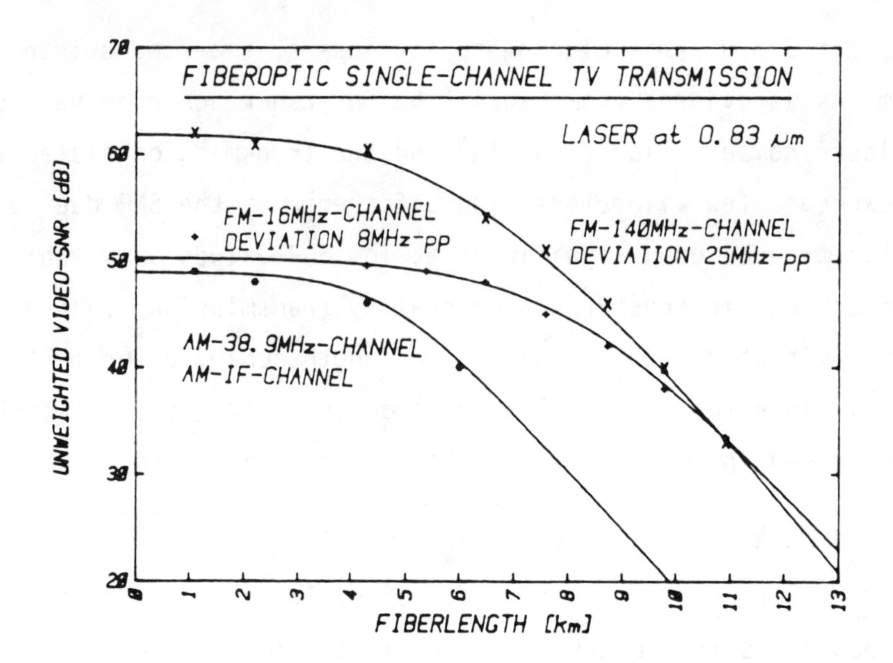

Fig.2 Calculated (full lines) and measured (points) dependence of the video SNR (unweighted) on fiber length. A single TV channel was transmitted using AM or FM and a laser diode at 0.83 µm as light source

Experimental results and discussion

Fig.2 shows the measured SNR in the TV channel as a function of fiber length for the laser system. Only one TV channel was transmitted, either with AM-IM or FM-IM. The modulation signal was 6 dBm electrical into the laser transmitter, corresponding to 3.8 mW peak-peak optical power from the laser or an optical modulation index of 0.67. The experimental results are given along with the calculated curves from equs.(1) to (6). By applying FM to the carrier, the contribution of the laser nonlinearities and noise is reduced and a significant increase in transmission distance over AM is achieved. On the other hand, the transmission distance with FM is strongly dependent on the frequency deviation. The experimentally determined video SNR for a FM-TV-channel at 140 MHz with a deviation three times larger as compared to the 16-MHz-channel is in good agreement with the theoretical prediction.

Two AM-TV-channels were transmitted over 4.4 km with a video SNR (unweighted) >40 dB and intermodulation distortions in the channels below -65 dB. Differential gain and differential phase changed by only ± o.1% and ± 0.1° respectively after the insertion of the fiber link. Four TV channels were transmitted simultaneously by AM with the aforementioned quality over 1.1 km fiber length. When the three FM channels at 16, 90 and 132 MHz are transmitted simultaneously, the SNR decreases with distance as in Fig.3 (power into the transmitter 0 dBm, opt. modulation index 0.4).

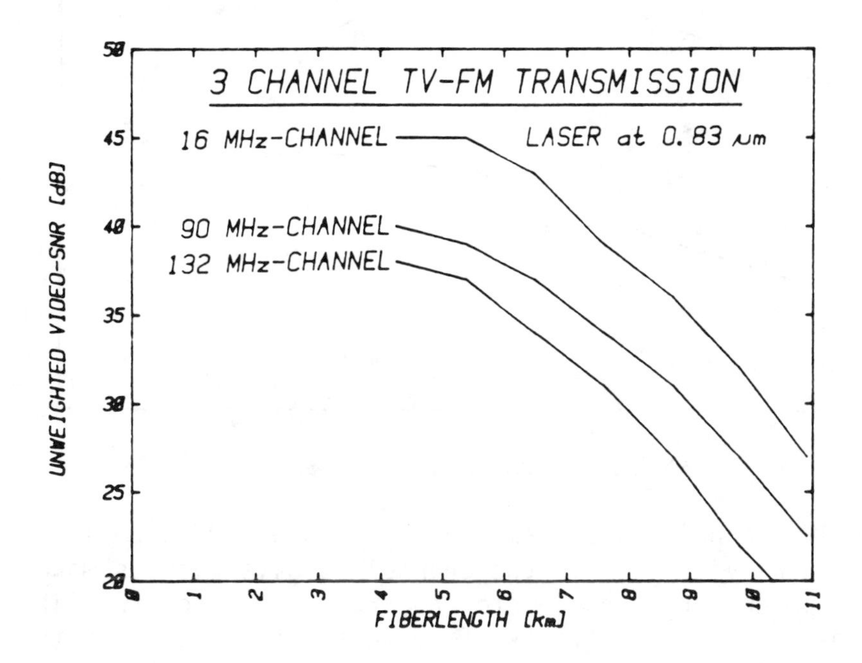

Fig.3 Simultaneous transmission of 3 TV channels using FM and a 0.83 µm laser source

As Fig.3 shows, high quality three channel transmission can be achieved with small deviation FM modems for fiber lengths up to 7, 4.3 and 3 km for the 16, 90 and 132 MHz channel respectively. The measured alterations in diff. gain and phase after insertion of a 4.4 km fiber length were within ± 0.5% and ± 3°.

The influence of the non-linearities of the laser transfer characteristic on the SNR for the three channel FM-IM transmission can be seen from Fig.4. Here the SNR is given as a function of modulation power per channel for the fiber lengths 5.4 and 8.7 km. Beginning at low modulation powers, the SNR first increases with increasing power at the rate of 1 dB per dBm because of the increasing optical power falling onto the detector. As the modulation power continues to increase, the SNR reaches a maximum and then decreases again because at high modulation depths the laser produces third order intermodulation products that fall into the channels.

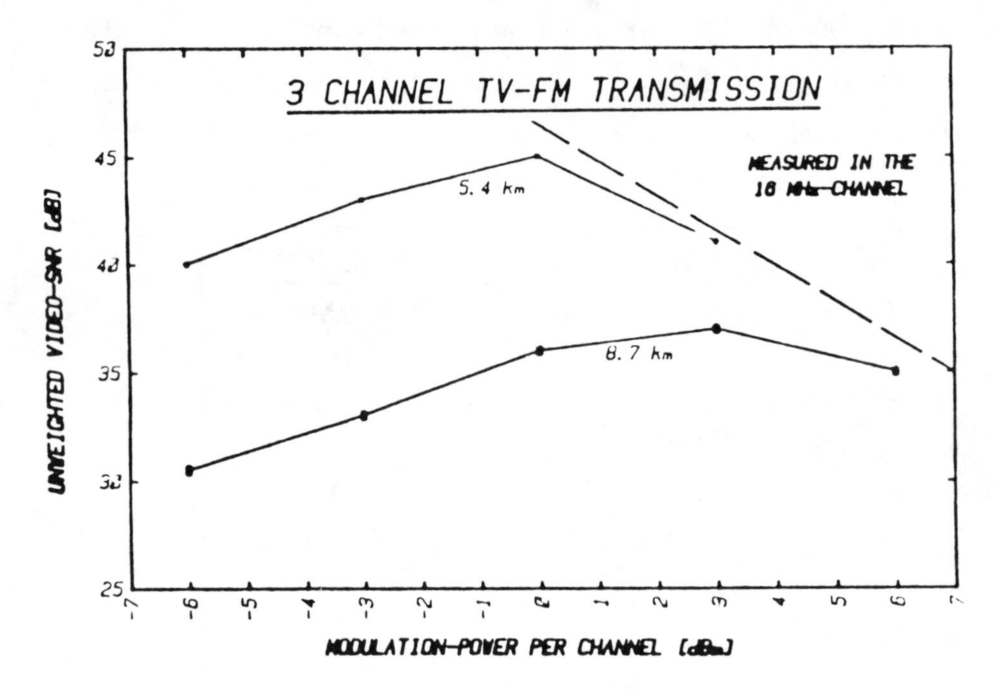

Fig.4 Three TV channel transmission using FM and a 0.83 μm laser source. The video SNR in the channel is given versus modulation power per channel into the transmitter for two fiber lengths. Due to the effects of noise and distortion of the light source, an optimum modulation power exists for a given fiber length.

For longer fiber lengths this maximum is reached at higher modulating powers than for shorter lengths. This occurs because at longer lengths the effect of the detector noise on the SNR is more pronounced and the consequences of the non-linearity can be perceived only at a higher modulation power. The rate of decrease of the SNR at higher modulation powers is near the theoretical value of -2 dB per dBm.

Comparing the results obtained with the 0.83 μm laser system and the 1.3 μm LED system, we found that the transmission distances reached for a picture quality given by a video SNR > 40 dB (unweighted) and an intermodulation distortion less than -65 dB are roughly comparable. The most serious drawback of the LED is the low optical power coupled in the fiber. This is to a certain degree compensated by the low attenuation of the fiber. However, the LED has to be driven at high modulation coefficients (practically 100%) in order to obtain more optical signal power at the detector. The consequence is high intermodulation distortion which limits the number of TV channels.

Table I Transmission lengths for TV channels. Signal-to-noise ratio (unweighted) in the TV channels better 40 dB, intermodulation distortion less than -65 dB

Number of channels per fiber	Transmission lengths				
	Laser at 0.83μm			LED at 1.3μm	
	AM	FM		AM	FM
	b=5MHz	ΔF=4MHz	ΔF=12.5MHz	5MHz	4MHz
1 channel		9.5km	9.7km (140MHz)	7.3km	10km
2 channels	4.4km	7km		3km	8.3km
3 channels		4.4km			
4 channels	1.1km				

Table I shows the results obtained with the two systems in more detail.

Conclusion

We have demonstrated theoretically and experimentally the capabilities of fiberoptic multichannel television transmission by using amplitude or frequency modulation (vestigial side band) of the TV carriers, commercial diode lasers at 0.83 μm and light-emitting diodes at 1.3 μm as light sources and multimode graded-index (50/125 μm) fibers. With the laser system two TV channels were transmitted over a fiber without repeater over 4.4 km with AM at 40 dB video signal-to-noise ratio in the channels (unweighted) and intermodulation distortion better than -65 dB. At the same picture quality, the two channels were transmitted over 7 km by using low cost, small deviation (4 MHz) FM modems. With the 1.3 μm LED system, the two channels were transmitted over 3 km with AM and over 8.3 km with FM. Finally, 4 TV channels were transmitted over 1.1 km using the 0.83 laser system. Our analysis show the influence of light source noise and linearity on the transmission distance for a given signal quality. At the present state-of-the-art of the laser diodes and LED's two high quality TV channels per fiber can be transmitted over several kilometers. The picture quality is expected to increase further as more linear light sources become available.

Acknowledgement

The authors wish to thank the Swiss PTT for supporting the project.

References

[1] H.P.Berger et al., Electr.Lett.17, pp.844-845, 1981.

[2] G.Morgensen, Optical and Quantum Electr.12, pp.353-381, 1980.

[3] H.Lange, Proc.of the 12th Intern.Television Symposium and Techn. Exhibition, Montreux, pp.283-297, 1981.

[4] H.P.Berger et al., Proc.8th Europ.Conf.on Optical Comm., pp.430-434, 1982.

[5] H.Jäckel, Ph.D.Thesis, ETH-Zurich, No.6447, 1980.

[6] R.Welter et al., Proc.8th Europ.Conf.on Optical Comm., pp.365-370, 1982.

[7] R.Bouillie et al., Proc.2nd Europ.Conf.on Optical Comm., pp.135-142, 1976.

[8] L.E.Weaver, IEE Monogr.Ser.9, 32 ff, Peregrinus, 1971.

[9] A.B.Carlson, Communication Systems, McGraw Hill, 1975.

AN EXPERIMENTAL DIGITAL VIDEO SWITCHING ARCHITECTURE

Vlack, D.   Lehman, H. R.

AT&T Bell Laboratories
Naperville, Illinois 60566

SUMMARY

This paper describes some new results related to an experimental video switching system. Three new aspects of such a system are discussed, and experimental results are detailed: a new system architecture (dual-star), a new crosspoint (PSECL), and a new network structure (Richards'). Modulation philosophies are also discussed.

1. INTRODUCTION

A wideband switching system has evolved from earlier studies [1] toward a very high-capability system. This paper discusses several new architectural aspects of this system, and describes results to date. In particular, a new digital video switch capable of conveying encoded conventional video signals and with the potential of being able to switch compressed high-definition video will be described. A new, rearrangeably nonblocking broadcast network will be discussed, as will results of work on a potentially inexpensive video PFM modulation scheme.

2. SIGNIFICANT NEW ARCHITECTURAL ELEMENTS

Three significant new architectural elements have arisen since the last-reported status [2]: the system architecture, the switching element, and the switching fabric form and utilization.

2.1 THE SYSTEM ARCHITECTURE

The system architecture has evolved from a "star" system switching and transporting digital voice/data and analog baseband video over direct fiber loops from the central office to a "dual-star" system utilizing intervening remote nodes and switching and transporting digital voice/data and digital video (see Figure 1). The economics of digital fiber suggest this approach, which provides a better distribution of costs among the elements necessary to serve a group of customers. The dual-star approach also provides a good partitioning of technology, allowing high-performance components to be used sparingly, and allowing low-performance componentry to be used within its limits.

The remote node, growable to a capacity of several hundred subscribers, contains a video switch, and its control resides in the central office. Video programming for broadcast purposes is conveyed at high time division multiplexed digital rates and multiple wavelengths over the feeder fiber for selection by a video switch. Two-way video is switched much like traditional voice. Data and signaling are time multiplexed for high-rate transmission.

Several possible methods for loop distribution transmission are under study. Three fibers, two fibers and one fiber, all employing wavelength-division multiplexing to a greater or lesser degree, offer solutions. Thus, for example, for a two-fiber approach, three downstream channels can be supported on one fiber, using three wavelengths, and two upstream. channels can be conveyed on the other, using two wavelengths. The downstream channels in this case could be comprised of two video signals and a data channel consisting of time division multiplexed voice, data and message signaling, and the upstream channels could consist of a video signal and a data channel.

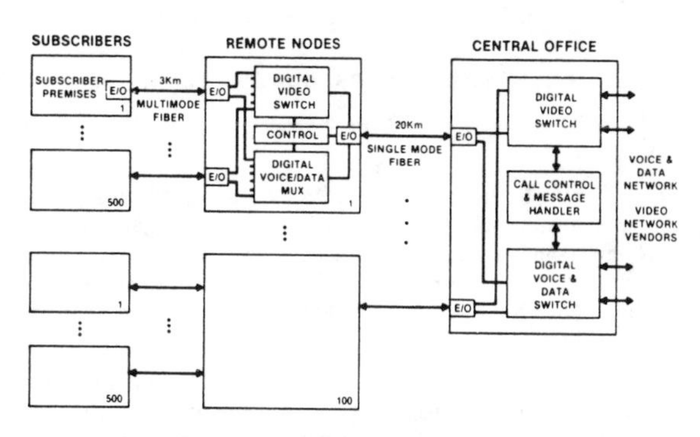

The Star-Star Network

FIGURE 1

2.2 A NEW SWITCHING ELEMENT: PSECL

For over twenty years, there has been one form of logic that has displayed the highest speeds at the lowest impedance level: Current-Mode Logic alias Emitter-Coupled Logic (ECL). In straining for ever higher speed, the two major opponents are parasitic capacitance and parasitic inductance. In the kinds of physical embodiments characterizing typical logic, the former predominates. It is therefore advantageous to utilize a type of logic that switches current rather than voltage, keeping voltage changes at a minimum. Such is the nature of ECL.

Unfortunately, as in many other endeavors, haste makes waste, and very high speed ECL logic is known for being highly dissipative. High dissipation unfortunately limits the density of devices which can be integrated upon a semiconductor chip, the number of chips that can be placed upon a circuit card, and the number of cards that can be placed in a card shelf of given size.

The speed that ECL logic can provide is a virtual necessity for a switch intended to have the capability to convey wideband signals in

Reprinted from *International Switching Symposium*, Florence, Italy, 1984.

pulse form, especially if it is to be sufficiently flexible to be able to accommodate future formats such as high-definition television. The high dissipation militates strongly against ECL usage in the high densities necessary for the required low costs. The situation is therefore one ripe for invention. How can the speed and off-chip drive capability of ECL be attained without the attendant dissipation?

One answer has been found in what has been dubbed Power-Switched Emitter-Coupled Logic (PSECL). Figure 2 depicts the philosophy of this device. High performance ECL utilizes an active current sink at the apex of the emitters of the switched transistors. The constant current drawn by this circuit is responsible for the constant, high dissipation of conventional ECL circuits. The PSECL circuit does not allow current to be drawn constantly in this manner. Rather, it causes a crosspoint to be selected via enabling its current sink, and deselects by cutting off the current sink. Thus, while a selected crosspoint dissipates as heavily as any high-performance ECL gate, an unselected crosspoint dissipates no power at all. Since selected crosspoints are always in the minority in a circuit-switched space division array in practical usage, the *average* dissipation per crosspoint is low.

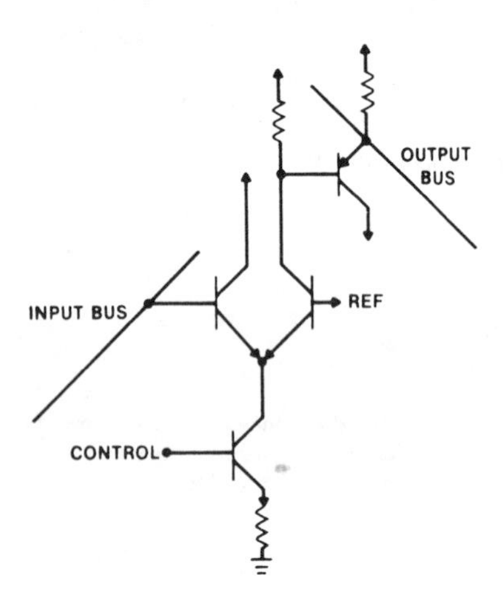

PSECL Crosspoint

FIGURE 2

By way of example, in a square (i.e., $N \times N$) array, N^2 crosspoints would be dissipating full power with conventional ECL gates, but *at most* N crosspoints would be dissipating power if the array were fashioned of PSECL crosspoints. The average dissipation per crosspoint for a PSECL array of such dimensions therefore has an upper bound of 1/N that of an ECL gate array.

It will be recognized that the state of the PSECL gate when its current sink is cut-off is unusual: neither of the upper two transistors is on, and the signals on the two collectors are no longer complements of each other. With the connections shown, however, the bus functions as an OR gate, and an unselected crosspoint does not load the bus since it is placed into a high-impedance state.

2.3 NETWORK PHILOSOPHY

The switching problem faced in this system is substantially different from that found in traditional telephony. Thus, conventional switching arrangements are generally not applicable.

The next section discusses those system aspects of wideband switching that differentiate it from other communication forms. Following that, a new network architecture is described that generally solves all of the problems encountered.

2.3.1 The Nature of the Switching Problem One of the ways that subscribers will utilize a video switching capability will be to access and view video source material for entertainment purposes. This has several implications for switching.

First, many of the video sources will be considered as "broadcast" sources for entertainment video viewing. These sources must always be accessible if their usage is to be consistent with that ordinarily assumed for over-the-air TV channels. Thus, the network that connects the sources to the subscribers must be (1) non-blocking and (2) capable of connecting any source to any or all subscribers simultaneously. Traditional telephone networks provide neither capability. Blocking is generally allowed as an economic consideration and connections are generally made pairwise, between a single inlet terminal and a single outlet terminal.

The need to be able to access a broadcast source at any time is considered a compelling enough reason to require a non-blocking network. However, even if it were not, another consideration would likely make blocking intolerable. Most of the accesses to video sources are for entertainment purposes. Such sessions tend to have much longer holding times than telephone calls. Queuing delays are directly related to server holding times. If blocking were allowed, delays would occur, and their length would be unacceptable.

Many subscribers will want more than one termination on the network (so that more than one video program can be viewed simultaneously on different sets). In addition, many "low runner" video sources will probably be accessible on the network. These two factors increase the size of the network greatly. The fabric of a wideband network tends to be expensive because of the high speeds required and crosstalk problems. Thus, efficient use of crosspoints is essential.

2.3.2 A New Solution A non-blocking broadcast network can be provided in a straightforward fashion via a single-stage rectangular switch. The required number of crosspoints equals the product of the number of video channels (inlets) and subscriber terminations (outlets).

Three-stage Clos networks can be designed with a sufficient number of middle-stage switches to provide the non-blocking broadcast function. If the network is large enough, fewer crosspoints will be required than that needed for a single-stage rectangular switch.

The only known way to reduce the crosspoint count further is to allow rearrangements; i.e., move existing connections when temporary blocking occurs. Rearrangeable networks can be designed for both pairwise connections and broadcast or multi-connections.

A totally new method due to Richards has been applied [3] to provide the non-blocking broadcast function in a rearrangeable network. The method is significantly more efficient in crosspoint usage than any other known published solution. In its simplest manifestation, only two stages are required as opposed to three with other approaches.

Remarkably, the new method requires fewer crosspoints than a single stage switch for a network as small as twelve inlets and five outlets. Of course, its efficiency increases as the network grows larger. Crosspoint count can be reduced even further in large networks by replacing specific individual switches with two-stage networks.

The price paid for crosspoint savings is the need to occasionally rearrange existing connections. However, various steps can be taken to minimize the probability of such events. These include such techniques as judicious inlet assignments, load balancing outlets, and adaptive path hunts. The result is an extremely efficient network that rarely has to be rearranged. A rearrangement occurring in a two stage network having a few hundred inlets would usually only require one connection to be moved, and almost never more than two or three.

In a later section some other aspects and additional details of this new switching invention will be discussed.

2.4 MODULATION METHOD

Conceptually, analog baseband switching represents the simplest way of switching video signals. However, an analog baseband signal has a very low immunity to noise, crosstalk and nonlinear distortions, so it is difficult to build a very large switching array employing an analog baseband technology. A much better performance can be obtained with digital modulation, but at the present time digital codecs are too expensive for universal application.

From the point of view of switching, for signals in AM form, linear amplifiers must be integrated within the switch matrix to enable broadcast fanout power. Digital crosspoints of the form described innately embody fanout capability within each crosspoint, supporting the use of digital or pulse-analog modulation techniques.

2.4.1 DPCM Differential Pulse Code Modulation (DPCM) is perhaps economical for application in the Central Office-Feeder-RN transport and switching due to subscriber sharing of the costs. It is thus a good candidate for use over the longest portion of the transport range.

2.4.2 PFM Pulse time modulation schemes (pulse position, pulse width, and pulse frequency) represent an attractive alternative to the baseband and digital switching for use in the loop and distribution part of the transmission plant. These schemes require relatively inexpensive modulators/demodulators, exhibit high immunity to nonlinear distortions, and can be implemented using simple digital and analog components. Among all pulse time modulation schemes, pulse frequency modulation (PFM) is probably the easiest to implement (since it does not require sampling), and it was therefore studied as a modulation method for wideband switching experiments. PFM has been used in several experimental and commercial fiber optic video transmission systems [4, 5, 6, 7], and might be considered as a contender for local distribution of video signals in the future wideband telecommunication systems. Compatibility between digital loop transmission and digital switching formats in such systems is another possible feature of PFM.

A PFM signal is comprised of a train of fixed-width pulses, the frequency of which is proportional to the modulating baseband signal. As can be seen in Figure 3, the spectrum of a sinusoidal modulated PFM signal contains a component of the modulating frequency f_m and clusters of sidebands centered around the average pulse frequency carrier, f_c and its harmonics [8]. Hence, the baseband signal can be recovered by means of a low-pass filter. A block-diagram of a PFM switched video system is shown in Figure 4. The pulse frequency modulator generates a train of pulses the frequency of which is controlled by the modulating signal. Noise present in the transmission channel (a link between two regeneration points) corrupts the transmitted pulses, so that the output of the channel represents a sum of the original pulse train and noise. Crosspoints in a PFM switch reconstitute the received signal via distributed hard limiting, producing a clean, full amplitude, sharp-edged replica. Noise present during the pulse rise (fall) time at the input of a regenerator causes an error in the position of the regenerated pulse, which translates into noise in the demodulated signal. At the end of the transmission path the regenerated pulse train is applied to a low-pass filter which recovers the original baseband signal.

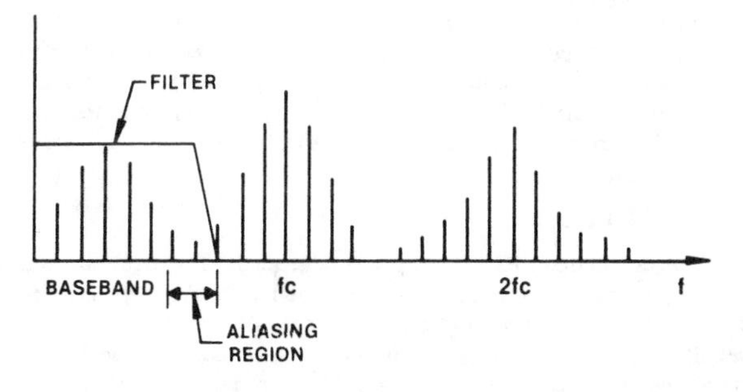

Spectrum of a PFM Signal

FIGURE 3

It should be noted that a PFM signal can be transmitted over a channel not only as a train of fixed width pulses, but also as a fixed duty cycle (50%) square waveform. The square waveform carrier requires smaller channel bandwidth than straight PFM. In the case of a square waveform carrier, the last regenerator not only regenerates the transmitted waveform, but also converts it into a train of fixed-width pulses via a simple digital quadrature detector. The pulses can be generated either for each transition in the square waveform, or only for the negative or positive transitions. The double edge regeneration doubles the carrier frequency and the frequency deviation, and therefore increases SNR, as will be seen shortly. A combination of square waveform and double edge demodulation was used in the experimental PFM system.

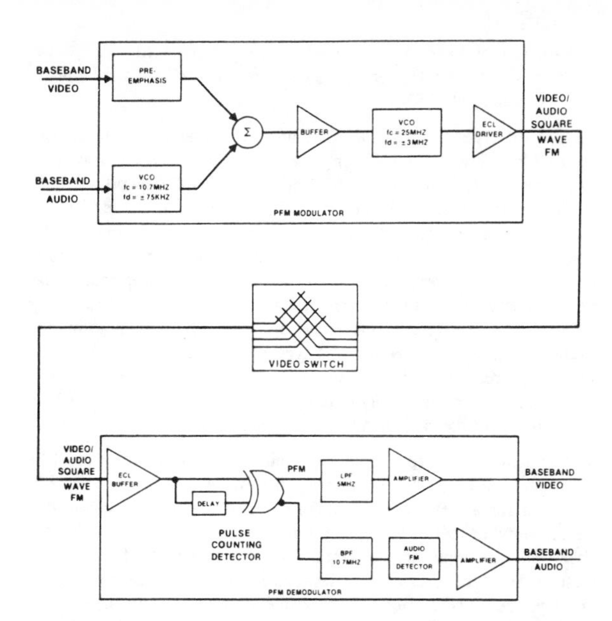

Block Diagram of a PFM System

FIGURE 4

The major source of noise in a switch is crosstalk between channels. Two basic cases of crosstalk coupling can take place in a PFM switch. In the first case, a channel will be subject to interference from several other channels driven by different modulators. This situation is typical for point-to-point connectivity. Experiments demonstrated that the resultant crosstalk in this case has a random noise-like appearance and crosstalk contributions from different sources add as uncorrelated voltages. Signal-to-random noise ratio for video signals is defined as a ratio of the amplitude of the luminance signal to the rms value of noise. For the flat field test signal (which is used for noise measurements in television) the amplitude of the luminance signal $V_S = A\Delta f\tau$, where A = pulse amplitude of the PFM carrier, Δf = the frequency deviation corresponding to the luminance amplitude, and τ = pulse duration. Then SNR (unweighted) can be written as follows

$$SNR = 20 \log \frac{V_S}{V_N} = 10 \log \frac{3}{8\pi^2 f_{cmax} f_m t_R^2} \left(\frac{\Delta f}{f_m}\right)^2 \left(\frac{A^2}{N}\right), \quad (1)$$

where V_N = rms noise voltage, $V_N = \sqrt{N}$, N = total noise power at the input of the regenerator, f_{cmax} = carrier frequency corresponding to the peak white, and t_R = 0% to 100% rise time defined for a trapezoidal representation of a real pulse waveform [9]. Experiments with an ECL-based PFM system demonstrated that expression (1) actually gives values that are 1 to 3 dB better than measured SNR.

In analyzing expression (1) it should be remembered that depending on crosstalk coupling conditions, N is a function of the rise time. For example, crosstalk power resulting from interference between two short conductors may vary as $1/t_R$ or $1/t_R^2$. On the other hand, crosstalk between two long conductors (for which round trip delay is greater than the pulse rise time) is virtually independent of the rise time [10]. It follows from this discussion that the most effective way of improving SNR is to increase the frequency deviation. The maximum frequency deviation is limited by the linear region of the "frequency vs. voltage" characteristic of the PFM modulator and by the upper cut-off frequency of the channel. Finally, the carrier frequency should be kept as low as possible. The lower limit on f_{cmax} is determined by the aliasing distortion, which occurs when the sidebands of the pulse repetition frequency penetrate into the baseband portion of the recovered signal (see Figure 3).

Additional improvement in random signal-to-noise ratio in PFM can be obtained by adding a pre-emphasis network at the input of the modulator and a corresponding de-emphasis network after the low-pass filter. For example, the CCIR pre-emphasis network can be used for this purpose [11]. A combination of standard video noise weighting and the CCIR network gives approximately a 14 dB advantage over unweighted SNR [12].

The second type of crosstalk coupling, correlated crosstalk, occurs when all interfering channels carry the same signal. This case is typical for broadcast connectivity. SNR in this case will depend not only on the relative magnitude of crosstalk, but also on the difference between the carrier frequencies of the affected and interfering signals. Since correlated crosstalk has a more defined appearance than random crosstalk, the peak value seems to be a more appropriate measure for such crosstalk than the rms value. A detailed analytical study of correlated crosstalk is presently underway, but the preliminary results indicate that SNR for this type of crosstalk can be approximated by expression (1) if rms noise values are replaced by peak values and an experimentally determined correction factor of 4 dB is subtracted from the right-hand side of the equation.

As was mentioned before, one of the most attractive features of PFM is its high immunity to nonlinear distortions. This is especially important for a PFM switch. Major sources of nonlinear distortions are attributable to the analog portion of a PFM system (modulator, fiber optic-preamplifier interfaces). In the switch itself the major cause of these distortions is reflections in conductors carrying high-speed pulses. The reflections occur in long transmission lines if the termination is not matched to the line's characteristic impedance. For a fixed propagation delay in a long line, the result of superposition of the incident and reflected waveforms depends on the frequency of a PFM signal, and therefore will affect linearity of the demodulated signal.

3. RESULTS TO DATE

Results obtained in incorporating the new concepts, techniques and devices into a wideband laboratory system have been been very favorable.

3.1 100k ECL Prototype Video Switch

3.1.1 Requirements Requirements were established for the switch such that it would be able to handle coding schemes and transmission rates expected in the future without a redesign. It was to be used for pulse analog encoding schemes at approximately 25 MHz rates, later for PCM-type encoding schemes at rates from 45 to 100 Mbps, and even later for compressed PCM-type encoding of high definition television signals at bit rates approaching 300 Mbps. Other requirements were that the switch must be capable of broadcast switching, it must be non-blocking, use only off-the-shelf parts, and use only convection cooling. Lastly, concerning signal degradation, the video switch should not add gain distortion, and should have an unweighted signal to noise ratio greater than 60 dB.

A 24X32 switch array based on 8x8 modules was implemented. Three different types of boards make up the switch array: (1) a Receiver/Driver board which terminates up to eight inputs and fans the signals out to four pairs of switch boards; (2) a First Stage Switch board which is a sparse array comprising the first stage of the two-stage rearrangeably non-blocking array described in this paper, and has 24 inputs and 16 outputs, and; (3) a Second Stage Switch board which completes the two-stage design, and has 16 inputs and 8 outputs. Provision was made for either coaxial cable or fiber optic inputs and outputs for flexibility in interfacing.

Each stage of the switch array is implemented using the partitioning architecture principles presented in [1]. The basic architecture is shown in Figure 5, which illustrates how the output signal of a stage is isolated from unused input signals by two gates. Thus extra gates (crosspoints) are used to achieve greater crosstalk isolation.

Small distributed microprocessors, linked to a larger central processor, control the switch crosspoints. An 8-bit microprocessor controls the crosspoints on each board, and all microprocessors receive commands over serial links from a 16-bit microprocessor-based switch controller board. This control board has a local terminal port, and a data link port for communication with the call processor. The switch control interfaces with the outside world, does path hunts through the switch network, and instructs the switch board controllers to set or clear particular crosspoints. It is also responsible for maintenance functions such as testing the switch network paths and diagnosing errors.

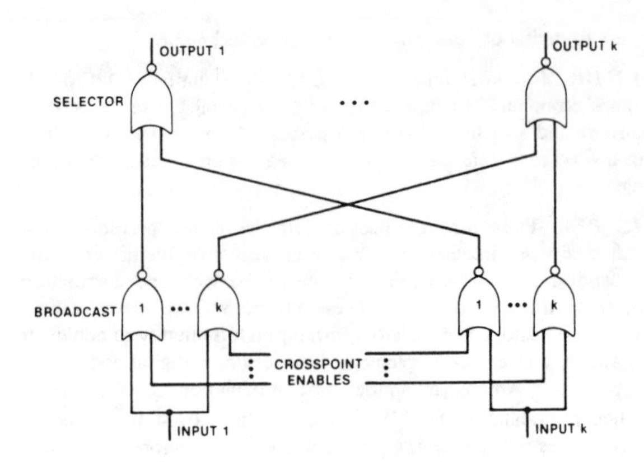

Wideband Structure for Extended Crosstalk Performance k x k 2-Tier Broadcast/Selector Arrangement

FIGURE 5

3.1.2 Performance The ECL switch can switch signal rates from DC to over 150 Mbps. The RMS signal to crosstalk ratio of the baseband signal recovered after traversing four stages of the switch in preemphasized PFM form was measured and exceeded the unweighted signal to noise specifications, and differential phase and gain were unaffected by the switch. The switch does not measurably degrade PCM or DPCM signals, whose performance is therefore limited only by quantizing noise.

Rearrangement adds an interesting problem to signal integrity. If one path is disconnected and then another is established the signal is absent for the time taken to reconnect the path. The switch could be designed to change a selected set of paths all at once which would cause at most, a glitch as wide as the latch gate delay variation. As a solution, consider the following: an alternate path could be set up in parallel with the one being reconfigured, and then the old one could be torn down (make before break). The present design ORs the last stage signals together, so the resultant signal is identical to the original. Then the first path is torn down and no glitches occur at all.

There is one problem, however. The different paths through the switch traverse different gates (with statistically different delays) and are of different lengths. Thus when ORed together, the resultant signal is altered. For PFM, a negligible phase shift due to the demodulation method occurs during the few hundred microseconds it takes to change a path. The amount of maximum delay is constant, so a maximum bit rate can be calculated for PCM type coding schemes such that no bit errors will occur. For this switch the maximum rate is about 150 Mbps.

3.2 PSECL Chip

An integrated circuit PSECL chip was fabricated. The array size was four inputs by four outputs (16 crosspoints), and the chip included I^2L control logic for storing crosspoint data. Control data is input serially with a data clock, and is double buffered. Sixteen bits of data, one bit for each crosspoint, are shifted into the input shift register and are latched in parallel, all at once, with a Load signal. The array and control is shown in Figure 6. The chip dissipates less than 250 mW with four crosspoints on, but the crosspoints are implemented with low power transistors and are only capable of 100 Mbps.

The circuit was also modeled with a computer simulation program to compare predicted and observed performance. The results of these methods agreed very closely. The maximum bit rate for one input signal to all four outputs (the worst case) is approximately 100 Mbps, and the worst case crosstalk for a recovered baseband signal was measured at >59 dB below the signal, the limit of the measurement apparatus. (It should be emphasized that this is for PFM signals.)

3.3 SOME DETAILS OF THE BROADCAST NETWORK

Past designs of non-blocking broadcast networks were primarily based on three-stage Clos networks, and were concerned with determining the minimum number of middle-stage switches required. However, the original Clos network was conceived as a non-blocking approach for a pairwise connection network; i.e., one in which any inlet is connected to at most one outlet, and vice-versa. In such a network, no connection state ever requires an inlet to have more than one appearance on the middle-stage switches.

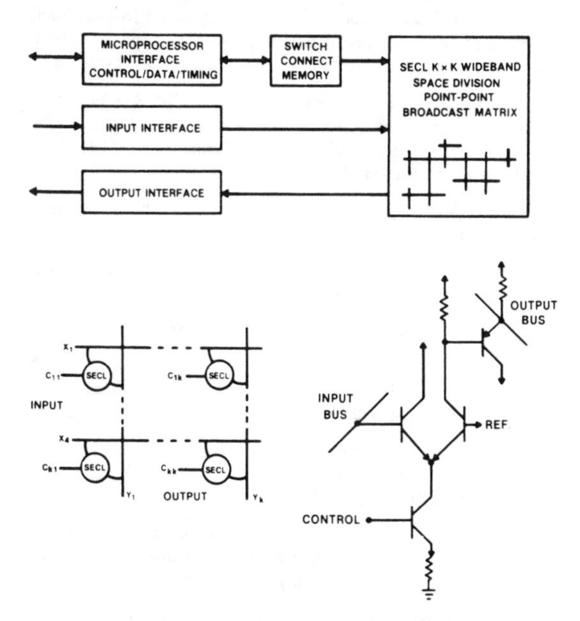

Experimental 4 x 4 PSECL Switch LSI Chip

FIGURE 6

In a broadcast network, an inlet may be connected to a number of outlets simultaneously, which might require the inlet to have multiple appearances on second-stage switches. This is the reason why more crosspoints are required to make a broadcast network non-blocking.

This difference, between pairwise and broadcast connection networks, has been used advantageously in the design of a more efficient network [3]. The method is based on a heretofore unaddressed generalization of Hall's theorem of a system of distinct representatives [13]. A set of switches is partitioned into overlapping subsets. The method of partitioning determines the value of a parameter k, such that any k subsets have k distinct representatives. (Hall's theorem deals with the case that k equals the cardinality of the given family of subsets.)

The result is a two-stage rearrangeable broadcast network. Design parameters may be chosen to minimize the required number of crosspoints for any number of inlets and outlets. In larger networks, specific individual switches may be replaced with two stage networks to further reduce the number of crosspoints. Figure 7 illustrates the required number of crosspoints per outlet as a function of the number of inlets for various broadcast networks. Two of the curves on the figure correspond to two- and three-stage networks as described in this paper. For comparison, two additional curves are included which show results from Masson and Jordon [14]. Both of these curves are for three-stage networks (the minimum configuration of their approach) and reflect values of 2 and 6 for the ratio of the number of outlets to the number of inlets. (The crosspoint count per outlet is essentially independent of the number of outlets with the new method.) The Masson and Jordon method was the most efficient known. The new approach is clearly superior. Both two- and three-stage networks require fewer crosspoints than Masson and Jordon.

Path hunts and rearrangement algorithms for this network are straightforward. The probability of rearrangements is expected to be small and can be made smaller through various design techniques. For example, a case was considered where no particular action or strategy was directed at reducing rearrangements. The network was provided with 169 inlet channels and 70% of the outlets were assumed to be busy. The probability of a rearrangement was estimated to be approximately 0.006.

It should also be noted that it is always possible to perform a "make before break" on a connection which must be moved; i.e., the inlet can be momentarily simultaneously accessed from both the old and new switches during the process of moving the connection. This is important in applications that cannot tolerate a loss of information resulting from a momentary open connection.

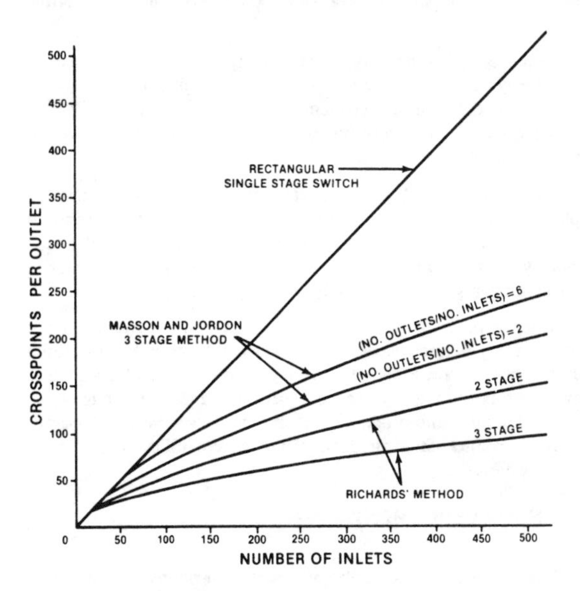

Crosspoint Count Comparisons of Rearrangeable Broadcast Networks and Rectuangular Single Stage Switch

FIGURE 7

The process of growing the number of inlets and/or outlets on a network is similar but not identical to the initial design. One reason is the desire to leave the existing configuration intact, so as to minimize labor and disruption of service. Another desire is to minimize the amount of unused fabric that results from anticipation of growth.

The broadcast network described here grows in accordance with these objectives. Inlets and outlets may be added without having to reconfigure existing interconnections or replace existing switches. Module sizes can be chosen to minimize unused fabric and, when these modules reach their capacities, it is possible to add stages or replica networks to continue the growth without disturbing the existing network. These are very important practical considerations. Networks without these attributes may be severely limited in their applications.

Since pairwise connections are special cases of broadcast connections, they can obviously be implemented on a broadcast network. However, this is only of practical interest if the crosspoint cost of the broadcast network is competitive with the cost of a pairwise connection network. A three-stage rearrangeable network, as described here, generally requires fewer crosspoints than a strictly non-blocking Clos pairwise connection network. Thus, it would be immediately competitive, except for the need to rearrange. However, in this application, the rearrangement process has essentially already been paid for. Furthermore, this network, like many others, exhibits a crosspoint cost function whose second derivative, with respect to the number of inlets, is negative. This means that the incremental cost of inlets continually decreases as the number of inlets increases. Thus, it is likely that the cheapest way to provide pairwise connection terminals is to simply add them to the existing broadcast network.

3.5 NETWORK ARCHITECTURE

The overall impact of the above-described new techniques upon the video switch network architecture is significant.

The use of digital crosspoints fabricated as logic gates frees the designer from concern for the network insertion loss. Since the crosspoints all return the signal to a standard level, the path length (measured in crosspoints traversed) is of no consequence relative to amplitude. The function of buffering and broadcast fanout is distributed amongst the crosspoints so that a separate facility for this purpose is unnecessary.

The high speed of the crosspoint control is conducive to the adoption of a rearrangeably nonblocking philosophy which saves many crosspoints in comparison to a network nonblocking in the strict sense.

The unclocked, pulse-responsive nature of the switch fabric makes it usable for conveying a variety of modulation scripts, including pulse frequency modulation for economical implementation, DPCM for bandwidth-reduced transmission of video, PCM, and a bandwidth-reduced form of High-Definition video.

The network fabric may also be adapted for a variety of transmission applications as a clear-channel "pipe." For example, it can serve in applications such as the transmission of high-definition digitized radiology images from a hospital to a centralized diagnostic center at virtually arbitrary rates (effectively DC to 300 Mbps).

Given such a potent network, a system can be built around it that can serve a variety of user needs. With a wideband transmission medium such as optical fiber, a digital signalling means, a control designed to grow in proportion to the network size, and means for interfacing with vendor wideband programming offerings, a system with capabilities by far transcending the goals of the current ISDN specifications may be achieved.

4. ACKNOWLEDGEMENTS

The authors thank their colleagues who contributed to the work reported in this paper. Special acknowledgement is extended to B. E. Briley, A. I. Drukarev, J. A. Hiltner, and G. W. Richards who, in addition, assisted in the preparation of the paper. The authors would also like to acknowledge the work done on the PSECL chip by R. L. Pritchett and P. C. Metz.

5. REFERENCES

[1] R. E. Cardwell and H. R. Lehman, "Experimental Wideband Switching System Capability," *Proceedings of the 10th International Switching Symposium*, Montreal, 1981.

[2] Ibid.

[3] G. W. Richards and F. K. Hwang, "A Two-Stage Rearrangeable Broadcast Switching Network," To be published.

[4] D. J. Brace, and D. J. Heatley, "The Application of Pulse Modulation Schemes for Wideband Distribution to Customers Premises", Sixth European Conference on Optical Communication, University of York, UK, 16-19 September, 1980.

[5] T. Kanada, K. Hakoda, and E. Yoneda, "SNR Fluctuation and Nonlinear Distortion in PFM Optical NTSC Video Transmission Systems," *IEEE Trans. Commun.*, vol. COM-30, pp.1868-1875, August 1982.

[6] "The WAVELINK Fiber Optic Communications System," *Data Sheet*, The Grass Valley Group, Inc., 1982.

[7] "Model FS 100/FS 101 High Quality Fiber Optic Video Systems," *Data Sheet*, Optical Information Systems, Inc., 1982.

[8] P. F. Panter, *Modulation, Noise, and Spectral Analysis*, McGraw-Hill, 1965.

[9] A. I. Drukarev, "Noise Performance and SNR Threshold in PFM," To be published.

[10] Bell Telephone Laboratories Staff, *Physical Design of Electronic Systems*, vol.1, *Design Technology*, Prentice Hall, Inc., Englewood cliffs, N.J., 1970.

[11] "Pre-emphasis Characteristics for Frequency Modulation Radio-Relay Systems for Television", Recommendation 405-1, CCIR XIV-th Plenary Assembly, vol. IX, Kyoto, 1978.

[12] "Signal-to-Noise Ratio in Television", Report 637-1, CCIR XIV-th Plenary Assembly, vol. XII, Kyoto, 1978.

[13] P. Hall, "On Representatives of Subsets," *J. London Math. Soc., 10, 1935, pp. 26-30.*

[14] G. M. Masson, "Upper Bounds on Fanout in Connection Networks," *IEEE Transactions on Circuit Theory*, 20, 1973, pp. 222-229.

Part II
Direct Broadcast Television from Satellites

BROADCASTING television directly from satellites to homes, apartments, or businesses is a very attractive service for areas where over-the-air VHF/UHF reception is either poor or lacking in variety or where cable distribution facilities are absent. In fact, even today there are thousands of privately owned earth stations receiving television from low power satellites on an informal basis, i.e., the owner of the satellite does not necessarily know of or approve of the receptions. Thus, it is not surprising that many government and private entities are planning direct broadcast satellites (DBS) for television.

A frequency band for DBS was allocated by the World Administrative Radio Conference near 12 GHz. Furthermore, channels having bandwidth 27 MHz have been assigned to the various countries which plan to offer DBS. Frequency modulation of the radio frequency carrier will be used to overcome transmission noise.

As of this writing, most services being proposed are still in a state of rapid flux. Thus, it is not possible to give a lasting description of the systems which are likely to evolve. However, the technical questions are fairly apparent, even though they are not resolved. For example, some proposals are for high satellite power, e.g., 200 W/channel, in which case the number of channels per satellite is limited. Others suggest a medium power satellite with more channels, but with a more limited service area on the ground.

Some systems will transmit standard composite television signals. Others plan to send separate color components. The leading contender for component transmission is the time compression multiplexing method [1], called "MAC" by the Independent Broadcasting Authority of the United Kingdom. Encryption is also desired by any provider who plans to charge a fee for service. Although earth stations would benefit if all carriers used the same method, no such standard encryption scheme seems to be in the offing. Many other details also remain to be worked out.

The first four of the papers in this part describe the technical aspects of various systems proposed for different countries. Other descriptions may be found from papers listed in the following bibliography. The remaining two papers of this part describe earth station designs, a topic of crucial importance to the technical and economic viability of DBS television distribution.

BARRY G. HASKELL
Associate Editor

BIBLIOGRAPHY

[1] T. S. Robson, "A compatible high fidelity TV standard for satellite broadcasting," pp. 181-199, this book.

[2] J. Georgy, "The French broadcasting satellite system TDF1," in *Proc. ICC84*, Amsterdam, The Netherlands, May 14-17, 1984, pp. 1084-1087.

[3] O. S. Roscoe, "Technical and economic models of a DBS system for Canada," in *Proc. ICC84,* Amsterdam, The Netherlands, May 14-17, 1984, pp. 1088-1091.

[4] Y. Nakamwra, "The operational broadcasting satellite system for Japan," in *Proc. ICC84,* Amsterdam, The Netherlands, May 14-17, 1984, pp. 1094-1097.

[5] J. L. Tejerina, "The Spanish TV satellite project," in *Proc. ICC84*, Amsterdam, The Netherlands, May 14-17, 1984, pp. 1107-1110.

[6] O. S. Roscoe, "Satellite broadcasting in Canada," *SMPTE J.*, vol. 91, pp. 1142-1147, Dec. 1982.

[7] A. Lafferty, "Using the wider TV channel bandwidth of DBS," *Commun. Int.*, vol. 9, pp. 30-31, Dec. 1982.

[8] E. E. Reinhart, "Three is not enough: Why the US specified four DBS service areas," *Satellite Commun.*, vol. 6, pp. 28-39, Dec. 1982.

[9] N. L. Cohen and G. R. Stone, "DBS platforms: A viable solution," *Satellite Commun.*, vol. 6, pp. 22-27, Dec. 1982.

[10] W. P. Marberg, "The problem of interference in direct broadcast satellite systems," in *Proc. Comput. Network. Symp.*, Gathersburg, MD, Dec. 10, 1982, pp. 18-27.

[11] J. F. Clark, "Proposed U.S. broadcasting-satellite systems," in *Proc. Globecom '82, IEEE Telecommun. Conf.*, vol. 2, Miami, FL, Nov. 29-Dec. 2, 1982, pp. 845-848.

[12] E. E. Reinhart, "National service requirements—Planning methods and system parameters for the 1983 Broadcasting-Satellite Planning Conference," in *Proc. Globecom '82, IEEE Global Telecommun. Conf.*, vol. 2, Miami, FL, Nov. 29-Dec. 2, 1982, pp. 829-834.

[13] "Direct broadcast satellite services for the United States," in *Proc. Nat. Telesyst. Conf., NTC '82, Systems for the Eighties,* Galveston, TX, Nov. 7-10, 1982.

[14] R. Miller and R. F. Buntschuh, "RCA direct broadcast satellite configuration," in *Proc. Nat. Telesyst. Conf., NTC '82*, Galveston, TX, Nov. 7-10, 1982.

[15] V. Bhaskaran, L. I. Bluestein, and M. A. Davidov, "Use of threshold extension in DBS system design," in *Proc. Nat. Telesyst. Conf., NTC '82*, Galveston, TX, Nov. 7-10, 1982.

[16] D. W. Maki, T. A. Milford, and M. A. Hreha, "TVRO design considerations," in *Proc. Nat. Telesyst. Conf., NTC '82*, Galveston, TX, Nov. 7-10, 1982.

[17] B. C. Fisher, "DBS regulatory policies: The FCC gives the green light," in *Proc. Nat. Telesyst. Conf., NTC '82*, Galveston, TX, Nov. 7-10, 1982.

[18] D. G. Thorpe, "Direct broadcast satellites the Canadian experience," *NTG-Fachber*, vol. 81, pp. 260-262, Oct. 1982.

[19] H. Khakzar, "Critical observation of determined radio-satellite system values at the WARC 1977 conference," *NTG-Fachber*, vol. 81, pp. 31-38, Oct. 1982.

[20] C. B. Wooster, "Direct broadcasting by satellite," *Commun. Broadcast*, vol. 8, pp. 27-34, Sept. 1982.

[21] P. A. Ratliff and A. Oliphant, "Consideration of improved quality sound and vision for satellite broadcast services in the UK," in *Proc. Int. Broadcast. Conv.*, Brighton, England, Sept. 18-21, 1982, pp. 174-180.

[22] M. D. Windram and D. K. W. Hopkins, "Satellite broadcasting-modulation methods and the FM channel," in *Proc. Int. Broadcast. Conv.*, Brighton, England, Sept. 18-21, 1982, pp. 169-173.

[23] D. Fashold, L. Heichele, M. Lieke, H. Pecher, and A. Ruthlein, "Transmitting antenna of the DBS-satellite TV-Sat," *Nachrichtentech. Z.*, vol. 35, pp. 592-595, Sept. 1982.

[24] J. Hutchon, "Corporation argues with authority where wider bandwidths don't narrow choices," *Int. Broadcast. Syst. Oper.*, vol. 5, June 16-18, 1982.

[25] P. Powell, "DBS transmission standards for the UK-proposals from the IBA and the BBC," *Int. Broadcast Eng.*, vol. 13, July 24-25, 1982.

[26] H. Krath, "Programmes from space: Details of TV-Sat and TDF-1," *Funkschau*, pp. 46-49, Sept. 17, 1982.

[27] T. B. McCrirrick, "The extended PAL system for DBS," *Electron Power*, vol. 28, pp. 581-583, Sept. 1982.

[28] T. S. Robson, "Why IBA says Mac for Europe," *Electron Power*, vol. 28, pp. 578-580, Sept. 1982.

[29] O. S. Roscoe, "Satellite broadcasting in Canada," *Bksts. J.*, vol. 64, pp. 414-419, Aug. 1982.

[30] P. Hawker, "Satellite broadcasting using MAC signals," *Middle East Electron.*, vol. 5, pp. 28-30, July-Aug. 1982.

[31] "A 450-W output multiplexer for direct broadcasting satellites," *IEEE Trans. Microwave Theory Tech.*, vol. MTT-30, pp. 1317-1323, Sept. 1982.

[32] K. J. Johnson, "A 12 GHZ downconverter for direct broadcast satellite ground station," in *Proc. IEEE Int. Conf. Commun., The Digital Revolution*, vol. 1, Philadelphia, PA, June 13-17, 1982.

[33] D. H. Miller and P. G. Ackerman, "A one kilowatt class direct broadcast satellite," in *Proc. IEEE Int. Conf. Commun.*, vol. 1, Philadelphia, PA, June 13-17, 1982.

[34] R. G. Gould, R. G. Hupe, and E. E. Reinhart, "Domestic broadcasting-satellite systems: The need for a common standard and the case for block allotment planning," in *Proc. IEEE Int. Conf. Commun.*, June 13-17, 1982.

[35] C. Kermarrec, P. Harrop, C. Tsironis, and J. Faguet, "Monolithic circuits for 12 GHZ direct broadcasting satellite reception," presented at IEEE 1982 Conf. Microwave and Millimeter-Wave Monolithic Circuits, Dallas, TX, June 18, 1982.

[36] D. Goodman, "Video 1990," *Radio Electron.*, vol. 53, pp. 77-80, Jan. 1982.

[37] "TV satellite broadcasting, An introduction," *Radio Electron. World*, vol. 1, pp. 69-71, Mar. 1982.

[38] D. L. Durand, "Direct broadcasting satellite system: Home equipment terminal characteristics," *Comsat Tech. Rev.*, vol. 11, pp. 241-247, Fall 1981.

[39] J. E. Withworth, "Direct broadcast satellite system: DBS/FS frequency sharing," *Comsat Tech. Rev.*, vol. 11, pp. 255-265, Fall 1981.

[40] D. L. Durand, "Direct broadcast satellite system: UP-link and ground control facilities," *Comsat Tech. Rev.*, vol. 11, pp. 248-254, Fall 1981.

[41] E. R. Martin, "Direct broadcasting satellite system: Satellite characteristics," *Comsat Tech. Rev.*, vol. 11, pp. 227-240, Fall 1981.

[42] E. E. Reinhart, "Direct broadcasting satellite system: The satellite television corporation (STC)," *Comsat Tech. Rev.*, vol. 11, pp. 202-214, Fall 1981.

[43] L. M. Keane, "A direct broadcast satellite system for the United States," *Comsat Tech. Rev.*, vol. 11, pp. 195-201, Fall 1981.

[44] C. Derieux, "TDF-1 French broadcasting satellite," *Aeronaut Astronaut*, no. 91, pp. 31-37, 1981.

[45] E. R. Martin and J. E. Whitworth, "Systems and technology aspects of a direct broadcast satellite service for the United States," in *Proc. IEEE 1981 Nat. Telecommun. Conf.*, vol. 1, New Orleans, LA, Nov. 29-Dec. 3, 1981.

[46] M. R. Freeling and L. Schiff, "Technical standards for direct broadcast satellite systems," *RCA Rev.*, vol. 42, pp. 408-423, Sept. 1981.

[47] S. E. Whythe and L. W. Steines, "European satellites—Evolution and potential," *Commun. Int.*, vol. 8, pp. 40, 43-44, 47, Nov. 1981.

[48] J. F. Arnaud, "The Franco-German direct broadcasting satellite project," in *Proc. Int. Conf.*, Liege, Belgium, Nov. 24-26, 1980, pp. 173-180.

[49] U. Gartelius, "Alternative modulation methods for the Nordsat project," in *Proc. Int. Conf.*, Liege, Belgium, Nov. 24-26, 1980, pp. 185-190.

[50] K. G. Freeman, "Direct broadcast satellite receivers," in *Proc. Clerk Maxwell Commemorative Conf., Radio Receivers and Associated Systems*, Leeds, England, July 7-9, 1981, pp. 247-256.

[51] B. F. Fabis, "Some aspects of the German TV-Sat-system," *ACTA Astron.*, vol. 8, pp. 775-783, July 1981.

[52] F. Forde Plude, "A direct broadcast satellite Delphi study: What do the experts predict?" *Satellite Commun.*, vol. 5, pp. 32-33, 35-36.

[53] W. L. Pritchard and C. A. Kase, "Getting set for direct-broadcast satellites," *IEEE Spectrum*, vol. 18, pp. 22-28, Aug. 1981.

[54] V. Sardella and K. Degnan, "Satellite broadcasting to homes: Service options, policy issues and costs," *Telecommun. Policy*, vol. 5, pp. 84-101, June 1981.

[55] D. Moralee, "The DBS dilemma," *Electron Power*, vol. 27, pp. 447-451, June 1981.

[56] J. W. B. Day, N. G. Davies, and R. J. Douville, "The applications of lower power satellite for direct television broadcasting," *ACTA Astron.*, vol. 7, pp. 1417-1431, Dec. 1980.

[57] J. G. Chambers, "An evolutionary approach to the introduction of direct broadcasting satellite service," presented at NTC'80, IEEE 1980 Nat. Telecommun. Conf., Houston, TX, Nov. 30-Dec. 4, 1980.

[58] C. A. Billowes, P. G. Bowers, and E. G. Rose, "Low power direct broadcasting satellite trials in Canada," *Bksts. J.*, vol. 63, pp. 128-130, Feb. 1981.

[59] L. G. Ludwig, "Satellite system for direct broadcast of television," in *Eascon '80 Rec., IEEE Electron. Aerospace Syst. Conv.*, Arlington, VA, Sept. 29-Oct. 1, 1980, pp. 103-108.

[60] "Receivers evolving for TV-by-satellite," *MSN Microwave Syst. News*, vol. 10, pp. 82-92, Sept. 1980.

[61] B. Salkeld, "OTS-broadcasters experiments," in *Proc. Colloquium Satellite Broadcasting*, London, England, Nov. 20, 1980.

[62] C. A. Billowes, P. G. Bowers, and E. G. Rose, "Low-power direct broadcasting satellite trials in Canada," in *Proc. Int. Broadcast. Conv.*, Brighton, England, Sept. 20-23, 1980, pp. 287-290.

[63] O. S. Roscoe, "Direct broadcast satellites—The Canadian experience," *Satellite Commun.*, vol. 4, pp. 22-23, 27, 29-32, Aug. 1980.

[64] A. Kinpara, "Rainfall attenuation characteristics of SHF waves for satellite broadcasting," *J. Inst. Telev. Eng. Jap.*, vol. 33, pp. 120-125, Feb. 1979.

[65] C. Dosch, "Propagation measurements in Munich using the 11.6 GHZ Beacomn of the Sirio satellite," *Alta Freq.*, vol. 48, pp. 363-367, June 1979.

[66] I. P. Shkarofsky and H. S. Moody, "Performance characteristics of antennas for direct broadcasting satellite systems including effects of rain depolarization," *RCA Rev.*, vol. 37, pp. 279-319, Sept. 1976.

[67] T. B. McCrirrick, "Extended Pal system for DBS," *Electron Power*, vol. 28, pp. 581-583, Sept. 1982.

[68] V. Sardella and K. Degnan, "Satellite broadcasting to homes," *Telecommun. Policy*, vol. 5, pp. 84-101, June 1981.

[69] J. F. Clark, "Direct broadcasting satellite: Black hole or bonanza?" *RCA Eng.*, vol. 27, pp. 82-88, July-Aug. 1982.

[70] K. G. Freeman, "Direct broadcast satellite receivers," *Radio Electron. Eng.*, vol. 52, pp. 127-133, Mar. 1982.

[71] J. E. Whitworth, "DBS/FS frequency sharing," *Comsat Tech. Rev.*, vol. 11, pp. 255-265, Fall 1981.

[72] L. M. Keane, "Direct broadcast satellite system for the United States," *Comsat Tech. Rev.*, vol. 11, pp. 195-201, Fall 1981.

[73] C. B. Wooster, "Engineering satellite ground stations," *Electron Power*, vol. 27, pp. 618-622, Sept. 1981.

[74] G. Bjontegaard and T. Pettersen, "Shaped offset dual reflector antenna with high gain and low sidelobe levels," presented at Int. Conf. Antennas and Propagat., Heslington, York, England, Apr. 13-16, 1981, IEEE Conf. Pub. 195, pp. 163-167.

[75] R. R. Bowen, C. A. Billowes, and R. J. Douville, "Development of a Canadian broadcasting satellite system at 12 GHz," in *Conf. Rec. Int. Conf. Commun., ICC '80*, vol. 3, Seattle, WA, June 8-12, 1980, pp. 51.3.1-5.

[76] F. Mancrief, "Japan strives for communications leadership," *Microwaves*, vol. 19, pp. 42-58, Jan. 1980.

DIRECT BROADCAST SATELLITE SERVICES FOR THE UNITED STATES

Nicholas J. Marzella*

Satellite Television Corporation
Washington, D.C.

ABSTRACT

This paper describes the major components and considerations relating to direct broadcasting satellite (DBS) services within the United States. Included is a discussion of the satellite and earth receiving technology which form the basis of a DBS system. Specific reference is made to Satellite Television Corporation's (STC's) DBS system which is presently planned as a subscription television service to be implemented beginning in 1986.

INTRODUCTION

On December 17, 1980, Satellite Television Corporation (STC) applied for authorization from the U.S. Federal Communications Commission (FCC) to provide the first direct broadcast satellite service in the United States.

In its application to the Federal Communications Commission, STC requested authority to implement a direct broadcasting satellite system for provision of service to an area approximating the Eastern Time Zone of the United States as a first phase in its ultimate plans to provide nationwide service; this paper provides a description of the complete STC DBS system. At the time of this writing, the FCC was still considering STC's application and STC was in the process of choosing a satellite contractor.

Over the past decade significant progress has been made in demonstrating the feasibility of broadcasting television programs by satellite to receiving terminals whose simplicity and cost make their use practical at individual residences. Such reception has been demonstrated with the joint U.S./Candian experimental Communications Technology Satellite (CTS) and Japan's Broadcasting Satellite for Experimental Purposes (BSE).

Direct broadcasting satellites are dissimilar from the majority of presently operating satellites in one major respect. Whereas, so-called fixed satellite systems have only a modest number of ground terminals (typically less than 100) direct broadcasting satellite systems will have ground terminals numbering in the millions. A direct consequence of this discrepancy in ground terminal population is a shift in economic burden from the earth segment to the space segment. That is, a fixed satellite system will place complexity and cost in the earth segment while the direct broadcasting satellite system will place complexity and cost in the space segment. This economic trade-off which seeks to minimize the ground segment cost of the direct broacasting satellite system drives satellite channel output powers to on the order of 200 watts for a U.S. time-zone sized coverage area while a fixed satellite's output per channel is well under 10 watts for a similar sized coverage area.

Thus, a satellite with a fixed power capability which could carry 20 fixed service channels would only be able to carry only a single direct broadcast channel.

PLANNED SERVICE AND SYSTEM DESCRIPTION

STC's broadcast service will provide three channels of television throughout the contiguous United States (CONUS) and to the major populated areas of Alaska and Hawaii. As currently conceived, STC's television programming will include movies, plays, concerts, nightclub acts, opera, dance, sports and public affairs and educational programs. Certain broadcast materials will be targeted at narrow segments of the public which are not now adequately served; pay-series and pay per-program attractions catering to small but intense public demand are also expected to be offered.

*Manager, System Analysis

Reprinted from *The National Telesystems Conference*, 1982, pp. A 3.5.1-A 3.5.6.

A combination of system economics, maturity of technology, system performance criteria and viewer preference has lead STC and others to configure systems based on 4 satellites serving 4 service areas each of which is approximately time-zone sized. Thus, a DBS system will be able to match the specific scheduling preferences of the entire U.S. developed over the past 40 or so years of terrestrial television transmission. Such service areas sizes coupled with the use of 200 watt traveling wave tube amplifiers (TWTAs) will be capable of providing an equivalent isotropically radiated power (EIRP) of approximately 57 dBW at the edge of coverage.

Typical video link performance objectives for DBS systems call for a carrier-to-noise (C/N) ratio of 14 dB and a signal-to-noise (S/N) ratio (weighted) of 42 dB for all but 0.25 percent of the average year. Satellite EIRP requirements, then, depend on the rain climatic zone in which a home receiving terminal is located. In addition, threshold performance is typically specified in order that below threshold operation is limited to, on the order of, 4 hours per year (0.05 percent of the average year).

While most areas of the U.S. can be provided with the first of these performance objectives (C/N & S/N) using receiving antennas with diameters of about 0.75 meters, threshold considerations will dictate that some locations utilize somewhat larger antenna diameters.

STC's system design allows all locations to use a maximum antenna diameter of 0.75 meters to meet the first criteria with larger antennas used for threshold reasons only.

The major aspects of STC's nationwide system are depicted in Figure 1. Four operating satellites are spaced 20° apart along the geostationary arc (115°W, 135°W, 155°W and 175°W longitude). Since a separation of approximately 15 degrees will allow frequencies to be reused between satellite serving identical or adjacent service areas, each of these satellites could be identical except for the antenna beam forming network which will necessarily be service area dependent. The satellites, each independently programmable, will serve areas within CONUS which are approximately the size of the time zones, designated Eastern Service Area (ESA) for the Eastern Time Zone, Central Service Area (CSA) for the Central Time Zone, etc. Satellite downlink transmissions will be in the 12 GHz Broadcasting Satellite Service (BSS) band with uplinks planned for the 17 GHz band.

Satellite broadcast transmissions will be received at individual residences by equipment composed of outdoor

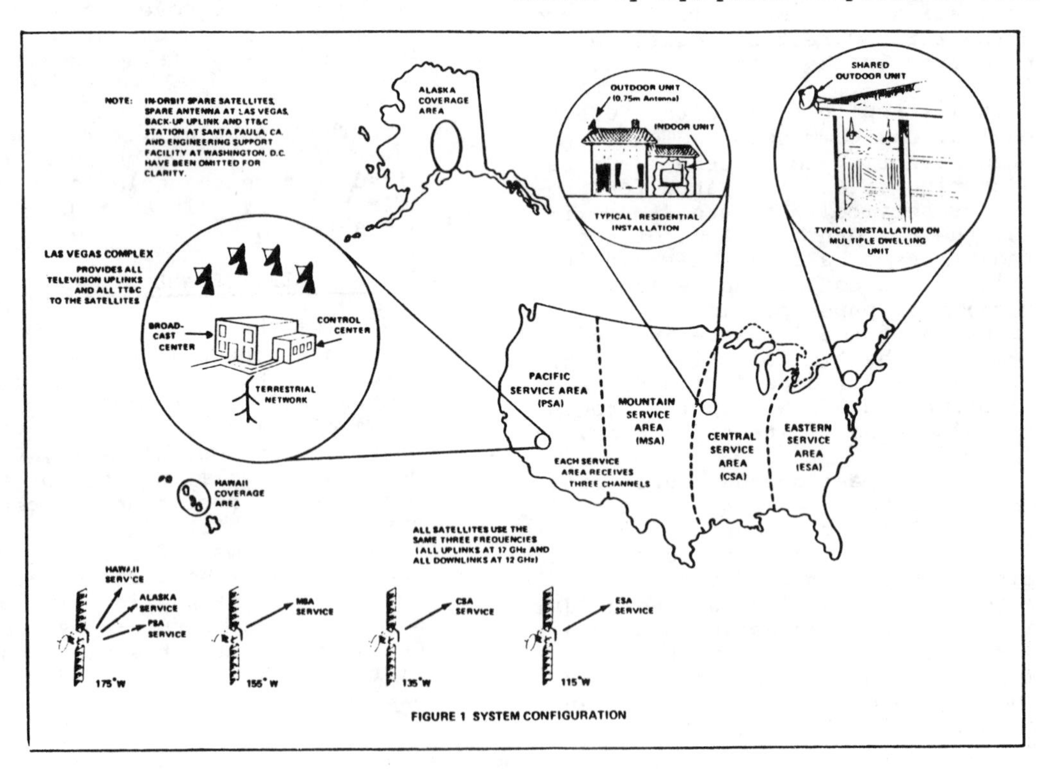

FIGURE 1 SYSTEM CONFIGURATION

and indoor electronic units as shown in Figure 2. Typically, the outdoor unit (ODU) consists of a 0.75 meter diameter receiving antenna and associated microwave electronics; in some areas of the U.S. antennae of 0.6 and 0.9 meters are used to equalize grade of reception and to minimize costs. The antennae may be mounted on residence rooftops, sidewalls, gable ends or at ground level. A cable connects the ODU of the home equipment to an indoor unit (IDU) which amplifies, demodulates, descrambles, and remodulates the receive signal to allow compatible video reception by a conventional television set.

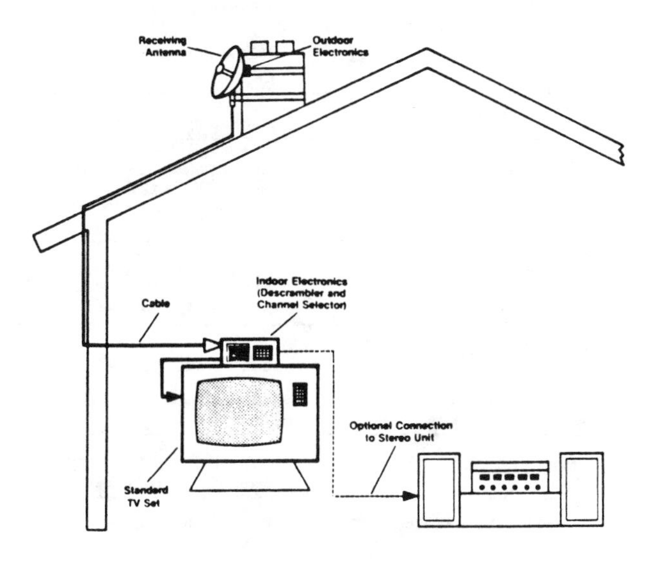

FIGURE 2 TYPICAL RESIDENTIAL INSTALLATION

STC intends to utilize PAM-D class satellites.* Each satellite is expected to generate prime DC power of approximately 1700 watts at end of life (over 2000 watts at beginning of life.) Three operating TWTAs will illuminate a shaped beam satellite transmitting antenna. Each TWTA has an RF output power of about 185 watts at end of life (215 watts at beginning of life). The transmit antenna patterns are tailored to the contours of the areas served for transmission efficiency reasons and to minimize, to the extent practicable, any signal radiated over foreign territories.

*PAM-D is the smallest booster available for carrying a spacecraft from the Shuttle orbit to geosynchronous transfer orbit. A PAM-D satellite can also be launched on a shared ARIANE III launch vehicle.

The choice of satellite orbital locations is dependent on several factors besides the need to accommodate frequency reuse: such as provisions of acceptable eclipse times (1 am or later) and elevation angles.

In addition to operating satellites, in-orbit spare satellites can be expected in a nationwide DBS implementation. STC intends to utilize 2 spares stationed close to the satellites serving the ESA and PSA, e.g., 115.05°W and 175.05°W longitude. Locating spares in close proximity to these operational satellites will allow restoration of service within minutes to the highly populated areas of the East and West coasts if a satellite malfunction occurs. Failure of the CSA or the MSA satellite will require repositioning of one of the spares, a process that would take several days to a week.

The receive coverage of a DBS satellite is highly dependent on the types of service envisioned; for example, some DBS operators will prefer to be able to access the satellites from anywhere within the U.S while others, such as STC, will prefer receive coverage areas which cover a smaller portion of the United States than the transmit coverage area.

STC's satellites will have a receive coverage area that extends from Los Angeles to Las Vegas. A transmitting ground station, which is planned for a location near Las Vegas, includes both a Broadcast Center and a System Control Facility. It will have four operating and one redundant antennae to feed and control the satellites reliably (the spare satellites will be controlled by the same antennae used to feed and control the satellites adjacent to them).

SATELLITE CHARACTERISTICS

Many of the system parameters of STC's DBS system were chosen on the basis of iterative tradeoff analyses which considered satellite constraints. Specifically, the size of the transmit service areas, the EIRP requirements and the number of channels have significant impact on satellite characteristics. These factors were combined with information regarding spacecraft hardware technology to yield the type of satellite described herein.

STC will procure spacecraft using specifications based predominantly on performance (e.g., EIRP, and receive figure of merit (G/T)) and not on design requirements. Therefore, spacecraft contractors will have considerable flexibility in the design of the satellites. For the purposes of characterizing the system, STC has synthesized a baseline spacecraft design and elements of this baseline spacecraft are described herein.

All satellites will be essentially the same. A capability will exist by ground command so that each satellite can be reconfigured to serve either of two services areas, provided it is located at specific orbital locations. The satellite design used to serve the ESA from 115°W longitude can be reconfigured by ground command and relocated to 135°W longitude in order to serve the CSA and vice versa; similarly the satellite used to serve the MSA can also be used to provide service to the PSA, and vice versa. The only difference between these two types of satellites is in antenna beam forming networks. For the nationwide implementation then, two operational satellites and a spare of one type will be used for ESA/CSA service, and two operational satellites and a spare of a slightly modified type will be used for PSA/MSA service.

A summary of the baseline satellite characteristics is present in Table 1. The following paragraphs identify the most important elements of the satellite.

The satellites will be designed to be compatible with two different launch vehicles: the National Aeronautics and Space Administration's (NASA's) Space Transportation System (STS, also referred to as the Shuttle), using a PAM-D booster to carry the spacecraft from the Shuttle's orbit into geostationary transfer orbit, and a shared launch (i.e., an STC satellite and another satellite launched together) on the ARIANE launch vehicle.

Spacecraft mass and size are limited by launch vehicle constraints. The STS/PAM-D is capable of placing a maximum mass of approximately 1247 kilograms into geosynchronous transfer orbit. The ARIANE Type III launch vehicle will be capable of launching two satellites simultaneously, each weighing approximately this amount. The spacecraft mass at the beginning of life will be approximately 650 kilograms.

A typical block-diagram of the communications subsystem of the ESA/CSA satellite, excluding redundancy, is presented in Figure 3. The satellite

TABLE 1

Satellite Characteristics

Mission	Television Broadcast for Individual Reception
Launch Vehicle	STS/PAM-D and ARIANE
Initial Mass on Station	650 kilograms
Satellite Mission Life	7 years
North-South Stationkeeping	± 0.1 degree or better
East-West Stationkeeping	±0.1 degree or better*
Prime Power	1700 watts end of life
Redundancy	100% all active electronic elements
Stabilization	Spin or body stabilized
Broadcast Channels	Three standard video - 16 MHz Bandwidth. Two (Alternative) HDTV Channels 28 MHz and 100 MHz Bandwidth
Emission Designators	Standard Video - M16 F5/9 HDTV 1 - M100 F5/9 HDTV 2 - M28 F5/9
Eclipse Capability	None
Receive Service Area Transmit Service Areas	L.A./Las Vegas *See* Figure 1
Frequency Bands	Transmit: 12.2 to 12.7 GHz Receive: 17.3 to 18.1 GHz
Minimum EIRP per Channel	Varies locally commensurate with rain attenuation statistics and range loss. Ranges between 58.2 dBW and 55.1 dBW.
Minimum Broadcast RF Output Power per Channel	185 watts
Saturation Flux Density per Channel	-88.1 dBW/m² at 17.3 GHz
Satellite Communications Subsystem G/T	7.7 dB/K
TT&C EIRP 4 GHz Band 12 GHz Band	 0 dBW minimum 15 dBW normally, 10 dBW during mispointing condition
Total TT&C RF Power 4 GHz Band 12 GHz Band	 2 watts Normal operation: 100 milliwatts Mispointing condition: 10 watts
Polarization	Transmit: RHCP for ESA, LHCP for CSA, RHCP for MSA, LHCP for PSA Receive: Orthogonal to transmit

*Stationkeeping of each in-orbit spare relative to its nearby operating satellite will be adequate to maintain nominal intersatellite spacing of 0.05°.

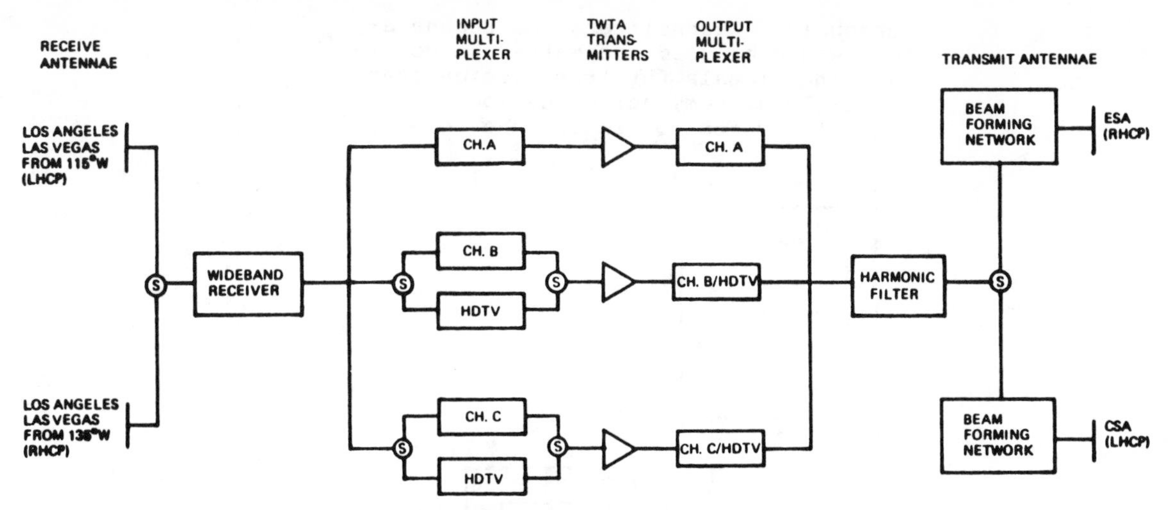

FIGURE 3 REPRESENTATIVE COMMUNICATIONS SUBSYSTEM BLOCK DIAGRAM (ESA/CSA SPACECRAFT)

design for MSA/PSA service would differ only in the receive and transmit antenna beam forming networks. The block diagram depicted shows a single conversion transponder, with the receiver translating the 17 GHs uplink signals directly to the 12 GHz downlink frequency band.

The beam forming networks are part of the antenna assembly, which provides shaped beams to CONUS service areas. Compared to simple beams of circular or elliptical cross section, shaped beams provide more efficient use of RF power because their patterns can be tailored to closely follow irregular contours; this minimizes the amount of RF power transmitted outside the service area and consequently enhances the net RF power falling within the service area.

HOME EQUIPMENT CHARACTERISTICS

The individual home reception equipment consists of three basis elements: a receiving antenna with its supporting mount, an outdoor microwave unit and an indoor unit. The receiving antenna taken with its universal mount and the microwave unit comprise the outdoor unit. The IDU and ODU are connected by a cable. ODU mounting will vary depending on home architecture and the extent of foliage or structural blockages in the direction of the satellite. Figure 4 shows a block diagram of the home equipment including the principal elements of the IDU.

The antenna and microwave unit are basically broadband devices which may be operated across the full BSS operating freuency band and would be suitable for operation with any conceivable 12 GHz U.S. DBS system implementation. The microwave unit will contain the necessary amplifiers, mixers, filters, and oscillators to down convert the receive 12 GHz signal to an intermediate frequency, nominally in the 800-1300 MHz range. Final specifications will be determined after further detailed discussions with hardware suppliers.

The IDU, located in proximity to the viewer's conventional television set, will contain the necessary electronics and controls to allow subscriber channel selection, FM demodulation, descrambling and AM remodulation. Normal channel reception is expected to be either on VHF channel 3 or 4. The IDU provides a standard NTSC formatted signal with the usual adjustments of color, hue, volume and tone retained in the viewer's TV set.

Multiple dwelling units (e.g., apartments, condominums) will be provided with service similar to the individual residences through use of a slightly different outdoor unit and distribution system. The configuration will provide scrambled service to each dwelling resident and uses IDUs at each set location identical to those used for individual residence reception. A somewhat larger receive antenna and higher gain IF distribution amplifier is expected for multiple dwelling unit installations.

High quality link performance can also be provided for cable head-ends also through the use of a larger receiving antennae. Head-end receiving equipment would down convert, demodulate and descramble all channels simultaneously. Channelizers and regenerators would be used thereafter to condition the signals for transmission over the cable system, using the local cable security system to protect the service.

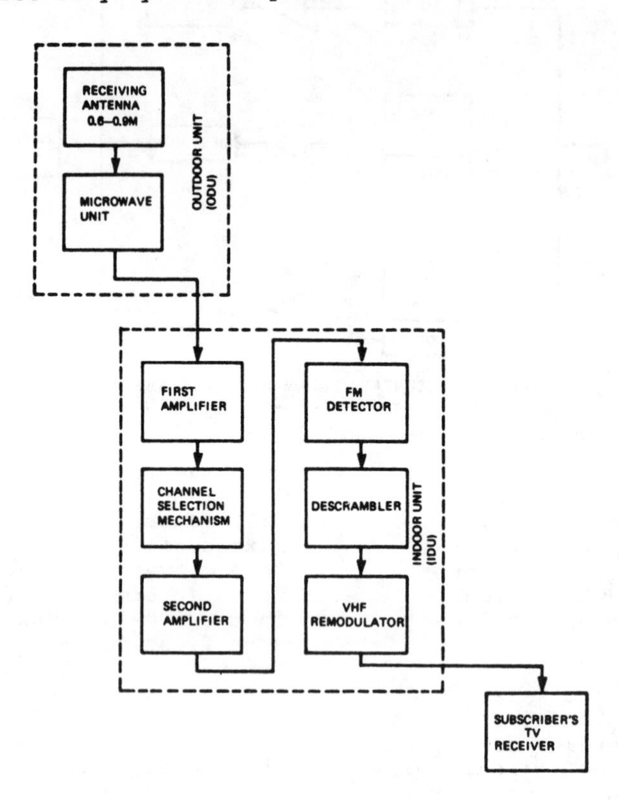

FIGURE 4 HOME EQUIPMENT BLOCK DIAGRAM

DEVELOPMENTS IN SYSTEMS FOR DIRECT BROADCASTING BY SATELLITE AT 12 GHz

by G.J. Phillips
BBC Research Department, Kingswood Warren

1. INTRODUCTION

The purpose of this paper is to present an up-date on interesting aspects of direct broadcasting by satellite (DBS) since the review given two years ago (1), rather than to provide a basic background paper (2). It will, however, consider a range of important topics affecting plans for the 11.7 to 12.5 GHz frequency band, namely the transmission system, satellite technology, receiver design, feeder link and cable distribution. In so doing it will review the important choices that need to be made and (where possible) indicate the decisions or likely decisions. Clearly some of the technical issues cannot be considered in detail, and other papers in the Symposium will provide more information.

2. TRANSMISSION SYSTEM

In view of the planned start of DBS transmissions in 1985 in W. Germany and France, and in the U.K. and Italy in 1986, probably to be followed in a short period by Sweden, Luxembourg, Switzerland and Austria, the transmission system for Western Europe countries must now be in the final decision stage if the receiver industry is to provide receivers in time and at a reasonable cost. It will be recalled that the ITU Plan prepared in 1977 had the following features.

- transmission channels 27 MHz wide, though planning allows for 19.18 MHz channels spacing.
- planning was based on service areas achieving a carrier-to-noise ratio (C/N) of 14 dB.
- the assumed transmission system was conventional video (PAL or SECAM colour in Europe) transmitted by FM with an FM subcarrier to provide a single analogue sound channel in the same way as terrestrial transmissions.

It is now agreed in member countries of the EBU that the transmission system should use digital sound and provide up to eight 15 kHz sound channels. At the time of writing there is a good prospect that a unified EBU 625-line transmission standard can be adopted on the basis

Reprinted with permission from *13th International Television Symposium*, Montreux, Switzerland, 1983, Systems Sessions, vol. 2, pp. 50–60.

of time-compressed components for video plus a phase-shift-keyed signal; the latter provides a data burst in every television line, for multiple sound channels. The features of such a system, which has been developed from various proposals, are as follows*.

Video components

The existing so called composite colour systems (PAL or SECAM) transmit the basic luminance or brightness video signal Y plus two chrominance signals U and V which are carried on a subcarrier near 4.43 MHz. Since the Y signal normally has at least 5 MHz bandwidth, the signals are band shared with a limited effective bandwidth for U and V of the order of 1 MHz. The band sharing results in cross-colour (colour patterns in fine detail) and cross-luminance (dot structure in the picture, usually confined sharp edges in the picture by the use of a 4.43 MHz notch filter in the receiver). The multi-plexed analogue component (MAC) system, developed in the U.K. by the IBA, (3) separates Y, U and V in time. In the form under EBU consideration, the transmission within the 64 µS television-line period consists of the chrominance signal (either U or V) followed by the luminance signal. Time compression by a factor of 3 for U or V, and 1.5 for Y, confines the duration of the video signal to about 52 µS leaving about 10 µS for digital sound transmission, allowing some room for transitions.

Modulation system

While the video signal is transmitted by FM in much the same way as envisaged in the original 1977 plan, the digital burst mentioned above is transmitted by the 2-4 PSK system, as developed in Norway.

In this system the main 12 GHz carrier is advanced 90° in phase for a digital "1" and retarded 90° in phase for a digital "0". It permits the use of a simple receiver demodulator in which the received signal is multiplied by a delayed version of itself, the delay being

---

* Space does not permit a review of various interesting alternatives that could well be used, including compatible adaptations of existing colour systems (4) and the use of subcarrier techniques to carry multiple sound (5a). One must pay tribute to the intense work undertaken, particularly to those workers who may have to see their proposals discarded in the interests of a unified standard.

one digital clock period and arranged to give a further 90° phase shift. The polarity of the product at the output varies according to the binary character. A clock rate of 20.25 MHz is adopted and both the video and digital components can be readily transmitted in the 27 MHz-wide DBS channel. The complete signal conforms to the 1977 Plan requirements that it should cause no more co- or adjacent-channel interference than the system assumed in the Plan. For transmission purposes it must be remembered that amplitude limiting will occur in the saturated travelling-wave-tube amplifier in the satellite and tests have included simulation of this.

Sound multiplex

The EBU agreed in 1981 on a digital sound coding standard for satellite broadcasting - namely that a 32 kHz sampling frequency would be used for 15 kHz audio channels. 14 bits per sample linear coding would be used and either transmitted directly or bit-reduced to 10 bits by near-instantaneous companding (6).

An important choice has to be made as to how the digital transmission is organised to carry multiple sound. The available mean bit rate is about 3 Mbit/s in the system just described. The two original alternative concepts of continuous multiplexing and packet multiplexing (5b) have been refined in stages. The first has been modified on a "structure map" principle. According to this Sveriges Radio proposal a small amount of extra data is sent to instruct the receiver on the number and time structure of the sound channels being sent. It allows flexibility in the forming-up of the multiplex so that specific numbers of linear or companded sound channels may be transmitted, leaving a variable amount of capacity for data or other uses. The second concept, from CCETT in France, now uses greater-length data packets transmitted in a regular manner, but still able to deal with alternative sound coding.

One remaining difference is that somewhat more "instructional" data, in total, is still required in the packet system (at the start of each packet) than in the structure-map system, but conceptually they have come closer. It is hoped that a final proposal based on a regularised packet system can be adopted by the EBU. The efficiency is such that with about 200 bits per burst in each television line (about 3 Mbit/s mean rate) a choice can be made to transmit 8 companded 15 kHz channels, or 6 linear (14-bit) channels on the basis of one parity bit per sample for error detection and concealment. Alternatively 6 and 4 channels respectively can be sent with three additional bits per sample for improved error correction. Also the bits that might be used for one 15 kHz channel can instead be used for two 7 kHz channels.

An interesting development, that has contributed to the efficiency, is that the 3-bit scale factor word required for each batch of 32 samples in the companded mode can be sent within the parity bits* this reducing the total bit rate per channel to 32 times 11 or 352 k bit/s.

It has also been realised that such a scale factor word can easily be sent with linear as well with companded coding. This is not of course required for digital expanding when transmitting the linear mode but it allows the receiver to reduce the effect of bit errors under poor receiving conditions. This is achieved by looking for received samples that are apparently outside the range of amplitudes signalled by the scale factor word, thus detecting the most audible types of error that have failed to be detected by the parity check. The error performance of linear coding is then made at least as good as for companded coding whereas, without this refinement, it suffers a disadvantage of greater audibility of errors.

3. SATELLITE TECHNOLOGY

The initial satellites for national services in Europe will not be capable of simultaneous transmission of more than two or three of the five channels which can be used. There will be some variations because of the different prime contractors involved. The function of the main transponders of the satellite will be to receive the incoming channels (in the 17.3 to 17.7 GHz or 17.7 to 18.1 GHz frequency range) and transmit on the appropriate frequency in the 12 GHz band.

The French satellite TDF-1, from Aerospatiale and Thomson-CSF, under the EUROSATELLITE consortium, will have a wide-band mixer, intermediate frequency amplifier (4.7 to 5.1 GHz), and second mixer to the 11.7 to 12.1 GHz band before separating the signals into separate channels for FET and travelling wave tube (250 watts) amplification at the output frequency. TDF-1 will have separate microwave dishes for reception and transmission, the former having a circular beam of 0.7° width and the latter providing the required elliptical beam.

---

* The 3-bit word indicates the multiplying factor to be used for a 32-sample block, and each sample is sent by an 11 bit word, including the parity bit. By identifying three groups of 9 samples, any one group can be transmitted with either positive parity in every 11 bit word or negative parity. A "majority vote" in the receiver gives the sign of parity, and hence both the scale factor word and the basis for rejecting (and concealing by interpolation) individual samples containing an error.

The U.K. satellite UNISAT due for launch in 1986 is from British Aerospace and GEC-Marconi, within the United Satellites joint company. It will employ a single frequency change after some amplification in the 17 GHz band. Channel filters will then separate the signals at 12 GHz to feed four separate amplifier chains of which any two can be powered. Each contains a 200 watt TWT output stage and can be switched to operate on any of the channels 4, 8, 12 or 16. Initial operation for the two BBC services will be on channels 4 and 16. Unlike the French satellite the U.K. satellite will adopt a single microwave dish with dual-frequency feeds for reception and transmission of the broadcast signals. It will also carry an independent communications payload for business services and television relay in the 14/12 GHz bands, with six channels of which four will normally be powered simultaneously.

The W. German TV-SAT from MBB and Telefunken will employ basically the same design as TDF-1 as the German companies are also within the EUROSATELLITE consortium, and the satellites are being produced as a joint venture.

4. RECEIVER DESIGN

Although there is an important part to be played by cable systems in distributing satellite television programmes, the justification and success of DBS will arise from its ability to provide a service to anyone who has access to a simple receiving antenna, bearing in mind that a large proportion of the population may not have access to a suitable cable system for many years.

There will therefore be a wide demand for the individual receiving installation, provided of course that cost can be kept to an acceptable level. We will consider separately the antenna, the outdoor frequency-changer unit, and finally the indoor unit or integrated receiver.

Antenna

The 1977 ITU Plan ensures that the five channels for any one national service are confined to one half of the 12 GHz band and have the same polarisation. Services of neighbouring countries may occupy the other half of the band, differ in polarisation, or come from a different orbit position, although the plan attempted to minimise these differences where common viewing would be widespread. Thus a simple installation can obtain all services of at least one country, while progressive complexity or duplication of equipment is necessary as more foreign services are required - subject to adequate signal level being available.

The first antennas are likely to employ parbolic reflectors between 0.7 and 0.9 metres in diameter but the longer term aim should be a flat panel antenna designed to fit neatly on the roof or vertical wall, and work towards this has been in progress for some time. As development proceeds it will be feasible to have duplicate panels, multiple feeds or electronic steering to meet requirements for alternative polarisation or orbit position.

Outdoor unit

A unit close to the antenna is required to act as the first amplifier and frequency changer. There have been significant developments in gallium arsenide field-effect transistor technology so that amplifier, oscillator and mixer stages can be realised. While present designs use separate FETs for each function, the design can already be seen for a monolithic Ga As device which will be produced, once the demand justifies it, in order to reduce the cost in large-scale production.

What is of greatest significance is that the noise figure that can be achieved, even over 800 MHz tuning range, is about 4 dB compared with 8 dB envisaged in 1977.

However, a problem remains with the choice of the intermediate frequency. The preferred approach hitherto has been to employ a fixed oscillator to translate the 800 MHz band to, say, 950 - 1750 MHz; this will be the frequency range at which the signal is fed from the outdoor unit to the indoor equipment. Owing to increasing use of both airborne and ground radar/navigation devices in the region of 1200 MHz there could be difficulty up to ten kilometres or more from airports in achieving adequate screening to avoid interference. It now seems likely that a cheaper and more satisfactory installation would result if 1400 - 1800 MHz were employed (or 1400 - 2200 MHz if translation of the complete 800 MHz band is required).

Indoor unit or integrated receiver

We can see three main categories of indoor installation

- a tuner/adaptor to produce a standard television signal to enter the VHF/UHF tuner of a present-day television receiver.

- a tuner/adaptor to produce component signals (R, G, B or Y, U, V) which a significant proportion of receivers will be able to accept by 1985/6, in accordance with EEC and IEC international standards.

- a new type of television receiver incorporating a tuner unit for satellite as well as terrestrial television which will accept the signal from the outdoor unit directly.

There will be a decreasing proportion of viewers in the first two categories as time progresses and there may also be a proportion of viewers who will prefer their receiving equipment in separate units (monitor, tuner, video cassette, disc) by analogy with audio units.

Regarding the tuner unit itself, the use of a second intermediate frequency near 125 MHz should be suitable. Normal types of discriminator (pulse counting, quadrature detector) preceded by an amplitude limiter can be used, but an attractive alternative is a phase-locked loop discriminator which, when used without an amplitude limiter, gives threshold extension. This means a reduction of noise in the picture on weak signals when the carrier-to-noise ratio is less than about 11 dB. There is also the possibility of using a phase-locked loop directly at the first intermediate frequency because it has certain self-selective properties, but the performance may not be satisfactory when receiving a signal that is weaker than others in the band.

The post-discriminator circuit will of course have some additional complexity if time compressed component signals are to be received. It is expected that large-scale production will justify the use of digital processing to achieve the required time expansion. Because the digital sound burst employs a 20.25 MHz clock frequency which is locked to the television line frequency, it is attractive to sample the incoming video signal at this frequency. Digital samples for one television line are fed into a store from which the chrominance is read out at 6.75 MHz and the luminance at 13.5 MHz, in order to obtain the required parallel feeds. Because chrominance U and V signals are sent only in alternate lines, the U (or V) from the previous line must be kept in store and read out a second time.

A further question is automatic frequency control (AFC). The proposed transmission system uses pre-emphasis (i.e. progressive increase in gain at higher modulation frequencies) but, for optimum performance, this is less than in conventional FM microwave television links and it is necessary to transmit with d.c. coupling for the video signal applied to the modulator. This is to prevent deviations outside the channel when there are extreme values of mean picture brightness. Accurate AFC therefore requires reference to the effective centre frequency of the transmission rather than the mean frequency. This may be obtained from the reference "zero chrominance" level provided for clamping purposes just before the transmission of the chrominance signal.

If an adaptor is required to feed a conventional receiver via the tuner, then remodulation on a carrier with PAL or SECAM coding will be required. With the second type of adaptor, for video connection to the receiver, not only will the remodulator be unnecessary but the picture quality will reap the benefit of the new system and be free of cross-colour and other composite-signal limitations.

Finally we must consider the sound signal. Because it is proposed that two audio channels should be used with the television programme for stereo sound, it will suit many viewers if the indoor unit provides audio outputs that can be connected to an existing stereophonic sound system, but newer receivers will provide, as an alternative, internally fitted stereophonic channels and loudspeakers.

It was indicated in the Section on the transmission system that the digital burst in the proposed EBU standard would employ 2-4 PSK modulation. At the present state of development it is recommended that receivers should employ a separate delay-line type of demodulator for this signal. In this case it can be preceded by an optimum filter of about 20 MHz bandwidth, independently of the video chain. However, other possibilities are a simplified receiver with a common discriminator, or a more advanced receiver in which the sound threshold for weak signals can be extended, e.g. by using phase coherent detection (although the transmission system is such that only 1 to 1.5 dB improvement would result from this) or using digital processing based on the difference in the characteristics of digit errors and a normal sound programme.

In conclusion it may be said that with a basic design of receiver the proposed EBU system should result in excellent vision and sound quality under normal receiving conditions. On weak signals the sound quality is still good when the picture is rather poor. Also fairly simple techniques are available for out-of-service-area reception which can improve both vision and sound thresholds by a few decibels, if required.

5. FEEDER LINK AND CONTROL STATION

In considering the operation of a satellite for direct broadcasting it is possible to separate the activities of control and of feeding the programme signals to the satellite. In many cases it will be convenient to use completely separate Earth stations.

Control of the satellite calls for flexibility and agility of the terminal, and specialist knowledge firstly to monitor various functions (station-keeping and pointing being obvious examples) and secondly to exercise control. Particular care is needed for example, around periods of eclipse, when the satellite is shadowed from sunlight by the Earth or Moon; without appropriate switching of certain functions at the right time, mal-functioning or even permanent damage could result. The control station could well be at a centre responsible for a number of different types of satellite.

On the other hand a feeder link station to send the broadcast programmes to the satellite can be a compact purpose-built station, relatively simple in concept and located at a broadcasting centre. Here a different specialisation of monitoring high quality television and maintaining continuity of programmes is required. Much of the material, especially from film or tape recordings, would originate from studios or equipment at the centre. This short review will be limited to certain aspects of feeder-link requirements.

Discussions in EBU suggest that the Earth station EIRP in the 17.3 to 18.1 GHz band should be sufficient to provide in the up-link a carrier-to-noise ratio (C/N) at least 24 dB for 99.9% time in the worst month of the year. This allows for the fact that, in contrast to the planning figure for the down-link of 14 dB C/N used in the 1977 Plan to cover worst-case, edge-of-beam receiving conditions, many viewers will have a down-link C/N greater than 20 dB for substantial periods of time (i.e. in dry weather, near beam centre and with improved receivers). Such viewers could therefore enjoy low-noise pictures provided the uplink can maintain a comparable standard.

A typical uplink budget calls for 80 dBW EIRP to meet this requirement, but the figure shows some variation with the size of the satellite receiving beam, climate at the Earth Station etc. It is generally felt that, for planning purposes, it is best to unify the EIRP of all Earth stations at an adequate figure of around 83 dBW. This could be the basis of a workable up-link plan in which some fine adjustment of some EIRPs can be made to optimise the margins of protection against mutual interference. A formal frequency plan for feeder links is unlikely before the first session (in 1985) of the ITU Conference on the geostationary satellite orbit, and interim agreements will have to be reached to cover the first broadcasting satellites.

A 1.5 kW klystron amplifier has been developed for the feeder-link band, and this would permit the use of a reflector size down to about 5 m. The control of mutual interference depends on good stability of the EIRP, but large reflectors (sizes up to 11 m diameter with about 0.1^{o} beam width have been proposed (7)) are undesirable as there could be gain fluctuations of as much as 2 or 3 dB due to effects of wind or non-uniform solar heating. A reflector of some 5 to 6 m diameter (0.2 to 0.25^{o} beamwidth) has the advantage that tracking would not be essential, provided that the satellite can be stabilised in position to about 0.05°. This possibility would simplify the design of an Earth station intended solely as a feeder link station. In conditions under which a second (spare) satellite had to operate on one channel, while other channels were carried on the primary satellite, even a steerable beam could not optimise its position for all channels and could be at a disadvantage if the beam were narrow.

6. CABLE DISTRIBUTION SYSTEMS

The advent of satellite broadcasts poses certain difficulties for cable distribution if their benefits (such as those of high quality digital sound) are to be maintained in the distribution system.

For smaller systems it may be attractive to distribute the FM satellite signals at the same frequencies (at UHF) as those recommended for the first intermediate frequency of an individual receiver. This would mean the receivers or receiver adaptors could be identical for individual reception and cable distribution.

For larger systems, distribution of standard terrestrial PAL or SECAM signals by remodulation may be used but, if a time-compressed component system is used, such a remodulator may be somewhat expensive and would not exploit the improved vision and sound quality. Studies suggest that a v.s.b. amplitude modulated signal could carry the satellite video modulation signal and that a separate carrier might convey the sound signal as a continuous 3 Mbit/s stream derived from the digital portion of the satellite signal. The complete signal could be accommodated in about 14 MHz of spectrum so that it would not occupy more space that two adjacent channels in a conventional 7 or 8 MHz channel system. It is difficult without further experimental work to judge whether FM or AM distribution would be preferable if circumstances permit a choice. It has been pointed out* that FM distribution can benefit from the greater robustness of the FM signal compared with VSB AM; thus the signal could be distributed at a lower level, and a higher relative level of intermodulation products could be allowed. These factors may compensate for the higher cable losses in the UHF band. It is not known without further experiment whether, in the distribution of time compressed signals with digital sound in bursts at 20.25 Mbit/s, the

---

* See CHAPLIN, J.G. and FROMM, H-H. "European Satellite broadcasting : the case for an FM television receiver" IEE, London, Colloquium Digest No. 1983/22 (Paper 10) on "Better television by satellite".

sound would be sufficiently robust to withstand the reflected signals at the levels that would normally be encountered in a cable distribution system.

A possible disadvantage of the FM system is that if satellite programmes of more than four countries are to be distributed, there would probably be a need for parallel coaxial feeders into each home because an 800 MHz range of frequencies can only distribute 20 FM channels. This could, however, be as cost-effective as other methods and simply requires a selector switch added at the receiver input.

7. CONCLUSION AND ACKNOWLEDGEMENT

In this paper it has only been possible to highlight a few topics of interest, some of which have been the subject of recent very rapid development. It is hoped - and there are encouraging signs - that the intense activity of the last few months will lead to high-quality television and sound transmissions in Europe, to a standard that is considerably better than would have resulted from the more conventional methods of transmission envisaged at the time of the 12 GHz Plan in 1977.

The author wishes to thank the Director of Engineering of the British Broadcasting Corporation for permission to publish this paper.

8. REFERENCES

1. SUVERKRUBBE, R.: "Projects for direct satellite broadcasting at 12 GHz" 1981 Montreux Television Symposium Record pp 19-47.

2. PHILLIPS, G.J.: "Direct broadcasting from satellites". Proceddings IEE, 1982 Vol. 129 Pt. A No. 7 pp 478-484

3. LUCAS, K. and WINDRAM, M.D.: "Direct television by satellite : desirability of a new transmissior standard" IBA Experimental and Development Report 116/81.

4. OLIPHANT, A.: "An extended PAL system for high quality television" BBC Research Department Report 1981/11.

5. CCIR, 1982 Plenary Assembly, Vol. XI Part 2, Reports (a) 632-2, (b)954.

6. CAINE, C.R., ENGLISH, A.R. and O'LAREY, J.W.H.: "NICAM 3: near instantaneously companded digital transmission system for high quality sound programmes". Radio and Electronic Eng. 1980 Vol. 50 pp 519-530.

7. KEANE, L.M.: "A direct broadcast satellite system for the United States", COMSAT Tech. Rev. 1981 Vol 11 pp 195-201.

Technical Aspects of Satellite Broadcasting in Japan

Koichi YABASHI
NHK(Japan Broadcasting Corporation)

1. Introduction

The Medium-scale Broadcasting Satellite for Experimental Purpose(BSE) was launched in 1978, and various experiments concerning satellite broadcasting by using it were carried out until 1982. Through these experiments, the practicability of satellite broadcasting has been fully proved. At present, NHK is making preparation for the launching of the first operational broadcasting satellite BS-2, which is scheduled to be launched in 1984 by means of N-II launch vehicle and serve two channels of NHK television program. The succeeding spacecrafts BS-3 are planned to be launched before the end of the life of BS-2, around 1988, so as to continue its service. Conceptual design of BS-3 had commenced already in late 1982.

This paper describes the technical aspects of BS-2 spacecraft and its launch vehicle N-II, and then the technical standards of the DBS transmission in Japan and the development of ground receiving equipments.

2. BS-2 Spacecraft

The space segment of BS-2 system consists of two spacecrafts, BS-2a and BS-2b, one for actual operation and another for the spare in orbit. These spacecrafts are now under the construction by Toshiba Co. with General Electric Co., which is a major subcontractor, in reflecting the experience obtained in the BSE program. The spacecraft BS-2 has almost the same scale and ability as BSE, but is modified its design a bit so as to conform to the technical standards established at WARC-BS in 1977 and extends the design life on orbit to 5 years. The configuration of it consists of a fixed antenna platform, North/South equipment panels, two rotating solar array panels for power generation and so on. The spacecraft is three-axis stabilized.

The communication subsystem is equipped with three 100W TWT amplifiers, two for operation and one for redundancy, and has the capability to transmit two channels of color television signals. In order to reduce the spacecraft weight, helix type TWTs are employed instead of coupled cavity type TWTs of BSE. NHK has been engaged in R & D works on TWT for DBS. According to result of them,

Reprinted with permission from *13th International Television Symposium*, Montreux, Switzerland, 1983, Equipment Innovations Sessions, vol. 1, pp. 191-200.

in future, helix type TWT with power of up to 200W or some more would be available and optimum for DBS because of its light weight.

The antenna subsystem on BS-2 has been developed by Toshiba Co. It has the function to receive and transmit 14/12 GHz band two channel TV programs and TT&C signals, and is composed of an elliptical reflector and a composite three-horn primary radiator so as to cover effectively all Japanese territory which consists of four main islands and remote islands including Okinawa and Ogasawara 500km and 1,500km south-west and south-east of main island respectively. The experiments to investigate the foot print of the BSE radiation showed the daily slight transformation of its pattern. This transformation was guessed to come of antenna surface thermal distortion caused by sunlight. In order to reduce the thermal distortion of the BS-2 antenna surface and the spacecraft weight, the reflector is made of graphite-epoxy, consequently the weight of all antenna subsystem is achieved about 10kg. The antenna is an offset parabola with three feed horns and radiates right hand circular polarized wave, so as to provide low spillover radiation outside the Japanes territories and comply with the WARC-BS technical standards.

Regarding DBS transmitting power, BS-2 spacecraft has the capability to transmit 100W for one channel at the output of TWT-Amplifiers. This output power necessitates an installation of ground receiving antenna of one-meter-diameter dish within main islands of Japan. From the point of view for the popularization of satellite broadcasting, however, the smaller dish antenna is desirable for DBS ground reception, so that the radiation power of 200W or more should be employed in the next generation of BS-2.

3. N-II Launch Vehicle

The BS-2 spacecrafts will be launched onto Japan's assigned orbital position of 110 degrees east longitude in early 1984 and mid 1985 from NASDA (National Space Development Agency) Tanegashima Space Center, which is located in Tanegashima island 100km south of main island, by means of N-II launch vehicles.

N-II was developed by NASDA as the successor of N-I launch vehicle, which was developed, based on the USA Thor-Delta, and had the capability to launch a satellite weighing about 130kg onto geostationary orbit. N-II has the total length of about 35 meters and the fairing diameter of about 2.4 meters, and is able to launch a geostationary satellite weighing about 350kg. The first and second stage of N-II are drived with liquid engines and nine strap-on boosters fitted on the first stage, and controled by an inertial-guidance control system in which acceleration and angular velocity data detected by sensors are calculated to decide the position and direction of the vehicle in the space and the control signals are sent to the engine system so as to follow the orbit planned in advance. N-II has been already in operational usage, that is, the Geostationary Meteorological Satellite Second(GMS-2) in Aug. 1981 and the Communication Satellite Second(CS-2) in Feb. 1983 were launched successfully onto each geostationary orbital position as planned.

Table 1 Specification of N-I, N-II and H-1

Specification \ Vehicle	N-I	N-II	H-I
Total length (m)	33	35	40
Diameter (m)	2.4	2.4	2.4
Weight (t)	90	135	140
Capability of launching geostationary satellite (kg)	145	350	550
Year of first launching	1975	1980	1985

With respect to the succeeding launch vehicle after N-II, at present H-1 launch vehicle is under development by NASDA, which has the capability of launching a geostationary satellite weighing 550kg, and the first test flight will be made in 1985. While BS-2 spacecrafts provide two 100W channels for NHK TV program, the Broadcasting University and commercial broadcasters are considered for possible candidates of the users of the BS-3, so that the number of TV channel for BS-3 could be three or four. Moreover, the transmitting power of more than 200W is needed for BS-3 as mentioned before. Therefore, the boost of H-1 launching capability is desirable. The specification of the Japanese launch vehicles is shown in Table 1.

Japan was allocated eight TV channels, all of which are in same orbital position, at WARC-BS in 1977, so NASDA has the future plan to develop a large launch vehicle which is able to put a satellite weighing up to two tons to broadcast 8 channels of program from one satellite.

4. The Technical Standards for Television Broadcasting by Satellite

With regard to technical standards to be applied to satellite television broadcasting in Japan, the Advisory Committee on Radiowave Technology to the Ministry of Posts and Telecommunications recommended adoption of frequency modulation, in line with the decision of the WARC-BS, 1977, for 525-line NTSC television signals with 4.5MHz video band-width, and digitally modulated subcarrier system for accompanying television sound signals, which leads to four high quality sound channels plus data or two studio quality sound channels plus data with a video signal under the big contribution by NHK. Fig. 1 shows the spectrum of video signal and a digital subcarrier.

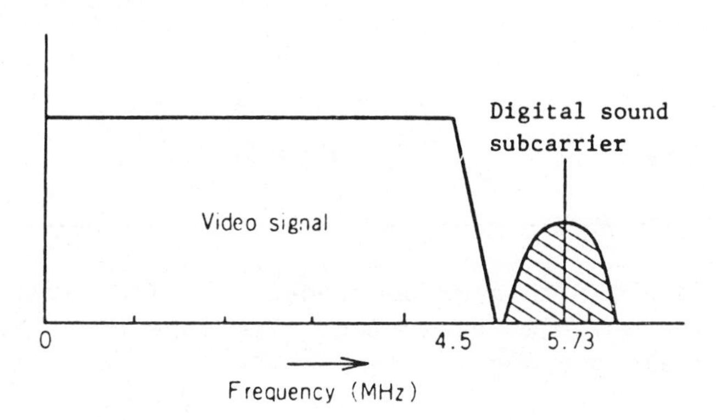

Fig. 1 Spectrum of the video signal and a digital sound sub-carrier

While discussing the technical standards, the Committee deliberated taking into account following items.

- Basic requirements
  - The standards must conform to the decision at WARC-BS, 1977.
  - The system should be capable of transmitting not less than four sound channels.
  - The receiving quality of picture and sound should be better than terrestrial TV and FM.

- Desirable Requirements
  - The compatibility with the conventional television system.
  - The availability of several additional data channel for the diverse usage in the future.
  - Inexpensive receivers.

4.1 The technical standards for video signal

The frequency modulation of 525-line NTSC television signal adopted as the technical standard for video signal achieves better quality picture and the compatibility with terrestrial television system. The specification of the standard is shown in Table 2.

Table 2 Transmission parameters of the video signal

Signal standard	525 lines M/NTSC
video bandwidth	4.5 MHz
Deviation	17 MHzp-p (including sync.)
Polarzation of modulation	Positive modulation for the video
Emphasis	CCIR Rec. 405-1
Energy dispersal	600 kHzp-p deviation 15 Hz triangle wave

In the case of 14dB C/N in a ground receiver, video S/N on a display are achieved 49dB, which is equivalent with more than Grade 4 in the five scale quality grade.

4.2 The technical standard for television sound signal

The committee adopted a single digital subcarrier system for the sound/data transmission for the reasons in following.

- It is possible to transmit four sound channels with resonable capacity for data.
- Flexibility in compliance with future needs for data broadcasting and others can be met.
- Home receivers will be made available for operational system of BS-2 which will be launched in February 1984.

Taking into account various services to be conceived in the future such as facsimile, teletext and data/code broadcasting, two sound transmission modes, Mode A and Mode B, were selected in following.

- Mode A can transmit four monophonic sound signals, which have 15kHz band-width, and data to meet requirement for multiple sound channels.
- Mode B can transmit two very high quality sound signals, which have 20kHz band-width, to meet requirement for receiving highest quality obtainable.

The transmission parameters of sound signal is shown in Table 3.

Table 3 Transmission parameters for sound in television broadcasting via satellite in Japan

Mode	A	B
Coding		
Sound signal bandwidth	15 kHz	20 kHz
Sampling frequency	32 kHz	48 kHz
Sound coding and companding	14/10 bits, near instantaneous companding (5 ranges)	16 bits linear attached range codes
Sound emphasis	50 μs + 15 μs	50 μs + 15 μs
Multiplexing		
Bit rate	2,048 Mbit/s	
Number of sound channels	4	2
Additional data capacity	480 kbit/s	240 kbit/s
Frame format	See Fig 4, 5	See Fig 4, 6
Error correction		
- sound and data	BCH SEC DED (63, 56)	
- range code	BCH, SEC, DED + Hamming (7, 3)	
- control code	majority decision	
Modulation		
Subcarrier frequency	5.727272 MHz	
Deviation of the main carrier by the subcarrier	±3.25 MHz	
Modulation scheme	4ϕ - DPSK	

In this system, the digital subcarrier frequency is set at 8/5 times of nominal color subcarrier frequency, and this digital subcarrier is 4ϕ-psk modulated by pulse stream of 2.048 Mbit/s bit rate, which includes error correction pulses and control pulses as well as PCM sound and data signals. The frame format for Mode A and Mode B is shown in Fig. 2. The transmission capacity of data channel in Mode A and Mode B are 480 Kb/s and 240 Kb/s respectively.

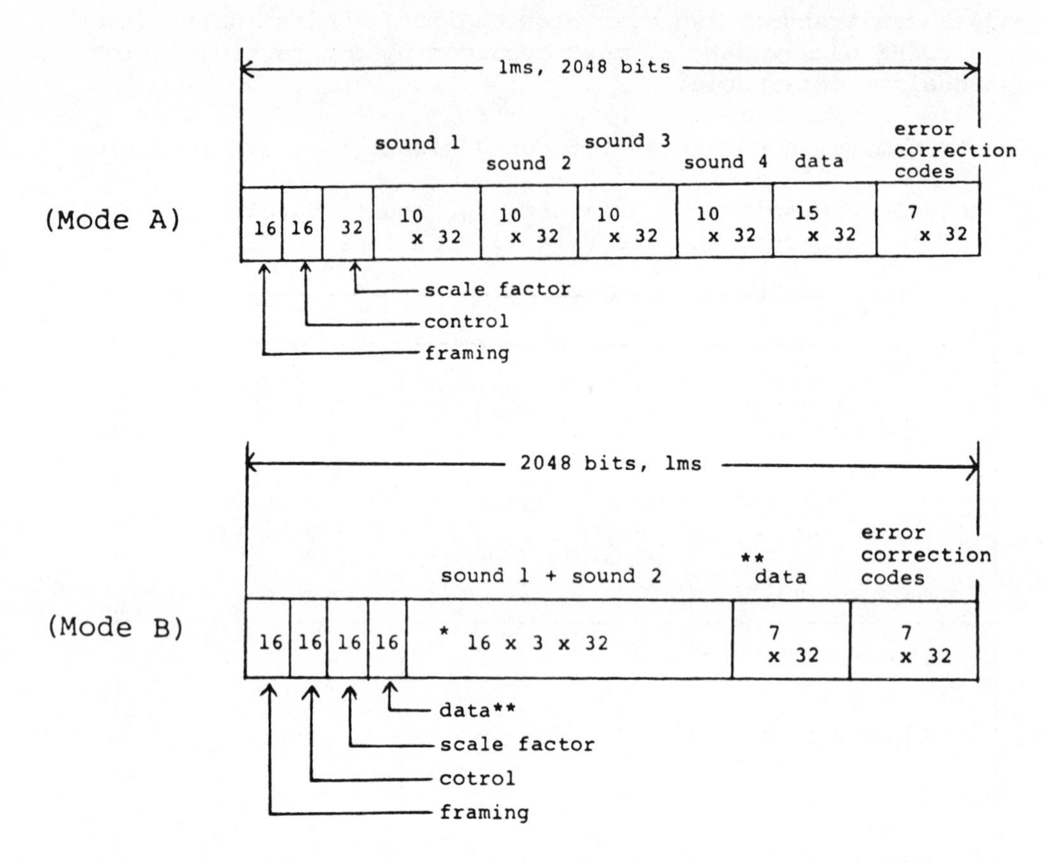

Fig. 2 Frame Format

The satellite television system mentioned before can provide better quality of picture and sound than the existing broadcasting services, and has good compromise between them. Furhtermore, it has the capability to provide new broadcasting services, so that this system is considered to be suitable system for several decades.

Recently, the introduction of MAC(Multipled Analogue Component) system into DBS was adopted in the UK and is being discussed in some other countries. While MAC system would provide better picture quality than conventional television system in the future, it requires the modification of existing equipment for program production, recording and so on, so that the introduction of MAC system would need some lead time to prepare such modification. Moreover, receivers for MAC system might be expensive because of A/D converters and memories composed in it, and picture display with RBG input for MAC signal has not been popular yet so much.

5. HDTV

HDTV provides new expression in the field of video production by means of higher resolution on wide screen, and it could be expected to become new generation of television in the coming century and contribute to creation of broadcasting culture. Since 1968, NHK has been engaged in the study in HDTV, and adopted a tentative technical standard for HDTV signal as shown in Table 4. Following the standard, TV camera, VTR and receivers for HDTV has been manufactured, and telecine system for 70-mm movie film using laser beam has also been developed.

Table 4 Tentative standard of the High-Definition Television System

Number of lines	1125
Aspect ratio	5 : 3
Interlace ratio	2 : 1
Field frequency	60 Hz
Video bandwidth	
Luminance	20 MHz
Chrominance	5.5 - 7 MHz

Transmission system is another item to be studied for realization of HDTV satellite broadcasting. The band-width of HDTV picture signal is about four or five times that of conventional television picture signal, and in case of frequency modulating HDTV signal directly, RF band-width of this signal occupies approximately 100MHz. Even if employing band-width compression technique, it is extremely difficult to transmit HDTV signal with one DBS channel occupying 27MHz band-width, prescribed at WARC-BS, 1977.

For the realization of HDTV service within this frequency band, however, two systems are considered to be possible to transmit the signal. One is to transmit the signal with two DBS channels, and another is to extend the band-width of one DBS channel, for example, 38MHz for one HDTV channel. Both of them need the development of band-width compression technique suitable for each system. Furthermore, the former has a problem to reduce available channels into half number, and the latter requires the study which proves for the signal not to give interference to other channel reception.

It should be noted that WARC-BS, 1977, has allocated the frequency band-width of 27MHz for one DBS channel for ITU Regions 1 and 3 based on the existing television system, taking no account of new HDTV services. If, in future, the demands for new television sytem such as HDTV or new broadcasting service grow up all over the world, it is expected that appropriate measures, which makes those new broadcasting service possible and contribute to effective usage of the DBS frequency band, might be taken.

6. Ground receivers and receiving technology

NHK has a department, which takes charge of the developmental work and guidance for viewers of broadcasting reception, and has also been studying DBS receiving technology, including receiver itself. The experiments for the reception under various environmental conditions had been carried out by NHK, and the results of those experiments, such as the influence by snowfall and wind pressure on dish antennas, are summarized in following.

Since dish antenna for DBS is usually set at outdoors, snowfall on it affects signal reception. In order to develop the antennas suitable for snowy area, NHK has been studying antenna shape and coating material on its surface. According to the results of the experiment, an off-set type dish antenna showed better performance than ordinary dish antenna because its reflector stands more vertically, and an off-set antenna covered with simplified radome of teflon material showed better reception. As for the coating material slippery for snow, acryl paint showed good result.

Dish antenna is apt to be affected by wind because of its large reflector, so it should be installed firmly. Since most of the roofs of wooden houses in Japan, however, are not so strongly built, the firm installation of dish antenna on a roof might be very difficult. For the study of influence by wind, NHK carried out an experiment by using a wind tunnel so as to get technical data of wind pressure on dish antennas. As an example of this experiment, maximum wind pressure on one-meter diameter antenna was equivalent to about 100kg load in the wind of 40m/s velocity.

It is well recognized that ground receiving equipments are very important part in satellite broadcasting system. For the spread of satellite broadcasting, the receivers must have low noise figure(NF) and high sensitivity. In the last several years, a large amount of effort has been devoted to the development of GaAs FET Amplifier in SHF band, and achieved NF of about 2.5dB. GaAs monolithic integrated circuit technology appears to lead to efficient fabrication for large scale production. After the Committee recomended the digitally modulated subcarrier system for sound and data transmission of satellite television, the manufacturers in Japan have been making big effort to introduce IC into each circuits of 4ϕ-psk demodulation, PCM signal processing, D/A converter and so on in receivers, so as to simplify those circuits and reduce the receiver price. The fabrication of DBS ground receivers is under way to supply them by the commence of BS-2 satellite broadcasting scheduled in spring of 1984. As for the diffusion of satellite broadcasting receivers in Japan, it is expected to be more than 500,000 by the end of BS-2 life and about 10,000,000 at mid of 1990's. A domestic receiving system comprises a dish antenna, a low-noise receiver front-end and indoor unit containing channel selector and demodulation stage and a television monitor. New broadcasting services such as HDTV, PCM Stereophonic Sound and so on might be realized through DBS in future, so additional equipment would be needed for the reception. Fig. 3 shows an example of a typical domestic receiving system comprising receivers for various new broadcasting services.

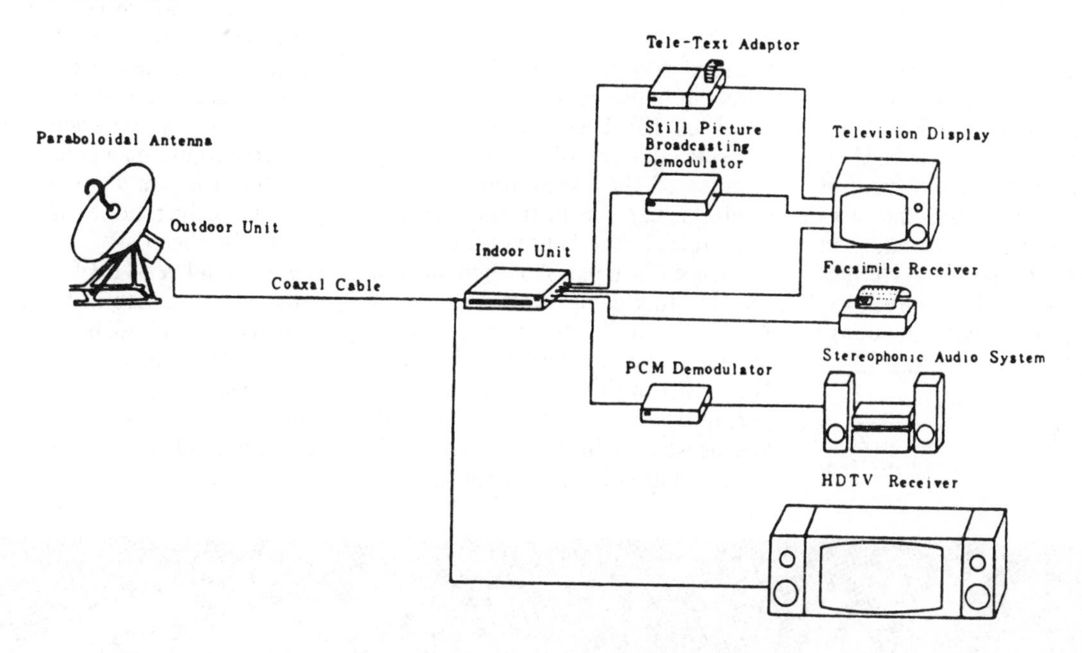

Fig. 3 Domestic receiving system expected in future

7. Conclusion

NHK will commence satellite broadcasting service in 1984 by using the BS-2, and the succeeding satellite BS-3 will be launched around 1988 for the continuity of the service. Many other countries moreover will also start satellite broadcasting service in the middle of 1980's, and the number of their services will increase remarkably in the next decade.

As for DBS services, the first step would be for conventional television, but in the next generation new broadcasting services, such as HDTV, PCM stereophonic and so on, are expected to be included. NHK has been engaged for a long time in R & D work on satellite broadcasting as well as new broadcasting services, and desires to cooperate with other countries in those fields so as to contribute to the progress in broadcasting technology.

Satellite Broadcasting in Canada

By O. S. Roscoe

Television broadcasting from satellites directly to low-cost individual home receivers will be widespread in many countries before the end of the decade. Canadian field trials have provided valuable experience with direct broadcasting satellites (DBS) operations and have increased awareness of the potential for such technology. A comprehensive study plan leading to the introduction of DBS service in Canada as early as 1983 is now under way.

Direct broadcasting to individual home receivers has emerged as the dominant new application of satellite technology for the 1980's. Canada pioneered in the development of this technology, first with demonstrations and experiments beginning in 1976 with the high-powered Hermes satellite (Fig. 1), and later with extended field trials using the modestly powered ANIK-B satellite (Fig. 2). Direct broadcasting satellite (DBS) television has been receiving an increasing amount of publicity lately because of the expectation of such service as early as next year. Several companies in the United States have announced firm plans to begin satellite broadcasting using Telesat Canada's ANIK-C satellite, which is scheduled to be launched by the space shuttle in November of this year. The same opportunities are there for Canadian broadcasters to make use of ANIK-C for direct broadcasting. In particular, the Pay TV licenses granted last month by the Canadian Radio, Television and Telecommunications Commission represent new broadcasting initiatives which could use ANIK-C for direct broadcasting as well as for distribution to cable head-ends.

Perhaps it would be worthwhile to define DBS service. A DBS transmits a signal of sufficient strength so that it can be received with a low-cost receiver ($500 or less when in quantity production) using an antenna which is 1 meter or less in diameter. With expected receiver noise figure performance in the 3 dB range, satellite power should be 54 dBW or greater for satisfactory service in those parts of Canada subject to heaviest rainfall conditions. This would make direct broadcasting satellites very large in comparison to present-day domestic telecommunications satellites. Typically, four to six times as much pri-

Presented at the SMPTE Montreal/Toronto/Rochester Miniconference, April 23–25, 1982 by O. S. Roscoe, Dept. of Communications, Government of Canada, Ottawa, Ontario. This paper was received on June 16, 1982.

Figure 1. Hermes satellite in space.

Reprinted with permission from *SMPTE J.*, vol. 91, Part I, pp. 1142–1147, Dec. 1982.
Copyright © 1982 by the Society of Motion Picture and Television Engineers.

Figure 2. Artist's conception of ANIK-B in orbit.

mary power would be required for a satellite providing eight channels of television into each of two quarter Canada beams, and two such satellites would be required to serve all of Canada. For ITU Region 2 — the Americas — the spectrum for transmission will be bounded somewhere between 12.1 and 12.3 GHz and 12.7 GHz, with the lower boundary to be decided at the 1983 Regional Administrative Radio Conference. Expectations are that this lower boundary will be set at 12.2 GHz.

ANIK-C falls short of the above definition by a significant degree. However, as is explained later, it is quite capable of providing acceptable service with receiver technology available today if the antenna size is increased to about 1.2 m, and if smaller rain margins are adopted.

Direct Broadcasting Satellite Experiments and Field Trials

The Hermes satellite, transmitting in the 12 GHz band with a boresight e.i.r.p. (effective isotropically radiated power) of 59.6 dBW, was the forerunner of the high-powered direct broadcasting satellites which proliferated in this decade (Fig. 3). Its principal communication parameters are shown in Table 1. Tests with Hermes demonstrated that good-quality reception could be achieved with high-performance receivers at e.i.r.p. values much below the maximum possible with the satellite.

Rapid advances in receiver technology were also being made, so that it was possible to consider an extended field trial of direct-to-home television using the ANIK-B satellite, which has an e.i.r.p. at 12 GHz ranging from 51 dBW to 46.5 dBW within the coverage area of each of its four beams. ANIK-B coverage is illustrated in Fig. 4. According to tests with Hermes, the e.i.r.p. of 49.5 dBW or greater within the inner contour of each beam would be sufficient to provide a very satisfactory TV picture in clear weather to a terminal with a G/T* of 13 dB/°K, using a 1.2 m antenna. A similar grade of service could be achieved in the outer contour ring in which e.i.r.p. dropped to 46.5 dBW if G/T was 16 dB/°K. This could be achieved by in-

* Earth station figure of merit (antenna gain to temperature ratio).

Table 1 — HERMES and ANIK-B and ANIK-C Parameters

	HERMES	ANIK-B		ANIK-C
Frequency Band (GHz)	14/12	6/4	14/12	14/12
Number of Transponders	2	12	4	16
Channel Bandwidth (MHz)	85	36	72	54
EIRP (dBW) Beam Edge	47 and 57	36	47	47
Downlink Beam	2–2½° Steerable Spot Beams	All Canada	4-2° Spot Beams	4-1° × 2° Spot Beams
Design Life (Years)	2	7		10

creasing the antenna diameter of the terminal to 1.8 meters.

ANIK-B, owned and operated by Telesat Canada, is a hybrid satellite, providing 12 transponders for Telesat's operational service at 6/4 GHz, and 4 transponders for experimental service at 14/12 GHz. The principal characteristics of ANIK-B are also listed in Table 1. Most of ANIK-B's 14/12 GHz capacity is released by the Department of Communications for a variety of pilot projects on specialized services which have the potential of becoming operational services.

The key to holding receiver antenna size as small as possible while using ANIK-B in a direct broadcasting mode lay in a design approach intended to minimize signal-level requirements. First, a video signal-to-noise ratio of 42 dB was considered to be appropriate for individual reception. Receiver bandwidth was reduced slightly below that required for broadcast standards. Margins were reduced drastically, allowing only for clear weather conditions. Typical received carrier-to-noise ratio (CNR) achieved on the inner contour using the 1.2 m antennas (or on the outer contour with a 1.8 m antenna) was 4 dB above the static threshold, which is defined as the point where the fm demodulator performance curve deviates from a linear extrapolation of the high performance portion by 1 dB. This is illustrated in Fig. 5. At this level a high-quality picture could be expected, as indicated in Table 2, which shows picture quality as a function of received CNR. Of course, degradation of the received video signal would be caused by rainfall which, in most of Eastern Canada, could be expected to cause attenuation of 3 dB or greater for approximately eight hours each year. Attenuation of 8 dB or greater, causing a complete loss of picture, occurs typically for about 48 minutes each year, during very heavy thunderstorm activity.

To carry out a pilot project demonstrating DBS service, 100 low-cost receiving terminals were purchased, 50 with 1.2-m antennas and 50 with 1.8-m antennas. The receivers are fully tuneable over 500 MHz, and nominal receiver noise figures are 4.5 dB. The terminals consist of three components; the antenna, a feed and outdoor electronics unit, and an indoor electronics unit. Table 3 gives details. In addition to these low-cost receivers, ten fixed-frequency receive-stations using 3-m antennas were acquired for use in fringe areas and cable head-end applications.

The pilot project is being carried out in two of ANIK-B's beams, with half of the receivers located in the center-east beam and half in the west beam. In the center-east beam, one program channel has been transmitted since September, 1979, using the full power of the transponder, for 94 hours weekly. The programming is supplied by the Ontario Educational Communications Authority, as a participant in the trial.

Table 2

dB relative to Static Threshold	Picture Quality
4	No threshold noise
2	Threshold noise just starts to appear on color bars. Not generally noticeable on pictures except those having wide deviation components.
0	Significant threshold noise on color bars, noticeable on picture.
−2	Large amounts of threshold noise on color bars, significant on picture.
−4	Large amounts of noise on all pictures, partial or complete loss of sync.

Static Threshold CNR = 9.0–9.8 dB.

In the portion of the trial being conducted in the west beam, two program channels are being transmitted with one transponder. The resulting power sharing and tube output back-off (total 4.5 dB) require that the 1.8-m receivers be used to achieve satisfactory picture quality. The program providers are the Canadian Broadcasting Corp. and BCTV Ltd., a CTV affiliate. Up to 154 hours of programming have been transmitted weekly since December, 1979.

In both areas, receivers are located in a wide variety of situations, from individual private homes to mining and logging camps, to small communities with local distribution systems such as cable or a low-power transmitter. In general, usage typified the mixed form of reception that could be anticipated in an operational satellite broadcasting system. The method of installation has varied widely, from firm attachment to buildings, to simply being placed in the ground and stabilized with materials at hand. This variety in quality of installation probably typifies an eventual consumer situation.

A program to evaluate performance was carried out by both the Department of Communications and the

Figure 3. Flight model of Hermes Communications Satellite as shown in the final stages of construction in August, 1975.

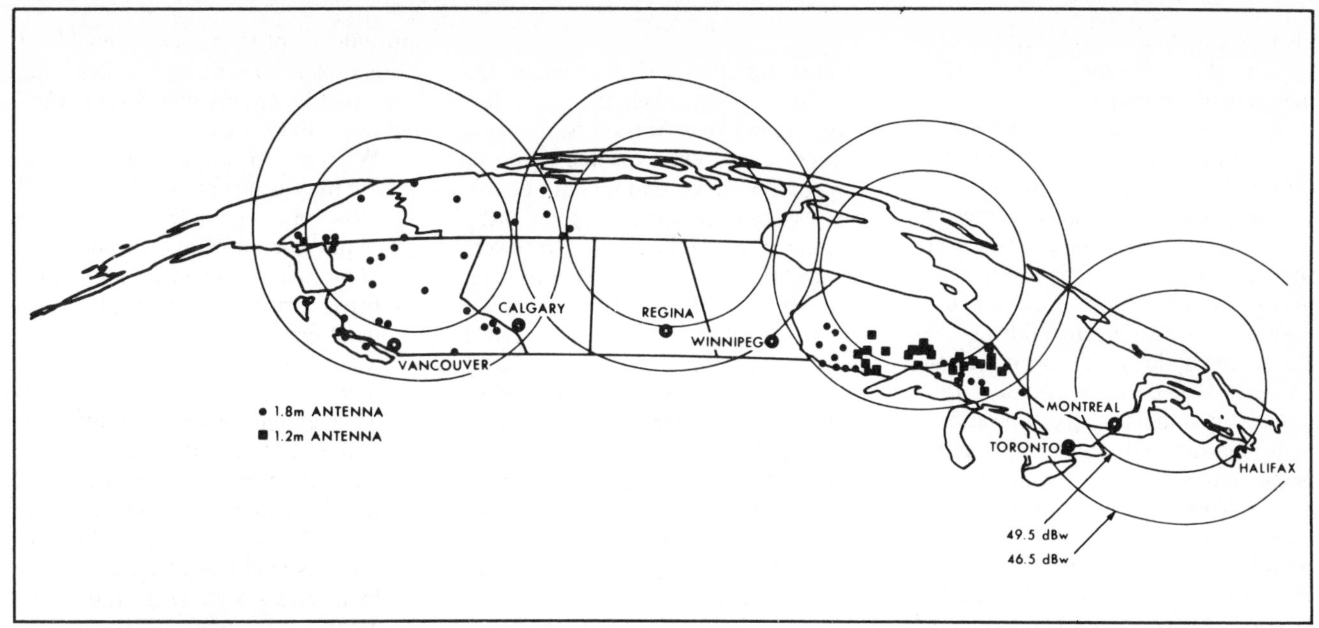

Figure 4. ANIK-B e.i.r.p. contours and low-cost earth terminal locations.

participating broadcasters. The evaluation involved both technical and subjective dimensions and was designed to provide information on the public acceptability of the service, as well as on the performance and reliability of the terminals in an operational environment and in the hands of the general public. Analysis of the results of the ANIK-B trials are ongoing, but a number of conclusions can be drawn:

a) The TVRO terminals operate satisfactorily in a wide variety of environmental conditions.

b) The size of antenna is not a serious factor in any installation. However, use of the 1.8-m antenna did require special consideration of transportation to some areas, and account had to be taken of the greater wind loading, since its area is about 2¼ times that of the smaller antenna. While a 1.2-m antenna could be readily mounted on the average roof, reinforcement is necessary to support the installation of the larger dish.

c) Although the margins for the reception of signals are small, the tendency is always to expand the reception area beyond that planned by the designers, rather than to contract it.

d) No user complained about or noted specifically the degradation or failure of reception during rainfall. It is expected that users will readily accept occasional signal deterioration to the receiver threshold or even below.

e) The satellite signals are usually at least as good and often better than any others available at the selected locations.

Table 3 — Specifications for Low Cost Earth Terminals

Antenna size	1.2m	1.8m
Overall G/T	13.0 dB/K	16.5 dB/K
Video SNR*	42 dB	
Noise figure	4.5 dB	
Tuning range	11.7–12.2 GHz	
RF bandwidth	18 MHz	
Video bandwidth	4.2 MHz	
Peak video deviation	6 MHz	
Audio subcarrier frequency	5.14 MHz	
Peak subcarrier deviation of main carrier	1 MHz	
Peak deviation subcarrier	60 KHz	
Peak energy dispersal deviation	200 KHz	

* At a C/N 4 dB above static threshold.

Implications of the ANIK-B Field Trials

An important and immediate implication of the successful experience with the ANIK-B direct broadcasting field trial is the possibility of using ANIK-C as a direct broadcasting satellite. Table 1 shows the similarity of its characteristics with those of the 14/12 GHz transponder on ANIK-B. The first ANIK-C is scheduled for launch in November,1982, and will be ready for service in January, 1983, with a second satellite available for service in June, 1983. By transmitting one television signal per travelling wave-tube amplifier, and using ANIK-C's half-Canada beams, up to eight programs can be made available. This should be an adequate number to encourage those consumers without access to cable TV to purchase receivers with 1.2-m antennas necessary for reception.

Of course, the availability of an attractive assortment of channels is an important prerequisite for market acceptance of DBS service. Pay TV and commercially-supported popular-appeal channels should be available, as well as CBC and publicly supported educational/instructional channels. Benefits would include extension of services to all Canadians, market build-up prior to the establishment of a higher-powered and more costly DBS system, opportunity for the broadcasting industry to adapt gradually to this new technology, and a head start for Canadian DBS receiver manufacturers against foreign competition.

Other implications concern the features of a specifically designed system for direct broadcasting service for Canada. The technical and operational experience gained as a result of the field-trial program has provided confidence in the performance expectations that can be achieved with an operational direct-broadcasting satellite. A satellite signal is very steady with only occasional deterioration occurring, predominately because of rain attenuation which has readily determined statistics. The nature of picture deterioration because of the FM signal falling to or below receiver threshold also appears to be less objectionable to viewers than low-level signal performance with terrestrial AM television broadcasting, permitting link design with less margin than engineers traditionally like to use.

There has been a steady progression from operation with large margins in the initial experiments to low margins in the later trials, with important economic implications. In addition, receiver technology has been demonstrated to be evolving rapidly, while low-cost manufacturing prospects have been maintained, giving far better performance than was assumed at the 1977 World Administrative Radio Conference that dealt with the broadcasting satellite service.

These factors indicate that a satisfactory service could be introduced with a DBS system employing a space segment considerably lower-powered than the high e.i.r.p. baseline used at the 1977 broadcasting WARC. This is particularly important for countries such as Canada where a relatively small population must be served. For example, a system providing eight television channels in each of four beams covering Canada has been estimated to cost over $600 million (1981 Canadian dollars) if it provided an edge-of-coverage e.i.r.p. of 58 dBW, whereas a system providing an edge-of-coverage e.i.r.p. of 54 dBW has been estimated to cost $200 million less.

This amount is significant at the start of service, when there are few receivers in place. Market penetration of receivers can be expected to follow a normal S-curve, and it could be five years, or past the half-way point of the lifetime of the satellite, before receivers are installed in significant numbers. Since a 54 dBW edge-of-coverage e.i.r.p. would result in a signal above the receiver threshold for 99.9% of time in Canada, assuming a receiver of up-to-date technology and using a 1-m antenna, this would constitute quite acceptable first-generation service. In later generations, after market penetration approaches saturation, it would be much easier to support higher e.i.r.p. systems.

Planning for a Dedicated DBS System for Canada

The DBS field trial has increased awareness of the technology in Canada among public and commercial broadcasters, education authorities, government agencies, and the general public. This has brought out numerous non-technical as well as technical issues concerned with the introduction of this technology into the broadcasting environment (perhaps in the late 1980's). Canada has a well-established broadcasting industry, and there is natural concern on the part of the broadcasters and the government that introduction of DBS service should not be disruptive to existing broadcasting. Specifically, among questions requiring examination are:

- What are the requirements? Who are the customers for DBS service? How can regional and language requirements be accommodated? How many channels and what beam arrangements would best meet needs?
- Should DBS be supported commercially, through subscription, by government, or by a combination of these? How much would viewers be willing to pay for improved television signals delivered in this way? What new, non-television services could be offered feasibly by a DBS system (e.g., radio, teletext)?
- What impact would DBS have on commercial broadcasters, particularly those serving areas of small population and meeting local needs?
- What regulatory changes would be necessary to accommodate DBS-delivered services?
- What technical parameters should be Canadian DBS system have, keeping in mind the Canadian market and the vast area to be covered?
- What industrial impact, from a hardware and software viewpoint, could be expected from implementation of a DBS service?
- What are the economics of providing television with a DBS throughout the entire country?
- What institutional arrangements would be most suitable for implementing a DBS service, and how can those currently in satellite services and broadcasting be best involved?
- What impact can be expected from unavoidable spillover from U.S. DBS satellites?

These questions have resulted in the Department of Communications undertaking a comprehensive, multidisciplinary study program covering requirements, socio-economic and policy/regulatory, as well as technical topics. General objectives of the study program are to:

- Develop a strategic plan for the possible introduction of a DBS system for Canada.
- Develop requirements and technical, economic, institutional, and policy information about a DBS system for Canada so that informed decisions can be taken re-

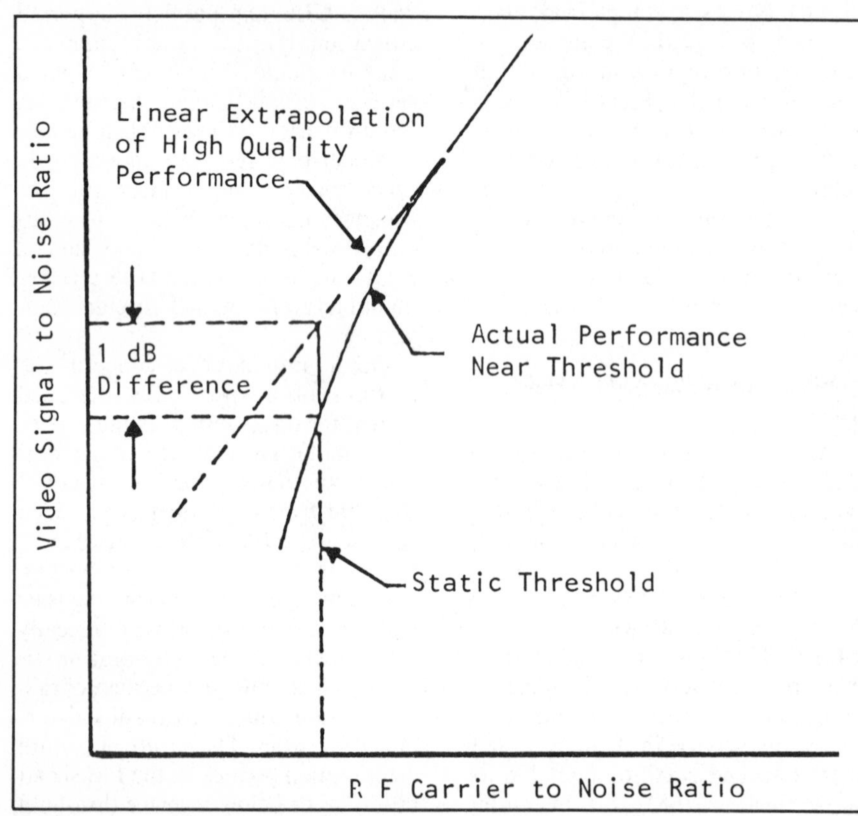

Figure 5. Graphic definition of static threshold.

garding the implementation of a DBS service.

- Document Canada's requirement for spectrum for direct broadcasting for the Regional Administrative Radio Conference which will allocate spectrum for the broadcasting-satellite service to the Region 2 countries.

The studies are scheduled to be completed at the end of 1982 with the publication of a general report discussing all of the above issues and outlining technically a system meeting Canadian needs as economically as possible. Public reaction to the report should supplement the information contained in it to permit formulation of a policy regarding implementation of a DBS system for Canada.

Some results are already available. Eighty-four percent of Canadian TV households, some 6 million, have access to television via cable distribution systems, which carry some 15 separate channels, and about 4.5 million subscribe to cable. The remainder, some 1.6 million homes, are in rural areas and for the most part are too dispersed for cable delivery to be economically viable. About 1 million of these homes have three or fewer channels of TV and could be considered as underserved, with about 260,000 people, or 80,000 households, receiving no television. Thus there is a substantial disparity in service between rural and urban Canada, as illustrated in Fig. 6. It is not surprising, therefore, that a market-demand survey conducted among rural

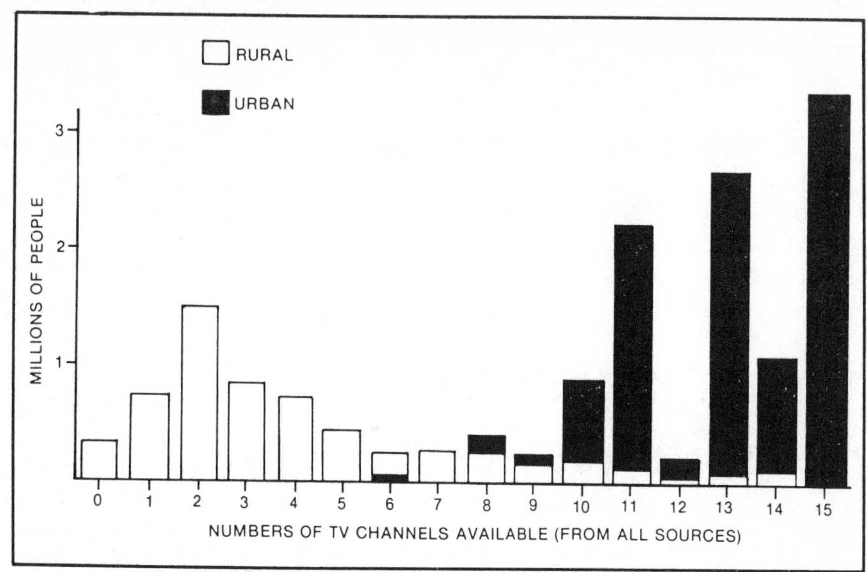

Figure 6. Disparity of service between rural and urban Canada.

Canadians has yielded the results shown in Table 4, which shows the market and expected penetration vs. the price of a DBS receiver. Of particular interest is the indication of the very rapid adoption of the technology. Within three years of the start of service, more than half of the potential purchasers at each price level would buy DBS receivers.

Clearly, the Canadians who would benefit most from a DBS system are those 1.6 million households, yet this is too small a market to support a DBS system on a commercial basis. A possible answer is to increase the potential market by making DBS programming available for carriage on cable, thereby considerably increasing potential viewers. This leads to the observation that programming on a DBS should be somewhat different from that already available, so that it is attractive to both the cable and direct-to-home markets. One could also conclude that new national or regional program offerings should use DBS as the delivery technology.

A specific case in Canada at the present time is Pay TV, which will be introduced next year. If delivered as a DBS signal, it could be equally available to rural subscribers using individual home terminals as well as to urban subscribers served with community antenna- and cable-distribution systems. Newly developing regional education networks also fall into this category. When full penetration of the individual home-receiver market is achieved, extra revenue from it should compensate for the difference in cost between using a DBS and a conventional satellite for distribution.

Table 4 — Long-Term Demand Forecasts for Improved Television Service through Satellite Technology

Penetration (in Thousands) vs. Price Level

YEAR	$400	$600	$800
1	298	197	151
2	361	260	210
3	323	256	220
4	220	189	171
5	124	112	105
6	62	58	55
7	29	28	27
8	14	13	13
9	6	6	6
10	3	3	3
11	1	1	1
12	1	1	1
Total Households Purchasing After 11 Years (in Thousands)	1,441	1,121	961
Potential Market	1,442	1,122	962

Conclusions

Television broadcasting from satellites directly to low-cost individual home receivers will be widespread in many countries before the end of the decade. Canadian field trials have been very valuable in providing experience with DBS operations and in increasing awareness of the potential of DBS technology. They have demonstrated that satisfactory service can be achieved with minimum satellite-link margins, a conclusion which has important implications for satellite costs and the economics of DBS service. The field trials are being complemented by a comprehensive study-program which will culminate in a strategic plan for introducing DBS service in Canada. Such service could well start as early as 1983 by making use of ANIK-C in the interim before a dedicated system could be launched.

References

1. Davies, N. G., Day, J. W. B., Jelly, D. H., and Kerr, W. T. "CTS/HERMES Experiments to explore the Applications of Advanced 14/12 GHz Communications Satellites," presented at the International Astronautical Federation 29th Congress, Dubrovnik, Yugoslavia, October, 1-8, 1978.
2. Bowen, R. R., Billowes, C. A., Day, J. W. B., and Douville, R. J. "The Development of a Canadian Broadcasting Satellite System at 12 GHz," International Conference on Communications. Conference Record 51.3.1-51.3.5.
3. Roscoe, O. S., "Planning for a Canadian Direct Broadcasting Satellite System," presented at the XXXI Congress of the International Astronautical Federation, Tokyo, Japan, September 21-28, 1980.
4. Billowes, C. A., Bowers, P. G., Rose, E. G., "Low-Power Direct Broadcasting Satellite Trials in Canada," presented at the International Broadcasting Conference, Brighton, England, September 20-23, 1980.
5. Roscoe, O. S., Davies, N. G., Kerr, W. T., "Canadian Experience with Satellite Direct Broadcasting and the Implications for the Future," presented at the Fourth Jatel Conference, Quito, Ecuador, Nov. 16-21, 1981.

Indexing Terms: Television receivers, Satellites, broadcasting

Direct broadcast satellite receivers

K. G. FREEMAN, B.Sc., C.Eng., MIERE*

Based on a paper presented at the IERE Conference on Radio Receivers and Associated Systems held in Leeds in July 1981

SUMMARY

Broadcasting of television and radio programmes from satellites to individual receivers is now technically possible. For Europe, allocations of up to five channels in the 12 GHz band to each country have been authorized and several countries plan to implement services in the next few years. The required characteristics of receivers for individual reception such as sensitivity, selectivity, signal processing, etc. are reviewed and it is shown that these can be satisfied by available technology.

* *Philips Research Laboratories, Redhill, Surrey RH1 5HA.*

1 Introduction

The feasibility of broadcasting direct to the home from a geostationary satellite was recognized at least two decades ago and by the mid-1960s the receiver industry was already considering the implications. In 1971 the World Administrative Radio Conference (WARC) allocated a number of frequency bands for satellite broadcasting, either exclusively or on a preferentially-shared basis with other services. Of particular interest for Europe (Region 1) was the allocation from 11·7 to 12·5 GHz, which offered multi-channel capability free from interference to or from existing terrestrial u.h.f. services. The WARC in 1977 made detailed allocations of channels and other parameters for broadcasting in this band—not merely to permit orderly implementation of satellite broadcasting but also for the benefit of other services permitted to share the band. Table 1 shows these allocations as modified by a subsequent WARC in 1979.

Table 1

Satellite broadcast allocations

Band (GHz)	Frequency range (GHz)	Region*	Notes
0·7	0·62– 0·79	All	Television only, subject to interference constraints
2·5	2·5 – 2·69	All	Community only
12	11·7 –12·1	1, 3	
	12·1 –12·2	All	
	12·2 –12·5	1, 2	
	12·5 –12·7	2, 3	Community only in Region 3
	12·7 –12·75	3	Community only
23	22·5 –23	All	
42	40·5 –42·5	All	
85	84–86	All	

* Region 1 Europe and Africa, Region 2 North and South America, Region 3 Asia and Australasia.

Although there was (and perhaps still is) a widespread misconception that satellite broadcasting would enable the individual to receive an almost unlimited selection of programmes from space, technical (and political) considerations strongly favoured a system of national satellites providing total coverage of individual countries with a limited number of programmes. A prime factor was the decision that programmes should be capable of individual reception anywhere in a country—although in urban areas community reception via cable distribution systems (often already existing) was likely as a convenient means of cost-sharing. Another consideration was frequency re-use without interference, so that each of the many countries involved in the European plan could be allocated enough channels to make a new service and new receiving equipment worthwhile.

2 Allocations and Transmission Parameters

In planning for satellite television broadcasting at 12 GHz a number of assumptions were made—in

Reprinted with permission from *Radio Electron. Eng.*, vol. 52, pp. 127–133, Mar. 1982.
Copyright © 1982 by the Institution of Electronic and Radio Engineers.

particular the use of existing 625-line colour television signal standards (with video signal base-bandwidth of about 5 MHz). Considerations of likely satellite power, receiver sensitivity, aerial directivity and permissible co-channel interference showed that for a practicable system of essentially national services conventional amplitude modulation was precluded. A plan was therefore evolved assuming wideband frequency modulation, but other systems, such as digital modulation, are not excluded. Eventually, after considerable computer modelling, a scheme providing five 27 MHz wide channels for each country was arrived at (although some groups of adjacent countries have opted for some channel sharing). Although allocated primarily for television, one or more channels may, if desired, be used by each country for other purposes, such as multiple programme sound broadcasting. However, parameters for such use have yet to be agreed.

Table 2

Probable 12 GHz television standards

baseband video	standard 625-line
baseband audio	f.m. subcarrier
12 GHz carrier modulation	f.m.
carrier deviation	13·3 MHz p-p
pre-emphasis	CCIR Rec. 405–1
signal bandwidth	~ 27 MHz
energy dispersal	600 kHz p-p

In arriving at the parameters for television broadcasting the following was assumed (Table 2):

(1) Wideband f.m. would be used with 13·5 MHz p-p deviation for a 1 V composite video signal at ~ 1·5 MHz (pre-emphasis according to CCIR Recommendation 405–1 being assumed). With a single subcarrier sound signal, having a peak-to-peak deviation of 2 MHz, this would give a signal bandwidth of about 27 MHz. The possibility of two (or more) television sound subcarriers for stereo or a second language commentary is, however, not precluded.

(2) Co-channel (and hence image) and adjacent channel protection ratios would be 31 dB and 15 dB respectively.

(3) The receiving system would have a nominal G/T of 6 dB/K—equivalent to a 0·9 m diameter receiving aerial of agreed polar response with a front-end noise figure of 6 dB and a 2 dB allowance for pointing and other losses.

(4) The received flux should yield good pictures within the service area even after allowance for satellite and receiver aerial pointing errors, some transmitter power loss with life and propagation losses due to rain, etc. (which were assumed to be < 3 dB for 99% of the worst month of the year).

(5) Satellite station-keeping and aerial pointing accuracy would each be $< \pm 0{\cdot}1^\circ$. Satellite locations would be chosen to minimize co-channel interference and to ensure that loss of solar power during the nocturnal satellite eclipses at the spring and autumn equinoxes would not occur until well after midnight when transmissions would have ceased.

(6) The service area 'footprint' of each satellite would be limited as far as practicable to the shape of the receiving country.

Table 3

12 GHz allocations—Region 1

channel spacings	19·18 MHz
(signal bandwidth	27 MHz)
number of channels	40
channels/country	5
national 'spread'	360 MHz
polarization	circular (L or R)
nominal power flux at Earth's surface	-103 dBW/m^2
satellite locations	$+5^\circ$, -1°, -19°, -25°, -31°, -37°

Based on these assumptions the following plan was arrived at (Table 3):

(1) The design flux (99% worst month, etc.) at a receiving installation at the edge of the service area is -103 dBW/m^2. For a geostationary satellite in a 42 000 km radius equatorial orbit this is equivalent to an e.i.r.p. of about 62 dBW. For a satellite with a 1° beam (~ 1·5 m dish), suitable for the UK, this corresponds to a satellite r.f. power of around 250 W, which is now practicable.

(2) A limited number of satellite locations have been chosen so that in many cases signals can be received (albeit at lower quality) from satellites serving adjacent countries without aerial re-pointing. For most large West European countries satellites are located at 19°W with the exception of those of the UK, Eire, Portugal and Spain which are at 31°W.

(3) Although slightly less convenient for the receiver, circular polarization (left- or right-hand) will be used since it eases satellite aerial design (and, incidentally, receiving aerial alignment).

(4) Channel spacing is set at 19·18 MHz which yields 40 channels in the 800 MHz band. Note that with a *signal* bandwidth of 27 MHz some overlap appears to occur, but adjacent channels are not allocated to the same or adjacent service areas and the necessary protection is allowed for.

(5) As stated, most countries have been allocated 5 channels, but there are some exceptions, e.g. the Scandinavian countries by arrangement have 3 national channels each and 2 common channels.

(6) For most countries the spread of allocations is limited to 360 MHz in the top or bottom half of the band, with guard bands 3 channels wide. This simplifies receiver design.

(7) To reduce potential interference to terrestrial fixed services energy dispersal will be used consisting of an additional 600 kHz p-p triangular deviation of the carrier at a sub-multiple of field frequency, e.g. 25 Hz.

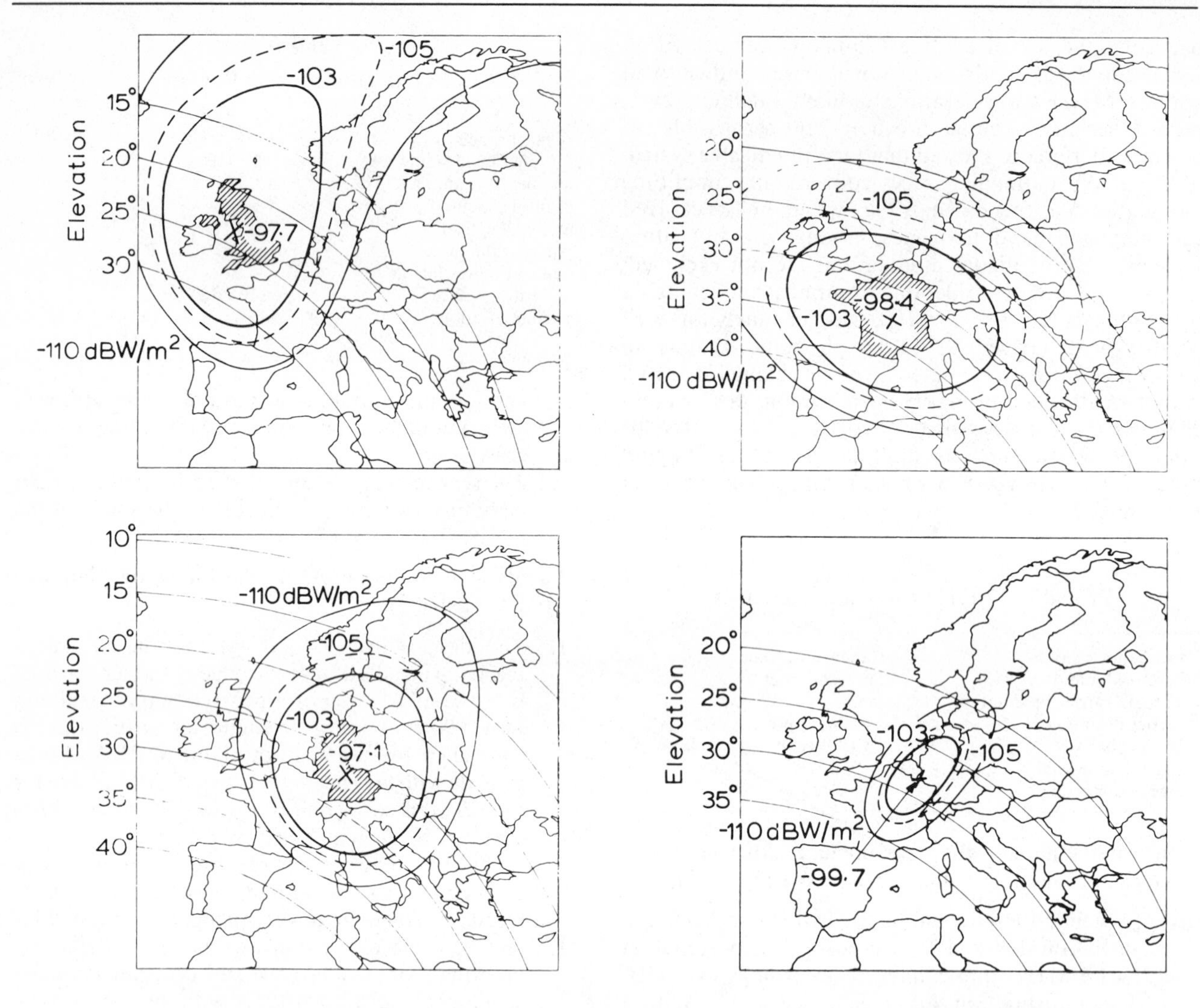

Fig. 1. Parameters and footprints for satellite coverage of the United Kingdom, France, West Germany and Luxembourg

	Parameters						
	Beam	Polarization	Channels	Peak e.i.r.p.	Orbital position	Orientation	Antenna peak gain
United Kingdom	1·8° × 0·7°	1	4, 8, 12, 16, 20	65·2 dBW	31°W	142°	43·0 dB
France	2·5° × 1·0°	1	1, 5, 9, 13, 17	64·0 dBW	19°W	160°	40·4 dB
West Germany	1·6° × 0·7°	2	2, 6, 10, 14, 18	65·7 dBW	19°W	147°	43·6 dB
Luxembourg	0·6° × 0·6°	1	3, 7, 11, 15, 19	63·1 dBW	19°W	0°	48·7 dB

Figure 1 summarizes the characteristics of the UK, French, West German and Luxembourg allocations.

3 Broadcast Satellite Plans

Within the next few years a number of European countries are expected to implement a 12 GHz satellite television service. West Germany and France, in particular, are already committed to launching full service satellites in 1983/4. These will probably have two accompanying sound channels and signals from the latter will be capable of reception with good quality in at least the southern part of the UK. Luxembourg, Switzerland and the Nordic countries are also known to be seriously considering introducing services. In the UK, with use of our fourth u.h.f. channel only recently authorized and the future use of the now obsolescent 405-line v.h.f. channels not yet resolved, there seems no urgency to do anything—but it is nevertheless argued that we should not fall behind. However, in Europe as a whole, it is clear that in only 2 or 3 years' time a rapidly growing receiver market is likely and the industry needs to be ready.

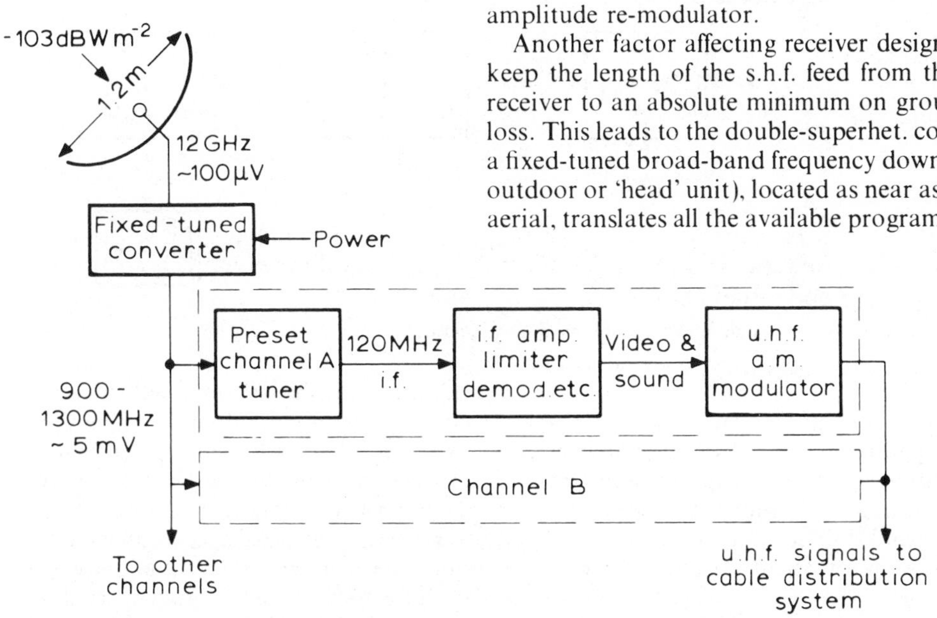

new satellite signals. However, since most households already have at least one receiver designed only for the existing service, the initial requirement will be for one or more add-on units capable of converting the s.h.f. f.m. signal to a u.h.f. a.m. signal suitable for the standard receiver (see Fig. 3). This is an unwelcome complication, but already proposals exist for fitting future standard receivers with a video and audio signal interface socket and this would at least remove the need for a u.h.f. amplitude re-modulator.

Another factor affecting receiver design is the need to keep the length of the s.h.f. feed from the aerial to the receiver to an absolute minimum on grounds of cost vs. loss. This leads to the double-superhet. concept, in which a fixed-tuned broad-band frequency down-converter (the outdoor or 'head' unit), located as near as possible to the aerial, translates all the available programme channels to a lower intermediate frequency, typically in the region of 1 GHz. Low-cost, low-loss cables for the downlead are then readily available. A second (indoor) unit associated with each standard television receiver then provides channel selection and signal demodulation to yield the required composite video colour signal and one or more subcarrier sound signals, which may then be suitably processed for feeding to the standard receiver.

An important advantage of this double superhet. approach, using a common outdoor unit and separate indoor units, is that, in the increasingly common situation where two or more receivers are used in the same household, the facility for viewing different national programmes simultaneously is retained. Indeed, in many European countries it will even be possible, without equipment modification, simultaneously to receive (albeit at lower quality) programmes of adjacent countries whose signals have the same polarization and satellite position and fall within the same half of the 12 GHz band. Only when one or more of these criteria is not met will it be necessary to have more complex receiving equipment.

Fig. 2. Outline of typical community television receiving system.

4 Television Receiving Systems

The form of receiving equipment required for satellite television reception will obviously depend on whether environmental and other circumstances compel or favour individual or small or large community systems. Of these the first is of greatest interest and poses the most difficult task for the manufacturer, since large quantities will be required having adequate performance at an economic price. Attention will therefore be concentrated on such receivers, since for large communities, particularly where cable distribution systems already exist, the problems are much less severe. For these, although multiple (and more sensitive) receivers will be required for simultaneous multi-channel signal reception and reprocessing (to v.h.f. or u.h.f. a.m.) for distribution, the greater cost can be shared between a large number of viewers. A typical scheme is outlined in Fig. 2. More modest versions could be used for small communities such as a large block of flats or a small housing estate.

Let us therefore examine in more detail the requirements of an individual receiving system for a single household, assuming frequency modulation of the 12 GHz carrier by a conventional composite video colour signal (e.g. PAL) accompanied by one or more f.m. sound subcarrier signals.

In the longer term we can expect there to be composite receivers on the market capable of receiving both conventional u.h.f. a.m. terrestrial broadcasts and the

5 Individual Satellite Television Receiver Components

Having discussed the general requirements for a satellite television receiver the various sub-units will be considered in rather more detail. (See Fig. 4.)

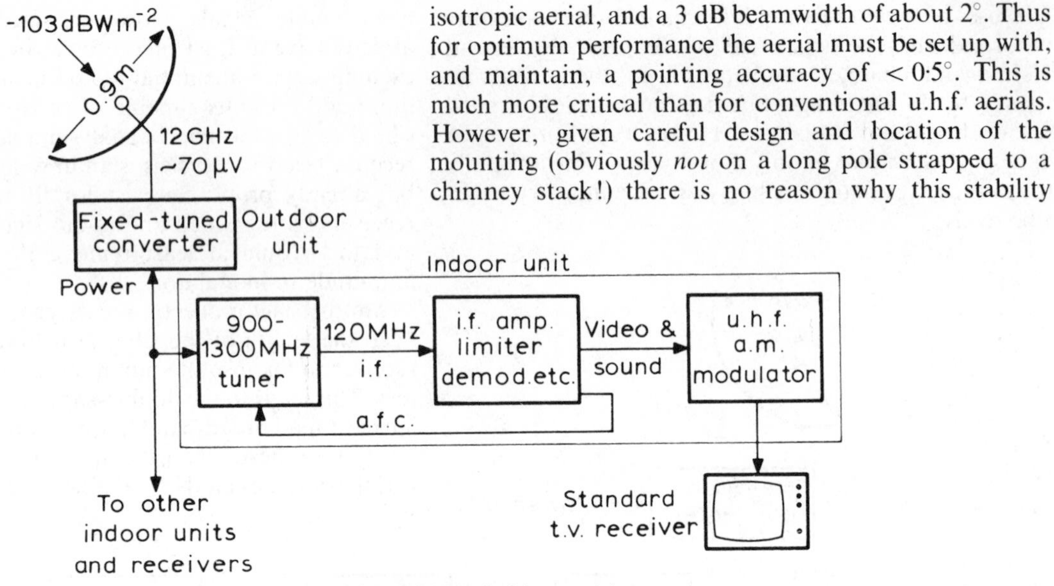

Fig. 3. Outline of typical individual television receiving system.

isotropic aerial, and a 3 dB beamwidth of about 2°. Thus for optimum performance the aerial must be set up with, and maintain, a pointing accuracy of < 0·5°. This is much more critical than for conventional u.h.f. aerials. However, given careful design and location of the mounting (obviously *not* on a long pole strapped to a chimney stack!) there is no reason why this stability should not be maintained under all reasonable environmental conditions. As regards initial alignment, experience has shown that given a simple compass and spirit-level (which might be an integral part of the aerial assembly) it is possible to provide initial mechanical alignment sufficiently accurately for a signal to be obtained from the satellite and then to use the signal itself for final precise alignment. It should be noted that the required aerial elevation and bearing will depend on the latitude and longitude of the receiving location relative to the satellite position. For the UK the elevation will range from about 28° in the Scilly Isles to about 17° in the Shetlands and the bearing from about 27°W of S in the Outer Hebrides to about 39°W of S in Kent.

It is generally assumed that the most likely form of aerial will be a prime feed or Cassegrain parabolic reflector with a waveguide feed from the focus. Depending on quantities and relative economics the reflector could either be pressed from metal, or moulded in, for example, glass reinforced plastic (fibre-glass), with an embedded conducting mesh, or a suitably-protected metallic surface coating. In either case, to achieve the required gain and polar response (Ref. 1 and Fig. 5) the profile must be accurate to within ±1 mm.

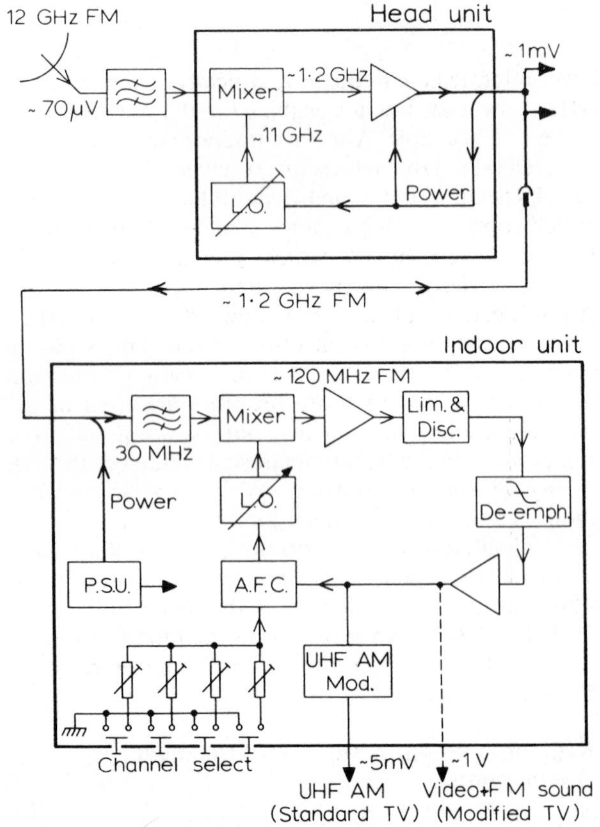

5.1 The Aerial

In the system planning an individual receiver aerial equivalent to a parabolic dish of 0·9 m diameter was assumed as being a reasonable compromise in terms of gain, directivity (including side-lobe response), required pointing accuracy and stability and, of course, cost. Such an aerial has a gain of about 37 dB relative to an

Fig. 4. Double superheterodyne 12 GHz f.m. converter for individual reception.

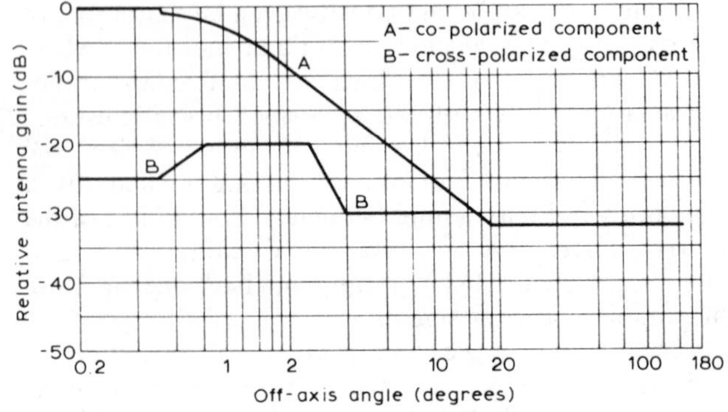

Fig. 5. Individual receiver aerial response.

However, provided they can meet the specification other solutions are not excluded. For example, some years ago the BBC suggested a Fresnel reflector.[2] This is essentially a sliced/compressed paraboloid which offers the interesting possibility of the choice of an offset angle of maximum response. There is also continued speculation about the prospect of arrays of small printed dipoles on a large dielectric panel. However these are still a very long way from yielding the required gain and directivity.

5.2 The S.H.F. Feed

The s.h.f. feed to the head unit has the additional functions of providing rejection of image frequency signals, suppression of local oscillator power radiation into the sky and selection of the appropriate signal polarization—if necessary with conversion to linear polarization for interfacing to the input circuit of the outdoor unit. Possible arrangements are a short circular waveguide feed, with the outdoor head unit located near the focus of the main reflector, or a Cassegrain (dual-reflector) arrangement with the head unit located behind the reflector.

In principle the problems of image signal rejection and oscillator radiation suppression can most simply be solved by choosing an 'oscillator-low' configuration. In this case the necessary filtering is readily achieved by virtue of the fact that a suitably chosen feed diameter will have a high-pass characteristic. This would be satisfactory in most locations but in some situations high-power radars in the 10 GHz band could be a potential source of interference. The 'oscillator-high' configuration may therefore be preferable in this respect, but involves extra complication, such as the insertion of iris diaphragms into the waveguide to yield the required bandpass response. Selection of the required polarization, and its conversion to linear polarization suitable for the head unit if required, may be achieved without too much difficulty, for example by insertion of appropriately dimensioned and spaced posts into the waveguide. However, where it is desired to be able to select either of the two circular polarizations at will by means of a switch, the feed will become much more complex and expensive.

5.3 The Head Unit

This unit converts and amplifies the band of received signals from 12 GHz (at a level $\simeq$ 70 μV) down to a first i.f. low enough (e.g. $\simeq$ 1 GHz) for feeding to the indoor unit via conventional coaxial cable (at a level of $\sim$ 5 mV). For reasons associated with the design of the indoor unit tuner it is desirable to restrict the bandwidth to around 400 MHz. The 5 channels assigned to most countries have therefore been arranged to fall in either the upper or lower half of the 800 MHz total band from 11·7 to 12·5 GHz. To minimize the risk of interference from other services, especially u.h.f. television, the first i.f. band is likely to be chosen to be 900–1300 MHz.

Until relatively recently, in the absence of low-cost s.h.f. pre-amplifiers, prototype down-converters have consisted of a direct mixer input, using Schottky barrier diodes, followed by i.f. amplification. One possible solution based on this approach uses conventional microstrip technology, in which distributed-line circuits are evaporated on alumina or a similar dielectric substrate, and incorporates a balanced mixer.[3] It is now known that with this arrangement it is difficult to achieve a noise figure much below 6 dB. A second solution developed by Konishi[4] consists of a single-ended mixer on a planar-in-waveguide configuration and has yielded noise figures approaching 4 dB.

To drive the mixer requires 10–20 mW of local oscillator power at s.h.f. Assuming automatic frequency control can be applied to the second local oscillator in the indoor unit, this s.h.f. oscillator must be stable to 1 or 2 MHz over an outdoor unit temperature range which in some parts of Europe can be 60°C or more. An obvious solution is to use a v.h.f. crystal oscillator followed by varactor or step-recovery diode multiplication, but the arrangement usually preferred has been a Gunn transit-time device in a low-cost waveguide cavity with dielectric temperature compensation (e.g. by means of a TiO_2 post) to obtain the required stability.

However, the last few years have seen rapid developments in low-noise GaAs field-effect transistors so that all-transistor head-units, with discrete-transistor s.h.f. preamplifiers, mixers, i.f. amplifiers and dielectric-resonator-stabilized s.h.f. oscillators, having a noise figure of less than 4 dB are now technically possible.[5,6] Given sufficient market impetus these should come down in price to an acceptable level. In the meantime, work is in progress aimed at monolithic integration of such units. In all cases it is a simple matter to provide power for the head-unit via the i.f. downlead.

It should be noted that whilst the 2–3 dB improvement in head-unit noise figure offered by these developments, as compared with the figure assumed originally for the planning, is not essential, it does offer a greater margin above threshold against propagation-fading and alignment errors, as well as the prospect of better reception of signals primarily intended for neighbouring countries.

5.4 The Indoor Unit

The remaining signal processing necessary to yield the desired satellite television programme will be provided by an indoor unit, which in the early stages of a service will be a separate unit associated with each standard receiver. This will incorporate a number of functions.

First, a 900–1300 MHz tuner of essentially conventional design, with automatic frequency control of the local oscillator, but having an output bandwidth of $\sim$ 27 MHz, will be used to select the required channel and convert it to the final i.f., which seems likely to be chosen to be around 120 MHz. Since the incoming f.m. signal will have better immunity against interfering signals than conventional a.m. (cf. Sect. 2) the requirements on image and adjacent channel rejection will be less stringent. The main i.f. selectivity will, however, need to have a carefully designed flat response over a 27 MHz bandwidth with a steep roll-off beyond, together with a good phase response, to keep intermodulation distortion products down to an acceptably low level. With conventional lumped LC elements this would require at least a four-pole filter configuration, but an attractive alternative is the acoustic surface wave filter, which can yield the desired response with only a single component.

The following functions of i.f. amplification and frequency demodulation can already be realized using available integrated circuits and the design of suitable mass-production custom i.c.s should present no real problems. After demodulation the necessary h.f. de-emphasis can be applied together with removal of the energy dispersal waveform. It is expected that in a domestic installation this latter function can be adequately performed by a simple d.c. restorer. (For community systems, however, more sophisticated techniques such as keyed-clamping, may be desirable.) After filtering the composite video colour signal and accompanying intercarrier sound signal(s) are available for further processing as necessary. From this point also can be derived the automatic tuning control for the indoor unit.

As indicated above, beyond this point the form of signal processing required will largely depend on the input facilities available in the domestic television receiver. For existing receivers designed only for u.h.f. input of an a.m. video signal with only a single sound carrier signal it will be necessary to convert the sound subcarrier deviation, etc. to the standard form (e.g. by demodulation and remodulation) and to remodulate this and the video signal to provide the required u.h.f. signal. If the satellite channel programme also has more than one accompanying sound signal it will also be necessary to select the one desired.

None of these processing functions poses any fundamental difficulties but they will obviously increase cost and could degrade the overall performance. Clearly, the planned availability in the near future of standard receivers with a composite video and audio input socket will offer useful advantages in both these respects.

In the longer term, of course, when the new satellite services become more generally available we can expect to see the additional satellite-reception indoor functions combined in an optimized cost-effective way in a dual purpose (satellite/terrestrial) television receiver.

6 Sound Broadcast Reception

The 12 GHz broadcast plan allows the use by each country of one or more of its channels for multiple-programme sound broadcasting. So far no standards for this purpose have been laid down but Ref. 7 lists some possibilities. An important factor affecting the choice is the resulting receiver complexity. For example, in a domestic installation it seems desirable—if not essential—to use at least the same aerial and outdoor unit as for television. This probably precludes the use of wideband (~ 600 kHz) simple f.m. systems, which although efficient from the point of view of satellite power and the number of programmes possible in a 27 MHz channel, require very high receiver local oscillator stability. At some cost in power efficiency and number of programmes, it therefore seems preferable to multiplex the different programmes in some way prior to frequency modulation of the 12 GHz carrier. If this is achieved within the 5 MHz video baseband the existing satellite television indoor unit could then also be used to recover this multiplexed signal at the video output point. It would then be a relatively simple matter to retrieve the required programme by de-multiplexing and demodulation, either in a separate unit or as an extra part of the existing television indoor unit. Possible methods include analogue f.m. in frequency division multiplex (f.m./f.m.) or digital signals in time division multiplex (e.g. p.c.m./f.m.). In view of the general move towards digital techniques and the likely inter-modulation problems with f.m./f.m., systems of the latter type would appear more attractive, but studies are not yet complete.

7 Conclusions

Satellite television broadcasting services will be introduced by a number of European countries in the next two or three years. The requirement for mass-produced equipment for individual reception of such signals has been shown to be realizable and development of low-cost solutions is now in progress. Although not yet resolved, solutions also exist for multiple sound broadcasting which are largely compatible with those for television, thus offering maximum economy for dual-purpose equipment in the home.

8 References

1 CCIR Report 810, 'Broadcasting—satellite service (sound and television)—Reference patterns and technology for transmitting and receiving antennae—Kyoto 1978, Vol. XI, pp. 271ff.

2 Millard, G. H., 'Satellite Broadcasting—An Aerial Design for Domestic Reception in the 12 GHz Band', BBC Report RD 1970/27.

3 Freeman, K. G., 'Experimental direct broadcast reception of 12 GHz television signals from the Canadian Communications Technology Satellite', *The Radio and Electronic Engineer*, **47**, no. 5, pp. 234–6, May 1977.

4 Konishi, Y., '12 GHz band f.m. receiver for satellite broadcasting' *IEEE Trans. on Microwave Theory and Techniques*, **MTT-26**, no. 10, p. 720, October 1978.

5 Dessert, R., Harrop, P., Kramer, B. and Vlek, T., 'All-f.e.t. front-end for 12 GHz satellite broadcasting reception', Proceedings of 8th European Microwave Conference, 1978, p. 638.

6 Dessert, R. and Harrop, P., '12 GHz f.e.t. front-end for direct satellite t.v. reception', Eurocon 80, Stuttgart, 1980.

7 CCIR Report 215–4 'Systems for the broadcasting satellite service (sound and television)', Kyoto 1978, Vol. XI, pp. 163ff.

Manuscript first received by the Institution in May 1981 and in final form on 8th October 1981

(Paper No. 2130/Comm 333)

DESIGN OF 12 GHz SATELLITE BROADCASTING RECEIVER TO NEW TV STANDARDS

D. Garrood
GEC-McMichael

The WARC 1977 proposals for DBS in Europe set out a comprehensive plan to provide high quality Television signals to domestic viewers. This was to be acheived via a series of geostationary satellites using receive terminals with a small antenna (a reflector of about 0.9 metres diameter).

A number of assumptions were made under the WARC plan with respect to the receive equipment in order to allow the overall system configuration to be defined. Prime among these were:

1) G/T of the receive terminals should be 6 dB
2) The noise figure of the front-end amplifier which would be readily available at consumer price levels would be 9 dB.
3) The antenna would have to be approximately 0.9 m in diameter as a result of the noise figure assumed.
4) A carrier to noise ratio of 14 dB would be required to obtain an acceptable picture.

Since 1977 however, technology has moved on in a number of ways, in particular with respect to the noise figure of the front-end amplifier. Already noise figures of 3 dB are commonplace and some further improvement can be expected before the service becomes operational in 1986. Also thresholds of FM demodulators may now be as low as 8 dB, thus the 14 dB C/N minimum assumed is perhaps an overkill for the domestic viewer as a worst case figure.

ANTENNAE

With such low noise amplifiers the antenna temperature takes on greater significance and further improvements in amplifier performance have less and less effect on the overall system temperature.

The system temperature (To) is given by:

$$To = T(antenna) + T(rx)$$

where T(rx) is the equivalent noise temperature of the receiver.

The 9dB noise figure assumed in WARC 77 is equivalent to a temperature of 2013 Kelvin whereas a noise figure of 3dB achievable now corresponds to only 288 Kelvin. The antenna temperature is of the order of 50 Kelvin consequently, as the receiver noise figure is reduced the advantage gained in overall system temperature reduces correspondingly. This diminishing return is clearly seen in Figure 1. The graph assumes that the antenna temperature remains constant.

Reprinted with permission from *13th International Television Symposium*, Montreux, Switzerland, 1983, Equipment Sessions, vol. 1, pp. 281-294.

Even if a further improvement in low noise technology were to be achieved it will have relatively little effect on the design of DBS receiver terminals.

The 6 dB improvement in the front end amplifier noise figure roughly translates to a 6 dB improvement in G/T, where G = net gain of the system ahead of a given reference point and T = the equivalent system temperature at the same reference point. This point is normally taken to be the output feed flange of the antenna. In this case G = antenna gain and T = antenna equivalent temperature. This 6dB improvement should in principle provide a higher picture S/N, but consumer quality demodulators, decoders and displays are likely to limit the degree of perceived enhancement. Consequently, the 0.9 metre diameter antenna may possibly be reduced to a diameter of 45 cms. to achieve the same perceived picture S/N, drastically simplyfying antenna installation and mounting problems.

This simplistic approach does however ignore some important considerations:

1) With a 9 dB receiver noise figure the antenna contributes only 2.5% of the total system noise whereas with a 3 dB system the antenna accounts for 15%. The antenna temperature is heavily dependant on the weather in its beam. It is easily possible for poor weather conditions to degrade the antenna noise temperature to 150 Kelvin or more. With a 9 dB system such an increase would raise the antenna contribution from 2.5% to 7% which would result in a G/T degradation of .2 dB but with a 3 dB noise figure system the percentage increases from 15% to over 34% ausing a G/T degradation of over 1.2 dB exactly when the maximum signal attenuation occurs.

2) Furthermore, reducing the diameter of the antenna broadens the beamwidth which degrades the performance in two ways:

Firstly the broader beam results in poorer sidelobe performance resulting in more of the beam seeing the earth which, at approximately 300 Kelvin, is very much hotter than the 4 Kelvin of deep space.

Secondly the beamwidth of the antenna is broadened beyond the 2 degrees assumed by WARC 77, raising the possibility of co-channel interference from satellites in adjacent orbital positions.

However, the situation is not so unsatisfactory as it first appears. There are no plans at present for these potentially interfering channels to be used and in the event that a channel were adopted it remains unlikely to cause interference simply due to the concentrated footprints of the majority of the allocations. Further, with the satellite manufacturing capacity of Europe virtually saturated for the foreseeable future it is likely that we would get at least three years warning of potential trouble by which time many early subscribers to DBS would be prepared to up-grade to a larger reflector if interference was experienced.

The radiation pattern which a 0.5 meter antenna could produce is shown in Figure 2. The nulls in the pattern are offset from the boresight by an angle primarily dependant upon the antenna diameter. Consequently, it is possible to choose the diameter such that interference from the regularly spaced satellites fall into one of these insensitive nulls. With all European DBS satellites located at intervals of 6 degrees an antenna diameter of slightly less than 50 cms. will place any potential interfering signal safely in a null.

An alternative approach is to retain a 2 degree beamwidth, as envisaged at WARC 77, but only in the plane of the geo-synchronous arc. The beamwidth in the North-South plane being significantly broader to take advantage of the reduced aperture requirement. This may, of course, be achieved with an elongated antenna oriented with the major axis lying in the plane of the geo-synchronous arc.

OUTDOOR ELECTRONICS

This unit comprises a low noise amplifier, mixer, local oscillator and IF amplifier. A block diagram of a typical solution is shown in Figure 3.

The low noise amplifier uses gallium arsenide field effect transistors (GaAs FET) on a thick film hybrid circuit which also carries the mixer.

The local oscillator is fixed tuned in order that the outdoor electronics may be kept as simple as possible. The effects of wind, rain and temperature variations together create a relatively hostile environment for the outdoor unit. In the interests of reliability therefore the sophisticated tuning circuitry will be retained indoors. The disadvantage of this approach is the wide bandwidth required of the first IF stage. Under the WARC plan the DBS transmissions will all fall in the band 11.7 -12.5 GHz. All the broadcasts from any one country will lie in either the bottom 400 MHz of the band or the top 400 MHz. Therefore to receive only the national channels a bandwidth of only 400 MHz is required at the IF. If all foreign broadcasts are also to be received the bandwidth must be increased to 800 MHz.

The oscillator must be sufficiently stable to ensure that the selected signal remains within the range of both the final IF bandwidth and the demodulator over a temperature range of at least 50 degrees centigrade. This may be achieved by using a dielectric resonator as the tuning device although many alternatives exist.

Some gain is provided in the outdoor unit at the first IF to ensure an adequate signal strength is received at the indoor unit at the other end of the coax downlead. The amount of gain in the outdoor unit and hence the signal level chosen for this coaxial transmission path is to some extent defined by the level of interference received on the downlead from other terrestrial transmissions using the same band as the IF. This consideration also affects the choice of IF frequency. A de facto standard is emerging which assumes the IF to be between 0.9 and 1.7 GHz.

In a large number of areas however there are high power radar systems operating in this band and the possibility of moving the IF to a higher band, around 2 GHz, is under consideration.

INDOOR UNIT

The functions performed in the indoor unit are as follows:

second downconversion
second IF processing
demodulation
decoding
audio demultiplexing
channel selection
decryption
(remodulation)

The UK DBS service at least, and probably most of the European services, will adopt the new colour encoding system - MAC (Multiplexed Analogue Components) - in which the luminance components are transmitted serially with the chrominance components in a Time Division Multiplex (TDM) system. This is in contrast to the PAL, NTSC and SECAM systems in which the components are simultaneously transmitted. These are essentially Frequency Division Multiplex (FDM) systems and since they share the same spectrum, suffer from crosstalk between luminance and chrominance components, an effect which is totally absent with a TDM system.

Furthermore, the audio channels are to be transmitted digitally using PSK (Phase Shift Key) modulation. Up to 8 audio channels being multiplexed into the digital signal.

These systems offer the potential for a significant improvement in the quality of the signals received by the consumer but require considerable development to be carried out on the "indoor unit". Two separate demodulators are needed a, frequency demodulator for the video signal and a PSK demodulator for the audio, a completely new MAC decoder, sophisticated audio demultiplexing circuitry and a decryption system, all to be made at an acceptable price.

A block diagram of the indoor unit is shown in Figure 4.

Second IF

The second local oscillator is used for the channel selection by downconverting the required channel to the chosen second IF. Contenders for the second IF range from 70 MHz through 134 to 140 MHz or higher.

Demodulation

After the second IF strip the signal is split into two. One path goes to the video frequency demodulator and the other to the audio QPSK demodulator. The demodulated output will be a burst of 20 megabits/sec. This is time expanded to form a nominal 2 megabits/sec multiplexed audio/data stream. This data stream is then demultiplexed into the separate audio channels before passing through digital to analogue converters restoring the original analogue audio signals. Packet multiplexing will provide up to eight separate channels within the 2 Mbit/sec digital signal, each providing 15 KHz bandwidth.

A variety of techniques are available for demodulation of the video signal one of the simplest being the pulse counting discriminator (Figure 5). FM demodulators exhibit a well defined knee in their output S/N against input C/N characteristic. At C/N values below this knee the output signal degrades very rapidly and quickly becomes unviewable. If this threshold can be lowered acceptable pictures are obtained for a correspondingly lowered C/N, in fact a 2 to 3dB improvement may be acheived.

All threshold extension demodulators (TED) work on the principle of reducing the effective noise bandwidth in which the demodulator has to work. This may be done in a number of ways:

1)Tracking Filter
An electronically tunable filter is placed ahead of the demodulator and is controlled by the demodulated signal (Figure 6). By tracking the input signal with a narrower filter than the overall RF bandwidth the total noise power presented to the demodulator is correspondingly reduced.

2)Frequency Feedback
In this technique the demodulator has a fixed filter preceeding it of a narrower bandwidth than the input signal (Figure 7). In order to pass this wideband signal through a narrow filter the deviation of the received signal is first reduced by mixing it with a signal derived from the output of the demodulator. The high frequency components in the baseband signal cause the largest deviations of the carrier and these would fall outside the bandwidth of the new filter. However the demodulated output is fed back to the local oscillator in such a way as to reduce the frequency deviation of the received signal in order that it passes through the filter. Thus all the carrier power is presented to the demodulator but the noise power is reduced by the filter.

3)Phase Lock Loop
A phase lock loop is effectively a tracking filter and may be designed to track the incoming signal. This it does by exactly recreating the frequency modulation it is receiving in its own voltage controlled oscillator. The voltage which controls this oscillator is of course the original baseband signal. Threshold extension is made possible by reducing the bandwidth of the loop filter.

Decoding

The vital process in the MAC decoder is the time expansion of both the luminance and chrominance components of the signal. Two techniques have been considered to achieve this, although both rely on shift registers which clock one line of information into the shift register quickly and clock it out more slowly. Coding and storage of the signal in digital form provides a flexible method provided that analogue to digital and digital to analogue converters can be produced at a sufficiently low price.

Video A to D converters using current chips cost many hundreds of pounds. An alternative exists in the form of charge coupled devices (CCD). These shift registers store the information in analogue form. The signal is sampled at a rate defined by the input clock and each sample voltage is used to generate a charge which is stored in a discrete location in the device. Under control of the output clock all the individual charges are shifted along to the next charge location and finally out as analogue voltages to form the continuous retimed video waveform.

Decryption

Attemps to date to encrypt analogue signals have met with only limited success. Really effective encryption has only been performed with digital signals.With a digital signal the method of encryption generally takes the following form.

A defined serial data stream is generated in a key generator circuit. This sequence is then modulo-2 added to the wanted digital signal to form an apparently unrelated digital signal. A receiver authorised to receive the signal will generate the same key sequence and having achieved synchronisation with the transmitting key generator will remove the code from the received data to restore the original signal. By choosing more or less complex key generator systems the level of encryption required may be achieved to the extent that the distribution of the key material becomes the weakest point in the system.

Encryption of the audio signal presents no difficulty and manipulation of the data and sync signals could be used to provide some protection of the picture.

Encryption standards are the subject of a number of studies currently in progress and these must lead to acceptance of a common encryption standard as part of a European standard for MAC to allow the maximum choice of programme to the viewer with the minimum of added expense.

Display Interfaces

If advantage is to be taken of the new MAC coding technique and superior transmission channels, the domestic television receivers currently in use cannot be used since most can only accept signals which are encoded in PAL or SECAM, rarely both, with vestigial sideband amplitude modulation on a UHF carrier. The output of the basic DBS receiver will be baseband video in component form together with baseband (stereo) sound.

Domestic televisions are now being launched which are capable of accepting component inputs. DBS is not of course the only factor which is demanding these inputs. Cable standards for the UK may well be based upon a form of MAC and many home computers have component outputs.

It must however be possible for DBS transmissions to be received on conventional television sets, at least in the first instance since baseband input types will still be rare in 1986. An adaptor module may easily be fitted to the indoor unit of the DBS receiver which will re-encode the components, this time into PAL or SECAM and then remodulate this signal together with the apropriate sound onto a UHF carrier. The signals will by then have been encoded and decoded and remodulated and demodulated an extra time, and all this in domestic quality electronics. Viewers are unlikely to be impressed by the quality of satellite television pictures on these sets.

It is therefore in the interests of everyone that a common display interface standard is agreed for Europe so that television receiver manufacturers can feel confident that their investment in the production of component form baseband input receivers is justified by a widespread acceptance of them as soon as possible.

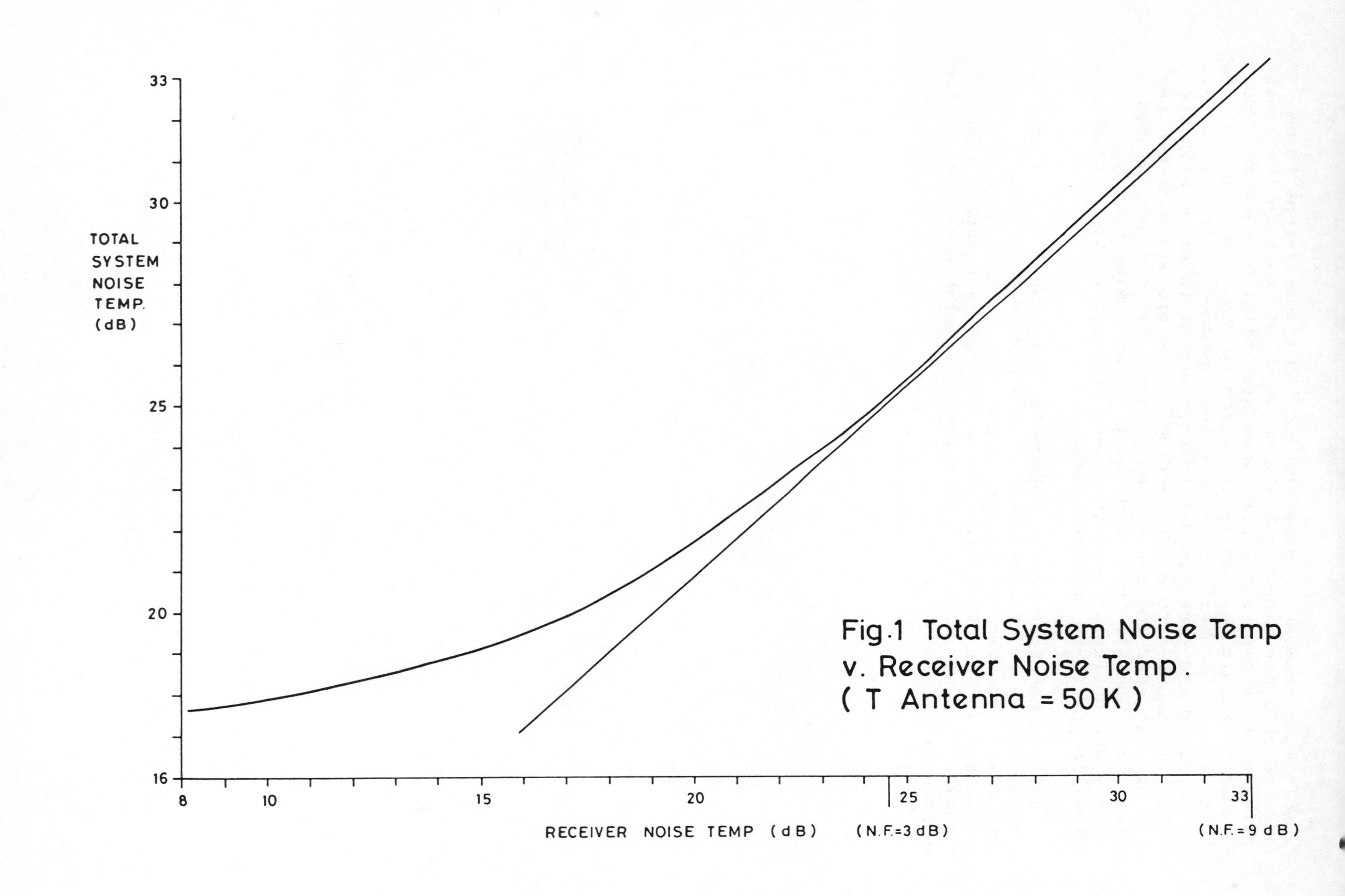

Fig.1 Total System Noise Temp v. Receiver Noise Temp. (T Antenna = 50 K)

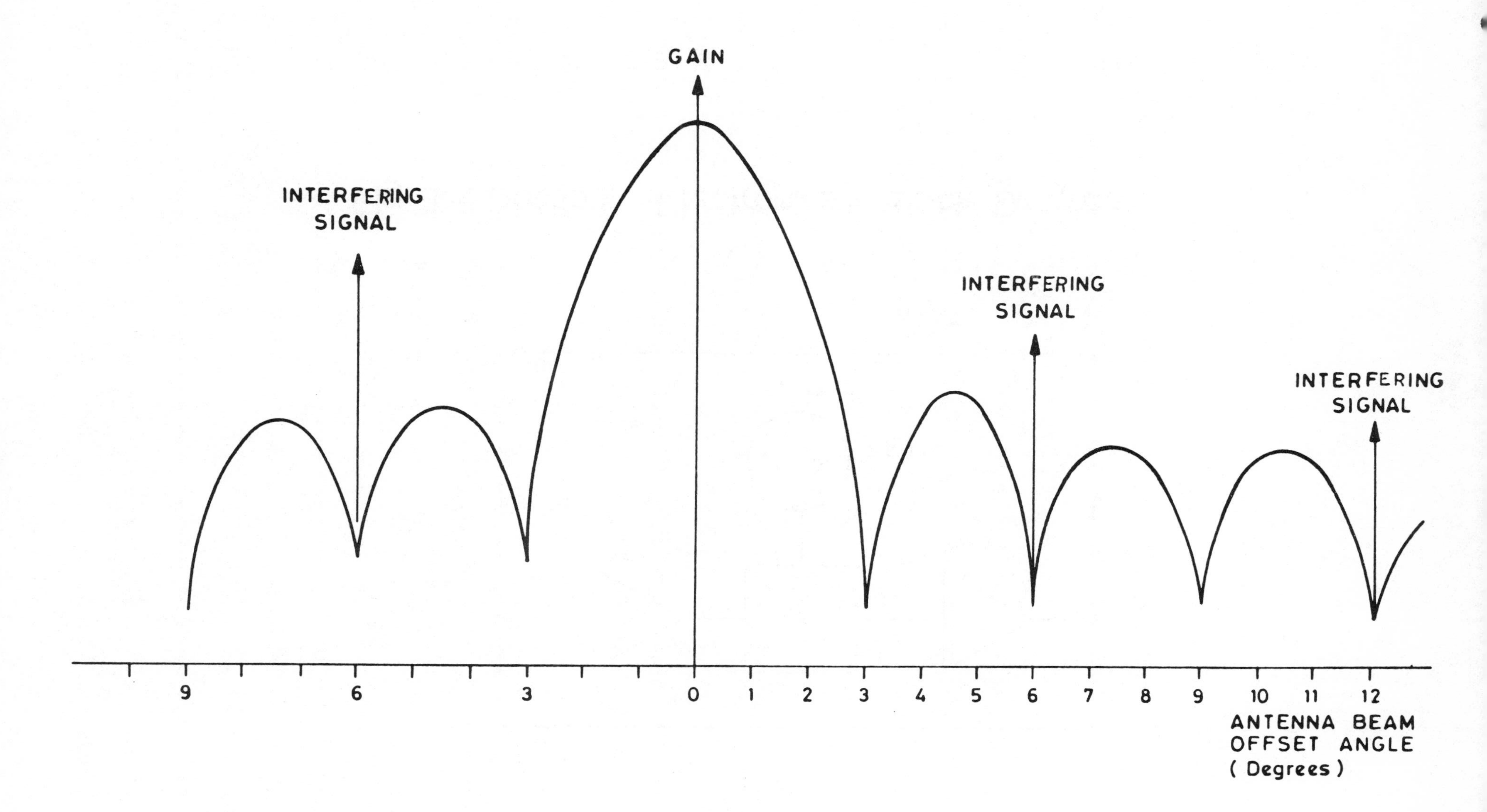

Fig 2 Radiation Pattern of 0·5m Diameter Antenna

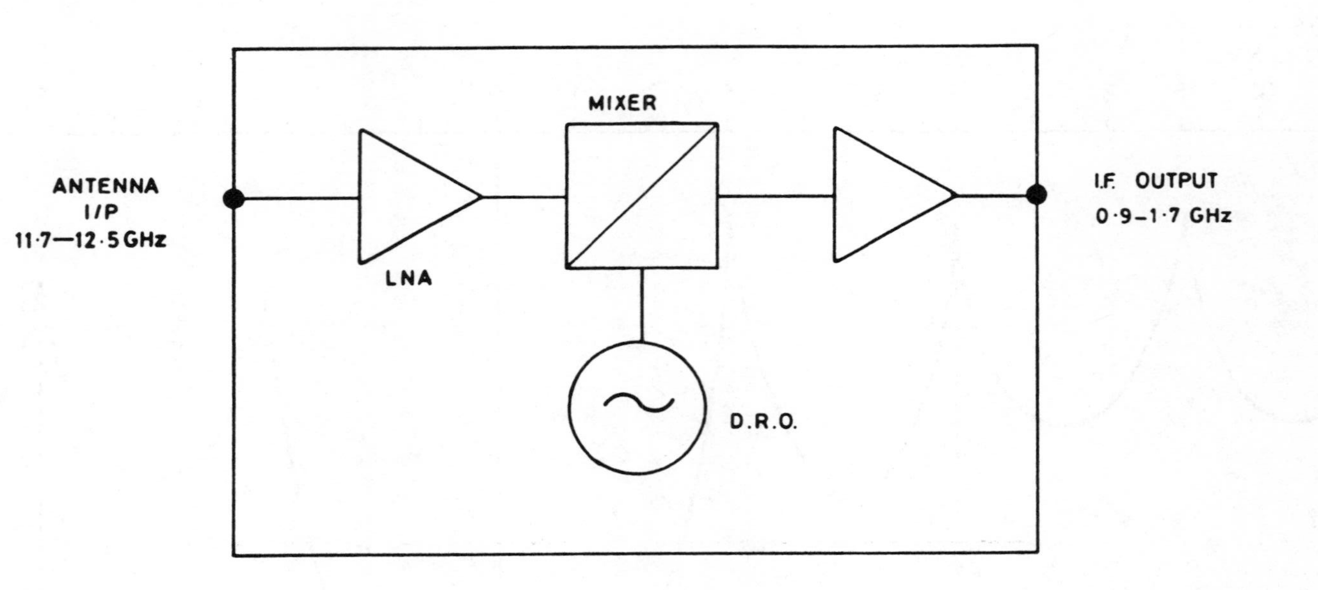

Fig. 3—Outdoor Electronics—Block Diagram.

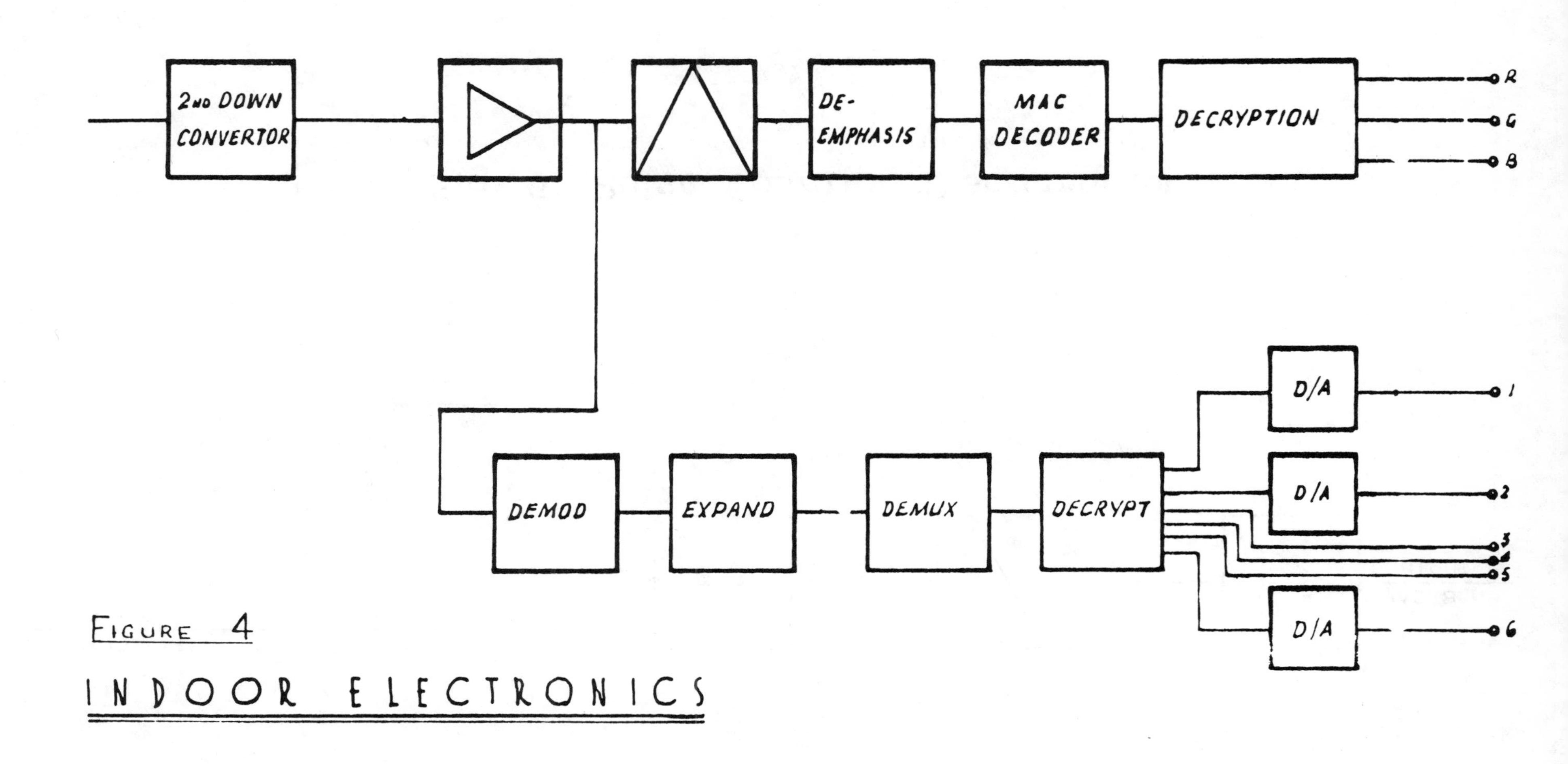

FIGURE 4

INDOOR ELECTRONICS

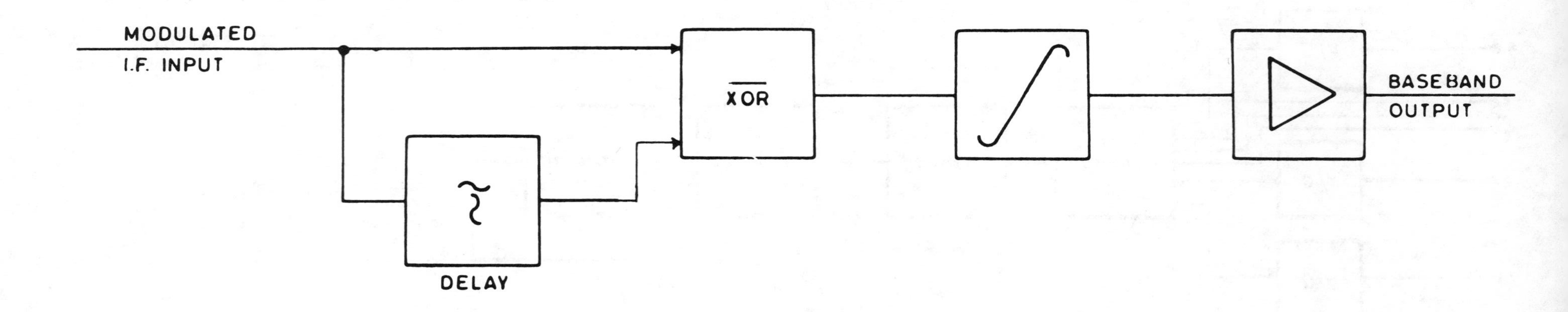

Fig 5 Pulse Counting Discriminator

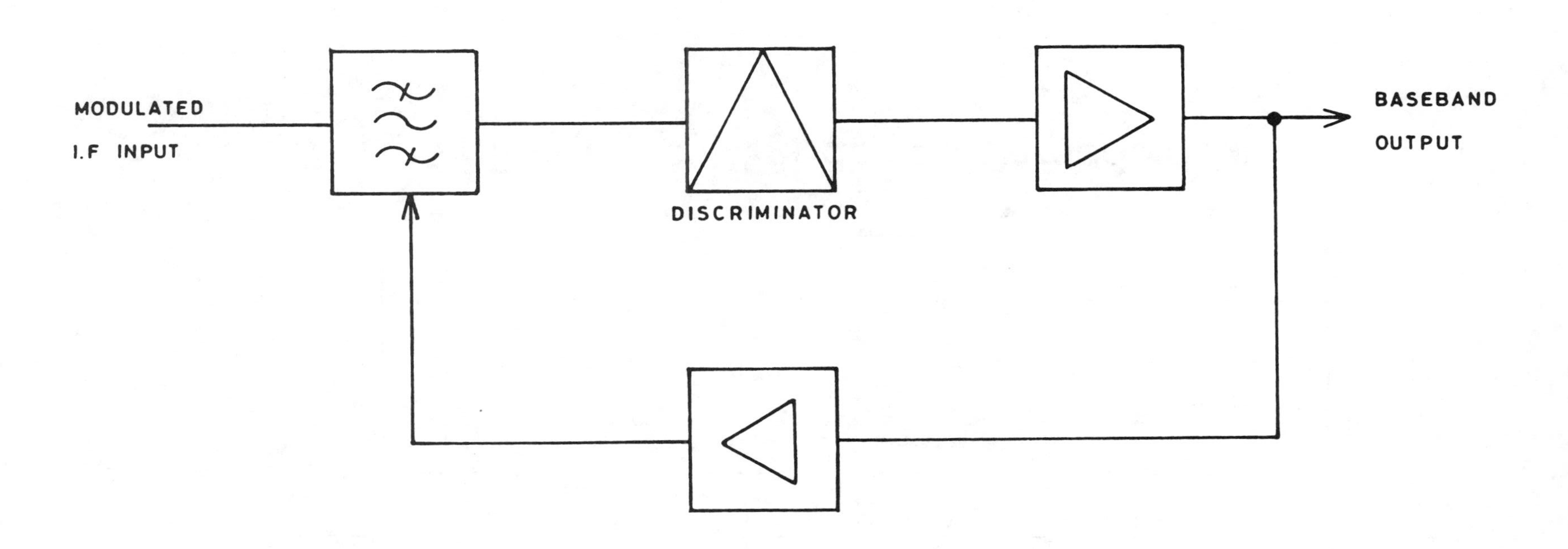

Fig. 6 Tracking Filter Demodulator

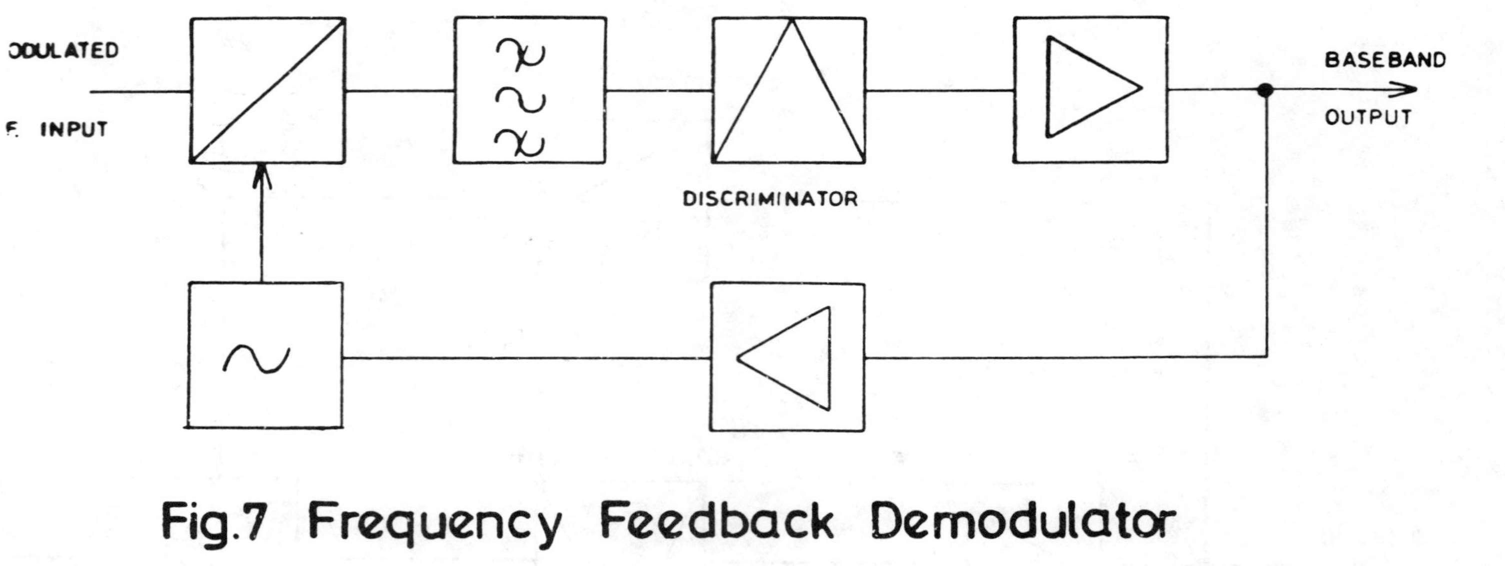

Fig.7 Frequency Feedback Demodulator

Part III
Advanced Television Systems

TELEVISION systems have come a long way from the monochrome images generated in the 1930's and 1940's. Full color pictures emerged in the 1950's and became a commercial reality in the 1960's. The decade of the 1970's was concerned with improvements in color images and the removal of visual artifacts. This naturally led the way to improved definition television systems, enhanced definition televison systems, and high definition television (HDTV) systems. These are indeed the advances of the 1970's and 1980's.

Improved definition TV refers to incremental improvements made within the existing television system, for example, the use of comb filters to reduce chrominance/luminance cross-talk. Enhanced definition TV consists of techniques applied within the current scan and display standards to effect a noticeable improvement in picture quality. Examples of such techniques include separate transmission of the luminance and chrominance signal to eliminate interference or predictive circuits used to increase the color resolution. High definition TV, on the other hand, wishes to improve the image quality by a quantum jump and is not necessarily bound to existing standards to any degree. HDTV systems strive to increase both the horizontal and vertical resolutions by approximately 2:1 such that they approach 600 lines along the horizontal and 500 lines in the vertical. They also seek to increase the aspect ratio to at least 5:3 (in some cases, e.g., landscape scenes, 2:1 is desired). In addition, it is important that an HDTV image exhibits little or no cross-luminance interference, cross-chrominance interference, and chroma-to-chroma cross-talk. Although somewhat orthogonal to video itself, stereophonic sound is normally considered to be a vital part of any large-screen HDTV system. Finally, it should be noted that the transmission bandwidth is typically constrained to 10–20 MHz for HDTV.

From a historical point of view, it is interesting to observe that throughout the 1970's there has been an ongoing effort in the HDTV area in Japan. The outstanding contributor here is Fujio and the scores of references in his early papers, found in the chronological bibliography at the end of this part, attest to this constant effort. However, the 1980's unmistakably mark the opening of the flood gates for HDTV systems throughout the world. Not only does one find an HDTV system originating from the Japan Broadcasting Corporation (NHK), but also from the British Broadcasting Corporation (BBC) and the Independent Broadcasting Authority (IBA) in the United Kingdom. In the United States, HDTV systems have been proposed by the Radio Corporation of America (RCA), the Columbia Broadcasting System (CBS), and Scientific Atlanta Corporation, to name a few. There have been contributions and system proposals from such notable organizations as Philips Research Laboratories, the University of Dortmund, AT&T Bell Laboratories, and Bell Northern Research (BNR) Corporation. And the list continues to grow.

Preceding the introduction for the seven papers found in this part, a brief note is in order regarding the attached bibliography. This is by no means an exhaustive list of contributions covering advanced television systems to date. It does not include many of the very good technical company reports that exist, nor does it include the many papers associated with general local conferences and conventions. The bibliography does consist of several dozen papers that were considered as candidates for this part and were drawn from notable journals and international conferences specifically addressing this topic. Both this bibliography and references found in the seven papers in this part can act as an excellent springboard for the reader interested in advanced television systems.

The lead paper in this part is entitled "High Definition Television Systems: Desirable Standards, Signal Forms, and Transmission Systems" by T. Fujio. This is an extensive article detailing the efforts and reporting the results of studies into HDTV systems in Japan since 1970. The primary factors included here are picture format (such as size and aspect ratio), scanning standards, and transmission signal characteristics. Considering psychophysical factors, such as the spatial frequency response of the human visual system, and a battery of subjective experiments, a provisional HDTV standard is proposed using 1125 scan line, a 5:3 aspect ratio, 2:1 line interlace, a 60-Hz field rate, and a luminance bandwidth of 20 MHz. Part of these standards include the definition of a wide-band and a narrow-band chrominance signal having 7 MHz and 5.5 MHz of bandwidth, respectively.

In discussing HDTV signal form, Fujio includes both a time-compressed integration scheme, using time division multiplexing (TDM), and a composite scheme, using frequency division multiplexing (FDM) and interleaving. The tradeoffs for each are included, but no consideration for compatibility with any existing standard is mentioned. The last major part of this paper contains a presentation of various transmission formats. Included here is a frequency modulation (FM) format using 100 MHz of radio frequency bandwidth for a 30-MHz composite baseband video signal. Also noted is a digital transmission scheme using 4 bit/sample differential pulse code modulation (DPCM), resulting in 250 Mbits/s and 100 MHz of bandwidth with four-symbol phase shift keying (PSK). The overall conclusion in this paper is that HDTV is a reality on the horizon if we are willing to expend the bandwidth and are not interested in compatibility. This conclusion is certainly supported by the results presented in this paper.

The second paper in this part is another marked contribution to the field of advanced television systems, by B. Wendland, entitled "Extended Definition Television with High

Picture Quality." Perhaps the most important contribution here is to show that line decimation and interpolation techniques along with digital signal processing could be advantageously applied to achieve high definition resolutions and still maintain compatibility with respect to line rate and the number of lines per frame. An excellent presentation of picture scanning theory, with progressive scanning, shows how vertical aliasing can be avoided with optimum prefiltering. Also covered is the effect of interlace scanning (with orthogonal and offset sampling) in the horizontal direction. One underlying point to be noted here is that to achieve true interlace compatibility, frame stores must be used to convert between progressing and interlace scans. However, the results are well worth the cost: a flicker-free picture without line interlace degradation such that the Kell factor is driven to unity. Some practical results of this theory are also presented for a scaled down HDTV system and they do confirm the theory quite well.

This paper concludes with an introduction to a multiplexed analog component (MAC) HDTV format for direct broadcast satellites (DBS) proposed by the IBA. Here, time compression of audio, luminance, and chrominance is proposed, followed by time division multiplexing of these signals. This scheme calls for alternate line transmission of the two color components and allows a wide range of aspect ratios, due to the compression. It also allows for good crosstalk protection. The costs are lack of direct compatibility and increased transmission bandwidth (due to the time compression).

The next paper utilizes some of the techniques found in the first two papers in proposing an HDTV system that is strictly compatible with National Television System Committee (NTSC) standards. This paper is authored by T. S. Rzeszewski and entitled "A Compatible High-Definition Television System." The proposed HDTV system here, called split-luminance split-chrominance (SLSC), would use 10 MHz of baseband bandwidth, where the lower portion contains an exact NTSC signal. The upper channel contains the high-frequency luminance and chrominance signals. Utilizing line decimation/interpolation techniques, this system is capable of 483 lines of vertical resolution. Relying on 7.5 MHz of luminance bandwidth, it can achieve 600 lines of horizontal resolution. By splitting both the luminance and the chrominance and employing frequency interleaving, as well as alternate line transmission of the chroma components, much of the current crosstalk between signal components is eliminated. This present version of SLSC allows room for teletext and multichannel sound but does not include aspect ratio improvements. The paper suggests that compatible transmission can easily be achieved using two 6-MHz channels and vestigial sideband amplitude modulation (AM). The concluding section of this paper has an interesting comparison for a number of HDTV systems, including resolution, bandwidth, compatibility, and crosstalk.

The next few papers present a broader approach to advanced TV systems, where concepts for enhanced video are included with suggested HDTV approaches. The fourth paper in this part is called "Compatible Systems for High-Quality Television" by R. N. Jackson and M.J.J.C. Annegarn. The premise presented here is that sophisticated signal processing can be employed, within the current standards, to reduce or eliminate artifacts such as large area flicker, line flicker, and luma/chroma crosstalk. An example of receiver processing that is discussed is a first-order discrete-time low-pass filter, where the filter delay element is a frame delay. Although this requires a frame of memory, it allows true temporal filtering of each picture element. A system is presented that filters four signals: the low-frequency luminance, the high-frequency luminance, and two color difference signals. The critical point is that the cutoff frequency of each filter is adaptively adjusted via motion detectors. Finally, these signals are converted from 50- to 100-Hz field rate to eliminate flicker.

Also contained in this paper is a proposal for a high-quality TV system that uses 1249 lines, 2:1 interlace, a 50-Hz field rate, and a 5:3 aspect ratio. To maintain compatibility it is suggested that one channel carry a compatible signal and a second channel carry the wide-screen and high-frequency information. To achieve a high vertical resolution the authors assume that they can oversample and then use line decimation/interpolation techniques. But they also add an interesting raster transcoding scheme and point to the need for movement adaption. Overall, the article favors an approach of compatible evolution to full HDTV quality and notes techniques that are within reach using current technology.

The fifth paper continues in the vein of enhancing the quality of existing TV by exploring the question of conversion between different scanning standards. This paper is authored by C. P. Sandbank and M. E. B. Moffat and is entitled "High-Definition Television and Compatibility with Existing Standards." It starts out by discussing four main impairments in standard TV: large-area flicker, interline flicker, line crawl, and the visibility of individual lines in a static raster, particularly on large screens. The authors discuss the result of up-converting from the normal 625 line, 2:1 interlace, 50-Hz field rate format to a series of improved quality formats. Employed here are two interpolation algorithms (basically with and without spatial and temporal filtering), four fields of storage, and extensive digital signal processing. The system described was designed to act as a testbed for monochrome images and a complete report is given on the effect of the conversions on the existing impairments.

Imbedded in this paper is a form of HDTV—an extended phase alternation line (PAL) system—that uses a split-luminance coding approach and frequency multiplexes the high-frequency luminance into an upper channel. This system would be aimed at direct broadcast satellites and occupies slightly more than 10 MHz of baseband bandwidth. In keeping with the main theme, the authors conclude this contribution by analyzing the effects of down-converting from HDTV standards to the current line and field rates. The main conclusion is that the choice of line standards is not critical in conversion, but field frequency values applied to conversion are very critical to maintain good picture quality.

The next to the last paper in this part is by T. S. Robson and titled "A Compatible High Fidelity TV Standard for Satellite Broadcasting." It is a rather lengthy article but includes some of the basic principles important in HDTV research. The initial part of this paper is concerned with conversion between a DBS

FM transmission format and the standard AM TV format, as DBS is considered as a primary transmission avenue for HDTV. This is followed by a discussion of current image impairments and techniques that can be exploited to improve picture quality.

With the idea of DBS FM transmission, the author gives a brief description of the MAC HDTV system. The MAC system discards direct compatibility for much reduced crosstalk and extensive freedom by using time compression and TDM techniques. The last part of this article covers general sampling theory, including pre- and post-filtering, while focusing on the vertical frequency domain and the temporal frequency domain. The concepts of oversampling and interpolation are included here and the entire discussion leads to possible methods of implementing HDTV systems like the MAC system.

The concluding paper in this part on advanced TV systems is written by a veteran in the television field, C. W. Rhodes, and entitled "An Evolutionary Approach to High Definition Television." The paper is written with a conversational tone and includes thoughtful discussions on such HDTV items as aspect ratio, bandwidth, interlace versus progressive scanning, vertical resolution and spatial aliasing, large area flicker, and motion portrayal with different types of scanning. In all cases, problems are pointed out and suggestions made to stimulate solutions.

The last part of this paper focuses on color considerations, where a TDM technique of line sequential time-compressed chrominance components is presented. The author notes that this relieves the crosstalk problems and eliminates dot crawl. He also notes that for good color image quality, the chrominance resolution should be at least one-quarter of the luminance resolution. The question of compatibility is included at the end of the paper, where transcoding between different formats becomes the important process. The primary purpose of this entire paper is really to suggest methods in which the camera and the picture display can operate on different scanning rates and structures in future HDTV systems. The admission here is that improved picture quality is certainly possible through massive digital processing of the video signal. I believe that this article is a catalyst for HDTV researchers to solve the existing problems related to image quality and produce an evolutionary, if not revolutionary, advanced television system.

JOSEPH L. LOCICERO
Associate Editor

BIBLIOGRAPHY

[1] T. Fujio, "A study of high-definition TV systems in the future," *IEEE Trans. Broadcast.*, vol. BC-24, pp. 92–100, Dec. 1978.

[2] D. G. Fink, "The future of HDTV," *SMPTE J.*, vol. 89, pp. 89–94, Feb. 1980.

[3] ——, "The future of HDTV," *SMPTE J.*, vol. 89, pp. 152–161, Mar. 1980.

[4] T. Fujio, "A universal weighted power function of television noise and its application to high-definition TV system design," *IEEE Trans. Broadcast.*, vol. BC-26, pp. 39–48, June 1980.

[5] T. Fujio, J. Ishida, T. Komoto, and T. Nishizawa, "High-definition television system—signal standard and transmission," *SMPTE J.*, vol. 89, pp. 579–584, Aug. 1980.

[6] T. Fujio, "High definition wide-screen television system for the future-present state of the study of HD-TV systems in Japan," *IEEE Trans. Broadcast.*, vol. BC-26, pp. 113–124, Dec. 1980.

[7] B. Wendland, "High definition television studies on compatible basis with present standards," in *Proc. 15th Annu. SMPTE Television Conf., Television Technology in the 80's*, Feb. 1981, pp. 151–165.

[8] T. Fujio, "The NHK high-resolution wide-screen television system," in *Proc. 15th Annu. SMPTE Television Conf., Television Technology in the 80's*, Feb. 1981, pp. 166–176.

[9] K. Hayoshi, "Research and development on high-definition television in Japan," *SMPTE J.*, vol. 90, pp. 178–186, Mar. 1981.

[10] T. Fujio, "High definition television systems: Desirable standards, signal forms, and transmission systems," *IEEE Trans. Commun.*, vol. COM-29, pp. 1882–1891, Dec. 1981.

[11] C. P. Sandbank and M. E. B. Moffat, "High-definition television and compatibility with existing standards," in *Proc. 16th Annu. SMPTE Television Conf., Tomorrow's Television*, Feb. 1982, pp. 170–185.

[12] C. W. Rhodes, "An evolutionary approach to high definition television," in *Proc. 16th Annu. SMPTE Television Conf., Tomorrow's Television*, Feb. 1982, pp. 186–197.

[13] T. S. Robson, "A compatible high fidelity TV standard for satellite broadcasting," in *Proc. 16th Annu. SMPTE Television Conf., Tomorrow's Television*, Feb. 1982, pp. 218–236.

[14] R. N. Jackson and S. L. Tan, "System concepts in high fidelity television," in *IBC 82 Conv. Rec.*, Sept. 1982, pp. 135–139.

[15] K. G. Freeman, G. de Haan, J. M. Menting, and D. W. Parker, "Experimental work towards high fidelity television," in *IBC 82 Conv. Rec.*, Sept. 1982, pp. 140–143.

[16] B. Wendland, "On picture scanning for future HDTV systems," in *IBC 82 Conv. Rec.*, Sept. 1982, pp. 144–149.

[17] I. Childs, "High definition television and its alternatives," in *Proc. HDTV Colloq.*, Oct. 1982, pp. 1.2-1–1.2-12.

[18] T. Fujio, "High-definition television of NHK," in *Proc. HDTV Colloq.*, Oct. 1982, pp. 1.3-1–1.3-11.

[19] T. S. Robson, "Television systems for the future," in *Proc. HDTV Colloq.*, Oct. 1982, pp. 1.4-1–1.4-6.

[20] C. W. Rhodes, "Towards the implementation of a compatible HDTV system in North America," in *Proc. HDTV Colloq.*, Oct. 1982, pp. 1.5-1–1.5-21.

[21] K. H. Powers, "Compatibility aspects of HDTV," in *Proc. HDTV Colloq.*, Oct. 1982, pp. 1.6-1–1.6-17.

[22] A. G. Toth, "Networking aspects of high-definition television," in *Proc. HDTV Colloq.*, Oct. 1982, pp. 2.1-1–2.1-13.

[23] B. Wendland, "Signal processing for compatible HDTV-systems, first results," in *Proc. HDTV Colloq.*, Oct. 1982, pp. 2.3-1–2.3-10.

[24] E. Dubois and B. Prasada, "Digital coding and processing of high-definition television signals," in *Proc. HDTV Colloq.*, Oct. 1982, pp. 2.5/2.6-1–2.5/2.6-16.

[25] R. N. Jackson and S. L. Tan, "System concepts in high fidelity television," in *Proc. HDTV Colloq.*, Oct. 1982, pp. 2.7-1–2.7-5.

[26] D. G. Fink, "International aspects of HDTV standardization," in *Proc. HDTV Colloq.*, Oct. 1982, pp. 5.3-1–5.3-12.

[27] C. W. Rhodes, "New chrominance and luminance components for multiplexed component video signals in HDTV systems," in *Proc. 17th Annu. SMPTE Television Conf., Video Pictures of the Future*, Feb. 1983, pp. 11–31.

[28] G. R. Southworth, "A high definition still frame television system," in *Proc. 17th Annu. SMPTE Television Conf., Video Pictures of the Future*, Feb. 1983, pp. 44–56.

[29] B. Wendland, "Extended definition television with high picture quality," in *Proc. 17th Annu. SMPTE Television Conf., Video Pictures of the Future*, Feb. 1983, pp. 57–71.

[30] M.J.J.C. Annegarn and R. N. Jackson, "Compatible systems for high quality television," in *Proc. 17th Annu. SMPTE Television Conf., Video Pictures of the Future*, Feb. 1983, pp. 72–82.

[31] R. N. Jackson and M.J.J.C. Annegarn, "Compatible systems for high-quality television," *SMPTE J.*, vol. 92, pp. 719–723, July 1983.

[32] G. R. Southworth, "A high-definition still-frame television system," *SMPTE J.*, vol. 92, pp. 834–842, Aug. 1983.
[33] T. S. Rzeszewski, "A compatible high-definition television system," *Bell Syst. Tech. J.*, vol. 62, pp. 2091–2111, Sept. 1983.
[34] B. Wendland, "Extended definition television with high picture quality," *SMPTE J.*, vol. 92, pp. 1028–1035, Oct. 1983.
[35] B. G. Haskell, "High-definition television (HDTV)—Compatibility and distribution," in *Proc. IEEE Global Telecommun. Conf.*, Nov. 1983, pp. 1070–1075.
[36] R. K. Jurgen, "The problems and promises of high-definition television," *IEEE Spectrum*, pp. 46–51, Dec. 1983.
[37] B. G. Haskell, "High-definition television (HDTV)—Compatibility and distribution," *IEEE Trans. Commun.*, vol. COM-31, pp. 1308–1317, Dec. 1983.

High Definition Television Systems: Desirable Standards, Signal Forms, and Transmission Systems

TAKASHI FUJIO

Abstract—**Since 1970, the Japan Broadcasting Corporation (NHK) has been making efforts in studies of future high-definition television systems in response to the expected demands of a future postindustrial society. The standards and signal systems and transmission parameters appropriate to high-definition TV are discussed. Several TV standards and the desired signal-to-noise ratios that will fit in with various viewing conditions from the viewpoint of the characteristics of the human visual system are described. Also described are the transmission primaries of the system and the signal parameters being used for a 1125-scanning-line system, which is the tentative standard chosen by NHK. In November 1978 and March 1979, transmission tests of the high-definition TV system were carried out via the broadcasting satellite "Yuri" in Japan.**

I. INTRODUCTION

AT NHK (the Japan Broadcasting Corporation) research into high-definition wide-screen television started in 1968. Wide ranging studies have been carried out by NHK on the establishment of desirable picture quality and signal standards of such broadcasting systems. We have developed new equipment from high-resolution TV cameras to wide-screen displays and transmission systems which we have tested via the experimental broadcasting satellite, "Yuri."

Part of the present state of high-definition television has already been reported in the CCIR (International Radio Consultative Committee) Report 11/452 [1], [2].

Manuscript received January 5, 1981; revised July 23, 1981.

The author is with the Technical Research Laboratories, Nippon Hoso Kyokai, Tokyo, 157, Japan.

This paper principally describes results obtained from studies of desirable TV standards, signal systems, and transmission systems for high-definition television.

II. HIGH-DEFINITION TELEVISION (HD-TV) STANDARDS

A. Viewer Requirement for Future Television Systems

In order to fix the parameters of a future television system, the following primary factors must be considered.

1) Picture format (standard picture size and aspect ratio).

2) Scanning standards (number of scanning lines, interlace ratio, and frame frequency).

3) Signal standards (signal type, bandwidth, and required S/N ratio).

To make these factors clear many tests were performed in respect to the items in Table I to seek the viewer requirements for future television systems. In these tests, 70 mm movie films, simulation films, and experimental high-resolution television systems were used.

B. Television Standards Viewed from the Human Visual System

The spatial frequency characteristics of the human visual system are shown in Fig. 1 and they are similar to those of a low-pass filter. The required number of scanning lines and required horizontal picture elements can be obtained by dividing the vertical and horizontal viewing angles of the TV picture

Reprinted from *IEEE Trans. Commun.*, vol. COM-29, pp. 1882–1891, Dec. 1981.

TABLE I
ITEMS OF VIEWER REQUIREMENTS AND HD-TV STANDARDS

Factors	Items			HD-TV system
Viewing	Desirable viewing distance			HD-TV Standard
Picture aspect	Aspect ratio			
	Picture size			Quality Factor (Q)
	Contrast and brightness			
Picture information	Still picture	Y	n	
			f_b	
		C	Transmission primaries	Preferable color
			f_W, f_N	Signal Standard
		Y ⇄ C cross talk		
	Moving picture	Smoothness, Tracking		Frame frequency,
		Desirable resolution		Display system
	Desirable S/N			Broadcasting system, Size

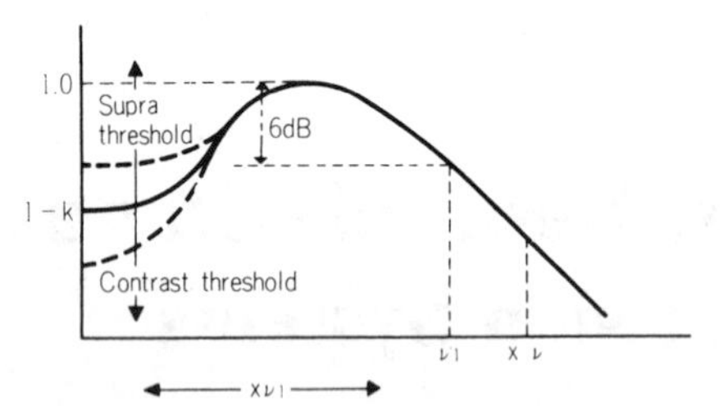

Fig. 1. Spatial frequency response of the human visual system.

by the minimum viewing angle at which the human visual system can resolve. So once the viewing distance of a television picture has been determined, general standards for a desirable television system may be defined.[1] The number of scanning lines n and signal bandwidth f_b required by the human visual system are given by the following equations.

$$n = \frac{4x_0\nu_1}{KK_1\eta_v}\tan^{-1}\frac{1}{2d}$$

$$f_b = \frac{8x_0{}^2\nu_1{}^2 f_p}{KK_i\eta_v\eta_h}\tan^{-1}\left(\frac{1}{2d}\right)\tan^{-1}\left(\frac{a}{2d}\right) \qquad (1)$$

where

x_0 = limiting spatial frequency (normalized value)
ν_1 = cutoff frequency for the human visual system (9.30 c/° arc)
K = Kell factor (=0.70)
K_i = interlace factor
η_h = effective horizontal scan rate
η_v = effective vertical scan rate
a = aspect ratio
d = viewing distance divided by picture height.

In (1), x_0 is the upper limit of the spatial frequency x shown in Fig. 1, and a desirable value of x_0 which is used for television systems designed by the author is 1.50.[2]

[1] Absolute picture size will exert influence on the psychological effect, such as size constancy of the picture, but with regard to the physical characteristics of the human visual system, if a viewing distance expressed by the multiple of picture height is determined, the required general standards of television could be defined.

[2] This value is determined by the perceiving power [1] of the visual system.

TABLE II
NUMBER OF SCANNING LINES REQUIRED AS A PARAMETER OF VIEWING DISTANCE

Viewing distance (dH)	4H	3.3H	3H	2.5H	n	525*2	951	1.351	1.601
n(lines)	935	1.125	1.241	1.481	dH	7.2H	3.9H	2.8H	2.3H
f_b(MHz)	11.0	16.0	19.3	27.4	f_b(MHz)	2.8	11.6	22.8	32.1
$(f_b)_{max}$ 〃	13.5	19.0	24.0	34.0	$(f_b)_{max}$ 〃	3.4	15.0	28.0	40.0
f_0(MHz)	4.5	6.5	7.9	11.4	f_0(MHz)	1.1	4.7	9.4	13.3
Visual angle*1 (deg.)	23.5°	28.3°	31.0	36.9°	Viewing angle (deg.)	10.7°	23.9°	33.7°	39.7°
Standard	Interlace ratio (Ri) 2 : 1, Frame frequency (f_p) 30Hz Aspect ratio (a) 5 : 3								

f_0 = cutoff frequency of noise weighting function
H = picture height
*1 Visual angle in horizonal direction viewed from optimum viewing distance of the system
aspect ratio 4 : 3

The quantity K_i is a factor representing the deterioration rate of the vertical spatial frequency response and is determined by the line-interlace scanning ratio. For a television system having a line-interlace ratio of 1:1 (noninterlace scan), K_i is 1.0, and for the system having 2:1, K_i is 0.6-0.7 at a field frequency around 60 Hz [1], [3]. Thus, the number of scanning lines required for high-definition television having an interlace ratio of 2:1, as a parameter of viewing distance, will be as shown in Table II.

The values of n and f_b in the table show the relation between the desirable number of scanning lines and the luminance signal bandwidth when the bandwidth is given. Conversely, values of $(f_b)_{max}$ (three times the f_0 for each system) are desirable signal bandwidths when the number of scanning lines is given. The cutoff frequency f_0 in the noise weighting function in such television systems is also shown in Table II.

A viewing distance for TV or movie pictures with rapid movements that produces no visual fatigue is about 4 H [4]-[6], but viewers tend to move closer if movement is not so rapid. The present author considers that the picture quality of future television should be good enough for a viewing distance of 3 H.

Starting from this assumption, the number of scanning lines and the signal bandwidth for HD-TV can be calculated [1], [7]. They work out to about 1240 lines as shown in Table II.

C. Picture Quality

To study the desirable level of picture quality for a future television system, a wide range of subjective evaluation tests using color transparencies [3], [4], [8]-[11] and experimental high-resolution television systems [13]-[17] were conducted. Subjective tests on the effect on picture quality of the number of scanning lines, the signal bandwidth, the balance of two-dimensional resolution (horizontal and vertical), the noise impairment produced by different spectra and the screen size have been conducted by projecting 4 X 5 in or 8 X 10 in color transparencies made by computer processing and a simulator onto a screen.

Tests on the line-interlace effect and how line structure affects picture quality were carried out by means of an experimental high-resolution television system.

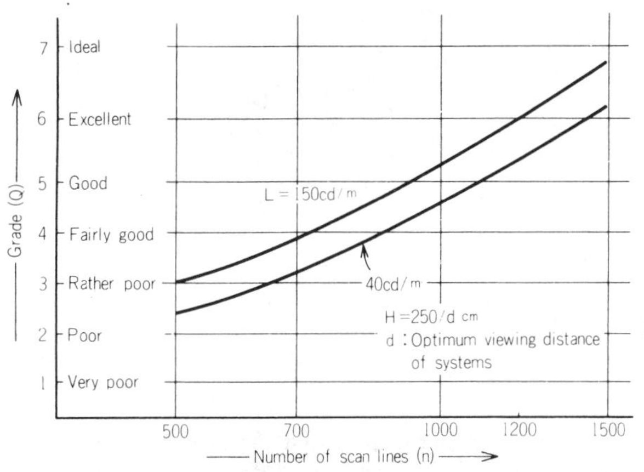

Fig. 2. System performance (quality factor Q) in various TV systems.

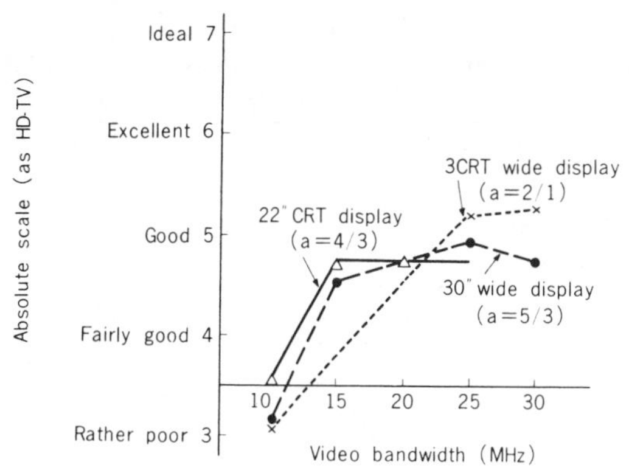

Fig. 3. Picture quality as a function of signal bandwidth.

A series of such tests produced the following results.

1) When the screen area was doubled, the subjective value of the picture quality was improved by about one grade [9], [10] on a 7-grade scale because of increased "realism" and the stronger impression produced by a bigger picture.

2) With a larger picture size, a wider aspect ratio 5:3 or 2:1 was preferred, and picture sizes of 1.4 X 0.8 m or 2.0 X 1.0 m are desirable for realistic and powerful picture representation [9], [11], [12].

3) A television system with 1500 scanning lines and a 1.7 X 1.0 m screen size would produce a subjective picture quality better by about 3.5 grades than that of a standard 525-line television picture displayed on a conventional 25-in display [8]-[11].

4) The quality of a picture displayed with a fairly large imbalance of the two-dimensional resolution depends only on the lower resolution because of the masking effect.

Through these tests, the author induced a quality factor Q defined by the primary factors of the number of picture elements (number of scanning lines), the display picture size, and the picture brightness. The value of Q gives the fineness of picture quality, and contains such effects as reality, presence, and the visual impact of large display pictures.

Fig. 2 gives the value of Q for television systems with various numbers of scanning lines and when the picture displayed in each system is viewed [21]. The picture size is derived from a fixed viewing distance of 2.5 m, as the picture height H is equal to $(2.5/d)$m. From Fig. 2, it will be found that a television system matched to a viewing distance of 3 H will produce a subjective picture quality better by about 3 grades than that of a conventional 525-line television system.

D. Scanning and Signal Standard

In order to study the effects of line-interlace scanning and signal bandwidth, an experimental HD-TV system with 1125 scanning lines and a high-resolution monochrome television system with a variable number of scanning lines [13], [14] up to 2125 were both used.

1) It was found that a television picture having n scanning lines and a 2:1 line-interlace ratio has the same subjective picture quality as that of a transparency having 0.6–0.7 n scanning lines [1]. This result agrees well with other results [3], [15] obtained from tests for interlace effects on picture quality using an experimental television system with variable numbers of scanning lines and interlace ratios.

2) The best line-interlace ratio is 2:1 [3], [15]. The improvement in picture quality produced by interlace ratios higher than 3:1 is less than that of a 2:1 interlace.

3) In a television system with a 2:1 interlace and a 30 Hz frame frequency, 1600 scanning lines are sufficient to obtain satisfactory picture quality [15] with no line flicker.

4) In an HD-TV system with 1125 scanning lines and a 4:3 aspect ratio, a desirable signal bandwidth is about 15 MHz[3] (Fig. 3) [16]. In this case, picture quality deterioration is not perceptible even at a viewing distance of three times the picture height.

5) In a 1125-line system with a 5:3 aspect ratio, desirable bandwidths for the wide-band chrominance component and the narrow-band chrominance component are 7 and 5 MHz, respectively [17].

E. Provisional Standard for an HD-TV

At NHK the development of camera equipment and a display for an HD-TV system with 1125 scanning lines started in 1970, together with studies of desirable standards for future television. Based on the results of these studies, provisional HD-TV standards were specified for further study without any change in the number of scanning lines as shown in Table III.

Desirable bandwidths are shown in Table IV for the luminance and chrominance signals in the systems described in

TABLE III
PROVISIONAL STANDARDS FOR AN HD-TV PROPOSED BY NHK

Number of scanning lines	1,125
Aspect ratio	5 : 3
Line interlace ratio	2 : 1
Field repetition frequency	60 Hz
Video frequency bandwidth	
Luminance (Y) signal	20 MHz
Chrominance (C) signal	
Wideband (C_W)	7.0 MHz
Narrowband (C_N)	5.5 MHz

$$\begin{pmatrix} Y \\ Cw \\ Cn \end{pmatrix} = \begin{pmatrix} 0.30, & 0.59, & 0.11 \\ 0.63, & -0.47, & -0.16 \\ -0.03, & -0.38, & 0.41 \end{pmatrix} \begin{pmatrix} R \\ G \\ B \end{pmatrix}$$

[3] All the displays used in these tests have an ability to reproduce the television signal with 30 MHz bandwidth (−6 dB) so the saturation characteristics in Fig. 3 can be regarded as that of the human visual system.

TABLE IV
TRANSMISSION SIGNAL BANDWIDTHS FOR VARIOUS HD-TV SYSTEMS

	n(lines)	935	1,241	1,481	951	1,125*2	1,351	1,601
Luminance signal f_Y (MHz)		14	24	34	15	19	28	40
Chrominance components	f_W (MHz)	5	8.5	12	5.5	7.0	10	14.5
	f_N (MHz)	4	6.5	9.5	4	5.5	7.5	11
	f_C*1 (MHz)				4.7	6.5	9.4	13.3

*1 Bandwidth of line-sequential chrominance signal
*2 Provisional used system at NHK

Table II. When a luminance signal bandwidth of 20 MHz has been secured as shown in the table, it seems that NHK's experimental HD-TV with 1125 scanning lines is acceptable for an imaging system matched to a viewing distance of three times the picture height.

III. HIGH-DEFINITION TELEVISION SIGNALS

A. Transmission Primaries and Their Signals

A study was also made into what transmission primaries are best for a high-definition color television system. For effective color-image information transmission, it is desirable to choose transmission primaries as luminance and chrominance primaries.

As to what transmission primaries should be chosen for a future television system, the following studies are being carried out [18], [19]:

1) Colorimetric design, using a new color primary system in which the color rendition range in the blue and purple boundary region of conventional television systems has been improved, and a signal form for various stimulus values.

2) Transmission primary signals for various chromaticity axes obtained from the examination of chromaticity responses in the human visual system.

3) Transmission primary signals weighted to the color stability of a certain chromaticity gamut in which the visual system is extremely sensitive to chromaticity error.

The author performed a colorimetric design for television systems using the UCS chromaticity diagram, and considers that chrominance primaries for broadcast television systems should be chosen corresponding to orthogonal axes on this diagram.

From such a viewpoint, several transmission primary systems suitable for a future high-definition television system are examined [18], [19].

The spatial frequency response for chromaticity [20] of the human visual system is as follows.

1) The visual system is sensitive to chromaticity in the direction of 625 and 492 nm of dominant wavelength.

2) It is less sensitive in the direction of 565 and 440 nm. The wide-band axis, narrow-band axis of color perception, and color axes for transmission are shown on the UCS (uniform chromaticity scale) chromaticity diagram in Fig. 4. The chrominance components C_W and C_N shown in Table III are proposed transmission primary signals.

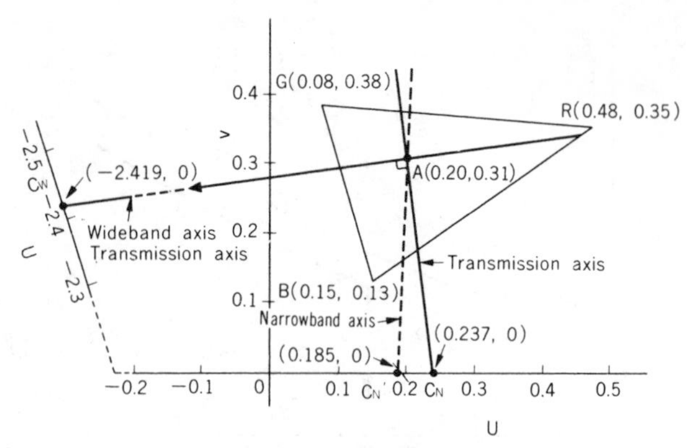

Fig. 4. Wide-band and narrow-band chromaticity axes and transmission primaries.

B. HD-TV Signal Form [19], [21]

In a high-definition TV signal, two different signal forms will be considered; one a composite system, in which transmission primary signals Y and C are frequency-division multiplexed in their spatial and temporal frequency domains as in conventional television signals; the other a TCI (time compressed integration) system in which the Y and C component signals are time-division multiplexed every one or two line scan periods after time compression. However, these systems are not equally adaptable to broadcasting systems. This depends on the modulation systems (such as AM, FM, analog, and digital) because of their different physical features.

In the composite system, signal processing, especially the decoding process at the receiving end, can be simplified and the efficiency of frequency utilization for the baseband signal can be high.

On the other hand, the TCI system has a positive feature in that balance of transmission quantity between the Y and C components—balance for noise impairment affecting the luminance and chrominance components of the reproduced image—is not strongly dependent on the broadcasting system, but fairly complicated processing is required for the system, such as time compressing and expanding of the signals.

In the FM transmission of an HD-TV signal, a Y-C separate transmission system in which the luminance signal and the line-sequential chrominance components are transmitted via separate FM channels is desirable from the transmission efficiency point of view.

C. Composite Signals

In a composite system, the author envisions three different signal forms (IYS, OYS, and OYC) in which the carrier chrominance signal is multiplexed in various frequency portions of the luminance signal as shown in Fig. 5.

The IYS (inside-Y subcarrier) system is one in which the signal multiplexes the carrier chrominance signal into the Y signal bandwidth as in a conventional television system. In this system, frequency utilization in the baseband signal is efficient, but it is difficult to remove enough of the cross interference between the Y and the C components.

In addition, it is difficult to satisfy the two conditions of effective encoding of the transmission primary signals with efficient bandwidth utilization, and freedom from crosstalk

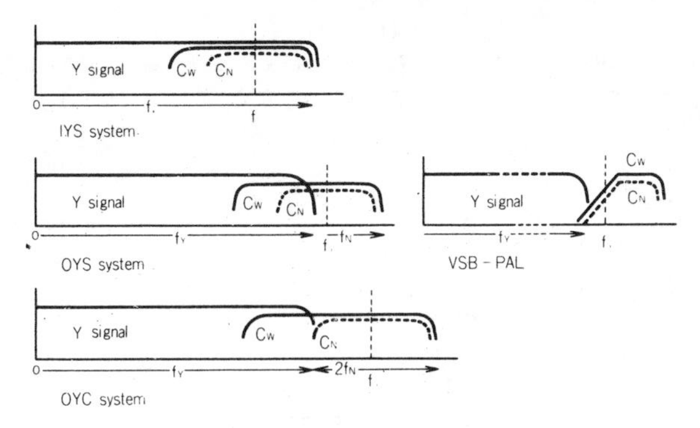

Fig. 5. Frequency spectrum composite signals.

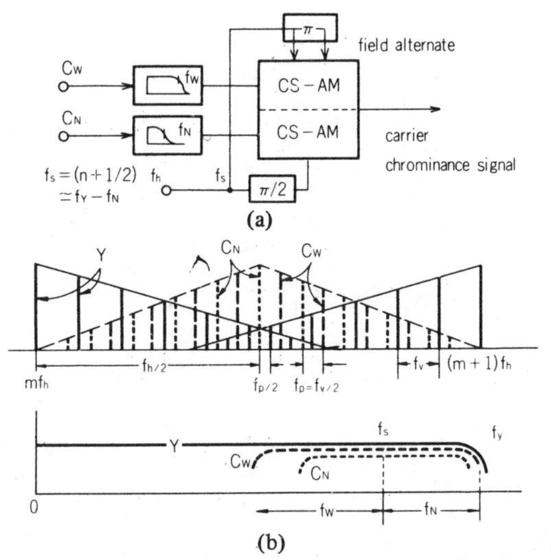

Fig. 6. Phase alternation by field (PAF) system. (a) Signal processing. (b) Frequency spectrum.

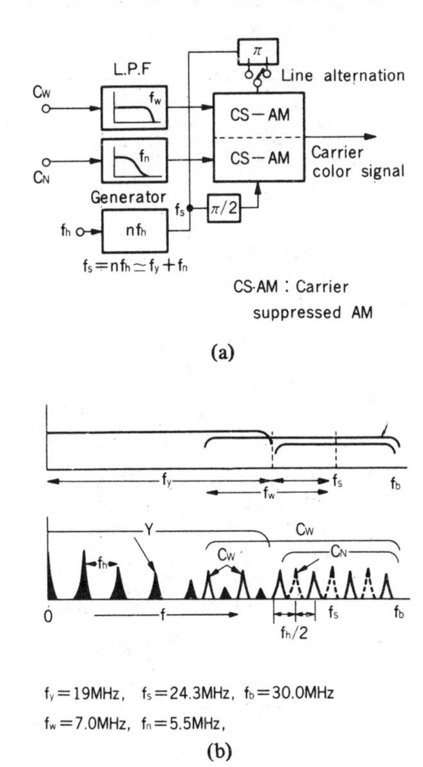

Fig. 7. The HLO-PAL system. (a) Block diagram of the chrominance modulator. (b) Frequency spectrum.

between color difference signals caused by degradation of the characteristics of the transmission channel. In order to make these two compatible, the author envisions a PAF [19] (phase alternation by field) system, of which a block diagram of the signal processing for encode and the baseband signal spectrum is shown in Fig. 6. The PAF system will be favorable in the near future when frame memory will be practical in the TV set.

A subcarrier dot and cross-color interference, which gives an unstable feeling in the reproduced image, are apt to appear when the IYS system is used.

For these types of interference, the OYS (outside-Y subcarrier) system, in which the color subcarrier lies outside the luminance signal bandwidth, as shown in Fig. 5(b), is effective. However, this system cannot be used effectively with frequency division multiplexing of the high frequency component of the luminance signal Y_H and the two chrominance components in the two-dimensional spatial frequency domain.

As to a system free of such defects, an OYC (outside-Y chrominance) system, which has a spectrum as shown in Fig. 5(c), is preferable. In this system, the efficiency of frequency utilization will be less, but an excellent signal can be formed with respect to interference between the Y and C signals, color resolution, and balance and stability of image for deterioration of the characteristics of the transmission channel.

The author believes that the IYS and OYS signals are preferable for broadcasting systems which perform the aim within allowable or grade 4 for noise impairment. For a broadcasting system in threshold noise impairment, the HLO-PAL signal [19] [which is an OYC system, as shown by its signal spectrum in Fig. 7(b)] was used in the design of a broadcasting service, and is now undergoing various transmission tests for a high-definition television. As this signal system is free from interference between the luminance and chrominance components and the interchrominance signal, good broadcasting service with stable picture quality can be achieved.

D. TCI Signals

By digitizing the television signal, signal processing such as time compression and time expansion will become easy, and if this is to be introduced into television sets the broadcasting service may be provided with a TCI signal in which the Y and C components are time-division multiplexed on each period of one or two scan lines after each signal is time compressed.

In this system, the TCI-LC (time compressed integration of line color signals) signal, which compresses and multiplexes the Y, C_W, and C_N signals, and the TCI-LSC (TCI of luminance and the line-sequential chrominance component) signal, which multiplexes Y and the line-sequential chrominance signal,[4] are taken into consideration as shown in Fig. 8.

[4] Two chrominance components for the TCI-LSC system and the Y-C separate system which transmit the chrominance components line sequentially are not necessarily used as C_W and C_N of which bandwidths are different. In this case, it will be possible to choose one pair of chrominance signlas corresponding to the orthogonal chromaticity axes on the UCS chromaticity diagram.

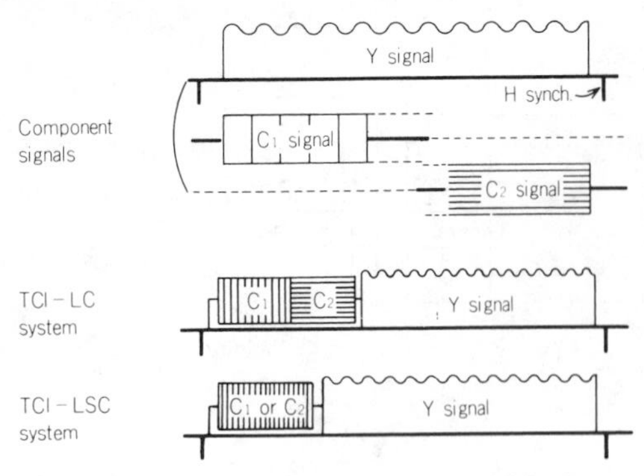

Fig. 8. Time compressed integration (TCI) systems.

TABLE V
RELATIVE VALUE OF LUMINANCE NOISE POWER FOR FM NOISE

System		Noise power	Noise power ratio
Composite system		$\frac{n_0}{3} f_Y^3$	0^{dB}
TCI system	L C	$\frac{n_0}{3} (f_Y + f_W + f_N)^2 f_Y$	+4.0"
	L S C	$\frac{n_0}{3} (f_Y + f_C)^2 f_Y$	+2.2"

$f_Y = 19.0^{MHz}$ $f_W = f_N = f_C = 5.5^{MHz}$

The efficiency of frequency utilization of the baseband signal of these systems may not become higher than that of the composite signals. Thus, these systems may not always be suitable for such broadcasting systems as VSB-AM if effective frequency utilization is required as a first priority.

However, these signals are superior to the composite signals for FM broadcasting and magnetic video recording from the point of view of transmission efficiency or of lower noise impairment.

E. Comparison of Signal Systems

Table V shows the relative value of luminance noise interference for triangular noise in various signal systems [21], derived on the supposition that the high-definition television signal is transmitted by FM, using the NHK experimental HD-TV system with 1125 scanning lines. Table V gives a comparison of picture noise deterioration values obtained by adding the chroma channel noise to the luminance channel noise in various signal systems [21].

In composite signals, when the signal is being formed so as to decrease crosstalk interference between the *Y* and *C* components in the order of IYS, OYS, and OYC, the chroma noise interference is greater in the case of FM transmission. Accordingly, noise impairment appearing on the picture depends only on the chroma noise as shown in Table VI.

The luminance channel noise of the TCI system is greater than that of the composite system in FM transmission as shown in Table V. However, the rate of noise impairment affecting the *Y* and *C* components does not depend greatly on the transmission system, so that the TCI system is superior to the composite system for FM broadcasting service from the point of view of transmission efficiency.

In the FM transmission of the TCI-LSC signal, chroma noise is much more apparent on the picture than luminance channel noise, while in the TCI-LC signals, luminance channel noise

TABLE VI
NOISE INCREASE BY ADDING CHROMA NOISE FOR FM NOISE

System		Rate of noise increase by adding chroma noise (Triangular noise)	
		General color scene	Saturated color scene
Composite	I Y S	-5.6^{dB}	-8.0^{dB}
	O Y S	−8.1"	−10.7"
	O Y C	−9.8"	−12.6"
T C I	L C	−2.1"	−3.6"
	L S C	−3.6"	−5.5"

$f_Y = 19.0^{MHz}$, $f_W = f_N = f_C = 5.5^{MHz}$
$(f_S)_{IYS} = 13.^{MHz}$, $(f_S)_{OYS} = 19.5^{MHz}$, $(f_S)_{OYC} = 24.3^{MHz}$
* : inclnde loss by two line repetition

TABLE VII
REQUIRED *S/N* RATIO FOR HLO-PAL FM TRANSMISSION

n(lines)	951	1,125	1,351	1,601
Signal bandwidth (f_c)	22 MHz	30 MHz	42 MHz	60 MHz
Desired S N ratio (weighted)	53 dB (for threshold of detectability)			
Weighting value	11.4 dB	12.0 dB	12.5 dB	13.2 dB
Required S N ratio (unweighted)	41.6 dB	41.0 dB	40.5 dB	39.8 dB
De-emphasis effect	2.8 dB			

Expected S N improvement by de-emphasis processing

will be dominant. The TCI-LC signal will not always be better than the TCI-LSC signal from the point of view of FM noise impairment.

For efficient FM transmission of the signal with low power and narrow bandwidth, the *Y-C* separate transmission system [22], [23] in which the luminance signal and line-sequential chrominance signal are transmitted through individual channels is desirable. In this system, the amplitude peak after pre-emphasis is suppressed to an extent of negligible picture degradation in order to make a large frequency deviation [24], [25]. Thus, a high signal-to-noise ratio can be achieved with a low transmitter power [23], [26].

IV. TRANSMISSION OF HIGH-DEFINITION TELEVISION SIGNALS

A. Desired Signal-to-Noise Ratio

By using the single weighting function for noise affecting the television signal which the author proposed (see [26]), *S/N* ratios required for the HD-TV signal of various standards were also studied [26]. Table VII shows the desirable *S/N* ratio for threshold noise impairment required for FM-transmitted HLO-PAL signals, and Table VIII shows those for *Y-C* separate FM transmission.

The *Y* and *C* signals are preemphasized. The crossover frequency f_t for the *Y* signal is 0.8 times the cutoff frequency f_0; and the f_t for the chrominance signal is 0.25 f_0. The gain in the high-to-low frequency range is 12 dB [26].

B. FM Television Broadcasting in SHF and EHF Bands [27]

The most economical and practical system for the proposed HD-TV service may be transmission by satellite broad-

TABLE VIII
REQUIRED *S/N* RATIO FOR *Y-C* SEPARATE FM TRANSMISSION

n(lines) / Y or C	951 Y	951 C	1,125 Y	1,125 C	1,351 Y	1,351 C	1,601 Y	1,601 C
Signal bandwidth	15 MHz	4.7 MHz	19 MHz	6.5 MHz	28 MHz	9.4 MHz	40 MHz	13.3 MHz
Desired S/N ratio (weighted)	53 dB (for threshold of detectability)							
Weighting value	12.7 dB	8.8 dB	13.4 dB	9.5 dB	14.2 dB	10.3 dB	14.8 dB	10.9 dB
Required S/N ratio (unweighted)	43.3 dB	47.2 dB	42.6 dB	46.5 dB	41.8 dB	45.7 dB	41.2 dB	45.1 dB
De-emphasis effect *	9.9 dB for Y signal, 10.1 dB for C signal							

Expected S/N improvement by de-emphasis processing

TABLE IX
TENTATIVE PARAMETERS FOR HD-TV SATELLITE BROADCASTING SYSTEM

Parameters	Composite colour signal transmission	Y-C separate transmission Y	Y-C separate transmission C	Composite color signal transmission	Y-C separate transmission Y	Y-C separate transmission C
Carrier frequency	22.8 GHz			42 GHz		
Video bandwidth	30 MHz	20 MHz	6.5 MHz	30 MHz	20 MHz	6.5 MHz
Type of modulation	FM	FM	FM	FM	FM	FM
Radio frequency bandwidth	100 MHz	75 MHz	25 MHz	100 MHz	75 MHz	25 MHz
Video signal-to-noise ratio* (unweighted)	40.9 dB	42.6 dB	46.5 dB	40.9 dB	42.6 dB	46.5 dB
Carrier-to-noise ratio (99.5% of time)	25.6 dB	17.3 dB	19.8 dB	25.6 dB	17.3 dB	19.8 dB
Receiving antenna diameter	1.6 m	1.6 m	1.6 m	1.0 m	1.0 m	1.0 m
Receiver noise temperature	800 k	800 k	800 k	1100 k	1100 k	1100 k
Atomospheric attenuation (99.5% of time)	6 dB	6 dB	6 dB	14 dB	14 dB	14 dB
Required e.i.r.p. at the edge of service area (−3 dB)	74.7 dBW	65.2 dBW	62.8 dBW	88.1 dBW	78.6 dBW	76.3 dBW
Satellite transmitting antenna beamwidth	1.0°			1.0		
Satellite transmitter power	3.2 kW	360 W	210 W	70.8 kW	8.0 kW	4.6 kW
		570 W			12.6 kW	

* S/N ratio for threshold of noise detectability

casting and reception by common-use antenna or by home antenna. At present, 11.7-12.2 GHz, 22.5-23 GHz (for Region 3 only), 40.5-42.5 GHz, and 84-86 GHz bands are reserved for satellite broadcasting services. In the 42 GHz band and the 85 GHz band rain attenuation is very large, so the use of the 23 GHz band looks promising.

Typical satellite link parameters are listed in Table IX based on the required signal-to-noise ratios given in Table VIII. The radio frequency bandwidth is 100 MHz, and the time fraction of signal-to-noise ratio which does not exceed the threshold of noise detectability is 0.5 percent during the worst month of the year. It can be seen from Table VIII that the satellite transmitter power is 600 W/° coverage for the *Y-C* separate transmission system mentioned in an earlier section. This transmitter power is likely to be achieved by the late 1980's.

The use of the 42 GHz band may be impractical since 20 times the power required in the 23 GHz band would be needed. This is because of large rain attenuation in the 42 GHz band.

From November 1978 to October 1979, HD-TV transmission experiments via the experimental broadcasting satellite "Yuri" were performed three times using the *Y-C* separate transmission system shown in Fig. 9. The experiments were performed with the cooperation of the Japanese Ministry of Posts and Telecommunications.

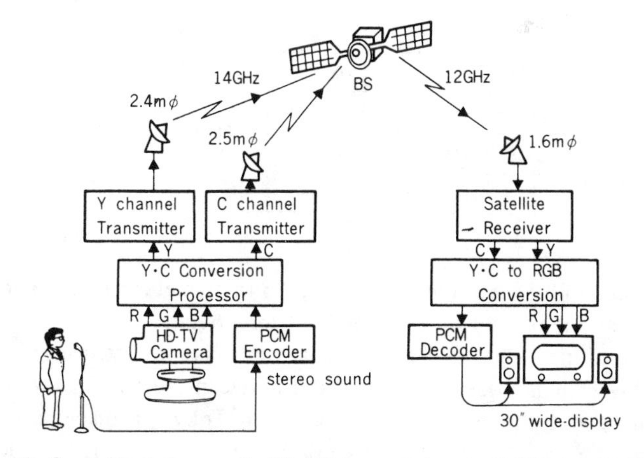

Fig. 9. A block diagram for HD-TV transmission via 12 GHz band BSE.

The satellite link parameters of the experiment are shown in Table X [28], and the audio signal standard is shown in Table XI. The experiment showed that the *S/N* ratio of the received picture was approximately equal to the threshold of

TABLE X
Y-C SEPARATE TRANSMISSION PARAMETERS VIA THE EBS AT 12 GHz BAND

Parameters	Y channel	C channel
Carrier frequency	12.0875 GHz	11.9625 GHz
Video bandwidth	20 MHz	6.5 MHz
Type of modulation	FM	FM
Frequency deviation	40 MHz	12 MHz
Radio frequency bandwidth	80 MHz	25 MHz
S/N ratio (unweighted)	42.5(38.6)dB	44.5(40.6)dB
C/N ratio (99% of time)	16.7(12.8)dB	22 (18.1)dB
Receiving antenna diameter	2.5m (1.6m)	
Receiver noise temperature	660 K	
Satellite transmitting antenna gain	37 dB	
Satellite transmitter power	100W	100W

TABLE XI
SPECIFICATIONS OF PCM AUDIO SIGNAL

Signal bandwidth	30 Hz ~ 15 kHz
Sampling frequency	33.75 kHz
Type of coding	Instantaneous compression from 14 to 12 bits (7 segment)
Number of parity bits	1 bit
Error concealment	Previous sample replacement or interpolation with succeeding sample
Transmission bit rate	12.15 Mb/s (equivalent)

TABLE XII
PARAMETERS FOR 38 GHz BAND FM TRANSMISSION

Radio frequency bandwidth		110 MHz	
Baseband signal	Bandwidth (f_b)	30 MHz	Threshold for detectability of S/N
	Desired S/N ratio	41 dB	
S/N improvement by FM		12 dB	
Desired C/N ratio		29 dB	
Transmitting power		400 mW	Transmitting ant. 40cmϕ
Receiving power		−37.5dBm	Receiving ant. 40cmϕ
C/N ratio		40 dB	
Allowable rain margin		11 dB	

detectability when received with a 2.5 m diameter antenna and was about grade 4 for a 1.6 m diameter antenna.

In June 1980, an HD-TV signal transmission test using the 38 GHz band was also carried out between the NHK Laboratories and the Broadcasting Center in Tokyo, a distance of about 8 km. For the test, an IMPATT power amplifier of 400 mW output was developed at the NHK Laboratories.

The link parameters of the experiment are shown in Table XII [2]. The experiment was continued for three days, and rain attenuation of about 6 dB was recorded during a moderate rainfall.

C. Digital Transmission [27]

It is important to consider not only FM transmission systems but also digital transmission systems, especially for

TABLE XIII
PARAMETERS FOR DIGITAL TRANSMISSION OF HD-TV

Parameters	Digital transmission (DPCM)
Bit rate	250 Mb/s
Type of modulation	4ϕ PSK
Radio frequency bandwidth	150 MHz
Bit error rate	10^{-5}
Carrier-to-noise ratio (99.5% of time)	18.8 dB
Receiving antenna dia.	1.6 m
Receiver noise temp.	800 K
Atmospheric attenuation (99.5% of time)	5 dB
Required e.i.r.p. from satellite at the edge of service area (−3 dB)	69.70 dB
Satellite transmitting antenna beamwidth	1°
Satellite transmitter power	1,010 W

future broadcasting systems. Considerable bit-rate reduction can be expected utilizing digital signal processing techniques such as intraframe coding (DPCM and block coding) and interframe coding. DPCM of 4 bits/sample (average rate) provides good picture quality. The total bit rate is 250 Mbits/s, including a sound channel. Digital satellite transmission parameters of the DPCM coded signal are shown in Table XIII. 4ϕ PSK modulation of bandwidth 150 MHz is used (where the product of bandwidth B and pulse spacing T equals 1.2), and the bit error rate is expected to be 10^{-5}, which can be improved to 10^{-8} by using an error-correction code. The carrier-to-noise (C/N) ratio required to keep the bit error rate (BER) under 10^{-5} is 18.8 dB, in which a total carrier-to-noise ratio degradation of 6 dB is assumed.

The degradation in carrier-to-noise ratio is caused mainly by intersymbol interference due to band limiting and the nonlinearity of transponders. In this case, the required satellite transmitter power is 1010 W, which is not large compared to FM transmission.

If interframe coding is adopted, a smaller bit rate can be utilized, and this will result in the reduction of the transmitting bandwidth and transmitter power.

D. Optical Fiber Transmission

Optical fibers are very suitable as transmission lines for broad-band HD-TV signals because of their wide bandwidth and low loss characteristics. We have developed several optical fiber transmission systems. In the first experimental system, a baseband HD-TV HLO-PAL signal modulates the light intensity output of an LED [29].

In this system which uses an LED with 0.82 nm wavelength and optical fiber with an attenuation loss of 3.0 dB/km, it is possible to transmit the HD-TV signal over 3 km without repeaters.

In recent years, an optical fiber system using a light source with a long wavelength of 1.3-1.5 μm has been studied actively and in detail. A transmission test with a laser diode (LD) modulated by the HLO-PAL baseband signal of HD-TV was conducted and it was found possible to transmit the signal over 20 km without any repeater [30].

A new system in which a laser diode is used with a transmission capacity of 500 MHz for long-distance trunk line transmission or distribution with a community antenna reception system from a broadcasting satellite has been developed [31]. In this system, frequency-modulated signals of HD-TV and VHF standard television signals were transmitted by frequency multiplex through graded-index optical fibers.

V. DEVELOPMENT OF HD-TV APPARATUS

In order to facilitate further study into HD-TV systems, NHK is making every effort to develop high-definition equipment for the 1125 scanning line system: for example, various practical equipment should be developed, and experimental broadcasting tests by using this equipment should be done. At present, various experimental equipment has been developed with the cooperation of some companies, such as TV cameras, Laser Telecine for 70 mm movie film, a color encoder and receiver for satellite FM transmission, display devices, and high resolution color monitors (see [2]). This equipment will be used effectively to help the study of HD-TV broadcasting systems to progress.

Currently, we are working on the development of a new videotape recorder and network system, as well as gas-discharge display panels for use in future large wall-type screen television receivers.

VI. CONCLUSION

Since 1970, NHK has been carrying out wide-ranging studies of high-definition television systems, including the development of equipment that can satisfy the social demands of future postindustrial societies. This paper mainly deals with desirable standards viewed from the characteristics of the human visual system, signal systems, and transmission parameters for broadcasting. With regard to transmission primaries and signal standards, a description is given of a 1125-line system, which NHK is considering as a tentative standard and using for various experiments. Consideration is also given to the parameters necessary for transmission with this system. Experiments show that a high-definition television should have more than 1000 scanning lines, so that a broadcasting satellite would be the most effective method for the distribution of television signals throughout Japan without the deterioration of picture quality. For a satellite broadcasting service, 11.7-12.2 GHz, 22.5-23.0 GHz (for region 3), 41-43 GHz, and 84-86 GHz bands are reserved, but because the rain attenuation in the 40 and 80 GHz bands is considerable in comparison to the 22 GHz band, it will be difficult to use it.

With regard to the standards for high-definition television systems, it will be necessary to exchange views, opinions, and information on a worldwide basis in order to establish unified international standards.

When it seems likely that the image-oriented society of the future will be formed on the basis of a progressing television technology, it is clear that there is a need for the development of higher precision and higher quality television technologies. It is expected that standards will be unified and established in order to promote the rapid development of high-definition television systems.

ACKNOWLEDGMENT

The author would like to express his gratitude to Dr. R. Takahashi, Director General of Engineering, K. Yabashi, General Managing Director, S. Shigeta, Director of the Technical Research Laboratories of NHK, and K. Hayashi, Director of the Broadcasting Science Research Laboratories, for their guidance and valuable suggestions during this research.

REFERENCES

[1] T. Fujio, "A study of high-definition TV system in the future," *IEEE Trans. Broadcast.*, vol. BC-24, Dec. 1978.
[2] ——, "High-definition, wide-screen television system for the future," *IEEE Trans. Broadcast.*, vol. BC-26, Dec. 1980.
[3] T. Nishizawa, "Visual effect of line interlace scanning," VVI46–9, ITE Japan, Sept. 1971.
[4] T. Ohtani, T. Fujio, and T. Hamasaki, "Subjective evaluation of picture quality for future high-definition television," *NHK Tech. J.*, vol. 28, no. 4, 1976.
[5] T. Ishida and T. Taneda, "70 mm motion picture and HD-TV," in *6th Conf. Image Tech.*, 5–3, 1975.
[6] ——, "Subjective quality tests on motion picture using 70 mm movie films," VV112–1, ITE Japan, Sept. 1975.
[7] T. Fujio, "High-definition TV system viewed from the response of the visual system," CS79–62, IECE Japan, July 1979.
[8] T. Hamasaki, T. Fujio, and T. Ohtani, "Subjective evaluation test of high-definition TV picture using simulated color slides," *NHK Tech. Rep.*, vol. 16, no. 10, 1973.
[9] T. Nishizawa and I. Yuyama, "A simulator and subjective assessments for a high-definition TV system," IE–75–96, IECE Japan, Jan. 1976.
[10] S. Sato and T. Kubo, "Picture quality of a high-definition television," NHK Lab. Note, Sept. 1979.
[11] T. Fujio and T. Nishizawa, "Film simulation for HD-TV picture and subjective test of picture quality," *NHK Tech. Rep.*, vol. 18, no. 11, 1975.
[12] T. Ohtani and T. Kubo, "An investigation of shape of screen for a high quality television system," *NHK Tech. Rep.*, vol. 14, no. 5, 1971.
[13] T. Mitsuhashi, "A 2125 lines high definition B&W CRT display," *NHK Tech. Rep.*, vol. 18, no. 11, 1975.
[14] T. Saito, "A high definition television camera—2125 lines real time scanning," *NHK Tech. Rep.*, vol. 18, no. 11, 1975.
[15] T. Mitsuhashi, "A study of the relationship between scanning specifications and picture quality," NHK Lab. Note, no. 256, 1980.
[16] H. Kusaka and T. Nishizawa, "A 1125 line high-definition three-vidicon color camaera," *NHK Tech. Rep.*, vol. 16, no. 10, 1973.
[17] J. Ishida, "The choice of bandwidths for the chrominance signals in 1125-line HD-TV," in *Nat. Conv. IECE*, no. 1620, 1976.
[18] T. Fujio, "The transmission primary signals and their transmission for a high-definition television system, CS75–79, IECE Japan, Sept. 1975.
[19] ——, "Prospect of high-definition TV system—Its signals and transmission," IE76–78, IECE Japan, Feb. 1977.
[20] H. Sakata and H. Isono, "The chromatic spacial frequency characteristics of human vision," in *Nat. Conv. ITE Japan*, 1–5, 1973.
[21] T. Fujio, "High-definition television for future imaging system," IE80–84, IECE Japan, Dec. 1980.
[22] T. Fujio and S. Nakamura, "Y–C separate transmission system of the standard color TV signal and transmitting parameter," in *1968 Joint Conv. Four Inst. Elec. Eng.*, no. 1980, Japan.
[23] T. Fujio, "Shortcoming of present color TV system and prospect of high-definition TV system," IE76–9, IECE Japan, May 1976.
[24] CCIR doc., CMTT/59 (4/219), 1963–1966, Japan.
[25] K. Hayashi, "Positive-sync. and nonlinear-emphasis processing for satellite FM transmission of television signals," *NHK Tech. Rep.*, vol. 8, no. 7, 1965.

[26] T. Fujio, "A universal weighted power function of television noise and its application to high-definition TV system design," *IEEE Trans. Broadcast.*, vol. BC-26, June 1980.

[27] T. Fujio, J. Ishida, T. Komoto, and T. Nishizawa, "High-definition television system," *SMPTE J.*, vol. 89, Aug. 1980.

[28] T. Komoto, J. Ishida, M. Hata *et al.*, "Y–C separate transmission of HD-TV signal by BSE," CS79–183, IECE Japan, Nov. 1979.

[29] J. Kumada and T. Mitsuhashi, "Optical fiber transmission for high-definition TV signal," NHK Lab. note, no. 228, Oct. 1978.

[30] K. Asatani, T. Miki *et al.*, "Fiber-optic transmission of HD-TV signals," CS79–147, IECE Japan, 1979.

[31] T. Komoto and J. Ishida, "FM transmission of the HD-TV signal through graded-index optical fiber," in *Nat. Conv. IECE*, no. 2248, 1980.

Extended Definition Television with High Picture Quality

By Broder Wendland

Present-day television systems do not offer optimum picture quality, particularly in new applications such as interactive videotex and still picture transmission, and where larger and brighter screens are used. There is a growing demand for improved resolution of details and absence of line crawl and flicker. This article proposes concepts for compatible improvements in picture quality, using the line numbers of present standards, by means of digital signal processing. Improved resolution through use of optimum pre- and postfiltering, the timeplex techniques of the MAC system, and the new FM channels with higher bandwidth are discussed.

Present-day TV systems do not offer a picture quality that will meet the requirements of future systems and applications. This is especially true for picture reproduction on large screens or in connection with new services such as teletext or computer graphics. Therefore, about 12 years ago, the Japanese (NHK) started experiments with high-line-number TV systems and created an 1125-line system with a luminance bandwidth of about 20 MHz.[1]

This system, with its 1125 lines and 60-Hz field frequency, is incompatible with the TV systems presently used in Europe that have 625 lines and 50-Hz field frequency. This incompatibility with the numerous standard receivers can be overcome only by expensive scan converters.

To use the 1125-line system, owners of the conventional receivers would have to spend a lot of money with no resulting improvement in picture quality. This is discussed in the 1980 report of the SMPTE HDTV Study Group.[2] The Study Group concluded that "... An HDTV system compatible with the existing domestic services ... is not feasible by any means known to or envisaged by the Study Group, in view of HDTV aspect ratios and bandwidths." This is due to the following restrictions:

1. Narrow bandwidth for luminance and chrominance,
2. 4:3 aspect ratio, and
3. Cross color and cross luminance.

However, with the application of highly sophisticated signal processing at the transmitter and receiver end, it is now possible to improve the picture quality within a given standard. This concept was proposed two years ago[3] and implemented at the University of Dortmund.

Following the concepts of optimum pre- and postfiltering[3] and the time-plex techniques,[4,5] there is a possibility for an enhanced system. The concept involves new FM channels with higher bandwidth, giving improved resolution in the horizontal and vertical directions, using the line numbers of present standards. With the time multiplex technique, it is further possible to use different aspect ratios for different kinds of receivers.

Picture Scanning Theory

In television systems, picture scanning is usually done line by line with line interlace 2:1 to conserve bandwidth, which provides a reasonably flicker-free picture. However, line interlace produces heavy aliasing in moving parts, bad vertical resolution, line flickering, and line crawl, etc.[3,6,7] To show how error-free scanning of frames can be performed, progressive scanning will be discussed first.

Progressive Scanning

With progressive scanning, it is assumed that the camera still works with

Presented at the Society's 17th Annual Television Conference (paper No. 17-4) in San Francisco on February 4, 1983, by Broder Wendland, Universitat Dortmund, Dortmund, Germany. This article was received January 14, 1983, and also appears in *Video Pictures of the Future*, published 1983, SMPTE.

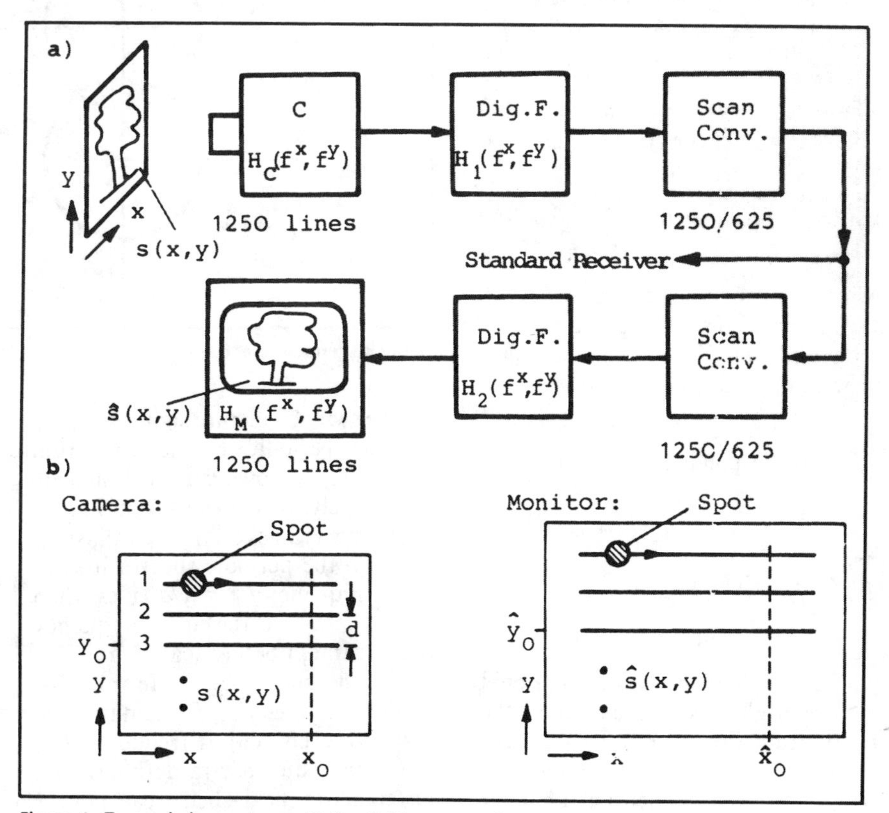

Figure 1. Transmission system with flat-field reproduction.

Reprinted with permission from *SMPTE J.*, vol. 92, pp. 1028–1035, Oct. 1983.
Copyright © 1983 by the Society of Motion Picture and Television Engineers.

25 frames/sec. To get flicker-free monitoring, digital frame stores are used. Figure 1a shows a picture transmission system containing a high-line-number camera with progressive scanning. The output signal of the camera is two-dimensionally band limited by a digital filter $H_1(f^x, f^y)$ and then transformed by scan conversion (digital frame store) into a 625-line signal with line interlace 2:1. This transmits a signal that is compatible with standard receivers.

However, for a high-line-number monitor, the video signal is retransformed by a scan converter and an interpolating two-dimensional filter $H_2(f^x, f^y)$ into a high-line-number signal. As will be shown, this process allows error-free scanning at the camera and, by flat-field reproduction at the monitor, an improved picture resolution in the vertical direction. At the transmission channel, the video signals are compatible with common receivers.

The progressive scanning scheme is shown in Figure 1b, and the point spread function $b(x,y)$ of the camera and its two-dimensional Fourier transform $\underline{B}(f^x, f^y)$, in Figs. 2a and 2b. With the assumption of a circular modulation transfer function (MTF) of the camera, the −40-dB level of $\underline{B}(f^x, f^y)$ gives a circular border for the so-defined cutoff frequency $f_c{}^r$ of the camera. If the δ-distribution series

$$\sum_{i=-\infty}^{\infty} \delta(y - id) = \perp\!\perp\!\perp_d(y) \quad (1)$$

where $\delta(y - id)$ represents a line singularity, we get the equivalence

$$\perp\!\perp\!\perp_d(y)$$

which corresponds to

$$f_s{}^y \cdot \perp\!\perp\!\perp_{f_s{}^y}(f^y)\delta(f^x) \quad (2)$$

where d is the line distance, and $f_s{}^y = 1/d$, the spatial sampling frequency in the y-direction. Then the line scanning process becomes

$$b_\perp(x,y) = b(x,y) \perp\!\perp\!\perp_d(y)$$

which corresponds to

$$\underline{B}_\perp(f^x, f^y) = \underline{B}(f^x, f^y) * f_s{}^y \times \perp\!\perp\!\perp_{f_s{}^y}(f^y)\delta(f^x) \quad (3)$$

$$\underline{B}_\perp(f^x, f^y) = (1/d) \sum_{\nu=-\infty}^{\infty} \times \underline{B}(f^x, f^y - \nu/d) \quad (4)$$

where $b(x,y)$ is the two-dimensional picture brightness (at the target of the pickup tube in the camera), $b_\perp(x,y)$ is the scanned picture signal, and $\underline{B}(x,y)$ is the two-dimensional Fourier transform of $b(x,y)$.

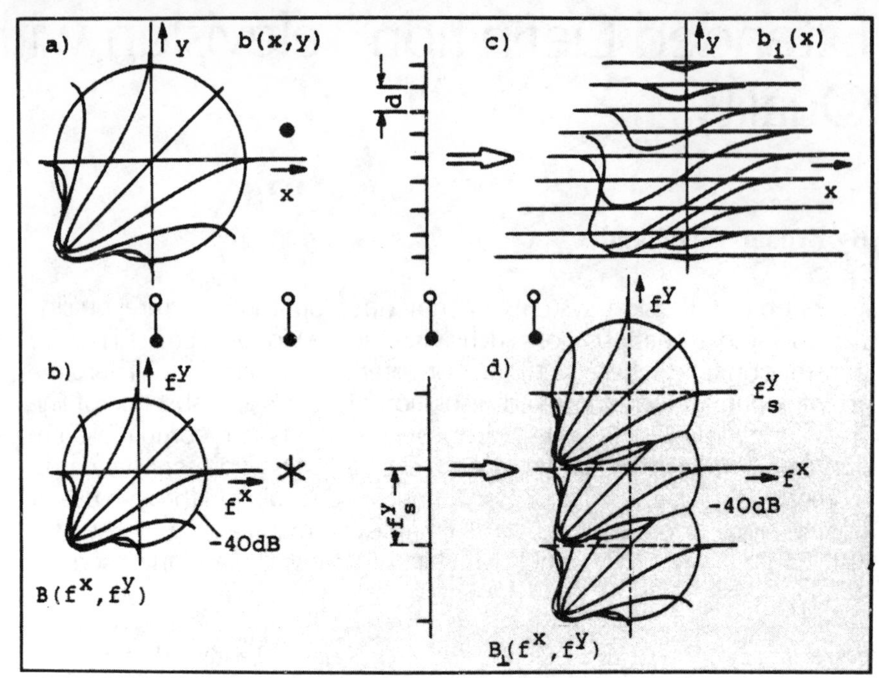

Figure 2. Picture signal $b(x,y)$, sampled signal $b_i(x)$ and their spectra.

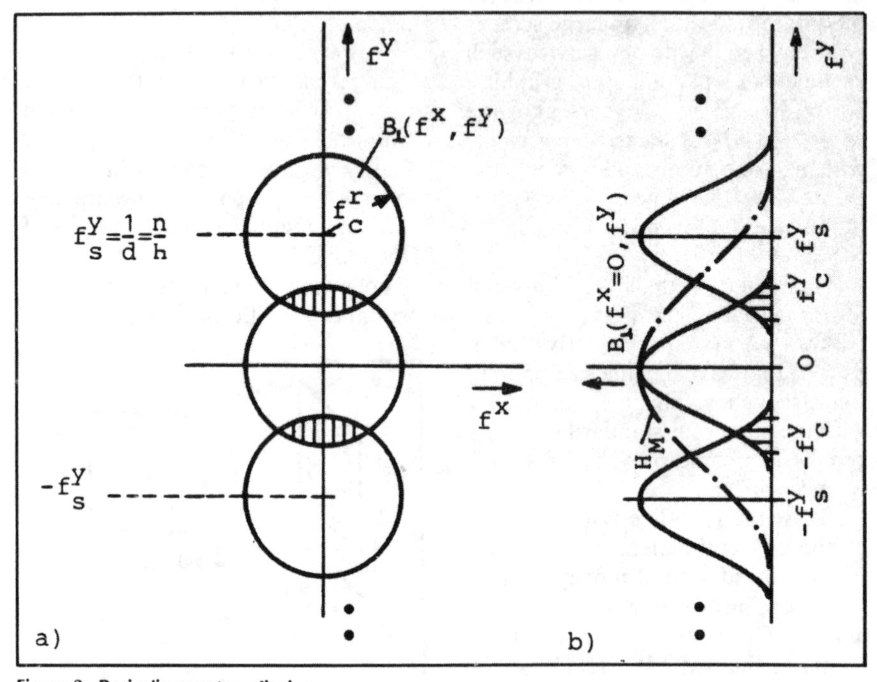

Figure 3. Periodic spectra, aliasing.

The scanned picture signal $b_\perp(x,y)$ and its periodic Fourier spectrum $\underline{B}_\perp(f^x, f^y)$ is shown in Figs. 2c and 2d, respectively. It is obvious from Fig. 2d that there is an overlapping (aliasing) between the periodic spectra if sampling frequency $f_s{}^y = 1/d$ is less than $2f_c{}^r$, where $f_c{}^r$ is the cutoff frequency of the camera before scanning.

This is shown again in Fig. 3a, where in the f^x-, f^y-domain the −40-dB levels (cutoff frequency $f_c{}^r$) of the periodic spectra define the interference in the dashed areas. For the spatial frequency $f^x = 0$, the periodic spectrum $B_\perp(0, f^y)$ is shown in Fig. 3b. With the transfer function $H_M(0, f^y)$ of the monitor, the alias components (dashed areas) and parts of higher order spectra are shown on the monitor screen. The moirés and irritating line structures are visible, both of which cause a reduction of vertical resolution. The design of an optimum camera/lens system involves a choice between high resolution with heavy aliasing (moirés), or no aliasing and bad resolution. Usually, in line interlace systems, fairly high resolution is chosen, accompanied by aliasing

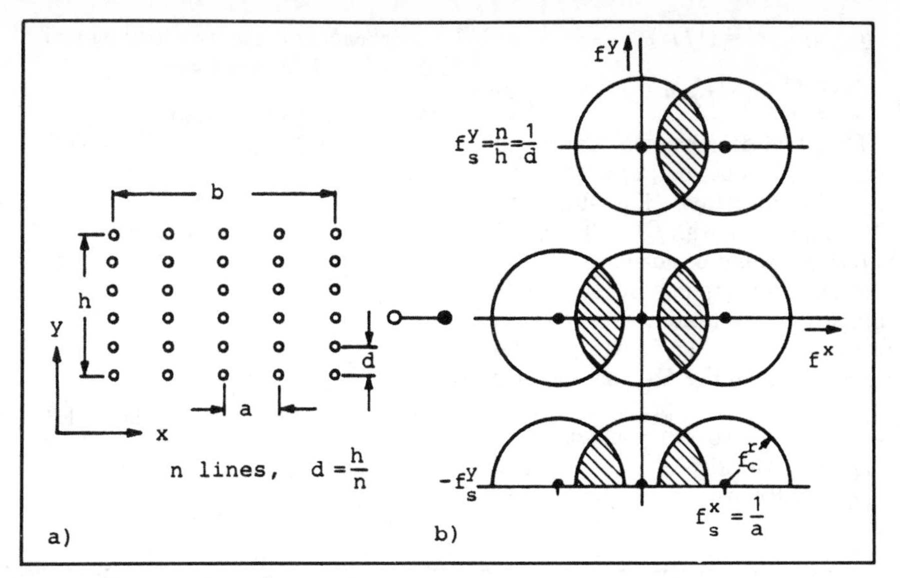

Figure 4. Orthogonal sampling, spectra.

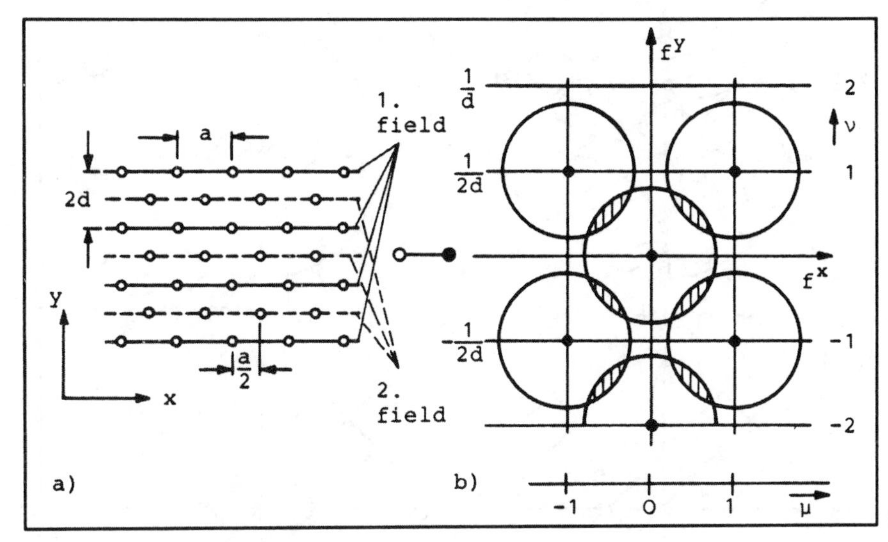

Figure 5. Offset sampling, spectra.

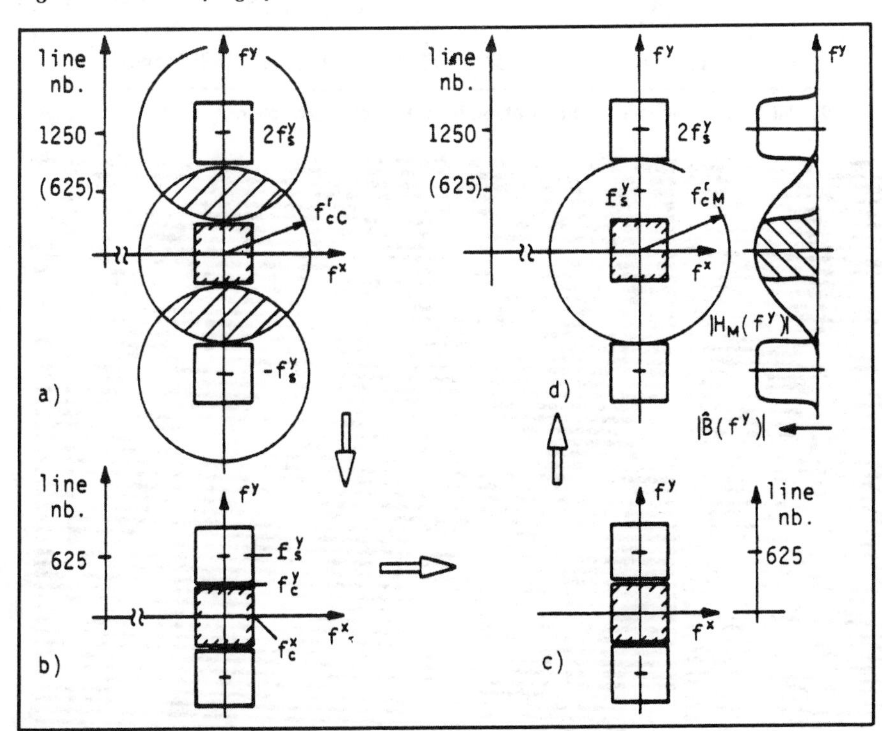

Figure 6. Alias-free scanning and flat-field reproduction.

effects. (In practice the line interlace effects dominate the aliasing effects.) The degradation of vertical resolution is quantitatively described by the Kell factor $k_1 = 0.64$ (for progressive scanning).

Interlace Scanning

All present-day TV systems make use of interlaced scanning to overcome large area flickering. With interlace scanning, there are two sequences of fields, $b_{\perp 1}(t)$ and $b_{\perp 2}(t)$, in which each field has double-spaced line structures with line distance $2d$, whereas the scanning patterns of field 1 and field 2 are in offset position. A total frame with n lines is described by two fields with $n/2$ lines each.

For a given three-dimensional picture signal $b(x,y,t)$, we have for two fields, according to one frame, the signals $b_1(x,y)$ and $b_2(x,y)$ with their corresponding Fourier spectra $\underline{B}_1(f^x,f^y)$ and $\underline{B}_2(f^x,f^y)$, respectively.

With respect to Eqs. 1 and 2, line interlace scanning gives for a first field:

$$b_{1\perp}(x,y) = b_1(x,y) \cdot \perp\!\perp\!\perp_{2d}(y) \tag{5}$$

which corresponds to

$$\underline{B}_{1\perp}(f^x,f^y) = \underline{B}_1(f^x,f^y) * \times (1/2d) \perp\!\perp\!\perp_{1/2d}(f^y)\cdot\delta(f^x) \tag{6}$$

and for a second field:

$$b_{2\perp} = b_2(x,y) \perp\!\perp\!\perp_{2d}(y-d) \tag{7}$$

which corresponds to

$$\underline{B}_{2\perp} = \underline{B}_2(f^x,f^y) * (1/2d) \perp\!\perp\!\perp_{2d} \times (f^y) e^{-2\pi j f^y d} \cdot \delta(f^x) \tag{8}$$

with contributions only for

$$f^y = \nu/2d,\ \nu = \pm 1, \pm 2 \ldots,$$

In total, line interlace scanning gives for a full frame from Eqs. 6 and 8

$$\underline{B}_{\perp}(f^x,f^y) = \underline{B}_{1\perp}(f^x,f^y) + \underline{B}_{2\perp}(f^x,f^y) = (1/2d)\left\{\sum_{\mu} \underline{B}_1(f^x, f^y - \mu/2d) + \sum_{\nu} (-1)^{\nu} \underline{B}_2(f^x, f^y - \nu/2d)\right\} \tag{9}$$

In the case of a still picture, where

$$\underline{B}_1(f^x,f^y) = \underline{B}_2(f^x,f^y) = \underline{B}(f^x,f^y)$$

we get from Eq. 9

$$B_{\perp}(f^x,f^y) = (1/d) \sum_{\mu} \underline{B}(f^x, f^y - 2\mu/2d) \tag{10}$$

because every second (odd) number of ν cancels the spectrum in Eq. 9.

Thereby, we get the same periodic spectrum as for progressive scanning (Eq. 4). This is true for still pictures and exact integration over two fields. However, in practice neither will still pictures be very interesting nor do we have very good integration over two fields with the human eye. Therefore, line interlace suffers from some deficiencies, including:

- Heavy field-aliasing effects by scanning with half the line number in every field.
- Incomplete cancelling of field aliasing by the human eye (25-Hz flickering).
- Bad resolution of vertical moving parts because of busy edges and 25-Hz flickering on horizontal contours with high contrast.
- Residual aliasing effects in vertically moving areas because of changing spectra $B_1(f^x,f^y) \neq B_2(f^x,f^y)$ in Eq. 9 (critical velocity $v_{cr}{}^y \approx 0.2$ pel/frame).
- Residual aliasing effects in horizontally moving areas because of noncompensating field aliasing.

Sampling Concepts

Progressive Line Scanning

A two-dimensional sampling scheme is shown in Fig. 4a, in which a progressive scanned video signal $b_\perp(x,y)$ is sampled in x-direction. With an orthogonal sampling pattern and sampling distance a, we get for the sampled video signal

$$b_{\perp\perp}(x,y) = [b(x,y) \perp\perp\perp_d(y)] \times \perp\perp\perp_a(x) \quad (11)$$

which corresponds to

$$\underline{B}_{\perp\perp}(f^x,f^y) = (1/ad) \times \sum_\mu \sum_\nu B(f^x - \mu/a, f^y - \nu/d) \quad (12)$$

The two-dimensional sampling process causes periodic spectra in the f^x- and f^y-directions (Fig. 4b). With a basic spectrum with $f_c{}^x \leq \frac{1}{2}a, f_c{}^y \leq \frac{1}{2}d$, there would be no aliasing.

Another sampling scheme with samples in offset position in adjacent lines is shown in Fig. 5a with its Fourier spectrum in Fig. 5b. This sampling scheme can be interpreted as the superposition of two orthogonal sampling schemes with video signals $b_{\perp\perp 1}$ and $b_{\perp\perp 2}$, each consisting of every second line.

For each of the two orthogonal sampling schemes, we get

$$b_{\perp\perp 1} = [b(x,y) \perp\perp\perp_{2d}(y)] \times \perp\perp\perp_a(x) \quad (13)$$

and

$$b_{\perp\perp 2} = [b(x,y) \perp\perp\perp_{2d}(y-d)] \times \perp\perp\perp_a(x - a/2) \quad (14)$$

with their Fourier transforms

$$\underline{B}_{\perp\perp 1} = (\tfrac{1}{2}ad) \sum_\mu \sum_\nu \underline{B}_1 \times (f^x - \mu/a, f^y - \nu/2d) \quad (15)$$

$$\underline{B}_{\perp\perp 2} = (\tfrac{1}{2}ad) \sum_\mu \sum_\nu \underline{B}_2 \times (f^x - \mu/a, f^y - \nu/2d)(-1)^{\nu+\mu} \quad (16)$$

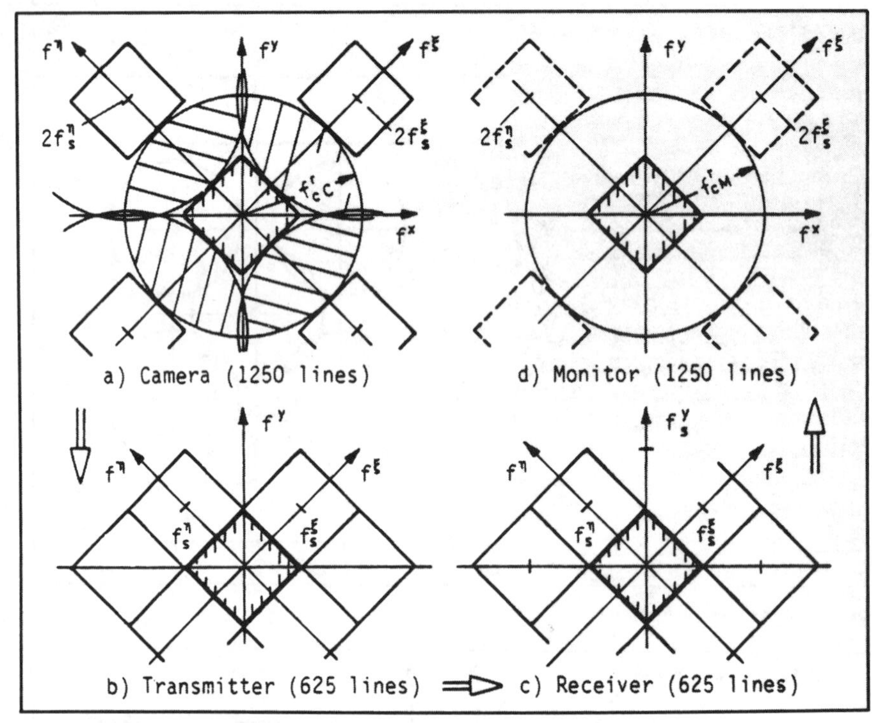

Figure 7. Flat-field reproduction with offset sampling and scan conversion.

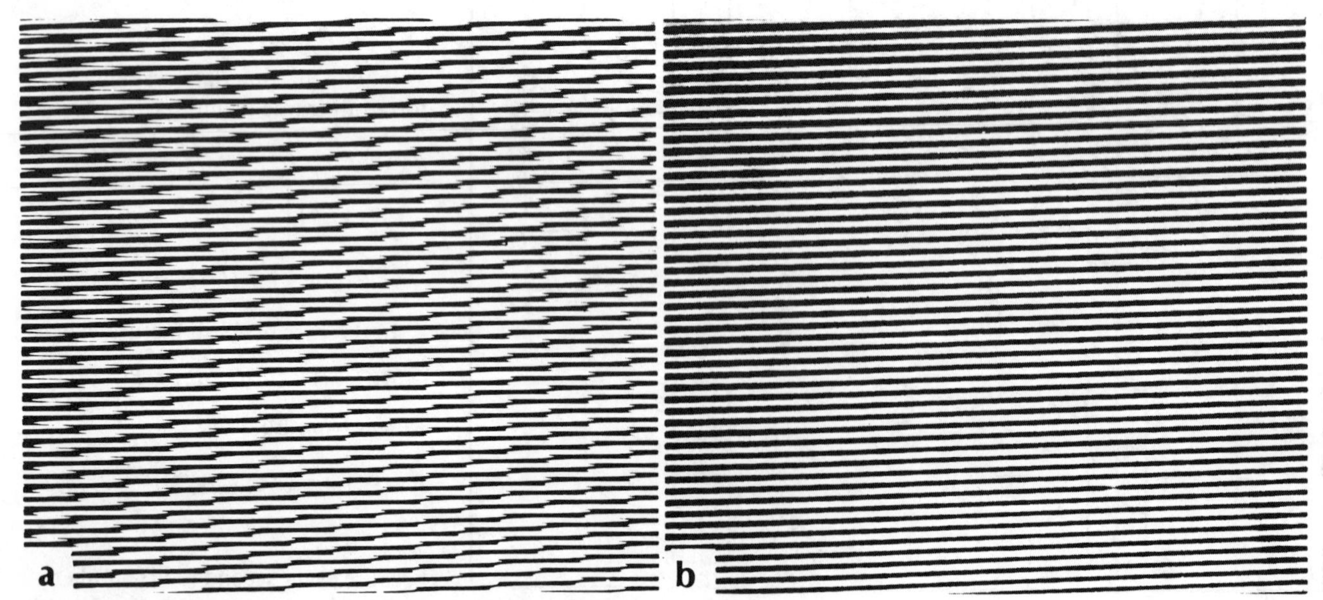

Figure 8. "801/100 TV lines": (a) Reference picture; (b) Horizontal and vertical filtering.

In the case of progressive scanning, $\underline{B}_1(f^x, f^y)$ is the same as $\underline{B}_2(f^x, f^y)$, and we get, for a full frame out of the superposition of Eqs. 15 and 16

$$\underline{B}_{\perp\perp} = (1/2ad) \sum_{\mu} \sum_{\nu} \underline{B}\,(f^x - \mu/a,\; f^y - \nu/2d)(1 - (-1)^{\nu+\mu}) \quad (17)$$

with repetitions of the basic spectrum $\underline{B}(f^x, f^y)$ only for $\nu + \mu = 2k$, $\pm k = 1, 2$. These periodic spectra in diagonal direction are shown in Fig. 5b.

Line Interlace Scanning

In the case of line interlace scanning, we will have different spectra $\underline{B}_1(f^x, f^y)$ and $\underline{B}_2(f^x, f^y)$ in moving areas of the picture. Thereby, we get additional residual spectra between the periodic spectra in Fig. 5b or 4b which cause aliasing effects. The sensitivity is found to be about 0.2 pel/frame for visibility in x- or y-direction.

Compatible Improvements by Pre- and Postfiltering

With a digital picture frame store at the receiver end, it is possible to reproduce a progressively scanned picture with a higher frame rate, resulting in a flicker-free picture without degradation by line interlace. So the degradation factor $k_2 \approx 0.65$ of line interlace is overcome.[1]

Moreover, with a picture frame store and a planar interpolating postfilter at the high-line-number monitor with flat-field reproduction, it is possible to get further improvements in picture quality.

Error-Free Scanning and Sampling

As it has been pointed out, vertical resolution is distorted by aliasing effects by the scanning process and by the line structure of the monitor. But in the case of a high-line-number camera in Fig. 1a, which has $2n$ lines/frame (whereas n lines have to be transmitted), we can tolerate some areas of aliasing (as shown in Fig. 6a). With proper vertical and horizontal filtering, we can transmit this error-free signal with 625 lines (Figs. 6b and 6c). After transmission, the video signal can be retransformed by a vertical interpolating filter into a high-line-number signal with 1250 lines. With the monitor transfer function $\underline{H}_M(f^y)$ in Fig. 6d out of this error-free signal, the picture can be reconstructed without irritating line structures by a flat-field monitor. With this technique, the picture quality can be improved and vertical resolution is increased within the same standard and for a given bandwidth, and the Kell factor is overcome.

Diagonal Pre- and Postfiltering

With proper planar filtering in the diagonal direction, resolution can be further improved in the horizontal direction by offset sampling, pointed out earlier and shown in Figs. 5a and 7. Again, we start with a high-line-number camera (1250 lines) in Fig. 7a, and by prefiltering in f^ξ- and f^η-direction, we get a basic area that is bordered by a diagonally oriented square. This basic spectrum can be transmitted with half the line number without any error (Figs. 7b and 7c).

At the receiver end, after analog or digital transmission, the signal is retransformed by an interpolating digital filter into a high-line-number signal. This high-line-number signal can be reconstructed at the monitor screen without irritating line structures (flat field).

This concept gives increased horizontal resolution. With a spatial sampling frequency $f_s{}^x$, an error-free sampling with an orthogonal sampling scheme needs a cutoff frequency $f_c{}^x \leq 1/2 f_s{}^x$. With offset sampling, using a diagonal prefilter (Fig. 7a), the horizontal cutoff frequency for error-free sampling can be twice as high, that is, $f_c{}^x = f_s{}^x$. This improvement of resolution in horizontal direction is accompanied by some loss in diagonal direction. However, the human eye is somewhat more sensitive to vertical and horizontal structures than to diagonal structures. In addition, there is a higher probability for horizontally

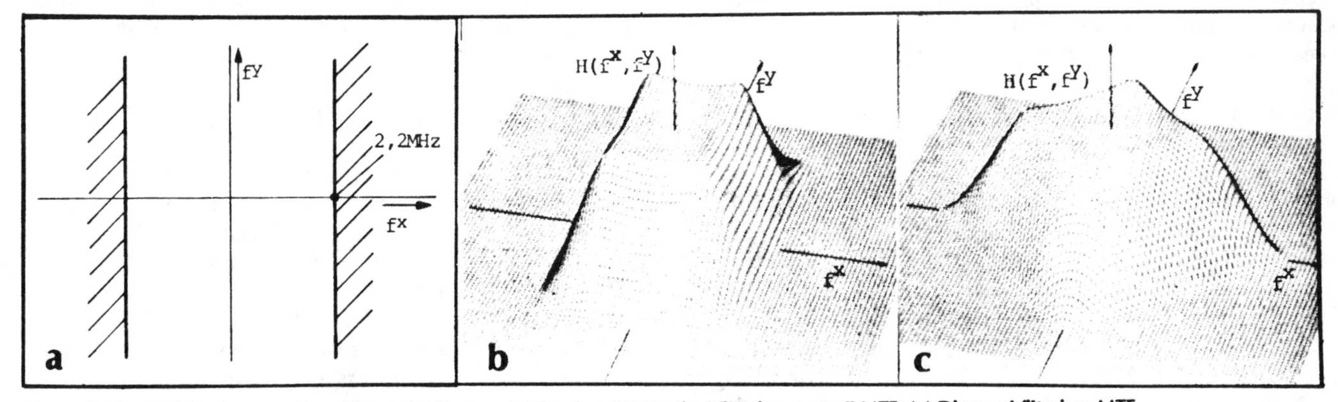

Figure 9. Total MTF of pre- and postfilters: (a) Horizontal filtering; (b) Vertical filtering, overall MTF; (c) Diagonal filtering, MTF.

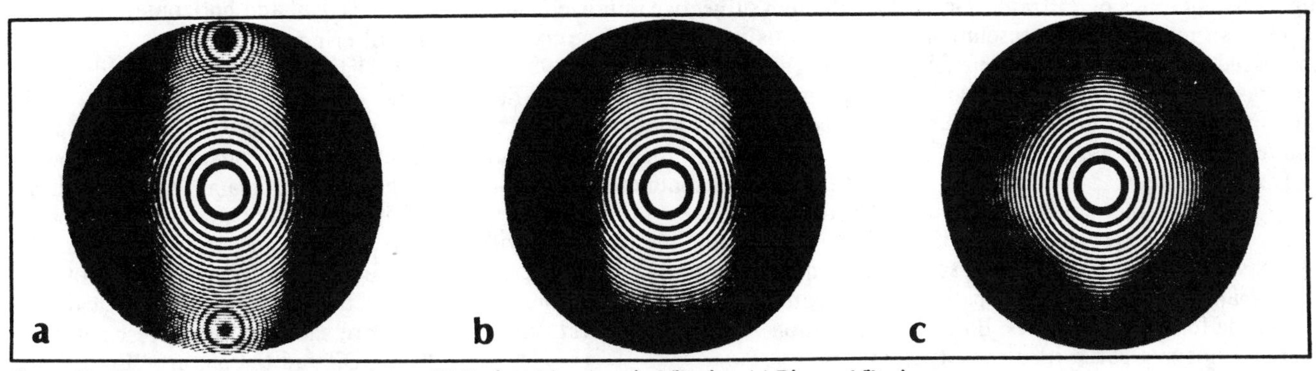

Figure 10. "Zone plate": (a) Reference picture; (b) Horizontal and vertical filtering; (c) Diagonal filtering.

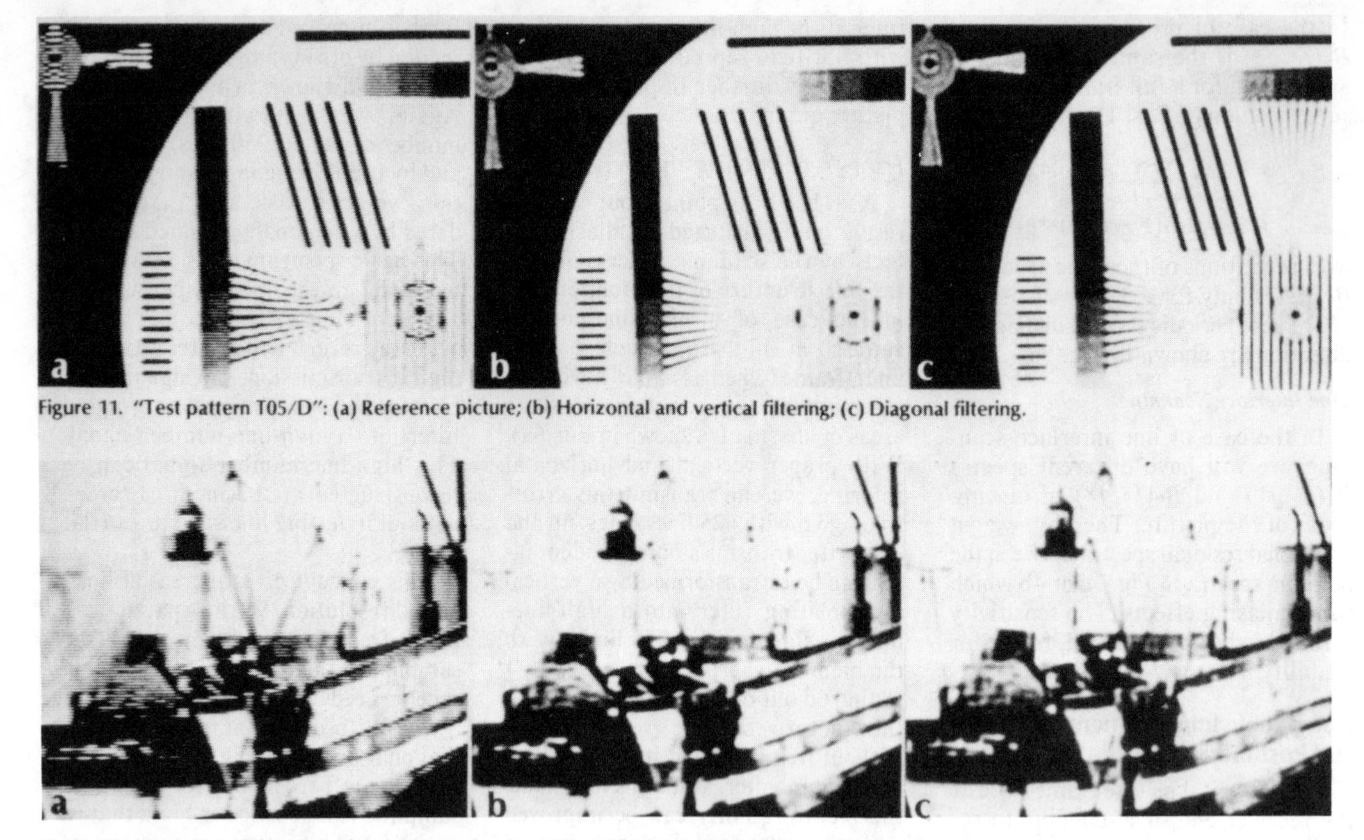

Figure 11. "Test pattern T05/D": (a) Reference picture; (b) Horizontal and vertical filtering; (c) Diagonal filtering.

Figure 12. "Boats": (a) Reference picture; (b) Horizontal and vertical filtering; (c) Diagonal filtering.

and vertically oriented structures in pictures.

Noise Figure

It is very important that the video signal be planar filtered *before* it is sampled in offset position.[8] Otherwise there will be low-frequency noise caused by subsampling noise, transforming high-frequency noise into the low-frequency domain. However, if the signal-to-noise ratio of a camera is given for a system with cutoff frequencies $f_c{}^x, f_c{}^y$, diagonal filtering reduces this to a quarter of the original noise power and thereby increases the signal-to-noise ratio.[7]

Concepts for Avoiding Moving Blur

Although line interlace was proposed to overcome large-area flickering for frame rates of 25 frames/sec, there is an improved time resolution compared with movies and their 25 pictures/sec. This is true for horizontally moving parts, but as it has been pointed out, this is not true for vertically moving areas.

So it is a further improvement if time resolution is working in both directions equally with 25 frames/sec accompanied by an increased spatial resolution for lower velocities. But, if there is a moving part with a velocity higher than about 2 pel/frame, the line scanning may be switched over to line interlace to overcome beginning moving flutter.

To avoid moving blur at the corresponding receiver, the interpolating postfilter switches from intraframe interpolation at low velocities to intrafield interpolation at higher velocities. This concept is discussed in more detail in Refs. 9 and 10.

First Practical Results

The proposed video system with progressive scanning and proper pre- and postfiltering offers error-free scanning in the sense of an ideal two-dimensional sampling system.[3,11] First practical results are:

1. Cancelling line interlace by "double speed" picture reconstruction out of a picture frame store gives a fairly high subjective improvement in picture quality (no more line crawl or 25-Hz flickering).

2. Alias-free scanning and flat-field reproduction is found to be very effective, especially for picture details with horizontal contours of high contrast.

3. Offset sampling combined with diagonal planar pre- and postfiltering gives an impressive improvement of resolution for fine vertical structures.

Some of these first results will be demonstrated by examples of unprocessed and processed pictures as shown in Figs. 8 through 12.[11]

In Fig. 8, an example is chosen, wherein a horizontally oriented line raster with 80 lines is scanned by just 100 TV lines. As the Kell factor is about 0.64 only 64 lines could be properly reconstructed with 100 TV lines. So without any pre- and postfiltering in Fig. 8a, the horizontal line raster cannot be recognized because of irritating artifacts. But, with proper pre- and postfiltering (Fig. 8b), there is a fine reproduction of the original raster.

In Fig. 9, the different total modulation transfer functions of the pre- and postfilters are shown for:

- Conventional scanning without vertical filtering (312 lines) (Fig. 9a).
- Vertical and horizontal pre- and postfiltering (Fig. 9b).
- Diagonal pre- and postfiltering (Fig. 9c).

With the "zone plate" (Fig. 10), the different steps of processing are shown in the Fourier domain. There is no alias left in the case of two-dimensional pre- and postfiltering, but an increased horizontal resolution in the case of diagonal filtering. Further practical results are shown with the test pattern "T05/D" and the picture "Boats" in Figs. 11 and 12.

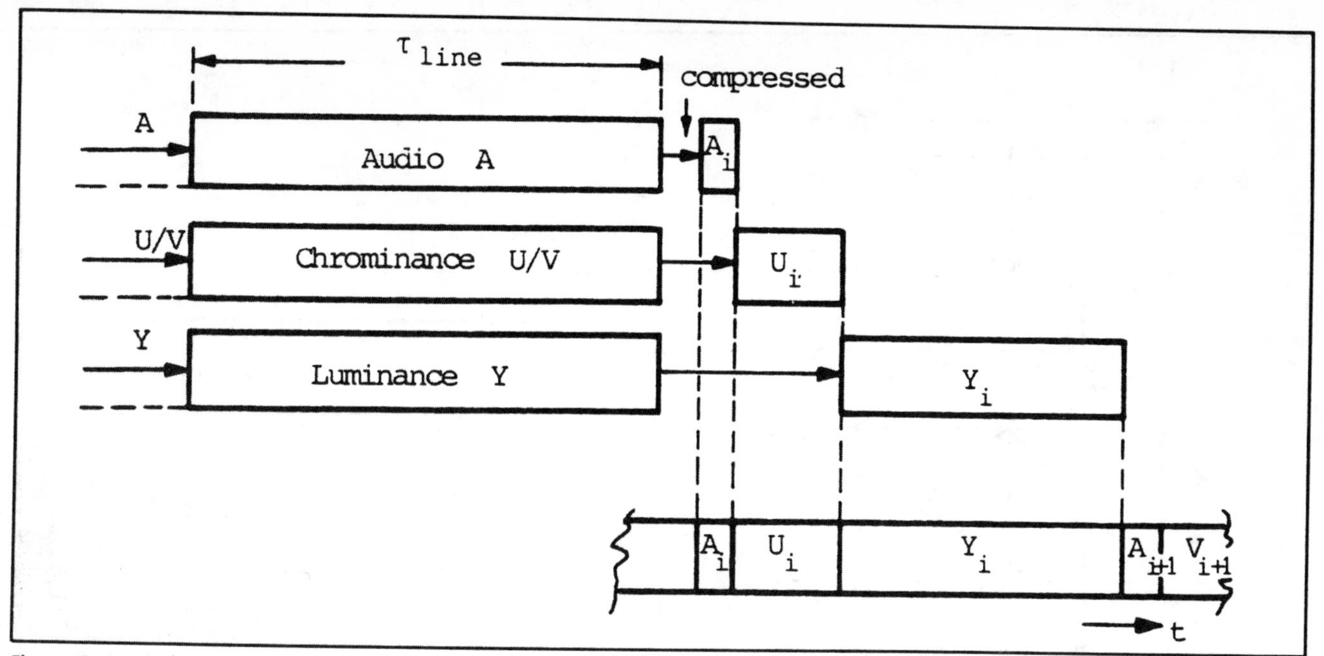

Figure 13. MAC/timeplex coding scheme, aspect ratio 5:3.

Enhanced Picture Quality by Timeplex Techniques

The start of direct television broadcast by satellite (DBS) in the mid-1980's could bring a new era in television broadcasting. This new service comes at a time of rapidly growing developments in the fields of digital signal processing and very large scale integration (VLSI) techniques for implementing them. There is also a growing interest in improved picture quality. Consequently, the IBA (U.K.) proposed the multiplexed analog component (MAC) television format for DBS, which keeps the luminance and chrominance components separated by a method of time division multiplexing.[5] This concept, which is also approved for FM channels of VTRs and which is called timeplex,[4] offers immediate improvements in picture quality, including:

- No more cross color because of time-sequential transmission of luminance and chrominance components.
- Very good signal/noise performance of both luminance and chrominance for FM channels with triangular rising noise versus frequency.
- Good bandwidth utilization.

In addition, with timeplex techniques, some more interesting improvements are possible:

- Compatible resolution improvement in x- and y-direction, depending on the bandwidth, given by the transmission channel.
- Selection of aspect ratios for different receivers by receiver type.

Additional options of the extended timeplex technique are discussed below.

Aspect Ratio Selection

The SMPTE Study Group concluded that compatibility is not feasible in view of aspect ratio and bandwidth. This is true, at least as far as bandwidth is concerned. It has been shown how the resolution bounds depend on the given transmission bandwidth. Nevertheless, by proper pre- and postfiltering, there are some limited quality improvements compared with the picture quality of a conventional TV system.

On the other hand, with new transmission channels such as satellite channels, new coding schemes and wider bandwidth could be applied, resulting in higher picture quality without any cross color and a wider aspect ratio.

Reference 5 describes the timeplex system called MAC-C, proposed by the IBA (U.K.), which works with time-compressed components (Fig.

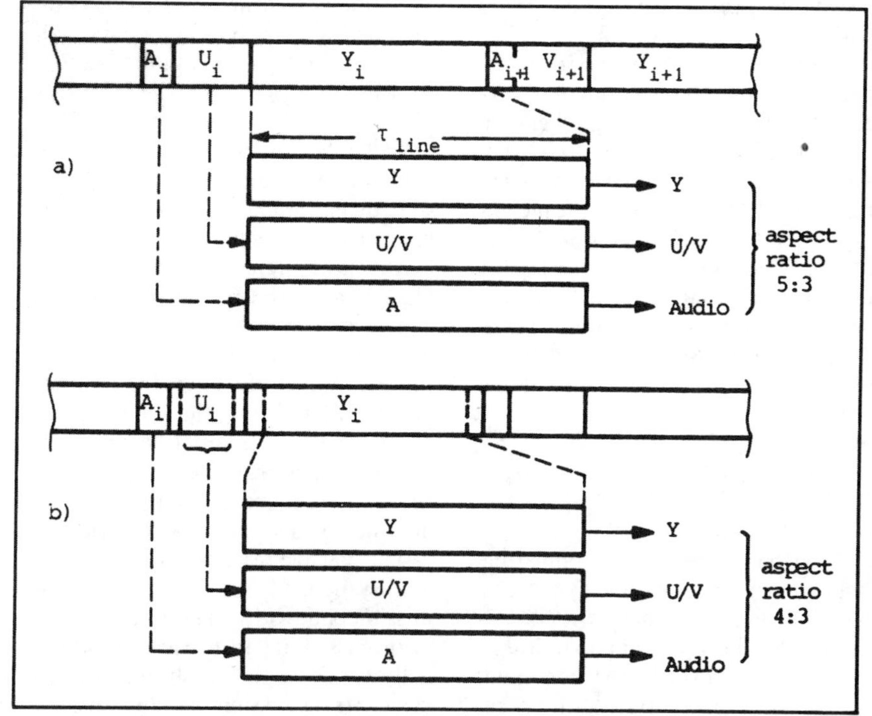

Figure 14. MAC/timeplex decoding scheme with variable aspect ratio.

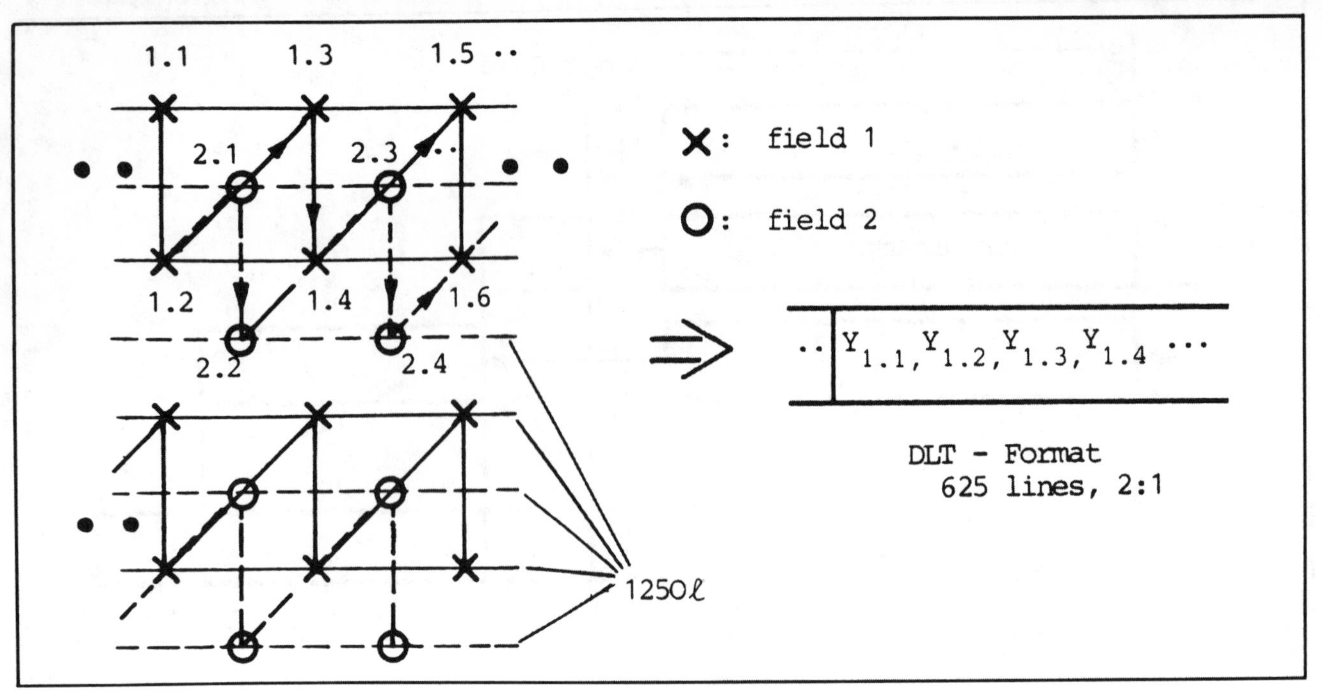

Figure 15. Dual-line timeplex (DLT) coding scheme.

13). The audio components A, the color components U or V, and the luminance component Y are individually time compressed and then transmitted line by line (U and V line sequentially).

At the receiver end, the incoming timeplex signal (Fig. 14a) is stored, decompressed, and transformed from time division multiplexed to space division multiplexed signals Y, U/V, and audio A. Assuming this coding and decoding is done for an aspect ratio of 5:3, in Fig. 14a we get the fully decoded signal components for a receiver with an aspect ratio of 5:3.

In the case of a receiver with an aspect ratio of 4:3 (conventional receiver), the timeplex decoding scheme in Fig. 14b gives a correctly decoded Y and U/V signal for just this aspect ratio, if a definable part of the incoming Y and U/V signals is properly decompressed. Obviously, with this timeplex decoding scheme, a new aspect ratio of 5:3 (2:1) could be introduced at the transmitter, without loss of compatibility with conventional receivers.

Compatible Improvement of Resolution

As it has been shown, compatible improvements are possible by pre- and postfiltering, but they are strictly limited by the given bandwidth. For increased bandwidth, the horizontal transmission resolution can be increased directly, but vertical resolution will be determined by the line number of the basic system.

However, with a dual-line timeplex format (DLT) the vertical resolution can be increased. This concept is principally shown in Fig. 15. For a given field 1 (field 2: dashed lines), there may be samples of brightness at the shown points 1.1, 1.2, 1.3, . . . of two adjacent lines of a high-line-number system. But with time multiplexing, these samples have twice the sampling frequency in a timeplex format (Fig. 15), and a compatible dual-line timeplex format results. During the line time of a 625-line system, all samples of a high-line-number system are available.

Conclusion

Compatible improvements are possible by pre- and postfiltering in combination with picture frame stores. With these signal processing techniques and with timeplex techniques, a new dual-line timeplex format is possible. This new format contains all the advantages of a high-line-number system. In addition, the DLT format can be decoded compatibly with selectable aspect ratio for standard receivers. Standard resolution can be obtained by a simple low-pass filter.

The possible resolution for improved quality depends on the given transmission bandwidth. These results will be discussed in a future article, which will also show the resolution bounds versus transmission bandwidths for different aspect ratios.

Acknowledgments

These studies originally were stimulated by the German Ministry for Research and Technology (BMFT). The author would like to thank the BMFT for supporting this project. Furthermore, the author is thankful to Dr. Schröder, Dr. Coy, and others for interesting discussions and results from their work.

References

1. T. Fujio, "A Study on HDTV-Systems," *IEEE Trans. on Broadcast.*, Vol. BC-24, No. 4, Dec. 1978.
2. D. G. Fink, "The Future of HDTV," *SMPTE Journal*, Vol. 89, No. 2, Feb. 1980 and Vol. 89, No. 3, March 1980.
3. B. Wendland, "HDTV Studies on Compatible Basis with Present Standards," in *Television Technology in the 80's*, Society of Motion Picture and Television Engineers, New York, 1981, pp 124–131.
4. G. Brand, G. Müller, H. Schönfelder, K.-P. Wendler, "Timeplex-ein serielles Farbcodierverfahren für Heim Videorecorder," *Fernseh und Kinotechnik*, 34. Jahrgang, Heft 12, Dez. 1980, S. 451–458.
5. Windram, et al, "MAC—A Television System for High Quality Satellite Broadcasting," IBA, Report 118/82, Crawley Court, Winchester, Hants, U.K.
6. G. J. Tonge, "The Sampling of Television Images," IBA, Report 112/81, Winchester, Hants, S021 2QA, U.K.
7. B. Wendland, "Zur Theorie der Bildabtastung," *ntz-Archiv*, Bd. 4, Heft 10, 1982.
8. B. Wendland, "Zur Theorie der Bildabtastung," Bericht des Lehrstuhls für Nachrichtentechnik, Universität Dortmund, 1982.
9. B. Wendland, "Entwicklungsalternativen für HDTV-Systeme," *ntz-Archiv*, Bd. 4, Heft 10, 1982, S. 285–293.
10. D. Uhlenkamp, E. Güttner, "Verbesserte Wiedergabe von Norm-Fernsehsignalen," *ntz-Archiv*, Band 4, Heft 10, 1982, S. 313–322.
11. H. Schröder, H. Elsler, "Planare Vor- und Nachfilterung für Fernsehsignale," *ntz-Archiv*, Band 4, Heft 10, 1982, S. 303–312.

A Compatible High-Definition Television System

By T. S. RZESZEWSKI*

(Manuscript received March 11, 1983)

A compatible HDTV system with the potential of having significantly better picture quality than the present NTSC color TV system is proposed. It realizes an increase in horizontal and vertical resolution and has considerably less crosstalk between the composite signal components compared to the NTSC signal. The increased resolution will allow a display as large as present home projection televisions with a sharper-looking and more detailed image than is possible with the present NTSC system. Also, the elimination of crosstalk adds to picture quality. A large screen size together with improved image quality should provide the user with a feeling of realism and involvement. This system also allows the use of more detailed graphics and more text per page for new services such as teletext.

I. INTRODUCTION

A compatible high-definition television (HDTV) system capable of producing an image quality significantly better than NTSC is proposed. Viewing the pictures produced by this system on a large screen should result in an increased sense of realism over the present NTSC television. This new system uses a split-luminance and split-chrominance (SLSC) type of transmission. The areas where the primary benefits are expected from this HDTV system over the National Television System Committee (NTSC) system are:

1. Increased horizontal resolution,
2. Increased vertical resolution, and
3. Less crosstalk between the components of the signal.

Several different approaches to HDTV can offer excellent quality. However, there are quality-independent attributes that will aid the acceptance of this type of service. In particular, compatibility with present NTSC receivers and a bandwidth of no more than twice the present 6-MHz channel for broadcast are critical.

Many different forms of compatibility are discussed today. However, as used here, compatibility means receiver compatibility. The signal must be able to feed an HDTV and an NTSC TV simultaneously and be received on the NTSC receiver with substantially the same quality picture that those sets presently realize, while the HDTV receiver realizes all the benefits, such as increased resolution.

Spectrum usage is another problem with some HDTV approaches. The NHK (Japanese) system uses 30 MHz of baseband.[1] This bandwidth is so large that this type of service could not be broadcast in the same manner as the present broadcast service, and would leave this service with fewer delivery systems. Of course, video tape or cable are still possible delivery systems. Direct broadcast by satellite (DBS) is also possible; however, the prime allocations that are not affected by the weather are likely to be used for DBS of NTSC in the near future.

Some other proposed systems have preserved some aspect of compatibility.[2,3] They retain the scanning format, but change the encoding such that a present receiver will not operate correctly.

A format is proposed here that uses a 10-MHz baseband composite signal that can be transmitted as a vestigial sideband, amplitude-modulated signal in a bandwidth of 12 MHz. Also, an NTSC receiver will operate with the same quality as at present when receiving this signal. The NTSC receiver must be tuned to the lower 6-MHz portion of the 12-MHz spectrum. The price that is paid for this compatibility is a reduction in the number of broadcast or CATV channels to approximately one half the present NTSC allocations. However, VHF stations are normally spaced at least every second channel allocation apart in a given location for broadcast, and UHF channels are spaced even further apart. Therefore, the impact on the present broadcast service is likely to be small. Also, new systems with bandwidth requirements like the NHK system would produce at least a six-to-one reduction in the number of channels compared to an NTSC service.

II. A COMPATIBLE APPROACH

2.1 System description

This new system is built around the NTSC baseband spectrum shown in Fig. 1. The NTSC baseband signal is modulated for broadcast as a vestigial-sideband, amplitude-modulated signal as shown in Fig. 2. The composite HDTV baseband is illustrated in Fig. 3. Note that there is approximately 0.75 MHz of spectrum left over for other

* Bell Laboratories.

©Copyright 1983, American Telephone & Telegraph Company. Photo reproduction for noncommercial use is permitted without payment of royalty provided that each reproduction is done without alteration and that the Journal reference and copyright notice are included on the first page. The title and abstract, but no other portions, of this paper may be copied or distributed royalty free by computer-based and other information-service systems without further permission. Permission to reproduce or republish any other portion of this paper must be obtained from the Editor.

Reprinted with permission from *Bell Syst. Tech. J.*, vol. 62, pp. 2091–2111, Sept. 1983. Copyright © 1983, AT&T.

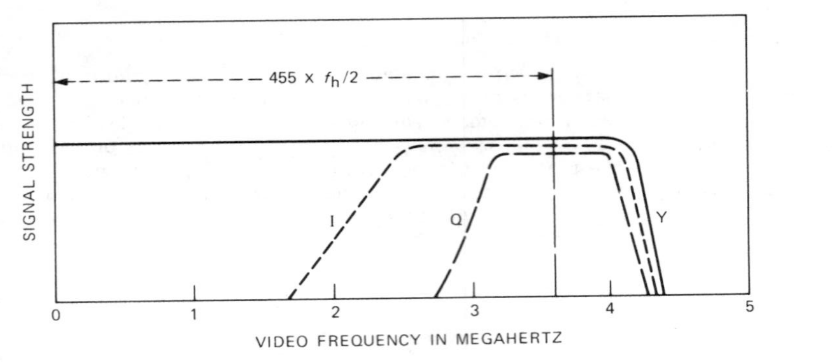

Fig. 1—NTSC luminance and chrominance bandwidth allocation.

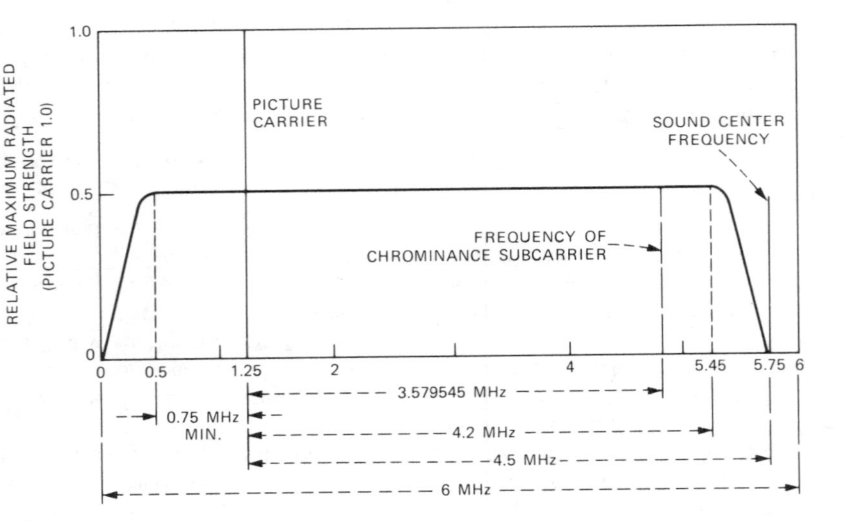

Fig. 2—Idealized picture transmission amplitude characteristic for NTSC.

services such as teletext and/or multichannel sound. The composite HDTV baseband signal could be modulated as a vestigial-sideband, amplitude-modulated signal for broadcast, as illustrated in Fig. 4. The selectivity of the NTSC receiver will reject the additional high-frequency portion of this signal at least as well if not better than an adjacent NTSC station.

The composite signal of Fig. 3 is obtained by starting with a 1050-line progressive scan source of wide-bandwidth red, green, blue (R, G, B) signals. The technique for improved vertical resolution is described

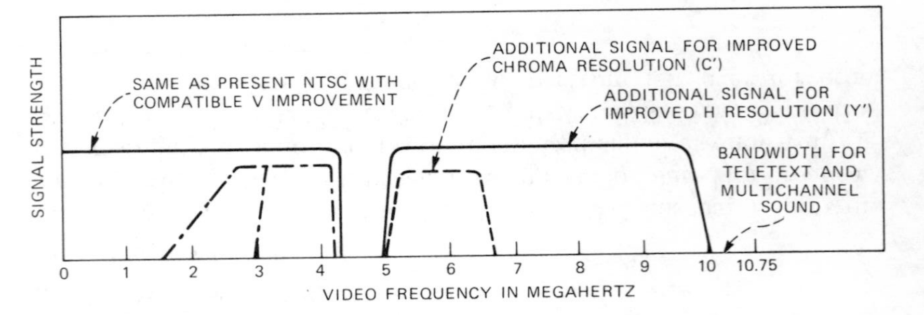

Fig. 3—SLSC HDTV baseband spectrum.

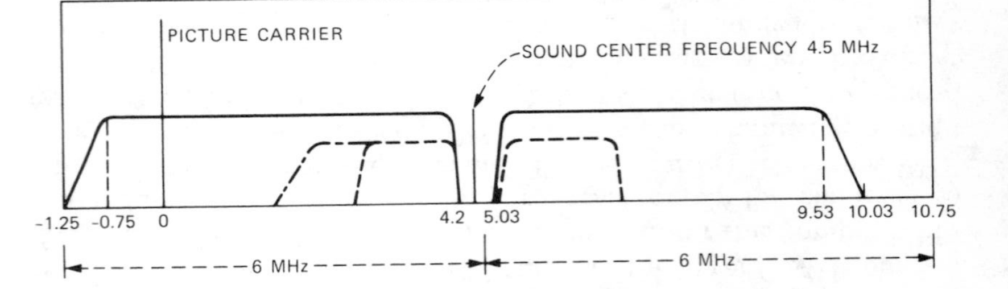

Fig. 4—Idealized broadcast picture transmission amplitude characteristic for SLSC HDTV.

in the literature.[4] That technique will be described in greater depth later. For now, it is sufficient to say that the wideband R, G, B signals are filtered and converted to a 525-line signal by a scan conversion that deletes every second line to obtain a 525-line signal suitable for transmission. Next, the wideband R, G, B signals are matrixed into a Y, I, Q format. This process is illustrated in Fig. 5. The NTSC encoder provides the appropriate selectivity and processing to create a standard NTSC signal that occupies the lower 4.2 MHz of the baseband spectrum in Fig. 3. The additional signal for improved horizontal resolution (Y′) that occupies the frequency region of approximately 5 to 10 MHz in the baseband is processed by the High Frequency Luminance Encoder. The High Frequency Chrominance Encoder creates the signal C′.* These signals are added to the signal obtained from the NTSC encoder to produce the composite HDTV baseband signal of Fig. 3. A detailed block diagram that elaborates on the encoding of these extra signals will be given later.

* A method of recreating the high-frequency components of the composite NTSC signal that were filtered out at the transmitter has recently been developed.[5] If this technique is proven to be acceptable for this application, the additional signal for extra chrominance resolution (C′) is not needed.

The system description in this paper utilizes the anti-aliasing filtering of the R, G, B signals in the encoder, as shown in Fig. 5. Consequently, the interpolation filtering in the decoder will also be performed on R, G, B as described later. An alternative approach that may have some advantages is to transform the R, G, B signals into a Y, I, Q format before performing the anti-aliasing filtering. In that case the I and Q signals will be filtered to a smaller bandwidth before the anti-aliasing filter. This approach simplifies two of the anti-aliasing and interpolation-filter designs and requires less frame-store memory. It will be considered further in the section on system alternatives.

Figure 6 contains the general block diagram of the decoder. Not shown are the tuner, IF, and video detector that are required to select and demodulate the HDTV signal to a baseband signal if the composite signal is modulated for broadcast; these functions will be described later. In the decoder, the three portions of the HDTV signal (the low-frequency signals obtained from the NTSC composite signal, the additional high-frequency chrominance signal, and the additional high-frequency luminance signal) are selected and processed in parallel. A detailed block diagram of these portions will be given later. The additional chrominance signals (Ih and Qh) are added to the chrominance produced by the low-frequency decoder to get I and Q signals that have a 2-MHz bandwidth each. These signals are then matrixed to produce 2-MHz color-difference signals. They are then added to the full 7.5-MHz-bandwidth luminance signal to produce R, G, B signals. The wideband luminance, Y, results from the addition of Yl and Yh.

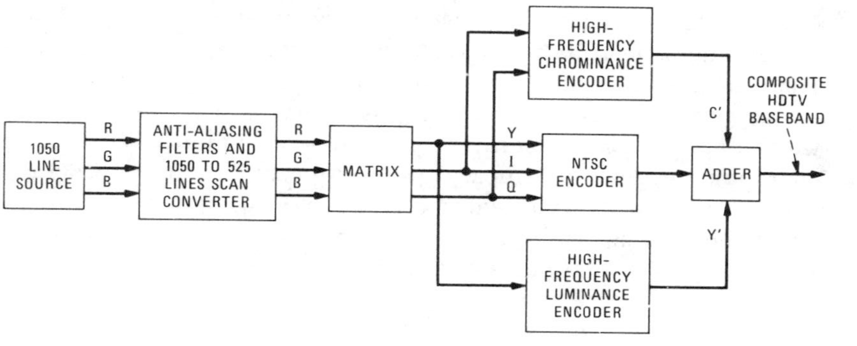

Fig. 5—SLSC HDTV encoder.

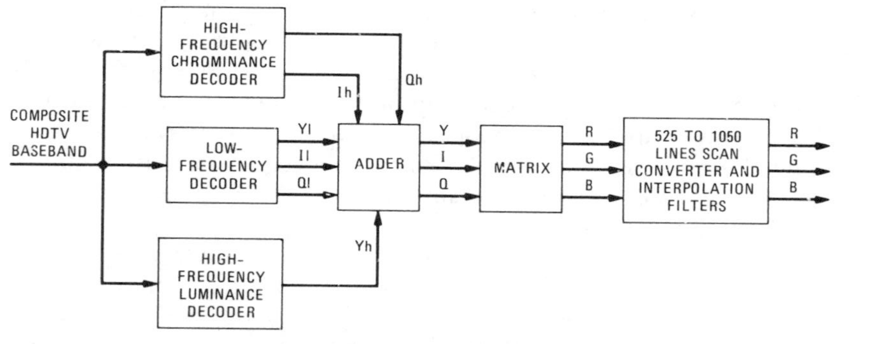

Fig. 6—SLSC HDTV decoder.

However, before being fed to the display, the signals must be processed to realize the increased vertical resolution. The process consists of a scan conversion back to 1050 lines from the interlaced 525-line transmission format. The monitor may be either a 1050-line progressive scan or interlaced scan at the discretion of the manufacturer. However, progressive scan is needed to realize the full potential of this system. Interpolation filtering is performed in the scan-conversion process, and the high-definition R, G, B signals are fed to the display device.

2.2 Vertical resolution

An analysis of the resolution capability of the NTSC signal is helpful to assess the improvements that HDTV will realize. Resolution is expressed in terms of vertical and horizontal equivalent TV lines. The vertical resolution tells the number of horizontal lines alternating between black and white that can be resolved in the TV image. It is tempting to equate this to the total number of scan lines minus the lines in the vertical interval that are not used for display. Unfortunately, the scanning process that changes the image into an electrical signal in the camera and then reassembles the image on the display is really a sampling process. It is well known that sampled signals must first be bandlimited or aliasing will occur. It is the aliasing and the replicated spectra that further reduce the vertical resolution. Equation 1 below expresses the actual vertical resolution in a TV picture by the addition of a Kell factor (k) that takes this extra loss into account.

$$Rv = (Nt - 2Nv)k. \tag{1}$$

Rv is the vertical resolution; Nt is the total number of scan lines (525 for NTSC); Nv is the number of lines in the vertical interval (21 for NTSC); and k is the Kell factor. The Kell factor normally ranges between 0.6 and 0.7 for TV systems and results in an Rv range between 290 and 338 lines for NTSC TV. The method of vertical resolution improvement employed for the SLSC HDTV system allows the vertical resolution to approach the full 483 lines of the active video; the Kell factor approaches unity.[4]

The modulation transfer function (MTF) in the camera and the display are analogous to the frequency response in linear system theory. It can be adjusted by shaping the electron beam in the camera and a CRT display. The contour of the scanning spot can be thought of as a two-dimensional impulse response commonly called the point-spread function. A narrow scanning spot in the vertical direction means a wide vertical spatial frequency spectrum and aliasing, and a wide scanning spot means overlapping of adjacent lines and low-pass filtering in the vertical direction (defocusing). In NTSC, the scanning spot is adjusted to compromise between aliasing and defocusing. Anti-aliasing (prefiltering) on a 1050-line progressive-scan source signal and interpolation (postfiltering) that eliminates replicated spectra aid the compromise. These filters can be used together with more lines at the camera in a compatible fashion to increase the vertical resolution.[4]

2.3 Horizontal resolution

The horizontal resolution is also expressed in terms of lines (equivalent vertical lines) that are the same width as the horizontal lines used to determine the vertical resolution above. There are two lines per cycle of video bandwidth. In other words, in the time—and therefore the horizontal space—it takes to display a cycle of the highest-frequency signal that will pass through the system bandwidth, the system will produce two lines on the display, one white and one black. The width of the lines is the same as for the vertical resolution, and the 4-to-3 aspect ratio is taken into account. Equation 2 below can be used to determine the horizontal resolution of a television system per unit of video bandwidth (Rh').

$$Rh' = 2Ta/AR. \tag{2}$$

Ta is the total active time for a horizontal line, and AR is the aspect ratio (these are 53.5 microseconds and 4/3 respectively for the NTSC system). This results in approximately 80 lines/MHz. Most NTSC receivers have at least 3 MHz of bandwidth, yielding a minimum of 240 lines of horizontal resolution. However, the total system luminance bandwidth can be 4.2 MHz if a comb filter is used in the NTSC receiver. This results in 336 lines.

The horizontal resolution is increased for the HDTV system proposed here by adding extra bandwidth to the baseband luminance signal. Figure 7a shows the 7.5-MHz bandwidth that is allocated to luminance in this compatible system. The same scanning format for transmission as NTSC is used; therefore, eq. (2) above indicates that the system will produce 80 lines/MHz. The 7.5-MHz bandwidth will

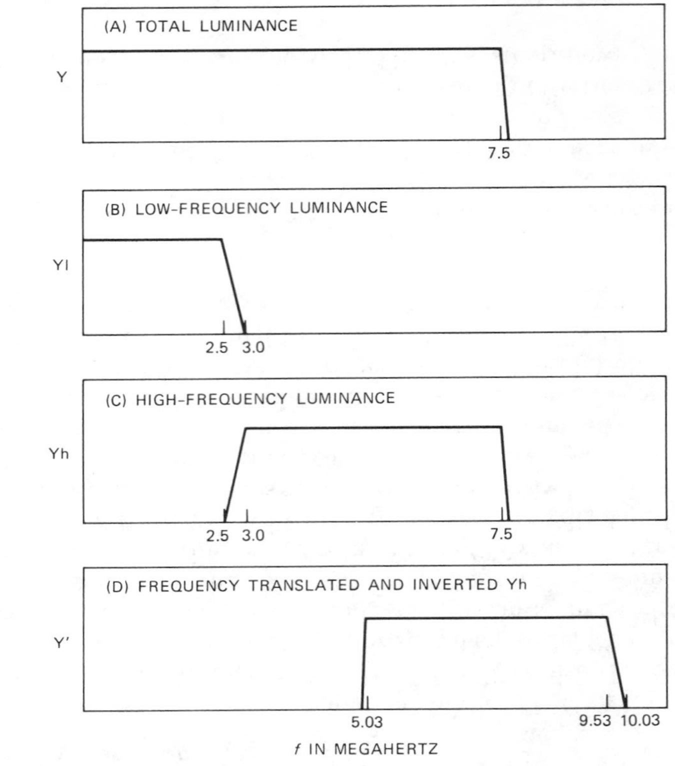

Fig. 7—Luminance (Y) processing in SLSC HDTV.

result in 600 lines of horizontal resolution.* This 7.5-MHz-bandwidth luminance spectrum feeds (1) an NTSC encoder that only uses the lower 4.2 MHz, and (2) a circuit that processes the high-frequency luminance detail (2.5 to 7.5 MHz) in a separate parallel channel.

Figures 7b through 7d illustrate the high-frequency luminance processing. Luminance (Y) is split up into a low-frequency portion (Yl) and high-frequency portion (Yh), each with a controlled roll off between 2.5 and 3 MHz (see Fig. 7c). If these two signals are added together, they will result in the original spectrum (Y) with a 7.5-MHz bandwidth. The high-frequency luminance will be processed so that it can be multiplexed with the NTSC baseband signal into a new, compatible HDTV baseband signal. This is accomplished by reversing

* This will result in 24 percent more horizontal resolution than vertical resolution for this system. There is the possibility of limiting the horizontal resolution to equal the vertical and simplifying the system. The amount of benefit produced by allowing greater horizontal resolution than vertical can be investigated when the system is implemented.

the frequency sense of Yh and translating it up in frequency to the location illustrated in Fig. 7d as Y′.

Note that the spectrum for Yh is cut off at a frequency below 2.5 MHz. Only the lower portion of the NTSC signal, which is substantially free from chrominance, is used for low-frequency luminance information (Yl).* Use of this lower cutoff frequency reduces the cross luminance problems that trouble the present NTSC system. An optimum cutoff frequency should be determined based on experiment since the higher the cutoff frequency the greater the horizontal resolution. The value indicated in Fig. 7 is a likely choice.

Figure 3 shows the complete baseband selectivity that defines the HDTV spectrum. Notice the additional signal for improved chrominance resolution (C′). Additional luminance resolution requires extra associated chrominance resolution and therefore additional chrominance bandwidth to appreciate the full improvement. The additional chrominance signal, C′, is time multiplexed between the Ih signal and the Qh signal components on alternating scan lines. However, unlike SECAM, the C′ signals are modulated as a single-sideband signal. The frequency of the carrier is selected to be a multiple of the horizontal frequency to minimize the crosstalk between this additional signal and the high-frequency luminance information, Y′. The purpose is to interleave the luminance (Y′) and chrominance (C′). The next section includes an explanation of the interleaving.

2.4 Detailed block diagrams

Figure 8 is a detailed block diagram of the encoder shown in Fig. 5. The method for increasing the vertical resolution mentioned previously is used,[4] so that a 1050-line progressive-scan source (camera) feeds an anti-aliasing filter. The R, G, B signals are all processed in the same fashion, each one is passed through a scan converter that converts from 1050 lines in a progressive-scan format to 525 lines in an interlaced format for compatible transmission. These signals are then matrixed into a Y, I, Q format.

The additional high-frequency signal for improved chrominance is time multiplexed to carry the Ih and Qh signals on alternate horizontal lines. However, it is not frequency modulated as in SECAM; rather, it is single-sideband amplitude modulated. This gives a spectrum that tends to cluster at multiples of the horizontal frequency and odd

* Yl will be completely free from any cross-luminance produced by the Q component of the color signal. If the I component is allowed the present specification of 1.5-MHz bandwidth, it can cause some cross-luminance. However, there is some question of whether that bandwidth should be transmitted since there are no consumer receivers that make use of this extra bandwidth; consequently, the extra bandwidth of the I signal can only cause cross-luminance problems in present NTSC receivers.

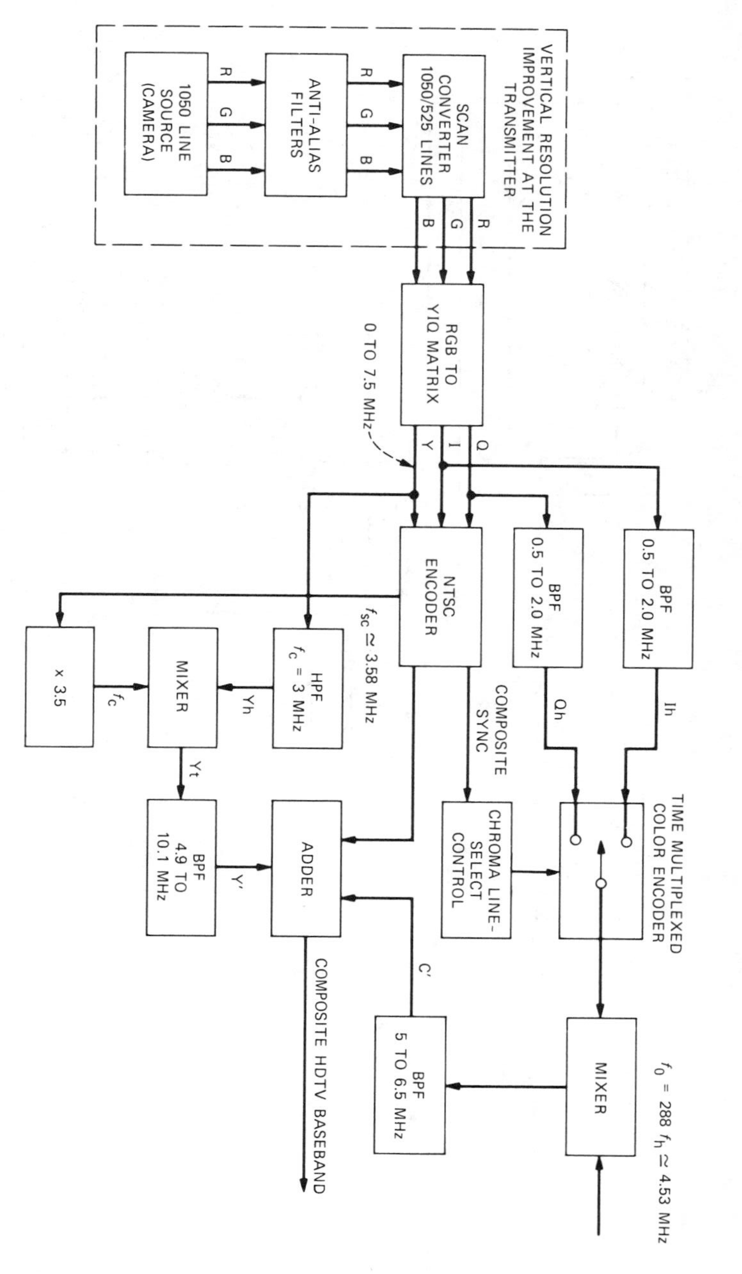

Fig. 8—Detailed block diagram of SLSC HDTV encoder.

multiples of half the horizontal frequency. The outputs of the two bandpass filters that select the Ih and Qh signals and the composite synchronization signal from the NTSC encoder feed the time-multiplexed color encoder (switch) that provides the additional signal for improved chrominance resolution, C′. The switch connects the Ih signal and then the Qh signal to the mixer on alternate scan lines. A carrier frequency, f_0, is the second input.*

$$f_0 = 288 f_h \sim 4.53 \text{ MHz}. \tag{3}$$

A tone burst of frequency f_0 could be inserted into the vertical interval for phase reference at the receiver. The bandpass filter at the output of the modulator selects only the sum signal, C′, that is in the frequency range of 5 to 6.5 MHz.

The NTSC encoder functions as it would normally. It provides the composite synchronization signal to the line-select control for the high-frequency chrominance processing described above, and the color subcarrier for the high-frequency luminance processing.

The translated and frequency-inverted high-frequency luminance signal is formed by first high-pass filtering the 7.5-MHz luminance to produce Yh. When Yh is applied to the mixer illustrated in Fig. 8, a double-sideband suppressed carrier signal is created. The carrier input to the modulator is f_c.

$$f_c = 3.5 f_{sc} = 3185(f_h/4) \sim 12.53 \text{ MHz}. \tag{4}$$

The NTSC color subcarrier, f_{sc}, is available from the NTSC encoder and can be used to derive f_c as indicated by eq. (4). However, a tone burst of f_c would be inserted into the vertical interval for phase reference at the receiver. The bandpass filter on the output only allows the frequency-inverted lower sideband, Y′, to pass. Y′ has a quarter-line frequency offset as can be seen from eq. (4). Consequently, f_0 is made equal to an exact multiple of f_h in eq. (3) so that the time-multiplexed signal C′ will interleave with Y′. This signal, Y′, is added to the output of the NTSC encoder and to the output of the high-frequency chrominance encoder, C′, to produce the composite HDTV signal of Fig. 3. Audio information is added just prior to transmission; this is the conventional manner of adding audio information to the NTSC broadcast signal. Additional subcarriers could be added to the baseband of the composite HDTV signal for multichannel sound or teletext as indicated in Fig. 3.

A detailed block diagram of the decoder is shown in Fig. 9. The decoder input comes from the video detector(s); more will be said

* This is just one of several strategies for selecting f_0. f_0 could be made 300 $F_h \sim 4.72$ MHz; then a low-level tone of this frequency can be inserted in the SLSC baseband.

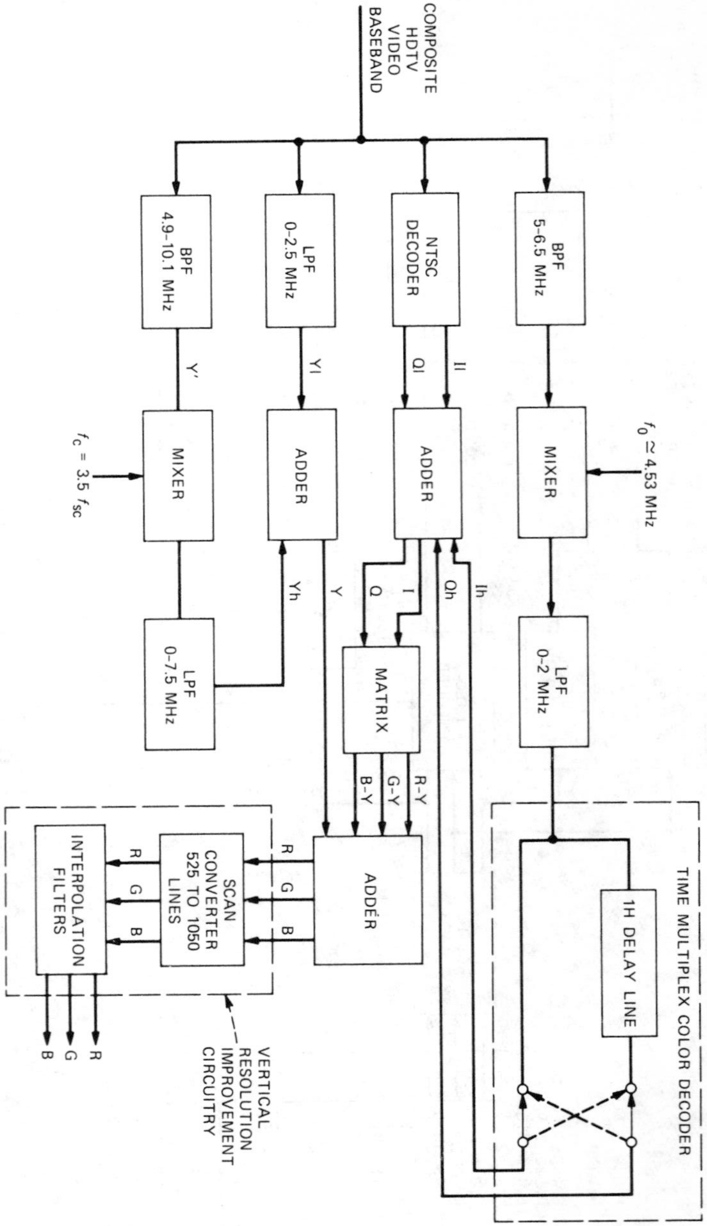

Fig. 9—Detailed block diagram of SLSC HDTV decoder.

about the receiver circuitry between the antenna and the decoder in Section 3.1. Since the composite HDTV signal was made up of three separate parts, each of these parts must be decoded. The high-frequency chrominance is decoded by first selecting it from the composite signal via the 5- to 6.5-MHz bandpass filter. This signal feeds a single-sideband demodulator that consists of a mixer and a bandpass filter that extracts the 0.5- to 2-MHz high-frequency color signal. However, this signal is still time multiplexed at the single-sideband demodulator output, so this output is fed into a time-multiplexed decoder to obtain simultaneous I and Q signals. The time-multiplexed decoder consists of a delay line that provides the storage of one horizontal line of color information, and the appropriate switches indicated in Fig. 9. The time-multiplexed decoder functions such that the present Ih (Qh) signal and the previous line Qh (Ih) signal are both present on the outputs providing the simultaneous Ih and Qh signals as outputs.

The NTSC decoder provides its normal I and Q color signals, called Il and Ql to emphasize that they convey the low-frequency portion (0 to 0.5 MHz) of the HDTV color signal. (Virtually all NTSC decoders produce I and Q signals of 0.5 MHz bandwidth in spite of the fact that the I channel should have extra bandwidth.) Il and Ql are added to Ih and Qh to form the complete 2-MHz I and Q signals. The high-frequency luminance, Y′, is passed by the 4.9- to 10.1-MHz bandpass filter and fed to the mixer. The carrier input is 3.5 f_{sc}, derived from the color subcarrier provided by the NTSC decoder with phase coherence obtained from a tone of f_c in the vertical interval. The demodulator output is filtered by a low-pass filter with a 7.5-MHz cutoff. The resultant signal, Yh′, which occupies a 2.5- to 7.5-MHz spectrum, is added to the 0- to 2.5-MHz low-frequency luminance signal obtained by passing the composite HDTV spectrum through a low-pass filter with a 2.5-MHz cutoff. The adder outputs the luminance signal, Y. Y, I, and Q are matrixed to provide color-difference signals and then added to Y to output R, G, B signals to the circuit that provides the vertical-resolution improvement. This circuit consists of a scan converter that inserts extra lines, converting the interlaced 525-line transmission standard to a progressive or interlaced 1050-line format along with interpolation filtering for each of the R, G, B signals. The resulting R, G, B signals feed circuitry that drives the picture display.

2.5 System alternatives

This system is flexible in that several variations are possible depending upon the needs and possibilities that surface during testing. The vertical resolution is limited to a maximum of 483 lines because the technique for vertical improvement can only promise a maximum resolution equal to the maximum number of active scan lines in the transmission standard. The actual resolution could be slightly less than this because the anti-aliasing filter and the interpolation filter may reduce the vertical frequency response somewhat. The maximum horizontal resolution that corresponds to 7.5 MHz of bandwidth is 600 lines. This resolution can be achieved by using a comb filter at the receiver (covering the 5- to 6.5-MHz portion of the baseband signal shown in Fig. 3), or by not transmitting C′ and using the inferred highs processing to restore wideband chrominance signals at the receiver.[5] Both approaches result in less vertical resolution (483 lines) than horizontal resolution (600 lines).

Alternatively, a horizontal resolution equal to the vertical resolution could be chosen, thus simplifying the system. If 480 lines of horizontal resolution were chosen, only 6 MHz of luminance bandwidth is required. The high-frequency luminance of Fig. 3 need not overlap the high-frequency chrominance. Therefore, a comb filter is not needed to realize the 480 lines of horizontal resolution while using a C′ signal. An additional simplification is possible; the carrier, f_c, used to translate the high-frequency luminance spectrum in Fig. 8 can be three times the color subcarrier, 3 f_{sc}, rather than three and one half times that frequency.

Less extensive changes are possible in the interest of optimizing the system. For example, the 2.5-MHz cutoff of the low-pass filter that forms Y1 (illustrated in Figs. 6 and 9) could be reduced to 2 or even 1.5 MHz in order to completely eliminate any possible crosstalk between low-frequency chrominance and luminance. Alternatively, the cutoff could be increased to 3 MHz if it is shown that the extra resolution is a more important benefit than the penalty of a small amount of extra crosstalk. The 2.5-MHz cutoff and many other parameters given are reasonable choices that may change somewhat once the system is tested. Another possibility is to bandlimit the I channel in the NTSC encoder to 0.5 MHz since virtually no NTSC decoders use more than 0.5 MHz at this time and the extra I channel bandwidth can only cause crosstalk in NTSC consumer receivers. With equal-bandwidth (0.5 MHz) I and Q channels in the NTSC encoder, the 2.5-MHz cutoff illustrated in Fig. 7 will result in no cross luminance from Ih and Qh.

A change in the encoder and decoder (illustrated in Figs. 5, 6, 8, and 9) is possible to simplify the system by taking the output of the 1050-line source in Figs. 5 and 8 and transforming to a Y, I, Q signal format immediately. Then the I and Q signals could be bandlimited to 4 MHz (twice the 2-MHz transmission bandwidth) rather than 15 MHz for R, G, B signals (twice the 7.5-MHz Y bandwidth) for a source with a 30-Hz frame rate. Anti-aliasing and interpolation filter implementation in these channels is simplified because of the lower-frequency operation. Framestore memory is also minimized because of the smaller bandwidth of I and Q.

III. DELIVERY SYSTEMS

3.1 AM broadcast

The front end of the HDTV set consists of a tuner, IF amplifier, and video detector. These functions can be handled in three basic ways, as illustrated in Fig. 10. Figures 10a and b represent the case where the two 6-MHz channels are adjacent to each other, as illustrated in Fig. 4.

Figure 10a illustrates a single wideband tuner and IF amplifier that is wide enough to pass the entire HDTV spectrum. The IF amplifier may be centered around the same IF frequency presently used, or a different center frequency may be chosen. Figure 10b uses a wideband tuner, but splits the spectrum up into two parts: an NTSC IF to receive that portion of the spectrum that is the same as the NTSC signal, paralleled by a second IF specially designed to receive the extra portion of the HDTV signal. After video detection in each channel, the two baseband spectra are added together to produce the baseband spectrum of Fig. 3.

A more versatile but complicated approach is illustrated in Fig. 10c; shown are two separate tuners for each portion of the spectrum, the NTSC portion and the extra portion. After processing by the respective IF amplifiers and detectors in each channel, the two portions of the baseband spectrum are added back together to obtain the spectrum of Fig. 3. If the arrangement of Fig. 10c is used, the extra 6-MHz spectrum for HDTV need not be located at the upper 6-MHz band shown in Fig. 4. It could be a totally separated channel of 6-MHz bandwidth. If the additional HDTV information is carried in the 6-MHz channel directly above the NTSC channel, 40.75 to 35.75 MHz is a likely IF frequency range for the additional IF. However, if the additional HDTV information is one or several channels away, another frequency may be chosen.

3.2 Cable

Modern CATV systems usually use a tree structure that has the capability of handling up to 50 or more NTSC TV channels per cable. Since present channel allocation is 6 MHz, any additional bandwidth requirement results in a minimum of two channels per HDTV signal. This two-to-one reduction in stations would be more acceptable to the CATV industry than HDTV systems requiring more than two 6-MHz channels per signal. It is almost certain that a higher-bandwidth system would cause the CATV industry large problems, since it would

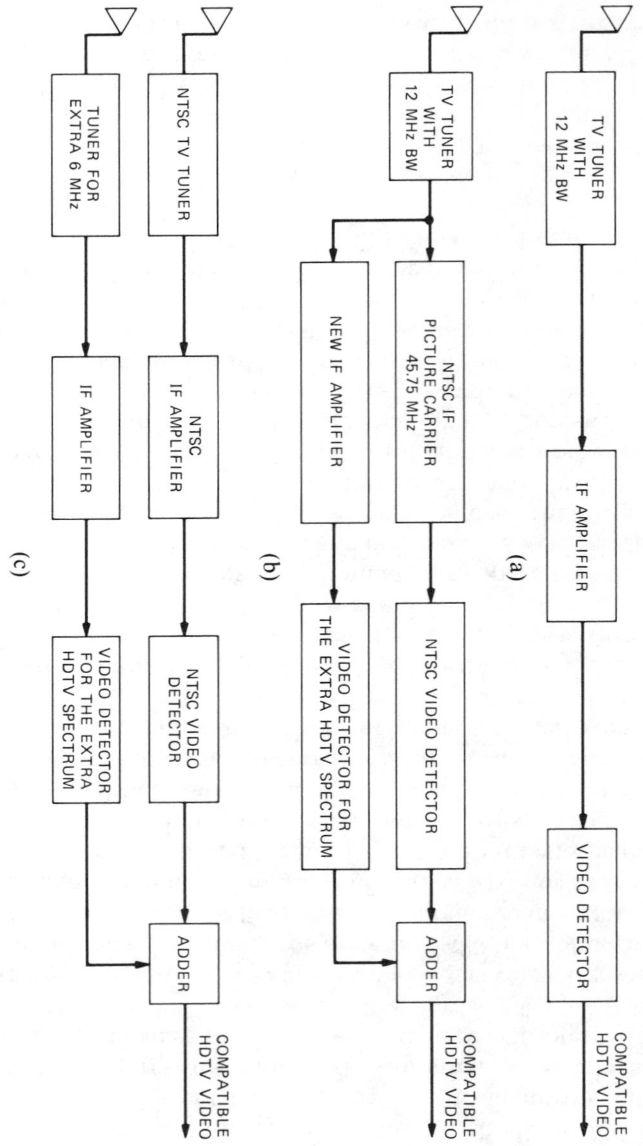

(a)

(b)

(c)

Fig. 10—Front-end alternatives for SLSC HDTV. (a) Single IF amplifier approach. (b) Single-tuner multiple IF amplifier approach. (c) Multiple-tuner approach.

severely limit the amount of programming that could be carried by a CATV system.

Switched video cable systems that have the ability to provide video on demand are just beginning to surface. It is very difficult to switch extremely wideband video; therefore, a reasonably compact HDTV spectrum is needed for this new type of cable system.

This compatible HDTV system has been carefully designed to occupy a 10-MHz baseband bandwidth. It will fully utilize two 6-MHz channels for transmission if desired. Therefore, it should have a minimum negative impact on present and future cable systems.

3.3 Satellite

Direct satellite broadcast to the home is a delivery system that is just about to become important. The modulation format and the frequency allocation of the service will eliminate compatibility with the present NTSC receivers. NTSC compatibility may lose some importance for this delivery system. However, bandwidth is always at a premium for satellite systems; thus the compact baseband as shown in Fig. 3 is very important. Also, the ability to modulate the HDTV signal into two NTSC channels could be advantageous.

3.4 Prerecorded

Even when the prerecorded media such as video tape and video disc are used as a delivery system, bandwidth is still important. It is much easier to record a 10-MHz baseband signal than a 30-MHz signal. Consequently, a tape recorder for the 10-MHz signal should be more economical. There may be a considerable problem with making a video disc unit that handles a 30-MHz baseband signal and still has sufficient playing time per disc.

IV. COMPARISONS

4.1 Vertical resolution comparisons

The vertical resolution, Rv, of a TV system is given by eq. (1). For the current NTSC system, the following is obtained:

$$Rv = (525 - 2x21)0.65 = 313 \text{ lines}. \quad (5)$$

This assumes a nominal Kell factor of 0.65. If the same Kell factor and 21-line vertical interval is applied to the NHK system, the result is 704 lines. For the Dortmund system using USA scanning standards and the proposed split-luminance, split-chrominance (SLSC) compatible system, the result should be the same as the NTSC with a Kell factor approaching unity.[4] Therefore, the vertical resolution should approach 483 lines. These comparisons together with modifications of the BBC and IBA systems for the USA scanning standards are shown in Table I.[2,3] (Note that there are actually two Dortmund systems described in the literature.[4] The system considered here is the diagonal sampling approach that provides both increased vertical and horizontal resolution over the present standard. The other system does not provide increased horizontal resolution.)

4.2 Horizontal resolution comparisons

Table I also contains a comparison of some HDTV systems with respect to horizontal resolution, crosstalk, and bandwidth requirements adapted to the NTSC environment where appropriate. The NTSC system is used as a reference. The horizontal resolution of the NTSC system ranges from 240 lines to 336 lines. The NHK system should be capable of approximately 30 lines/MHz based on eq. (2). This predicts a horizontal resolution, Rh, of approximately:

$$Rh = (30 \text{ lines/MHz})(20 \text{ MHz}) = 600 \text{ lines}. \quad (6)$$

The Dortmund system approach applied to the NTSC system should have the ability to reproduce the same horizontal resolution as vertical. Further, the vertical resolution should approximate the number of active scan lines. Therefore, it should result in 480 lines.[4]

4.3 Crosstalk

Crosstalk is a very important factor in the image quality of a video system. There are three types of component crosstalk in present television standards. They are cross luminance (crosstalk of chrominance into luminance), cross color (luminance crosstalking into chrominance), and chrominance to chrominance crosstalk. Cross luminance, also called dot crawl, normally shows up in the NTSC picture at the edges of color areas. The high-frequency subcarrier signal appears in the luminance channel as a dot pattern crawling up the edges of color areas. Cross color is most obvious when the scene contains a detailed pattern such as a striped shirt or tweed suit. It is

Table I—HDTV comparison chart

System	V Resolution (k = 0.65)	H Resolution	Min. No. NTSC Channels When Broadcast	Compatible	Crosstalk
NTSC	313	240–336	1	Yes	Bad–Small
NHK	704	600	—	No	Small
Dortmund*	483	480	1	—	—
BBC*	313†	336	2	Can Be	None
IBA*	313†	336	2	No	None
SLSC	483	600	2	Yes	Small–None

* The values are adapted to the USA scanning standard.
† This value can be made 483 by applying the vertical improvement technique.

seen as a color pattern over the area of detail that obviously does not belong, and destroys the ability to see the detail in that area. Chrominance-to-chrominance crosstalk is less obvious than the other two because it is correlated with the scene. It results in color distortions.

The first two types of crosstalk mentioned above can produce significant picture degradation in the NTSC system unless the luminance bandwidth is sacrificed or a comb filter is used. However, a comb filter tends to produce some luminance degradation of its own. It reduces vertical resolution by averaging two or more lines at a time and rejects diagonal lines in the picture. This situation can be improved for luminance by only comb filtering the high-frequency luminance where the actual crosstalk frequency components occur. Further, a comb filter will produce a dot crawl on horizontal edges of saturated colors in a picture that is not present without it.

As shown in Fig. 7, the low-frequency luminance (Yl) used by the SLSC HDTV system is the lower 2.5 MHz of the NTSC luminance. This portion of the NTSC luminance is not interleaved with any of the Q component of the NTSC color signal. There is some interleaving of the I component of the NTSC signal and the low-frequency luminance, Yl, of the HDTV signal. However, there is some question of whether the full frequency range of the I signal should be transmitted for NTSC receivers since there are no consumer receivers that make use of it, and it can only cause crosstalk in the present NTSC receivers.

Cross luminance in the high-frequency luminance (Yh) can also be minimized. The upper 1.5 MHz of the luminance—the 6-MHz to 7.5-MHz region of the original luminance signal—interleaves with the high-frequency chrominance. This upper end of the luminance ends up occupying the 5- to 6.5-MHz region of the frequency inverted and translated spectrum Y′ in Fig. 3. This region can be comb filtered—rolled off at 6 MHz (480 lines of resolution)—or simply allowed to pass with the small amount of cross luminance mentioned above.

The last alternative is possible with only a small amount of degradation because of the high-frequency nature of the cross luminance and the fact that the chrominance that is talking into the luminance will be down in level compared to the signal that is producing cross luminance in the NTSC system (it is only the 0.5- to 2-MHz region of the chrominance that will contribute to crosstalk here). Since this relatively low-level chrominance will also be producing a much smaller dot pattern in the luminance than the NTSC color subcarrier would cause in the NTSC system, the crosstalk should be much less obvious.

Cross color can be reduced in several ways. A comb filter could be used to remove that portion of the luminance that interferes with the low-frequency chrominance. Alternatively, use can be made of the fact that a portion of the luminance that represents 2.5 to 4.2 MHz of the NTSC luminance is repeated in the high-frequency luminance portions of the composite HDTV signal. This portion of the high-frequency luminance is free of any interleaved chrominance. However, the corresponding portion of the NTSC part of the signal contains chrominance interleaved with the luminance. These two corresponding portions of luminance can be subtracted, leaving an uncontaminated chrominance signal. Also, the luminance above 3 MHz could be rolled off in the NTSC part of the baseband as another way to eliminate cross color without any degradation to the vast majority of NTSC receivers that do not use a comb filter. Cross color into C′ can be reduced with a comb filter.

4.4 *Encoding errors*

There are a number of color deficiencies in the NTSC color standard that degrade the quality of the reproduced image.[6-8] The crosstalk problems mentioned in the previous section are a part of these deficiencies. Encoding errors are another facet of these problems. The study of encoding errors or distortions in the NTSC system is an involved topic that will only be briefly mentioned here. The problems center around the fact that part of the luminance is carried by the chrominance when color is broadcasted, and it is aggravated by the nonlinear gamma characteristics of the system.[6] The transmitter corrects for a nominal gamma of 2.2 by raising the signal to the 1/2.2 power so that the overall response is linear. The result is that saturated (vivid) colors lose details. Also, there may be substantial errors in the transients between certain colors (usually complementary colors produce the worst transients). With wideband color signals, there will be fewer errors. The luminance is still carried partly by the chrominance, but the chrominance is now wideband and it degrades the end result less.

V. SUMMARY

A split-luminance, split-chrominance (SLSC) HDTV system has been described that is NTSC compatible and uses a 10-MHz baseband signal. This baseband signal can be modulated to produce an amplitude-modulated, vestigial-sideband signal in a 12-MHz bandwidth for broadcast. Alternatively, this compatible signal can also occupy two separate 6-MHz channels. Compatibility and bandwidth conservation are two of the most important attributes of this HDTV system. A compatible system is likely to penetrate the market place much more rapidly than a non-compatible system because of the huge investment in NTSC equipment. The two most common delivery systems are broadcast and CATV; both are very sensitive to the bandwidth requirements of a new system.

The present NTSC channels are 6 MHz. Conventional channel allocation forces additional bandwidth to come in increments of 6 MHz. A two-channel, 12-MHz requirement is the largest bandwidth that these systems can reasonably accommodate for this new HDTV service. Therefore, this new compatible system has been designed to occupy a 12-MHz bandwidth when using the present broadcast format of vestigial-sideband amplitude modulation.

The important parameters of this new compatible HDTV system are a horizontal resolution potential up to 600 lines, a vertical resolution potential up to 483 lines, and less crosstalk between the individual components of the color signal compared to the NTSC system. Further minimization of color-encoding errors may be possible by using recently developed processing techniques.[5]

VI. ACKNOWLEDGMENTS

The author gratefully acknowledges Joseph LoCicero for many useful discussions that aided his understanding of much of the HDTV area. The author expresses his gratitude to Bruce Briley, Kristin Kocan, and Harvey Lehman for their review of this paper.

REFERENCES

1. T. Fujio, "High-Definition Wide-Screen Television System for the Future-Present State of the Study of HD-TV Systems in Japan," IEEE Trans. Broadcasting, *BC-26* (December 1980), pp. 113–24.
2. C. Sandbank and M. Moffat, "High-Definition Television and Compatibility with Existing Standards," *Tomorrow's Television*, Scarsdale, New York: SMPTE, 1982, pp. 170–85.
3. T. Robson, "A Compatible High Fidelity TV Standard for Satellite Broadcasting," *Tomorrow's Television*, Scarsdale, New York: SMPTE, 1982, pp. 218–36.
4. B. Wendland, "High Definition Television Studies on Compatible Basis with Present Standards," *Television Technology in the 80's*, Scarsdale, New York: SMPTE, 1981, pp. 151–65.
5. D. Richman, "Color Television Receiving System Utilizing Inferred High Frequency Signal Components to Reduce Color Infidelities in Regions of Color Transitions," U.S. Patent 4,181,917, Jan. 1, 1980; "Color Television Receiving System Utilizing Multimode Inferred Highs Correction to Reduce Color Infidelities," U.S. Patent 4,183,051, Jan. 8, 1980; and "Color Television Receiving System Utilizing Inferred High Frequency Signal Components to Reduce Color Infidelities in Regions of High Color Saturation." U.S. Patent 4,245,239, Jan. 13, 1981.
6. K. McIlain and C. Dean, *Principles of Color Television*, New York: John Wiley, 1956, Chapter 11.
7. W. Gibson and A. Schroeder, "Color Television Luminance Detail Rendition," Proc. I.R.E., *43* (August 1955), pp. 918–23.
8. D. Livingston, "Reproduction of Luminance Detail by NTSC Color Television Systems," Proc. I.R.E., *42* (January 1954), pp. 228–34.

Compatible Systems for High-Quality Television

By R. N. Jackson and M. J. J. C. Annegarn

The feasibility of introducing a picture memory into domestic receivers at an acceptable cost opens new ways to approach the improvement of television picture quality. Until very recently, straightforward improvement of the picture memory opens up at least two alternative approaches (described in outline in a previous paper by Jackson and Tan[1]). First, if normal transmission were left unchanged, a combination of memory plus picture processing could achieve a greatly improved picture. With the alternative starting point of a large-screen picture of very high quality, processing could retain as much compatibility as possible. This article gives more detail about work already done along both these lines and also suggests some new possibilities.

The use of existing transmission standards (NTSC, PAL, SECAM) has the advantage of compatibility with existing home equipment, so that the introduction of such a system presents few problems. Within the existing standards, we believe it may be possible to increase the amount of information to be displayed by a factor of two.

Improved Picture Quality using Existing Standards

One can appreciate this improvement, however, only if all artifacts such as area flicker, line flicker, cross-color, and cross-luminance are also eliminated or greatly reduced.

The main processing equipment has to be situated at the receiver. But it should also be permissible to use processing at the transmitter, provided that the transmitted signal itself remains within the system specification. It is convenient to consider these two cases separately, beginning with the case of receiver processing, which could produce an immediate improvement over existing transmissions.

Receiver Processing

In our laboratories, we have used receiver processors which employ the same memory for the simultaneous reduction of several artifacts. Although the experiments have been carried out using PAL signals, the basic method is applicable to NTSC and other standards. Initial demonstrations given at the International Broadcast Convention in September, 1982, showed that, while some problems concerning movement adaptivity remained to be solved, the results were very promising and a significant picture improvement could already be achieved.

Although full details of our latest progress in this field will be given in special papers published elsewhere,[2,3] a brief summary of the technique is appropriate here.

The method is based upon the use of first-order temporal low-pass recursive filters as shown in Fig. 1. Such filters are normally used for the reduction of noise. But they also reduce, effectively, the temporal frequencies of cross-color and cross-luminance components, leaving the average value of both artifacts — which happens to be zero — at the output.

Thus the two processes of noise reduction and cross-color/cross-luminance reduction are carried out simultaneously.

This strategy is very effective for still parts of the picture but can fail when motion is present. For this reason, movement detectors are included in the equipment which vary the Kell (K) factor of the filter, depending on the picture content. The precise form of the movement detectors has proved to be a critical feature of the design and is still the subject of experimental work.[3]

Figure 2 shows the complete processor block diagram (omitting the movement detectors). The luminance- and color-difference signals are obtained from a PAL-Simple decoder, the luminance signal being split into a low-frequency and a high-frequency component. The low-frequency luminance filter reduces noise, while the high-frequency filter reduces both noise and cross-luminance, and the filters in the R-Y and B-Y channels reduce both noise and cross-color. In this way, separate control of the respective K factors is achieved, taking into account the fact that motion may exist in one domain while not being present in another.

For display, the signals are converted to a 100-Hz field rate. This eliminates large-area flicker effects. Two modes of operation are incorporated in the equipment. If A is the first field of an input picture and B is the second field, one mode displays the sequence AA,BB which removes large-area flicker and causes no

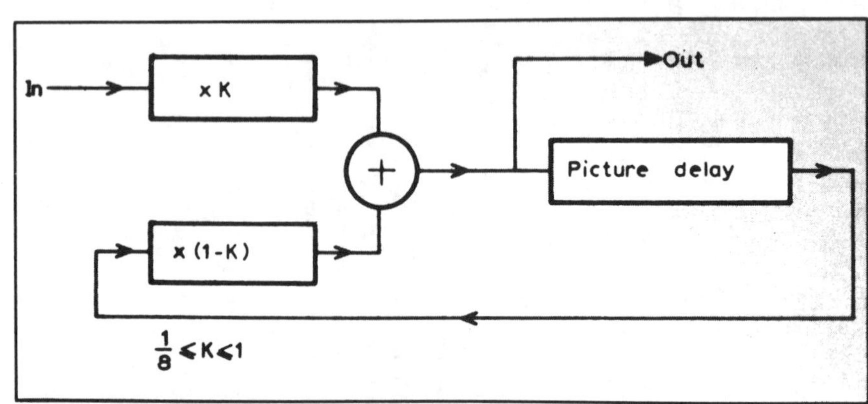

Figure 1. Principle of the recursive filters.

Presented at the Society's 17th Annual Television Conference in San Francisco (paper No. 17-6) on February 4, 1983, by Richard N. Jackson, Philips Research Laboratories, Surrey, U.K., and Dr. M. J. J. C. Annegarn, Philips Research Laboratories, Eindhoven, The Netherlands. This paper also appears in *Video Pictures of the Future*, published 1983, SMPTE.

Reprinted with permission from *SMPTE J.*, vol. 92, pp. 719–723, July 1983.
Copyright © 1983 by the Society of Motion Picture and Television Engineers.
This paper reprints information originally reported in a lecture to the SMPTE 17th Annual Television Conference in February 1983. Further work is reported in Reference 8.

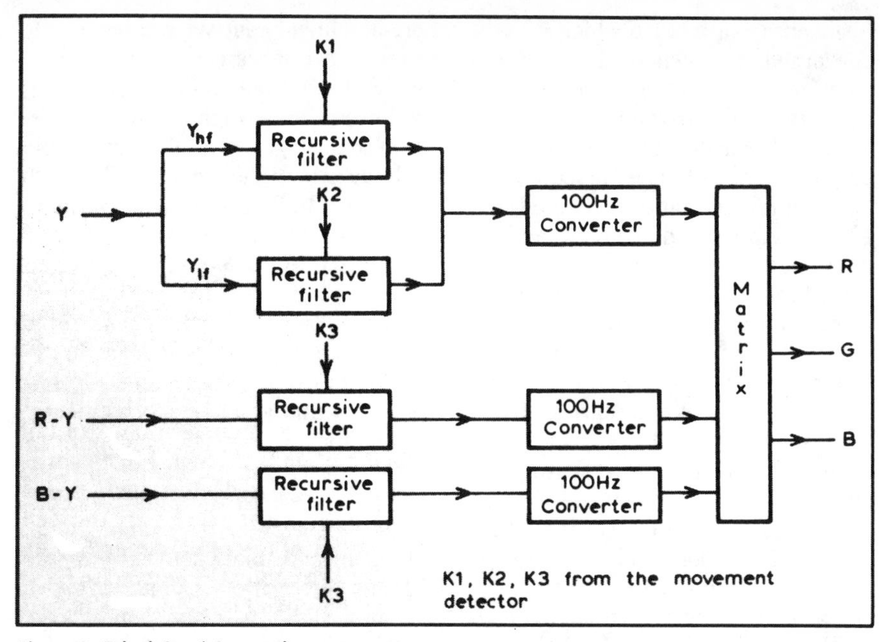

Figure 2. Principle of the receiver processor.

movement problems but does not reduce interline flicker. The second mode displays the sequence *AB,AB* which reduces both large-area flicker and interline flicker but can cause disturbing effects on moving scenes. Investigations have already shown that a movement-adaptive switching between these modes is desirable.

The total amount of storage required by such a processor is clearly important. In principle, large-area flicker reduction can be achieved using only a single field of memory, although the read and write cycle-speeds of this memory are high. A two-field memory has advantages with respect to speed and complexity of memory address, and it is shown in Ref. 2 that such a memory can readily provide a signal suitable for the recursive filter function mentioned above. Thus, it is possible to reduce large-area flicker, noise, cross-color, and cross-luminance in a single picture memory, provided this is based on component signals rather than encoded signals. For interline flicker reduction combined with large-area flicker reduction, a further field of memory is required.

The performance of a processor, such as that described, has to be seen to be believed. Many of the effects, such as flicker and cross-color, are dynamic and cannot be illustrated on paper. To give some idea of one aspect of the performance, Fig. 3a shows a portion of a circular zone plate pattern centered on the color subcarrier frequency (4.43 MHz) before processing, while Fig. 3b shows the same area after processing.

Apart from the obvious reduction of the cross-color rings, it should be noted that the receiver is displaying the full luminance resolution available. This is made possible because we have used a specially developed shadow mask tube having a 500-μ mask pitch which provides improved resolution compared to normal domestic receiver displays, together with freedom from moiré effects and good brightness. The combined effect of the elimination of flicker, plus the resolution and cross-effect improvements illustrated here, add up to a very acceptable picture.

Studio Processing

The addition of some processing at the TV studio side could be used to provide additional improvement without the need for a change in the transmitted standard. One possibility is the further improvement of cross-color performance.

Theoretically, it is impossible to remove all cross-color at the receiver. It is possible, however, to detect in the studio those situations where the receiver processor will fail. By using the same processing algorithm on the composite signal at the studio as that used at the receiver (e.g., by using a receiver similar to that described above) and by comparing the output with the *RGB* signal, then a signal could be derived to indicate when receiver errors were inevitable.

In such cases — and only in such cases — appropriate two-dimensional filtering of the studio signal would be applied. The combined effect of such studio and receiver processing would be to eliminate cross-color completely, while preserving the full available horizontal resolution on all except some (probably rare) moving scenes.

Once line-flicker has been improved by using processing at the receiver, it then becomes worthwhile to consider vertical-resolution enhancement techniques at the studio. This is an area where many workers believe potential improvements can be made, although work is as yet at an early stage.[4,5,6] We are also studying such techniques, the essence of which is to originate the picture signals at a higher-than-normal line number and to transcode these to 625 lines using vertical filtering. It has been stated that such techniques could lead to an effective improvement of K factor from the normal value of 0.65 to about 0.9.

HQTV: Compatible High-Quality Picture Transmission

In Ref. 1, the case was argued for a new TV concept having as a target a picture about 1 m^2 in area giving twice the normal viewing angle. Such a picture should contain about twice as much information in both horizontal and vertical directions as present standards. In addition to this, a widening of the aspect ratio would give a more pleasing picture and a better film compatibility: 5:3 has been suggested. Some people refer to this concept as High-Definition TV (HDTV). Because we think other visual improvements must also be included, we shall call it High-Quality TV (HQTV).

Display and Pickup Equipment

It is a prerequisite of any such concept that there should be a display system capable of producing pictures of the required size and quality and pickup devices capable of generating the signal sources. Accordingly, our laboratories have studied both of these problems in some detail and have built experimental equipment with which to test the feasibility of the proposal.

Initially, the equipment has been designed to operate on a standard of 1249 lines, 2:1 interlace, 50-Hz field rate. This was chosen primarily because it allows easy conversion to normal European 625-line standards and thus aids the study of standards interchange and vertical filtering already referred to. It is also a sufficiently high standard to challenge the relevant technologies and explore their limits. We do not regard this choice of line number as immutable (see below), but it has provided a useful starting point. Full descriptions of the equipment have already been given by Freeman et al.[7], so only a summary is given here.

Projection Display

The concept outlined above calls for a display of dimensions about 50 in. wide by 30 in. high (1.3 m × 0.8 m) with a resolution of 1000 picture elements and a brightness in excess of 50 fL (170 cd/m). While the use of a direct-view display technique cannot be ruled out for the future, we concluded that, at present, the specification can be met only by some form of TV projector. We have, therefore, used three 5-in. (130-mm) CRTs operating at 50 kV. By adopting gimbal-mounted electro-magnetic focus coils and beam astigmatism correction, we can generate rasters with a 75-μm spot-size at peak currents of about 300 μA.

Consideration of the characteristics of the screens available at the start of the design period and their effect on brightness and color shift led us to choose in-line forward projection with a 7-ft (2.2-m) throw onto a 5× gain screen. This led to the requirement for an $f/1$ optic system having a magnification of 13× and good response at a spatial frequency of 6.5 line pairs/mm at the cathode ray tube.

Although Schmidt systems with aspheric correctors offer very high apertures and excellent axial resolution, their corner resolution and size made them unsuitable for our application. We therefore commissioned 150-mm, $f1$, 9-element refractive lenses, which have been measured to yield axial and corner modulation transfer functions (MTFs) of 60% and 30% respectively.

To ensure a picture which may be converged everywhere to within a fraction of a pixel, we use separate correction coils driven by digital programmed waveforms under microprocessor control. The corrections at 16 × 16 image points are stored in a 12-bit memory and are read out and interpolated in real time. The unexpected requirement for a 12-bit word length arose when we discovered that the eye has the extraordinary ability to detect deviations of the order of 0.1 min of arc at the boundaries of two vertical interpolated regions. This is equivalent to $^1/_{10}$ of the raster line pitch.

Preliminary results confirm that 25-MHz resolution can be achieved on 1249-line standards. Operation of the projector on alternative standards such as 625 lines, 2:1 interlace, 100-Hz field rate has also been demonstrated.

Camera

As one source of high-quality pictures, we have developed a 1249-line camera giving 25-MHz *RGB* output signals. This has been based on the well-known Philips LDK3 camera. The camera uses three 30-mm 45XQ "frog-head" Plumbicon tubes, scanned with a 25-mm diagonal raster. Specially developed assemblies contain deflection, alignment, and focus coils which are arranged in two parts: a long coil around the deflection region, and a large-diameter coil around the tube head to minimize edge-effects. For good corner resolution, dynamic focus is also applied to the tube collectors. Vertical out-of-green contour correction has been incorporated using digital delay lines.

Input capacitance and noise pickup are minimized by mounting the pre-amplifier input stage directly onto the deflection assembly. This stage, which lowers the input impedance, is followed by a high-frequency pre-emphasis stage. The total maximum voltage gain of the pre-amplifier is 120 dB with a 3-dB bandwidth of 35 MHz.

Finally, image registration and correction are achieved by a combination of analog and digital techniques. Coarse correction is achieved by the analog circuit, and a digital system based on 14 × 11 matrix points provides fine vertical and horizontal correction.

Telecine

As an additional source we chose 35-mm cine film. Compared with 70-mm film, this offered a broader range of program material and is also available in several aspect ratios — e.g., 4:3, 5:3 wide screen, and 2.35:1 CinemaScope. The smaller 16-mm format has, of course, inadequate resolution. In our telecine, the line scanning is performed using 1728-element P^2 charge-coupled device (CCD) line-sensors. Three of these are mounted, together with a prism color-splitter to provide the three color signals at a data rate of 64 MHz.

Compared to a camera-telecine, the CCD scanner gives relatively small registration problems, and the advantage over a flying spot scanner is the absence of conflict between resolution (spot size) and signal-to-noise ratio (tube light output). There is no lag and no EHT requirement — leading to good reliability. Any small nonuniformities of the sensor output signals may be corrected by means of fast-operating line memories.

Figure 4 shows a photograph of the

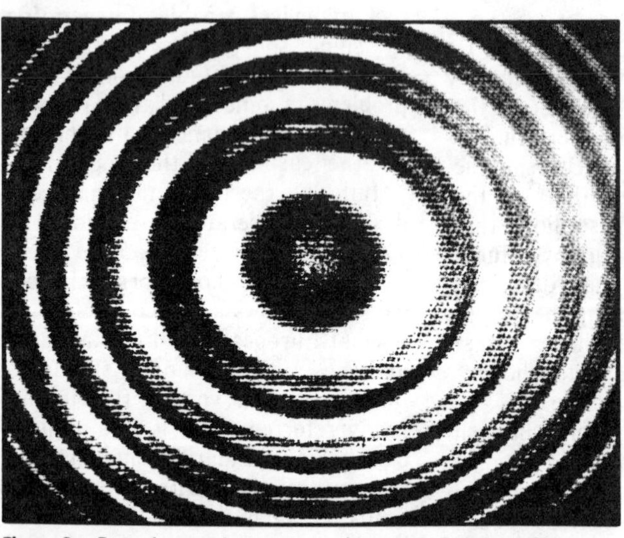

Figure 3a. Part of zone plate pattern with normal PAL decoding.

Figure 3b. Part of zone plate pattern after processing.

Figure 4. 1249-line HQTV telecine equipment.

complete telecine equipment. For vertical scanning, the film is driven continuously via its sprocket holes, the sprocket-wheel servo being locked to the line-synchronizing pulses to obtain a positioning better than one HQTV line. A conversion memory is incorporated to change the non-interlaced video from the line sensors to the 2:1 interlace pattern chosen for our initial experiments.

Transmission Methods

If we were to transmit signals relating to the full HQTV specification by a method directly analogous to the existing signal transmission, without any special processing, then we would require 5× the normal bandwidth — about 25 MHz. We consider this to be unacceptably large for any medium and quite impossible within any existing channel allocations, right up to 12 GHz. One of the underlying assumptions of our work is that, within a reasonable time span, it will be possible to provide one or more fields of storage at each end of the TV chain. This implies that it will no longer be necessary to use identical scanning standards for the studio, the transmission, and the display.

The advantage of this new flexibility is that we may well be able to process signals so as to eliminate unwanted artifacts, to reduce redundancy, and to provide for a compatible evolution from today's standards to those of the future. An attractive possibility is to use (only) two standard channels for one HQTV transmission, one of which carries a signal compatible with existing TV receivers.[1] By way of example, we will describe a hypothetical system of this nature.

A picture of 5 × 3 aspect ratio, having about one million picture elements and designed for the 1-m^2 screen, presents three problems: (a) the aspect ratio differs from existing transmissions; (b) the number of scanning lines will also be different; (c) the required bandwidth is very large.

One method of dealing with problem (a) is indicated in Fig. 5. One channel might carry a 4 × 3 aspect ratio conventional signal representing the center portion of the 5 × 3 original. The second channel would then carry the wide screen edge information, plus the high-frequency components of the center portion. We have verified, experimentally, that edge-joining of signals transmitted via separate transmitters can, indeed, be successfully carried out by using storage techniques in association with vertical interval test signals.[1]

This technique does not solve problem (b). In order to consider this problem in more detail, let us first study the requirements for vertical resolution. The first fact to note is that, for a given picture repetition rate and assuming equal horizontal and vertical resolution, the bandwidth required in a TV signal is proportional to the square of the number of scanning lines. Thus, it is desirable to try to achieve the minimum number of lines in the raster, consistent with adequate vertical resolution. If we assume that the K factor in existing conventional systems is about 0.65, then the present vertical resolution (European Standard) is

$$575 \text{ (active lines)} \times 0.65 = 374 \text{ picture elements}$$

The height of the HQTV picture will be approximately double that of the standard picture, so

$$\text{HQTV vertical resolution} = 748 \text{ picture elements}$$

Let us further assume that, by oversampling at the studio together with down-conversion/up-conversion and appropriate filtering, a K factor of 0.87 can be achieved. Then the HQTV

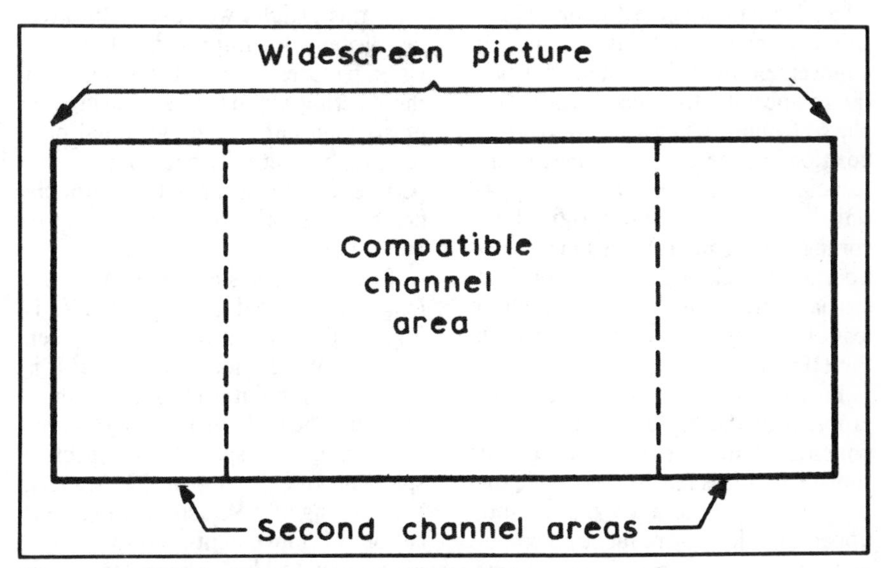

Figure 5. Spatial separation of two-channel HQTV picture.

raster requires that

748 (elements) ÷ 0.87 = 860 active lines

Allowing a similar blanking ratio to the existing system (8%) leads to the conclusion that

the HQTV raster should have

$$860 \div 0.92 = 935 \text{ lines}$$

This number is very close to 1.5 × the number of lines in the normal (European) TV standard, so the choice of 937 lines seems promising for HQTV. We may now consider a picture store in the TV studio which is replenished every 40 ms by a 937-line picture having 2:1 interlace, (Fig. 6a). One compatible way to transmit this is to send the total information over a longer period of time than normal, as shown in Fig. 6b. Here the suggestion is to transmit 20-ms fields, each having 312.5 lines, so that in each 40-ms period a 625 line raster is transmitted. In each field, the lines transmitted are the nearest available lines from the 937-line raster, as indicated by the unbracketed numbers. It can be seen that some lines repeat at 40-ms intervals while others repeat at 80-ms intervals.

For still pictures, the HQTV receiver acquires all the necessary information in the correct spatial position. The normal receiver, however, will produce a slightly degraded picture, since some of the information will be displaced by ⅓ of a line interval from its true position. We suggest a maxim for TV engineers: "Compatibility always means degradation of the existing standard!"

When movement occurs in the scene, a compromise has to be made, since some of the lines are displaced by 20 ms from their correct temporal position. To counteract this, we envision a movement-adaptive system in which the temporally displaced lines are replaced by lines which are in the correct position in time but are derived by interpolation from the nearest two spatial lines in the appropriate field (bracketed numbers). This means deliberately exchanging temporal and spatial information and accepting a loss of resolution on moving scenes in the HQTV picture.

Finally, we have to consider the problem of the high-frequency components. Our previous work has shown[1] that it is feasible to divide the baseband signal of a TV picture into upper and lower frequency components and to transmit these via two separate channels. If we adopted this strategy to increase the horizontal resolution of the 4 × 3 portion of the picture transmitted as outlined above by a factor of 2:1, then the high-frequency components of that portion would completely fill the second channel. Since we do not need to maintain raster compatibility with existing standards in the second channel, however, we may omit the repeat fields in that channel (Field 3 of Fig. 6b). This gives us a spare channel capacity of 25% for transmission of the edge information which, together with savings which can be made in signal blanking, provides theoretically adequate total capacity.

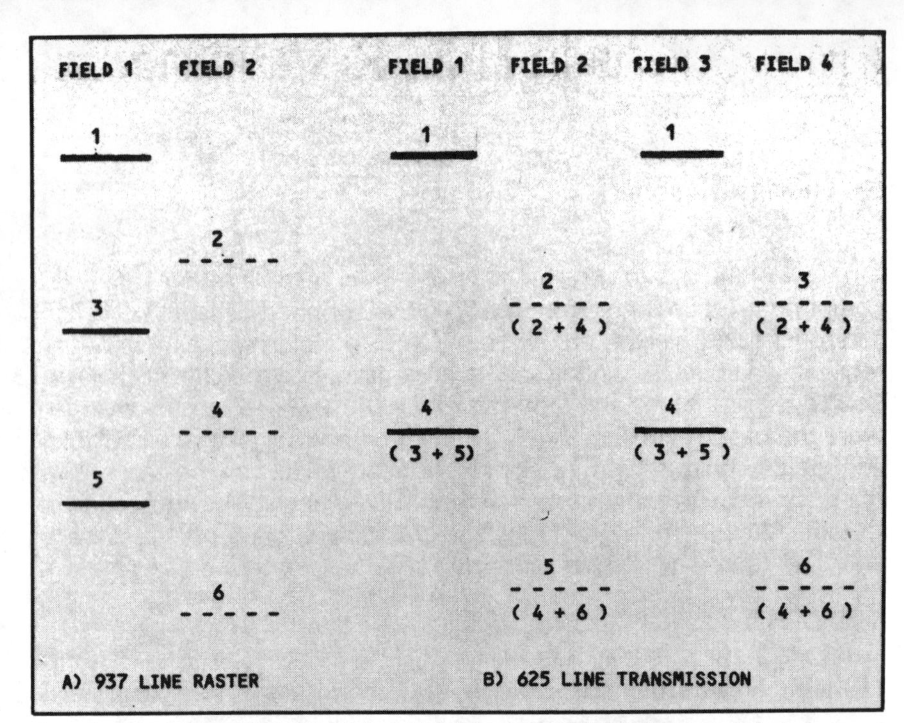

Figure 6. Compatible HQTV transmission: (a) 937-line raster, (b) 625-line transmission.

Conclusion

In this article, we have postulated two systems for improvement of television pictures. The first, starting with the existing standards and using sophisticated but realizable signal processing, has actually been built in experimental form and has publicly demonstrated a major improvement in picture quality.

The second system, aimed at the longer term goal of full HQTV, is highly speculative and has not been realized. We do not pretend that it is other than a pointer toward possibilities. Still, the main features of this type of system, viz., spatial and spectral splitting and joining of TV signals, transcoding of TV rasters, and application of movement-adaptive techniques, have all been shown feasible in principle, either by direct experiment or by simulation.

It is not our purpose here to propose a new standard. Nevertheless, we do wish to repeat the message we have given in previous papers from our laboratories:

1. Significant improvements in quality are possible within the limitations of existing signal standards.
2. We favor the approach of compatible evolution, so that upgrading of the existing standards can proceed over the next decade as we approach the full HQTV goal.
3. The introduction of full HQTV standards will require considerable time and money. The industry should, therefore, conduct sufficient detailed experiments and use its customary ingenuity to find the optimum solution before deciding on a new standard.

References

1. R. N. Jackson and S. L. Tan, "System Concepts in High Fidelity Television," IBC 82 Convention Record, Sept. 1982, pp. 135–139.
2. M. J. J. C. Annegarn, "Signal Processing in PAL Receivers." Paper read at the 1983 Picture Coding Symposium, Davis, Calif., March 1983.
3. M. J. J. C. Annegarn, "Movement Detection in Composite PAL Signals," Ibid., 1983.
4. B. Wendland, "On Picture Scanning for Future HDTV Systems," IBC 82 Convention Record, Sept. 1982, pp. 144–149.
5. G. J. Tonge, "Extended Definition Television through Digital Signal Processing," Ibid., pp. 148–151.
6. A. G. Constantinides and D. A. Jenkins, "A Technique for Three Dimensional Filter Design." Paper to be read at the 1983 Picture Coding Symposium, Davis, Calif., March 1983.
7. K. G. Freeman, G. de Haan, J. M. Menting, and D. W. Parker, "Experimental Work Towards High Fidelity Television," IBC 82 Convention Record, Sept. 1982, pp. 140–143.
8. S. L. Tan and R. N. Jackson, "Flexible Coding and Decoding for Extended Television," IBC 84 Convention Record, Sept. 1984, pp. 107–110.

High-Definition Television and Compatibility with Existing Standards

By C. P. Sandbank and M. E. B. Moffat

This paper details the technological considerations involved in bringing high-definition television to television receivers already in use, and in achieving high-definition compatibility with existing standards. It lists means by which conventional 625/50 signals may be improved for display on high-definition receivers, and discusses recent investigations into many different scanning standards.

The current interest in high-definition television (HDTV) is somewhat surprising, since during the last 20 years there have been no significant changes in display technology to herald the availability of consumer products giving a large area, high-resolution, bright display, and which are capable of doing justice to a picture scanned with more than 625 lines. However, the interest in new TV standards must be seen in the light of the fact that, displays apart, during these same 20 years there have been fundamental changes in most other aspects of television technology. Solid-state devices have led to the introduction of digital techniques in the studio, as well as for recording and distribution. The use of both satellites and optical fiber cable has been established as means of transmitting video directly to the home viewer. New technology for storing video on tape or disk has appeared in the consumer area.

It is to be expected, therefore, that attention should be turned to the possibility of new high-definition TV standards. If one assumes that there is an element of the "chicken and egg" situation with respect to the display, attention to new standards may stimulate the development of this device. Also, there are some industries, e.g., film production, visual aids, simulators, etc., which are not limited by the absence of improved domestic display devices and can take immediate advantage of many of the other recent advances in video technology.

Presented at the Society's 16th Annual Television Conference (paper No. 16-26) Nashville, Tennessee, February 5, 1982, by C. P. Sandbank, Research Department, BBC, U.K. This paper was received January 12, 1983.

For the broadcaster, the possibility of a new standard presents a greater challenge. If future developments suggest further changes in standards, these could be introduced fairly readily, for example, into the process of electronic cinematography. Not so for the broadcaster, who has to maintain faith with every generation of consumer purchasing equipment to a new standard. Furthermore, the broadcaster's degrees of freedom to increase definition are limited by the available radio spectrum and pressure for more channels. Thus it becomes the first responsibility of a broadcaster (particularly a public service broadcaster) to ensure that new technology is fully exploited to improve the current standard. At best, this should be done in a manner available to all viewers with existing equipment, and next best, in a manner compatible with current equipment, but allowing the improvements to be obtained by the purchase of new equipment.

Accepting the fact that in the near future there will be new HDTV standards, and that these will be applied without some of the constraints, such as bandwidth limitation, which apply to many aspects of broadcasting, it is important that their relationship to present TV standards be considered when these new standards are being established. This is necessary so that HDTV source material can readily be used for transmissions to existing standards. It will also assist in bridging the quality gap between material derived from broadcast sources limited by consideration of bandwidth, etc., and from sources not subject to these limitations when these are displayed on future HDTV screens.

Improved Display of Conventional 625/50 Signals

The conventional 625-line, 50-field, 2:1-interlace scanning standard can be displayed well on present-day receivers, to the satisfaction of all but a very small minority of viewers. However, HDTV will be associated with the advent of larger, high-brightness display equipment; and the display of conventional 625/50 signals on such equipment may have to be improved to bridge the quality gap between existing and HDTV sources.

The four main impairments to be minimized are as follows:

1. Large-area flicker: this occurs at 50 Hz and is perceived primarily in peripheral vision. Thus it is aggravated, for a given brightness, as the screen size increases or the viewing distance decreases. It is also a function of display brightness: as the brightness increases, the response time of the eye is

Table 1 — Some Possible Display Standards Which Have Been Compared on the Test Bed, Using Various Interpolation Algorithms.

Lines per Picture	Fields per Second	Interlace Factor	Scan Frequency in kHz
625	50	2:1	15.6
1250	50	2:1	31.2
625	50	1:1	31.2
625	100	2:1	31.2
1250	100	4:1	31.2
625	100	1:1	62.5
1250	100	2:1	62.5

Reprinted with permission from *SMPTE J.*, vol. 92, pp. 552–561, May 1983.
Copyright © 1983 by the Society of Motion Picture and Television Engineers.

reduced and flicker becomes worse.

2. Inter-line "twitter": this is a flicker, referred to here as twitter to distinguish it from large-area flicker. It occurs at 25 Hz and is caused by the 2:1 interlace. It arises in areas of high-amplitude vertical detail with spatial frequencies approaching the Nyquist limit of 312.5 cycles per picture height. Perception of twitter is a function of display contrast because of an ocular property similar to the brightness/response-time relationship. It may also be dependent on viewing distance or size of display, because of peripheral vision, and on the scanning spot-size in the studio source.

3. Line crawl: this is an unavoidable effect resulting from the interlaced scanning process; it is most readily seen when vertical motion in a scene forces the eye to scan the image at or near a critical rate of 11.5 sec per picture height, which is the number of active lines per picture divided by the number of fields/sec (i.e., 575/50). When line crawl is apparent, the viewer unfortunately perceives a 312.5-line picture. It is also unfortunate that the rate of vertical rolling of captions in television programs frequently emphasizes line crawl.

4. Static raster: the individual lines of a picture can be seen when viewed on modern, high-brightness displays at moderately close viewing distances. This impairment would be aggravated by increases in screen size and brightness, if the size of the scanning-spot in relation to picture size were maintained or reduced. With most pictures, however, the overall subjective impression is usually dominated by line crawl, so that a 312.5-line picture is perceived more frequently than the 625-line raster.

The possibility of including complex digital signal processing in future consumer products suggests that these impairments can be reduced by up-converting the 625/50 signal to a new standard with higher line and/or field rates.[1] Such an up-conversion, using a ratio of 1:2 for simplicity, is considered here for use in receivers. Wendland[2] and Lucas and Windram[3] have also suggested a ratio of 1:2 in their proposals for extended-definition television based on present scanning standards. Kraus[4] has published results for a 1:2 up-conversion in field frequency to 100 Hz for conventional 625/50 signals, to avoid large-area flicker in domestic TV receivers. For 66-cm (26 in.) picture tubes with mean and peak brightness of 200- and 750-candela/m^2 (58 and 219 fL) respectively, he asserted that a field frequency of at least 75 Hz was needed, so that 100 Hz gave a comfortable margin.

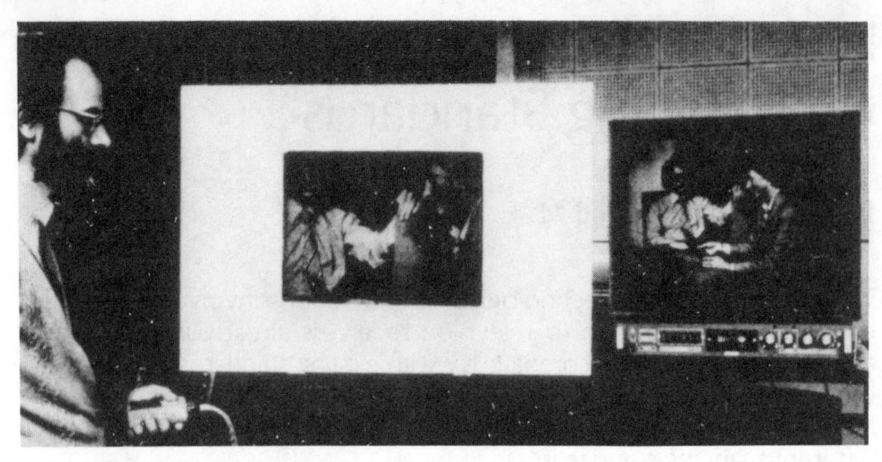

Figure 1. Experimental test bed showing a simulated large-screen display.

A flexible test bed has been developed at the BBC Research Department to allow a wide range of possible scanning standards to be examined. These are obtained by up-conversion from a normal 625/50 signal. It can be seen from Table 1 that, to examine some of the higher standards, line-scanning frequencies of 62.5 kHz are required. In the test bed, which is shown in Fig. 1, this is facilitated by a technique whereby part of the original image signal is processed to become double sized (linearly), filling the whole of the monitor screen. This represents 25% of a large-screen display, twice the linear size of the monitor. The method also allows a standard CRT display to be used to simulate the visual effect of a large, bright, high-resolution display with an area greater than that possible with CRT technology.

The digital standards converter comprises four fields of storage and two interpolators (two to ease high-speed processing difficulties), together with appropriate analog interfaces and synchronizing pulse generation; with this arrangement, a wide variety of interpolation algorithms was investigated in monochrome, as outlined below. Color experiments were deferred, since many important factors can be examined in monochrome with simpler instrumentation.

A 1250/50/2:1 raster is obtained by doubling the line frequency to 31.250 kHz. Large-area flicker is unaffected, but inter-line twitter may be slightly improved by the interpolation which produces the extra 625 lines. The twitter frequency remains constant at 25 Hz. Line crawl is present at more speeds than it is for the 625/50 standard, its lowest rate being 23 sec per picture height, but is less visible because of the finer structure. The 2:1 static-raster improvement is dominated by line crawl, but the crawl has a 625-line structure.

A 625/50/1:1 raster also involves doubling the line frequency. Again, large-area flicker is unaffected. Inter-line twitter and line crawl are eliminated for static pictures, although some twitter may be introduced on moving pictures by the intra-field interpolation used under these conditions. The static raster is unaffected.

A 625/100/2:1 raster is obtained by doubling both the horizontal and vertical scanning rates. Large-area flicker is effectively eliminated. Inter-line twitter is raised to 50 Hz, thereby virtually eliminating its visibility. Intra-field interpolation, however, may re-introduce some image-dependent twitter. The slowest rate of line crawl is quickened to 5.75 sec per picture height, and the static raster is unaffected. This standard is depicted in Fig. 2. Note that in Figs. 2–6 the y axis represents the vertical position on the screen and the x axis represents time. Thus, the horizontal scanning lines should be regarded as perpendicular to the plane of the paper.

The 1250/100/4:1 raster is obtained by doubling both the line and field frequencies, as for 625/100/2:1. Large-area flicker is eliminated and twitter is potentially so, depending again upon interpolation. The static raster is also improved, but a 312.5-line structure appears when line crawl is observed, and the visibility of line crawl is directly affected by the sequence chosen for the 4:1 interlace structure. There are two types of 4:1 interlace structure, progressive and scrambled, and they are depicted in

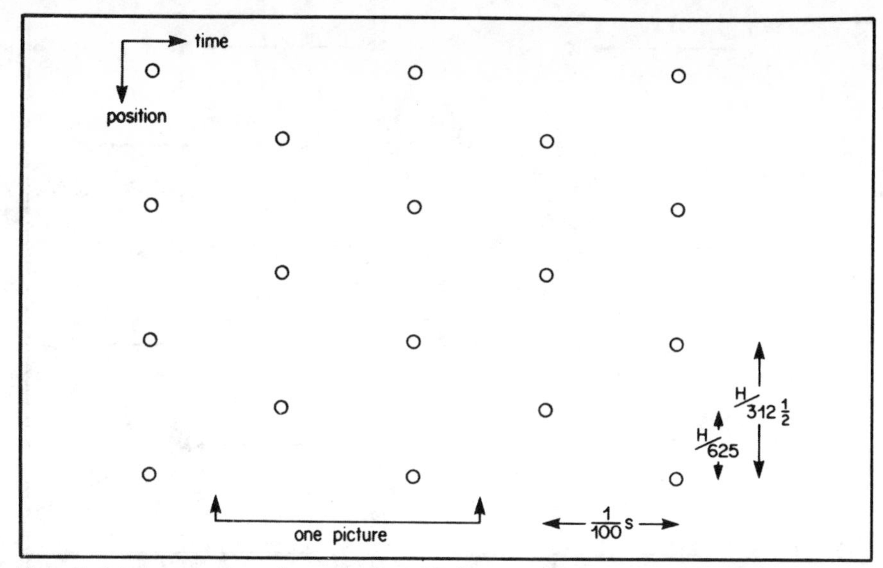

Figure 2. Interlace structure of scanning lines: 625 lines per picture: 2:1 interlace, 100 fields/sec (H is picture height).

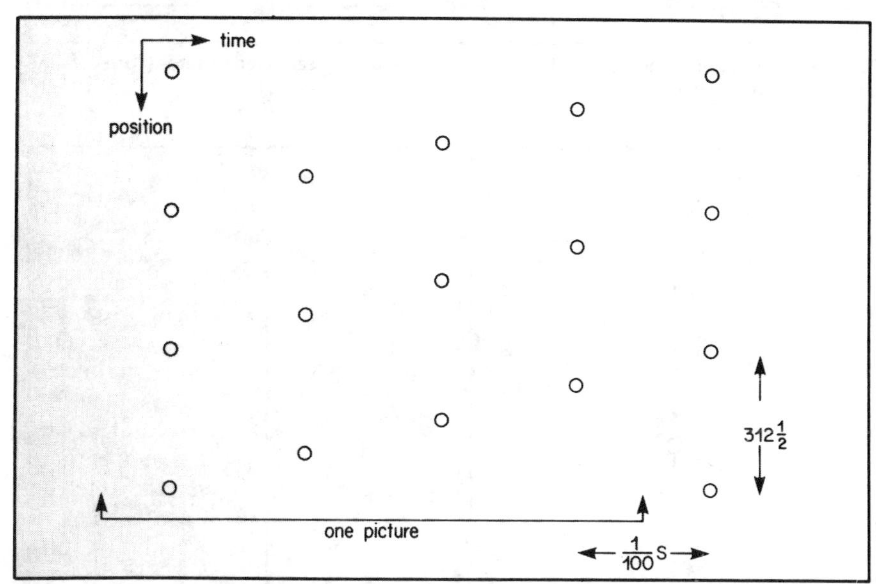

Figure 3. Interlace structure of scanning lines: 1250 lines/picture, 4:1 interlace, 100 fields/sec, upward progression of scanning lines.

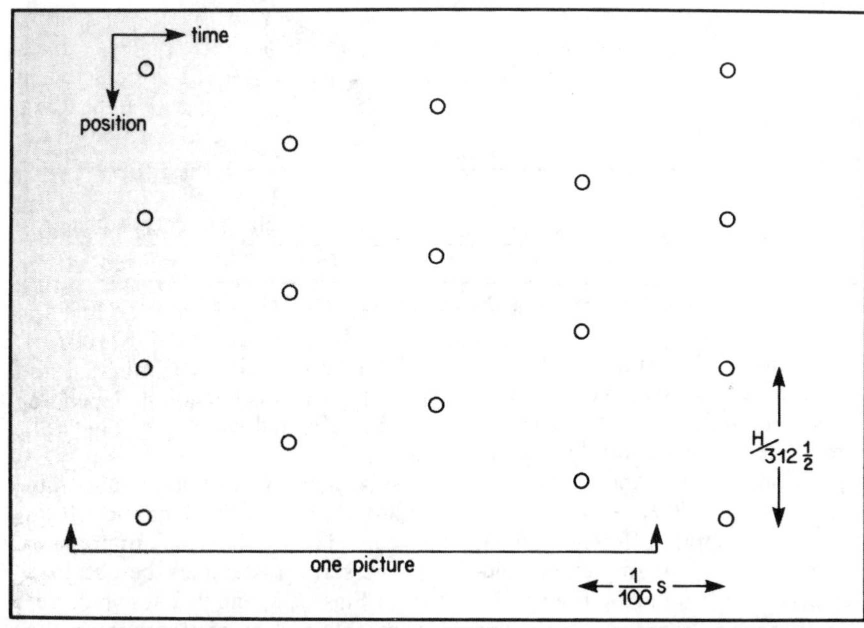

Figure 4. Interlace structure of scanning lines: 1250 lines/picture, 4:1 interlace, 100 fields/sec, scrambled sequence of scanning lines.

Figs. 3 and 4. The upward progression in Fig. 3 is at a rate of 11.5 sec per picture height, coupled with a much faster (less visible) downward crawl at a rate of 3.83 sec. Obviously, there is also an option for a downward progressive structure at the 11.5-sec rate and its associated fast upward crawl at a rate of 3.83 sec could reduce the impairment of upward-rolling captions. The line crawl of the scrambled structure in Fig. 4 has equal upward and downward rates and is less visible than that of the progressive structures; it is similar to but less visible than that of the 625/50 standard. The presence of the 312.5-line structure re-introduced by line crawl tends to diminish the advantage of the finer static raster.

A 625/100/1:1 raster is obtained by doubling the field frequency and quadrupling the line frequency. Large-area flicker, inter-line twitter, and line crawl are eliminated (although, again, inter-line twitter may be re-introduced by intra-picture interpolation). Static raster is unaffected.

A 1250/100/2:1 raster is obtained by doubling the field frequency and quadrupling the line frequency. Large-area flicker is eliminated and inter-line twitter can be greatly reduced if suitable line and field interpolation is used. This time, however, an improvement in the static raster is exchanged for a reintroduction of some line crawl. The slowest line crawl is at 11.5 sec per picture height. Its 625-line crawling pattern makes it less visible, and the benefit of the 1250-line static raster can be appreciated.

The wide range of interpolation algorithms that were examined subjectively could be broadly divided into two classes, A and B. Class A algorithms did not use any spatial or temporal filtering and were better suited to stationary images. Class B algorithms used spatial filtering and yielded better movement portrayal than Class A at the expense of poorer vertical resolution.

Figure 5 shows a Class A algorithm applied to a 625/100/2:1 structure. This illustrates the Class A feature that each line of an up-converted (output) field, marked as a cross, is taken from the nearest 625/50/2:1 (input) field in which a line, marked as a circle, can be found in the correct vertical position. The particular algorithm Type A2 shown in Fig. 5 is an extrapolation. It yields an output sequence of fields in which "time reversal" of input fields occurs, thereby causing judder (periodic horizontal jumps) on moving images from television cameras. With televised film, however, each frame is simply shown twice and is correctly phased. No time reversal occurs and there is no judder other than that normally experienced with film.

It can also be seen in Fig. 5 that a Class A algorithm conserves the sequence of odd and even fields. This causes the inter-line twitter frequency to be doubled, rendering the twitter invisible on virtually all stationary images.

Figure 6 shows a Class B algorithm applied to a 625/100/2:1 structure, illustrating the Class B feature that each output field is generated from the nearest input field. As a result, interline twitter at 25 Hz remains visible, depending upon the image being displayed. With the particular algorithm Type B¼ shown in Fig. 6, the introduction of flicker due to interpolation was avoided by effectively subjecting all fields to the same processing. Motion portrayal was found to be good, but not perfect. A horizontally moving vertical edge became serrated, or comb-like, corresponding to the displacement of odd and even fields, for edge movements at about 5 sec per picture width. The same combing effect can be seen on film televised in the usual way, but at edge movement rates of about 10 sec per picture width. At higher edge movement rates, a somewhat different impairment occurs, similar to the judder seen on conventionally televised film; this is fringing, where the moving edge is seen as two distinct images. Apparently, for combing and fringing, the eye transforms the temporal errors into spatial errors.

Several algorithms, aimed at combining the good vertical resolution properties of Class A with the good movement portrayal of Class B, were examined experimentally. Unfortunately, no satisfactory hybrid was found. Therefore, attention is being turned toward an adaptive interpolator, including a movement detector to control selection between a Class A algorithm and a Class B. The video noise reducers used on BBC Television networks are based upon a BBC Research Department design incorporating a movement detector. The experience gained with those noise reducers showed that it is possible to achieve satisfactory movement detection to control an adaptive process. (Incidentally, a noise reducer is another example of instrumentation that will improve large-screen displays).

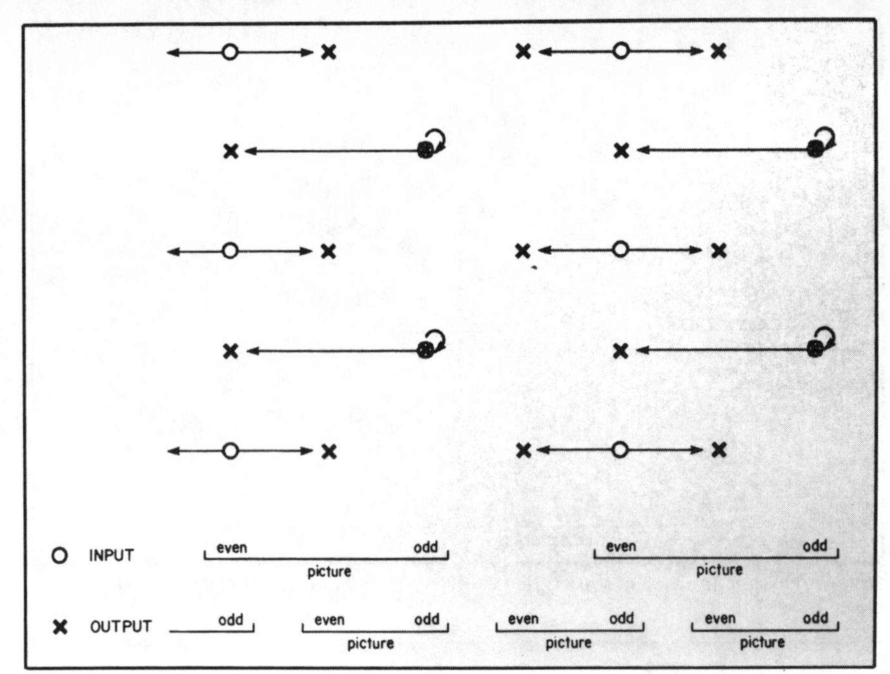

Figure 5. Type A2 algorithm applied to a 625/100/2:1 structure: repeated pictures (correct frame phase for film).

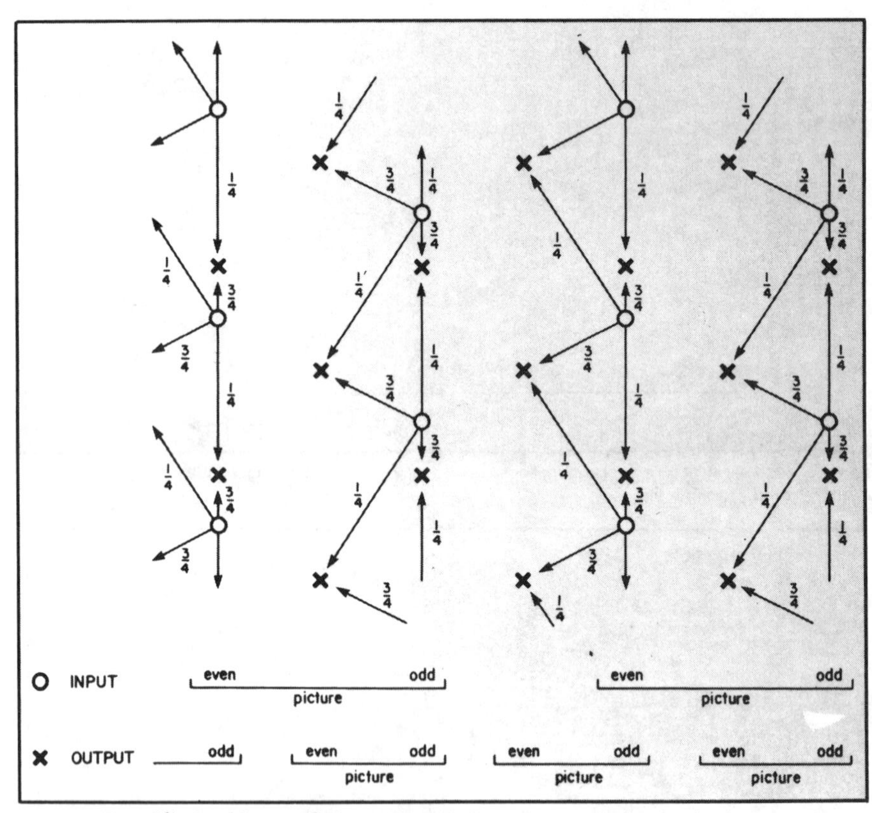

Figure 6. Type B¼ algorithm applied to a 625/100/2:1 structure: all pictures line interpolated.

One proposal for a movement detector that would be suitable for interpolation involves measuring the ratio $(D_T - D_S)/(D_T + D_S)$, where D_T is a temporal difference signal generated from two samples, one picture-period apart, and D_S is a spatial difference signal from two samples one line-period apart.[1] The ratio would be +1 for pure motion and −1 for pure stationary vertical detail; it is unlikely to be affected much by random noise.

It is, of course, difficult to illustrate photographically the dynamic affects discussed above, but some impression of the static raster may be obtained from Figs. 7, 8, and 9. These pictures were obtained by photographing the left-hand monitor in the equipment

Figure 7. Central 25% of a simulated large-screen display of the 625/100/2:1 standard with a 625/50/2:1 input signal.

Figure 8. Central 25% of a simulated large-screen display of the 1250/100/2:1 standard with a 625/50/2:1 input signal.

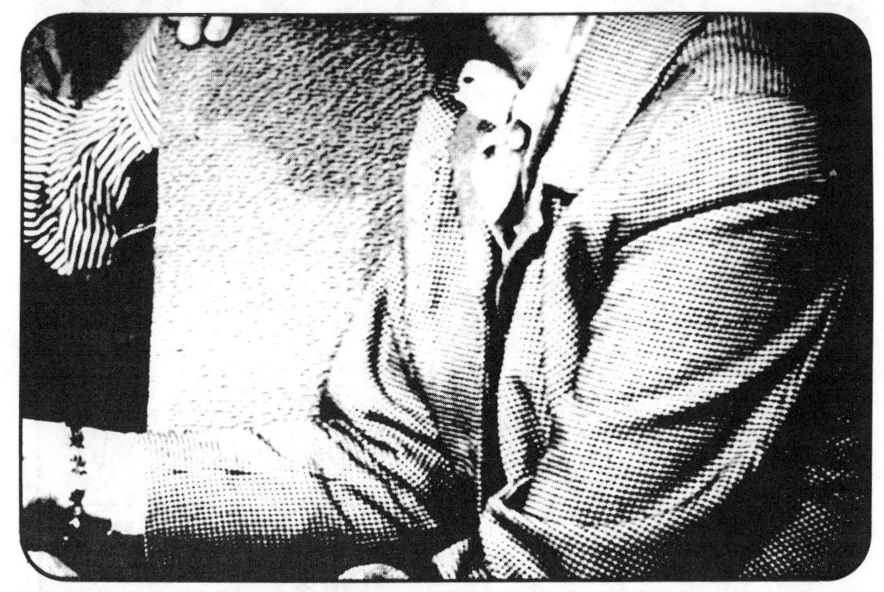

Figure 9. Central 25% of a simulated large-screen display of the 1250/100/2:1 standard from a source operating on the equivalent to a 1250 line studio standard.

shown in Fig. 1. The image is 25% of the area of the EBU test slide known as "Young Couple," which was designed to include a large degree of fine detail for the purpose of resolution and cross-effect tests. The full image is seen on the right-hand monitor in Fig. 1.

Figure 7 is a photograph of the simulated large-screen display using a 625/100/2:1 raster. About 300 picture lines appear on the monitor representing the central 25% area of the display. Figure 8 is a similar picture, but with the test-bed set to stimulate a 1250/100/2:1 display so that, in this case, about 600 lines appear on the monitor. At viewing distances around three times picture height, there is a significant improvement, due to the fine static raster when the number of lines on the simulator is increased to 1250. The finer raster also produced another subjective improvement which might be observable from the photographs. Areas of fine texture, such as the back of the chair and those parts of the sleeve subject to prominent spatial aliasing, appeared to have a higher resolution when displayed with 1250 lines. However, any such improvement is, of course, limited by the fact that the source was the same for the experiment recorded in Fig. 8 as that used to produce Fig. 7.

Thus, starting with a conventional 625/50/2:1 signal, many of the impairments of this standard can be overcome by converting to a different standard in the receiver. The work reported above provides useful guidance on the best process to use for improving the displayed picture. The optimum choice depends very much on the nature of the display device. For displays with the characteristics of a CRT with a large, bright image viewed with significant areas in peripheral vision, doubling the field rate to 100 Hz might give the most effective improvement of signals originated at 625/50/2:1. However, with a different type of display (perhaps one using a matrix with inherent memory), quite different raster parameters and interpolation algorithms might be appropriate.

Another factor which may determine the choice of raster in the display device is the possibility that the establishment of electronic cinematography and other non-bandwidth-limited means of distributing TV programs may lead to the use of a high-definition standard in the studio well before such signals could be broadcast directly. Figure 9 is a photograph of the screen using the same display conditions as used to produce Fig. 8, but the input signal is produced by sampling the picture with a raster equivalent to a 1250-line studio standard.

This experiment serves to illustrate that, for signals originating in the studio and transmitted at a 625-line standard, conversion in the receiver to a higher line standard may not be worthwhile. In order to obtain true high definition, it is necessary to sample, transmit, and display at a higher standard than that used at present.[5] However, it may be possible to take an input of the type used to produce Fig. 9 and pre-filter this for transmission at the 625-line standard and then post-process the signal in the receiver to recover some of the increased resolution of the original.

This involves compromises such as those between diagonal and horizontal resolution, as well as between spatial and temporal resolution. Thus, under circumstances where the compromises associated with the pre-filtering of a high-definition signal for transmission by a lower standard are subjectively acceptable, conversion to a higher standard in the receiver may well be worthwhile. In addition to providing data for the development of new displays, the results reported above are also relevant to the choice of new HDTV standards and their relationship to the existing standards.

Improved Decoding of Conventional PAL Signals

The 625/50/2:1 CCIR System I PAL television standard used in the U.K. has a video bandwidth of 5.5 MHz, but it is difficult to take much advantage of this above 4 MHz with conventional PAL decoders because of color and luminance cross-effects. Proposals for overcoming such cross-effects have been made[6,7] and instrumented[8,9] at the BBC Research Department. A two-picture-store, line-locked digital PAL decoder[10] has displayed stationary 5.5-MHz bandwidth pictures with virtually no impairment due to composite coding, as depicted in Figs. 10 and 11. However, adaptive methods were necessary to reduce impairment on moving pictures containing saturated colors.

These adaptive methods operate by detecting the moving areas of the picture which produce impairments with the picture-based decoder, and changing the decoding method to a line-based movement mode. Thus, the decoded signals have slightly reduced

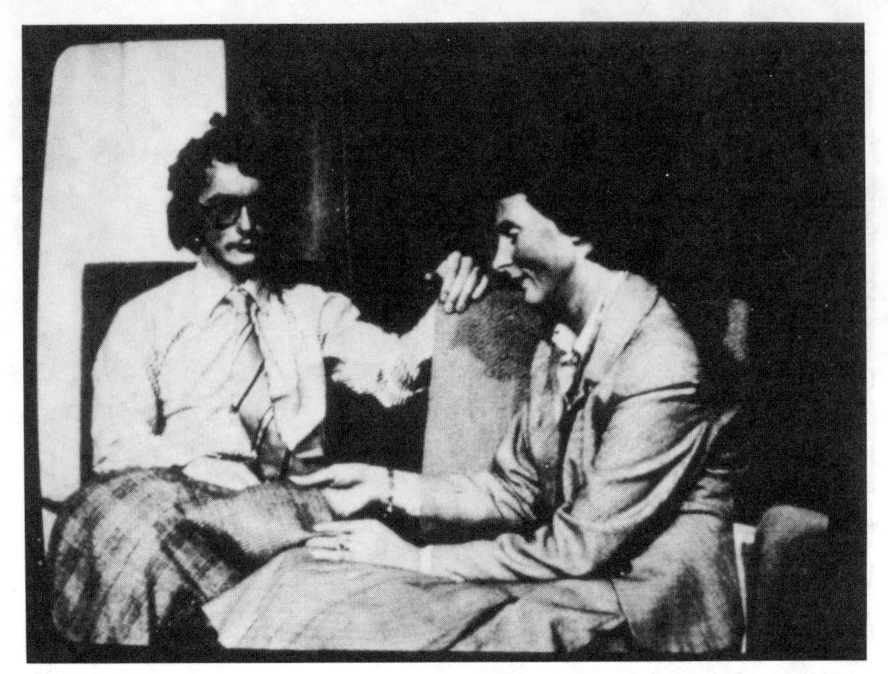

Figure 10. Color picture decoded with a conventional PAL decoder; green signal shown to highlight cross-effects.

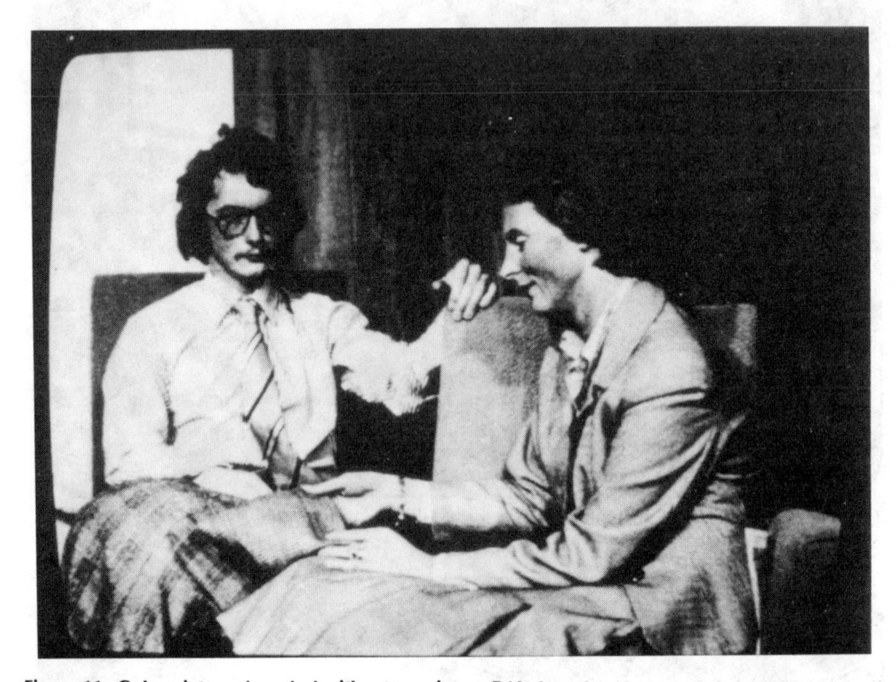

Figure 11. Color picture decoded with a two-picture PAL decoder; the green signal is shown.

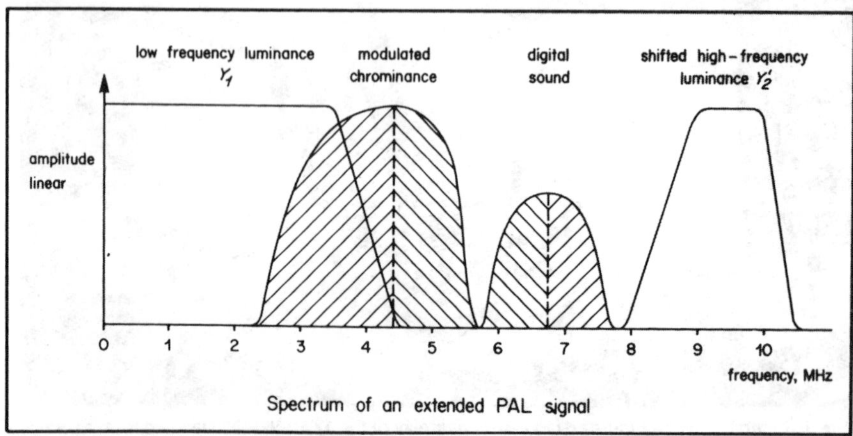

Figure 12. Spectrum of split-luminance coding for DBS.

resolution in moving areas, but are free from the more serious impairments caused by picture delays. Simple adaptive techniques can cause the decoder to use the wrong mode. This is usually the result of the adaptive circuit detecting moving, high-frequency luminance more readily than moving chrominance. It can also result from noise and distortion in the composite signal. However, these problems are largely overcome by careful filtering and by using a specially programmed, graded changeover between the still and moving-picture modes under the control of an optimized, non-linear law. By this means, then, it has been possible to obtain acceptable performance on rapid movement while retaining the nearly perfect performance of picture-delay decoding for still and slowly moving scenes.

This decoder was designed to provide an interface between conventional studios and those using the new component digital standard. The role of this type of decoding in connection with the evolutionary approach to higher quality TV has been discussed,[11] and its use in prototype domestic receivers has recently been demonstrated by Philips Research Laboratories.

Figure 13. Standard PAL signal with conventional decoder.

Figure 14. 6-MHz *RGB* signal coded and decoded by the split-luminance method.

Direct Broadcasting from Satellites with Compatibly Extended Definition

Several proposals have been made for coding luminance and chrominance signals and applying them to direct broadcasting from satellites (DBS) in such a way that the received picture quality would be better than conventionally decoded PAL or NTSC signals would give.[3,5] Within the BBC, techniques have been investigated experimentally with the emphasis on maximizing compatibility with current PAL-receiver hardware, so that the prospect of inaugurating a public DBS service in the mid-1980's would not be prejudiced by the lengthy processes required to establish new standards. Compatibility, however, need not mean that higher definition with enhanced receivers is ruled out. Certain methods, compatible with the use of conventional PAL decoders, are being explored which would not only virtually eliminate cross-color and cross-luminance in enhanced receivers, with consequent full use of the present 5.5 MHz video bandwidth, but would also have the potential for extended definition.

It is not within the scope of this paper to detail these methods, but it is appropriate to outline how one of them could offer enhanced definition. Figure 12 shows the spectrum of a split-luminance PAL-compatible coding method proposed for DBS by Oliphant.[12] The luminance spectrum, from about 3.5 MHz to 5.5 MHz, is shifted up to a band above the sound channels, leaving a luminance-free band for the PAL chrominance signal. This virtually eliminates cross-color in both enhanced and conventional receivers, at the expense of some slight loss of horizontal definition in the latter. The point at issue here is that, while retaining the conventional PAL subcarrier frequency for compatible chrominance modulation, the shifted high-frequency luminance band and the upper sideband of the chrominance signal could be widened. The result would be PAL-compatible higher definition transmission of YUV signals, which would do justice to the digital studio standard based on YUV sampling frequencies of 13.5, 6.75, and 6.75 MHz. Figures 13 and 14 compare a standard PAL signal with a 6-MHz *RGB* signal, coded and decoded by the extended system.

The above argument does not deal with the frequency modulation and radio-frequency spectrum problems associated with the WARC 77 plan for satellite broadcasting to Europe. However, experimental work with the split-luminance method showed that it is consistent with the WARC 77 plan, as well as entirely compatible with current PAL decoders. Also, it appears quite feasible to reconcile a significant extension of picture definition (i.e., beyond 5.5 MHz for luminance) with the WARC 77 plan, using a split-luminance or other method compatible with PAL decoders. The scope for the addition of energy in the 7–10 MHz modulation region can be assessed from the work reported on the effect of additional sound carriers on the protection ratios for adjacent channel interference.[13,14]

Further, methods of improving the display of 625/50/2:1 signals, discussed earlier in this paper, and pre-filtering techniques[2] could be applied to such DBS signals to complement methods for enhancing quality and extending definition described here. Additionally, the improved decoding methods described could be instrumented relatively simply with the spectrum shown in Fig. 12 to eliminate effectively any residual cross-effects. It is possible that some of the circuitry used for such decoding could be common to that required for the post-processing carried out to improve the display.

If time division multiplex[3] is used in place of the frequency division methods discussed above, separation between chrominance and luminance is achieved more readily, but at the expense of compatibility with existing receivers. Thus, the choice of enhanced transmissions for DBS would be influenced by the importance attached to compatibility with existing transmissions.

Improved Sources: Compatibility with Present Scanning Standards

Quite apart from the pre-filtering possibilities mentioned earlier, the prospect of converting HDTV source signals to the existing 625/50 and 525/60 interlaced scanning standards must be faced as HDTV programs will, to a greater or lesser degree, also be viewed via non-HDTV channels using existing standards. Compatibility between standards having differing aspect ratios is not considered in this section, although it is worth pointing out that if most of the horizontal scan period were devoted to the active line,[15] the aspect ratio would be nearly 5:3.

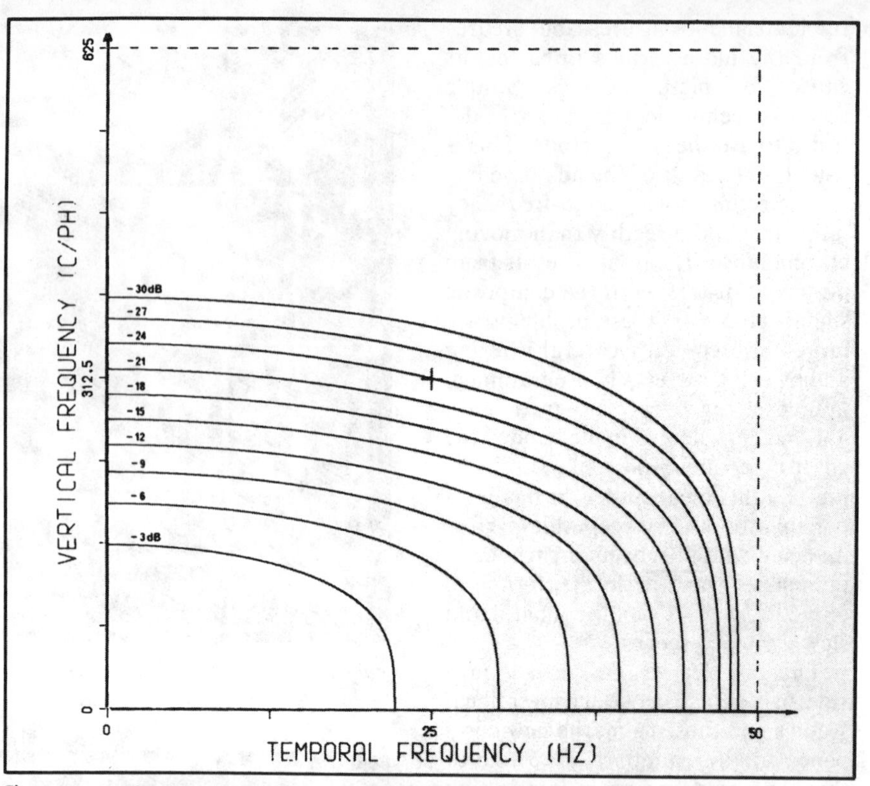

Figure 15. Vertical/temporal frequency characteristics of a typical television camera operated on the 625/50/2:1 standard.

The problem of line-standards conversion is not as great as field-standards conversion. To quantify this, consider first the relative amplitudes of the alias components which are indicated in Fig. 15 showing the vertical/temporal frequency characteristics of a typical television camera operated on the 625/50/2:1 standard. To understand Fig. 15, consider two examples. First, with a perfectly stationary image, the temporal frequency is zero and the vertical resolution loss of the camera, relative to its maximum value, can be seen to be 6 dB at around 200 cycles per picture height. Secondly, with a moving image such as a horizontally moving pattern of vertical lines, the vertical detail (frequency) is zero and temporal components, at around 30 Hz in the image, will be attentuated by 6 dB.

If the camera output is sampled at 625 lines per picture and 50 fields/sec, it can be seen from Fig. 15 that there will be virtually no vertical aliasing, as the characteristics are 20 dB down at 312.5 cycles per picture height, i.e., half the vertical sampling rate. On the other hand, temporal aliasing will predominate as the characteristic is only 4 dB down at 25 Hz, half the temporal sampling rate. It is this in-band temporal aliasing that makes perfect field standards conversion impracticable, whereas line standards conversion can be extremely effective because of the low degree of vertical aliasing that is usually encountered. It should be noted that, for a given field rate, temporal aliasing will become worse if horizontal resolution is increased; this is because the level of temporal components will increase in proportion to the horizontal components.

Having shown that the magnitude of alias components due to field rate conversion is greater than those due to conversion to a different line standard, it is now necessary to examine the nature of the temporal alias components.

Consider first the case of conversion from a high-definition standard of say, 1125 lines, down to one of 625 lines, when both systems use a 50-Hz field frequency. By tracing the progress of a particular temporal component at 15 Hz, shown in Fig. 16, it will be seen that, when the image is scanned at 50 Hz, an alias component appears at 35 Hz in addition to the true component at 15 Hz. When this is displayed on a CRT as indicated in Fig. 16c, it is modified by a process which is equivalent to a low-pass filter, largely due to the response time of the phosphors. We can regard the line standards conversion process to be equivalent to re-scanning the image on the CRT with a camera operating at the 50-Hz

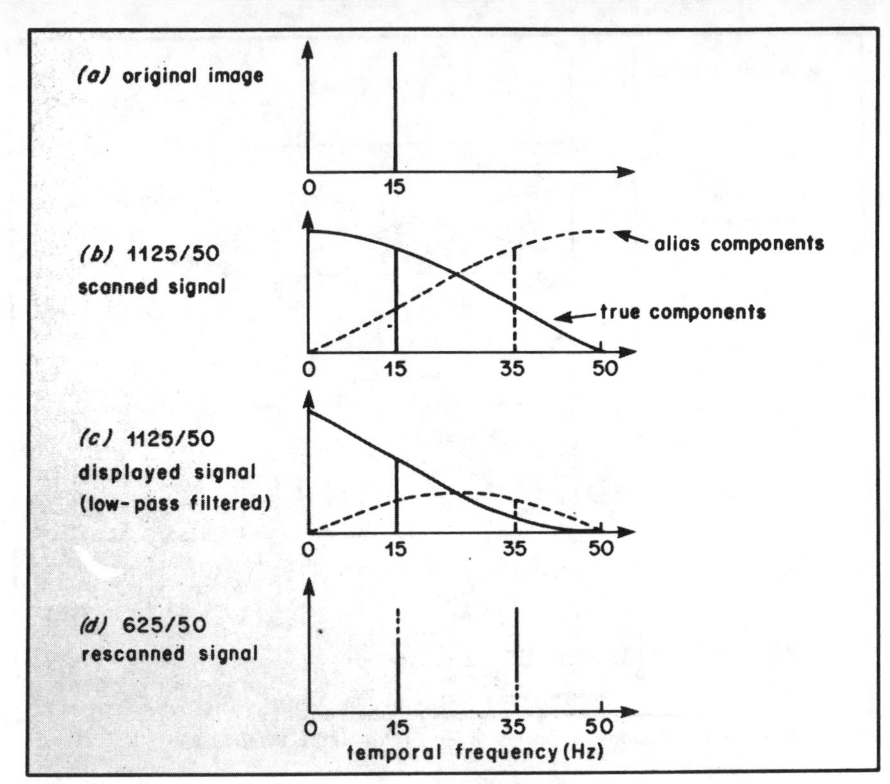

Figure 16a–d. Stages in the standards conversion process from 1125/50 to 625/50.

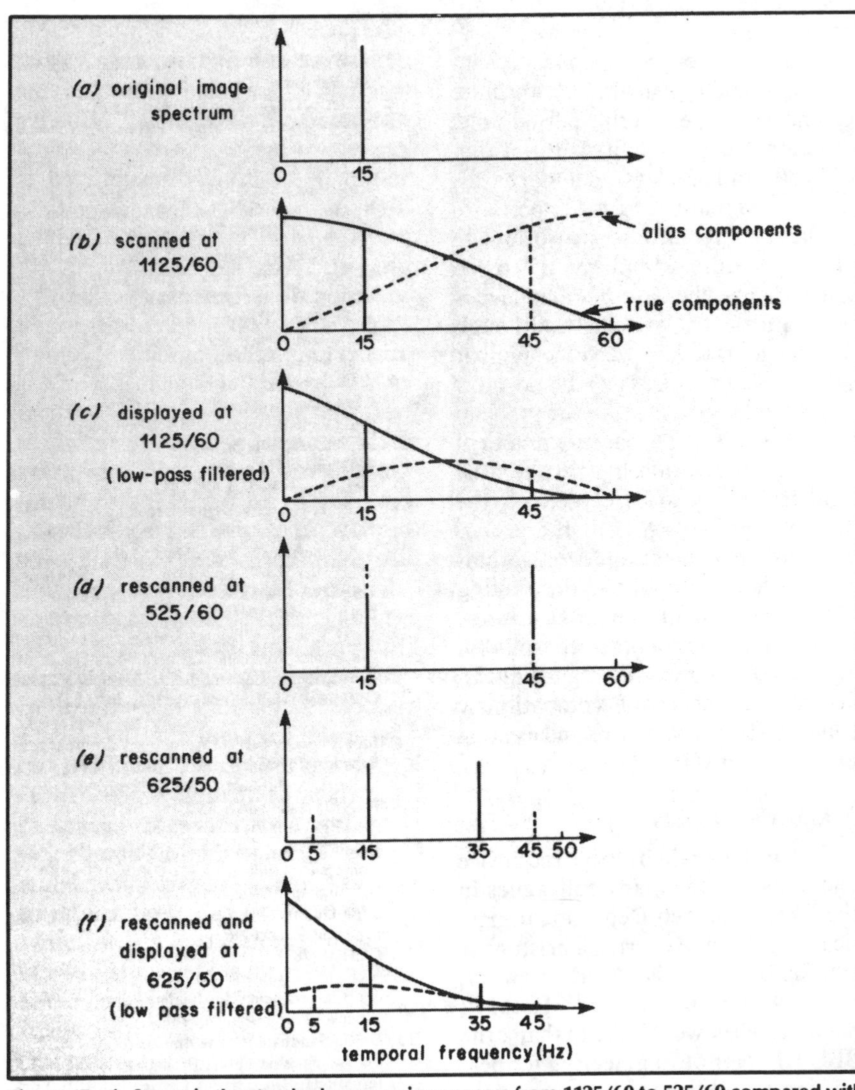

Figure 17a–f. Stages in the standards conversion process from 1125/60 to 525/60 compared with conversion from 1125/60 to 625/50.

field frequency, but at the lower line standard.

In fact, early standards converters used just such a principle. The resulting components are shown in Fig. 16d. The re-scanned 15-Hz component from Fig. 16c appears as a true component at 15 Hz and as an alias component at 35 Hz. The re-scanned alias component at 35 Hz in Fig. 16c appears in Fig. 16d at 35 Hz and again at 15 Hz. Thus, the resultant components in Fig. 16d are similar to those in Fig. 16b. Therefore, if the signal from such a standards converter is displayed, the temporal characteristics will be very similar to those indicated at Fig. 16c for the displayed signal which has not been standards-converted.

Consider now the case of down-converting from an input HDTV signal at 1125/60 shown in Fig. 17. The circumstances for conversion to 525/60, indicated in Figs. 17a, b, c, and d, are similar to the corresponding sequence in Fig. 16, with the alias components at 45 Hz instead of 35 Hz. However, if the displayed HDTV signal indicated in Fig. 17c is re-scanned at 50 Hz as shown in Fig. 17e, then additional alias components appear. When this signal is displayed as shown in Fig. 17f, it will contain alias components which are significantly different from those which would be present when the original signal is displayed as shown at Fig. 17c. Although Fig. 17f bears a resemblance to Fig. 16c, it is significant that there is a substantial 5-Hz alias component present in Fig. 17f which would give rise to noticeable movement judder not present in Fig. 17c or Fig. 16c.

Figure 18 considers the case of conversion from 1125/100 to 625/50. This illustrates the benefit of a simple relationship between the field rates, which is demonstrated by the similarity between Fig. 18e and Fig. 16c. No in-band alias components are present in Fig. 18e which are not in Fig. 16c.

Figure 18 also illustrates another important factor relevant to the readiness with which a high-definition standard may be converted to existing standards. Comparing Fig. 18e with Fig. 16c, it can be seen that the alias components are less significant in relation to the true components when a high frequency is used for the original scanning process. This suggests that if a frequency like 100 Hz were chosen for the field frequency of the HDTV standard, the temporal alias components produced by converting down to a 60-Hz standard would be significantly less than those which would be produced by converting from 60 to 50 Hz.

Thus, the desirability of minimizing residual impairments in programs intended for worldwide compatible distribution might suggest the use of a high field rate for a new universal HDTV studio standard, provided that the additional capital and running costs associated with such a field frequency are not excessive.

Conclusions

When considering a new standard, broadcasters must keep faith with the standards in existing domestic equipment. To that end it is important to maximize the fidelity obtainable with current television standards.

With the cost of sophisticated instrumentation steadily falling, it would be surprising if pre- and post-transmission processing of signals broadcast on existing standards could not be developed to enhance the picture fidelity available to the domestic viewer. However, research described in this paper suggests that, by doubling the field and line frequencies in a 625-line, 50-field receiver, a worthwhile quality improvement could be obtained even with existing transmissions, particularly when combined with adaptive de-coding methods.

Coding methods are being formulated for direct broadcasting from satellites (DBS) using frequency division or time division to separate luminance and chrominance in order to do justice to the inherent picture definition of the new digital YUV studio standard. One example, for extending PAL based on a split luminance method, has been outlined in the paper. It is shown how extended definition and elimination of cross-effects can be achieved while maintaining compatibility with conventional system I PAL decoders.

Further enhancements are possible by pre-filtering HDTV sources for DBS transmission within the constraints of the WARC 77 frequency allocations and by subsequent post-processing. This should allow viewing of DBS programs with a quality moving toward that obtained from material which is not subject to broadcast bandwidth limitations, but is, of course, still limited by the capabilities of the display device.

The use of HDTV with scanning rasters of more than 1000 lines will now increase slowly but surely. Although it may be a considerable while before high-definition signals can be distributed with a penetration approaching that of conventional TV, their use as a source for conversion to existing standards might become quite extensive in the interim period. The question of compatibility with existing 625/50 and 525/60 standards is, therefore, a matter of great importance if the opportunity of a worldwide HDTV studio standard is to be grasped. The choice of line standard is not critical in this context, although low integer ratios between the number of lines in the standards could have instrumental advantages. The problem requiring the most careful practical study is the relationship between the field frequency to be chosen for the HDTV standard, so that the probability of movement judder is minimized when converting to the existing 625/50 or 525/60 standards.

Overall, there are reasonable grounds for optimism regarding an acceptable degree of compatibility between HDTV standards and existing broadcasting standards.

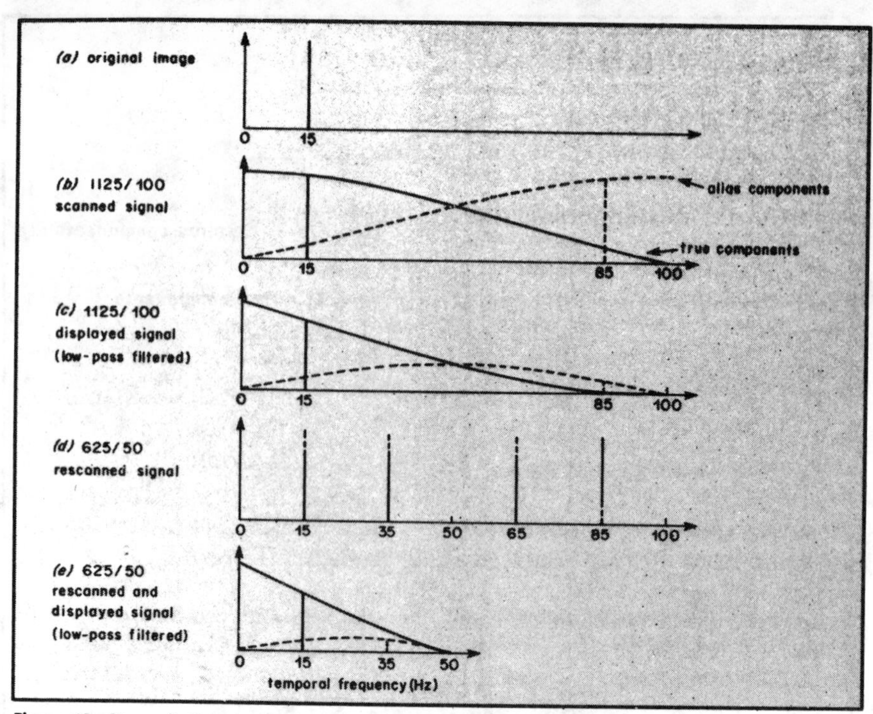

Figure 18. Stages in the standards conversion process from 1125/100 to 625/50.

Acknowledgments

The authors acknowledge the advice and assistance of many colleagues in the BBC Research Department, particularly Alan Roberts, Christopher Clarke, Ian Childs, John Drewery, Andrew Oliphant, and Richard Sanders. They would like to thank the BBC Director of Engineering for permission to publish the paper, and for helpful discussions.

References

1. UK Patent Specification GB 2050109A (May 1979).
2. B. Wendland, 1981, "High Definition Television Studies on a Compatible Basis with Present Standards," *Television Technology in the 80's*, pp 151–165, Society of Motion Picture and Television Engineers, New York, 1981.
3. K. Lucas and M. D. Windram, "Direct Television Broadcasts by Satellite: Desirability of a New Transmission Standard," IBA, Winchester, Experimental and Development Report 116/81.
4. U. E. Kraus, "Avoidance of Large-Area Flicker in Home TV Receivers," Proc. FKTG 1981 Convention, West Germany, 1981, pp 458–473.
5. K. Hayashi, "Research and Development of High Definition Television in Japan," *SMPTE J.*, March, 1981, pp 178–186.
6. J. O. Drewery, "The Filtering of Luminance and Chrominance Signals to Avoid Cross-Colour in a PAL Colour System," BBC Engineering, 104, September 1976, pp 8–39.
7. A. Oliphant, "Weston Clean PAL," BBC Research Department Report No. 1980/1.
8. C. K. P. Clarke, "Digital Decoding of PAL and NTSC Signals Using Field-Delay Comb Filters and Line-Locked Sampling," *Television Technology in the 80's*, Society of Motion Picture and Television Engineers, New York, 1981, pp 200–209.
9. M. G. Croll, "A Digital Storage System for an Electronic Rostrum Camera," International Broadcasting Convention, IEE London, 1980 Conference Publication No. 191, pp 252–255.
10. C. K. P. Clarke, "High Quality Decoding for PAL Inputs to Digital YUV Studios," International Broadcasting Convention, IEE, London, 1982, Conference Publication No. 220, pp 363–366.
11. R. N. Jackson and S. L. Tan, "System Concepts in High Fidelity Television," International Broadcasting Convention, IEE, London, 1982, Conference Publication No. 220, pp 135–139.
12. A. Oliphant, "An Extended PAL System for High-Quality Television," BBC Research Department Report No. 1981/11.
13. CCIR 1978–82, Document 10-11/1011.
14. V. Rajic, "Transmission of Two Sound Components in Frequency-Modulated Television," EBU Review (Technical), No. 166, Dec. 1977, p 300.
15. C. P. Sandbank, "The Impact of Some New Technologies on Future Broadcast Services," International Broadcasting Convention, IEE, London, Conference Publication No. 220, pp 4–9.

A Compatible High Fidelity TV Standard for Satellite Broadcasting

Tom S. Robson
Independent Broadcasting Authority
United Kingdom

Introduction

Direct broadcasting of television to the home from the satellite will be a reality by the middle of this decade in Europe. In the United States of America, satellite plans are not yet fixed and a number of proposals have been made including the use of High Definition Television, with higher numbers of lines and requiring a much wider bandwidth than that for existing television systems. In the USA therefore, the options for Satellite Broadcasting are still wide open. But what is the position in Europe, and are ideas now developing in Europe applicable to the USA?

Plans for direct television broadcast in Europe and throughout regions 1 and 3 of the world were established at the 1977 World Broadcasting Satellite Administrative Radio Conference (WARC 1977). These plans were based on the frequency modulation of a 12 GHz carrier by a composite NTSC, PAL or SECAM video signal with associated FM sound sub-carrier. In region 1, including Europe there are no further frequency bands available to the broadcaster below 40 GHz, a frequency which is unlikely to be technically suitable for direct reception for many years to come. Since the plans were adopted in 1977, there has been a growing interest in higher quality television signals; for example the use of component signals for programme production.

For this reason, in the Independent Broadcasting Authority we have been considering how Europe can avoid losing the last opportunity for perhaps twenty years or more to have a high fidelity television standard for satellite broadcasting. In this paper, we demonstrate that the possibility exists for a new television standard which retains compatibility with the existing line and frame frequency standards and with the allocated channel frequencies, while offering the possibility of higher fidelity TV, particularly for the larger screen viewer.

Such a system, when developed could form the basis of a European standard for a Satellite Reception Unit that could be used to feed PAL or SECAM receivers. This would make the satellite transmission independent of the terrestrial television standard being used.

2. SATELLITE BROADCAST STANDARDS IN EUROPE - The Case for Change

In Europe, the satellite band at 12 GHz is split into 40 channels of 27 MHz bandwidth with a 19.18 MHz channel spacing. Most countries are allocated five channels, and the situation for the United Kingdom is illustrated in Figure 1.

The use of conventionally coded composite signals was a natural assumption when the WARC '77 plans were devised. At that time (and indeed today) the majority of programmes were manipulated and stored on magnetic tape in the composite signal form. Also there remains the firm intention in Europe that prospective viewers of the satellite service should be spared the cost and inconvenience of purchasing new television

Reprinted with permission from *Tomorrow's Television*, Scarsdale, NY, SMPTE, 1982, pp. 218-236.
Copyright © 1982 by the Society of Motion Picture and Television Engineers.

receivers beyond the essential dish aerial and satellite RF converter and decoder unit as shown in Figure 2 where it is called the 'Indoor Unit'. Therefore the satellite signals will be interfaced initially to existing domestic UHF receivers which, necessarily, already contain either a PAL or SECAM decoder prior to the display unit. However, it should not be assumed because of this that the satellite signals are 'compatible'. They are not because:

i) they are frequency modulated not amplitude modulated
ii) they are in a completely different frequency band (12 GHz) with different channel bandwidth and spacing.

Hence, each domestic installation requires a satellite indoor unit, the minimum complexity of which is illustrated in Figure 3. However, the existing standards in Europe imply that, for many satellite broadcast users, increased complexity of reception equipment will be unavoidable if there is a requirement to receive the programme from a satellite that is aimed at a country with a different terrestrial standard from their own.

Whereas VHF and UHF broadcasts involve only limited overspill, the satellite plans involve significant overspill as shown in Fig. 4. The satellite broadcasts for any one country will cover major areas of other countries. Thereby, multinational reception will become an important factor. Unfortunately, under the existing proposals, viewers will be unable to receive broadcasts from other countries because of the additional 'incompatibility' between the PAL and SECAM signals (and indeed between the different versions of PAL), without additional PAL/SECAM or SECAM/PAL transcoding. Not only does this add to the complexity and cost but it also is likely to result in a degradation of technical quality.

It is of interest to note that despite the current differences in video coding standards, attempts are being made within the European Broadcast Union (EBU) to standardise the audio transmission of the satellite channels using a digital system If these efforts prove successful, they will imply, even for reception of national transmissions, trans-coding within the satellite indoor unit. This will be necessary in order to match the standardised audio format to the various specifications of existing terrestrial transmissions. For example, the UK uses a sound carrier of 6 MHz FM, Germany uses 5.5 MHz FM, and France 6.5 MHZ AM.

Therefore, the question of compatibility is a complex one, and relates to the sophistication of the satellite indoor unit. Compatibility with existing transmissions in any one country leads to international incompatibility and to diversity in the design of satellite reception unit.

Complexity is important only insofar as it affects the critical parameters of cost and convenience; the economies of scale may be equally important. Nevertheless, it has been assumed hitherto that these parameters would be optimised by an appropriate choice of either PAL or SECAM modulation, thereby avoiding unnecessary transcoding for reception of the national service.

Developments since 1977 give cause to doubt the validity of arguments in favour of composite coded video satellite signals. Firstly, the rapid advances of digital technology applied in television studio equipment have led to an agreement throughout the world on a standard for digital video signals. This standard is based exclusively on separate-component YUV (luminance, and colour-difference) signals rather than on a composite-coded signal such as PAL or SECAM. It has become clear that, in the future, programmes will be stored on magnetic tape in a high-quality YUV format.

A second development which has recently taken place is that television manufacturers in Europe are begining to introduce a direct separate component interface to the television receiver display as a standard fitting. Such an interface at the separate component level is the only one which can become common throughout Europe. Unlike PAL or SECAM broadcasts, seperate component broadcasts by any European nation could be received throughout Europe by any viewer with a suitable standard decoder. Certain multilingual populations (who do not normally possess a dual-standard receiver) would thereby gain new television channels. It might be thought that a common standard of this type is well-suited to a broadcast technique which, by its nature, crosses the national boundaries of Europe.

It is worth noting at this stage that just as for the studio application, because of the similarity in line frequencies of systems M, B, G, I, L etc, a suitably designed component coded signal for satellite broadcast could lead to a common standard world-wide, with differences only in the synchronisation timing.

Can we even consider any departure from PAL or SECAM in Europe?. Fortunately the architects of the WARC 1977 plan had the foresight to include the provision:
'The WARC '77 plans .... do not preclude the use of other modulation signals having different characteristics .... provided that the use of such characteristics does not cause greater interference than that caused by the system considered in the plan'.

The foregoing arguments do not establish the need to consider a separate-component form of modulation for direct satellite broadcasts; they merely indicate the feasibility of so doing should that prove desirable. Any departure from composite encoding would have to be based on one or both of the following propositions:

i) That the current proposals for NTSC/PAL/SECAM using frequency modulation are, in some respects, inadequate.
ii) That a departure from NTSC/PAL/SECAM modulation would provide an improved public service with better quality pictures and the possibility of High Fidelity pictures without significantly increased costs.

These propositions are examined in detail in the following Sections.

3. DEFECTS OF THE COMPOSITE CODED FREQUENCY MODULATED SIGNAL

The composite signals NTSC, PAL and SECAM in use today were developed some twenty or thirty years ago, and were appropriate to the technology available at that time and had to be compatible with the existing monochrome services. They do, however, introduce well-known picture impairments such as cross-colour and cross-luminance, although modern technology can be employed to reduce these defects (or rather, to trade them for less objectionable ones). The systems took into account the problems of noise and distortion which occur in AM transmissions.

However, they are less well adapted to the problems of noise, distortion and threshold effect which will occur in FM transmissions.

a) Noise

For any broadcast system, the required transmission power is defined by the signal/noise ratio which can be achieved at the limits of the intended service area.

In amplitude modulated signals, the power spectrum of the noise is approximately constant throughout the frequency range of the received baseband video signals as shown in Fig. 5a. In the composite-coded signals, the amplitude of the colour subcarrier was determined with due regard to creating a reasonable balance between the S/N ratios in the decoded luminance and colour-difference signals, as well as balancing the defects of cross-colour and cross-luminance. In FM systems, the noise voltage spectrum is triangular (Fig. 5b) increasing linearly from zero (at DC) to the edge of the band. The colour subcarrier with its sidebands, resides at the top of the band, and is therefore subjected to high levels of noise. Consequently, there is an imbalance between the S/N ratios which may be achieved in the luminance and chrominance signals which amounts to 11 dB (weighted)[1]. To reduce the imbalance, pre-emphasis is used. The agreed pre-emphasis network (CCIR recommendation 405-1) improves the chrominance S/N ratio by 2½ dB, and the luminance S/N by ½ dB[2]. Consequently an imbalance of about 8 or 9 dB remains in favour of the luminance signal. Experiments show that (in terms of weighted S/N ratios) we can accept more chrominance noise than luminance, although the degree of tolerance is highly picture dependent, and amounts to about 4 dB. Colour noise (particularly below 800 kHz) would therefore be the major source of picture impairment.

The effect of the mismatch between the FM noise spectrum and the requirements of the PAL subcarrier signal is such that the received pictures would be more noisy than would be the case with alternative signal formats. The problem would be less severe if it were possible to compensate by increasing the power transmitted by the satellite. Unfortunately, such a solution would be extremely difficult to apply because of increased satellite weight, reduced reliability and dramatically increased cost.

The mismatch of approximately 5 dB might not seem very great, but it implies that the performance of the channel as a whole is limited by the chrominance signal alone. It follows that, if we could increase the chrominance signal to noise ratio by 5 dB, we would achieve a reduction in picture noise which would otherwise require an increase in the total satellite power of a factor of 3.

The noise imbalance in composite coded signals cannot be overcome due to problems of overdeviation and distortion if the colour amplitude is raised. With the necessary assumptions of the WARC plan and for composite coded signals many viewers would receive a picture quality little better than grade 3½ in the CCIR 5 grade scale. Removal of the noise mismatch could offer the possibility of significant improvement in grade.

It has been suggested that improvements of technology will provide some improvement to the viewer. It is possible however that improvements in receiver noise performance will be used to permit the use of smaller dishes perhaps 70 cm in diameter.

b) Distortion

Distortion in an FM system arises from:

i) bandwidth limitations in the RF and IF stages of the receiver

ii) non-linearity in the FM discriminator

Composite coded signals contain high amplitude, high frequency signals due to the subcarrier, and these result in very high deviation in the FM channel with the possibility of distortion. The subcarrier also makes the signal more sensitive to distortion due to the possibility of intermodulation of the subcarrier with sound subcarriers which may be present. This leads to linearity requirements ~1% which imposes a severe constraint on receiver design.

c) Threshold Effect

A further phenomenon of importance in satellite reception is that of the FM threshold effect. This occurs at Carrier to Noise ratios below ~10 dB. Tests have shown this effect to be worsened by the presence of high levels of subcarrier, and that de-emphasis further increases the subjective impairments of threshold on the TV picture.

It is concluded that an improvement to the chrominance S/N ratio, a decrease in the susceptibility of the signal to non-linear distortions and a reduction in the effects of FM threshold would be advantageous.

4. AN IMPROVED SATELLITE SERVICE AT SIMILAR COST?

We now consider the options for an improved satellite service.

4.1) High Resolution Television Services

Since the WARC plan was made in 1977, there has been a growing interest in the possibility of high-resolution television services, capable of display on larger screens. Teams of engineers in a number of laboratories around the world (particularly in the USA and Japan) have been developing experimental equipments for the origination, distribution and display of such signals. The reason for the interest in this field arises from two main sources. Firstly, it is known that pictures viewed at distances of less than 3H (three times picture-height) provide an enhanced experience, by causing the eyes of the viewer to move to take in the whole scene. Until recently, it has been assumed that high-resolution pictures would demand 3-4 times the bandwidth of existing 525/625-line services; but broadcast systems are essentially limited in bandwidth due to the increasing demands on the radio frequency spectrum. This would not be of particular concern if it were forseeable that broadcasters could employ new areas of the radio spectrum to provide high-resolution transmissions. But, in Europe at least, all the useful frequencies up to 40 GHz have been allocated to non-broadcast services apart from the direct-broadcast allocation in the 12 GHz band. Transmissions in the 40 GHz region would not be very suited to television applications, due to their sensitivity to local weather conditions, particularly rain. For the 40 GHz band, assuming national coverage, and allowing a reasonable margin for fading during rainfall, a minimum satellite power of around 10 kW will be required.

The technology to achieve this will not be available for some years to come. Direct-broadcast transmissions in the 12 GHz band may be the last chance for broadcasters in Europe to provide a service with the potential for large-screen display.

The second source of interest in high-resolution systems is the advantage which may be gained from electronic production techniques in the film industry. The delays inevitable with chemical processing encourage the practice of multiple re-takes of each scene so as to be sure of having enough suitable material in editing. The facility of immediate inspection from electronic recorders improves efficiency and reduces costs. Electronic display is also seen as a possibility in small cinemas, but the main research effort is directed towards high quality transfer from magnetic tape to film as the final stage of the production process.

It is difficult to see these two applications (broadcast television and electronic film production) as requiring the same standard. For the film industry, a quality equal that of 35 mm film would be necessary; but, for broadcast television, the standard would be influenced by bandwidth limitations in transmission and questions of compatibility in domestic displays. It is likely that most European broadcasters would be attracted to an improved resolution service by satellite only if it was compatible with existing standards and providing the signals could be displayed on existing sets without a cost penalty to small-screen viewers.

The only reasonable prospect for an improved resolution television service in Europe is to make use of the 12 GHz WARC channels, and to do so using a signal which can also be displayed through an interface to existing receivers. The direct-broadcast channels are seen by many public service broadcasters as a way in which television can be brought to the small percentage of viewers unserved by existing terrestrial transmitters. It is essential that these viewers should not be burdened with additional cost as a result of using a signal which has the potential for large-screen viewing.

4.2 The 'All-Digital' Broadcast Signal

It has been suggested that an all-digital broadcast signal might provide the optimum solution for the WARC channel. However, estimates for the capacity of the WARC channel vary in the range 15-35 Mb/s, and (optimistically) an extended-definition picture (using frame-store bit-rate reduction methods) would require $\sim$100 Mb/s . Research work in progress will probably reduce this figure significantly, but the techniques involved at the moment offer no low-cost option for the small screen viewer. Therefore, the digital solution must be excluded due to the risk that the technology will not exist within the timescale envisaged for the new services.

4.3 Improvements to Existing 525/625 Line Television Broadcast Services

The constraints on bandwidth and bit rate, as mentioned above, are severe. A service which would use 3-4 times the bandwidth of conventional television reduces significantly the number of channels potentially available for satellite broadcasting if no channels have yet been planned, and is unlikely to be acceptable for the 12 GHz band in regions 1 and 3 which already have channels allocated.

In the Independent Broadcasting Authority, we have therefore been looking at methods of improving the definition of the existing composite coded television signals. The following impairments can be seen when conventionally displayed composite signals are viewed at a distance of 3H:

Cross-colour
Cross-luminance
Large-area flicker
Temporal aliasing
Vertical aliasing
Lack of vertical resolution
Lack of horizontal resolution

Cross-colour and cross-luminance may be excluded by using a separate component transmission system. Alternative methods such as 'clean' PAL are available; but, in general, these make almost impossible the elimination of some of the other impairments on the list.

Large-area flicker is a peripheral vision effect which is not merely a function of the signal, but also of the display device. It may be removed by displaying each frame twice within a single frame period or, more efficiently, by using a screen which continues to display information until refreshed.

Temporal aliasing is the effect which causes wagon wheels to appear to go backwards. Increasing the frame-rate will not reduce the effect, it will merely change the frequency at which it occurs. The right approach is to remove high temporal frequencies prior to display. There is no evidence to suppose that the temporal resolution afforded by the 50 or 60 Hz field-rate is insufficient.

Vertical aliasing (visibility of line-structure) is perhaps the worst defect in high-quality YUV signals displayed in the conventional interlace format. Again, it is the result of inadequate filtering rather than insufficient resolution.

This leaves the question of whether the vertical and horizontal resolution in 625-line YUV signals is sufficient to permit the use of large screen displays.

In the case of vertical resolution, we must carefully separate transmission format from the display format. For example, we find that a 625-line 2:1 interlace display is capable of only 156.25 cycles/picture height of alias free resolution, but that a 625 line sequential display (625 lines/field) is capable of 312.5 cycles/picture height even when the signals are derived from a conventional 2:1 interlace signal. Furthermore, it has been found[3] that a television picture having n scanning lines and a 2:1 line-interlace ratio has the same subjective picture quality as that of a line sequential picture having 0.6 to 0.7 n lines. This means that a 625 line sequential display has a subjective picture quality equivalent to a 951 line interlace display.

An improvement to horizontal resolution would follow from an increase in channel bandwidth, but this is not possible in Europe with the WARC channel due to consideration of interference and noise. However, alternative methods for increasing horizontal resolution are available.

Composite coded signals make inefficient use of the spectrum. If we consider 625 line PAL as an example, there is an obvious inefficiency as 24% of the signal is absorbed in sending blanking periods rather than useful picture information. The field-blanking period (~8%) is rather difficult to exploit, but the line-blanking periods (~16%) could be used to improve picture quality. A second cause of inefficiency arises due to the repetitive nature of the signal. Any line of a television picture is very similar to its adjacent neighbours, and a repetitive signal of this type produces a line-spectrum in the frequency domain. In fact, because there are small differences between adjacent television lines, the energy falls, not on discrete frequencies, but in a series of narrow packets occurring at regular intervals along the frequency axis.

The interval between packets is precisely equal to the repetition frequency of the signal, that is the television line frequency of 15.625 kHz. The spectral occupancy, therefore, varies in a predictable way, with much of the spectrum (between the packets) carrying little useful information.

The photographs of Fig. 6 show the poor spectral occupancy achieved by conventional signals. This property may be exploited to provide improved horizontal resolution without increasing the transmission bandwidth.

4.4 The Requirements of a New Satellite Broadcast Signal

An ideal signal for direct broadcast by satellite would have the following characteristics. It would be analogue, and compatible with display on a conventional 2:1 interlace 625 or 525 line receiver through a suitable interface. It would avoid the use of a high-frequency colour subcarrier which introduces problems of cross-colour, cross-luminance and susceptibility to FM noise and distortion. It would provide increased resolution by avoiding the spectral inefficiency of conventional composite signals. It would allow for the removal of vertical and temporal aliasing in the receiver by those who wish to pay for the necessary processing. A signal of this type would provide an extended-definition television service to the public without incurring additional costs for those who do not wish to use the extended-definition facility. Finally, such a signal should meet the bandwidth constraints of the WARC channel as already defined in Europe.

5 MULTIPLEXED ANALOGUE COMPONENT (MAC) SIGNALS

In this section much of the discussion is based on comparison with the 625 line PAL signal. Similar arguments exist for comparison with the SECAM and NTSC signals, but the figures are not included here.

From the foregoing discussion it will be clear that it should not be difficult to devise a signal which will improve on the quality of PAL in an FM channel. One method is described below, and variants of the scheme will be discussed in the remainder of this paper.

PAL signals normally achieve about 3.8 MHz of useful luminance bandwidth, higher frequencies being impaired by cross-luminance effects in typical decoders. 4.5 MHz of 'clean' luminance bandwidth will be assumed as a figure which would represent an improvement (or at least comparability) with the performance of PAL. Similarly, a figure of 1.3 MHz will be assumed for the colour-difference channels, with a vertical resolution equal to half that of the luminance channel.

If each line of the luminance signal is compressed in time from 52 µs to 40 µs, the transmission bandwidth increases proportionally. A luminance signal limited to a bandwidth of 4.5 MHz would then pass through a 5.85 MHz channel in this time-compressed form. At least 20 µs of the line would be available for the colour-difference signals, allowing a sequential colour-difference transmission of 2.2 MHz bandwidth (Fig 7). Filtering to 1.3 MHz bandwidth would be applied in the receiver to improve S/N ratio. For a signal with this format, calculations show that the weighted S/N ratio in the colour channels would be improved by 5 dB in comparison with pre-emphasised PAL, (the luminance S/N ratio being virtually unchanged). This technique of video component compression has been previously and independently proposed for the application of analogue video tape recording[4].

In addition to the noise improvements, cross-colour, cross-luminance and the effect of truncating the upper chrominance sideband would be eliminated. Moreover, the calculation does not take account of the fact that the new signal has no high-frequency subcarrier, and probably would be less sensitive to distortion. Therefore, the inefficiency caused by the mismatch between the PAL signal and the FM noise spectrum, together with the line-blanking period, provides sufficient margin to allow a separate-component transmission with improved S/N ratio. These factors alone portend a significant improvement (~5 dB) in the signal quality for the small screen viewer without incurring significant costs in the home satellite converter. The converter requires a CCD line-store for time decompression of the component signals as shown in Fig. 8. Preliminary experiments confirm that the noise improvement as measured subjectively at a viewing distance of 4H is approximately 4-5 dB. It is possible to use common compression ratios and detailed structure within the line in both 525 and 625 line countries opening the possibility for standards which are common worldwide other than for the synchronisation information.

However, the main cause of inefficiency in composite signals (namely the repetitious nature) would remain. A compatible extension of resolution for large-screen applications would be based on exploiting this latter redundancy through signal processing.

6 EXTENDED DEFINITION THROUGH SIGNAL PROCESSING

6.1 The General Theory

The analogue television signal represents a dynamic scene which has been sampled in two dimensions. It has been sampled vertically by the line structure, and temporally by the field structure. Any sampling process generates spurious spectral components which can give rise to aliasing - impairment of the baseband spectrum through intermodulation with the sampling frequency.

This process of aliasing is illustrated in Figure 9. For a sampling frequency f_s, a frequency component f in the baseband spectrum produces an intermodulation component at frequency (f_s-f). This gives rise to a repeat of the baseband spectrum centred on the sample frequency f_s, as shown. Any frequencies above fs/2 have 'aliased' components below fs/2. To avoid this process of aliasing, the baseband spectrum must be low pass filtered at or below fs/2 to remove frequencies exceeding fs/2 before sampling. This is illustrated in figure 10. Furthermore, to remove the repeat spectrum, a post filter with bandwidth less than fs/2 is required. Hence, to use the maximum capacity of a channel which involves a sampling process, pre and post filtering is required.

Although the television signal samples the scene both vertically and temporally, no such pre-filtering or post-filtering has taken place. Consequently, television pictures are impaired by high vertical frequencies, and by high temporal frequencies. In particular, vertical frequencies exceeding 156.25 c/ph (cycles per picture height), or temporal frequencies exceeding 12.5 Hz cause aliasing. To remove these impairments, it is necessary to introduce a vertical/temporal pre-filter and post-filter. It is only when we consider the necessary characteristics of these filters that we see the potential for increasing the vertical and temporal resolution.

The line-interlace structure, considered as a sample frequency, must be treated in two dimensions. The lines sample the scene vertically, but are also displaced along the temporal axis (i.e. between adjacent fields). Therefore, we must introduce the concept of a two-dimensional frequency plane as shown in Fig. 11. The interlaced line structure is represented by a point in this plane at 312.5 cycles/picture height, and 25 Hz temporal frequency. The baseband signal is centred on the point (0,0) and is unrestricted in extent, except by the sampling aperture (that is, the spot-size and persistence), there being no pre-filter either vertically or temporally. As in the one-dimensional case, the effect of the sampling process is to introduce a repeat baseband spectrum on the site of the sampling frequency. Clearly, vertical frequencies exceeding 156.25 c/ph or temporal frequencies exceeding 12.5 Hz, enter the baseband region unless pre-filtering is employed. Even if suitable pre-filtering is employed the repeat spectrum will be apparent in the display unless post-filtering is also applied - this is merely the visibility of line-structure and flicker familiar in all television pictures. The simplest arrangement for the removal of vertical and temporal aliasing would be to introduce separate filters in the vertical and temporal directions, as shown in Fig. 12.

This would avoid the aliasing which arises from overlap between the base-band and repeat spectra. However, half the frequency space would then be wasted. Therefore it is possible to double either the vertical or the temporal resolution without causing aliasing. Alternatively, a combined vertical/temporal filter can produce the result shown in Fig. 13, in which a peak resolution of 312.5 c/ph and 25 Hz is achieved. This characteristic approximates that of the human eye which can perceive high spatial resolution only in static scenes.

In order to produce this type of filter response, it is necessary to provide an initial scan which contains more lines per field than 312.5 (for example by doubling their number to 625 lines per field). This signal can then be pre-filtered (vertically or temporally) to remove those frequencies which cannot be carried by the 312.5 line interlace signal. Alternate lines are then discarded to produce the conventional interlace signal, which then carries enhanced vertical/temporal resolution (without aliasing) as indicated in fig.13.

This signal could be displayed on a conventional interlace monitor, which would then show improved aliasing characteristics. The line structure would remain visible however, due to the presence of the repeat spectrum at 312.5 c/ph/25 Hz. The full potential of the signal can be realised only if a post-filter is included before the display. The repeat spectrum is unavoidable in a 312.5 line interlace display, but the post-filter has the effect of interpolating the lines which were originally dropped, so as to recreate the 625 lines/field signal. Indeed, it may be shown that the functions of post-filtering and line interpolation are mathematically equivalent[5]. The procedure just described provides a transmission signal which may be viewed either on a conventional display, or with enhanced resolution by including the necessary filtering. It therefore represents a compatible extension to resolution, the costs of which are borne only by users who wish to pay for the improved picture quality. It cannot be simply applied to composite signals, because the post-filtering operation destroys the subcarrier phase relationships. It can, however, be applied separately to each component of a YUV transmission.

The compatible enhancement of vertical resolution arose through an optimisation of the filters associated with the vertical sampling (scanning) of the scene. It is interesting to consider whether a similar enhancement can be achieved in the horizontal direction by deliberately introducing a horizontal sampling process where none was necessary. This question has been considered in some detail[5] where it is found that an improved horizontal bandwidth can be achieved through a similar technique.

Suppose that a luminance signal is sampled at a frequency of, say, 9 MHz, using the 'field-quincunx' sample structure of Fig. 14. This sample structure must be analysed in three dimensions, the samples being displaced horizontally, vertically and temporally. Consequently, the repeat spectra generated by the sample structure occur in three-dimensional frequency space.

Suitable three-dimensional filtering can (in principle) then create an alias free (unity gain) region extending to 9 MHz horizontally 312.5 c/ph vertically and 25 Hz temporally, as shown in Fig. 15. As in the two-dimensional case, the source coding must be a 625 lines/field sequential scan signal, with alternate lines being dropped after the pre-filtering process, thereby creating a compatible field-interlace structure. The extended definition receiver would include a similar post-filter/interpolator prior to a 625 line sequential display.

6.2 Implementation of Extended Definition for MAC

PAL transmission typically achieves only 4 MHz of horizontal resolution, a little over 156.25 cycles per picture height vertically and 12.5 Hz temporally, and is impaired by aliasing, cross-colour and cross-luminance, and it is therefore quite unlikely that such a signal would ever be suitable for large-screen display.

For the MAC signal, in principle, impulses at a 9 MHz rate can pass through the proposed channel of 4.5 MHz bandwidth, without impairment, or through the 5.85 MHz channel when time compressed from 52 μs to 40 μs per line. It is less than obvious that the technique also improves the spectral efficiency of the signal, or indeed that the extended definition arises directly as a result of this improvement. It was indicated in Section 4 that conventional signals contain a series of gaps in the spectrum which carry only high-frequency diagonal information. When high-frequency diagonals are excluded, as in Fig. 15, these gaps become available to carry more useful information. The effect of the 3-D sampling process described is to deliberately alias useful high frequency information into these gaps in the spectrum. The original spectrum can then be truncated, in the safe knowledge that high frequencies are carried elsewhere in the signal. The advantage is illustrated in the case of two-dimensional processing by the spectrum shown in Fig. 6b. The post filter/interpolator restores the folded energy to its rightful position at high frequency, thereby recreating the original spectrum. It follows that, when such signals are viewed directly on a conventional line-interlace display (without the 3-D post-filter), an alias product is present on high-frequency gratings. However, experiments with unfiltered field-quincunx structures have been carried out, and it is thought that the impairment for the small-screen viewer should be significantly less disturbing than cross-colour in the PAL signal.

Similar techniques to those described above could be separately applied to the colour-difference channels. Investigation work is in progress to optimise the balance between the resolution achieved in the three component channels, taking account of the receiver complexity.

7. RECEIVER IMPLICATIONS

The effects of the MAC system proposals on receiver design are considered in the following two sections.

7.1 Conventional Displays

The simplest conception of the indoor unit of a Satellite Reception Unit (using FM composite modulation) is a s shown in figure 3. The desire for multiple sound channels in Europe has upset this simple scheme and there is a proposal for a single digital subcarrier for multiplexed sound and data. For this reason, the scheme of figure 3 must be abandoned, and a basic converter would appear as in Fig. 16. This modification is unlikely to increase cost significantly, because the necessary circuitry can be minimised through Large Scale Integration.

If a MAC (separate component) transmission system is considered, it is necessary to include additional circuits for baseband video processing, and, as an interface to most existing television sets to include circuits for encoding the signal into composite form as shown in Fig. 17. Both these operations, (including a video decompression store using CCDs) employ known technology and are amenable to large scale integration. Composite encoders already exist as single integrated components. Therefore, the additional costs involved should be relatively small. Against these cost increases, there are factors which will tend to decrease costs. Firstly the improved S/N ratio in the chrominance channel should, in many locations, allow the use of a smaller dish aerial. These locations will include all sites with sufficient protection ratio from adjacent satellites. Apart from being more acceptable from the environmental viewpoint, smaller dishes have the advantage of requiring a lesser pointing accuracy. This could be important for installations designed to capture the services of several satellites. Cost reductions will also occur on the periphery of the service area (or indeed outside it), where an increase in the dish size from one metre will be avoided. Secondly, the situation as shown in figure 17 would only be temporary. Many receivers in Europe are already being fitted with baseband video inputs, in which case the composite video encoder, FM sound modulator and UHF modulator can be dropped. The satellite converter.is then as shown in figure 18. As well as providing higher quality due to removal of cross-colour etc, which results from composite encoding, this receiver configuration would certainly be less expensive than the one currently proposed (Fig. 16).

Because of the closeness of line frequencies for 525 and 625 line systems and because the MAC signal can be used to give common structure within the line, this type of interface could be common world wide with differences only in the synchronisation signals to the display device.

7.2 High Fidelity Television Displays

An extended definition receiver would consist of two modules. A down converter (Fig 18) would provide YUV signals on the conventional interlace line-standard. These signals would then be processed by the high definition module which consists of a frame store filter/interpolator which regenerates the missing lines and provides signals for a 625 (or 525)-line sequential scan display (Fig. 19). As storage and display technology decrease in cost, an increasing number of users will be able to buy the compatible unit to provide extended definition on a larger display.

8 CONCLUSIONS

Satellite broadcasting has arrived on the scene at a time of rapid developments in television technology. Several factors strongly suggest that analogue component coded signals of the type described in this paper should be used for this new medium.

These factors include the following:

(1) Maximum benefit to the user in terms of both quality and cost will be achieved through a common world-wide standard for satellite broadcasts. The use of composite coded signals would represent a failure to achieve this objective. In view of the imminence of an international studio standard based on separate component signals, and on the development of separate-component interfaces to domestic receivers, this form of modulation could be adopted throughout Europe and hopefully the world.

(2) Composite coded signals are poorly matched to the characteristics of the FM broadcast channel, and result in levels of chrominance noise which are unnecessarily high. Forms of separate component transmission (in particular, Multiplexed Analogue Components) remove this problem,

eliminate cross-colour, cross-luminance and U/V crosstalk, and open the door to an extended definition service in the future. These advantages could be achieved without a significant increase in cost to the small-screen viewer; indeed, the receiver cost would actually decrease in the long term.

(3) The use of a MAC type system would enable broadcasters to provide a high fidelity television system without wasting the very scarce resource of spectrum which must result with systems using higher line standards. It would also provide a broadcast system compatible with existing displays, permitting the home viewer to make the choice of whether to watch programmes on a conventional display, or pay the extra cost and watch high fidelity pictures.

REFERENCES

Ref. 1. Macdiarmid and Allnatt
'Performance Requirements for the Transmission of PAL Coded Signal'
Proc. IEE, 125, 1978

Ref. 2 R. A. Harris
'OTS Repeater Breadboard, Programme of FM TV Tests, Preliminary Test Results'
ESRO Report, EBU Doc. Com. T.1974(N) 101

Ref. 3. Fujio, T, Ishida, J., Komoto, T., Nishizawa, T., NHK Laboratory Note no 239, 1979

Ref. 4. Brand, G., Schonfelder, H. and Wendler, K.P., 8 Jahrestagung der FKTG. P 209, 1980

Ref. 5. G J Tonge, 'The Sampling of Television Images'
IBA E & D Report 112/81.

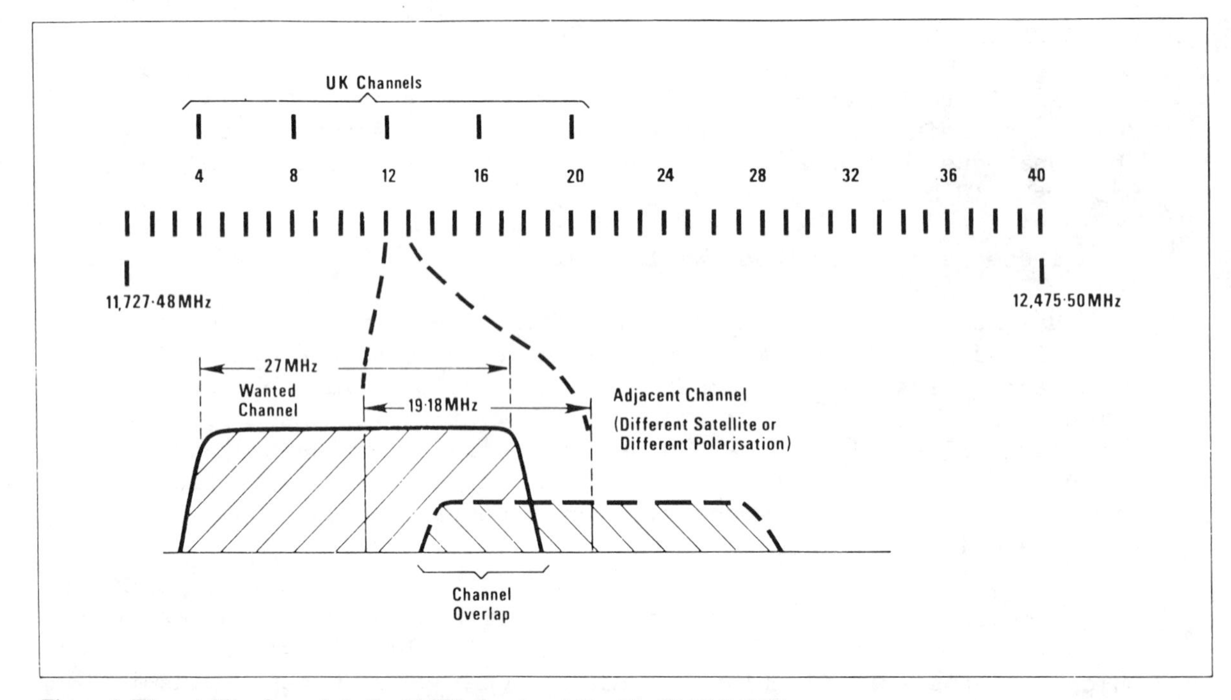

Figure 1. The satellite channels in the 12 GHz band as defined by WARC 1977.

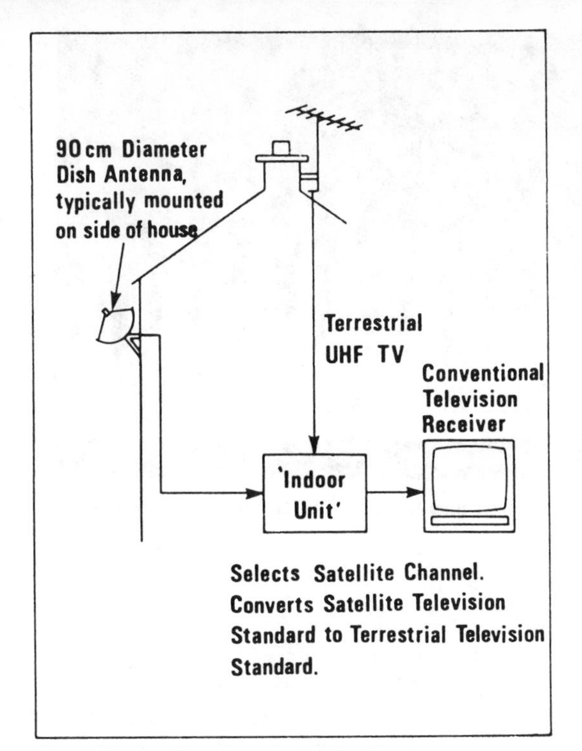

Figure 2. Satellite reception in the home.

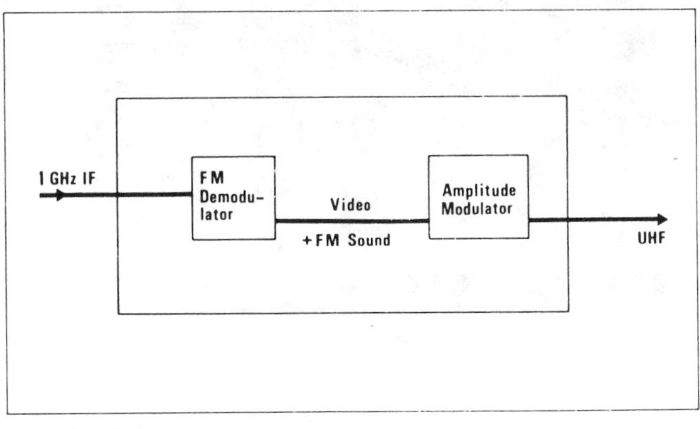

Figure 3. Satellite indoor unit.

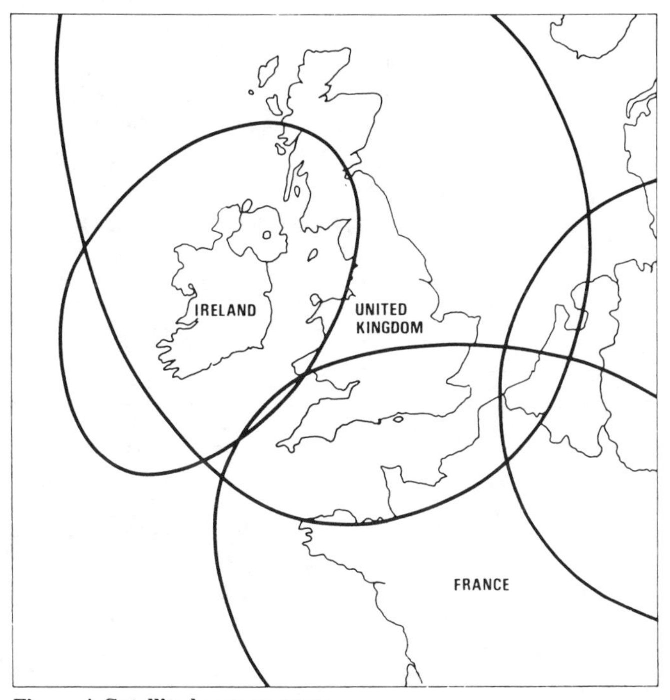

Figure 4. Satellite beam coverage.

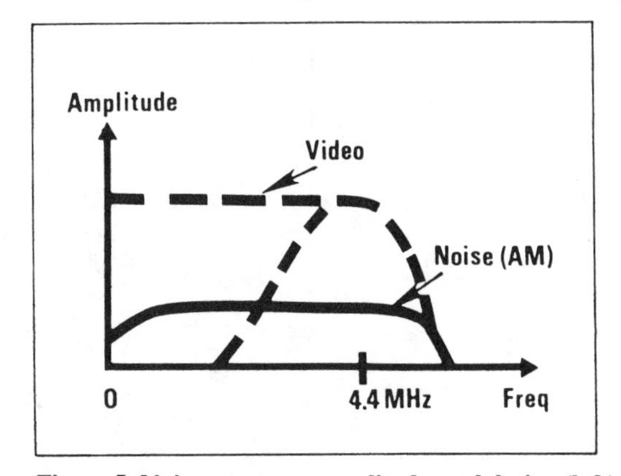

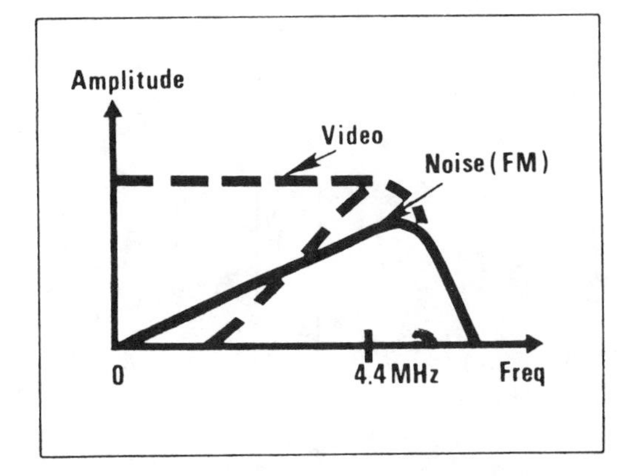

Figure 5. Noise spectrum–amplitude modulation (left), and noise spectrum–frequency modulation (right).

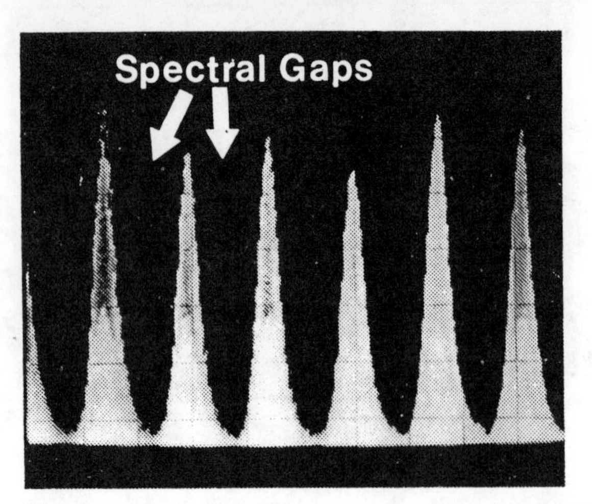

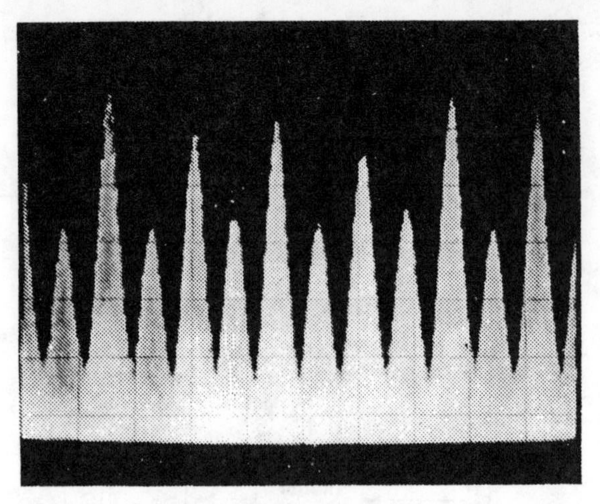

Figure 6. Poor spectral efficiency of conventional signals (left), and improved spectral efficiency through two-dimensional processing (right).

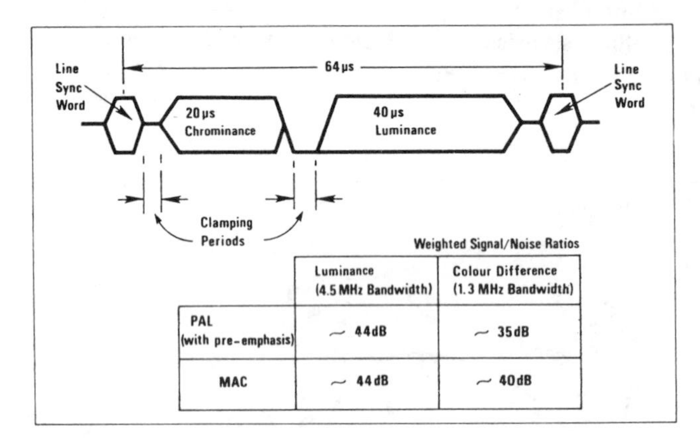

Figure 7. A multiplexed analogue component (MAC) signal.

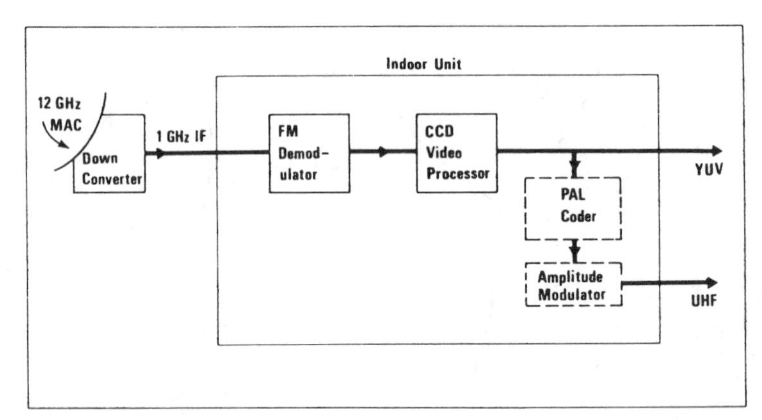

Figure 8. A MAC satellite converter.

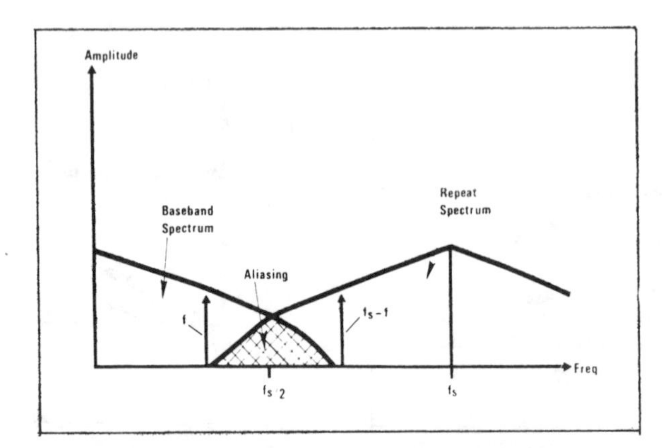

Figure 9. Repeat spectra due to sampling frequency.

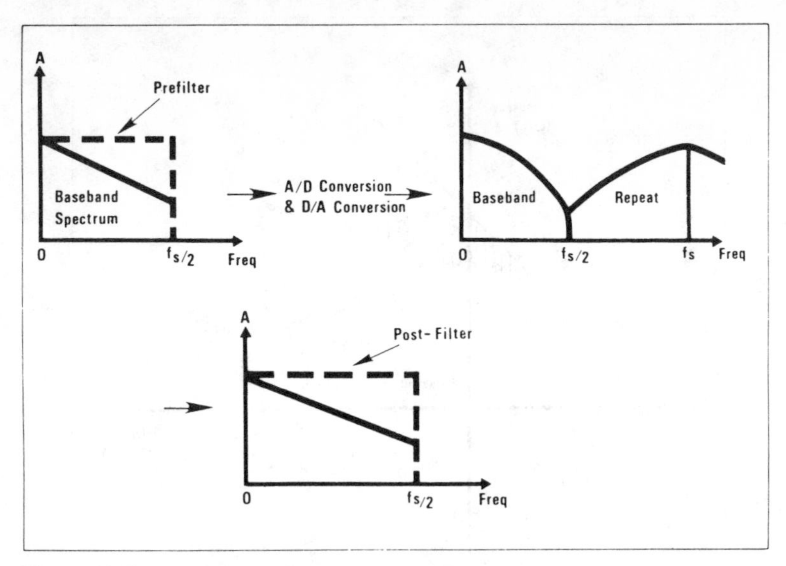

Figure 10. Pre-sampling and post-sampling filters.

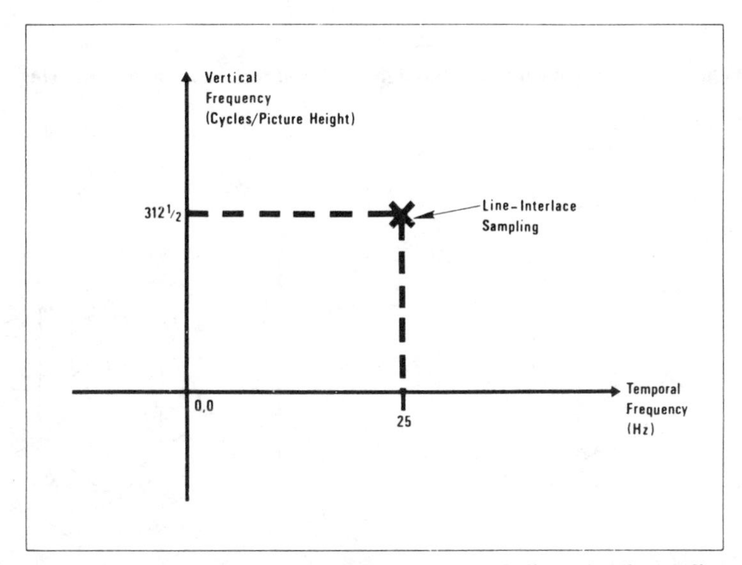

Figure 11. Line-interlace scan, considered as a 2-dimensional sampling frequency.

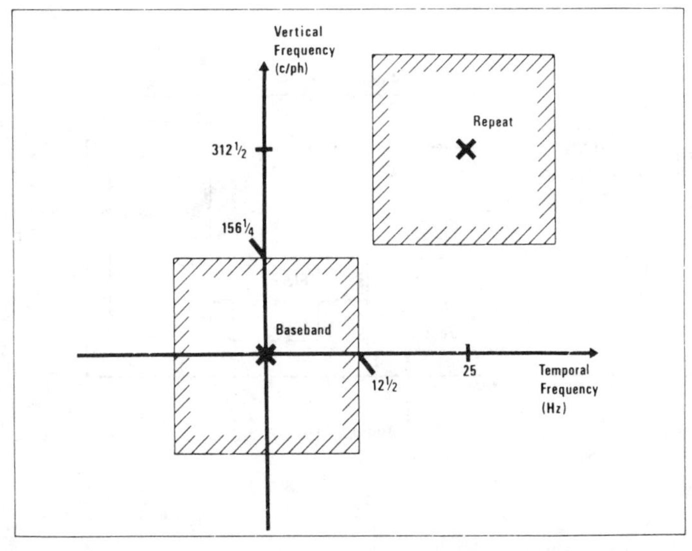

Figure 12. Alias free zone with separate vertical and temporal filters.

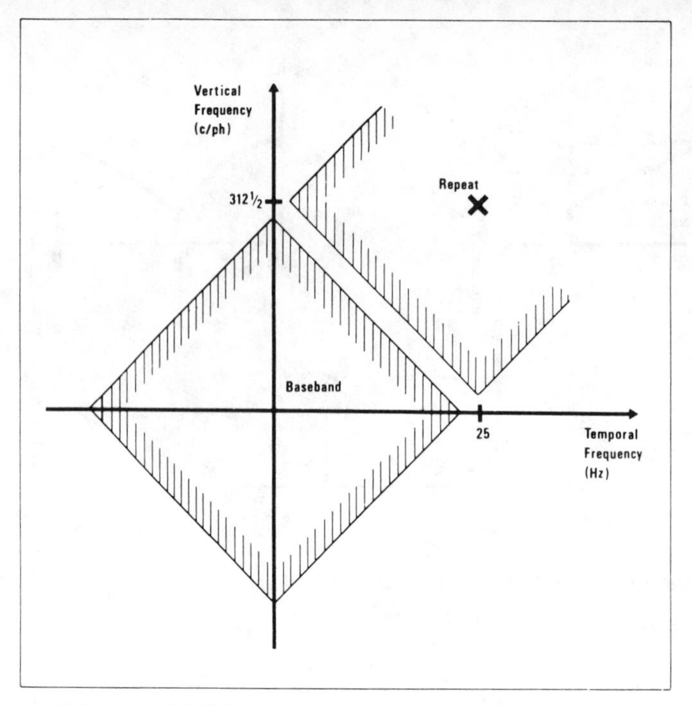

Figure 13. Improved definition through combined vertical/temporal filtering.

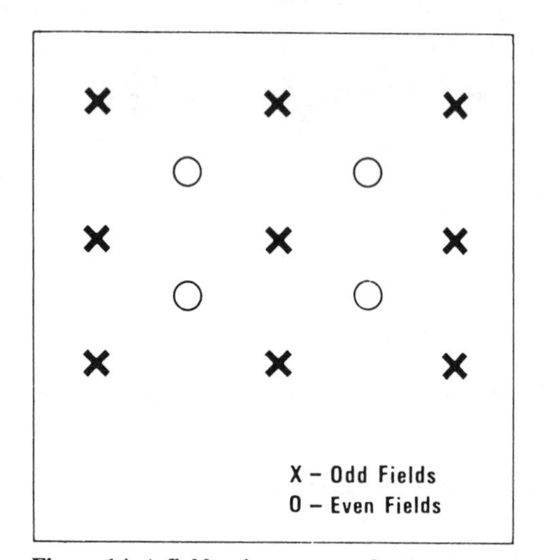

Figure 14. A field-quincunx sample structure.

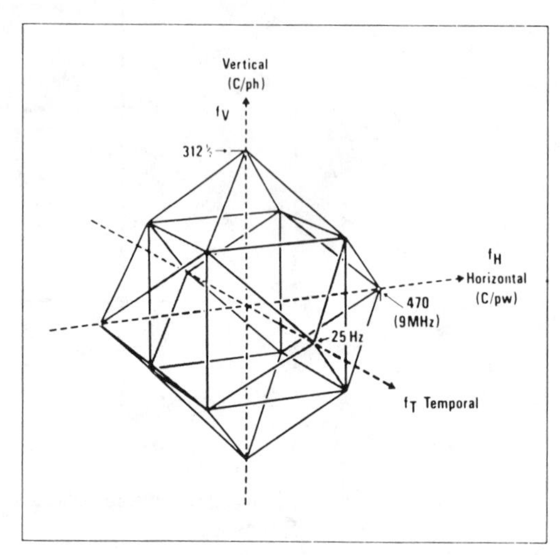

Figure 15. Spacio-temporal resolution with three-dimensional signal processing.

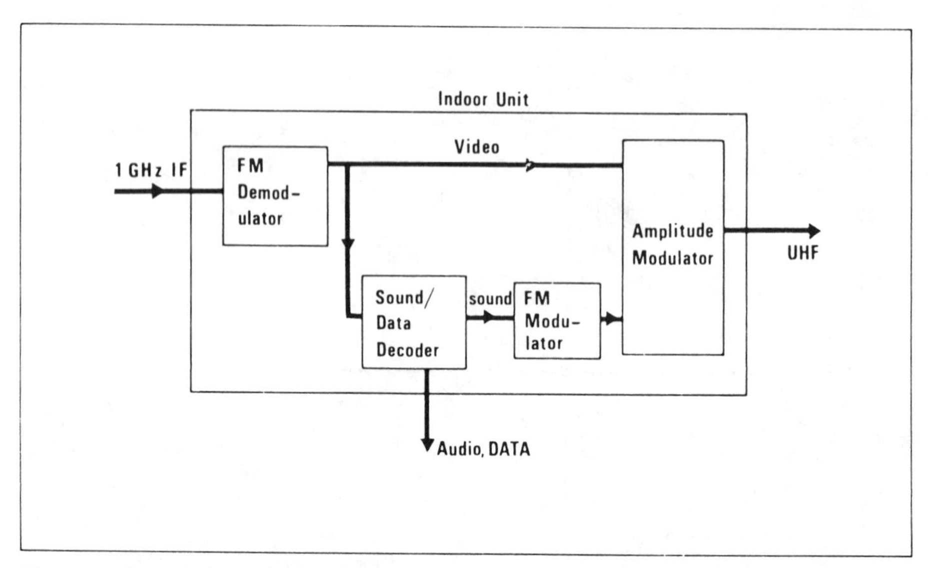

Figure 16. A single composite standard satellite converter with digital sound decoder.

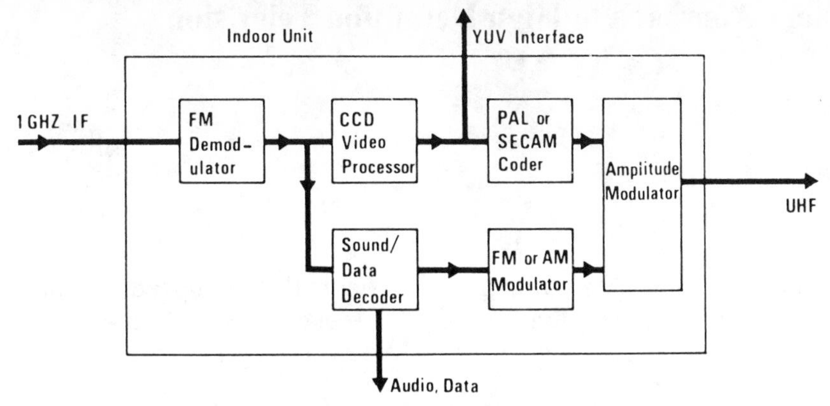

Figure 17. Multi-standard MAC satellite converter with digital sound decoder.

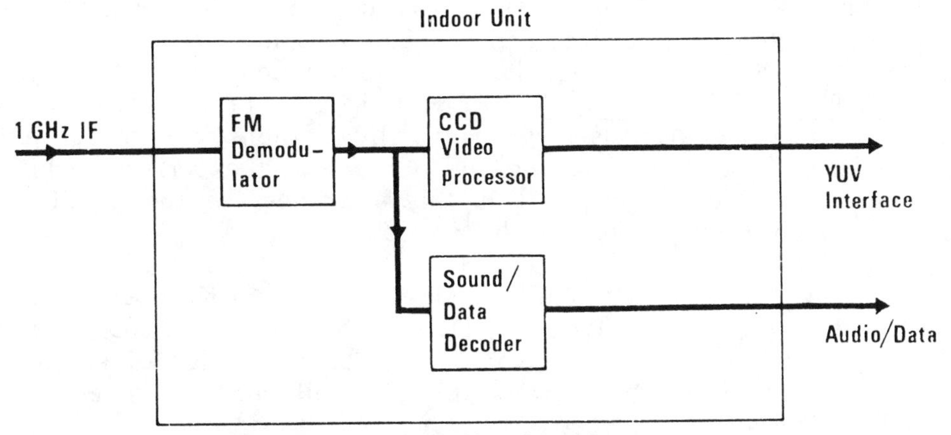

Figure 18. Universal MAC converter with YUV interface only.

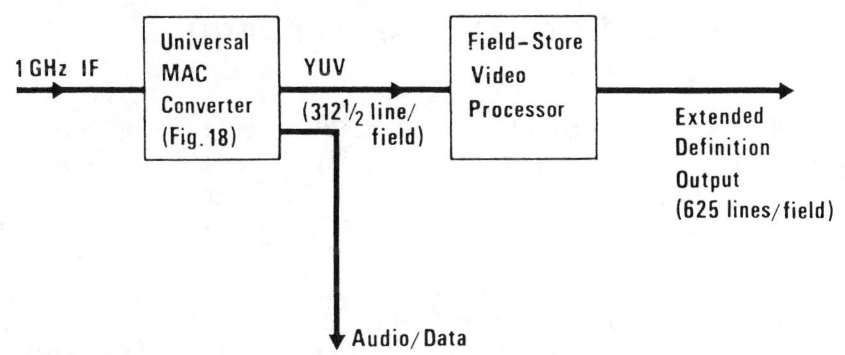

Figure 19. An extended definition MAC converter.

An Evolutionary Approach to High Definition Television

Charles W. Rhodes
Tektronix Inc.
Beaverton, Oregon

The number of lines per picture is generally accepted as an indicator of the performance of a television system. Fiqure 1 shows the number of lines per picture, with the approximate date at which each broadcast service was initiated.

Extrapolating the data in figure 1, it is not surprising that an 1125 line high definition televion system (HDTV) was demonstrated in the late 1970's. If the data in figure 1 were indicative of a continuing trend, such a service might have commenced by around 1980. Demonstrations of HDTV by NHK at the 1981 SMPTE WINTER TV CONFERENCE in San Francisco have increased broadcaster awareness that it is now technically possible to plan for a quantum leap in television picture quality, and the new viewing experience which HDTV, viewed on a large screen evokes. It is this awareness of the possibility of an HDTV system which is the reason for this session.

The brilliant work by Dr. Fujio and others at NHK, has given us much insight into the requirements for an HDTV system in terms of psychophysical research and beyond that, reduction to practice of the design of studio cameras, monitors, and other experimental work already well known through publications in the SMPTE JOURNAL.

ASPECT RATIO and BANDWIDTH

It has been demonstrated quite convincingly that roughly doubling the number of lines per picture affords the possibility of resolution comparable to 35 mm film images on large screen TV displays. This carries with it a heavy spectrum cost. This is due principally to the bandwidth of video being proportional to the square of the number of lines, and also due to a desire to increase the aspect ratio. The number of pixels resolved in the vertical direction is, of course, related to the number of active lines, while the number of pixels in the horizontal direction is a direct function of the aspect ratio. Thus, wide screen formats require more bandwidth.

The question of spectrum utilization cannot be ignored in any proposal for a broadcasting system today any more than when our present broadcast standards were developed. Thus, we must look into every aspect of a new proposal for an HDTV system to assess how efficiently it makes use of the frequency spectrum.

INTERLACED SCANNING?

All present broadcast systems make use of the interlace concept to conserve spectrum, while providing a reasonably flicker-free picture, as viewed at the usual viewing distance. While there was no alternative available then, it appears that with digital frame stores now available, and becoming cheaper still, the HDTV camert need not be interlaced. It is only in the display that

Reprinted with permission from *Tomorrow's Television*, Scarsdale, NY, SMPTE, 1982, pp. 186–197.
Copyright © 1982 by the Society of Motion Picture and Television Engineers.

interlaced scanning serves its purpose to reduce flicker. Alternate means for flicker suppression exist so it is possible that even at the display end, interlace may not always be used, at least were ultimate system capability is desired.

There are a number of advantages to not interlacing the camera's scanning pattern and there is always the possibility of providing interlace by digital means further downstream, within the studio.

One interesting possibility would be to scan the camera in 1/60 second, blank the beam for another 1/60 second, thus delivering into the interlacing unit 30 frames/second progressively scanned. By scanning the camera in 1/60 second, motion blur is reduced and so is the incorrect portrayal of horizontal motion.

Even more compelling reasons for the use of progressive scanning exist. Consider that in the 525 line system there are 483 active lines per frame, or 242 lines per field. In the vertical direction, the camera scanning process samples the image. The Nyquist criteria for sampling requires a minimum of two samples per cycle of the highest frequency being sampled. In the vertical direction then, the scanning process samples 242 times per field. Thus, spatial frequencies above 121 cycles per picture height are not permitted. That is, spatial frequencies above the Nyquist limit generate aliasing. In examining a resolution test pattern, we note the effect of aliasing in the fine vertical resolution wedge. While the detail perceived in this wedge may be mistaken for true resolution of fine detail, it is not. The aliasing products do not line up with the true detail, and they flicker. This is easily shown by photographing this wedge so that the film is exposed for somewhat less than the duration of one field. This is shown in figure 2. The conclusion is that with interlaced scanning, the 525 system cannot resolve more than 121 line pairs or cycles per picture height without aliasing.

The effect of interlacing two scanning fields is to partially cancel the aliasing products. This is shown in figure 3 where the test wedge was photographed with the film exposed over an interval of several frames. With film, the results are far better than with the human visual system due to the incomplete retention of the first field over the interval of a frame. This then represents a real limitation of interlaced scanning. When there is interfield motion, the cancelling effect illustrated in figure 3 does not occur. The aliasing components are at the wrong locations. This is a second failing of interlaced scanning.

The design of a camera and lens system involves a careful balance between providing too much resolution of the image upon the target, as read by the scanning beam, which results in aliasing, and too little resolution, producing a soft looking picture.

The lens/camera system is designed to suppress aliasing by pre-filtering the spatial frequencies before they are sampled by the beam. The spatial filter which removes alias components also reduces horizontal resolution because the lens/camera scanning beam comprise a two dimensional filter.

One may ask whether a one dimensional filter might be employed to suppress aliasing in the camera, without reducing horizontal resolution. Suppose the camera has sampled the image in twice as many places, say 483 lines per field. Now the alias-free resolution limit is 242 line pairs or cycles per picture height. To be sure, spatial frequencies above this value will cause aliasing, but these are removed by the finite MTF of the lens and camera scanning beam which is now chosen to filter out these much higher frequencies.

An example of one-dimensional filtering to suppress aliasing is to scan the camera with twice the number of lines per picture as are to be used to reproduce the picture. In the case of the 525/60 system, the camera horizontal scanning frequency would be 31.5 kHz which produces a 1050 line raster repeated 30 times per second. With the even number of lines per camera raster, the camera is being scanned progressively. The 1050 lines may be converted to a 525 line raster by digital signal processing. The 525 line picture which results is notably free of aliasing. Picture resolution can be significantly improved by improvements in the lens/camera tube MTF now that it need not be compromised. This is one step on the bridge to higher definition pictures.

DE-INTERLACING FOR DISPLAY

Just as it is possible to interlace a progressive scanned camera signal, it is also possible by digital signal processing, to de-interlace a video signal for display.

Consider, for example, that the 525 line 60 field interlaced signal is written into a frame store. During the first field, even numbered lines are written and field two is written in between these lines, on the odd numbered lines. Now the frame store contains a full frame of video in interlaced form. This can be read out line sequentially, This provides a non-interlaced (progressive scan) video signal at 30 frames/ second.

By doubling the read clock frequency, with respect to the write clock, the picture is read in half the time, 1/60 second. This permits re-reading the same frame twice. The picture repetition rate of 60 Hz. will not exhibit any inter-line flicker as does a 30 frame/second interlaced picture. This may be a substantial advantage with large screen pictures where inter-line flicker is quite objectionable. It may be of greater value when alpha-numerics are displayed. These, experience has shown, fare poorly on interlaced displays.

While this point may be trivial in entertainment, it is of great importance in educational applications. Admittedly, the use of a frame store at the display site is an expensive proposition today. Its cost is expected to decrease sharply and it would be reasonable to expect frame stores to be used in this fashion in a few years.

In the case of the 625/50 systems, it is interesting to speculate about the potiential for frame store de-interlacing of the picture and increasing the picture repetition rate to 75 Hz. This would provide a picture free of both inter-line flicker and large area flicker at brightness levels which are far beyond those presently attainable with known displays.

The point to be made is that improvements in the present broadcast system are possible with (massive) digital signal processing.

MOTION PORTRAYAL

Horizontal motion in interlaced television systems introduces a blurring effect as viewed at normal distances. This is perhaps an advantage as the eye cannot perceive detail in those parts of the picture which are in motion. This blur is due to the horizontal displacement of detail in the time interval of one field. Figure 4 shows a simple example; a black block (4a) which moves across the screen. If the block moves some number of pixels within the time duration of one frame, it will be depicted as being tilted along the direction of motion. Spatially adjacent lines, being one field apart, will stagger the edge in motion as in figure 4b. With progressive scanning, the tilt is doubled, however the staggered edge detail of the interlaced picture is absent (4c). At normal viewing distances, the staggered transition is not resolved and the blur is not objectionable, except in the case of a "frozen frame". This general advantage of interlaced scanning of horizontal motion, less tilt, is offset by the more serious problems with vertical motion.

Better portrayal of horizontal motion with progressive scanning is possible. Consider the camera to be scanned vertically in 1/60 second and that each raster has 1050 lines. The horizontal scanning frequency is 63 kHz. The camera beam is unblanked only for alternate frames. The resulting video frames have a duration of only 1/60 second and repeat at a 30 frame/second rate. Each frame is stored in a frame store. Two frame stores are required. When these are both read out in parallel, they are read out at one-half the rate at which they were written into. That is, the frames are now 1/30 second in duration and their rate is 30 Hz. These two parallel video signals are then averaged to form one frame. The result is depicted in figure 5. Motion blur exists where the two frames were not identical. This scheme would seem to provide improved portrayal of horizontal motion.

Actual application of the principle just described is a formidable task. Two extremely fast frame stores would be required in the studio. The video signal bandwidth in the path from camera to frame store is staggering. Nevertheless, this may be a practical proposal in the future. The greatest difficulty in its reduction to practice is that the picture reproducer (CRT) has a very non-linear transfer characteristic. Consider the situation shown in figure 5 again. The average of the 100 IRE white background and 0 IRE black block is 50 IRE. With an assumed display gamma of 2.2, this 50 IRE motion blur will be undesirably dark. The same would be true of a white object moving against a dark background.

This difficulty has a solution which results in significant increases in the complexity of the digital signal processing.

One solution is to average the two consecutive frames upstream of the gamma correction where the camera signals are still in linear form. The increased complexity of this is due to the need to digitize the linear video signals with perhaps 11 bits, thereby adding substantially to the size of the frame stores. The greater complexity lies in the A/D conversion.

With the lower frame rate of motion pictures, this kind of signal processing may be of interest in electronic cimematography.

COLOR CONSIDERATIONS

Since the development of the present color TV systems, their early major problems have become quite minor and problems of a more fundamental nature remain quite clearly as limitations to system performance. These are the result of an in-band chrominance subcarrier, regardless of how it is modulated.

An alternative to frequency domain multiplexing of chrominance into the luminance spectrum is time domain multiplex. The horizontal blanking interval represents 18% of the line period. Active line time to blanking time is 4.7:1. One chrominance component could be transmitted during one line blanking period and the other during the next blanking period. This would require time compression of the chrominance components before they are inserted into horizontal blanking. This is possible because the chrominance components can be filtered to occupy substantially less bandwidth than luminance. Time compression expands the signal spectrum in the same ratio as the time compression. Chrominance resolution should be at least one fourth that of the luminance signal. This is not possible with the amount of time compression readily available in the 525/60 standard. One possibility is to time compress the luminance signal to gain more chroma transmission time. For example; time compress luminance to occupy 48 uS which extends chroma transmission time + sync time to 15.6 uS. This time ratio is adequate for transmission of one chrominance signal of one fourth the luminance bandwidth, and a sync pulse. This is shown in figure 6.

Reducing the luminance transmission time reduces luminance bandwidth to 48/52.6 its present value. This gives adequate resolution, with freedom from cross-color and dot crawl artifacts. Color errors due to differential phase, and/or differential gain are avoided. Chrominance/luminance gain and delay distortions which are associated with composite coded signals do not occur.

These advantages of time domain multiplex are especially great where FM transmission is used, as in satellite transmission, and in analog VTR.

In higher line rate TV systems, the ratio of active line time to blanking time is reduced because of the problems in faster retrace of the CRT beam which can be translated into a power cost.

There may be no need to time compress the luminance signal in such systems. This would reduce complexity of the receiver. This is shownin figure 7.

Systems which transmit chrominance line sequentially have an inherent artifact which is worth mentioning. Interlaced systems will exhibit a flicker at half frame rate at vertical color transitions. Such an artifact may be substantially less significant in progressively scanned systems. Such systems should also be carefully examined for artifacts around moving objects.

In existing broadcast systems, 7.4% of the line time is allocated to horizontal sync pulses. In any new TV standard, this can be significantly reduced.

Phase-locked loops are universally used to provide noise immunity for the horizontal deflection of monitors and receivers. This makes it unnecessary to provide a sync pulse in every blanking pulse. Alternate blanking pulses would carry an adequate number of sync pulses. Because only the leading edge of the sync pulse contains the timing reference, it is thought that the sync pulse need not be as large a part of a line as in the present standards.

These economies of transmission time for sync can be used in a line sequential chrominance transmission system to identify the chrominance component to immediately follow. As a result, the now available space in the alternate blanking periods can carry more chrominance information. This would be used for the chrominance signal for which human vision has the greater acuity.

In the NHK proposal for an HDTV system, Dr. Fujio has identified two new chrominance axes, C_w, a wideband signal, and C_n. These are included in Table 1. C_w is reported to correspond closely to the maximum chromatic acuitance of human vision. In accord with the concept being put forth, C_w would be transmitted in those blanking periods not carrying a sync pulse.

The choice of chrominance transmission signals will be influenced by consideration of visual acuity and the visability of transmission noise. Table 1 shows the transmission noise for the recovered R-Y, B-Y and G-Y for different transmission signals. Research is needed to determine the relative sensitivity to noise for each color difference signal. Experience with NTSC suggests the differences are significant.

Important industrial uses for HDTV exist today. Electronic cinematography is perhaps the most exciting early possible application, both in capital investment and that it might become a starting point for HDTV broadcasting.

The 1050 line/30 frame (progressively scanned) camera and a wideband video recorder would be useful production tools for the production of programming to be aired as 525/NTSC. The primary advantage would be to increase the flexibility of post-production editing. Using a wider than normal lens, a greater field of view would be recorded. In editing, at least a 2:1 picture expansion with re-framing is possible without loss of definition. Reframing would be a transparent process. Improved chroma-keying would also result from the fast, clean keying edges available. Less extensive use of tight close-ups might differentiate the two products. Perhaps most important, these high definition programs would be archived in preparation for the eventual start of HDTV broadcasting.

Some of the image quality improvements of the HDTV production system would accrue directly as benefits to picture quality after the signal is transcoded to NTSC. These would include improved vertical resolution, free of vertical aliasing. Films produced by electronic cin ematography would also benefit.

Transcoding of HDTV to 525/60 NTSC will be an important process. It can be expected to be a routine operation in post-production facilities. This process is greatly simplified when there is a 2:1 ratio of number of lines per picture between the HDTV and the existing standard.

Reversed direction transcoding will be an important function of HDTV receivers. These should be able to display 525/60 NTSC in addition to HDTV pictures. Once the NTSC video is decoded into baseband components, simple line stores can convert the 525 line video to the normal scanning rates of the HDTV display. By twice-reading each line from the line stores, the video signal presented to the display is at its normal scanning rate. This would allow the display system to function at constant scanning rates and video signal bandwidth with considerable advantages over any display system which must operate on several different scanning standards.

Just how and when an HDTV broadcasting service to the public may begin in the U.S. cannot be foretold. It does appear that such a system, based on very extensive signal processing and a line rate twice that of 525/60 NTSC offers great and early promise.

The primary purpose of this paper is to suggest ways in which the camera and picture display should operate on different scanning rates and structures in a future HDTV system. This is in marked contrast to all broadcast systems. This possibility for improved picture quality depends on what we know is possible thru admittedly "massive" digital processing of the video signal. The burden of most of this processing can be assigned to the studio.

I should qualify these views as being my own. I am not speaking as a spokesperson for my employer. I do wish to acknowledge the contributions of collegues to numerous to mention here. I do wish to express my gratitude for the support of this work by both Mr. Tom Long and Charles Barrows of Tektronix Inc.

BIBLIOGRAPHY

Beiser, Leo. Laser Beam Techinques for Color Film Transfer. CBS Laboratories,Inc. Presented at Second Armed Forces Audio Visual Communications Conference, Washington D.C., November 5, 1969.

CCIR. Draft Revision of Report 801 - The Present State of High Definition Television. (question 27/11). Doc. 11/170-E, October 9. 1978-1982.

Dill, F. H. High Resolution NTSC Television System. IBM Technical Disclosure Bulletin, Volume 21, No. 5. October, 1978. Pages 2148 cf.

Fugio, Takashi, Dr. High Definition Television System-Signal Standards and Transmission. Presented by the author at the International Broadcasting Conference, Brighton, U.K., September, 1980. NHK Laboratories Note 239, August, 1979. Journal of SMPTE, Volume 89, August, 1980. Pages 579 cf.

----. NHK High Resolution Wide Screen Television System. Presented by the author at the 15th SMPTE TV Conference, February 1981, San Francisco, California.

----. High Definition Television of the Future, IEEE Transaction on Broadcasting, Volume BC-26, No. 4, December, 1980. Page 113 cf.

----. Present State of the Study of HD-TV Systems in Japan. Presented by the author at the IEEE Broadcasting Symposium, September, 1980.

----. Study of Component Coding Standards of Color Television Signal for Electro-Cinematography. NHK Laboratory Notes 259. January, 1981.

----. Study of High Definition TV Systems in the Future. IEEE Transactions on Broadcasting, December, 1978, Volume BC-24, No. 4, Pages 92 cf.

----. Universal Weighted Power Funtion of Television Noise and Its Application to High Definition TV System Design. IEEE Transaction on Broadcasting, June, 1980, Volume BC-26, No. 2, Pages 39 cf. Also NHK Laboratories Note 240, September, 1979.

Hayashi, Kozo. Research and Development of High Definition Television in Japan. Journal of SMPTE, Volume 90, No. 3, March, 1981. Pages 178 cf.

Isono, H. et. al. Subjective Evaluation of Apparent Reduction of Chromatic Blur Depending on Luminance Signals.

Lucas, K., Windram, M.D. Direct Television Broadcasting by Satellite - Desirability of a New Transmission Standard. IBA Experimental and Development Report 116/81. IBA, Crowley Court, Winchester, Hants, SO212QA, United Kingdom.

McIlwain, Knox. Requisite Color Bandwidth for Simultaneous Color Television Systems. Procedures of IRE, August, 1952.

Mitshashi, Tetsuo. Study of the Relationship Between Scanning Specifications and Picture Quality. NHK Laboratories Note 256. October, 1980.

Schade Sr., Otto H. Image Quality, A Comparison of Photographic and Television Systems. RCA Laboratories, Princeton, New Jersey 1975.

SMPTE Study Group on High Definition Television. Dr. Don Fink, Chairman. The Future of High-Definition Television Report. Journal of SMPTE. Volume 89, No. 2/3, February/March, 1980.

Sato, Shoichi and Kubo, Tokuji. Picture Quality of High Definition Television. 30 Inch Wide Screen Color CRT Display. NHK Laboratories, Note 241. October, 1979.

Schonfelder, Helmut, Timeplex Transmission System for Phonovision, Fernseh-und-Kino-Technik. Volume 33, No. 9, 1979. Pages 308 cf (in German).

Taneda, T., et. al. High Quality Laser Color Television Display. Journal of SMPTE Volume 82, No. 6, Pages 470 cf. June, 1973.

Tonge, G.J. Sampling of Television Images, Independant Broadcasting Authority (U.K.) Report 112/81.

Wendland, Broder. Alternative Lines of Development for Future TV Systems. Faculty for Information Transmission, University of Dortmund, W. Germany. Presented by the author at Fernseh-und-Kino-Technische Gesellschaft, Berlin on October 9, 1980.

----. Lines of Developement for Future Television Systems. Presented by the author at the 12th International Television Symposium at Montreux, June 3, 1981

----. HDTV Studies on Compatible Basis with Present Standards. Presented by the author at the SMPTE 15th Winter Conference, San Francisco, California, February 6, 1981.

Wengenroth, G. and Wendt, H. Die Optimale Bandbrette for Leuchlicke - und Farbdifferenzsignale. Post Office Research Department. Fernsch und Kino Technick 31 Jahrang, No. 11. Pages 393 cf.

Yamomoto, M., PhD. 1125 Scanning Line Laser Color TV Display. Central Research Lab, Hitachi Ltd. Hitachi Review, Volume 24, No. 2, 1975. Pages 89 cf.

PATENTS

Japanese Patent, Sho 44-13,538. Color Signal Transformation. Published June 17, 1969. Inventor: T. Kubota. Applicant: Sony Corp.

Japanese Patent Sho 51-48623. Transmission System of Color TV Signal. Published December 22, 1976. Inventor: S. Iwamura, et. al. Applicant: NHK.

Japanese Patent Sho 54-6843. Compatible TV Signal System. Published April 2, 1979. Inventor: K. Iinuma, et. al. Applicant: NEC. Related Sho 54-16696 of NEC.

West German Patent No 26 29 706. Procedure for the Transmission and/or the Recording of Colour TV Signals. Inventor: Robert Bosch GMBH.

Table 1 — Chrominance Transmission Noise

Transmission Signal Equation	Dynamic Range	Coding Gain	Decoder Matrix Gain (S/N dB)		
			(R-Y)	(B-Y)	(G-Y)
I = .60R -.28G -.32B	± .60	.83	1.15	1.33	0.34
			(+2.4 dB)	(+7.0 dB)	(-2.5 dB)
Q = .21R -.52G +.31B	± .52	.96	0.655	1.79	0.67
R-Y= .70R -.59G -.11B	± .70	.71	1.43	0	0.73
			(+3.1 dB)	(+5.1 dB)	(-1.9 dB)
B-Y=-.30R -.59G +.89B	± .89	.56	0	1.79	0.34
C_w = .63R -.47G -.16B	± .63	.79	1.57	0.48	0.63
			(+4.0 dB)	(+4.8 dB)	(-2.8 dB)
C_n =-.03R -.38G +.41B	± .41	1.22	0.18	1.66	0.35

Note 1: Coding Gain is the gain to normalize the Dynamic Range of the component in question to ± .50.

Note 2:

$$G-Y = -0.4627(C_n) - 0.4982(C_w)$$

$$B-Y = 2.0220(C_n) - 0.3799(C_w)$$

$$R-Y = 0.2153(C_n) + 1.2390(C_w)$$

Note 3: Decoder Matrix Gains, example of calculation:

$$R-Y = 0.2153\,C_n + 1.239C_w$$

$$= \frac{0.2153(1.22C_n)}{1.22} + \frac{1.239(0.79C_w)}{0.79}$$

$$= 0.1765(1.22C_n) + 1.568(0.79C_w)$$

0.1765 and 1.568 are the Decoder Matrix Gains (R-Y)

Note 4: S/N, example of calculation:

$$S/N = 20 \log_{10} [(0.1765)^2 + (1.568)^2]^{1/2} = 3.96 \text{ dB}$$

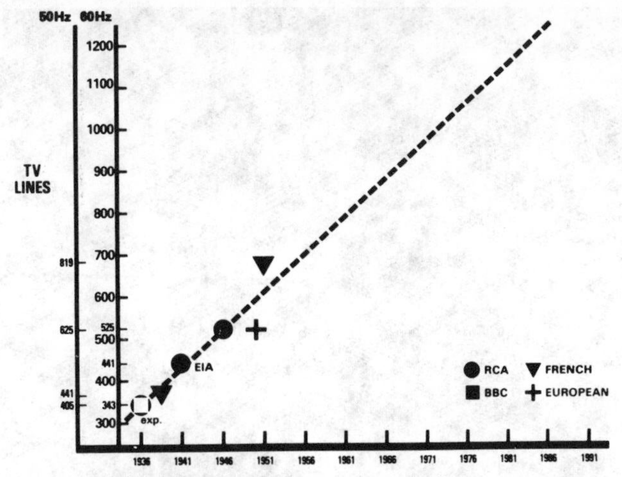

Figure 1. Number of lines per picture of TV systems with associated year of introduction.

Figure 2. Vertical aliasing due to sampling process of scanning. Shutter open less than 1 field.

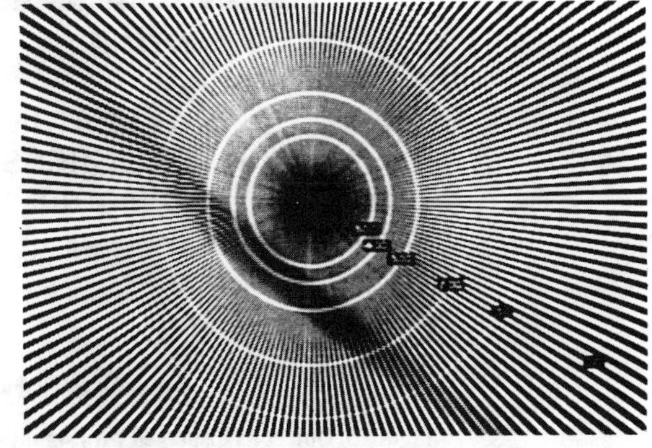

Figure 3. Reduction of vertical aliasing with integration of both fields. The shutter is open more than 1 frame.

(a)

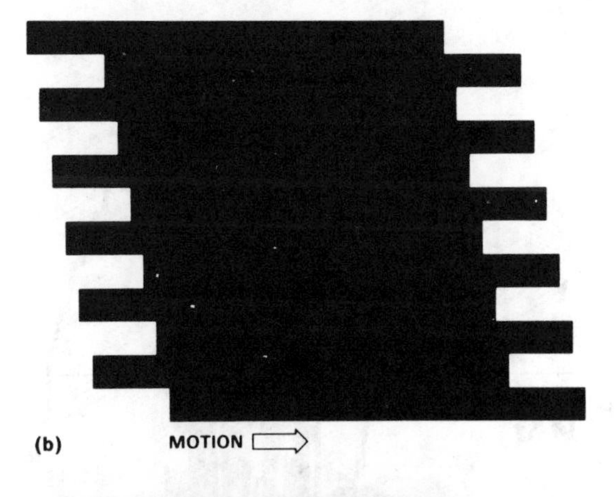

(b)

Figure 4. (a) Black block, stationary. (b) Object as displayed with interlaced scanning by camera. Note diagonal geometric distortion and image breakup field-to-field (1 field = $\frac{1}{60}$ sec, 1 frame = $\frac{1}{30}$ sec). (c) Object as displayed with progressive scanning by camera. Note increased diagonal geometric error (1 frame = $\frac{1}{30}$ sec).

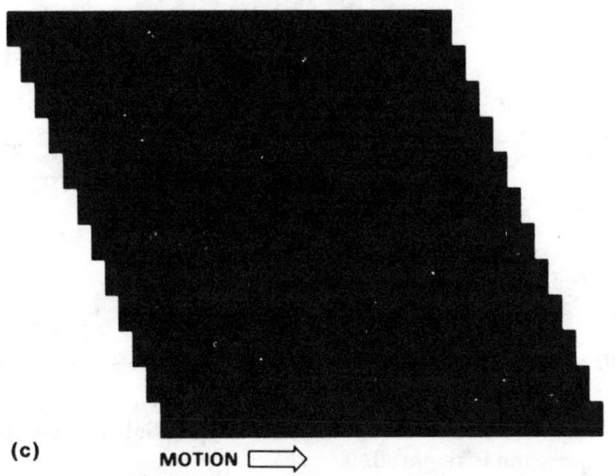

(c)

Figure 5. Object as displayed with signal averaged between two fast progressively scanned frames by camera.

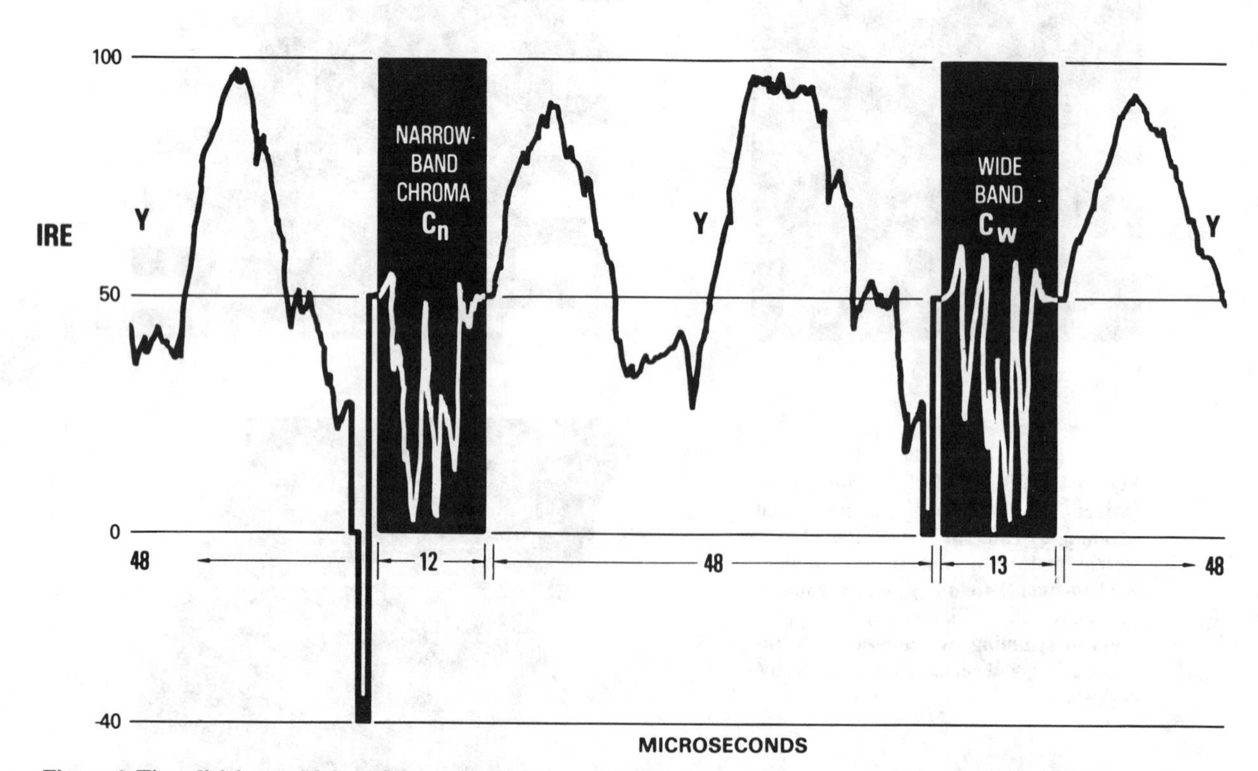

Figure 6. Time division multiplex of line sequential time compressed chrominance components (525 lines). Slight luminance time compression is required.

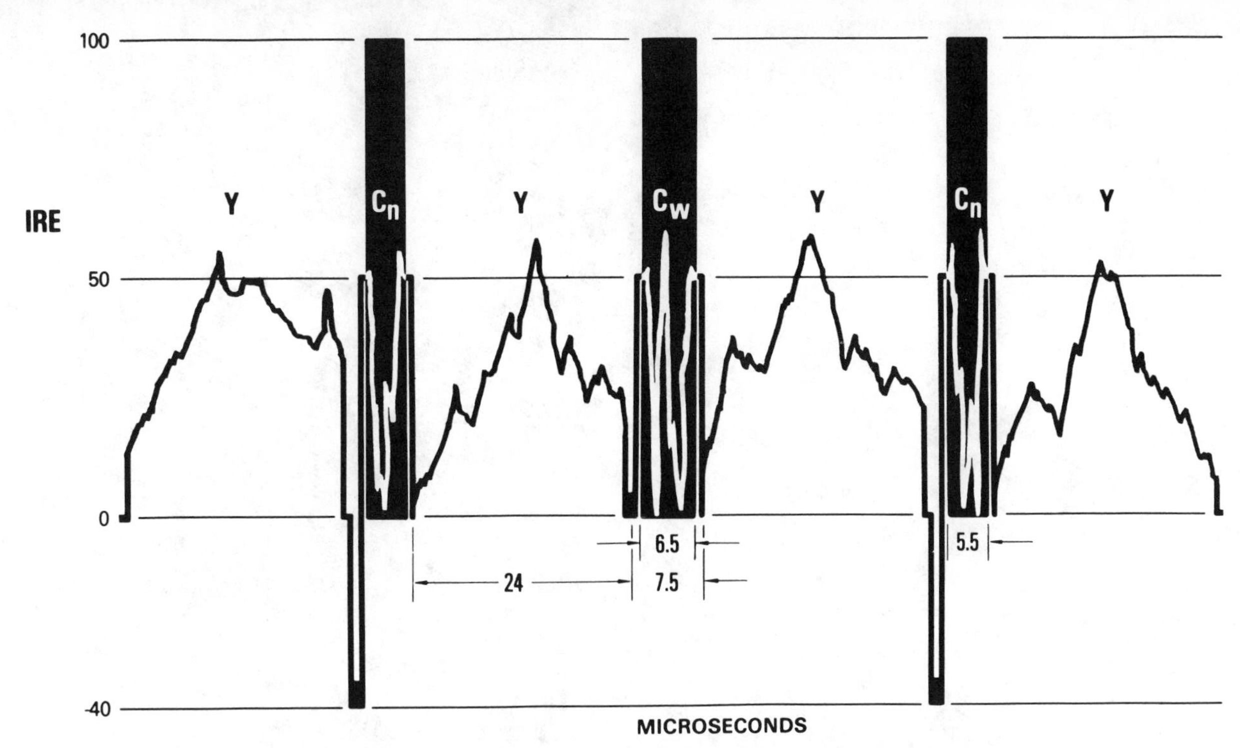

Figure 7. Time division multiplex of line sequential time compressed chrominance components (1050 lines).

Part IV
Digital Television

THE advent of VLSI technology has recently opened the door to the new field known as "digital television." Despite the recent boom of digital signal processing in various other fields, the high frequency of video signals is one of the main reasons for the late start of digital television. There are two aspects of digital television. On the one hand, one implies digitization and subsequent coding of a video signal for digital transmission purposes. On the other hand, we refer to digitization and further processing occurring in a television receiver whose input is an analog video signal. Both subfields are growing along with high-speed digital signal processing VLSI devices and digital memory technology. Major issues related to the ground work for both areas are addressed in the collection of papers that follow. An effort was made to equally cover both areas; however, the reader may find the papers relating to the first subject somewhat more analytical and tutorial in nature as opposed to those in the second area which are less involved. The reader will find, at the end of this part, a list of other papers on the subject, that we were unable to include because of space limitations. Next, we provide a short introduction for each of the two categories together with summaries of the related papers.

The advantages of digital transmission could be exploited by a controlled television network environment, where program material is determined by real-time customer request and video signal transport is switchable. Coding schemes that provide for efficient bandwidth utilization and quality reception have been developed. Conventional digital signal processors that are in use for voice-type applications are not fast enough to support algorithms for coding of video signals. Custom processors are being developed for specific coding arrangements while general-purpose digital signal processing devices operating at very high speeds are under study. High capacity digital memory devices that are being developed will also contribute to the realization of coding schemes that demand substantial memory. Our first paper, entitled "Digital Coding of Color Video Signals—A Review," constitutes a fine tutorial for those unfamiliar with this field. It sets the stage using analog television signal representation and has a good introduction to the basics of color, color television signals, and signal transformations; it then examines digitization and coding processes. The paper includes a good collection of video coding schemes under the two major categories of component and composite techniques and in relation to the three analog transmission schemes of NTSC, PAL, and SECAM. For further information on picture coding see the paper [1] which provides a wide coverage. Most of the coding schemes under consideration have been computer simulated and very few actually brought up in hardware. Video coding techniques related to existing products have been concentrated around schemes that provide very low bit rate digital video of relatively poor (conference) to fair (satellite transmission) picture quality. In the arena of consumer television broadcasting, a high picture quality at a reasonable bit rate is required.

Our second paper, entitled "Predictive Coding of Composite NTSC Color Television Signals," was selected as representative of the state of the art of high-quality digital video coding; high-quality performance and real-time implementability combined with a reasonable bit rate characterize candidates with a high potential for utilization with digital transmission. The limiting factors here are the speed of processing devices and the size and cost of memory devices. Picture quality is always a concern that varies with the application. Bibliography items [2]-[5] provide additional material related to differential and predictive coding techniques.

Digitization of the baseband video signal and subsequent processing, all taking place in a television receiver, has been of continuously growing interest. It is an expanding field with a lot of room for television receiver evolution and great challenges for those who will dare to revolutionize the consumer market by introducing exciting features. Some television manufacturers believe that replacement of most of the analog functions by digital signal processing will yield lower cost, higher set reliability, better picture quality, and facilitate maintenance and troubleshooting; it will also contribute to significantly reduce the bulk of the set itself and provide features unknown to today's analog sets. Other manufacturers' orientation is that digital television should exist as part of a home information system and as such should be designed accordingly; they see digital television more as a generalized/sophisticated computer terminal with probably digital video input and graphics capabilities, and as such they envision it appearing in the future when digital home information networks become a reality. Although there seems to be controversy in the industry on the consumer benefits of early versions of digital TV sets, certain TV manufacturers are proceeding with production as early as the end of 1984.

Our third paper, entitled "Feature IC's for Digivision TV Sets," describes a set of new feature IC's that can be part of a digital TV set. Beyond the basic VLSI devices that perform demodulation and signal decomposition, additional digital signal processing in customized VLSI provides for teletext processing, comb filtering improvements, and interlace free pictures. The reader is encouraged to review [6]-[9] for a better understanding of the notion, capabilities, advantages, and feature variety of digital TV. To digitally implement a variety of features and picture improvements, large amounts of digital memory are required.

Our fourth paper, entitled "Applications of Picture Memories in Television Receivers," considers the variety of features and picture improvements described in relation to charged-coupled memory devices (CCD's). High bit density and serial operation are key features of CCD's; reference [6] provides more information on CCD-MOS field memory device technology. Although dynamic RAM's currently are not cost effective, their random access operation and high data reliability are still attractive for certain classes of digital processing applications. Digital TV manufacturers are planning on future use of DRAM's without greatly affecting the overall TV set cost.

In summary, it is evident that the field of digital television is going through its beginnings with high expectations for the future. The notion of a TV set being the general means of interfacing with the network of tomorrow, where fiber technology is combined with digital video transmission and a variety of information banks, is a feasible one. Early digital TV sets with analog interfaces and capabilities covering the video reproduction is a first step in this direction. Additional papers which appeared in the proceedings of this year's International Conference on Consumer Electronics refer to the continuing work on features and indicate that the first marketable digital TV set will appear before the end of 1984. The reader is advised to search the August 1984 issue of the IEEE TRANSACTIONS ON CONSUMER ELECTRONICS for copies of some of these papers. Both fields of digital television have a lot to gain from the advent of VLSI technology in that it will vastly minimize the size and maximize the performance and functionality of both video coders and digital TV receivers.

DIMITRIOS P. PREZAS
Associate Editor

BIBLIOGRAPHY

[1] A. N. Netravali and J. O. Limb, "Picture coding: A review," *Proc. IEEE*, vol. 68, pp. 366-406, Mar. 1980.

[2] R. C. Brainard, A. N. Netravali, and D. E. Pearson, "Composite television coding: Subsampling and interpolation," *SMPTE J.*, pp. 717-724, Aug. 1982.

[3] R. C. Brainard and J. V. Scattaglia, "Programmable test-bed for composite television," *SMPTE J.*, pp. 906-911, Oct. 1982.

[4] T. R. Lei, N. Scheinberg, and D. L. Schilling, "Adaptive delta modulation systems for video encoding," *IEEE Trans. Commun.*, vol. COM-25, pp. 1302-1314, Nov. 1977.

[5] R. Wilson, H. E. Knutsson, and G. H. Granlund, "Anisotropic nonstationary image estimation and its applications: Part II—Predictive image coding," *IEEE Trans. Commun.*, vol. COM-31, pp. 398-406, Mar. 1983.

[6] T. Fischer, "What is the impact of digital TV?" *IEEE Trans. Consum. Electron.*, vol. CE-28, pp. 423-430, Aug. 1982.

[7] E. J. Lerner, "Digital TV: Makers bet on VLSI," *IEEE Spectrum*, pp. 39-43, Feb. 1983.

[8] M. Jacobsen, "Picture enhancement for PAL-coded TV signals by digital processing in TV receivers," *SMPTE J.*, pp. 164-169, Feb. 1983.

[9] S. Suzuki, Y. Kudo, M. Nakagawa, A. Yoshimoto, and T. Namioka, "High picture quality digital TV for NTSC and PAL systems," in *Proc. IEEE Conf. Consum. Electron.*, June 1984.

[10] M. J. M. Pelgrom, M. J. J. C. Annegarn, H. A. Harwig, H. F. Peuscher, L. C. M. G. Pfennings, J. G. Raven, A. Slob, J. W. Slotboom, and H. J. M. Veendrick, "A digital field memory for television receivers," *IEEE Trans. Consum. Electron.*, vol. CE-29, pp. 242-250, Aug. 1983.

Digital Coding of Color Video Signals—A Review

JOHN O. LIMB, SENIOR MEMBER, IEEE, CHARLES B. RUBINSTEIN, AND JOHN E. THOMPSON

(Invited Paper)

Abstract—This paper reviews the field of the efficient coding of color television signals. Because this paper is perhaps the first review on this topic, some background is given in the areas of colorimetry, visual perception of color and color television systems. We assume that the reader has some familiarity with luminance encoding techniques.

Coding techniques themselves are divided into two broad groups: component coding methods in which each component (usually three) is coded separately, and composite coding methods in which the composite television signal with its "color" modulated subcarrier is processed as a single entity. Both approaches are covered in detail.

The field is still growing, pushed primarily by the desire in the television area to find digital coding standards accepted by both broadcasters and carriers and suitable for use with NTSC, PAL and SECAM television systems. We discuss this aspect by comparing composite and component coding methods.

I. INTRODUCTION

THE digital coding of color video signals has received considerably less attention than the coding of monochrome video signals. However, given the current widespread proliferation of color television systems and the general preference for color pictures versus monochrome, it is obvious that the efficient coding of color picture signals is of prime importance. Broadcast color television systems make highly efficient use of the analog bandwidth to accommodate the increased information content of color signals. If the same relative efficiency is to be achieved in the encoding of color signals as has been achieved for monochrome, it is going to require a great amount of ingenuity.

The first attempt in digitally encoding a color signal most probably started in 1960 with the work of R. L. Carbrey [1] on applying PCM to a broadcast color television signal. A few papers were published on the subject during the following 11 years until 1971 when there was a marked increase which has persisted to the present and gives every indication of continuing. It therefore seems especially appropriate to review this field now while many new techniques are still being explored.

In laying the foundations of the present-day color television standards in the late 1940's, much study went into various background topics such as colorimetry and visual perception so as to match the resulting signal to the color fidelity requirements of the human observer. Further, additional studies have been made in the area of threshold color-difference perception and on the interaction between the "brightness" (luminance) and "color" (chrominance) components of the signal. To enable the reader to fully appreciate the various factors that bear on the encoding of color signals, we have gone into background material in some detail. Section II provides basic information in the area of colorimetry, laying a foundation for the colorimetric properties of color television. Section III describes the format of the three major color television systems in use in the world today, and some of the considerations that led to these standards. Section IV covers aspects of the work in color vision which bear on the problem of efficiently encoding color video signals. Final preparation for the coding sections is given in Section V where the statistical nature of color signals is described. Readers with familiarity in these background areas may wish to jump straight to Section VI although there is some work on color vision in Section IV that is perhaps not widely known to workers in the coding field.

We assume that the reader has some familiarity with basic waveform encoding techniques (see [2]). Coding of the luminance component per se will not be discussed in any detail. Of course, in coding the composite signal it is not feasible to divorce the coding of the luminance and chrominance components. Reviews of work on luminance encoding are available for readers who would like more background. Reference [3] covers the proceedings of a special conference on efficient picture coding, covering all aspects. References [4], [5] and [6] are special issues on signal processing which contain large sections describing video coding techniques.

From the first experiments in color encoding two somewhat separate approaches have been exploited. The first is to operate directly on the composite television signal, whereas the second divides the signal into three components, codes each component separately and then, after transmission, combines them again to form a composite signal. These two different approaches are examined in some detail in Sections VII and VI, respectively. Finally, in Section VIII we compare the implications of the composite and component encoding methods and conclude that, at least in the short- to medium-term future, both coding strategies will find important application.

We express a word of caution concerning the assessment of the performance of different encoding techniques. At the very minimum, such assessment requires the measurement of the bit-rate for a given picture quality. It is primarily the assessment of picture quality that is so variable. Picture quality depends on many factors, for example lighting conditions and monitor adjustments, the range of picture material that is presented, whether a single stored frame (or photograph) is being viewed, the amount and type of movement contained in the scene and the experience and expectations of the viewers.

Manuscript received December 7, 1976; revised May 10, 1977.

J. O. Limb and C. B. Rubinstein are with Bell Laboratories, Holmdel, NJ 07733.

J. E. Thompson is with the Post Office Research Center, Ipswich, England.

Reprinted from *IEEE Trans. Commun.*, vol. COM-25, pp. 1349–1384, Nov. 1977.

In the work on component coding described in Section VI, rates in the range of 1-2 bits/pel (picture element) have been found to give "good" quality pictures. But, in many cases, quality is based on single-frame simulations or on a small range of picture-material not representative of that handled by broadcast television. The situation will only be eased when all workers adhere to a common testing procedure (such as that recommended by the CCIR [7]) and use common picture material.

This difficulty in comparing picture quality, in turn, complicates the task of the authors in comparing different coding schemes. Even side-by-side comparison of picture quality would not provide definitive ratings since other factors such as error performance, cost, complexity and compatibility will all affect the final decision on the type of coder most suitable for a given application. Finally, the field of color coding is the subject of a great deal of current activity, and much of the research required to reach firm conclusions on the relative worth of various schemes has yet to be completed.

II. REPRESENTATION OF COLOR

A. Trichromacy

There is a long history associated with the science of color, and much of this history has been recently reported by MacAdam [8]. Those interested in the early work will also find an interesting collection of papers in a book by MacAdam [9] that spans a period of approximately 2000 years. The credit for the great advance in the science of color concerning the trichromacy of vision generally goes to Thomas Young for his advancement of the concept that the retina is composed of three sets of sensitive mechanisms—one for each of three principal colors [10]. Today it is generally accepted that there are three types of cones in the retina which mediate color vision at light levels greater than approximately 10 cd/m^2.

The natural counterpart to the trichromacy of vision is the trichromacy of color mixture. Maxwell studied color mixtures and demonstrated the first three-color projection in 1855, a description of which appears in the reprint of his 1857 paper [9]. Maxwell's work formed the basis for colorimetry—the technique of the measurement of color.

B. Colorimetry

Colorimetry is based on the premise that a relationship can be found between the physical stimuli and the visual sensation that arises from them. At the foundation of 3-color colorimetry lie a series of rules generally attributed to Grassman [11]. Two of the more important rules can be expressed as follows:

1. Any color stimulus can be matched (in appearance) by the additive mixture of three matching stimuli provided that no one of the three matching stimuli can be matched by the remaining two. This can be expressed as

$$(C) = \alpha(P_1) + \beta(P_2) + \gamma(P_3) \tag{1}$$

where (C) is a color that is matched by α units of color (P_1), β units of color (P_2), and γ units of color (P_3). The colors (P_1), (P_2) and (P_3) are conventionally called primaries.

2. The luminance[1] of a color mixture is equal to the sum of the luminances of the components in the mixture

$$L = L_1 + L_2 + L_3. \tag{2}$$

These rules have been extensively tested experimentally, and some discrepancies have been found (see Guth [12] for a review).

Wintringham [13] discusses the foundations of colorimetry in terms of a somewhat idealized set of experiments designed to determine what mixture of three primary colors would appear exactly like a spectral color. In these experiments an adjustable mixture of three well-chosen monochromatic lights act as the primaries. Subjects adjust the strengths of the three primaries to match each color in the series of spectral colors after first matching a white that has equal energy in all parts of the visible spectrum. The match to white represents one unit of each primary.

Note that we have not mentioned the absolute intensity or radiance of the spectral colors that are matched. This was intentional because the relative proportions of each of the three primaries is independent of the radiance over a wide range. Instead of referring directly to the number of units required to make the match, colorimetrists use normalized quantities called chromaticity coordinates expressed by the relations:

$$r = R/(R + G + B)$$

$$g = G/(R + G + B)$$

$$b = B/(R + G + B) \tag{3}$$

where we have changed the notation such that the units α, β and γ, which are called tristimulus values, have been replaced by the symbols R, G and B, respectively, to agree with the common usage of Red, Green and Blue primaries in these experiments. The results obtained by Guild [14] in his fundamental measurements of color mixture for 7 subjects are shown in Fig. 1.

The normalizing process we have just carried out to obtain the chromaticity coordinates has eliminated the radiance information, and we are left with only two pieces of information. The third dimension of color is obtained by a separate measurement of luminance. In a three-primary match to the reference white, the ratios of each of the three component luminances contributed by each primary to the total luminance are called the luminosity coefficients.

C. Color Transformations

A colorimetrist designates color by a graphical representation in a color space. This requires the data for a "Standard Observer" that would be representative of the mixture data of color normals. As noted in Fig. 1, Guild [14] had obtained

[1] Luminance is a quantity measured in a photometer. It is the photometric brightness of a uniform, small field.

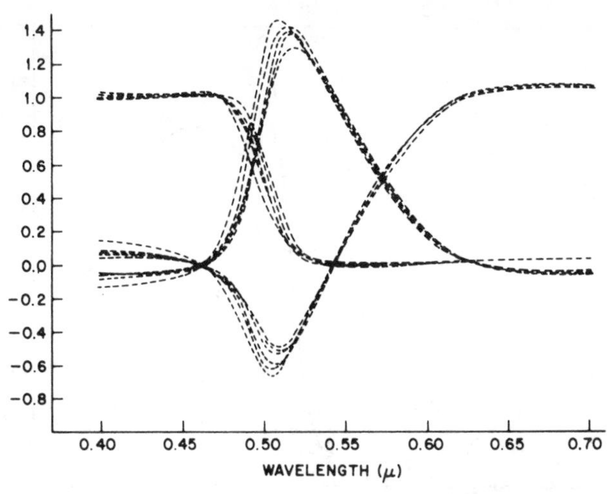

Fig. 1. The chromaticity coordinates versus wavelength of the spectral colors for seven observers using Guild's trichromatic colorimeter primaries, and the National Physical Laboratory (NPL) reference white (from Guild [14]).

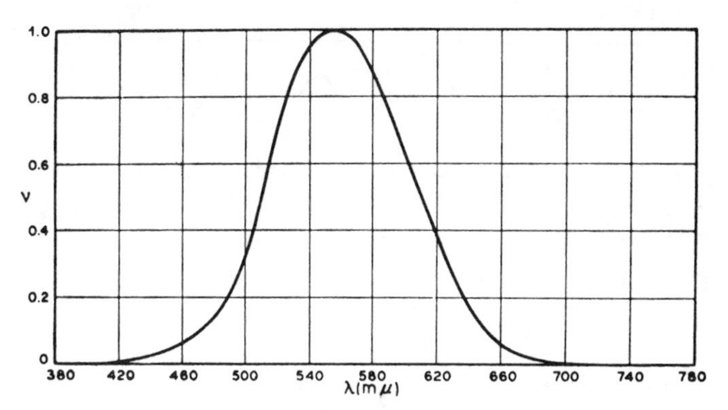

Fig. 2. Luminosity curve for the Standard Observer.[3] At 555 nm, one W is equivalent to 680 lumens (from Wintringham [13]).

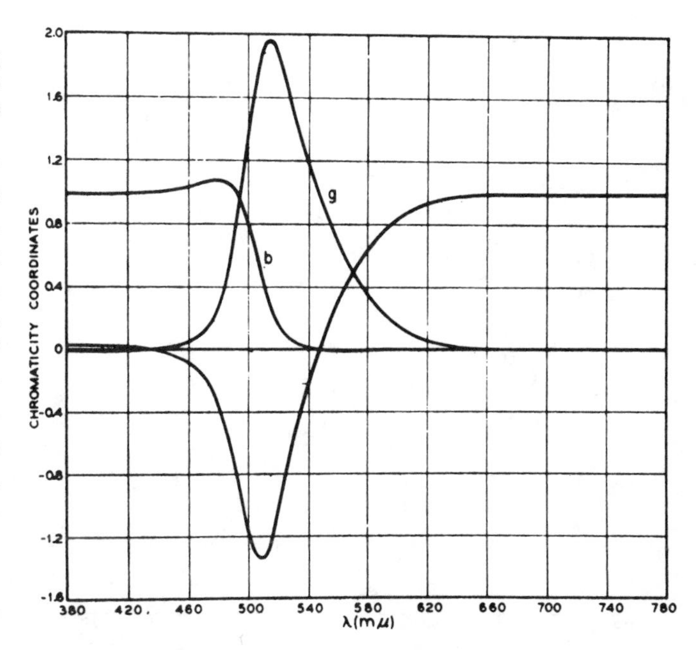

Fig. 3. Chromaticity coordinates of spectrum colors for the Standard Observer (from Wintringham [13]).[3] Primaries: 700.0 nm, 546.1 nm and 435.8 nm (NPL primaries). Reference white: equal-energy white.

data on 7 observers with a particular set of spectral primaries and a particular reference white. Wright [15] obtained data for 10 observers using different spectral primaries and did not normalize to the same white. In order to make use of the data of both Guild and Wright a transformation must be found between the tristimulus values of a color for two arbitrary sets of primaries. This has been solved by a number of workers for certain special cases. Wintringham [13] has treated the problem in a very general form in which the two reference whites are not the same. The result can be expressed in terms of a 3 X 3 matrix transformation.

D. Chromaticity Diagrams

The Commission International de L'Eclairage (CIE) in 1931 defined the color-matching data of the Standard Observer to be the mean of the data of Wright and Guild, and the equal-energy white was adopted as the reference white.[2] These specifications enable us to define the color mixture data of the Standard Observer (see, for example [13, 17]). The data indicate how much of each primary is needed to match spectral stimuli of equal radiance for the Standard Observer. The chromaticity coordinates of the spectral colors for the Standard Observer based on the National Physical Laboratory primaries are shown in Fig. 3. Corresponding tristimulus values for spectral stimuli of equal radiance are shown in Fig. 4. Chromaticity coordinates express fractions of a whole mixture. On the other hand, tristimulus values express how much of a primary is needed in a match to a given spectral color.

One form of graphical representation of this information on a chromaticity diagram using the r and g chromaticity coordinates is shown in Fig. 5. The spectral colors plot on the elongated horseshoe shaped curve called the spectral locus. The straight line connecting the two extremes of the spectral locus is called the line of purples. Note that the spectral locus extends outside the triangle formed by the three primaries which are, of course, located at (0,0), (0,1) and (1,0).

An important property of a chromaticity diagram concerns the calculation of the chromaticity of a mixture by a method analogous to a center of gravity system with the luminances of the components acting as weights. If two colors are additively mixed, then the chromaticity of the mixture lies on the straight line between the two chromaticities of the components. For a three-color mixture the chromaticity of the result lies within the triangle formed by the three component chromaticities. The extension of the spectral locus outside of the color triangle formed by the three primaries in Fig. 5 is a consequence of the necessity of adding one of the primaries to some of the spectral colors in order to carry out the match (equivalent to moving one of the terms from the right side of (1) to the left).

Suppose we wish to calculate the chromaticity of an object

[2] The CIE [16] had already adopted a standard relative luminous efficiency function V_λ shown in Fig. 2 for photopic ("normal daylight vision") conditions. The function represents the results derived from several different photometric methods of equating the brightness of spectral energy sources.

[3] The symbol "mμ" on the axis is outdated and has been replaced by "nm" in our text.

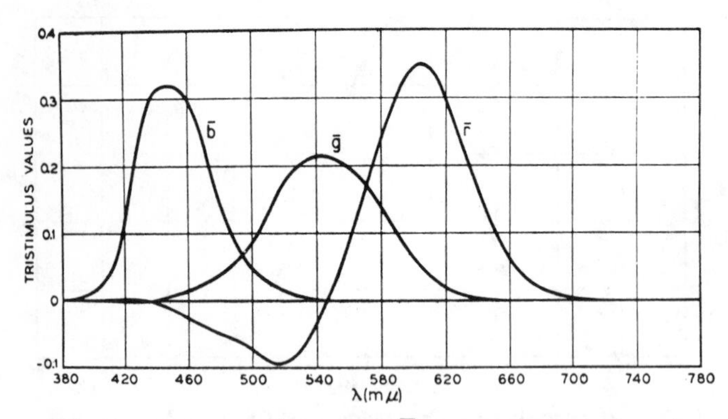

Fig. 4. Tristimulus values $\bar{r}$, $\bar{g}$ and $\bar{b}$ of spectral stimuli of equal radiance for the Standard Observer (from Wintringham [13]).[3]

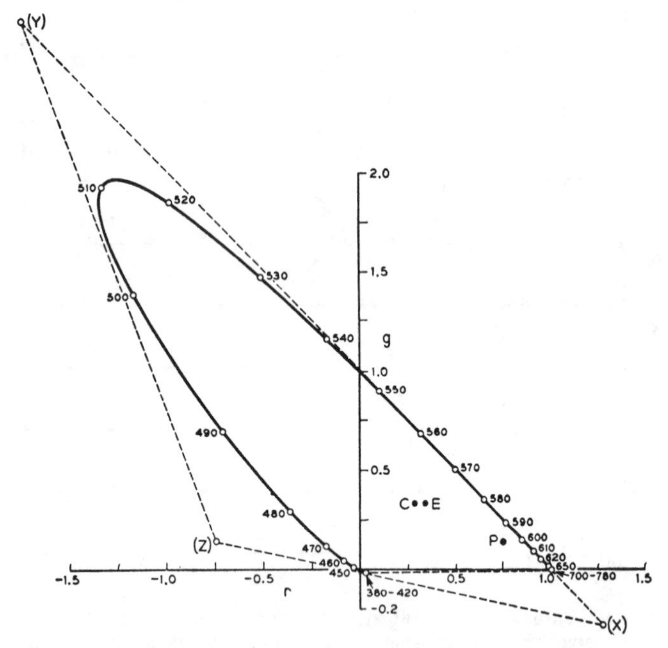

Fig. 5. The *rg* chromaticity diagram for the Standard Observer (from Wintringham [13]). The wavelengths (in nm) of the spectral colors appear on the horseshoe shaped locus. Point *E* represents equal-energy white, *C* represents Illuminant *C* which is a standard bluish-white source, *P* represents a specific color sample irradiated by Illuminant *C* and (*X*), (*Y*) and (*Z*) are the standard CIE nonphysical primaries discussed in Section II. D.

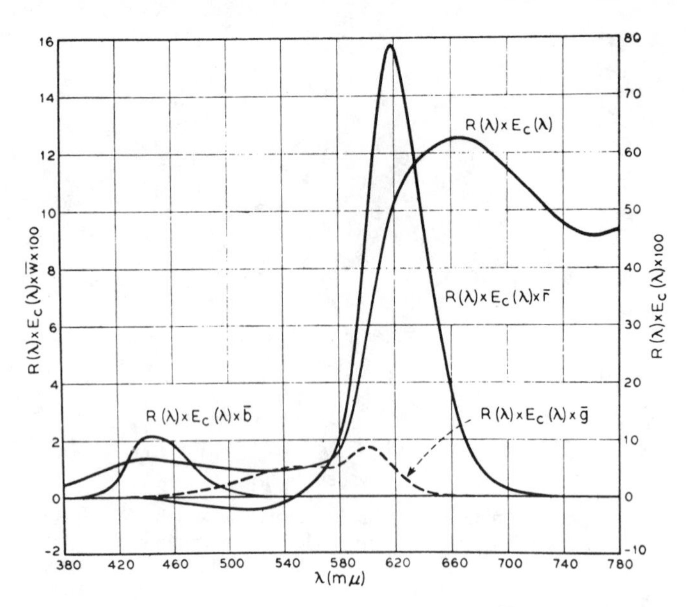

Fig. 6. Products of the tristimulus curves $\bar{r}$, $\bar{g}$ and $\bar{b}$ with the reflectance of a color sample, $R(\lambda)$, irradiated by Illuminant *C*, $E_c(\lambda)$ (from Wintringham [13]).[3]

that is illuminated by a light of a specific spectral distribution. The spectral distribution of the reflected light may, of course, be thought of as being composed of an infinite series of spectral colors. To determine how much of each primary is needed in the mixture, a product is formed of each of the tristimulus values with the spectral reflectance of the object as shown in Fig. 6. The areas under each curve, as obtained by integration, are the desired tristimulus values *R*, *G* and *B* for the sample. Chromaticity coordinates can be calculated using (3) and the point "*P*" is plotted in Fig. 5

In 1931 such calculations were commonly performed on desk calculators, and the negative lobes of the functions of Fig. 4 introduce negative product terms in which the negative sign is error prone with repetitive summing and differencing operations. It would be much better if there were no negative lobes, and it would be convenient if the quantities were zero over as large a range as possible. Calculation of the luminance of a color would be made much easier if the luminosity coefficients of two of the primaries were equal to zero. The luminance of a color would then be equal to the number of units of the other primary used in the match.

It was these considerations, among others, that led the CIE to propose a new set of primaries. The spectral locus is totally contained within the triangle formed by these new primaries denoted by (*X*), (*Y*) and (*Z*) (as can be seen in Fig. 5) implying that all spectral colors can be matched with a positive quantity of each primary. The use of such "nonphysical" primaries should be no cause for concern. For measurement purposes any real set can be used and the results can be transformed by a 3 × 3 matrix to the nonphysical set.

Luminosity information is obtained from the tristimulus value of the (*Y*) primary—the luminosity coefficients of the other two primaries are equal to zero. The resulting tristimulus values $\bar{x}$, $\bar{y}$ and $\bar{z}$ are shown in Fig. 7. Note the all-positive nature of the functions, and that $\bar{y}$ is identical to the V_λ curve of Fig. 2. The *xy* chromaticity diagram for the 1931 CIE Standard Observer is shown in Fig. 8. The equal-energy white (*E*) has the coordinates (1/3, 1/3) because it is the reference white for this system. All the color mixture properties that we have previously described for the *rg* diagram are valid for this diagram, however, the equations for color mixture are especially simple for this case. If we are given the chromaticities x_1, y_1, x_2; y_2 and their luminances L_1 and L_2 the chromaticity of the mixture is simply

$$x_3 = \frac{x_1(L_1/y_1) + x_2(L_2/y_2)}{(L_1/y_1) + (L_2/y_2)}$$

$$y_3 = \frac{L_1 + L_2}{(L_1/y_1) + (L_2/y_2)} \tag{4}$$

and the luminance is

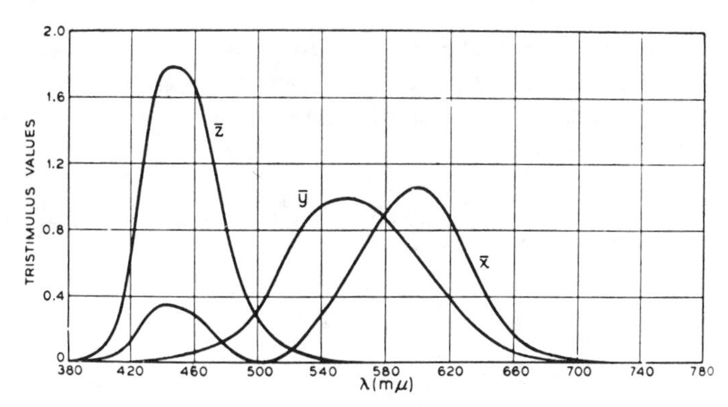

Fig. 7. Tristimulus values $\bar{x}$, $\bar{y}$ and $\bar{z}$ of spectral stimuli of equal radiance for the Standard Observer (from Wintringham [13]).[3]

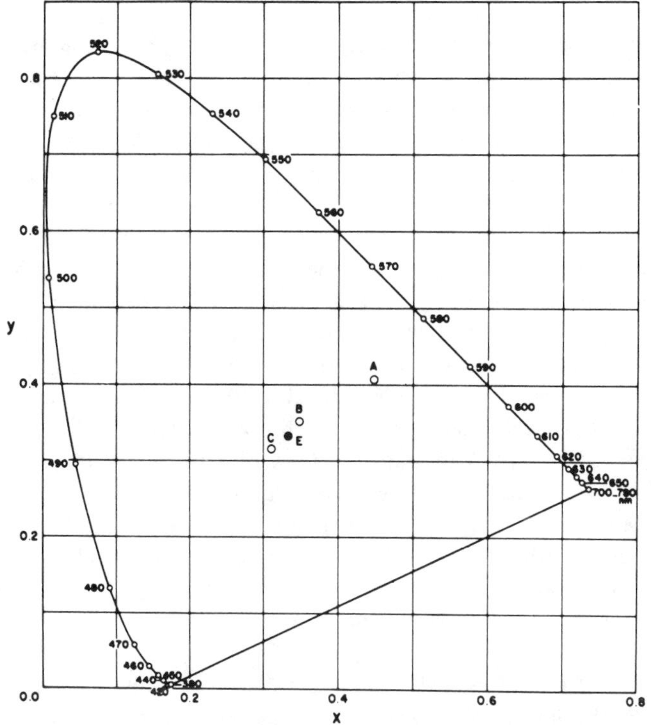

Fig. 8. 1931 CIE *xy* chromaticity diagram containing the spectral locus, line of purples, and the chromaticity locations *A*, *B*, *C* and *E* for CIE standard Illuminants *A*, *B*, *C* and equal-energy white (from Wyszecki and Stiles, Fig. 3.10 [17]).

$$L_3 = L_1 + L_2. \tag{5}$$

The 1931 Standard Observer was based on color matches using a 2° field. In 1964, the CIE defined another Standard Observer, but based on a 10° field. The data base was obtained from Stiles and Burch [18] and Speranskaya [19]. The location of certain wavelengths are somewhat shifted along the spectral locus in the 1964 $x_{10}y_{10}$ diagram as compared to the 1931 *xy* diagram as can be seen by examining Fig. 2.17 of [20].

It is important to realize that the chromaticity diagrams we have discussed pertain to color in the objective sense. They are based on color matches and not on color appearance. Information concerning the subjective color sensations of hue and saturation are not obtainable from a chromaticity diagram.

E. Uniform Chromaticity Diagrams

The *xy* chromaticity diagram is still in wide use today, but it has a major shortcoming for some practical applications. Color samples which have chromaticities which are equally distant from each other on the diagram are not equally different in appearance. MacAdam has determined the loci of chromaticities that are equally noticeably different from each of 25 representative colors for a constant level of luminance [21, 22]. Such loci are ellipses and they are shown on the *xy* diagram in Fig. 9. Notice how the differences vary over the diagram.

There have been many attempts to make a diagram in which equal distances correspond to equal differences in perception under the restriction that it be obtained by a linear transformation of the *xy* diagram [17].[4] However, the goal is impossible to attain strictly although improvements can be made. The 1960 CIE-UCS diagram [23] is one such diagram, and it is shown in Fig. 10 with a plot of MacAdam's ellipses. These ellipses tend to be more circular than those shown in Fig. 9. The transformation from *x, y* to *u, v* is as follows:

$$u = 4x/(-2x + 12y + 3)$$

$$v = 6y/(-2x + 12y + 3). \tag{6}$$

III. THE COLOR TELEVISION SIGNAL

A. Relation to Colorimetry

We can draw an analogy between a color television system and a colorimeter. The three phosphors of the receiver correspond to the three primaries of the colorimeter, and the camera taking filters correspond to the color mixture curves for these primaries. Each color in the scene before the camera must be matched by suitably controlling the light output of the receiver phosphors. This goal will be achieved if the light contributions from the receiver phosphors are adjusted to be equal to the tristimulus values appropriate to this sytem of primaries for each of the colors in the original scene. If we assume for the moment that the television system is linear, then the three signals from the camera should be proportional to these tristimulus values. As can be surmised from Section II, this is achieved by making the spectral sensitivity of the color filter in each of the three channels in the camera proportional to the corresponding color mixture curve of the Standard Observer for the receiver primary it controls. These relations were applied in the design of color television systems.

B. Format of the National Television System Committee (NTSC) Color Television Signal

The NTSC color television standards meet two basic requirements: (1) compatibility with existing monochrome receivers and (2) bandwidth containment of the color signal within the existing bandwidth for monochrome television.

[4] The latter condition is needed in order to retain the facility to carry out color mixture on the diagram by the method analogous to center of gravity systems.

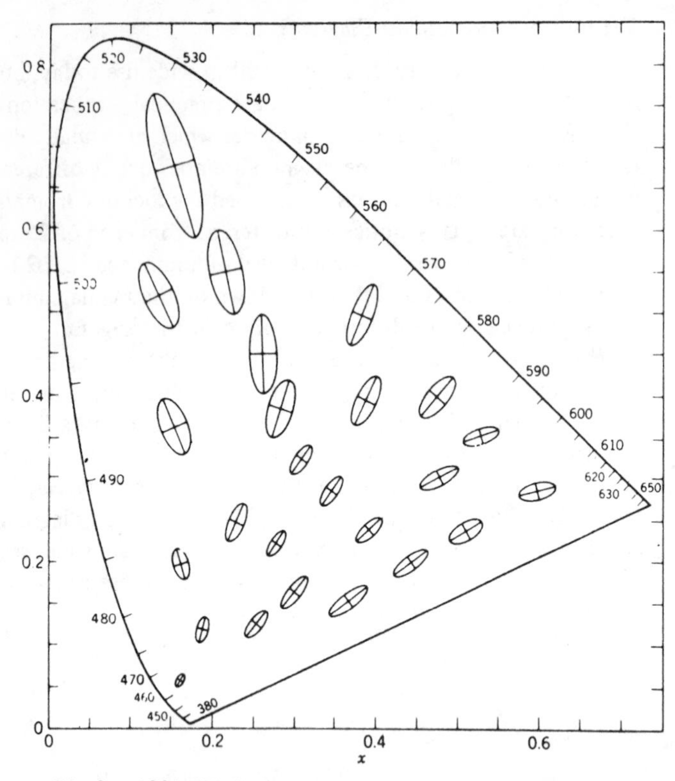

Fig. 9. 1931 CIE diagram showing MacAdam's ellipses enlarged 10 times (from Wyszecki and Stiles, Fig. 6.36 [17]).

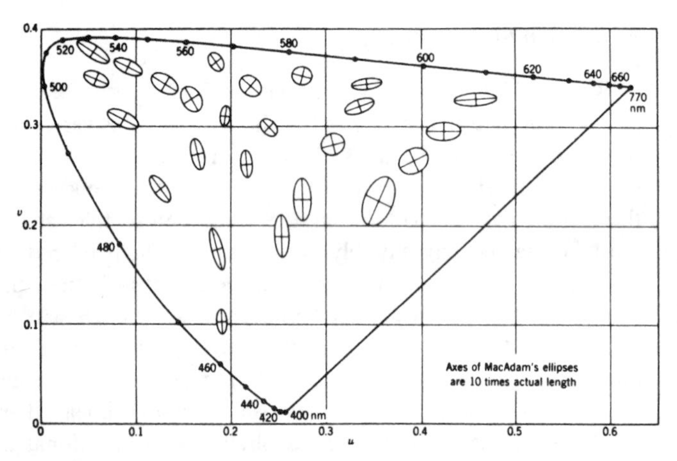

Fig. 10. 1960 CIE-UCS diagram showing MacAdam's ellipses (from Wyszecki and Stiles, Fig. 6.48 [17]).

This is achieved by transmitting a luminance signal representing the monochrome information and another signal comprising two "chrominance" components which supply the additional information needed to represent the chromaticities of the original scene. Frequency multiplex techniques are used to simultaneously transmit these signals in the bandwidth previously allocated for monochrome television. In order to make the most efficient use of the bandwidth of the channel the properties of the human visual system are utilized. The two chrominance components were chosen such that they could be accorded much less bandwidth than the luminance signal. The exact form of this signal will be described shortly.

In order to form the luminance signal, the chromaticities of the three receiver primaries and the condition for normalizing the three color signals must all be specified. The primaries were chosen such that there was a large gamut of chromaticities contained within the triangle formed by the three receiver phosphors used as primaries. The three color signals, E_R, E_G and E_B were adjusted to be equal on the reference white (Illuminant *C*). For the conditions specified by the NTSC, the luminance signal is

$$E_Y = 0.30E_R + 0.59E_G + 0.11E_B. \tag{7}$$

The coefficients in this equation sum to unity so that when Illuminant *C* is reproduced

$$E_Y = E_R = E_G = E_B. \tag{8}$$

Several advantages accrue if, instead of transmitting the signal E_Y and two of the three color signals E_R, E_G and E_B, the luminance signal is transmitted along with two "color difference" or "chrominance" signals such as $E_R - E_Y$ and $E_B - E_Y$. The receiver can recover the E_R, E_G and E_B signals by a particularly simple linear matrix operation. Variations in the two color difference signals do not affect the reproduced luminance. The same is true if the other two signals were chosen from E_R, E_G and E_B. This imperviousness of the luminance to perturbations in the other two signals is called the "constant luminance principle." A specific advantage of using color difference signals and normalizing to the reference white is that for achromatic signals both color difference signals are equal to zero. Variations in the relative strengths of the three signals do not affect the color balance of the reproduction of any level of grey. As will be pointed out in more detail in Section V, the majority of colors in most scenes have chromaticities that are close to the reference white. This reduces the amount of information that the color difference signals must carry.

Of special interest in terms of color television transmission is the ability of the visual system to detect spatial detail in the color difference signals. Pearson [24] gives a short account of most of the applicable early work. Briefly stated, the NTSC color difference signals denoted as *I* and *Q* were chosen to be

$$E_I = ((E_R - E_Y)\cos 33°/1.14) - ((E_B - E_Y)\sin 33°/2.03)$$

$$E_Q = ((E_R - E_Y)\sin 33°/1.14) + ((E_B - E_Y)\cos 33°/2.03). \tag{9}$$

This choice is near optimum for the constraints involved. The E_I signal carries orange-red to cyan information and the E_Q signal carries green to magenta information. The chrominance bandwidths assigned prior to forming the composite signal are as specified in Table I. The luminance bandwidth is a nominal 4.2 MHz.

The disparity in the relative apportionment of the bandwidth for the three transmitted signals results in a "mixed highs" system. For large area color the chromaticity is repro-

TABLE I
BANDWIDTH OF THE CHROMINANCE SIGNALS

NTSC Q-Channel

at 400 kHz less than 2 dB down

at 500 kHz less than 6 dB down

at 600 kHz at least 6 dB down

NTSC I-Channel

at 1.3 MHz less than 3 dB down

at 3.6 MHz at least 20 dB down

PAL (Systems B, G, H and I)
U- and V- Channel

at 1.3 MHz less than 3 dB down

at 4.0 MHz at least 20 dB down

duced correctly. As the area becomes smaller and smaller, first the E_Q signal and then the E_I signal is attenuated, and finally, only the high frequencies of the luminance signal remain.

In order to fit the chrominance signals into the luminance bandwidth they are amplitude modulated onto two color subcarriers of the same frequency (3.58 MHz) but shifted in phase by 90°. The exact frequency for the color subcarrier is chosen to be an odd multiple of half the horizontal line frequency so that the clumps of energy in the subcarrier fall between the clumps in the luminance signal spectrum, and so are of minimum visibility when viewed on a monochrome receiver. The form of the composite color signal is:

$$E = E_Y' + [E_Q' \sin(\omega t + 33^\circ) + E_I' \cos(\omega t + 33^\circ)]. \tag{10}$$

The two chrominance components in (10) can be thought of as modulating the amplitude and phase of a single subcarrier. The primes on the symbols refer to the gamma corrected voltages. Up to this point we have been assuming a linear system which is not the case in practice. For the recommended gamma correction of 2.2, $E_{R,G,B}' = E_{R,G,B}{}^{1/2.2}$ and

$$E_Y' = 0.30E_R' + 0.59E_G' + 0.11E_B'. \tag{11}$$

The chrominance components can be plotted on the 1960 CIE UCS diagram in terms of the normalized amplitude and subcarrier phase as shown in Fig. 11 (Townsend [25]). These loci for a gamma of 2.2 are not circles and straight lines as would be ideal.

C. PAL and SECAM

The Phase Alternation Line (PAL) color television system is in use in most of Western Europe, and the SECAM[5] system is in use in France, Iran, the Middle East and Eastern Europe. (See Carnt and Townsend [26] for a description of these systems.)

In PAL and SECAM the chrominance components are designated as

$$U = (B - Y)/2.03$$
$$V = (R - Y)/1.14 \tag{12}$$

where the factors 2.03 and 1.14 are the same as those appearing in the NTSC signal, but the cosine and sine terms are absent.

In the NTSC system, deviation by more than 600 kHz above the subcarrier frequency exceeds the nominal bandwidth (4.2 MHz). Crosstalk between the *I and Q* channels is avoided by restricting the Q bandwidth to 600 kHz which allows enough room for a double sideband (therefore no crosstalk), and confining sideband distortion to the I signal which begins to roll off around 1.3 MHz.

An alternative technique of avoiding U/V crosstalk is adopted in the PAL system whereby the phase of the V signal is changed by 180° between successive lines in the same field. Complementary crosstalk errors therefore occur on adjacent lines and can be reduced significantly by averaging in a delay line decoder. Similar bandwidths may therefore be allowed for the U and V signals (Table I) which are transmitted, as in NTSC, in phase quadrature (subcarrier is located at 4.43 MHz). However, since a subcarrier which is an odd multiple of half the line frequency (NTSC) would now cause the line alternating V component to crosstalk into vertically correlated luminance energy, the PAL subcarrier is chosen to bear a quarter-line offset. Alternation of the V component now results in spectral separation of the U and V signals between line harmonics, as in Fig. 12a as compared with the NTSC spectrum shown in Fig. 12b. An additional subcarrier offset of 25 Hz relative to a quarter-line harmonic gives appreciable improvement in compatibility (with monochrome reception) and crosscolor, but requires eight fields to complete the spatial subcarrier cycle. In addition to the advantage mentioned in respect to sidebanding, the alternation of the V component gives a degree of protection against phase errors (e.g., differential phase distortion) and multipath reception. These errors change the saturation but not the hue. As a results of averaging out phase errors, hue and tint controls are not required in the PAL receiver.

Immunity to gain and phase distortion of the chrominance is provided in the SECAM system where frequency modulation of two subcarriers is used to transmit the U and V signals individually on alternate lines, thus precluding the possibility of U/V crosstalk. Unfortunately, the system is still susceptible to differential gain distortion through the effects of gain errors on line-time nonlinearity on the luminance signal. It also imposes a tight tolerance on the slope of the differential phase/luminance characteristics, so that the requirements of SECAM and PAL in these respects are broadly similar. In SECAM the luminance signal is transmitted on every line, and a delay line in the receiver allows repetition of each chrominance signal so

[5] The name SECAM is not an acronym. It refers to the use of a sequential chrominance signal and a memory device (sequential à mémoire).

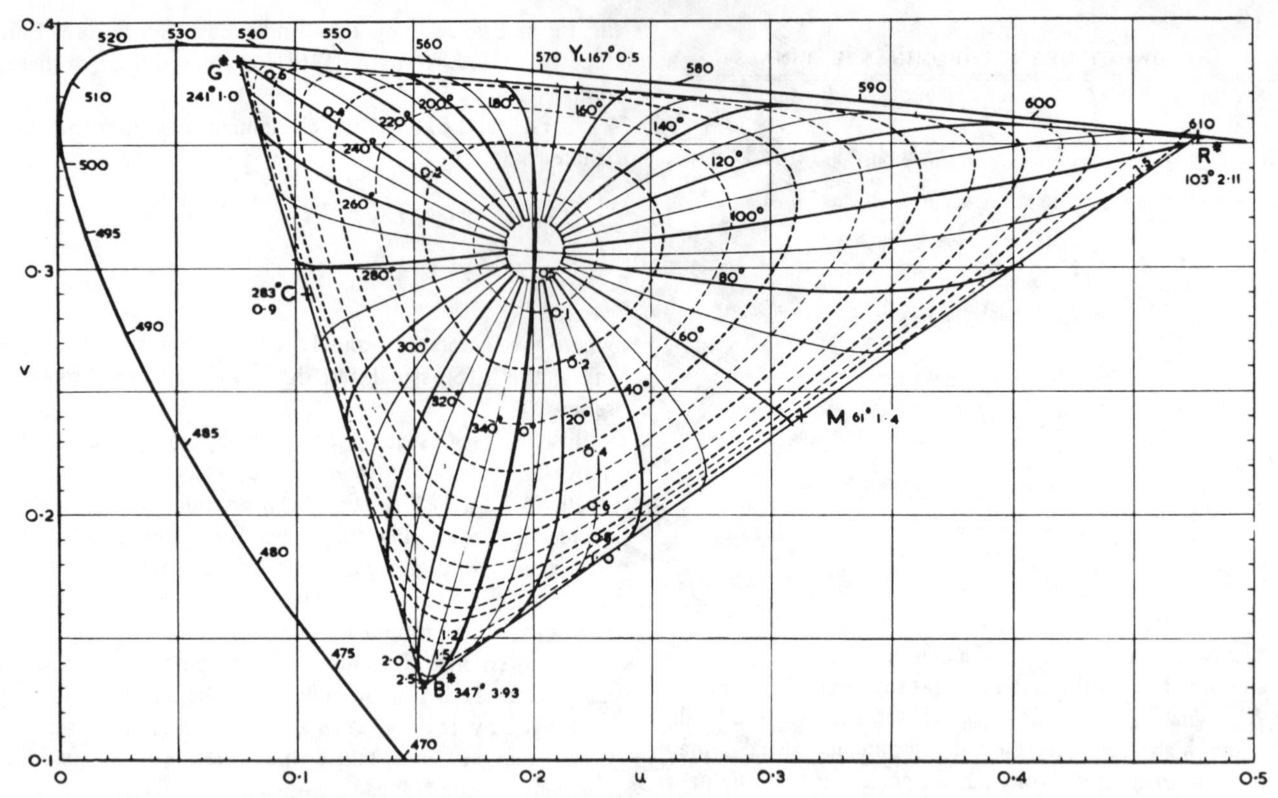

Fig. 11. Loci of constant NTSC subcarrier phase and normalized amplitude for gamma = 2.2 on 1960 CIE-UCS diagram (from Townsend [25]).

Fig. 12. Frequency spectra (neglecting 25 Hz offset) for: (a) PAL with 1/4 and 3/4 line offset, (b) NTSC with 1/2 line offset (from Carnt and Townsend, Fig. 2.7 [26]). The closest luminance line harmonic to the subcarrier frequency is 284 f_H (horizontal line frequency) because the U subcarrier is 283 3/4 f_H.

that Y, U and V signals are simultaneously available. The corresponding reduction in vertical chrominance definition is marginally offset by alternating the order of transmission of the U and V signals in successive frames. The U and V subcarriers are at 4.25 and 4.41 MHz respectively, and are reset in phase at the beginning of each line to reduce dot pattern visibility on monochrome receivers.

Unlike PAL and NTSC, SECAM transmits a subcarrier signal for grey. Mixing and crossfading of two signals are more difficult than with NTSC since fading reduces the luminance, but not the frequency-modulated chrominance. SECAM studio mixers, therefore, require separate luminance and color difference signals at least during mixing, and it is perhaps this inflexibility which has deterred the development of composite coding techniques for the SECAM form.

D. Component Representation

Just as for the analog coding processes of NTSC, PAL, or SECAM, the color signal could be presented for digital coding as a luminance component and two chrominance components. However, it is realized that the criteria for efficient filtering of components prior to analog assembly of the NTSC, PAL or SECAM signals are very different to those for defining bandwidths prior to a digital assembly of a component structure (perhaps involving sub-Nyquist sampling). This is currently a subject for study under a CMTT question.

In the remainder of this paper, the IQ and UV signal designations will be loosely applied to describe signals resembling those of the parent composite system. The terms $C1$ and $C2$ will be used to denote chrominance signals where the specific nature of the signals are not specified.

IV. PSYCHOPHYSICS OF COLOR VISION

A. Accuracy of Representation

Let us make the reasonable assumption that the intention is to encode a picture such that in appearance it is as close to the original as possible subject to any constraints put on the system such as cost and bit-rate. The essential feature underlying the assumption is that the pictures are being presented for appreciation by a human observer. This assumption is valid for situations such as conventional television. Recognizability, such as in target identification, is excluded. For television, and similar cases, we need to know how accurately to encode the color of each point in the picture. The reproduction of a color picture without encoding already represents a compromise with accuracy, namely: the choice of the objective that the color reproduction system must satisfy. There are six different types of color reproduction as defined by Hunt [27] (see Table II).

The objective that is generally used in color television is "colorimetric color reproduction" which requires only that the chromaticities of the original and reproduction be identical.[6] Although color television can theoretically attain this objective, it does not in practice. There are a number of reasons for this failure including: some chromaticities in the original scene lie outside the range of the receiver phosphors, the camera sensitivity curves are not exactly matched to the phosphors, improper alignment of all the equipment in the chain from the studio to the home receiver, and nonlinearity in the system.

Improper camera sensitivity curves present an interesting case study of the effects of nonstandardization. Modern receivers use different phosphors and a different reference white than those originally specified by the NTSC. The theoretical camera sensitivity curves must have negative lobes for any set of phosphors. Fig. 13 shows the required sensitivities for the NTSC phosphors. The sensitivities for real cameras, however, tend to approximate the positive portion of these curves, ignoring the negative lobes and thereby causing colorimetric errors in the system. This could be obviated by specifying that cameras use an all-positive set of sensitivities obtained from a linear combination of the CIE 1931 color matching functions shown in Fig. 7, and requiring a matrix at the camera to convert the signals to be appropriate for a standard set of phosphors. This procedure is followed in the United Kingdom. It is incumbent upon the receiver to incorporate any additional matrix to correct for other phosphors. It is interesting to note that, even with all these sources of error, color television does produce pleasing pictures, but it is capable of doing even better.

As is evident from even this sketchy account, there are many unanswered questions in reference to the accuracy required for color reproduction in color television. The considerations that are entailed in efficiently encoding the picture place additional importance on answering these questions. Fortunately, some of the information obtained from coding monochrome television signals is applicable here.

B. Learning from Luminance Coding

The effort to exploit the statistical redundancy in coding monochrome television signals has a long history. The corresponding effort to exploit the redundancy relative to the human observer has received relatively less attention, but it is equally important. In terms of color signals the surface has been barely scratched. Reference to much of the literature for the monochrome case is covered in a recent article by Limb [28] in which he considers two aspects of the general problem of determining the required accuracy of representation: "(1) the threshold of an arbitrary shaped stimulus presented against a uniform background and (2) the threshold of an element adjacent to a large change of luminance."

If one encodes the three components of a color signal, then the quantizer designs that have been tailored to the requirements of human vision for monochrome signals can be used to encode the luminance component. There are other properties of human vision that can be specifically exploited to efficiently encode color signals. These properties relate to aspects of color discrimination, threshold versus suprathreshold requirements, spatial resolution requirements of color signals, self-masking effects of a color component and the masking effect of one color component upon another. We do not have a complete understanding of human color vision and some of the above properties have not been fully measured, but this has not prevented their use in encoding color pictures. Some of these properties are discussed in the following sections and the coding aspects are discussed later in the paper.

TABLE II
CLASSIFICATION OF COLOR REPRODUCTION SYSTEMS BY GOALS

CATEGORY OF COLOR REPRODUCTION	REQUIREMENT
Spectral	Exact spectral match
Colorimetric	Chromaticities match for same illuminant
Exact	Chromaticities and luminance match for same illuminant
Equivalent	Colors match in appearance under different illuminants
Corresponding	Same as Equivalent with no lightness match
Preferred	Appearance is made to match subjective preference rather than original

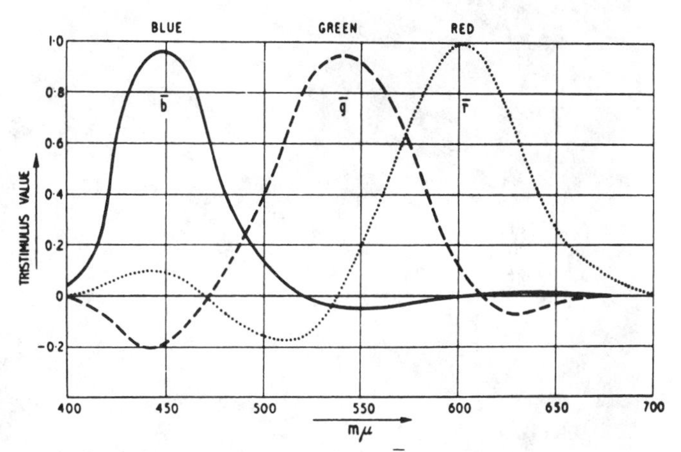

Fig. 13. Tristimulus values $\bar{r}$, $\bar{g}$ and $\bar{b}$ of spectral stimuli of equal radiance based on the NTSC phosphors and Illuminant C as the normalizing white (from Carnt and Townsend, Fig. 1.2 [26]).[3]

[6] Many television commercials employ "preferred color reproduction" by making the colors in the scene before the camera depart from their "real life" counterpart in a direction to make them more pleasing.

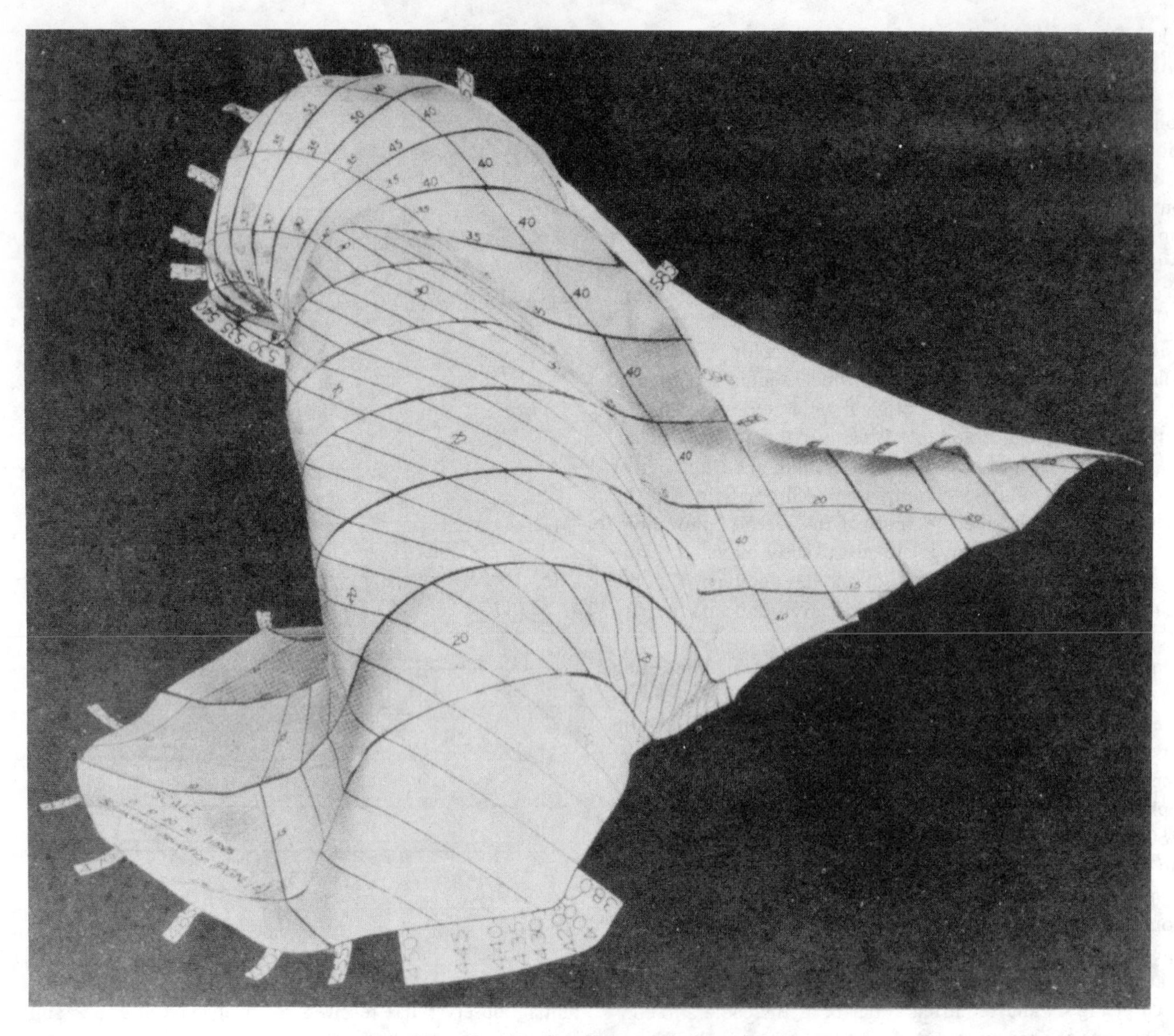

Fig. 14. UCS surface based on MacAdam's discrimination ellipses for observer PGN (from Wyszecki and Stiles, Fig. 6.39 [17]).

C. Color Metrics

It would be especially helpful in encoding color pictures to have a color space in which the sensitivity to errors is equal at all points in the space. One example of an approximation to such a space was previously discussed (see Section II.E). This area of knowledge is still the subject of intense investigation under the headings: color differences, uniform color scales, color order systems, line elements and color metrics. In general, these fields concern the subjective impressions of the human observer. In contradistinction to the terminology of dominant wavelength, excitation purity and luminance used in objective color measurements in colorimetry, the terms of hue, saturation and lightness are used to describe the attributes of color perception. Hue is the color name, such as red, green or blue. Saturation expresses the comparative difference of the color from a reference achromatic color.[7] Lightness is the attribute of an object color which permits it to be classed as equivalent to one of a series of achromatic color perceptions ranging from black to white.

[7] An achromatic color is a color not possessing a hue.

Many of the aspects of color metrics are discussed in Section 6 of the book by Wyszecki and Stiles [17]. The literature on line elements is extensive, and no attempt will be made to cover the general field. The importance of a line element in color theory lies in its specification of the distance between a pair of colors that are just noticeably different. For all such pairs, the line element is a constant whether or not the space is Euclidean. Much of the work is referred to in the published proceedings of the Helmholtz Memorial Symposium on Color Metrics held in 1971 [29], and there has been some recent work by Jain [30, 31].

As an example of the concepts involved, we will discuss one particular line element which has been used extensively. MacAdam's line element is concerned with just-noticeable differences (jnd) of pairs of colors having a luminance of 15 millilamberts [22]. It is based on his measurements of over 25,000 trials at color matching for a single observer, Perley G. Nutting, Jr. (PGN) [21]. The standard deviations of the trials are represented in the form of ellipses as mentioned in the section entitled Uniform Chromaticity Diagrams. The line element is

$$(ds)^2 = g_{11}(dx)^2 + 2g_{12}dxdy + g_{22}(dy)^2 \qquad (13)$$

where g_{ij} are constants for each ellipse and $ds = 1$ corresponds to the standard deviation of color matching (approximately 1/3 of a jnd). MacAdam has constructed a uniform chromaticity scale surface based on this line element [32], and the complicated Riemannian surface that results is shown in Fig. 14. It is impossible to flatten it to a plane without rupture which certainly makes it painfully difficult to find a plane diagram to approximate a uniform chromaticity scale surface.

In addition to line elements, there is a related method for assessing color differences through the use of color difference formulas. These formulas generally assume that color-perception space is Euclidean. The main body of data on which they are based are those of Wright, MacAdam, and Brown.[8] The data generally entail measurements involving small differences above threshold. An account is given in [17] along with an extensive Table of color difference formulas. Each formula is restricted to a specific set of viewing conditions, and either a small or large color difference. The one recommended by the CIE in 1964 is based on the three-dimensional U^*, V^*, W^* space of Wyszecki [33]. This system is an outgrowth of the 1960 CIE-UCS diagram. The values of U^* and V^* depend upon the value of W^* so as to take account of the apparent increase or decrease in saturation with lightness when the chromaticity is kept constant.

Two other formulas have been recommended for use by the CIE in 1976. They are called the $L^*a^*b^*$ formula and the $L^*u^*v^*$ formula [34]. The former is to be used for small color differences, and the latter is to be used for large color differences. The $L^*u^*v^*$ formula is intended to be an improvement over the CIE 1964 $U^*V^*W^*$ formula. The $L^*a^*b^*$ formula is based on a simplification of Adams-Nickerson space. Both formulas are described in detail in [34].

If one is interested in *chromaticity differences* without taking direct account of the effect of luminance, and if the restriction on requiring a linear transform of the 1931 CIE diagram is removed, then a two-dimensional representation that more closely approximates a uniform chromaticity diagram can be obtained. Such a diagram has been derived by MacAdam based on a nonlinear transformation of the CIE x,y coordinates and the color matching data of 14 normal observers [35]. It is shown in Fig.15 with MacAdam's ellipses for the same observer (PGN) as shown in Fig. 9 and 10. Any straight line in this diagram represents the series of colors involving the least number of jnds in the transition between the points at

[8] Color difference data are not easily obtained because of the time consuming and exacting nature of the experiments. There has been some more recent work by Wyszecki and Fielder [121, 122] among others.

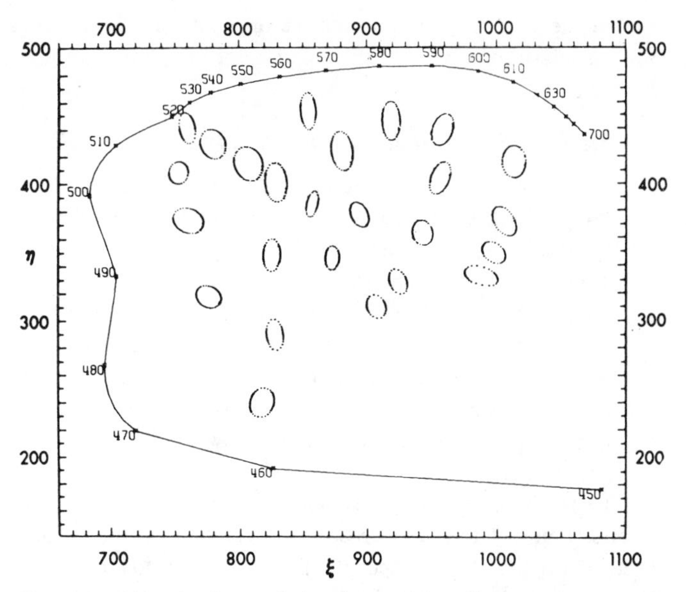

Fig. 15. MacAdam's geodesic chromaticity diagram shown with ellipses (enlarged 10 times) obtained from the variances of color matching by observer PGN (from MacAdam [35]).

the ends of the line. This diagram is not useful for predicting results of additive mixtures because of the nonlinear nature of the transformation.

It is not obvious that any of these previous formulas are applicable to the television viewing situation. Pazderak examined color differences in TV viewing, but under conditions where a subject's memory of color appearance was tested [36]. The sample colors were electronically inserted into a projected picture via a TV display. The colored areas within the display subtended 2° while the whole display subtended 11°. Each subject was shown the starting sample color which was then occluded for 3 seconds while it was changed. The sample was then exposed again during which time the subject had to judge whether the sample had changed color. This cycle was repeated until a change was detected. The median value of the difference in 10 sample colors amounted to 20 MacAdam units of color difference as defined in [17]. The samples ranged from a minimum of 9.6 units to a maximum of 36 units.

Dupont-Henius [37] used a TV display to evaluate color differences, but in his case the background was a uniform color rather than a color picture, and memory color was not a factor. The sample was displayed at one of six positions on the display, and the subject was required to determine if the sample was perceptible. An extensive body of data has been gathered, and large inconsistencies with MacAdam's data have been found.

DeMarsh and Pinney have proposed a color difference formula specifically for the television case [38]. It comprises a small modification to the CIE 1964 color difference formula reducing the contribution of the $(W^*)^2$ term by a weighting factor of 1/4.

Color order systems are another way to attempt to represent a perceptually uniform color space. The Munsell and DIN system are two examples [17]. The recent proposal of the Optical Society of America Committee on Uniform Color Scales for a regular-rhombohedral system of color sampling is another such system.

D. Threshold, Suprathreshold, Perceptibility and Acceptability

So far we have been discussing threshold differences. A study of suprathreshold color differences has been reported by Witzel *et al.* [39]. They had subjects vary one color attribute at a time to match a standard "color difference" between two achromatic samples. They were able to determine suprathreshold perceptual color ellipsoids about the six colors they investigated. The precision of the color-difference matching was a function of the size of the color interval of the reference colors and the ellipses tended to orient toward the nearest colorimeter primary.

The study of threshold differences is akin to the stimuli difference being at the just-perceptible level. Perceptibility, however, is not the same as acceptability. All color difference formula are intended to predict the perceptibility of color differences not the acceptability. For encoding purposes, acceptability might be the desired criterion. Attempts have been made to predict acceptability from color difference formulas, but this has not met with any great success [29]. Additional parameters, other than those associated with color vision, must be used to determine acceptability. These parameters are functions of the particular application for which acceptability is to be judged.

E. Masking Effects

The masking effects we are concerned with pertain to the influence of the signal surrounding a point in a picture on the perception of the signal at the point. In terms of the components of a color television signal our interest will be restricted to the effect of the masking of the chrominance signal by the luminance signal, and the masking of the chrominance signal by the chrominance signal.

The case of luminance masking of the chrominance signal was investigated by Hacking [40]. He measured the just perceptible degradation in the sharpness of the chromaticity component at a boundary between two colors. Optical projection was used to produce the color transitions. The luminance was maintained at an equivalent bandwidth of 3 MHz at the boundary at all times, and different amplitude luminance transitions were introduced with neutral density filters. The field of view contained a reference pair of colors in the upper half, and the same pair of colors in the lower half. Starting with identically sharp transitions (3 MHz) in the upper and lower half of the field, the sharpness of the chromaticity transition in the lower half was then degraded until the subject stated that he could just perceive a degradation or blur at the boundary between the lower two colors. The test was repeated for decreasing blur, and a threshold bandwidth was obtained from these settings.

The results obtained are shown in Fig. 16 for the nine color pairs investigated. The threshold for blur changes rapidly between luminance contrasts of 1:1 and 2:1 (or 1:2), and changes much more slowly for luminance contrasts which exceed 4:1 (or 1:4). The luminance contrast ratio has a large influence on the threshold as does the choice of the color pair. At a luminance contrast of 8:1 (or 1:8) the bandwidth requirement (inversely related to blur threshold) is down by a factor of about 3 from the 1:1 case. Thus, the change in the luminance signal partially masks the ability of the subject to detect the blur at the transition border.

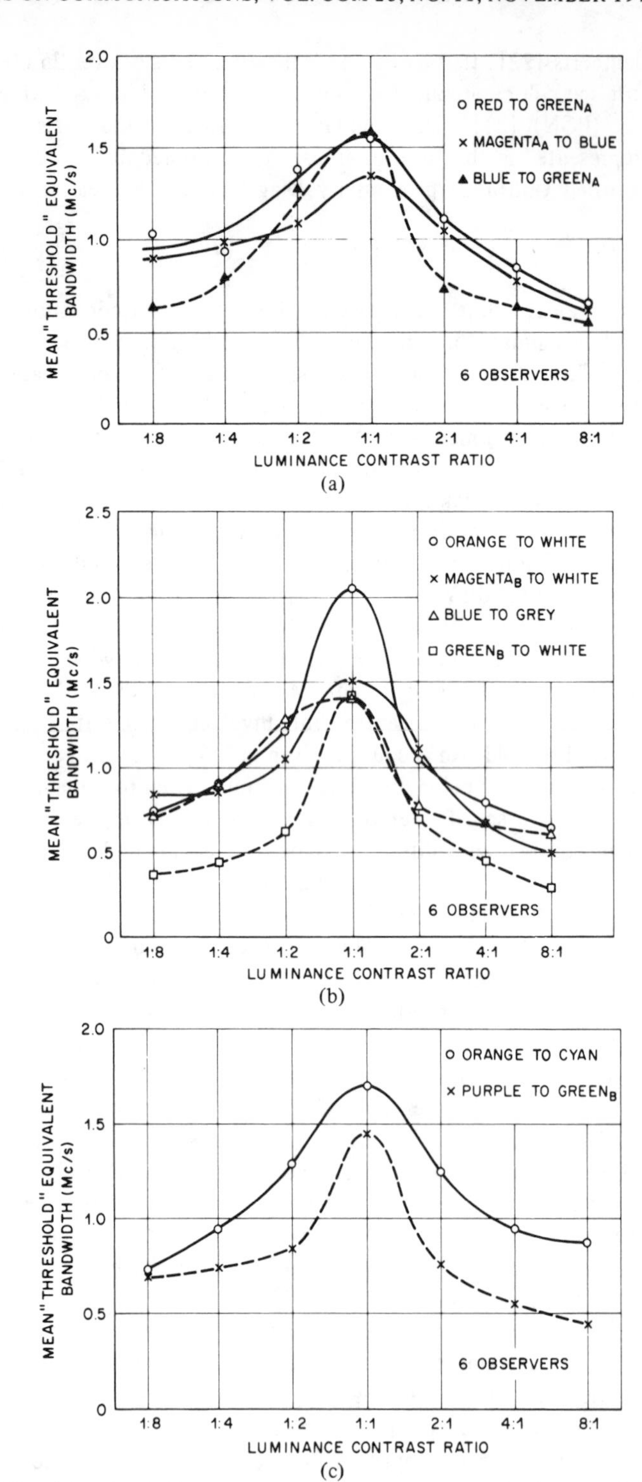

Fig. 16. Bandwidth versus luminance contrast ratio (expressed in terms of the first color for each pair) for: (a) transitions between saturated colors, (b) transitions between saturated and "white" colors, (c) transitions between orange to cyan and purple to green (from Hacking [40]).

A related investigation was carried out by Rubinstein and Limb [41]. These experiments used television signals, permitting direct adjustment of the chrominance signals to blur the transition between the two colors. For a perfectly sharp transition between two colors of different luminance, all three signal components would contain a jump at the transition point. To vary the sharpness, the subject changed the slope of a ramp in the chrominance signals at the transition. Control of the two ramps was ganged so that the width of the ramp region was the same for both chrominance components. This signal generation process is representative of color television systems that employ equal filtering in the chrominance channels.

The exact form of the chromatic path at the transition between the two colors was restricted to a straight line in the UCS diagram, but the traversal of the path varied from a case involving monotonic progression from one color to the other, to a case involving chromatic jumps. Fig. 17 illustrates this effect. The upper right quadrant contains curves for three of the color pairs studied with a 1:1 luminance contrast ratio. The ordinate represents the straight line drawn from the first named color to the second named color on the UCS diagram. Distance along this line is plotted on the abscissa in terms of the fractional distance from the first named color. The physical width of the transition region in the display is represented by the full extent of the abcissa. If the path plots as a straight line at 45° on this graph, then there is a uniform change of chromaticity in the transition region between the two colors.

The path for any luminance contrast ratio in the Hacking-type experiment is identical to the 1:1 luminance contrast ratio case. The paths for the 1:2 and 2:1 luminance contrast ratios are shown in the other three quadrants of Fig. 17 illustrating the chromatic jumps that occur for these ratios in contrast to the experiment of Hacking. In a further experiment [41] the transition region between the two colors was made achromatic and the subject controlled the width of this region. The results obtained support those of Hacking, and in addition they indicate that the exact form of the chromatic path has an effect on the threshold bandwidth.

The effect of blur on color discrimination thresholds has been investigated by Rentschler for the case of equiluminance in a bipartite field [42]. As the blur of the chromatic dividing line increased in a 2.5° visual field, there was an immediate rise in threshold by about a factor of two, followed by a leveling off. In a 12.5° field, the threshold is relatively independent of boundary blur over a wide range.

Luminance masking of the chrominance signal can be evidenced in other ways. Noise was added to the chrominance signal at all points in a portrait-type picture where the horizontal slope of the luminance signal exceeded a certain threshold (Netravali and Rubinstein [43]). Pictures impaired by such additive noise were then compared in subjective tests to a picture having white noise added to the luminance component. The white noise was added to the entire picture, and the subject controlled the amount in order to equate the two impaired pictures. Based on these measurements, visibility functions were obtained which express the relative visibility of unit noise added to all parts of the picture where the horizontal slope has a certain value. The visibility of this noise decreased with increasing size of the luminance slope. This same experiment was repeated using the chrominance slope as a control to add noise

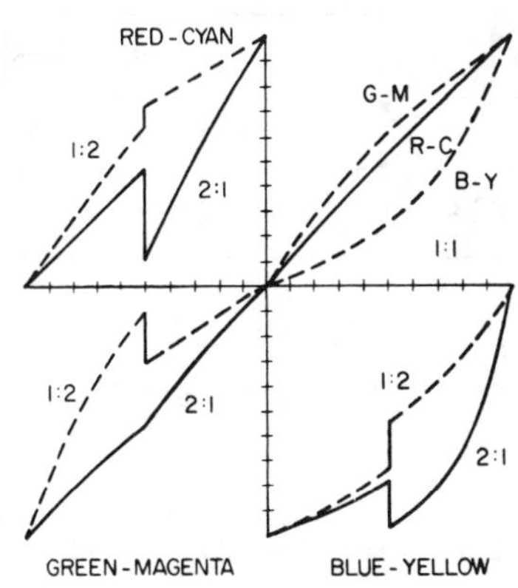

Fig. 17. Chromatic paths across a transition between various color pairs. The ordinate is normalized chromatic distance on the UCS chromaticity diagram and the abcissa is normalized spatial distance across the blurred transition (from Rubinstein and Limb [41]).

to the chrominance signal. The visibility of the noise decreased with increasing size of the chrominance slope. However, masking by luminance changes was much greater than masking by chrominance changes, but this conclusion may well be the result of the relative filtering of the luminance and chrominance signals.

F. Color Perception Models

There are a number of color perception models and we will briefly discuss some features of each before describing one in detail. The importance of having a model of color perception in relation to color signal encoding lies in the possibility of carrying out the coding in "perceptual space" (see Judd for a fuller explanation of the following summary [44]).

The Young-Helmholtz theory assumes the existence of three independent response mechanisms which mediate the color sensations. One is mainly sensitive to the long wavelengths, the second to the middle wavelengths, and the third to the short wavelengths. Young believed that the mechanisms yielded a red, green and blue sensation, respectively.

In the Hering-Hurvich-Jameson opponent-colors theory there are three opposing pairs of processes: black-white, yellow-blue and red-green. Light is absorbed by the photopigments in the receptors, and this absorption starts the three opponent processes which are directly responsible for our color sensations.

There are various models of color vision that combine the Young-Helmholtz three-components theory with the opponent colors theory. A three-stage model was formulated by Müller entailing a photopigment stage, a cone response stage, and an optic nerve stage. The first stage comprises the Young primaries, the second stage is of an opponent colors form and the third stage is the opponent colors formulation of Hering. Judd and Yonemura have expanded and further defined the Müller Theory. Adams' three-stage model proposed in 1923 was the first theory to contain a nonlinear transformation between the stages. The first and third stages are as in the Müller theory while the second stage comprises the 1931 CIE primaries.

A color vision model that has marked similarity to color television was described by Shklover [45]. His model takes account of a logarithmic transformation at the output of the receptors. A block diagram of the model is shown in Fig. 18.

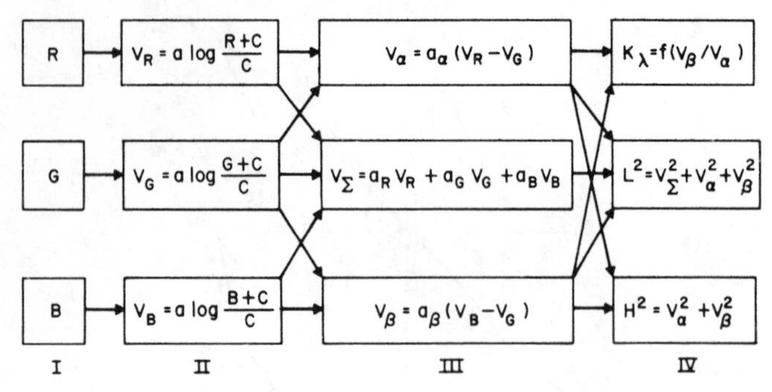

Fig. 18. Block diagram of Shklover's nonlinear model of the color vision process (from Shklover [45]).

The input radiation is received by three receptors in (I) whose sensitivities are those of the retinal cones. The output signals of the receptors are logarithmically transformed (II) and then matrixed to produce two difference signals and a signal akin to a luminance signal (III). Finally (IV), signals directly related to the percepts of hue (K_λ), saturation (H) and lightness (L) are formed. If we assume that the gamma correction at the camera approximates the log responses of Shklover's photoreceptors, then the expression for hue and saturation in Shklover's model is in close agreement with the signals used in color television if one equates hue with color carrier phase and saturation with color carrier amplitude. Various stages of the other models we discussed also are strongly related to color television signals. The remarkable feature is that the similarity between the electronic signal processing of a color television signal and these models of color vision was not deliberately chosen on the part of the NTSC or the investigators who proposed the models.

A "color television model" with gamma as a parameter has been used by Tannenbaum [46] to fit the hue and saturation (color carrier amplitude and phase) loci to the loci of constant hue and saturation of the Munsell system as plotted on the 1960 CIE-UCS diagram. In order to carry out such a fit, it was also necessary to allow a rigid rotation of the loci (derived from the model) about the reference white in the UCS diagram. The rotation can be interpreted as a linear transformation of the NTSC primaries to a set more closely related to the visual system. The fit for a gamma of 1.6 is very good. The model appears to have at least the same predictive capability as the Müller-Judd and Shklover models. A very similar color television model has been proposed by Frei to encode color pictures by transforming them into the "perceptual space" [47]. Frei's model attempts to take account of spatial resolution, simultaneous contrast and adaptation effects. Further refinements of Frei's model have been made by Faugeras [48].

V. SIGNAL STATISTICS

A. Comparison of Signal Transformations

There have not been a great number of statistics gathered on color signals, but we will briefly review what is known. Starting with an RGB signal, we will assume that each component has a range 0-1.0 unit such that equal quantities of R, G and B produce an achromatic mixture. The probability density function of the R, G and B signals varies from picture to picture (Fig. 19), but one could postulate that if an average were

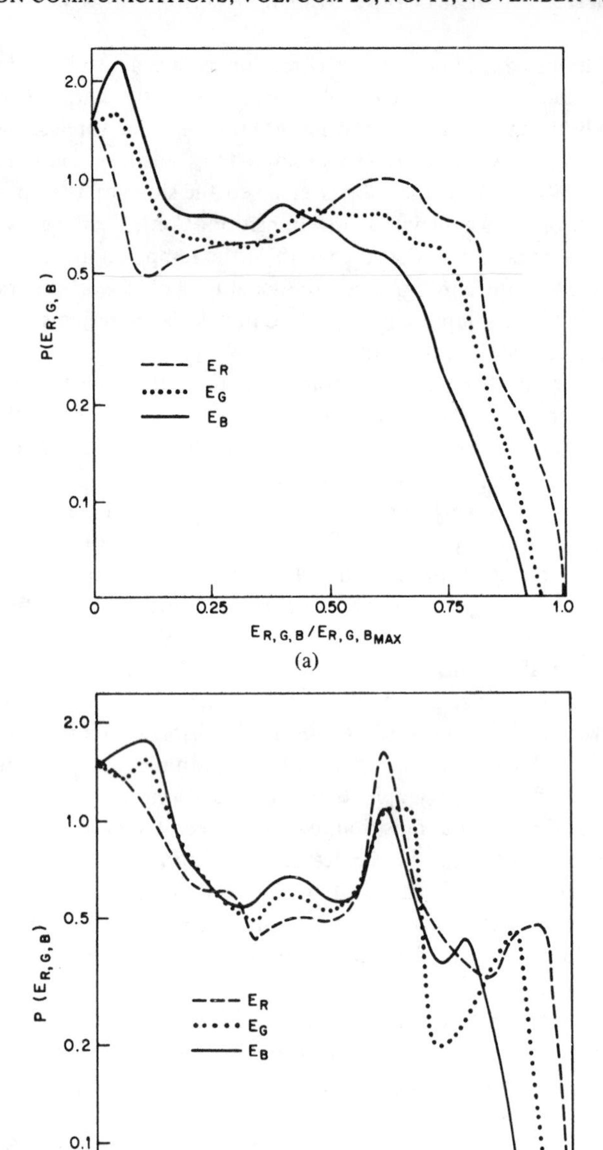

Fig. 19. Probability density functions of the Red, Green and Blue signals for a single picture: (a) Picture shown in Fig. 22d and (b) Another portrait-type picture.

taken over a large number of pictures, the resulting probability density would be much more uniform, just as one experiences with the luminance signal. Looking at the relation between a pair of components (Frei and Jaeger [49]), we see that they are highly correlated (Fig. 20). This result is not surprising since most natural scenes do not contain many saturated colors. The correlation between R, G and B increases as the gamut of colors decreases, and is obviously complete for an achromatic scene.

It is possible to transform R, G and B into three new components so as to eliminate the correlation between components. This is achieved with the Karhunen-Loeve (K-L) transformation (Habibi and Wintz [50]). Most of the signal energy is in one component and little is in the other two (referred to as energy compaction). Hopefully one could then code the

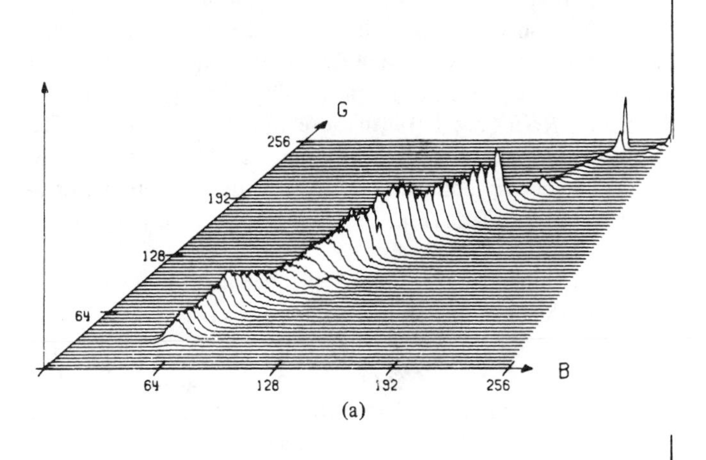

(a)

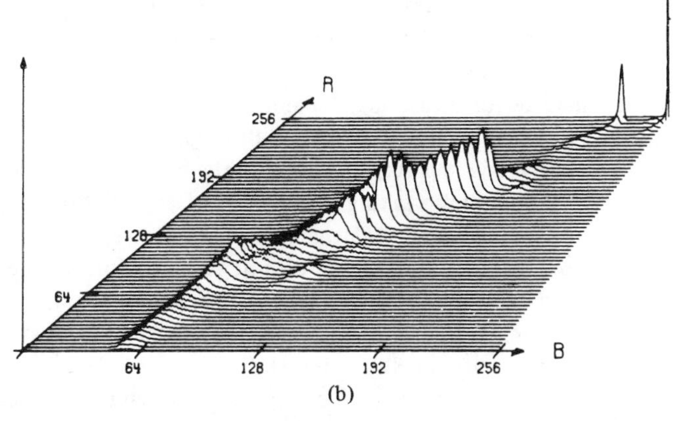

(b)

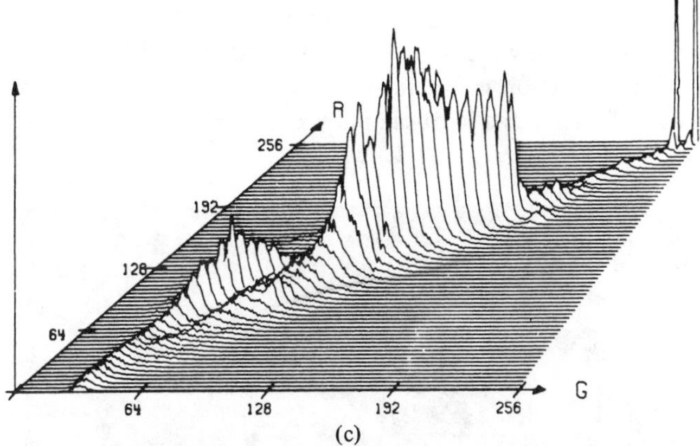

(c)

Fig. 20 Two-dimensional probability density functions of the three-dimensional *RGB* signal, for SMPTE Slide No. 2, showing the strong correlation between components. (a) *B* vs. *G*, (b) *R* vs. *B*, and (c) *R* vs. *G* (from Frei and Jaeger [49]).

TABLE III
COMPARISON OF THE ENERGY COMPACTION ABILITY OF THREE DIFFERENT COMPONENT REPRESENTATIONS ON TWO PICTURES

Picture	Coordinate System	Power (%) 1st Comp.	Power (%) 2nd Comp.	Power (%) 3rd Comp.
GIRL	RGB	45.14	35.41	19.45
	YIQ	78.32	17.54	4.14
	K1, K2, K3	85.84	12.10	2.06
COUPLE	RGB	51.55	31.09	17.36
	YIQ	84.84	13.81	1.35
	K1, K2, K3	92.75	6.46	0.79

high-energy component accurately and the low energy components less accurately so that the overall bit-rate is reduced in comparison, say, to coding each of the components of the RGB signal with high accuracy. This transform, which depends upon the correlation between components, will vary from picture to picture. The first component, denoted by *K*1, consists of nearly equal parts of the red, green and blue signals for the SMPTE Test Slide used by Pratt [51]. This apportionment probably stems from the somewhat balanced distribution of colors in the picture. This interpretation is supported in another study in which the K-L transformation was calculated for six pictures [52]. Again the *K*1 component consisted of almost equal amounts of *R*, *G* and *B* except for two of the pictures which were unusual in that they contained large flat areas of predominantly one color. The *K*1 component is similar to the luminance signal except that, if displayed alone, it would yield much brighter pictures in the blue areas. The energy compaction ability of the *K-L* transform is compared with that of the *RGB* signal and the *YIQ* signal in Table III [51, 53]. It can be seen that the *YIQ* signal provides almost as high an energy compaction as does a K-L signal and, of course, it leads to a compatible system.

Frei [47] obtains about the same energy compaction as the K-L transform using a two-stage process comprising a transformation from *RGB* to tristimulus *XYZ* followed by a log transform. However, direct comparison with unitary linear transformations is probably not relevant because of the non-linearity of the model.

From this point on we will assume we have a luminance signal and two chrominance signals and study the properties of just the chrominance components.

B. Range of the Chrominance Signal

In practice most scenes require a very much smaller range of chromaticities than those enclosed by the spectral locus. A television system can only reproduce chromaticities falling within the triangle formed by connecting the chromaticities of the three display-tube phosphors. That this range is not very restrictive can be appreciated from Fig. 21 where it can be seen that this range of chromaticities is larger than that encompassed by the range of all pigments, dyes and inks [54]. However, most normal scenes encompass but a small fraction of the total allowed range as shown in the plots of Fig. 22 for two specific pictures where the darkness of the location in the UCS diagram indicates the frequency of occurrence of that particular chromaticity in the accompanying diagram. The adjacent plot shows the same information plotted logarithmically; one can regard the light areas here as being close to the extreme range of chromaticities found in the picture.

Assuming the NTSC signals of (9), it is easy to see that not all chromaticities within the triangle of Fig. 21 are obtainable at all luminances. In fact, as the luminance reaches a value of 1 unit, the maximum range of obtainable chromaticities shrinks to a point centered on the white point. Consider, for example, a pure blue signal of maximum value 1. From (7) the luminance will be 0.11. Higher luminances can only be obtained by adding red or green to the signal which moves the chromaticity away from the blue corner of the triangle. The maximum range of achievable chromaticities at different values of luminance is shown in Fig. 23 for a linear system.

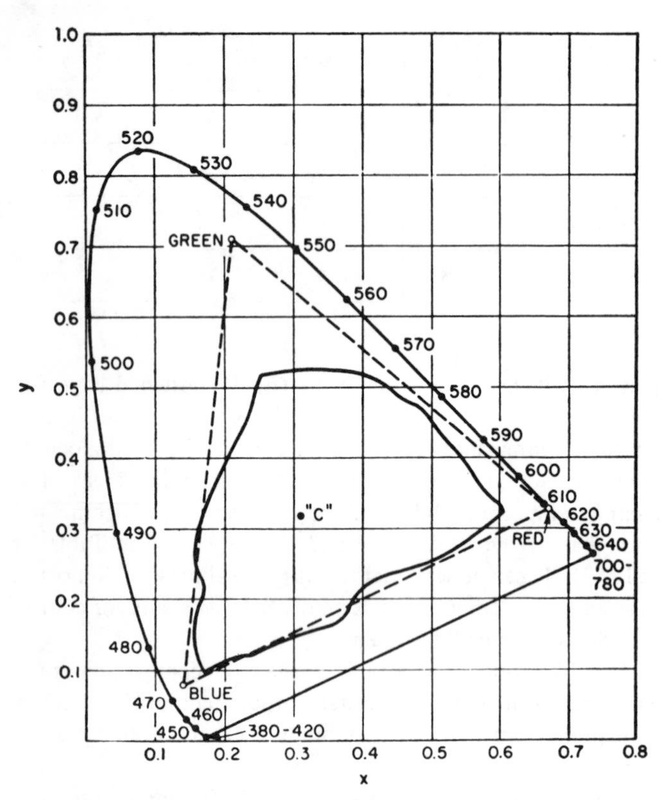

Fig. 21. CIE x, y diagram comparing the gamut of colors reproducible by the NTSC receiver (area lying within the dashed triangle) and the extreme purities of pigments, dyes and inks under Illuminant C (from Fink, Fig. 4.7 [54]).

It is perhaps more helpful to think in terms of the range of the chrominance signals rather than chromaticities because these are the signals which are frequently coded in practice. It is interesting to compare the probability density functions for the signals RGB (Fig. 19) with those of YUV. The plots in Fig. 24 are average results from six test pictures (Pirsch and Stenger [55]). The energy compaction effect of the YUV transformation is clearly evident from the reduced dynamic range of the signals relative to those in Fig. 19.

Davidse [56] has measured the probability distribution function, not of the color difference signals (I or Q) but of the amplitude of the color subcarrier which can be regarded as the root-sum-square of I and Q or, equivalently, of U and V. Fig. 25 gives a histogram of the subcarrier amplitude for 295 studio scenes and slides. Of course, individual pictures will vary greatly from this average. The average amplitude of the color subcarrier for this set of pictures is 11 percent of the maximum possible value.

One method of measuring the amount of redundancy in a signal is to calculate the entropy of the signal. Simple first-order entropy is a measure of the bit-rate required to code each sample separately, taking no account of any relationship between an element and its neighbors. The joint entropy of two or more signals indicates the bit-rate required to code the signals when they are processed jointly instead of separately. Rubinstein and Limb [52] and Pirsch and Stenger [55] have used joint entropies to study the statistical dependence between the components of a color signal. Measurement of the cross-correlation coefficient between components would be an alternative method of measuring statistical dependence, but by using an entropy measure, a quantity more directly related to bit-rate reduction is obtained. For six test pictures the average value of 11.22 bits/joint-sample was obtained for the quantity $H_S = H(Y) + H(U) + H(V)$ where Y, U and V were individu-

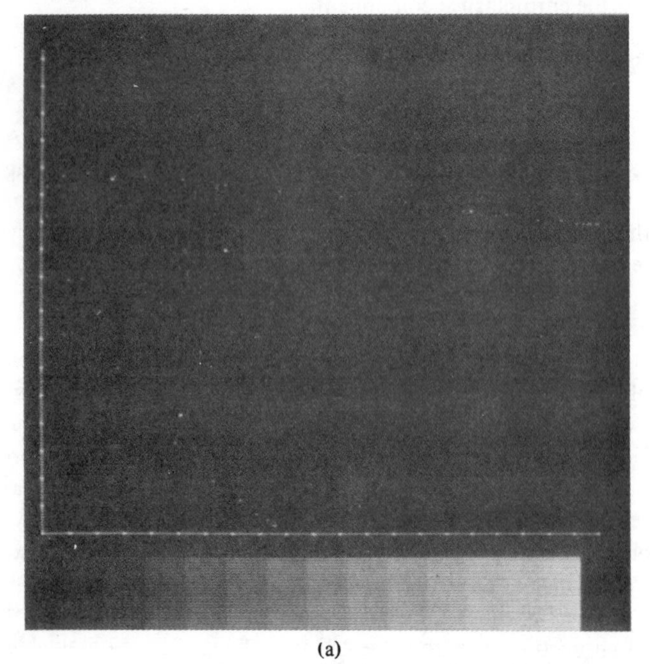

(a)

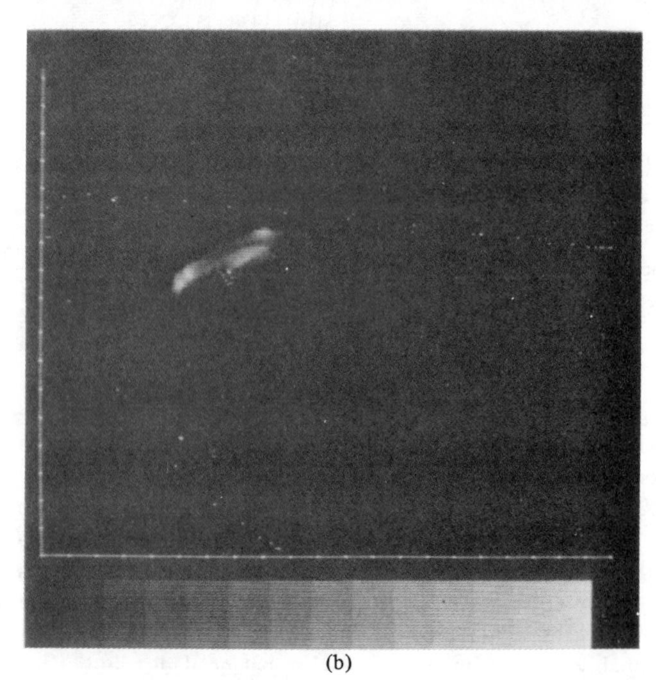

(b)

Fig. 22. Probability density plot of chromaticity for two pictures plotted in the CIE-UCS diagram: (a) and (b) Linear and logarithmic plots, respectively, for the picture shown in (c). (e) and (f) Linear and logarithmic plots, respectively, for the picture shown in (d). The lighter the area, the greater the probability density.

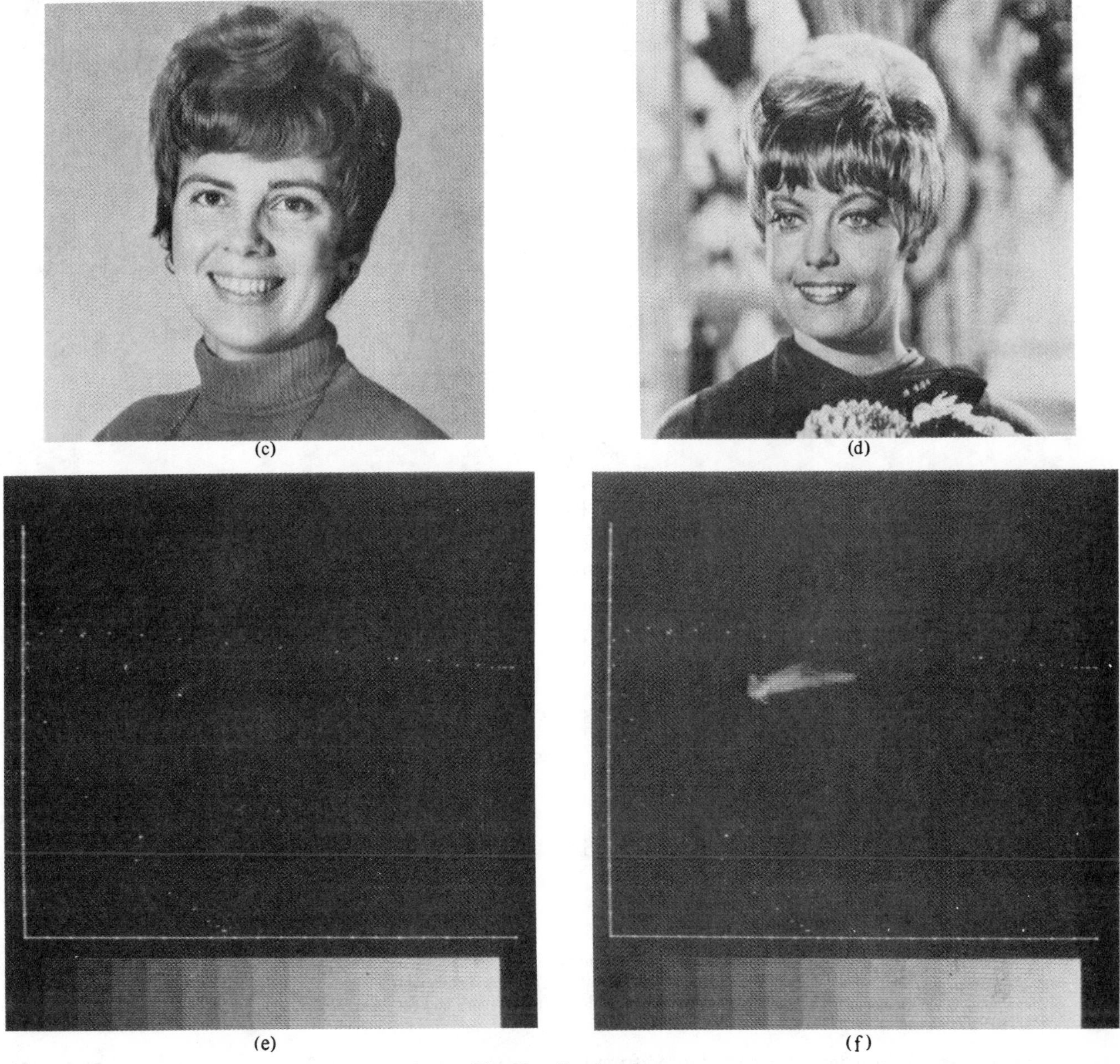

(c) (d)

(e) (f)

Fig. 22. Continued.

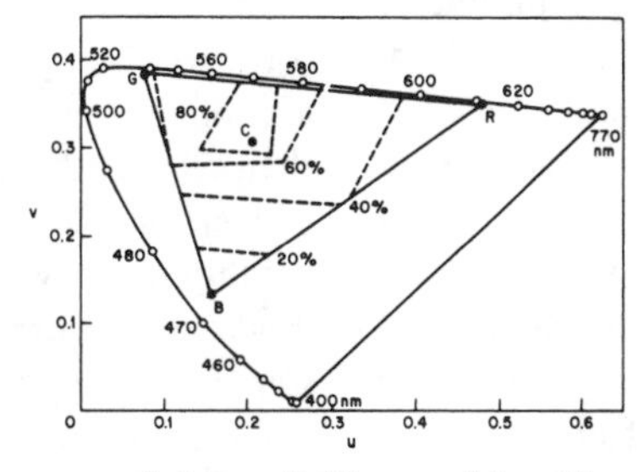

Fig. 23. The range of chromaticities possible with a television signal at a given luminance. The dashed contour lines indicate the range at 20, 40, 60 and 80 percent of maximum luminance. Below a luminance of 11 percent, all chromaticities within the triangle are achieveable.

ally quantized with an accuracy of 6 bits. This compares with a value of 9.04 bits/joint-sample for the quantity $H_T = H(Y,U,V)$ [55] (there was one chrominance sample per five luminance samples). Further measurements show that about one-half of this redundancy results from dependence between U and V and only about one-quarter each from dependence between Y and U and between Y and V.

The statistical dependence between *differential signals* of a DPCM coder rather than the PCM signals have also been studied. The results in Table IV show that for six pictures the

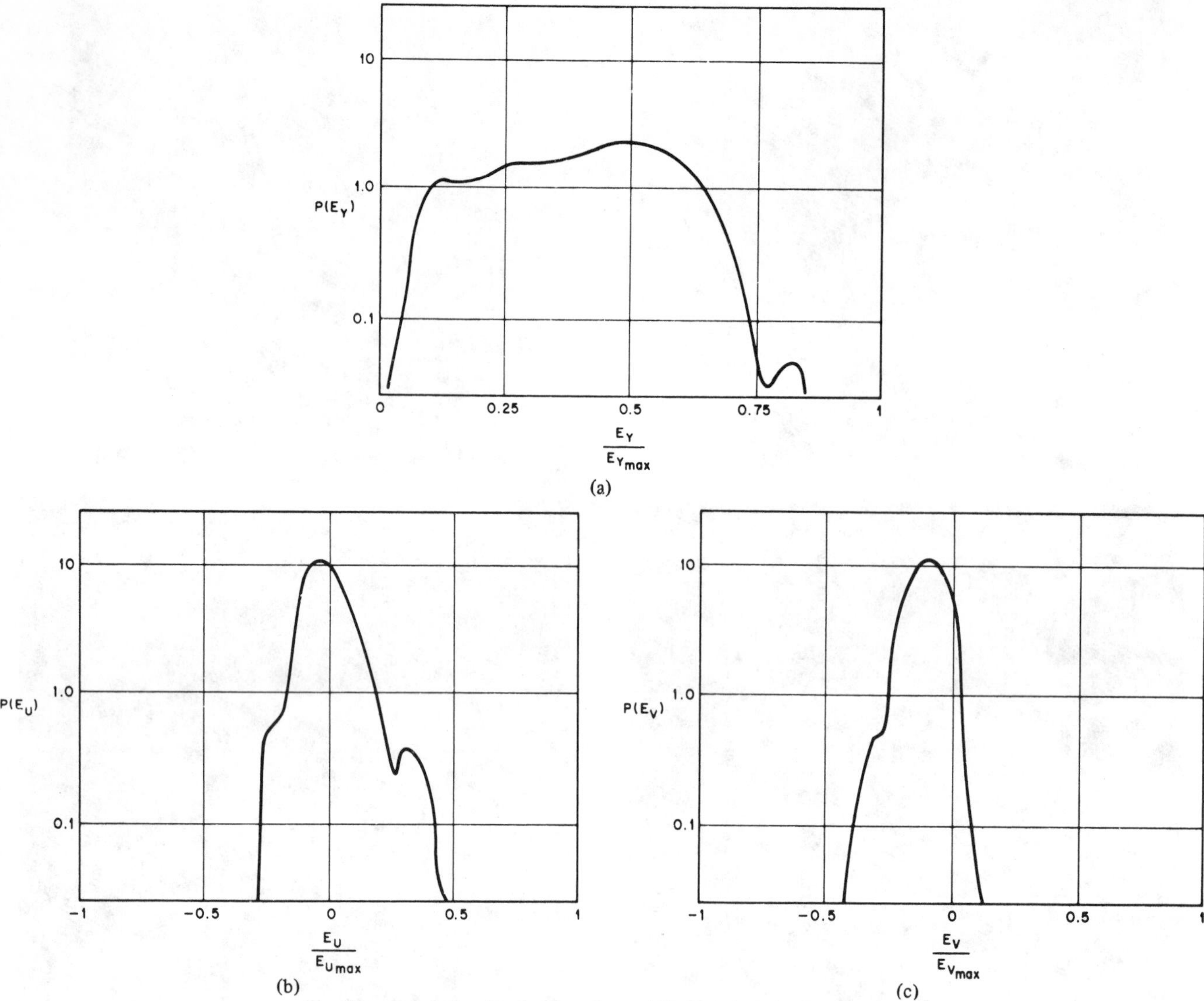

Fig. 24 Probability density functions of Y, U and V signals averaged across six different test pictures. The reduced dynamic range of the U and V signals is clearly evident (from Pirsch and Stenger [55]).

entropy of the differential chrominance signals (derived from I and Q signals sampled at 1/3 and 1/6 the rate of Y) $H(\Delta I)$ and $H(\Delta Q)$ are changed very little whether or not the amplitude of either the luminance signal or the differential luminance signal is known [52]. Measurements on Y, I and Q signals sampled at the same rate showed very little increase in statistical dependence. Measurements on YUV differential signals, using both a previous element predictor and an "optimized" predictor, generally support the results of Table IV [55], indicating that there is little *statistical* gain to be obtained by using knowledge of the *amplitude* of the luminance signal in differentially encoding the chrominance signal.

C. Other Chrominance Transformations

Assuming we desire to retain a luminance signal and two color difference signals, we could consider coordinate conversions other than the chrominance signals IQ and UV to obtain different variables which may be more suitable for coding purposes. As we have seen, a rotation of the UV signals by 33° yields the NTSC signals. The coordinates could be rotated so as to eliminate correlation between the components (K-L conversion), but there is no a priori reason why the K-L transform will be optimum. If the color difference signals could be accurately modeled by a two-dimensional Gaussian source, then all statistical dependence would be eliminated by the K-L conversion. Besides being grossly non-Gaussian, the joint probability density of the color difference signals is also multimodal and appears as a number of spikes and ridges which can sometimes be quite sharp (see Fig. 22). Alternatively, one could rotate the axes so as to find a minimum in the resulting bit-rate.

A difficulty in comparing different transformations is that in practice the optimal division of bandwidth (and sampling-rate) between the two components will vary depending on the angle of rotation; and for a critical comparison of different angles the bandwidth allocation should be optimized separately at each angle. Adding further complexity, one can investigate nonorthogonal transformations of the two color difference components; and the assumptions in making valid com-

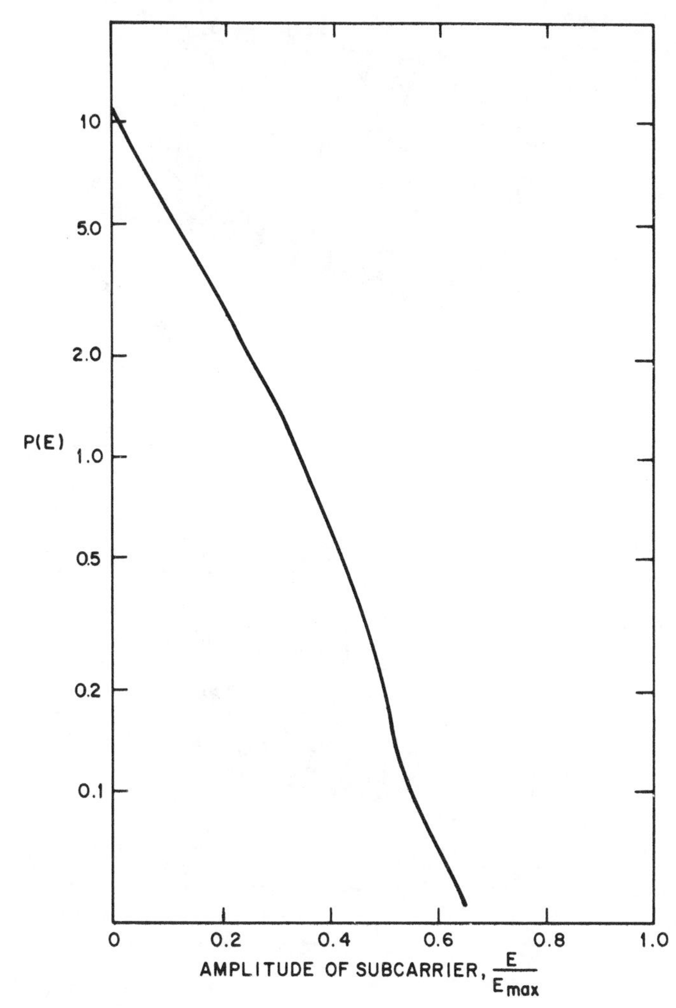

Fig. 25. Probability density function of the NTSC color subcarrier (see equation 10) (from Davidse [56]).

TABLE IV
COMPARISON OF ENTROPY IN BITS/SAMPLES OF CHROMINANCE SIGNALS GIVEN EITHER THE DIFFERENTIAL LUMINANCE SIGNAL OR THE UNCODED LUMINANCE SIGNAL

Picture	H(ΔI)	H(ΔI/ΔY)	H(ΔI/Y')	H(ΔQ)	H(ΔQ/ΔY)	H(ΔQ/Y')
A	2.17	2.09	2.06	1.56	1.53	1.47
B	1.73	1.62	1.60	1.30	1.26	1.15
C	1.58	1.47	1.41	1.16	1.11	1.02
D	1.75	1.66	1.63	1.28	1.24	1.11
E	2.26	2.05	2.01	1.75	1.62	1.57
F	2.08	1.72	1.87	1.34	1.26	1.26

parisons are more severely strained. A compression of the color space effectively occurs where the angle between the axes is acute leading to a diminution in the effect of errors in that part of the color space with a corresponding expansion where the angle is obtuse.

In spite of these problems, a comparison has been made of the entropy and redundancy resulting from different choices of component conversion [52]. The results are summarized in Table V for a set of six color pictures. If the same orthogonal transformation is used for all pictures, the transformation corresponding to a 23° rotation gives the lowest joint entropy values for most pictures, closely followed by the *IQ* signals. If

TABLE V
COMPARISON OF THE JOINT ENTROPY (H_T) AND THE REDUNDANCY ($R = H_S - H_T$) OF A Y, $C1$ AND $C2$ TRIPLET FOR DIFFERENT TRANSFORMATIONS OF $C1$ AND $C2$.

Picture		0°	23°	33°	K-L	Non Orthog.
A	H_T	6.46	6.31	6.38	6.31	
	R	0.34	0.20	0.22	0.20	
B	H_T	5.45	5.21	5.27	5.27	
	R	0.27	0.22	0.26	0.25	
C	H_T	4.83	4.70	4.64	4.80	
	R	0.25	0.24	0.32	0.18	
D	H_T	5.58	5.53	5.52	5.50	
	R	0.29	0.18	0.22	0.18	
E	H_T	5.94	5.78	5.88	5.77	5.31
	R	0.84	0.62	0.49	0.59	.44
F	H_T	5.50	5.33	5.26	5.26	
	R	0.88	0.68	0.52	0.51	

the orthogonal transformation is selected separately for each picture, the K-L transform is marginally better than any fixed rotation across all pictures, although in most cases one of the fixed rotations has a lower value. A nonorthogonal transform was investigated for one picture, and significantly lower entropies were obtained than for the orthogonal transforms. The reduction came not from a reduction in redundancy *between* the chrominance components, but rather from a reduction in entropy of the components themselves. The above findings take little account of the sensitivity of the human observer to various orientations of the color axes. More work relating statistical and perceptual factors is needed.

The primary conclusion to be drawn from these studies is that redundancy between the *amplitudes* of the differential signals of the components of a well chosen color signal, although present, is not very significant, and if exploited will not lead to any large reduction in bit-rate.

VI. COMPONENT CODING METHODS

The starting point for the coding of color components is usually an *R, G, B* signal or the components of a composite signal, e.g., *Y, I, Q*. These signals may be transformed to another space for digitization. This digitization may be carried out using a criterion that depends on measurements in a further space. The digitization may involve independent quantization of each component or, alternatively, multidimensional quantization by simultaneous treatment of more than one component. The resulting digitized signal is sometimes further quantized and statistically coded before being transmitted. This will be made clearer in the sections below.

A. PCM Encoding

If *R, G* and *B* signals are to be digitized, it is not necessary to quantize each with the same precision since they are not equally sensitive to added noise. For example, for a picture containing only one of the three primary colors displayed at a constant luminance, the threshold signal-to-noise ratio is 36 dB for blue, 41 dB for red and 43 dB for green (Mueller and Wengenroth [57]), although these figures vary somewhat with luminance. The difference in noise sensitivity of the color components is even larger for "normal" pictures. In the blue

TABLE VI
ENTROPY IN BITS/SAMPLE OF THE THREE COMPONENTS OF A PAL SIGNAL; THE CHROMINANCE SIGNALS ARE BANDLIMITED TO 1.0 MHz AND SAMPLED AT 2.0 MHz (FROM PIRSH AND STENGER [55])

COMPONENT	H(X)	H(X|A)	H(X|C)
Y	7.34	4.66	4.85
U	5.57	4.15	2.96
V	5.24	3.85	2.98

signal, noise is 10 dB less visible than in the red signal and 20 dB less visible than in the green signal (Bushan [58]). For quantization of the full bandwidth signals with no visible error, at least 4 bits are required for the blue signal, 5 bits for the red signal and 6 bits for the green signal [58]. The drawback of coding the components of the *RGB* signal is that relatively high spatial resolution is required by each component. Relative to an original bandwidth of 2 MHz for the green, the bandwidth of the red and blue signals could be reduced to 1.5 MHz and 1.0 MHz, respectively for typical scenes [58]. (It is not clear that even this reduction in bandwidth would be acceptable for pictures that were either predominantly red or predominantly blue.) This relative reduction is very much less than that possible with chrominance signals [24].

In der Smitten [59] partially overcame this problem of the increased *RGB* requirements by splitting the input into a low-band (0-1.25 MHz) *RGB* signal and a high-band (1.25-3.75 MHz) luminance signal. The RGB space was quantized three-dimensionally into 125 volumes. The boundaries of these volumes were determined by subjective experiment. The *RGB* signal would require 7 bits at a sample rate of 2.5 MHz, while the high-band signal would require perhaps 3-bit DPCM at 7.5 MHz, giving 17.5 + 22.5 = 40 Mbits/s. The *RGB* signal requires a relatively large fraction of the total bit-rate, and with just 125 colors some color-contouring is visible in the coded pictures.

Luminance and chrominance signals may be directly digitized (Gronemann [60]), and variable-length coding (such as Huffman coding [61]) can be used to further reduce the bit-rate. A little more reduction in bit-rate is achieved for the *U, V* signals than for the luminance signal because of the greater reduction in dynamic range of the chrominance signals relative to the luminance signal as previously seen in Fig. 24. Average values of first-order entropy are shown in Table VI for six different test slides (Pirsch and Stenger [55]). However, much greater reduction in entropy is achieved by exploiting knowledge of the previous sample in either the vertical direction or the horizontal direction. From Table VI we see that the conditional entropy using the homologous element in the previous line is significantly lower than the conditional entropy obtained using the previous element in the same line. This, no doubt, results from the fact that the horizontal spacing between chrominance samples is greater than the vertical spacing, interlace notwithstanding.

A number of studies have investigated the visibility of errors resulting from quantizing the chrominance signals. The effect of digitizing the NTSC *I* and *Q* signals, the PAL *U* and *V* signals and a hue-saturation type signal is shown in Fig. 26 (Frei,

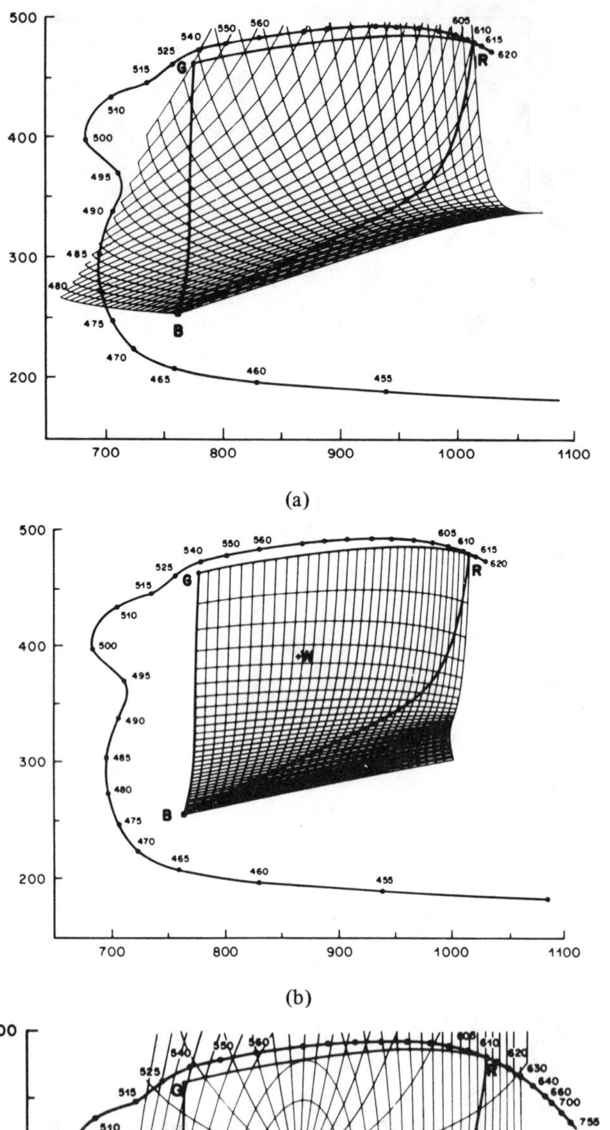

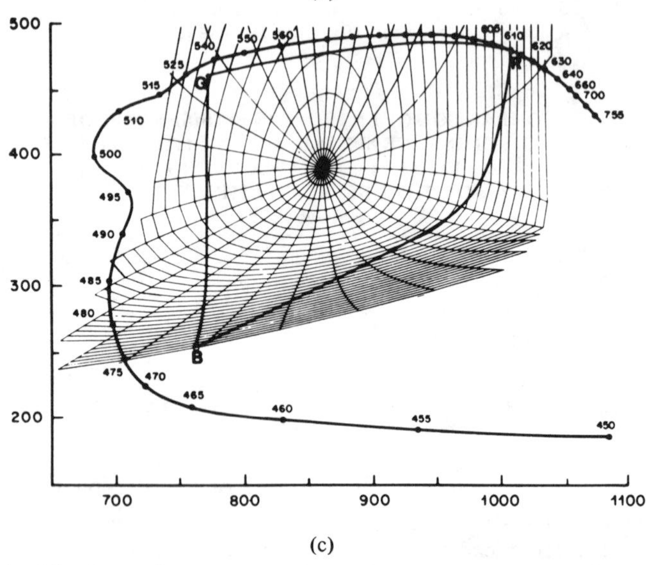

Fig. 26. Plots of 5-bit quantization grids mapped from three different color spaces onto MacAdam's geodesic diagram: (a) NTSC *I* and *Q* signals, (b) PAL *U* and *V* signals, and (c) hue and saturation signals derived from *U* and *V* signals (from Frei, Jaeger and Probst [62] and Frei [63]).

Jaeger and Probst [62, 63]). A rectangular quantization grid in each case is mapped into a new space in which equal distances represent more nearly uniform changes in perceived chromaticity. The space used is McAdams's Geodesic space (Section IV.C). The plot is given for one value of luminance, and no ac-

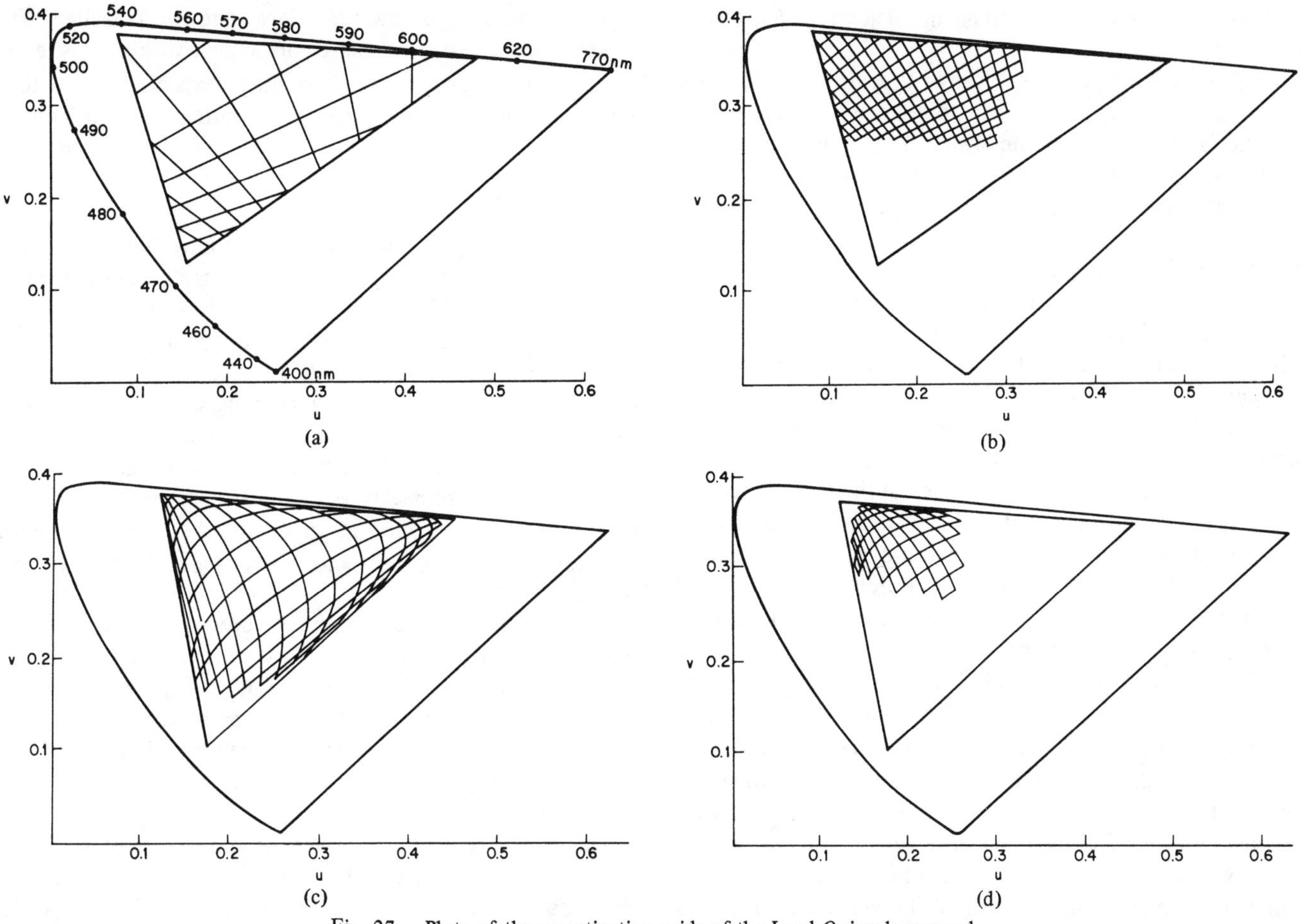

Fig. 27. Plots of the quantization grids of the I and Q signals mapped onto the CIE Uniform Chromaticity scale. Quantization is to approximately 20 levels: (a) and (b) Quantization without gamma correction at luminances of 10 percent and 50 percent of the maximum value, (c) and (d) The same as parts (a) and (b) but applied to a gamma corrected signal (from Marti [65]).

count is taken of gamma correction (Section III.B). None of the quantizations give very uniform areas in MacAdam's Geodesic diagram. In all cases there is considerable compression of the grid in the blue and purple area. The hue and saturation signals appear inferior to the chrominance signals.

Jain and Pratt [64] have uniformly digitized color signals in five different coordinate systems using a Geodesic measure of color difference to evaluate the error at the three primary colors and the three binary mixtures of these colors. A total of 12 bits was allocated to the three components, and different divisions of the bits between the three components were investigated. The *RGB* components gave the smallest errors primarily because the actual range of colors better filled the quantizing space.

When the quantization steps for I and Q are mapped onto the 1960 UCS CIE diagram rather than MacAdam's Geodesic diagram, the grid is again quite nonuniform with significant compression in the blue area (Fig. 27 a, b, Marti [65]). At higher luminances, less of the u, v space is occupied and the quantization in the remaining area is more perceptually uniform. Notice that the grid size is also finer, implying a dependence on Y. In practice the television signals are gamma corrected prior to transmission, and this nonlinear operation significantly affects the quantization process. In Fig. 27c, d [65] are shown the same quantization steps as appear in Fig. 27a, b, but assuming a gamma (=2.2) corrected signal. The grid is now somewhat less dependent on Y although still quite nonuniform. Logarithmic correction of *RG* and *B* signals can be used to remove the dependence of grid size on Y, and nonlinear pre- and post-processing of the chrominance signals can be employed before and after digitization to produce even more perceptually uniform quantization. Such a correction curve has been calculated by Marti based on errors in both UCS space and the CIE U^*, V^*, W^* space [65].

For a linear transmission system the luminance information is carried by the Y signal, and this is advantageous from the point of view of monochrome-color compatibility. An additional advantage is that errors in the chrominance signals are less visible than errors in the luminance signal; therefore, coding or transmission errors can be matched to the different sensitivity requirements. With gamma correction this independence of the luminance signal is partially lost. Consequently, the effect of an error in the chrominance signal on luminance as well as chrominance should be taken into account. For this reason Marti used a three-dimensional space, $U^* V^* W^*$, in deriving the nonlinear correction for chrominance.

In the methods described so far the luminance and chrominance components of the signal have been coded independently. However, this leads to certain inefficiencies [65, 66] in that:

(a) The range of possible values of the chrominance signal at a given luminance is less (and sometimes very much less) than the overall range of the signal. This reduction, as previously described, occurs at high luminances because of the maximum signal value permitted the individual components, and at low luminance through the requirement that $C1$ and $C2$ go to zero as the luminance signal goes to zero.

(b) The sensitivity of the human observer to perturbations in $C1$ and $C2$ varies throughout the three-dimensional space.

Now in order to digitize the signal components efficiently, two approaches are possible:

(a) transform the signal components to a new space which approximates a uniform chromaticity space, digitize uniformly and code; the procedure is then reversed at the receiver, or

(b) map the uniform quantization of some UCS space back to the original signal space and digitize and code in this space.

Stenger [66] has taken the latter approach and quantized the U and V signals of the PAL system as a function of Y, U and V by constructing volumes having approximately the same size when measured in $U^*V^*W^*$ space. Four different quantizations are employed covering four different luminance ranges. In each case about 1000 cells of different sizes are used. This variable, two-dimensional quantization need not be difficult to implement since the boundary regions can be stored in read-only memories for simultaneous processing of the two chrominance signals. In fact, the process is much simpler than converting the signals to another space. The penalty paid here is that since the cells are of odd shape a simple metric for the quantized data no longer exists, making it difficult to perform subsequent processing such as DPCM or transform encoding. The UV space may be divided up somewhat more restrictively by using two sets of straight lines. Lines within each set need not be parallel but may not cross within the dynamic range of the signals. The cells constructed by such a quantization can be ordered in the same way as a rectangular division of the color space, and hence coding with the usual types of algorithms is possible [66].

Solomon [67] has sought to exploit the factors discussed above, but using approach (a). Quantization takes place directly in the CIE UCS space, and a procedure called "luminance scaled chromaticity" is used to scale the chrominance signals at a given luminance, so that they completely fill the available quantizing range. With this scaling, four-bit quantization of the chromaticity components produces no objectionable chrominance-contouring or color errors.

Finally, the effect of transmission errors on a PCM coded signal has been studied by Pratt [68]. It is assumed that each bit of the digitized signal has the same independent probability of being corrupted (binary symmetric channel). General equations for the shift in luminance and shift in chromaticity (in uv space) and the variances associated with these shifts are derived and applied to various transformations: RGB, YIQ, Yrg and Yuv. The general effect of transmission errors is to shift the displayed luminance slightly toward a mid value. Similarly, chromaticity values are shifted towards the average value of the coded signals which is the white point for RGB and YIQ. For Yrg the shift is towards yellow, and for Yuv the shift is toward purple. Transmission errors appear to have least effect on the RGB signal which is perhaps expected, since these signal components have the most uniform distribution of energy.

B. DPCM Coding

Most studies of color-signal encoding have started with a luminance signal and two color components, although there are some exceptions [51, 58]. As previously stated, we are primarily concerned in this paper with the coding of the chrominance signals rather than the luminance signal. However, the overall efficiency of a coding technique will depend more on the efficiency with which the luminance signal is coded. Approximately speaking, for simple one-dimensional coding of normal video-telephone type pictures, between 2 and 3 bits/pel might be considered adequate for the luminance signal, whereas 1/2 to 1 bit/luminance pel is needed for both chrominance signals.

Delta- and DPCM coding may be used on the chrominance signals in a similar manner to its use in luminance coding [69, 70, 71, 43, 55]. Using simple DPCM with previous element prediction, good quality pictures can be obtained with as few as 7 levels for I and 5 levels for Q when the filtering and sampling for I and Q is reduced to 1/3 and 1/6 that of the luminance signal [70].

Dither can be used to improve the appearance of the DPCM encoding by breaking up color contours that may appear if the quantization is too coarse. However, because of the large size of a color sample, less visual filtering occurs at normal viewing distances; and hence the subjective improvement is rather marginal (Limb and Rubinstein [70]). The resulting signal may be coded with a variable-length code (Rubinstein and Limb [52]), and the reduction in bit-rate over that achieved with a constant-length code is comparable to that obtained for the luminance signal. For the 7- and 5-level signals previously mentioned, the resulting entropy of constant-length codes is 2.81 and 2.32 bits/pel, respectively, whereas for variable-length codes the corresponding figures range from 1.6 to 2.2 bits/pel for I and from 1.2 to 1.8 bits/pel for Q for six different head and shoulders pictures; the reduction would be less for pictures with a larger amount of detail.

A method for further reducing the chrominance bit-rate is to reduce chrominance resolution in the vertical direction. One digitally coded chrominance component can be transmitted at the end of each line of coded luminance signal, a format which is highly compatible with monochrome television [69, 72]. Better picture quality can be obtained if the missing chrominance information is reconstructed, not by repeating the previous line as done with SECAM, but by averaging the previous and next lines [70, 71]. This procedure requires an additional line of delay, not only for the chrominance signal but also for the luminance signal.

It is not necessary to build separate DPCM coders for the luminance and chrominance coders [70]. Fig. 28 illustrates a simple method for using a single DPCM coder for the three components. Assume that the analog chrominance information is first time-compressed. This is achieved by sampling and

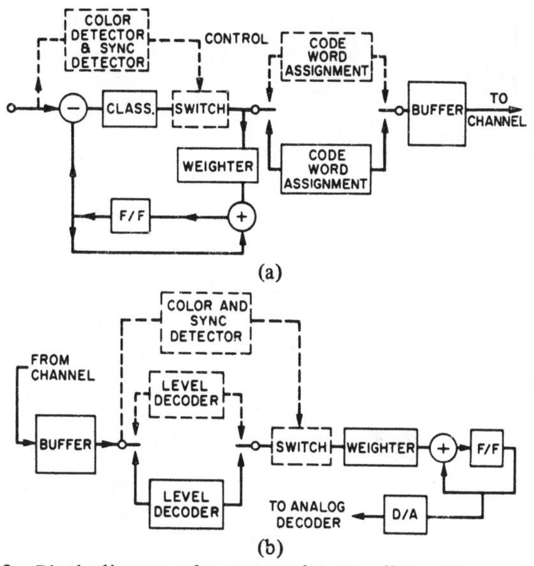

Fig. 28 Block diagram of a system for encoding a time-compressed color signal: (a) Coder, (b) Decoder. The chrominance signals use a subset of the quantizing levels used by the luminance signal (from Limb, Rubinstein and Walsh [10]).

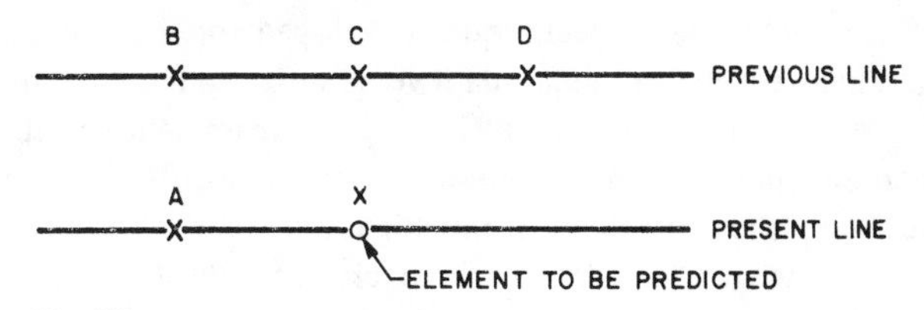

Fig. 29. Location of picture elements used in measurements of entropy.

TABLE VII
WEIGHT OF COEFFICIENTS FOR TWO DIFFERENT TYPES OF PREDICTORS, MINIMUM ENTROPY AND MINIMUM FOURTH POWER OF THE ERROR; SEE FIG. 29 FOR THE POSITION OF PICTURE ELEMENTS *A*, *B* AND *C* (FROM PIRSCH AND STENGER [55])

SIGNAL COMPONENT	PREDICTOR COEFFICIENTS MIN. ENTROPY			PREDICTOR COEFFICIENTS MIN. 4TH POWER OF ERROR		
	A	B	C	A	B	C
Y	7/8	-1/2	5/8	7/8	-5/8	3/4
U	3/8	-1/4	7/8	5/8	-1/2	7/8
V	3/8	-1/4	7/8	3/8	-1/4	7/8

storing the chrominance signal at, for example, 1/4 of the sampling rate used for the luminance signal. This signal is then read out of the store at the full luminance sampling rate during the horizontal blanking interval.[9] The only change in the basic DPCM loop is that a switch is inserted between the classifier and weighter, and the coarser quantization for the chrominance signals is obtained by inhibiting use of the outer quantizer levels. Variations abound: both input and output quantizer levels can be switched for better matching of quantizer characteristics to the properties of the signal; to avoid time-multiplexing, the input luminance and chrominance samples can be interleaved [72] using a more complex predictor in the feedback loop (luminance samples would be placed in one store and chrominance samples in another).

DITEC [71], a digital DPCM coder for the NTSC signal, operates at a bit-rate of 29.2 Mbits/s for the picture signal by DPCM coding with 5 bits/pel for *Y* and 4 bits/pel for *I* and *Q*, transmitted on alternate lines. The DPCM coders use uniformly spaced levels, switching between fine spacing in low-detail portions of the picture and coarse spacing at edges. One bit is used to indicate the scale being employed. Sampling rates as low as 6.02, 1.77 and 0.67 MHz for the *Y*, *I* and *Q* signals, respectively, would normally produce aliasing errors in all three signals. However, frequency interleaved sampling [73] is used on all components in the coder, and comb filters are used in the decoder to remove aliasing errors before reforming the composite signal. In *A-B* comparisons with white noise, the coded signal was judged to be better than an analog signal with a 56 dB peak-to-peak signal to weighted noise ratio. The main source of impairment in the coder appears to be loss of vertical resolution due to comb-filtering and line-alternating transmission of the *I* and *Q* signals.

[9] Depending on the chrominance resolution that is required, it may be necessary also to time-compress the luminance information by a small amount to make enough room for the chrominance samples.

OCCITAN is a computer controlled coding facility for studying the DPCM coding of signal components derived from a SECAM signal [74]. The sampling rate, quantizing characteristic and prediction method for the luminance and chrominance signals (*U* and *V* are coded in the same circuit on consecutive lines) may all be varied, as well as the comb-filter used to separate the components.

It was found that previous element prediction for the DPCM coder was most suitable since corresponding samples of the same chrominance component were too far removed vertically (because of the transmission of *U* and *V* on alternate lines). Interleaved sampling was used, but little reduction in sampling rate from the nominal Nyquist rate was possible without introducing degradation. A rate of 2.63 MHz was used. A 4-bit quantizer was necessary if edge-busyness was to be avoided. Total bit-rates in excess of 40 Mbits/s were required to produce an acceptable picture quality for television transmission (4 on the CCIR 5-point scale [7]). The chrominance signal accounts for approximately 25 percent of this rate.

Pirsch and Stenger [55] have studied two different methods for designing linear predictors for DPCM coders in which a weighted average of previous samples, *A*, *B* and *C*, is used (Fig. 29). One technique minimizes the entropy of the prediction error, whereas the other technique minimizes the fourth power of the prediction error. In both cases computer search techniques were used, and the coefficients were adjusted in steps of 1/8. The weights of the predictor coefficients for the two design methods are given in Table VII. It is immediately obvious that the predictors for the two chrominance signals, although quite similar, are quite different from the luminance predictor. In the chrominance predictors much greater weight is given to the homologous element of the preceding line than occurs for the luminance predictor. The error power for the chrominance coders is reduced by a factor of about 5 in going from a previous-element predictor to the optimized third-order predictor.

More sophisticated techniques developed for the encoding of luminance signals such as adaptive quantization are just now being explored as to their efficacy for chrominance signals. Other techniques such as multidimensional quantization and adaptive prediction still await investigation.

The design of luminance quantizers for DPCM coders has developed into quite an art (see for example, Sharma and Netravali [75]). The viewer's sensitivity to coding errors introduced at an edge are incorporated into the design procedure in different ways. Color quantizers for DPCM coders have been designed using a Lloyd-Max type method [76] to place the levels so as to minimize the mean square quantizing error. However, instead of using the probability function to do the minimization, a "visibility" function is used instead (Netravali and Rubinstein [43]). This visibility function is a measure of the visibility of noise at an edge in a particular picture as a function of the horizontal slope of the edge. The function is shown in Fig. 30a. The function decreases with increasing slope because, firstly, the eye is less sensitive to errors at large slopes and, secondly, a typical picture contains fewer picture elements with large slopes. The quantizers designed in this way (visibility quantizers) are less companded than minimum-mean-square-error quantizers, but the inner levels are more coarsely spaced. A quantizer was designed on the basis of measurements on a specific color picture and applied to a number of video-telephone type head-and-shoulders pictures. The DPCM encoder used simple previous-element prediction, and the chrominance signals were filtered and sampled at a quarter of the luminance values. With 13, 9 and 7 levels for the Y, $C1$ and $C2$ signals, respectively, high-quality pictures were obtained having first-order entropies in the range 1.3-1.8 bits/pel for the luminance signal and 0.4 and 0.3 bits/luminance pel for the $C1$ and $C2$ signals, respectively. The resulting entropy bit-rate was thus in the range 2.0-2.5 bits/pel [43].

As described in Section IV.E, changes in the luminance signal partially mask the ability of the human observer to discriminate small perturbations in the chrominance signal in the immediate vicinity of the luminance changes. This is illustrated in the visibility function of Fig. 30b. DPCM quantizers were designed to take advantage of this masking in the following manner [43]. Visibility functions of the $C1$ and $C2$ signals were experimentally measured (Section IV.E) not as a function of the slopes of the $C1$ and $C2$ signals but as a function of the slope, ΔY, of Y. The range of ΔY was then divided into four sub-ranges and the appropriate step size for each sub-range determined by referring to the visibility function. The chrominance coders switched to the appropriate uniform quantizer, depending on ΔY. The adaptive uniform quantizers were compared with nonadaptive uniform quantizers and found to give an advantage in entropy of 0.35 bits/chrominance pel in a total of 3.32 bits/chrominance pel. Although these figures demonstrate the advantage of luminance-adaptive chrominance coding, uniform quantization may not be the best method of exploiting the luminance-chrominance interaction.

C. Plateau Coding

Plateau coding attempts to break the chrominance signal into areas having approximately the same chrominance value (plateaus) (Limb and Rubinstein [77]).[10] These areas can then be represented by a single pair of values. Normally one would need to transmit the boundaries of the plateaus, but this could require a relatively large number of address bits. The

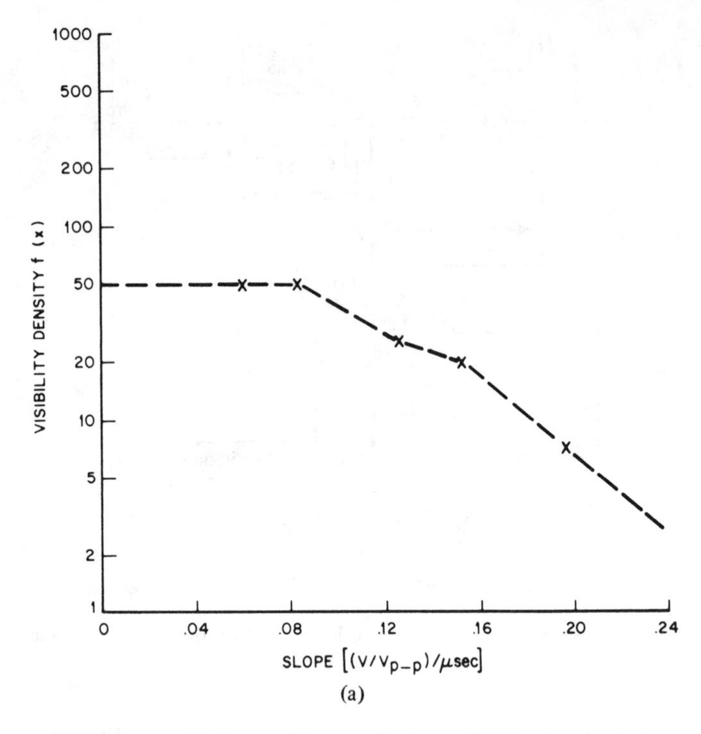

(a)

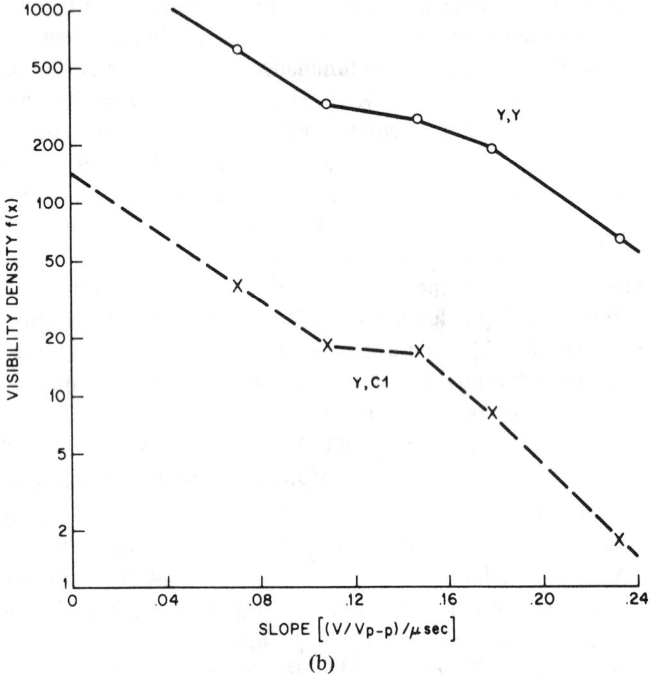

(b)

Fig. 30. Plot of visibility density function of an I-type chrominance signal against slope of a signal as measured through subjective experiment on the picture of Fig. 22c: (a) Visibility of the I signal as a function of the slope of the I signal. (b) Visibility of the I signal as a function of the slope of the Y signal. Also shown for comparison is the visibility density function of the Y signal (from Netravali and Rubinstein [43]).

[10] Ideally, the segmentation should be based on chromaticity rather than chrominance. However, chrominance signals are more easily obtained and more commonly used, and the technique works well in spite of the change in chrominance signal with luminance.

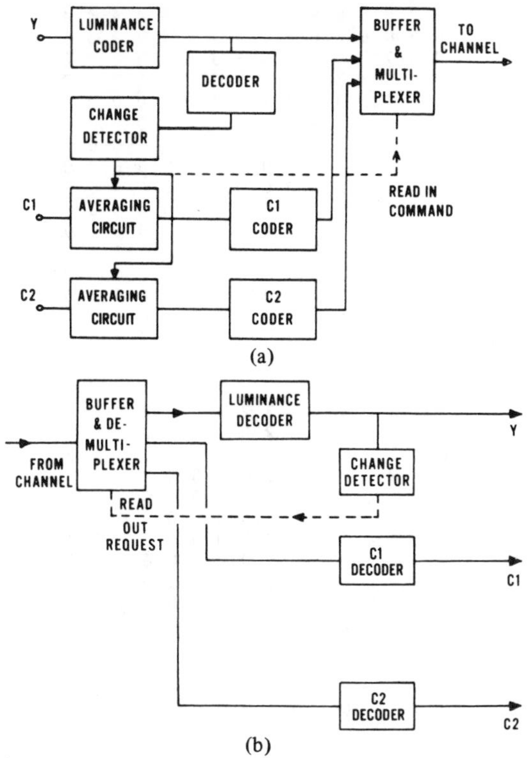

Fig. 31. Diagram of plateau coder. (a) Coder, (b) Decoder (from Limb and Rubinstein [77]).

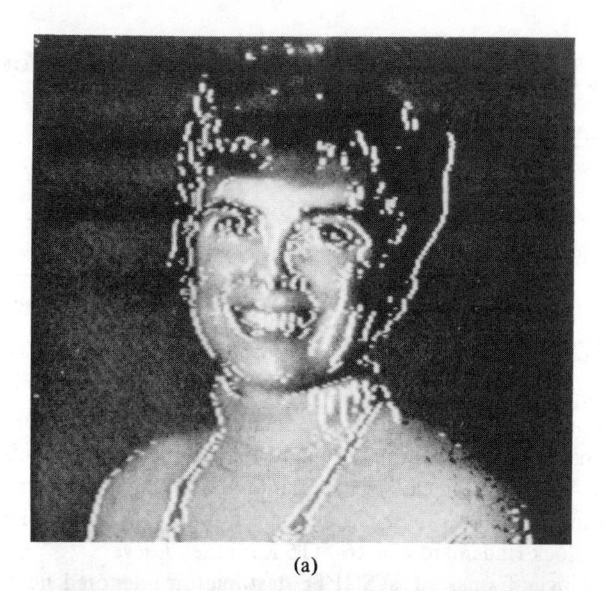

(a)

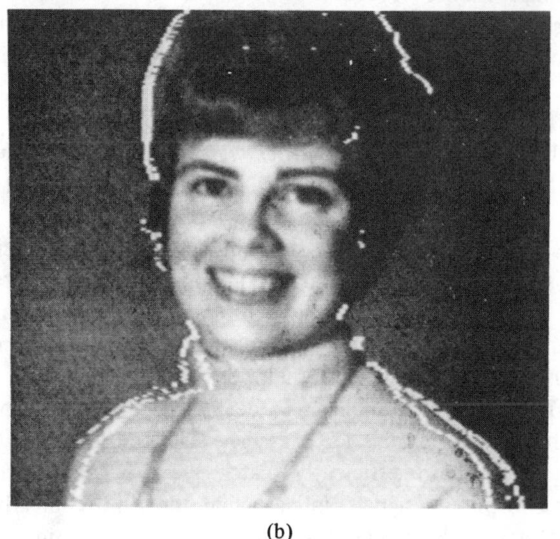

(b)

Fig. 32. Pictures showing significent changes marked by white dots: (a) Those points detected from changes in the luminance signal, (b) Those points detected from the chrominance error. The two sets of points clearly show the complementary action of the two methods (from Limb and Rubinstein [77]).

need for address bits is overcome by extracting the plateau boundaries from the luminance signal. Although changes in luminance do not necessarily indicate a change in chrominance (e.g., luminance texture falling within a single colored area), it is rare that a change in chrominance is not accompanied by a corresponding change in luminance. Thus by treating every luminance edge as signaling a possible chrominance edge, very few chrominance edges would be missed, albeit, many unnecessary luminance edges would also be included. This technique has been applied in one dimension, as shown in Fig. 31.

The omission of even a few chrominance changes in the one-dimensional simulation produced significant degradation in the picture. This may be overcome by examining the local chrominance error in the encoding operation. When this error exceeds some threshold, the plateau can be terminated. Normally this additional information would require an address to be transmitted; however, this can be obviated by forcing a change in the luminance signal just large enough to signal a chrominance boundary. These perturbations can be designed to have low visibility and will occur at a point in the picture where the chrominance signal is changing rapidly, and hence chrominance masking (although not as strong as luminance masking [43]) will diminish the visibility of the perturbation. Fig. 32a shows the change points detected by examing the luminance signal only (flagged by white markers). Fig. 32b indicates additional points detected by the chrominance detection circuit. Note the complementary nature of the two sets of points

The technique described in [77] is concerned with reducing the number of chrominance samples that needs to be transmitted. Netravali and Rubinstein [43] describe a method to efficiently PCM encode these samples for transmission. The best division of levels between $C1$ and $C2$ was calculated such that a measure of subjective distortion is minimized. Pictures coded in this manner were compared subjectively to pictures coded with other level assignments, keeping constant the number of bits required to code the plateaus. The calculated level assignments always gave the better quality picture.

Plateau coding gives high-quality reproductions on single pictures, but when applied to a sequence of pictures, variations of the change points from frame to frame can produce disturbing effects. By extending the algorithm to two dimensions, it may be possible to establish the plateau boundaries with less fluctuation, and of course even fewer bits would be required to specify the amplitudes of the plateaus.

D. Transform Encoding

As with DPCM encoding, transform encoding of the chrominance signals has largely followed techniques used successfully with monochrome coding (see the review by Wintz [78]). Pratt [51] studied the transform coding of various

TABLE VIII
RESULTS OF TRANSFORM CODING THE COMPONENTS OF A COLOR SIGNAL; ZONAL FILTERING AND PCM QUANTIZATION (FROM PRATT [51])

	COMPONENTS					
	RGB	YIQ	Yuv	YIQ	YIQ	YIQ
Transform	Fourier	Fourier	Fourier	Hadamard	Hadamard	K-L (Thresh.)
Quant.	R=1.25 bits G=1.25 bits B=1.25 bits	Y=3.0 bits I, Q=0.75 bits	Y=3.0 bits U, V=0.75 bits	Y=3.0 bits I, Q=0.75 bits	Y=3.0 bits I, Q=0.75 bits	Y=1.0 bits I, Q=0.75 bits
Bit-Rate Total (Bits/Pel)	3.75	3.75	3.75	3.75	3.75	1.75
MS Error Total	1.22%	0.63%	0.67%	0.75%	0.72%	0.69%

color signal representations, primarily the linear transformations *RGB, YIQ,* Karhunen-Loeve transformation (see Section V) and the nonlinear transformation *Yuv.* The types of transforms applied to these signals were Fourier, Hadamard, 16 × 16 block Hadamard and 16 × 16 Karhunen-Loeve.

Single frames of a SMPTE test picture, denoted here as "Girl," were coded using the method of zonal filtering in which all transform samples outside a central zone are discarded. Six-bit uniform quantization was applied to the remaining coefficients (see Table VIII for a summary of the results). The resulting bit-rate for all pictures was a total of 3.75 bits/pel. Another picture was processed using the Karhunen-Loeve transform on just the luminance signal with 10:1 threshold coding;[11] the resulting bit-rate is 1.0 bit for the luminance signal and 0.75 bits for the I and Q components, giving a total of 1.75 bits/pel. Apart from the RGB encoding, there appears to be little difference in the picture quality or measured mean-square-error of the different methods.

The slant-transform has also been applied to the encoding of chrominance signals [79, 53]. The slant-transform has been designed to more efficiently match the type of waveforms found in picture signals, in particular the almost linearly increasing and decreasing segments typically found in signals caused, for example, by shading of a smooth surface [80]. The slant-transform performs almost as well as the optimal Karhunen-Loeve transform in terms of compacting the signal energy into the low-order transform components.

The method of quantizing the coefficients was more complex than in the previously described study. The quantization of the transformed signal is designed to minimize the mean-square-error in the reconstruction. This is achieved by making the number of quantization levels assigned to the quantizer for each transform sample proportional to the expected variance of the sample. The levels of each quantizer are then positioned to minimize mean-square-error by the Lloyd-Max procedure. The best division of bits between the three color planes is determined by trial and error for typical color pictures by iterative search techniques. Typical bit assignments for the transform coefficients Y, I and Q are shown in Fig. 33 for a 16 × 16 block. The reduced spatial resolution and increased quantization error of the I and Q signals relative to the Y signal is evident from the bit assignments. High quality reconstructions were achieved by coding at a rate of 2 bits/pel.

[11] Ten percent of the transform components having the largest amplitude are transmitted together with a code to specify which components they are. The remaining 90 percent of the components are assumed to be zero.

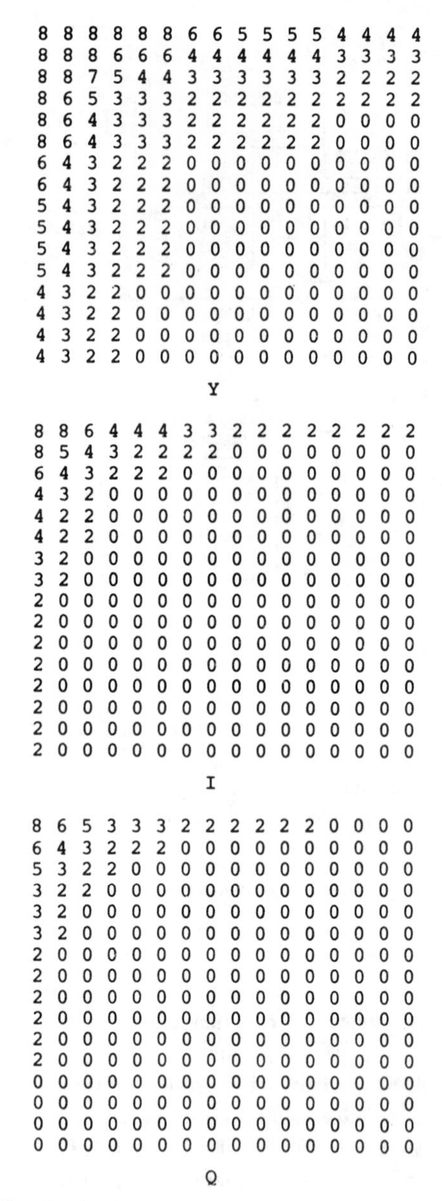

```
8 8 8 8 8 8 6 6 5 5 5 5 4 4 4 4
8 8 8 6 6 6 4 4 4 4 4 4 3 3 3 3
8 8 7 5 4 4 3 3 3 3 3 3 2 2 2 2
8 6 5 3 3 3 2 2 2 2 2 2 2 2 2 2
8 6 4 3 3 3 2 2 2 2 2 2 0 0 0 0
8 6 4 3 3 3 2 2 2 2 2 2 0 0 0 0
6 4 3 2 2 2 0 0 0 0 0 0 0 0 0 0
6 4 3 2 2 2 0 0 0 0 0 0 0 0 0 0
5 4 3 2 2 2 0 0 0 0 0 0 0 0 0 0
5 4 3 2 2 2 0 0 0 0 0 0 0 0 0 0
5 4 3 2 2 2 0 0 0 0 0 0 0 0 0 0
5 4 3 2 2 2 0 0 0 0 0 0 0 0 0 0
4 3 2 2 0 0 0 0 0 0 0 0 0 0 0 0
4 3 2 2 0 0 0 0 0 0 0 0 0 0 0 0
4 3 2 2 0 0 0 0 0 0 0 0 0 0 0 0
4 3 2 2 0 0 0 0 0 0 0 0 0 0 0 0

               Y

8 8 6 4 4 4 3 3 2 2 2 2 2 2 2 2
8 5 4 3 2 2 2 2 0 0 0 0 0 0 0 0
6 4 3 2 2 2 0 0 0 0 0 0 0 0 0 0
4 3 2 0 0 0 0 0 0 0 0 0 0 0 0 0
4 2 2 0 0 0 0 0 0 0 0 0 0 0 0 0
4 2 2 0 0 0 0 0 0 0 0 0 0 0 0 0
3 2 0 0 0 0 0 0 0 0 0 0 0 0 0 0
3 2 0 0 0 0 0 0 0 0 0 0 0 0 0 0
2 0 0 0 0 0 0 0 0 0 0 0 0 0 0 0
2 0 0 0 0 0 0 0 0 0 0 0 0 0 0 0
2 0 0 0 0 0 0 0 0 0 0 0 0 0 0 0
2 0 0 0 0 0 0 0 0 0 0 0 0 0 0 0
2 0 0 0 0 0 0 0 0 0 0 0 0 0 0 0
2 0 0 0 0 0 0 0 0 0 0 0 0 0 0 0
2 0 0 0 0 0 0 0 0 0 0 0 0 0 0 0
2 0 0 0 0 0 0 0 0 0 0 0 0 0 0 0

               I

8 6 5 3 3 3 2 2 2 2 2 2 0 0 0 0
6 4 3 2 2 2 0 0 0 0 0 0 0 0 0 0
5 3 2 2 0 0 0 0 0 0 0 0 0 0 0 0
3 2 2 0 0 0 0 0 0 0 0 0 0 0 0 0
3 2 0 0 0 0 0 0 0 0 0 0 0 0 0 0
3 2 0 0 0 0 0 0 0 0 0 0 0 0 0 0
2 0 0 0 0 0 0 0 0 0 0 0 0 0 0 0
2 0 0 0 0 0 0 0 0 0 0 0 0 0 0 0
2 0 0 0 0 0 0 0 0 0 0 0 0 0 0 0
2 0 0 0 0 0 0 0 0 0 0 0 0 0 0 0
2 0 0 0 0 0 0 0 0 0 0 0 0 0 0 0
2 0 0 0 0 0 0 0 0 0 0 0 0 0 0 0
0 0 0 0 0 0 0 0 0 0 0 0 0 0 0 0
0 0 0 0 0 0 0 0 0 0 0 0 0 0 0 0
0 0 0 0 0 0 0 0 0 0 0 0 0 0 0 0
0 0 0 0 0 0 0 0 0 0 0 0 0 0 0 0

               Q
```

Fig. 33. Typical bit assignments for the Y, I and Q transform coefficients for 16 × 16 blocks (from Chen and Pratt [79]).

Tescher [81] also assigned quantizing bits in proportion to the expected variance of the transformed sample, but did it adaptively, estimating the variance from adjacent, previously transmitted samples. In this way no additional decoding information need be transmitted, but there is a resulting increase in sensitivity to the effect of transmission errors. The R, G and B components for the whole picture were first individually Fourier transformed and then combined in a further three-point transform to obtain components akin to the luminance and chrominance signals. The resulting complex Fourier components were then converted to amplitude and phase signals before being coded. Pictures were coded at the rate of 0.55 to 1.2 bits/sample, but significant loss of detail is evident in the processed pictures.

Chen and Smith [82] adaptively encoded the signal by adapting to the properties of each transformed block, rather than continuously adapting over the whole picture. A cosine transform [83] was used which, at least for Markovian sources, gives energy compaction which is very close to that achieved by the Karhunen-Loeve transform. The activity within a block is measured by the energy of the signal with the DC component removed, and based on this quantity the blocks are divided into four equal groups. More bits may now be allocated to highly detailed blocks where they are needed, and fewer to blocks having low detail. The number of bits allotted to each transform sample of each class is calculated from the variance of that sample within each class in a similar manner to that described above. The bit allocation is based on a rate-distortion equation rather than just the variance of the sample.

For such a system it is now necessary to transmit more overhead data; for each block there are 4-bit maps for each of the three tables, three normalization factors to set the gain for each of the color components and a block classification word. All this information can be transmitted quite efficiently. If a rate 1/2 error code is used to protect just these overhead data, the total overhead data are but 16 percent of the total bit-rate.

When evaluated on single frames of "Girl" and "Couple" (another SMPTE test slide), excellent picture quality was obtained at 2 bits/pel total and good picture quality at 1 bit/pel. Results are also given for simulated bit error-rates of 10^{-4} and 10^{-2}

E. Frame-to-Frame Coding

There appears to be but one description of a frame-to-frame color coding system that operates on separate components (Iinuma *et al.* [84]). The incoming NTSC signal is converted into a new format in which the chrominance signals (R-Y and B-Y) are compressed and time multiplexed into the blanking region of the signal (see Section VI.B). The chrominance signals are subsampled by a factor of 5 relative to the luminance signal, and only one component is sent per line; at the receiver the missing lines are obtained by repeating the previous line. The resulting signal can now be processed by a frame-to-frame coder of the type normally used for luminance signals (see Candy *et al.* [85] for a review of frame-to-frame coding). A circuit detects when there is a significant change between a reference frame stored in a frame memory and the incoming picture signal. The difference between the two signals is quantized, coded and transmitted to a frame memory in the receiver. At the same time the local reference frame memory is being updated with the new quantized data. It is necessary to transmit address information so that the receiver can insert the correction signal at the correct point in the receiver frame memory. The resulting data generation rate is irregular and must be buffered to provide a constant transmission rate before being transmitted over a channel. Various techniques are available to reduce the information rate as the amount of movement in a scene increases. These techniques produce a gradual reduction in picture quality as the buffer fills, as opposed to a sharp cutoff in data rate and the resulting jerk in the picture that would occur without such control. The reduction in bit-rate is achieved by two broad classes of method: transmit only the more significant frame-to-frame changes, or, reduce the spatial or temporal resolution.

Because the chrominance components represent only 20 percent of the total signal, most frame-to-frame statistics of the color coder will not differ greatly from the corresponding luminance coder. However, in comparing frame-to-frame changes of the chrominance and luminance components, we see that the chrominance signal has only 10 percent of the frame-to-frame change even though it has 20 percent of the samples [84]. Such a comparison will, of course, be affected by the relative amplitudes of the luminance and chrominance signals; in this case the dynamic ranges were adjusted to be equal.

The frame-to-frame coder could be adjusted to operate at transmission rates in the range 6.3-16 Mbits/s. At 6.3 Mbits/s, large area movement, such as zooming and panning, produce visible edge effects due primarily to the spatial subsampling of the signal (spatial subsampling is a control mode used to prevent buffer overload). Buffer overload occurs for very active pictures, resulting in a certain amount of jerkiness. However, for video-telephone or video-conferencing use, where the camera is fixed and there is little motion, the picture quality may be quite acceptable. At a transmission rate of 16 Mbits/s, it is difficult to see degradation, and the picture quality is claimed to be approaching broadcast quality requirements.

VII. CODING OF THE COMPOSITE SIGNAL

We mentioned in the last section that hardware economies can be achieved by coding a time multiplex of component signals by a single DPCM codec. The multiplex might involve storing the chrominance components to place them in the (luminance) line blanking interval. Alternatively, luminance and chrominance can be sample-interleaved and separate stores can be used in a more complex predictor feedback loop. Extend this concept to interleaved linear sums of the luminance and chrominance samples, and we arrive at a coded signal which could equally have been obtained by encoding a *composite* signal at a sampling rate which is an integer multiple of the subcarrier.

Coding techniques which retain the signal in composite form will be reviewed more generally in this section, but first it is useful to consider certain aspects of basic sampling and quantization.

A. Sampling

Whereas for component signals considerable freedom exists for the choice of sampling frequency,[12] composite signals are best sampled at an integer multiple of subcarrier frequency. This is not a theoretical requirement for picture coding, as will become evident later; rather it helps to avoid intermodulation between sampling and subcarrier frequencies that occurs in the (nonlinear) process of A/D conversion. Operation is possible with minimal patterning near harmonics of the line rate for NTSC, and additionally near odd harmonics of half the line rate for PAL, and at simple fractional relationships with respect to the subcarrier (e.g., $8/3 f_{SC}$), but avoiding visible crawl with the 8-field cycle of the PAL subcarrier is particularly difficult.

Aside from coding considerations for transmission, there

[12] Typically, the sampling frequency for component signals should be an integer harmonic of the line rate to reduce the crawl of edge busyness, or an odd multiple of half-line rate for minimum aliasing with sub-Nyquist sampled images.

are good operational reasons for sampling at an integral multiple of the subcarrier, even for component signals, since conversion to the composite format can then be done conveniently in digital form. This could be of value (Baldwin [86]) for subsequent video tape recording or radio broadcast of the composite signal where the modulation to an intermediate frequency that is required can be done digitally. For the composite signal too, processing is much simplified since in the case of sampling at three times the subcarrier frequency, $3f_{SC}$ (the lowest multiple giving a sampling frequency greater than the Nyquist rate), either I or Q information (U or V) can be excluded from every third sample by aligning this sample with an appropriate subcarrier axis (i.e., by phasing every third sample to occur at a zero crossing of the I or Q subcarrier component). Furthermore, using the PALE technique for NTSC signals (Rossi [87]), whereby the sampling phase is reversed on alternate lines, the additional benefits of a vertically aligned (line synchronous) sampling pattern can be obtained for comb filtering or two-dimensional prediction. These benefits also accrue (for PAL and NTSC) with $4f_{SC}$ sampling, and now each chrominance component can be excluded from alternate samples by aligning a sample with one quadrature axis; a time division multiplex (TDM) signal results consisting of linearly matrixed luminance and chrominance (IQ or UV) signals. Simple two-dimensional digital filtering can then be performed (Weston [88]) to generate basic signal components at $2f_{SC}$ or f_{SC} sampling rates for coding or standards conversion (though with some U/V "crosstalk") while retaining ideal reversibility for subsequent regeneration of the composite (PAL) format.

Considerable bit-rate reduction can be achieved by sampling the composite signal at twice subcarrier frequency ($2f_{SC}$). For PAL signals, this corresponds to an odd harmonic of half-line frequency (neglecting the 25 Hz subcarrier offset) so that efficient sub-Nyquist sampling of the luminance component can be achieved with alias suppression by a comb filter. (Sub-Nyquist sampling has of course also been used for component signals [73]). For PAL signals, the sampling axis is aligned (Devereux [89]) at 45° and 225° relative to the U-axis, as in Fig. 34, and a post-sampling comb filter exploits the V-axis switch to restore a normal modulated subcarrier prior to D/A conversion. An additional pre-sampling comb filter reduces chrominance noise and cross-color and alias components on diagonal luminance edges, but both comb filters serve to attenuate high spatial-frequency luminance components in diagonal directions.

The principle of $2f_{SC}$ sampling can also be used with NTSC signals (Rossi [90]) by sampling at $2f_{SC} \pm 1/4f_H$ (horizontal line frequency). Again pre and post comb filters are used, but since they must now operate in the 1/4-line gaps they are constructed to average information from alternate lines. The combined filters therefore embrace at least 5 lines of the field, whereas for PAL in which adjacent lines are averaged and the 1/2 line gaps are combed, only 3 field lines are embraced. An additional disadvantage of NTSC is that the $1/4f_H$ sampling offset of f_S from f_{SC} causes any subsequent nonlinearity, such as in A/D conversion and picture coding, to generate noticeable hue and saturation changes on alternate lines. These color beats can be reduced by shifting the sampling phase by 90° on alternate lines.

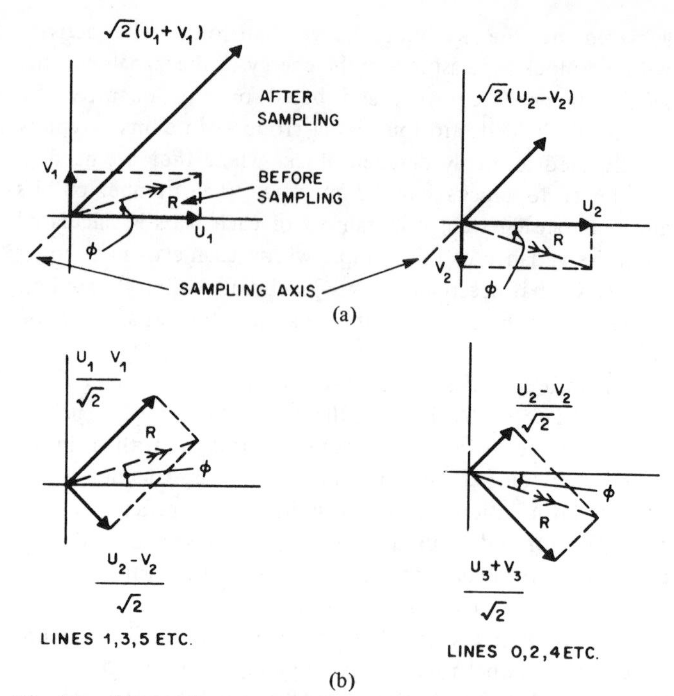

Fig. 34. Vector diagrams showing reconstruction of PAL subcarrier after $2f_{SC}$ sampling.
(a) Subcarrier vector (R) and its components before and after sampling. Note: the $\sqrt{2}$ factor arises from $1/\sqrt{2}$ coefficient through projection on to the 45° sampling axis, doubled by the addition of alias components from a double sideband sampled signal. (b) Addition of subcarrier components in comb filter (which averages across an exact line delay). Note: the 90° rotation of the "next line" contributions to the comb filter shown here results from the quarter line offset of subcarrier and line frequencies (25 Hz offset neglected) (from Devereux [101]).

A further disadvantage of the post comb filter which average lines to restore the normal PAL or NTSC subcarriers is the generation of flicker on horizontal boundaries between areas of different hue. The flicker, at 12.5 Hz (PAL) or 7.5 Hz (NTSC) is particularly noticeable with 100% saturated split-field color bars. Impairments due to comb filtering can be alleviated, though at high cost, by use of field delay filters.

B. Quantization

Under the most critical test conditions, it is generally accepted that uniform 8 bits/sample coding of a (gamma corrected) monochrome signal is adequate to avoid perceptible contouring although the presence of a dither signal allows less bits to be used. For a composite signal with subcarrier-asynchronous sampling, the subcarrier acts as a very efficient dither, but the remanent nonlinearity arising from the nonuniform dither probability density function (pdf) corresponds to a mechanism of chrominance-to-luminance crosstalk (Furman [91]). Beat patterns between sampling and subcarrier frequencies which may be generated by the quantizer nonlinearity are imperceptible at 8 bits/sample coding, though considerable patterning is often generated by other nonlinear parts of practical A/D converters (e.g., nonideal operation of the Sample and Hold circuit).

For synchronous sampling, however, the situation may be considered differently. At $3f_{SC}$, for example, a simple model consisting of a constant luminance signal added to a subcarrier of constant phase and amplitude (D'Amato [92], Thompson

[93], Felix [94]) shows that the maximum luminance quantizing error is still $\pm q/2$ (where q is the quantizing stepsize), but the subcarrier suffers differential gain or phase through the addition of an elemental subcarrier of peak-to-peak amplitude $4q/3$, and one of six phases determined by the phase relationship between the sampling and subcarrier frequencies. On a standard test signal (0.14 V peak-to-peak subcarrier on 0.7 V luminance ramp with the codec adjusted to accept 100% saturated color bars) an 8 bits/sample codec can generate peak values of ±4.5% and ±2.6° differential gain and phase [93]. Fig. 35 shows a vector scope plot of subcarrier amplitude and phase of such a test signal where the phase of the input subcarrier is swept through 360°. Translating the differential gain figure of ± 4.5% into a variation in color components for a 100% saturated color signal, we find that the demodulated U and V signals could show peak quantizing errors of ±1% and ±0.7% of maximum (100% saturated) amplitude, respectively. Peak-to-peak variations of 2% in the U signal are not perceived, however, since a change of only one quantal step in the luminance pedestal causes the subcarrier to cycle sample by sample through three states of differential gain of differing sign and magnitude [93] so that in practice a much reduced average quantity is invariably obtained (particularly for PAL where six states are averaged over two lines through the V-switch). On this basis, one should question the use of the terms differential gain and phase when they correspond to quite different impairments from those described in analog systems.

In practice, the most critical requirement of quantization applies to the luminance signal in areas where the subcarrier is of zero amplitude.

C. Predictive Coding

Although signals bearing subcarriers are incompatible with optimum predictive coding of the baseband signal (since the effect of the color subcarrier on a monochrome DPCM system is to maintain a continuous state of slope overload), it was realized in 1971 (Thompson [95]) that by modifying a DPCM encoder to differentiate across an integral subcarrier period, useful prediction of luminance could still be obtained. Most simply, the sampling rate is chosen to be n times the subcarrier frequency, and the nth-previous sample is used as the prediction. Alternatively, more complex prediction can be designed to match the bimodal spectrum of the composite signal so that simultaneous prediction of the luminance and chrominance channels is almost as efficient as could be obtained with separate encoding of the components. The use of a common quantizer restricts flexibility in principle, and in some designs (e.g., planar prediction of PAL) it is difficult to achieve efficient error decay, and at the same time obtain a predictor with near-unity gain both at low frequencies and at the subcarrier frequency.

Prediction of composite signals can be obtimized by recourse to the noise-feedback and pre-whitening filter analogy of Kimme and Kuo [96] whereby the frequency response of the predictor is made to fit the bimodal signal spectrum as closely as possible (Iijima *et al.* [97]). Viewed alternatively, just as Harrison's planar predictor [98] is effectively a cascade of two loops (Thoma [99]) which independently reduce the horizontal and vertical modes of redundancy of a monochrome signal, a number of loops can be cascaded to efficiently predict the composite color signal in one, two, or three dimensions

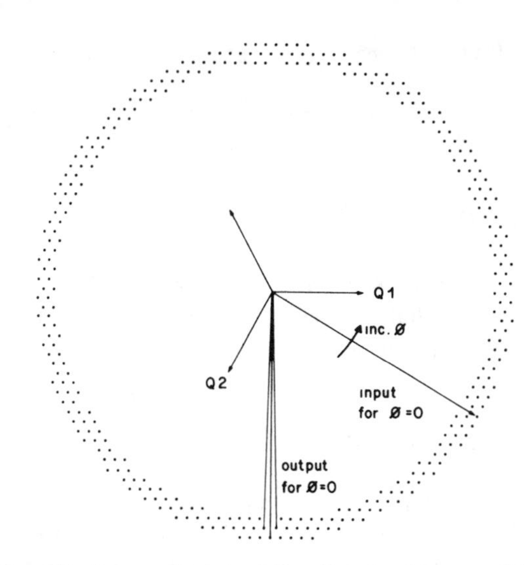

Fig. 35. "Vectorscope" plot of the discrete states possible at the output of an 8 bits/sample PCM codec sampled at $3f_{SC}$ for a (NTSC) subcarrier-on-ramp test signal. Subcarrier amplitude is 0.14 V peak-to-peak, and its phase (ϕ) is variable relative to the axis specified by one of the three sample positions per cycle. Although the ramp is 0.7 V (black-white), the coder is adjusted to (just) accept 100% saturated color bars. Apart from the constant 60° sampling shift of phase, the input vector differs from the polar coordinates of the nearest output state by maximum values of ±4.5% (amplitude) and ±2.6° phase (from Thompson [93]).

D. One-Dimensional (Intraline) Prediction

For intraline prediction, identical techniques may be used for both NTSC and PAL and most simply involve sampling at either $3f_{SC}$ (Thompson [100]) or $2f_{SC}$ (Devereux [101]) and using the third or second previous sample, respectively, as a prediction. Although the prediction distance is similar, the former clearly gives less quantizing error since three samples, instead of two, per period are available to follow rapid luminance transitions.

Transmission error propagation can be reduced by making the predictor slightly leaky, so that the effect of an error decreases exponentially. A better technique is that of hybrid DPCM (Devereux [102]). Whenever the prediction error exceeds a given threshold, an n bit/sample DPCM system reverts to $(n-1)$ bits/sample PCM and any transmission error is terminated. Additional ruggedness is obtained by switching to PCM whenever the input (8-bit) PCM signal coincides with one of the available $(n-1)$ bit PCM levels. Typically, an order of magnitude increase in error incidence can be permitted by this technique.

Relative to simple prediction across one period of the subcarrier, improved mid-frequency luminance performance can be obtained by higher order (slope) prediction, though with some sacrifice in the ability of the predictor to deal with transitions containing energy near half the sampling rate (since the spread of the prediction function is increased and embraces more of the signal history). With $3f_{SC}$ sampling, for example, by cascading two loops to transmit "the element difference of the third element difference," the prediction error, "$1 - P(z)$," is given by

$$1 - P(z) = (1 - az^{-1})(1 - bz^{-3}) = (1 - az^{-1} - bz^{-3} + abz^{-4}),$$

i.e.,

$$P(z) = az^{-1} + bz^{-3} - abz^{-4} \tag{14}$$

where $P(z)$ is the prediction function and z is the Z-transform operator such that z^{-1} represents a shift of 1 sample period. Therefore, using the nomenclature of Fig. 36, $P(z)$ could be restated as

$$\hat{S}_1 = aS_2 + bS_4 - abS_5. \tag{15}$$

Iijima and Ishiguro [97] termed such prediction "High Order DPCM" (HODPCM). They found that the best compromise between improved performance in predicting signal components below f_{SC} and reduced performance above f_{SC}, relative to the simple z^{-3} predictor, is obtained with $a = 1/2$ and $b = 15/16$. The latter value is specifically nonunity to allow error decay. In a subsequent paper, Iijima and Suzuki [103] generalized the prediction function in terms of sampling frequency, while imposing the condition $1 - P(z) = 0$ at both zero and subcarrier frequencies. The result was

$$P(z) = (1/2 - \alpha)z^{-1} + 3\alpha z^{-2}/2 + (1 - \alpha/2)z^{-3} - (1/2)z^{-4},$$

where

$$f_S = 2f_{SC}/\cos^{-1}[(1 + \alpha)/2]. \tag{16}$$

The case for $3f_{SC}$ is obtained with $\alpha = 0$ (without error decay, i.e., $b = 1$), but the generalization is particularly significant as it generates predictions which *do not require the sampling rate to be an integer multiple of the subcarrier frequency.* In particular, if values of α are chosen which yield predictor coefficients that are relatively easily realized, the NTSC predictors shown in Table IX are obtained. The last predictor in the table allows the bit-rate for a single codec meeting broadcasting standards to be reduced from 48.3 Mbits/s ($3f_{SC}$, 4.5 bits/sample) to 41.9 Mbits/s [97].

However, while recognizing the usefulness of the above results, the advantages of an integral relationship with the subcarrier were stated at the beginning of this section. In particular, for $2f_{SC}$ sampling we obtain the prediction error,

$$1 - P(z) = (1 - az^{-1})(1 - bz^{-2}),$$

i.e.,

$$P(z) = az^{-1} + bz^{-2} - abz^{-3}. \tag{17}$$

As for $3f_{SC}$, Ishiguro *et al.* [104] found best results for $2f_{SC}$ with a $= 1/2$ and $b = 15/16$.

Just as for monochrome signals (Thoma [99]), even higher order predictors may be designed to follow the *change* in slope. For $3f_{SC}$, we might consider

$$1 - P(z) = (1 - az^{-1})(1 - bz^{-1})(1 - cz^{-3}),$$

TABLE IX
PREDICTOR COEFFICIENTS FOR VARIOUS SAMPLING FREQUENCIES (NTSC) (FROM IIJIMA AND SUZUKI [103])

α	f_s (MHz)	$P(z)$
0	$3f_{sc}$ (10.74)	$(1/2)z^{-1} + z^{-3} - (1/2)z^{-4}$
1/4	10.01	$(1/4)z^{-1} + (3/8)z^{-2} + (7/8)z^{-3} - (1/2)z^{-4}$
3/8	9.66	$(1/8)z^{-1} + (9/16)z^{-2} + (13/16)z^{-3} - (1/2)z^{-4}$
1/2	9.3	$(3/4)z^{-2} + (3/4)z^{-3} - (1/2)z^{-4}$

i.e.,

$$P(z) = (a + b)z^{-1} - abz^{-2} + cz^{-3} - c(a + b)z^{-4} + abcz^{-5}. \tag{18}$$

Three examples of this were investigated by Dunne *et al.* [105]:

$$P_1(z) = z^{-1} - 1/3z^{-2} + z^{-3} - z^{-4} + 1/3z^{-5}$$

$$P_2(z) = z^{-1} - 1/4z^{-2} + z^{-3} - z^{-4} + 1/4z^{-5}$$

$$P_3(z) = z^{-1} - 1/2z^{-2} + z^{-3} - z^{-4} + 1/2z^{-5}. \tag{19}$$

In fact, only P_2 is physically realizable in the cascaded loop configuration ($a = b = 1/2$, $c = 1$), but the predictors were conceived in transversal form with P_2 and P_3 representing simpler hardware (binary coefficients) realizations of P_1. $P_1(z)$ was in turn conceived (referring to Fig. 36a) on the basis of predicting

$$\hat{S}_1 = S_2(\text{lum}) + S_4(\text{chrom}) \tag{20}$$

with the assumptions

$$S_2(\text{lum}) = S_2 - S_4(\text{chrom})$$

$$S_4(\text{chrom}) = S_4 - S_4(\text{lum})$$

$$S_4(\text{lum}) = 1/3(S_5 + S_4 + S_3).$$

Using a reflected quantizer designed to minimize rms error on luminance detail, they concluded from subjective tests that a 5 bits/sample codec based on $P_3(z)$ gave adequate quality for broadcast purposes, although on average it was only 0.1 of a grade (6-point Impairment Scale) better than z^{-3} prediction. Objective measurement of the unweighted quantizing noise using 5 bits/sample quantization showed larger differences (of questionable significance), showing an average of 2.7 dB improvement for the more complex codec, and being only 5 dB short of 8-bit PCM.

Although the above techniques are optimum for signals bearing phase modulated subcarriers where the required prediction distance is constant in areas of uniform hue, they can be applied very effectively to SECAM signals. For example, Devereux [106] found that a 5 bits/sample codec using z^{-3} prediction (prediction using the third previous element), which gave negligible distortion on all PAL signals sampled at $3f_{SC}$,

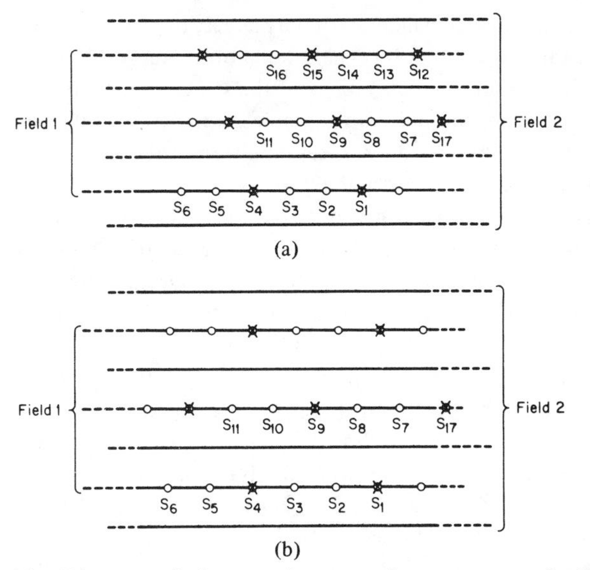

Fig. 36. Diagrams of picture points near the present sample ($S1$) which may be used for prediction. ○ Sample points; ⊗ Points of similiar subcarrier phase.

also gave negligible distortion on most types of picture material coded in the SECAM system. (The distortion was definitely perceptible, however, on color bars.)

This is borne out by computer measurements of Iijima and Ishiguro [97] of the effect of sampling frequency on the S/N ratio of HODPCM for NTSC signals. They found that deviations of 6 line-harmonics from $3f_{SC}$ had no effect on the S/N ratio of a coded subcarrier of 1 V p-p, and for half that amplitude there was no reduction over the calculated range ($\pm 8f_H$). It seems possible therefore that a simpler z^{-3} predictor could accommodate the full ±300 kHz deviation of the SECAM subcarriers with acceptable error.

Alternatively, adaptive prediction, as used for coding periodic speech components, might be devised for SECAM whereby a measure of instantaneous subcarrier frequency could be used to control predictor coefficients. However, there appears to be little interest in composite prediction in SECAM countries and developments in this direction seem unlikely.

Besides linear one-dimensional techniques, a nonlinear predictor (Devereux [102]) should also be mentioned. It uses z^{-3} prediction until large prediction errors occur, whereupon the system switches to z^{-1} for three samples. Subjective tests indicated a marginal preference of expert observers for the adaptive predictor over z^{-3} (approximately 0.1 of a grade for 5 bits/sample PAL signals on a 6-point Impairment Scale) because of reduced busyness at sharp luminance edges, but this was largely offset by increased noise in colored areas containing rapid changes in luminance. Subsequent measurement of the rms quantizing error (Devereux [107]) showed the advantage was 1.2 dB averaged over 5 pictures, compared with a 1.1 dB improvement over z^{-3} by an averaged two-dimensional predictor (see below).

Intraline prediction clearly has an advantage for broadcast signals with respect to compatibility of Vertical Interval Test Signals (VITS) and Teletext waveforms. However, for certain types of picture material there may be advantage in the reduced spatial spread of the two-dimensional predictor in spite of the more complex sampling techniques which may require (for PAL) accurate phasing of samples relative to the subcarrier.

E. Two-Dimensional (Intrafield) Prediction

Most of the two-dimensional techniques which have been developed for monochrome signals can be adapted to composite NTSC signals and PAL signals (with some additional complexity to cope with the V-axis switch). There is little point in using previous line information for SECAM signals, however, on account of the differing alternate-line subcarrier frequencies. Similarly it is difficult to apply two-dimensional coding techniques to $2f_{SC}$ -sampled PAL and NTSC signals since alternate lines of the sampled signal carry sum and difference signals of the chrominance components. Information from the second-previous line could be used (Thompson [100]), but the prediction distance is quite large.

For $3f_{SC}$ -sampling, Fig. 36a and 36b show the 625 and 525 rasters, respectively, and those points near the present sample (S_1) which may be used for prediction. The simplest two-dimensional predictor for NTSC signals (Thompson [108]) is perhaps $(S_4 + S_9)/2$ and gives considerable improvement since S_9 is much nearer to S_1 than S_4. Alternatively, as for monochrome signals, a less directional prediction is obtained by choosing a previous line point to the right of S_1 such as S_{17}.

The same principle can be applied to PAL signals (Devereux [107]), but consideration must be given to line-by-line reversal of the V-component (otherwise the prediction is only valid in areas of the picture where the V-component is zero). One possibility for $3f_{SC}$ sampling is to choose one sample out of three (e.g., S_1, S_9, etc.) to fall on the U-axis. Interchanging the intermediate sample (e.g., S_1, S_{11}), as required in the previous line contribution to the prediction, then effects a reflection of the subcarrier phasor in the U-axis and cancels the effect of the PAL switch. The cyclic three-state prediction algorithm is then

$$\hat{S}_1 = (S_4 + S_9)/2$$
$$\hat{S}_2 = (S_5 + S_8)/2$$
$$\hat{S}_3 = (S_6 + S_{10})/2. \tag{21}$$

On average, subjective test results showed the advantage of this predictor to be 1/3 bit for the same picture quality relative to third-previous sample prediction. (A similar advantage was measured by Dunne *et al.* [105] for their high-order intraline predictor.)

In the averaging predictors for both NTSC and PAL, the prediction distance of the same-line point is considerably in excess of the previous line contribution. In an attempt to improve performance for high frequency luminance components, Thompson [108] developed a technique for PAL and NTSC of previous sample prediction with a chrominance-only contribution from the previous line. The expected improvement was obtained by this "chrominance-corrected" prediction, but a bandpass filter and (for PAL) a low-pass filter and PAL modifier in the prediction loop are not easily realized by digital methods. In a later form (Thompson [109]), however, an all-digital planar predictor was derived. Conceptually, this predictor is obtained from the high order, one-dimensional predictor

$$P(z) = 1 - (1 - az^{-1})(1 - bz^{-3}) \tag{22}$$

by changing the last term of (22) to provide prediction in the vertical spatial dimension, that is, by replacing $(1 - bz^{-3})$ by $(1 - bz - n^{-3})$ where there are approximately n samples per line. For NTSC signals, a somewhat skew planar predictor results

$$\hat{S}_1 = S_2 - S_{10} + S_9. \tag{23}$$

For PAL signals, the V-axis switch is overcome by choosing one sample out of three (e.g., S_1, S_9, etc.) to lie on the V-axis and by cycling through a sequence of three states,

$$\begin{aligned}\hat{S}_1 &= S_2 + S_8 - S_9\\ \hat{S}_2 &= S_3 + S_{10} - S_{11}\\ \hat{S}_3 &= S_4 + S_9 - S_{10}.\end{aligned} \tag{24}$$

No formal subjective tests have been made for the planar predictor, but a direct comparison was made with the complex one-dimensional predictor of Dunne *et al* [105]. On average, the systems were of similar performance. The difference, which was only perceptible at 4 bits/sample coding or less, showed up as edge busyness on the fully saturated colored boundaries in the horizontal direction for the planar predictor and in the vertical direction for the one-dimensional predictor. The transmission error propagation for the planar predictor is severe, however, affecting the remainder of the afflicted field. Furthermore, for PAL signals the problem of balancing the predictor to achieve near-unity loop gain at dc and correct phasor prediction for the subcarrier in all three states of the prediction algorithm severely restricts the use of leaky prediction for rapid error decay.

Leaky predictors are more readily applied to two-dimensional coding of NTSC. In particular, Hatori and Yamamoto [110] describe simulation results of three intrafield predictors that use samples in the previous one or two lines. Using line-harmonic sampling near $3f_{SC}$, one predictor,

$$\hat{S}_1 = (3/4)S_2 - (1/4)S_8 + (3/4)S_{13} - (1/2)S_{14} + (1/4)S_9 \tag{25}$$

gave approximately 1.25 dB S/N improvement over one-dimensional HODPCM. This predictor also has relatively efficient transmission error decay.

F. Three-Dimensional Prediction

There has been relatively little published work on interframe coding of composite signals.

Ishiguro *et al.* [111] chose to sample the NTSC signal at a line-harmonic near $3f_{SC}$ and to apply z^{-3} prediction to the frame difference signal. In order to remove the subcarrier from the difference signal in stationary areas, however, it was first necessary to apply a reversible transformation to obtain, alternately, the sum and difference of the sampled lines. In areas of constant hue, the line sum is then essentially a luminance-only signal, and the line difference mainly the subcarrier. The line difference is then inverted frame by frame against the NTSC phase reversal. Although the A/D converter and frame store operate continuously at $3f_{SC}$, above a predetermined level of buffer occupancy (i.e., where there is a large amount of movement) subsampling is introduced by transmitting samples at $2f_{SC}$ coded by HODPCM ($P(z) = (1/2)z^{-1} + z^{-2} - (1/2)z^{-3}$). Under this condition, therefore, the system transmits the element difference of the second element difference of the frame difference of the line difference on alternate lines! With the aid of nonlinear filtering of the frame difference signal, it is hoped that broadcast quality NTSC signals might be transmitted at a rate in the region 16-32 Mbit/s.

Hatori and Yamamoto have computer simulated [112] an interframe planar algorithm for coding NTSC signals. Using a sampling rate which can be as low as desired, provided that it is an integral harmonic of line frequency, they propose the prediction

$$\hat{S}_1 = \bar{S}_8 + \bar{S}_1 - S_8 \tag{26}$$

where $\bar{\cdot}$ refers to the previous frame.

G. Transform Coding

It has generally been found for monochrome signals (Habibi [113]) that transform techniques tend to be less sensitive to variation in picture content and have considerable transmission error advantage over predictive coders. In addition, they are more efficient at low bit-rates. Shibata [114] has now shown that, even for high quality transmission of NTSC signals, transform coding may provide more efficient source encoding.

Sampling just below $3f_{SC}$ at a multiple of 8 times line rate, he used an 8-element slant transform in the horizontal direction and a 4-line Hadamard transform (which matches the line-by-line subcarrier phase reversal better than the slant transform) in the vertical direction. A highly flexible quantization scheme allows a differing number of bits and quantization characteristics to be chosen for each of the 32 transform coefficients. At an average rate of 3.1 bits/sample the picture quality obtained was, in the opinion of one of the authors (JET), comparable to that obtainable with 4 bits/sample, two-dimensional predictive techniques (Thompson [109]) on "natural" pictures although the predictive coder may give less edge busyness on certain critical test patterns. The complexity of the transform coder was, however, considerably greater than that of the predictive coder.

Less optimistic results have been presented by Clarke [115] for transform coding of PAL signals. A 32-element (intraline) Hadamard transform was used with bit reduction achieved by removing the more significant bits of certain coefficients (which are virtually never used), and by rounding the lesser significant bits of each coefficient. A single, nonuniform, quantization law was used for all coefficients, and the sampling rate was near $3f_{SC}$. Subjective tests on five pictures, using a 6-point Impairment Scale, showed 5 bits/sample DPCM (z^{-3}) to be consistently better than 5.5 bits/sample transform coding. More specifically, using 5 bits/sample coding with either technique, the advantage of DPCM ranged from 3/4 grade on a test card and detailed scene to 2.75 grades on monochrome alphanumeric characters. For a sampling rate near $3f_{SC}$, Clarke explored a technique in which 96 consecutive samples were divided into three blocks of 32 samples prior to transformation such that each block contained consecutive samples of the same phase (akin to the philosophy of z^{-3} prediction for DPCM). This technique permitted bit savings for pictures having saturated color by concentrating low-frequency luminance and chrominance energy into the same basis functions. Correspondingly, performance on high-frequency luminance components was poorer (e.g., alphanumeric characters).

The results of Shibata *et al.* [116] suggest that the slant transform would have been more efficient than the Hadamard, but their two-dimensional codec relies on the affinity of the vertical Hadamard transform for the line-by-line phase reversal of the NTSC subcarrier. This would presumably be inefficient for the PAL subcarrier, or for the SECAM signal which uses differing subcarriers on adjacent lines.

In terms of the transmission error advantage of transform coding, the system of Shibata [114] gives only slight degradation at an error probability of 10^{-5} and is "usable" (though annoying) even at 10^{-3}. For the more serious error pattern of the 32-element intraline Hadamard transform, Clarke [115] established that an error probability of 10^{-5} was definitely perceptible but not disturbing, while the threshold of perceptibility was around an error probability of 10^{-7}. DPCM, on the other hand, is more sensitive to errors by about two orders of magnitude.

VIII. COMPARISON OF COMPONENT AND COMPOSITE CODING

There has been considerable debate over whether it is better to code separately the components of a composite television signal or to code the composite signal itself. Many factors bear on the question, and only some of them relate directly to coding. The implications of such a decision could be far reaching, and we will consider some of the issues that are involved.

A. Compatibility

The existence of many different television standards in the world today complicates the problem of program exchange between countries having different standards. At present we have two levels of incompatibility: (a) at the level of the basic monochrome signal with respect to scanning rates and signal bandwidth and (b) in methods for forming the color signals and multiplexing together the luminance and chrominance components. The "opportunity" now arises for a further level of incompatibility, the method of digitizing the signals. The PAL and SECAM signals use essentially the same baseband components, and it would be possible to arrive at a common component encoding method. The degree of digital compatibility possible between the NTSC standard on one hand and the PAL and SECAM systems on the other appears rather limited because of the different line-rates and scanning methods.

The arguments for and against each method are rather involved and will not be discussed in full here (see [117, 118]). In summary, it appears that where the signal originates in composite form, and the system used by the receiving location is the same as that of the source, then coding in composite form could be the best solution—particularly during the period when digital techniques are being introduced and when the complete chain could contain several digital and analog sections in cascade. However, where the long-term goal is an all-digital system (from camera up to but not including the transmitter), then it is generally accepted that component coding is most desirable. But even here, in applications where transmission, storage or processing will subsequently be required with a phase-modulated subcarrier (i.e., NTSC or PAL), there are good reasons for initially sampling the component signals at a multiple of f_{SC} since this allows a more simple digital synthesis of the composite signal.

Much concern is generally expressed [117, 118] over degradation that may result in conversion from one format to another. This is indeed an important consideration in a system that may contain a number of interspersed analog and digital sections. However, where conversions are from one digital format to another and where the cost of transmission is high so that efficient coding techniques are used, coding degradations are likely to be of more concern than digital-to-digital format conversions.

Looking well into the future, one could envisage digital distribution: cable, satellite, or even VHF/UHF broadcast transmission. If such be the case, the possibility would exist for the distribution of higher quality pictures. By using component coding, it is possible to eliminate the shortcomings of frequency multiplexed color signals such as cross color effects and (for the NTSC signal) single-sideband distortion. In addition, the components could be upgraded with additional time-multiplexed components to give super-resolution should such a service become attractive. Considering a CCIR suggestion that the overall performance of systems employing digital, or digital and analog methods should not be worse than that of the equivalent analog system [119], the wisdom of overzealous compression either of the composite (i.e., sampling at $2f_{SC}$) or of component signals to fit a low hierarchical level might also be questioned.

B. Efficiency and Performance

According to the results surveyed in previous sections, to achieve the highest performance in the composite encoder, de-facto separation of the signal would be necessary to match coding requirements to the properties of each component and the relationships between them (for example, masking of one signal component by another). However, some of these properties are already recognized in existing analog composite designs, and consequently the resulting margin may not be so large.

For example, consider two approaches to coding in countries using the PAL system. Zschunke [120] claims to have achieved a quality rating between good and excellent on the CCIR five-grade scale from component coding using 10 MHz sampling with 3-bit DPCM for the luminance signal and 2 MHz sampling with 4-bit DPCM for an alternate-line transmission of the chrominance signals. By exploiting the line blanking interval, an average transmission rate of 30.9 Mbits/s is achieved. From the results of Devereux [107], using $2f_{SC}$ sampling of the composite PAL signal, a similar quality is claimed for 5-6 bits per sample DPCM with only a simple second-previous-element predictor. If more efficient prediction and quantization were to give a further saving of about 1 bit per sample, then, again exploiting the line blanking interval, it would still be possible to achieve transmission at the proposed European 34 Mbits/s transmission level.

Although composite coding, by definition, has the advantage of only transmitting one component, component coding methods do currently appear to give slightly lower rates for the same picture quality. There is no established basis for comparison, however, and considerable uncertainty exists in comparing ratings on different scales, under different viewing conditions and with different signals, parameters and criteria.

Small differences in efficiency may be more or less important depending upon the exact transmission facilities that are available.[13] For example, if a digital facility is available at the 3rd level of the digital hierarchy as adopted in the United States and Japan (44 Mbits/s), the ability to obtain a picture of high quality at 22 Mbits/s would be paramount, and a further small reduction would be of little significance. However, if the 3rd level digital hierarchy called for a rate of 34 Mbits/s, then a system that permitted operation at approximately 16 Mbits/s would be very attractive and a system that operated at 22 Mbits/s would have little advantage over a system operating at 34 Mbits/s.

In summary the *current* situation appears to be largely a matter of convenience. One approach that would find appeal among broadcasters, particularly in the near future, is to employ composite coding for distribution of signals within an area employing the same standard but to convert to component coding when changing from one standard to another. Further, considering the digital PAL/*YUV*/PAL converter described by Weston [88], a way in Europe is clear for composite and component systems to be compatible or even concomitant, rather than competitive.

ACKNOWLEDGMENT

We wish to thank the following people for their very helpful critiques of the manuscript: W. Frei, L. S. Golding, I. F. Macdiarmid, A. N. Netravali, D. E. Pearson, P. Pirsch and L. Stenger. Acknowledgment is made by John E. Thompson to the Director of Research of the British Post Office for permission to publish this paper.

[13]The CMTT [118] has expressed particular interest in coding methods leading to bit-rates at the hierarchical levels recommended by the CCITT or, where this would be too extravagant, at nonheirarchical levels formed from multiples or submultiples of recommended hierarchical levels.

REFERENCES

[1] Carbrey, R. L., "Video Transmission Over Telephone Cable Pairs by Pulse Code Modulation", *Proc. IRE,* Vol. 48, September 1960, pp. 1546-1561.

[2] Jayant, N. S., Ed., *Waveform Quantization and Coding,* IEEE Press, New York, 1976.

[3] Huang, T. S., and Tretiak, O. J., Eds., *Picture Bandwidth Compression,* New York: Gordon and Breach, 1972.

[4] Cutler, C. C., Ed., *Proceedings of the IEEE, Special Issue on Redundancy Reduction,* Vol. 55, No. 3., March 1967.

[5] Aaron, M. R., Ed., *IEEE Trans. on Commun. Technol., Special Issue on Signal Processing for Digital Communications,* Part I, Vol. COM-19, No. 6, December 1971.

[6] Andrews, H. C., and Enloe, L. H., Eds., *Proceedings of the IEEE, Special Issue on Digital Picture Processing,* Vol. 60, No. 7, July 1972.

[7] C.C.I.R., "Method for the Subjective Assessment of the Quality of Television Pictures," 13th Plenary Assembly, Rec. 500, 1974, Vol. 11, pp. 65-68.

[8] MacAdam, D. L., "Color Essays," *J. Opt. Soc. Am.,* Vol. 65, No. 5, May, 1975, pp. 483-492.

[9] MacAdam, D. L., *Sources of Color Science,* Cambridge, MA, and London, England: MIT Press, 1970.

[10] Young, T., "On the Theory of Light and Colors," *Philosophical Transactions of the Royal Society of London,* Vol. 92, 1802, pp. 20-71.

[11] Grassman, H. G., "Theory of Compound Colors," translated in *Philosophic Magazine,* Vol. 4, No. 7, 1854, pp. 254-264.

[12] Guth, S. L., Donley, N. J., and Marrocco, R. T., "On Luminance Additivity and Related Topics," *Vision Res.,* Vol. 9, May 1969, pp. 537-575.

[13] Wintringham, W. T., "Color Television and Colorimetry," Proceedings IRE, Vol. 39, October, 1951, pp. 1135-1172.

[14] Guild, J., "The Colorimetric Properties of the Spectrum," Phil. Trans. Roy. Soc. A, Vol. 230, 1931, pp. 149-187.

[15] Wright, W. D., "A Re-determination of the Trichromatic Coefficients of the Spectral Colours," *Transactions of the Optical Society,* Vol. 30, 1928-1929, pp. 141-164.

[16] *CIE Proceedings 1924,* Cambridge: Cambridge University Press, 1926, pp. 67-70.

[17] Wyszecki, G., and Stiles, W. G., *Color Science,* New York: John Wiley and Sons, Inc., 1967.

[18] Stiles, W. S., and Burch, J. M., "N. P. L. Colour-Matching Investigation Final Report (1958)," *Optica Acta,* Vol. 6, 1959, pp. 1-26.

[19] Speranskaya, N. J., "Determination of Spectrum Color Coordinates for Twenty-Seven Normal Observers," *Optics and Spectroscopy,* Vol. 7, 1959, pp. 424-428.

[20] Judd, D. B., and Wyszecki, G., *Color in Business, Science and Industry,* New York: John Wiley and Sons, Inc., 1965.

[21] MacAdam, D. L., "Visual Sensitivities to Color Differences in Daylight," *J. Opt. Soc. Am.,* Vol. 32, No. 5, May, 1942, pp. 247-274.

[22] MacAdam, D. L., "Specification of Small Chromaticity Differences, " *J. Opt. Soc. Am.,* Vol. 33, 1943, pp. 18-26.

[23] MacAdam, D. L., "Projective Transformations of ICI Color Specification," *J. Opt. Soc. Amer.,* Vol. 27, August 1937, pp. 294-299.

[24] Pearson, D. E., *Transmission and Display of Pictoral Information,* London: Pentach Press Ltd, 1975, pp. 184-186.

[25] Townsend, G. B., "The Correspondence Between Chromaticity and N.T.S.C. Chrominance," Int. TV Conf., London 1962, IEE Conf. Rept. Series 5, pp. 519-527.

[26] Carnt, P. G., and Townsend, G. B., *Colour Television–Volume 2,* London: Iliffe, 1969.

[27] Hunt, R. W. G., "Objectives in Colour Reproduction," *J. of Photographic Science,* Vol. 18, 1970, pp. 205-215.

[28] Limb, J. O., "Visual Perception Applied to the Encoding of Pictures," Proceedings of the SPIE National Seminar on Advances in Image Transmission Techniques, Vol. 87, 1976, pp. 80-87.
[29] Vos, J. J., Friele, L. F. C., and Walraven, P. L., *Color Metrics,* Soesterberg: AIC/Holland, 1972.
[30] Jain, A. K., "Color Distance and Geodesics in Color 3 Space," *J. Opt. Soc. Am.,* Vol. 62, November, 1972, pp. 1287-1291.
[31] Jain, A. K., "Role of Geodesics in Schrodinger's Theory of Color Vision," *J. Opt. Soc. Am.,* Vol. 63, August, 1973, pp. 934-939.
[32] MacAdam, D. L., "On The Geometry of Color Space", *Journal Franklin Institute,* 1944, Vol. 238, pp. 195-210.
[33] Wyszecki, G., "Proposal for a New Color Difference Formula," *J. Opt. Soc. Am.,* Vol. 63, 1963, pp. 1318-1319.
[34] Anonymous, "Technical Note: CIE Colorimetry Committee–Working Program on Color Difference," *J. Opt. Soc. Am.,* Vol. 64, June, 1974, pp. 896-897.
[35] MacAdam, D. L., "Geodesic Chromaticity Diagram Based on Variances of Color Matching by 14 Normal Observers," Applied Optics, Vol. 10, No. 1, January, 1971, pp. 1-7.
[36] Pazderak, J., "Color Differentiation in Television Images," *Slaboproudy Obzor,* Vol. 24, 1963, pp. 69-75.
[37] Dupont-Henius, G., private communication.
[38] DeMarsh, L. E., and Pinney, J. E., "Studies of Some Colorimetric Problems in Color Television," *J. SMPTE,* Vol. 79, April, 1970, pp. 338-342.
[39] Witzel, R. F., Burnham, R. W., and Onley, J. W., "Threshold and Suprathreshold Perceptual Color Difference," *J. Opt. Soc. Am.,* Vol. 63, No. 5, May, 1973, pp. 615-625.
[40] Hacking, K., "The Choice of Chrominance Axes for Colour Television," *Acta Electronica,* Vol. 2, 1957, pp. 87-94.
[41] Rubinstein, C. B., and Limb, J. O., "Color Border Sharpness," *Colour 73,* London: Adam Hilger 1973, pp. 377-380.
[42] Rentschler, I., "Color Perception Studies," *Optik,* Vol. 32, 1971, pp. 471-478.
[43] Netravali, A. N., and Rubinstein, C. B., "Quantization of Color Signals," *Proceedings IEEE* to be published, July 1977.
[44] Judd, D. B., "Fundamental Studies of Color Vision From 1860 to 1960," *Proc. N.A.G.,* Vol. 55, 1966, pp. 1313-1330.
[45] Shklover, D. A., "A New Uniform Color Space," *Colour 73,* Adam Hilger, London, 1973, pp. 312-319.
[46] Tannebaum, P. M., private communication.
[47] Frei, W., "Rate-Distortion Coding Simulation For Color Images," Proceedings of SPIE National Seminar on Advances in Image Transmission Techniques, Vol. 87, 1976, pp. 197-203.
[48] Faugeras, O., "Digital Color Image Processing and Psychophysics within the Framework of a Human Visual Model," Ph.D. Thesis, University of Utah, June 1976.
[49] Frei, W., and Jaeger, P. A., "Some Basic Considerations for the Source Coding of Color Pictures," Proc. Int. Conf. Commun., Seattle, Wash., June 11-13, 1973, pp. 48.26 – 48.29.
[50] Habibi, A., and Wintz, P. A., "Image Coding by Linear Transformation and Block Quantization," *IEEE Trans. on Commun. Technol.,* Vol. COM-19, February 1971, pp. 50-61.
[51] Pratt, W. K., "Spatial Transform Coding of Color Images," *IEEE Trans. on Commun. Technol.,* Vol. COM-19, December 1971, pp. 980-992.
[52] Rubinstein, C. B., and Limb, J. O., "Statistical Dependence Between Components of a Differentially Quantized Color Signal," *IEEE Trans. on Commun. Technol.,* Vol. COM-20, October 1972, pp. 890-899.
[53] Pratt, W. K., Chen, W., and Welch, L. R., "Slant Transform Image Coding," *IEEE Trans. on Commun. Technol.,*Vol.COM-22, August 1974, pp. 1075-1093.
[54] Fink, D. G., ed., *Color Television Standards; Selected Papers and Records of the National Television Systems Committee,* New York: McGraw-Hill, 1955.
[55] Pirsch, P., and Stenger, L., "Statistical Analysis and Coding of Color Video Signals," *Acta Electronica,* Vol. 19, No. 4, 1976, pp. 277-287.
[56] Davidse, J., "N.T.S.C. Colour-Television Signals; Measurement Techniques," *Electronic & Radio Engineer,* October, November 1959, pp. 370-376 and pp. 416-419.
[57] Mueller, J. and Wengenroth, G., "The Perceptibility of Random Noise in Color TV Pictures Using the N.T.S.C. System," *NTZ,* Vol. 16, No. 4, 1963, pp. 163-166.
[58] Bhushan, A. K., "Efficient Transmission and Coding of Color Pictures," M.S. Thesis, Massachusetts Institute of Technology, Cambridge, MA, June 1967.
[59] In Der Smitten, F. J., "Data-Reducing Source Encoding of Color Picture Signals Based on Optical Chromaticity Classes," *Nachrichtentech. Z.,* Vol. 27, 1974, pp. 176-181.
[60] Gronemann, U. F., "Coding Color Pictures," Ph.D. Dissertation, Elec. Res. Lab., Massachusetts Institute of Technology, Technical Report 422, June 1964.
[61] Huffman, D. A., "A Method for the Construction of Minimum-Redundacy Codes," *Proceedings of the IRE,* September 1952, Vol. 40, pp. 1098-1101.
[62] Frei, W., Jaeger, P. A., and Probst, P. A., "Quantization of Pictorial Color Information," *Nachrichtentech. Z.,* Vol. 61, 1972, pp. 401-404.
[63] Frei, W., "Quantization of Pictorial Color Information; Nonlinear Transforms," *Proc. IEEE* (Lett.), Vol. 61, April 1973, pp. 465-466.
[64] Jain, A. K., and Pratt, W. K., "Color Image Quantization," NTC 1972 Record, Dec. 1972, pp. 34D-1 – 34D-6.
[65] Marti, B., "Preliminary Processing of Color Images," CCETT, ATA/T/3/73, September 5, 1973.
[66] Stenger, L., "Quantization of TV Chrominance Signals Considering the Visibility of Small Color Differences," this issue, pp. 1393-1406.
[67] Solomon, R. D., "Color Picture Coding for Facsimile," M.I.T. Research Laboratory of Electronics, Progress Report No. 117, January 1976.
[68] Pratt, W. K., "Binary Symmetric Channel Error Effects on PCM Color Image Transmission," *IEEE Trans. on Information Theory,* Vol. IT-18, September 1972, pp. 636-643.
[69] Teacher, C. F., and Yutz, R. W., "Secure Color Video Techniques," Tech. Rep. RADC-TDR-64-339, Vol. 1, Rome Air Development Center, Comm. Techniques Branch, Griffiss AFB, NY, February 1965.
[70] Limb, J. O., Rubinstein, C. B., and Walsh, K. A., "Digital Coding of Color PICTUREPHONE Signals by Element-Differential Quantization," *IEEE Trans. on Commun. Technol.,* Vol. COM-19, December 1971, pp. 992-1006.
[71] Golding, "DITEC- -A Digital Television Communications System for Satellite Links," presented at the 2nd Int. Conf. on Digital Satellite Commun., Paris, France, November 28-30, 1972.
[72] Stenger, L., and Wengenroth, G., "Digital Coding and Transmission of Color Television Signals," *Nachrichtentech. Z.,* Vol. 24, 1971, pp. 321-325.
[73] Golding, L. S., Garlow, R. K., "Frequency Interleaved Sampling of a Color Television Signal," *IEEE Trans. on. Commun. Technol.,* Vol. COM-19, December 1971, pp. 972-979.
[74] Sabatier, J., "Le Codage Differential des Composantes du Signal de Television Couleur," *Acta Electronica,* Vol. 19, No. 3, 1976, pp. 245-253.
[75] Sharma, D. K., and Netravali, A. N., "Design of Quantizers for DPCM Coding of Picture Signals", to be published.
[76] Max, J., "Quantizing for Minimum Distortion," *IRE Trans. Inform. Theory,* Vol. IT-6, March 1960, pp. 7-12.
[77] Limb, J. O., and Rubinstein, C. B., "Plateau Coding of the Chrominance Component of Color Picture Signals," *IEEE Trans. on Comm.,* Vol. COM-22, June 1974, pp. 812-820.
[78] Wintz, P. A., "Transform Picture Coding," *Proc. IEEE,* Vol. 60, July 1972, pp. 809-820.
[79] Chen, W. H., and Pratt, W. K., "Color Image Coding with the Slant Transform," Applications of Walsh Functions, Proceedings of the Symposium (4th), Catholic University of America, April 16-18, 1973, pp. 155-161.
[80] Enomoto, H., and Shibata, K., "Orthogonal Transform Coding System For Television Signals," presented at the 1971 Symp. Application of Walsh Functions, Washington, DC, April 1971.
[81] Tescher, A. G., "The Role of Phase in Adaptive Image Coding," University of Southern California, Image Processing Institute Report 510, December 1973.
[82] Chen, W., and Smith, C. H., "Adaptive Coding of Color Images

Using Cosine Transform," 1976 ICC Convention Record, pp. 47-7 – 47-13.

[83] Ahmed, N., Natarjan, T., and Rao, K. R., "Discrete Cosine Transform," *IEEE Trans. on Computer,* January 1974, pp. 90-93.

[84] Iinuma, K., Iijima, Y., Ishiguro, T., and Kaneko, H., "Interframe Coding for 4 MHz Color Television Signals," Conference Record, ICC '75, pp. 23-26 – 23-30.

[85] Candy, J. C., Franke, M. A., Haskell, B. G., and Mounts, F. W., "Transmitting Television as Clusters of Frame-to-Frame Differences," *Bell Syst. Tech. J.,* Vol. 50, July-August 1971, pp. 1889-1917.

[86] Baldwin, J. L. E., "Sampling Frequencies for Digital Coding of Television Signals," *IBA Technical Review,* Vol. 9, September 1976, pp. 32-36.

[87] Rossi, J., "Color Decoding a PCM NTSC Television Signal," *J. SMPTE,* Vol. 83, No. 6, June 1974, pp. 489-495.

[88] Weston, M., "A PAL/YUV Digital System for 625-Line International Connections," *BBC Research Dept.,* Report 1976/24, September 1976.

[89] Devereux, V. G., "Digital Video: Sub-Nyquist Sampling of PAL Colour Signals," *BBC Report* 1975/4, January 1975.

[90] Rossi, J. P., "Sub-Nyquist Encoded NTSC Color Television," *J. SMPTE,* Vol. 85, January 1976.

[91] Furman, G. G., "Improving the Quantization of Random Signals by Dithering," RAND Corporation Memorandum RM-3504-PR, May 1963.

[92] D' Amato, P., "Distorsioni Della Crominanza in un Segnale PCM Video Lineare," *Elettronica E.Telecommuniczioni,* No. 3, 1972, pp. 111-118.

[93] Thompson, J. E., "Methods of Digital Coding for Television Transmission," *Royal Television Society Journal,* Vol. 15, September/October 1975, pp. 384-391.

[94] Felix, M. O., "Differential Phase and Gain Measurements in Digitised Video Signals," *J. SMPTE,* Vol. 85, February 1976, pp. 76-79.

[95] Thompson, J. E., "Digital Encoding System," UK Patent No. 1,344,312 August 27, 1971, US patent No. 3,921,204.

[96] Kimme, E. G., and Kuo, F. F., "Synthesis of Optimal Filters For A Feedback Quantization System," *IEEE Trans. Circuit Theory,* Vol. CT-10, September 1963, pp. 405-413.

[97] Iijima, Y., and Ishiguro, T., "Differential Pulse Code Modulation of NTSC Colour Television Signals," IECE (Japan) Comm. Systems Study Group Monograph, CS 73-44, July 26, 1973.

[98] Harrison, C. W., "Experiments with Linear Prediction in Television," *Bell Syst. Technical Journal,* Vol. 31, 1952, pp. 764-783.

[99] Thoma, W. "Schaetzwertrechner fuer die DPCM von Fernsehsignalen," NTG – Fachtagung, 'Signalverarbeitung', Erlangen, April 4-6, 1973.

[100] Thompson, J. E., "Differential Coding for Digital Transmission of PAL Colour Television Signals," Int. Broadcasting Convention, London, September 1972 (IEE Conf. Pub. 88, pp. 26-32).

[101] Devereux, V. G., "Digital Video: Sub-Nyquist Sampling of PAL Colour Signals," BBC Report 1975/4, January 1975.

[102] Devereux, V. G., "Digital Video: Differential Coding of PAL Signals Based on Differences Between Samples One Subcarrier Period Apart," BBC Report, 1973/7, June 1973.

[103] Iijima, Y., and Suzuki, N., "Experiments on Higher Order DPCM for NTSC Color Television Signals," NEC Central Research Laboratories Report, 1974.

[104] Ishiguro, T., Suzuki, N., Iijima, Y., and Kawayachi, N., "32 MB/s Higher Order DPCM of NTSC Color Television Signals," IECE (Japan) Comm. Systems Study Group, CS 75-69, September 10, 1975.

[105] Dunne J., Wilkinson, J. H., and Greenfield, R., "Design and Evaluation of an Experimental DPCM Equipment for Color TV Signals," *IBA Review,* to be published.

[106] Devereux, V. G., "Comparison of Picture Impairments Caused by Digital Coding of PAL and SECAM Video Signals," BBC Report 1974/16, April 1974.

[107] Devereux, V. G., "Digital Video: Differential Coding of PAL Colour Signals Using Same-Line and Two-Dimensional Prediction," BBC Research Dept., RD 1975/20, June 1975.

[108] Thompson, J. E., "Differential Encoding of Composite Color Television Signals Using Chrominance-Corrected Prediction," *IEEE Trans. on Commun. Technol.,* Vol. COM-22, August 1974, pp. 1106-1113.

[109] Thompson, J. E., "Data Compression of Composite Colour Television Signal Using Planar Prediction," Third International Conference on Digital Satellite Communications, Kyoto Japan, November 1975, INTELSAT/IECE/ITE, Pub. pp. 315-321.

[110] Hatori, Y. and Yamamoto, H., "Direct Intra-field Coding of NTSC Signals (DPCM)," National Conference Record of IECE (Japan), Paper No. 982, March 1976.

[111] Ishiguro, T., Iinuma, K., Iijima, Y., Koga, T., Azami, S., and Mune, T., "Composite Interframe Coding of NTSC Color Television Signals," National Telecommunications Conference, Dallas, Texas, December 1976.

[112] Hatori, Y. and Yamamoto, H., "Direct Frame to Frame Coding of NTSC Color TV Signals (Simulation Results)," Institute of Television Engineers (Japan) Monograph, September 25, 1975.

[113] Habibi, A., "Hybrid Coding of Pictorial Data," *IEEE Trans. Comm.,* Vol. COM-22, No. 5, May 1974, pp. 614-624.

[114] Shibata, K., Private Communication.

[115] Clarke, C. K. P., "Hadamard Transformation: Assessment of Bit-Rate Reduction Methods," BBC Report, 1976/28, October, 1976.

[116] Shibata, K., and Matsumoto, J., "Two-Dimensional Orthogonal Transform Coding for NTSC Color Television Signals," KDD Research and Development Laboratory, Tokyo, Japan.

[117] C.C.I.R., "Comparison of Different Methods for the Digital Coding of Television Signals," CCIR Study Groups, Period 1974-1978, Doc. CMTT/39-E.

[118] C.C.I.R., "Digital Transmission of Television Signals," 13th Plenary Assembly, Doc. 646, 1974, Vol. 12, pp. 208-209.

[119] C.C.I.R., "The Specification of Performance for Transmission Circuits which Employ Digital Methods," 13th Plenary Assembly, Rep. 645, 1974, Vol. 12, pp. 206-208.

[120] Zschunke, W., "Digital Transmission of TV by Satellite," Third International Conference Digital Satellite Communications, Kyoto, Japan, November 11-13, 1975, pp. 322-329.

[121] Wyszecki, G., and Fielder, G. H., "New Color-Matching Ellipses," *J. Opt. Soc. Am.,* Vol. 61, No. 9, September, 1971, pp. 1135-1152.

[122] Wyszecki, G., and Fielder, G. H., "Color-Difference Matches," *J. Opt. Soc. Am.,* Vol. 61, No. 11, November, 1971, pp. 1501-1513.

Predictive Coding of Composite NTSC Color Television Signals

By R. C. BRAINARD, A. N. NETRAVALI, and D. E. PEARSON

A study is reported of intrafield predictive coding of composite NTSC television signals sampled at four times the color subcarrier frequency (14.3 MHz). The choice of predictor was found to be governed by three constraints, which are described. Of several predictors which satisfied these constraints, it was determined that one of them performed the best. A procedure for designing quantizers for composite signals was developed based on a three-parameter piecewise-linear approximation to the quantizer error threshold function. Observations of picture quality were made using real-time processing on a mixed hardware/software experimental facility, from which it was concluded that between 5 and 6 bits/pel were required for broadcast quality. Preliminary results were obtained on the effect of composite predictive coding on the vertical interval test signals (VITS). Our observations indicate that the waveform distortion criteria applied to VITS give some indication of granular noise visibility (in the SNR measurements); these criteria, however, are too severe in their requirements for slope overload distortion and do not consider edge busyness, an important temporal impairment in predictive coding.

Introduction

New generations of transmission media which have a digital channel format are presently coming into being. Among these are lightwave and satellite systems.[1] One advantage of the digital format is that a group of messages representing voice or data may be replaced readily by a single bit stream representing a television signal. The term "fungible" is being used by engineers to describe this property of interchangeability. Since digital channels will be introduced initially in short sections as replacements for analog channels, future transmission of television signals will be part analog and part digital. Composite-coding techniques[2,3] are attractive in these circumstances since they do not require decomposition of the NTSC signal into its luminance and chrominance components and subsequent recomposition, a process which introduces some impairment.

Previous studies of composite NTSC television signal coding[4-7] have mainly used sampling frequencies exactly or approximately equal to $2f_{sc}$ or $3f_{sc}$, where f_{sc} is the subcarrier frequency. Our use of $4f_{sc}$ sampling follows the recommendations of the SMPTE Working Group on Digital Video Standards.[8] It results in simplifications in the weighting coefficients needed to combine composite-signal predictor pels (pixels or picture elements) so that the resultant subcarrier phase is the same as that of the pel being predicted.[9] However, such oversampling results in high bit rates; with PCM (pulse code modulation) encoding at 9 bits/pel the rate is 128.7 Mbit/sec.

Our work sought to reduce this bit rate by means of a simple intrafield predictive coder while maintaining full broadcast quality. This involved keeping any coding distortions strictly below visual threshold even on very critical pictures. We also made some limited observations of the effect of the coding on the vertical interval test signals (VITS), which are used to monitor linear waveform distortion in analog channels.

Predictive coding is also known as differential pulse code modulation (DPCM). In this technique the current sample to be encoded is predicted from the encoded values of previously transmitted samples. The error resulting from the subtraction of the prediction from the actual input value is quantized into a set of discrete amplitude levels. These levels are assigned to binary words and transmitted. The receiver generates the prediction in the same way in which it is accomplished at the transmitter and adds the quantized error to generate the output signal.

A paper presented at the 122nd annual SMPTE Technical Conference, November 9-14, 1980, New York, N.Y. AUTHORS: Bell Laboratories, Holmdel, N.J.

Reprinted with permission from *SMPTE J.*, vol. 91, pp. 245-252, Mar. 1982.
Copyright © 1982 by the Society of Motion Picture and Television Engineers.

The paper first explores the mathematical rules for forming predictors as linear combinations of pels with the correct resultant subcarrier phase. It is shown that the allowable combinations are: (1) pels having the same phase as the pel being predicted, (2) particular triads of other pels, and (3) "null" predictors consisting of differences between pels of identical phase. The triad combinations were found to be susceptible to limit-cycle oscillations — oscillations resulting from nonlinear characteristics.[10] This made it necessary to insert additional representative levels in the quantizer to prevent this. The additional "stability" levels tended to cause edge busyness, requiring still more repesentative levels to suppress this effect. Use of the triad groups was necessary if pels very close to the pel being predicted were used as predictors. Thus an interesting tradeoff was observed between prediction and quantization; some good predictors tended to require more quantization levels. The experiments were performed using a mixed hardware/software test facility capable of coding 4.2-MHz standard NTSC signals in real time.

Composite-Signal Prediction

The composite NTSC system M signal is given by

$$E(t) = Y(t) + I(t) \cos (2\pi f_{sc}t + 33°) + Q(t) \sin (2\pi f_{sc}t + 33°) \quad (1)$$

where $Y(t)$ is the luminance signal, and $I(t)$ and $Q(t)$ are the baseband chrominance signals formed from the gamma-corrected red, green, and blue camera signals.

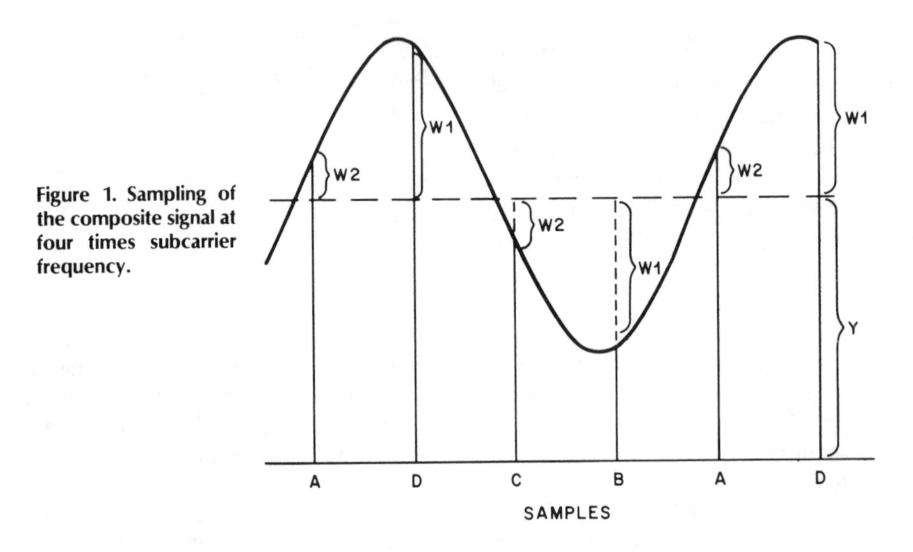

Figure 1. Sampling of the composite signal at four times subcarrier frequency.

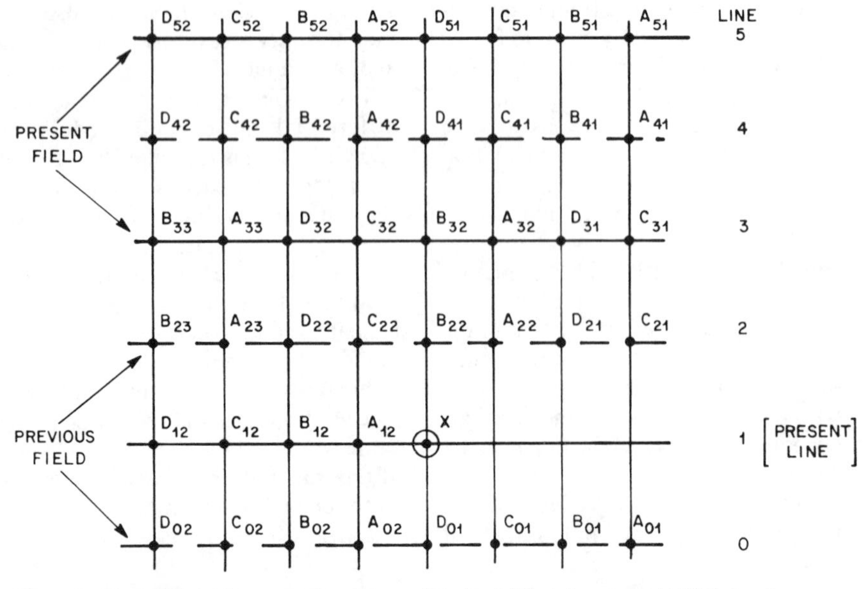

Figure 2. Map of the pels surrounding the predicted pel X in 4 f_{sc}-sampled NTSC signals.

With $4f_{sc}$ sampling there are four samples for every cycle of the subcarrier. It is assumed here that the phase relationship between the samples and the subcarrier is arbitrary (Fig. 1). A simplification results when the samples are taken at $2\pi f_{sc}t + 33° = n\pi/2$, where n is an integer, since in this case the sample sequence is $Y + I$, $Y + Q$, $Y - I$, $Y - Q$, $Y + I$, etc. In the general case, however, each composite-signal sample contains components from all three of the Y, I, and Q signals. Consider four samples taken sequentially and denoted by A, B, C, and D, as in Fig. 2, and the present sample X (having same phase as D) to be predicted. With $4f_{sc}$ sampling, due to 180° phase shift from line to line, any vertical column (Fig. 2) consists of either pels A and C, or pels B and D.

If the sampling takes place in an area of the picture where both the luminance and chrominance are constant, then from Fig. 1 it can be seen that the following four sample amplitudes result:

$$A = Y + W_2$$
$$B = Y - W_1$$
$$C = Y - W_2$$
$$D = Y + W_1 \quad (2)$$

where W_1 and W_2 are instantaneous subcarrier amplitudes whose values depend on the subcarrier amplitude and the phase relationship between the samples and the subcarrier. From Equation 2 we have

$$Y = \frac{1}{2}(A + C) = \frac{1}{2}(B + D) \quad (3)$$

$$W_1 = Y - B = D - Y$$
$$= \frac{1}{2}(A + C) - B = \frac{1}{2}(D - B)$$
$$= D - \frac{1}{2}(A + C) \quad (4)$$

$$W_2 = Y - C = A - Y$$
$$= \frac{1}{2}(A - C) = \frac{1}{2}(B + D) - C$$
$$= A - \frac{1}{2}(B + D) \quad (5)$$

Since X has the same phase as D, any linear predictor $\hat{X}$ for X can be represented as

$$\hat{X} = \hat{Y} + \hat{W}_1 \quad (6)$$

where $\hat{Y}$ predicts the luminance at X, and $\hat{W}_1$ the instantaneous subcarrier amplitude. From Equation 3 we have two possible elementary predictors for Y, namely

$$\hat{Y}_1 = \frac{1}{2}(A_{ij} + C_{ij}) \quad (7)$$

$$\hat{Y}_2 = \frac{1}{2}(B_{ij} + D_{ij}) \quad (8)$$

and from Equation 4 we have three possible elementary predictors for W_1, namely

$$\hat{W}_1 = \frac{1}{2}(A_{ij} + C_{ij}) - B_{ij} \quad (9)$$

$$\hat{W}_2 = \frac{1}{2}(D_{ij} - B_{ij}) \quad (10)$$

$$\hat{W}_3 = D_{ij} - \frac{1}{2}(A_{ij} + C_{ij}) \quad (11)$$

The A_{ij}, B_{ij}, C_{ij}, and D_{ij} may be selected from any of those shown in Fig. 2, but clearly the elements closer to X would be more desirable. Taking all possible combinations of Equations 7–8 and 9–11 yields only two elementary predictors for X, namely

$$\hat{X}_1 = D_{ij} \quad (12)$$

and

$$\hat{X}_2 = A_{ij} - B_{ij} + C_{ij} \quad (13)$$

Since the A_{ij}, B_{ij}, C_{ij}, and D_{ij} may be chosen in a number of different ways and formed into linear combinations to give complex predictors which satisfy Equation 6, a more general expression for $\hat{X}$ is

$$\hat{X} = \sum_{i,j} a_{ij}D_{ij} + \sum_{i,j} b_{ij}(A_{ij} - B_{ij} + C_{ij}) \quad (14)$$

with

$$\sum_{i,j} a_{i,j} + \sum_{i,j} b_{i,j} = 1$$

Finally we could add to Equation 14 any number of "null" predictors which are zero in an area of constant luminance and chromaticity such as Fig. 2, but which may give useful information about the slope of luminance or chrominance changes in an area of change. This converts Equation 14 to

$$\hat{X} = \sum_{i,j} a_{ij}D_{ij} + \sum_{i,j} b_{ij}(A_{ij} - B_{ij} + C_{ij}) + \sum_{i,j} c_{ij}(A_{ij} - A'_{ij}) + \sum_{i,j} d_{ij}(B_{ij} - B'_{ij}) + \sum_{i,j} e_{ij}(C_{ij} - C'_{ij}) \quad (15)$$

with

$$\sum_{i,j} a_{ij} + \sum_{i,j} b_{ij} = 1$$

for a unity-gain predictor. The notation A'_{ij}, B'_{ij}, and C'_{ij} is intended to indicate that these pels are chosen to be in the same neighborhood as A_{ij}, B_{ij}, and C_{ij} and have the same subcarrier phase. Although for simplicity of notation a common subscripting has been used in all the terms in Equation 15, the summations may extend over a different number of pels in each term. For example, a predictor that we have used is

$$\hat{X} = \frac{1}{4}D_{12} + \frac{1}{4}D_{32} + \frac{1}{4}D_{31} + \frac{1}{4}(A_{32} - B_{32} + C_{32}) \quad (16)$$

Here there are three D pels, one $(A - B + C)$ group, and no null predictor.

Equation 15 states the constraints on linear combinations of pels such that the prediction has the same resultant subcarrier phase as X has. It does not say how these pels are to be selected. As with monochrome predictors, the best pels to select are those which are close to X in space and time, in a combination which gives reasonable anticipation of edges of various orientation. With $4f_{sc}$ sampling the pels are spaced quite close together along any line. Calculation shows that B_{32} is 2.3 times further away from X than is A_{12} (Fig. 2).

In principle therefore, the use of same-line pels A_{12} and B_{12} is desirable on the grounds of closeness. However, to constitute a valid predictor they must be combined with some C pel, say C_{12} or C_{32}, into an $(A - B + C)$ group. With a sample interval of $1/(14.3 \times 10^6) = 70$ nsec, it is difficult with present device speeds to complete this arithmetic in a predictive loop which also includes a RAM (random-access memory) as a quantizer. Thus device speeds give additional constraints on the choice of pels for the predictor.

Quantization

In a predictive coding system the prediction error

$$e = X - \hat{X} \quad (17)$$

is quantized to give a quantized prediction error e_q. The quantization error q is then

$$q = e - e_q \quad (18)$$

Several different procedures have been suggested[11] for determining the allowable magnitude of q as a function of e. Quantizers for broadcast-standard composite color signals, however, have generally been adjusted by trial and error to give good subjective results.[5,12] Since there are a large number of variables in a broadcast-quality quantizer, trial-and-error optimization is not a straightforward process. We experimented therefore with an algorithmic procedure, explained below, which simplified the selection of representative and decision levels in the quantizer.

It is to be noted, firstly, that the effect of predictive quantization on a color signal is similar in kind to that for monochrome signals. If the quantization characteristic has steps near the origin which are too large, granular noise occurs. Steps in the central part of the characteristic give rise to edge busyness. Slope overload results from having a maximum representative level which is too small. These three types of distortions can occur as purely luminance variations or purely chromaticity variations or, more typically, in combination.

With component-color signals, which have one quantizer for each of the three components Y, I, and Q, the chrominance and luminance variations can be separately controlled. With composite signals, however, the Y, I, and Q signals are simultaneously quantized in a single quantizer. The quantization error is then distributed between the luminance and chrominance components in the demodulation process at the receiver.

Analysis of this process[13] suggests that procedures used to design threshold quantizers for luminance signals[14] can be used for composite-color signals, provided the function $T(e)$ linking the luminance threshold magnitude of quantization error $|g|$ to the magnitude of the prediction error $|q|$ is stretched horizontally. This amounts to forming a new threshold function

$$T'(e) = T(e/k), \quad k > 1 \quad (19)$$

In practice $T(e)$ is determined subjectively using a particular predictor. If a pel close to the predicted pel X is chosen as a predictor, quantization defects such as granular noise and slope overload are less visible than for a pel further away, such as D_{12}. As we have seen in the previous section, it is difficult to utilize pels close to X in $4f_{sc}$ sampled composite-signal prediction. This means that the threshold will be low, the stretched threshold will be low, the stretched threshold function $T(e)$ will be even lower, and a significantly larger number of quantization levels may be needed in the quantizer

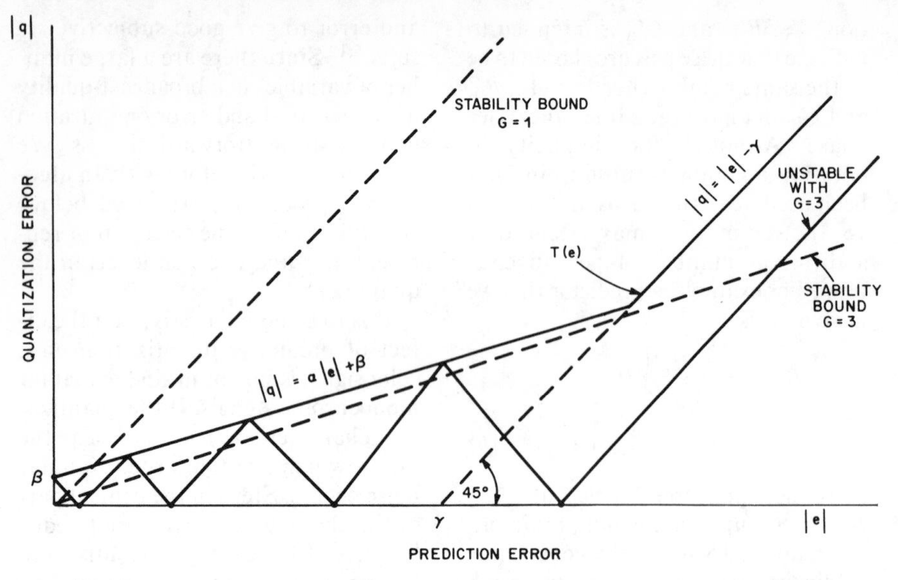

Figure 3. Approximation used for the threshold function $T(e)$. Also shown are the stability limits for $G = 1$ and $G = 3$.

than for a luminance-signal coder where the choice of predictor pels is not restricted.

Our experimental work involved parallel optimization of the predictor and quantizer, and this interaction between the two made the quantizer design more difficult. Rather than obtain $T(e)$ subjectively for all possible predictors of interest — a tedious process — it was noted that previous measurements[15,16] of $T(e)$ could be approximated by the three-parameter piecewise linear function illustrated in Fig. 3. Since $T'(e)$ is just a stretched or scaled version of $T(e)$, Fig. 3 also serves to describe $T'(e)$. Representative and decision levels are obtained as shown by constructing lines at 45° to the horizontal with alternate positive and negative slope.[16] The intercept parameter β determines the step size around the origin, and the slope parameter α determines the rate of increase of the step size. The parameter γ represents a value which must be equaled or exceeded by the largest representative level in the quantizer; it determines the slope overload. Figure 4 shows the positive half of a symmetrical 6-bit quantizer generated in this way with $\alpha = 1/31$, $\beta = 1.8$, and $\gamma = 245$, each expressed on a scale of 0 to 511, assuming initial 9-bit quantization. In cases such as this, where a definite number of representative levels is required in the quantizer, a few iterations of the procedure in Fig. 3 may be required, with slightly different values of the parameters α, β, and γ. It was found that this was best accomplished by fixing β and γ at subjectively determined values and altering α.

The procedure portrayed in Fig. 3 for calculating the decision and representative levels is, in practice, carried out with whole numbers, which results in some subtleties. Intervals between representative levels turn out always to be odd numbers (Fig. 4), and representative levels increase alternately as odd and even numbers. If an interval between two representative levels is chosen to be an even number, as for example

12		16	
11	13	14	18

and the quantizer produces subthreshold impairment, this can always be replaced by the more economical

12		17	
11	14	15	19

since the maximum quantization error for a prediction error of 14 remains 2.

A significant point in designing quantizers for composite NTSC signals is that while the total composite signal range is 166 IRE units, the luminance range is only 100 IRE units. If there is initial 9-bit (512-level) analog-to-digital conversion of the composite signal, the luminance signal is quantized to $512 \times 100/166 = 308$ levels as compared with 256 levels for normal 8-bit conversion of the luminance component signal. The quantization levels in a 9-bit composite predictive coder have therefore an effective spacing of 256/308 or 83 percent of the values with 8-bit luminance-component coding.

Stability Considerations

If quantization error q is made in a simple predictive code using a single

Representative level	Decision interval
1	0 2
4	3 5
7	6 8
10	9 11
13	12 14
16	15 17
19	18 20
22	21 24
27	25 29
32	30 34
37	35 39
42	40 44
47	45 49
52	50 55
59	56 62
66	63 69
73	70 76
80	77 83
87	84 91
96	92 100
105	101 109
114	110 118
123	119 128
134	129 139
145	140 150
156	151 162
169	163 175
182	176 189
197	190 204
212	205 220
229	221 237
246	238 511

Figure 4. A 6-bit composite symmetrical quantizer characteristic generated according to the procedure of Fig. 3, with $\alpha = 1/31$, $\beta = 1.8$, and $\gamma = 245$.

pel as a predictor, limit-cycle oscillations can occur[16] if

$$|q| \geq |e| \quad (20)$$

where e is the prediction error as in Equation 17. The oscillations occur because a quantization error q made in coding pel X affects future predictions. If the prediction $\hat{Y} = X_q$ is exact in the absence of the error q_X, i.e.,

$$Y - X = 0 \quad (21)$$

then the prediction error e_Y is composed entirely of the quantization error in X, i.e.,

$$e_Y = Y - \hat{Y} = q_X \quad (22)$$

But if the characteristic of the quantizer is such that $|q_Y| \geq |e_Y|$, then $|q_Y| \geq |q_X|$. Hence the quantization error at X results in an equal or greater quantization error at Y. This process can result in oscillations between $+q$ and $-q$ in some areas of the picture.

The use of a single $(A - B + C)$ group as a predictor is illustrated in Fig. 5, where A, B, and C may be selected from any pels reasonably close to X in Fig. 2. Extending the above argument, it is now possible for quantization errors q_A, q_B, and q_C at pels A, B, and C respectively to produce a total quantization error greater than any one. Thus, if

$$|q_A| = |q_B| = |q_C| = q \quad (23)$$

and

$$q_A = q_C = -q_B \quad (24)$$

then the error made in the prediction is $3q$. The worst-case quantization error gain G is 3 in this example, and therefore to insure stability.

$$|q| < \frac{|e|}{G} = \frac{1}{3}|e| \quad (25)$$

This result may be generalized to any linear prediction

Table 1. Worst-Case Gain for Selected Predictors

Predictor	Worst-Case Gain G
$\hat{X}_1 = D_{12}$	1
$\hat{X}_2 = \frac{1}{4}D_{12} + \frac{1}{4}D_{32} + \frac{1}{2}D_{31}$	1
$\hat{X}_3 = A_{32} - B_{32} + C_{32}$	3
$\hat{X}_4 = \frac{1}{4}D_{12} + \frac{1}{4}D_{32} + \frac{1}{4}D_{31} + \frac{1}{4}(A_{32} - B_{32} + C_{32})$	1.5

$$\hat{X} = a_1X_1 + a_2X_2 + \cdots + a_nX_n \quad (26)$$

$$\sum_{i=1}^{n} a_i = 1$$

for which the worst-case quantization error gain is

$$G = \sum_{i=1}^{n} |a_i| \quad (27)$$

The stability condition is then,

$$|q| < |e| \Big/ \sum_{i=1}^{n} |a_i| \quad (28)$$

Pirsch[17,18] has shown that this is a sufficient condition for stability, but that a lower bound holds with certain classes of predictors. Table 1 lists the value of G for some of the predictors we have used.

Equation 28 is shown in graphical form in Fig. 3 for values of $G = 1$ and $G = 3$. For $G = 1$ and $\beta \neq 0$, small-amplitude limit-cycle oscillations can occur but will typically not be visible. Of particular interest is the fact that with an $(A - B + C)$ predictor, for which $G = 3$, the quantization characteristic will be unstable if γ is too low; this effect was observed experimentally and is described later. It follows from expression 25 that the criterion is

$$\gamma > 511\left(1 - \frac{1}{3}\right) \text{ or } \gamma \geq 341 \quad (29)$$

The 6-bit quantizer shown in Fig. 4 therefore is not stable with an $(A - B + C)$ predictor and needs additional "stability levels" at ±341 to make it so. If $\gamma < 171$, another pair of stability levels needs to be inserted at ±171. We report in the next section on the effect of these stability levels on picture quality.

The difficulty in satisfying inequality 28 becomes more severe as the number of quantization levels is reduced. With 6-bit quantizers it is relatively easy to find an extra one or two pairs of levels for stability purposes. But with 3-bit, 4-bit, and even 5-bit quantizers, all the available representative levels are utilized in the lower part of the characteristic to minimize the visibility of quantization distortion, and it may be impossible to satisfy inequality 28 without a noticeable deterioration in picture quality. In this case, predictors with more D samples and fewer $(A - B + C)$ groups or null groups must be considered as a better trade-off.

Experimental Results

The experimental results are divided into three sections: (1) a description of the mixed hardware/software test facility, (2) selection of the predictor and quantizer for best subjective picture quality, and (3) modifications to the predictor and quantizer that are required to code vertical interval test signals (VITS).

Test Facility

The test facility consisted of a Digital Equipment Corporation PDP 11/40 computer linked to a Z80 local microcomputer which was used to change the predictor and quantization characteristic. Composite 525-line, 2:1 line-interlaced, 4.2-MHz NTSC color signals could be obtained from a camera or off-air. Back-lighted 8 × 10-in (203 × 254-mm) transparencies of various scenes were available as test pictures. Some of the transparencies were made by using the EBU* test

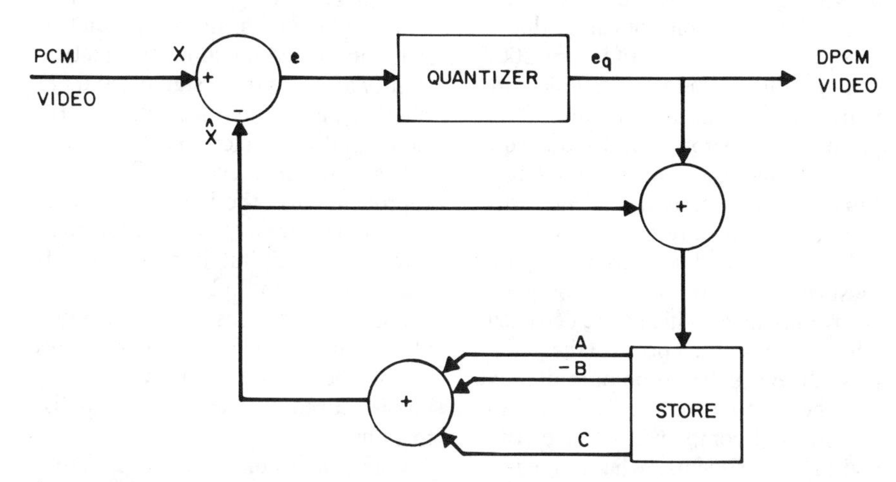

Figure 5. Composite-signal prediction using an $(A - B + C)$ combination of pels.

* EBU = European Broadcasting Union.

slides. Initial 9-bit analog-to-digital (A/D) conversion of the composite NTSC signal was accomplished using a Computer Labs A/D converter.

Predictor and Quantizer Selection

The method adopted in the selection of a suitable predictor and quantizer was one of visual observation by the authors based on critical pictures. This is perhaps less appealing than finding a minimum mean-square error predictor and combining it with a minimum mean-square error quantizer. The latter procedure, however, has been shown to give less than optimum subjective picture quality with component color pictures due to inappropriateness of the mean-square error criterion.

In composite-signal coding, predictor–quantizer interaction is of even greater significance because the predictors have to satisfy the phase-combination property given by Equation 15, and, in so doing, they tend to amplify quantization errors, due to the inclusion of $(A - B + C)$ and null groups. This amplification can make the predictor–quantizer combination unstable, as discussed, but even within the stability bound, a gain factor $G > 1$ can increase the visibility of quantization error.

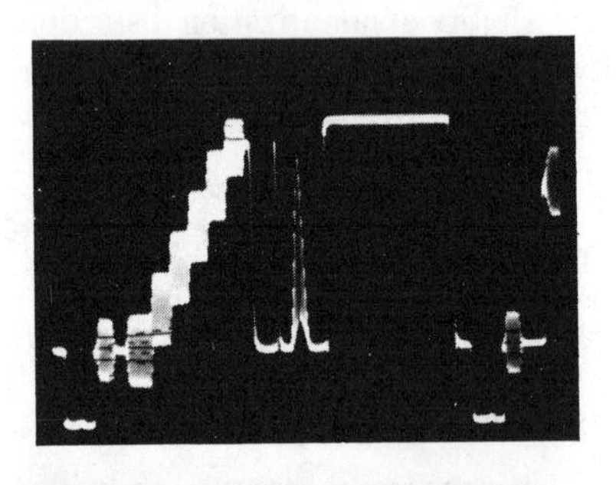

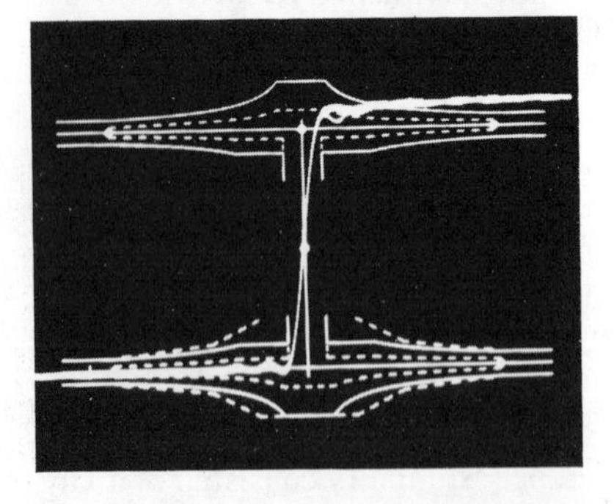

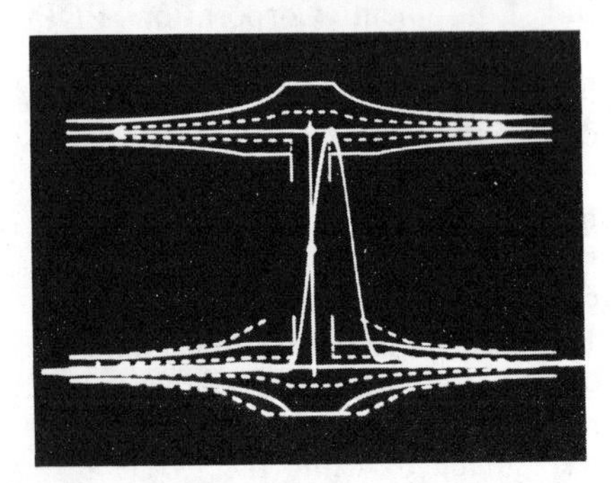

Figure 6. Predictive coding of the Vertical Interval Test Signal (VITS).

Of the many possible intrafield predictors which satisfy Equation 15, we concentrated attention on the four shown in Table 1, which showed promise in early trials. Observations were made with a number of quantizers ranging from 3 bits/pel to 6 bits/pel, designed according to the procedure in the *Quantization* Section. It was found that with $\hat{X}_3$, severe limit-cycle oscillations occurred with 3-bit and 4-bit quantizers, typically triggered by the first line of picture information or by the VITS in the field blanking interval. Even with a $4\frac{1}{2}$-bit (22-level) quantizer, transient breakup occurred at particular features involving both horizontal and vertical edges. We concluded that $\hat{X}_3$ was an unsuitable predictor to use with low-bit quantizers.

Of the remaining predictors, $\hat{X}_1$ and $\hat{X}_2$ each have a worst-case gain $G = 1$ and had no stability problems. Predictor $\hat{X}_4$ incorporated one $(A - B + C)$ group which raised G to 1.5, but gave improved interline prediction as a result. We observed no stability problems with it and found in general that it gave a slightly higher picture quality than $\hat{X}_2$, which was the next best predictor. We do not at present rule out the possibility that there may be more complex predictors giving a better performance than any listed in Table 1.

The three-parameter quantizer design procedure summarized in Fig. 3 proved to be convenient. In a rough approximation, the value of β determined the granular noise, α the edge busyness, and γ the slope overload. Limited observations indicated that $\alpha \simeq 1/15$, $\beta \simeq 2/511$, and $\gamma \simeq 200/511$ produced below-threshold impairment for one or another of the predictors in Table 1 and required a quantizer having between 5 and 6 bits. Further observations are needed with a wider variety of test pictures.

Slope overload appeared to be the most common form of distortion in the many quantizers which were designed. This may have been because the design procedure was based on data derived for the luminance signal in monochrome and component-color picture coding experiments, where the prediction pels can be closer to the predicted pel than is possible in composite coding. With the fourth-previous element D_{12} as a predictor, for example, the predictive loop responds rather slowly to an edge in the picture.

Experiments were conducted with artificial "stability levels" inserted in 5-bit and 6-bit quantizers at 171 and/or 341. This tended to cause edge busyness as the quantizer flipped between the stability level and the next representative level below it on a frame-by-frame basis. The cure was to insert a series of intermediate levels between the maximum representative level required on subjective grounds and the stability level, so that overall the representative levels incremented gradually. This, however, significantly increased the total number of levels in the quantizer. Eventually the practice of stability level insertion was abandoned in favor of finding predictors with lower worst-case gains. A further advantage of low-gain predictors, which we did not explore experimentally, is their potential for increasing the decay rate of transmission errors and therefore lowering their visibility.

Our overall conclusion was that broadcast-quality pictures could be produced with a quantizer having between 5 and 6 bits, using $4f_{sc}$ sampling. Thus a simple intrafield predictive coder, without allowance for blanking-interval removal or error correction, requires around 75–78 Mbit/sec to satisfy subjective broadcast criteria.

Vertical Interval Test Signals

We have carried out some preliminary experiments concerned with the prediction and quantization of the vertical interval test signals (VITS) normally inserted in the field blanking interval. Our aim was to discover two things: (1) how a simple predictive coder optimized for subjective picture quality affected these signals, and (2) whether small modifications to predictor and quantizer could yield improved waveform rendition. The latter approach skirts the issue of whether such signals are appropriate for measurement of digital impairments. It attempts only to find a means of providing as much transparency as possible on a digital link which constitutes a short section of an otherwise analog link impaired by linear waveform distortions.

Our conclusion was that different predictor and quantization characteristics were required for the VITS. In particular, the fourth-previous element D_{12} was a better predictor than any of the others in Table 1 because pels in the previous line yield no information at all about the VITS; therefore, they severely reduce the height of the pulse and bar due to slope overload and also produce a poor SNR (signal-to-noise ratio) in the central section of the bar where this measurement is normally made.[19]

The quantizer in Fig. 4 proved to be adequate for the pulse and bar section of the VITS. The height of both of these is 100 IRE units or 308/511 quantization levels. Observed on a waveform monitor, there was noticeable slope overload, edge busyness, and bar noise. Photographs of VITS are shown in Fig. 6. A modified asymmetrical 6-bit quantizer was designed (Fig. 7) which had complete transparency between $-1/511$ and $+1/511$ and could be reflected[20] about 256 to provide quantization levels around 308. There was one more level in the positive half of the characteristic than in the negative half.[21] Limited observations of both waveform distortion and subjective picture quality with this quantizer indicate that it may be possible to find a 6-bit compromise design which satisfies both waveform and subjective criteria.

Our observations to date indicate that the waveform-distortion criteria that are normally applied to VITS give some indication of granular noise visibility (in the SNR measurement), are too severe in their requirements for slope overload distortion, and do not consider edge busyness, an important temporal impairment in predictive coding.

-247	-228	-211	-194	-179	-164	-151
-512 -238	-237 -220	-219 -203	-202 -187	-186 172	171 158	157 145
-138	-127	-116	-107	-98	-89	-80
144 133	132 122	121 112	111 103	102 94	93 85	84 77
-73	-66	-59	-52	-47	-42	-37
76 70	69 63	62 56	55 50	49 45	44 40	39 35
-32	-27	-22	-19	-16	-13	-10
34 30	29 25	24 21	20 18	17 15	14 12	11 9
-7	-4	-1	0	1	4	7
8 6	5 3	2 1	0	1 2	3 5	6 8
10	13	16	19	22	27	32
9 11	12 14	15 17	18 20	21 24	25 29	30 34
37	42	47	52	59	66	73
35 39	40 44	45 49	50 55	56 62	63 69	70 76
80	87	96	105	114	123	134
77 83	84 91	92 100	101 109	110 118	119 128	129 139
145	156	169	182	197	212	229
140 150	151 162	163 175	176 189	190 204	205 220	221 237
246						
238 511						

Figure 7. Asymmetrical quantizer used in reflected form. The positive half is the same as Fig. 4; the negative half was generated with $\alpha = 1/29$, $\beta = 1.8$, and $\gamma = 245$.

Conclusions

A study is reported of intrafield predictive coding of composite NTSC signals sampled at $4f_{sc}$. The choice of predictor was found to be subject to several constraints: (1) linear combinations of pels had to be used which were either D samples (same phase as the predicted pel) or $(A - B + C)$ groups or null predictors; (2) several same-line pels were not suitable because they could not be formed into valid linear combinations in the short time period resulting from the 14.3-MHz sampling frequency; (3) $(A - B + C)$ groups gave rise to stability and edge busyness problems. Of several predictors which satisfied these constraints, one which used a single D element in the same line, two in the previous line, and one $(A - B + C)$ group in the previous line was found to perform best.

A procedure for designing quantizers for composite signals was developed based on a three-parameter piecewise linear approximation to the quantization error threshold function. A family of quantizers was generated in this way, with broadcast quality being obtained at between 5 and 6 bits/pel. With 6 bits/pel there was no visible impairment on critical pictures using very close inspection.

Limited observations were made of waveform distortion caused when coding vertical interval test signals (VITS). It was concluded that a different predictor and quantizer were ideally required for these signals to satisfy their performance criteria, but that at 6 bits/pel some compromise might be found which provided broadcast-standard subjective quality and met waveform distortion criteria.

Acknowledgment. The authors would like to acknowledge with gratitude helpful discussions with J. A. Bellisio, A. B. Larsen, and P. Pirsch.

References

1. I. M. Jacobs and J. Stauffer, "FT3-Metropolitan Trunk Lightwave System," *Proc. IEEE*, *68*, No. 10, 1286–1290, Oct. 1980.

2. J. E. Thompson, "Differential Encoding of Composite Color Television Signals using Chrominance-Corrected Prediction," *IEEE Trans. on Commun. Tech.*, *COM-22*, 1106–1113, 1975.

3. J. O. Limb, C. B. Rubinstein, and J. E. Thompson, "Digital Coding of Video Signals — A Review," *IEEE Trans. on Commun. Tech.*, *COM-25*, No. 11, 1349–1384, 1977.

4. J. P. Rossi, *"Sub-Nyquist-Encoded PCM NTSC Color Television," SMPTE J.*, *85*, 1–6, Jan. 1976.

5. Y. Hatori and H. Yamamato, "Predictive Coding for NTSC Composite Color Television Signals Based on Comb-Filter Integration Method," *Trans. IECE of Japan*, *E62*, No. 4, 201–208, 1979.

6. I. Dinstein, "DPCM Prediction for

NTSC Composite Signals," *Comsat Tech. Rev.*, *7*, No. 2, 429–446, 1977.
7. T. Ishiguro et al., "Composite Interframe Coding of NTSC Color Television Signals," *Proc. Nat. Telecomm. Conference*, Dallas, Texas, 1976, pp. 6.4-1 to 6.4-5.
8. R. S. Hopkins, Jr., "Report of the Committee on New Technology," *SMPTE J.*, *89*, 449–450, June, 1980.
9. H. Gharavi and R. Steele, "Predictors for Intraframe Encoding of PAL Picture Signals," *Proc. IEE*, *127*, Pt. F, No. 3, 205–211, 1980.
10. A. V. Oppenheim and R. W. Schafer, *Digital Signal Processing*, Prentice-Hall, Inc., Englewood Cliffs, N.J., 1975.
11. A. N. Netravali and J. O. Limb, "Picture Coding: A Review," *Proc. IEEE*, *68*, No. 3, 366–406, 1980.
12. V. G. Devereux, "Differential Coding of PAL Video Signals Using Intrafield Prediction," *Proc. IEE*, No. 12, 1139–1146, 1977.
13. D. E. Pearson, "Digital Processing of Signals in Communications," *IERE Proc.*, No. 49, 1981, pp. 413–426.
14. D. K. Sharma and A. N. Netravali, "Design of Quantizers for DPCM Coding of Picture Signals," *IEEE Trans. Comm.*, *COM-25*, No. 11, 1267–1274, 1977.
15. P. Pirsch, "Design of DPCM Quantizers for Video Signals Using Subjective Tests," *IEEE Trans. Comm.*, *COM-29*, No. 7, 990–1000, 1981.
16. J. C. Candy and R. H. Bosworth, "Methods for Designing Differential Quantizers Based on Subjective Evaluation of Edge Busyness," *Bell System Tech. J.*, *51*, No. 7, 1495–1516, 1972.
17. P. Pirsch, "Optimierung von Farbfernseh-DPCM-Systemen unter Berücksichtigung der Wahrnehmbarkeit von Quantisierungsfehlern," Doktor-Ingenieur Dissertation, University of Hannover, 1979.
18. P. Pirsch, Private Communication.
19. "IEEE Standard on Video Signal Measurement of Linear Waveform Distortion," IEEE Standard 511-1979.
20. H. G. Musmann, "Predictive Image Coding," in *Image Transmission Techniques*, edited by W. K. Pratt, Academic Press, New York, 1979, pp. 73–112.
21. F. Kretz, J. L. Boudeville, and P. Sallio, "Optimization of DPCM Video Coding Scheme Using Subjective Quality Criterions," *Proc. Conf. on Digital Processing of Signals in Communications*, Loughborough, England, IERE Conf. Publication No. 37, 1977.

Discussion After Presentation of Paper

JIM GASPER, MCDONNELL-DOUGLAS: I noticed that you labeled your sampling points *D*, *C*, *B*, and *A*, in that order. Is that because of different weights you place on the different sampling points?

BRAINARD: Thank you. I think I left out a couple of points in there. What I really intended to say was that we labeled the points in normal fashion from right to left and from bottom to top, because what we are interested in is history in generating our prediction. I have used here the same nomenclature we started with. There is nothing special about the *A*, *B*, *C*, and *D*; they are just four successive samples, and there is no intent that they should have any particular significance relative to the *I* and *Q* signals either.

FRED HODGE, 3M COMPANY: I noticed you evaluated the VITS with a separate predictor. Have you done anything with character generator signals in the video since they have a single-line transition that is kind of unusual; also, how about chromakey effects of the signal downstream after it has been predicted and transmitted?

BRAINARD: On the second part, on chromakeying, we have done nothing so far. Adding alphanumerics into a picture or mixing signals that result in a sharp horizontal transition gives a signal that is beyond the normal bandwidth of a television signal. . .

HODGE: Well, beyond the vertical bandwidth. . .

BRAINARD: Yes, we have one line that is very different from the previous line as in the case of the VITS. However, we know that the VITS is intended to be viewed on a waveform monitor. This is intended to enable easy measurement of degradations from an analog channel. In this case we have an edge in a picture. The sensitivity of our visual system to errors at an edge in a picture is rather low; such errors are not as visible. In a digital system where we have control we want to put the noise near an edge for concealment.

KEN DAVIES, CBC, MONTREAL: I noticed your prediction involves taking samples from two fields. Could you give us some idea of what kinds of evaluations you are doing of the results of motion defects?

BRAINARD: What I am talking about this morning is strictly infield (intrafield) using only the current line and the previous line. At a future time, I hope I can talk about also using the previous fields.

DAVIES: I am not entirely sure from your presentation whether you are attempting to get to 88 or 44 megabits per second (Mbit/sec) with this work?

BRAINARD: Well, what I was talking about this morning was limited to a 90-Mbit/sec system; for such a system, we can code the entire signal at four times the color subcarrier frequency ($4f_{sc}$) with 6 bits per sample using DPCM (differential pulse code modulation) coding. Now, to go to 45 Mbit/sec (and I was not talking about this today), in addition to transmitting only the active portion of the signal, one would have to do many other things to get a picture at the receiver with no visible degradation.

A. A. GOLDBERG, CBS: You started off with 9-bit encoding. What was the significance of that? And was it necessary to start off with something with a higher quantization resolution?

BRAINARD: We started off with 9 bits. In coding a monochrome television signal, if we consider 8-bit resolution as adequate, then for a composite color signal, which includes the chrominance information added to the luminance signal, we have a larger amplitude range, and 9-bit resolution would be adequate. That is why we started with 9 bits. When we reduce this to 6-bit DPCM for transmission we use a nonlinear quantizer. In smooth areas of the picture the prediction error is small and errors are relatively visible. Here we use small steps corresponding to 9-bit PCM resolution. At edges in the picture, where errors are less visible and prediction errors are larger, we use coarse steps to represent these prediction errors. The quantizer, when appropriately balanced, results in a picture with no visible degradation.

FEATURE IC'S FOR DIGIVISION TV SETS

R. Deubert
ITT Intermetall
Freiburg, West-Germany

With the introduction of digital signal processing in color TV sets further possibilities for the implementation of new functions are made available.

In the following, three new VLSI's will be described which can directly be implemented in the digital color TV concept of Intermetall. These are

- Video Processing Unit for NTSC featuring combfilter for Luminance and Chrominance
- Decoder for Teletext or Videotext signals
- Interlace free picture

To clarify the interface of the new VLSI's in the digital TV concept a brief remainder will be given on the basic digital TV chip set.

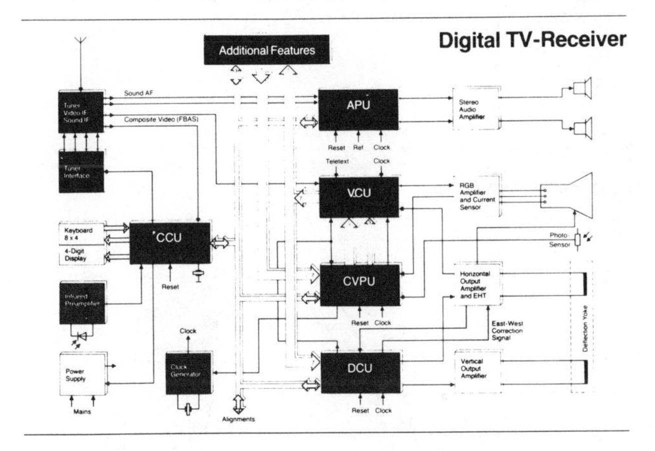

Fig. 1

Fig. 1 shows the block diagram of a digital color TV set.
The digital processing is performed by 5 VLSI's:

- the microcomputerbased Central Control Unit
- the Audio Processing Unit
- The Video Codec Unit performing all the analog to digital and digital to analog conversion for the video signals
- the Video Processing Unit
- the Deflection Control Unit

Manuscript received 6/17/83

As interface between the processors two types of digital bus systems are used:
A video bus system giving access to the composite video signal and to the processed luminance and color difference signals and a control bus system which is used to control the different processors. The clock rate of the video bus system is fourtimes color subcarrier frequency and the clock rate of the control bus is approx. 170 kHz. All new systems are designed to interface directly with the two bus system of the digital TV chip set.

NTSC Video Processor

The basic chip set shown in Fig. 1 can perform both PAL and NTSC processing having a color subcarrier trap in the luminance channel. However, combfilters are becoming more and more important for NTSC TV sets because they can significantly improve the quality of TV pictures. Up to now mainly glass delay lines are used to perform the function of combfilters and many external components are additionally required. Moreover the performance of the combfilters realized in this way is mainly determined by the characteristics of the glass delay line. The usage of a whide band digital delays offers the possibility to optimize the combfilter system. Therefore a special NTSC Video Processor is developed including combfilter. It is pin compatible with the Video Processor of the basic chip set and can be directly inserted in the TV PC board instead of the Video Processor.

Block diagram

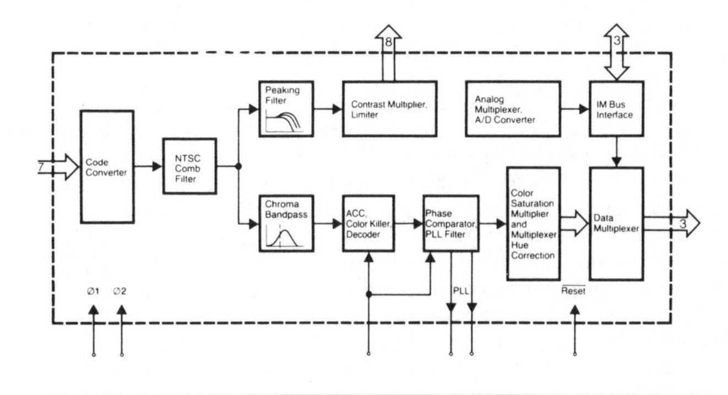

Fig. 2

It's internal block diagram is shown in Fig. 2

Reprinted from *IEEE Trans. Consum. Electron.*, vol. CE-29, pp. 237–241, Aug. 1983.

The chip combines all the functions necessary to process the luminance and the chrominance signals. The description will only concentrate on the blocks "NTSC Combfilter, Peaking Filter, Chroma Bandpass". See Fig. 3.

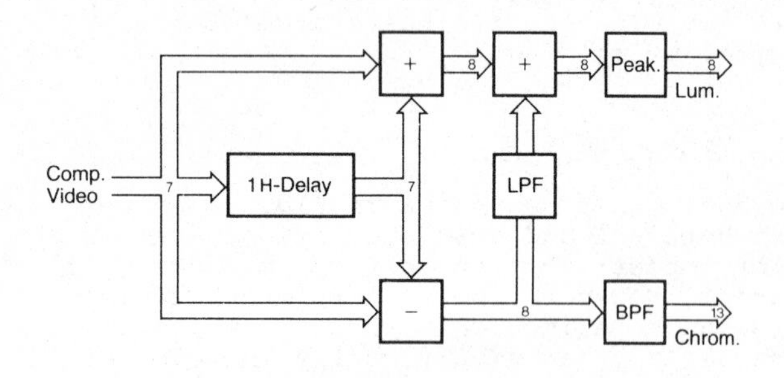

Fig. 3

The composite video signal having passed through the code converter is delayed by one horizontal line with the aid of a 910x7 bit DRAM. The delayed composite video signal is then supplied to an adder and a subtractor in which it is added to and subtracted from the non delayed composite video signal, thus generating the comb filtered luminance signal at the output of the adder and the comb filtered chrominance signal at the output of the subtractor. In order to regain the vertical resolution in the luminance signal, which is reduced due to the combfilter, low frequency parts of the comb filtered chrominance signal must be added back to the luminance signal by which the comb filter effect for the low frequency domain is eliminated. This is achieved with a low pass filter (LPF) and a second adder. The characteristics of the low pass filter determines the frequency domain in which the luminance signal is comb filtered. Due to the fact that the chrominance signal is now completely removed from the luminance signal the peaking filter can be optimized.

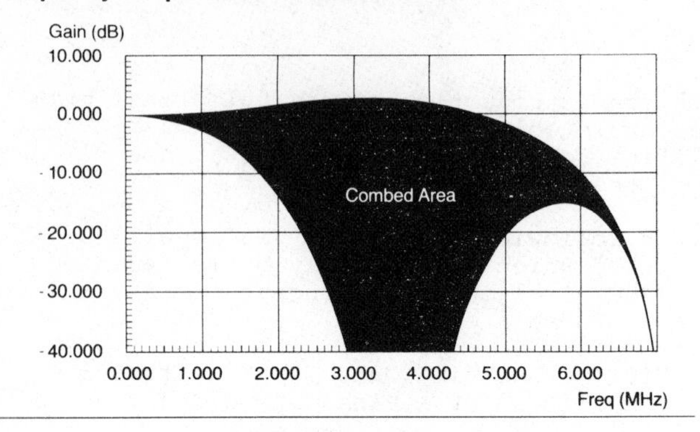

Fig. 4

Fig. 4 illustrates the overall frequency response of the luminance channel including the area in which the combfilter is active. The frequency response can be changed by +10 db and -6 db at a frequency of 3,5 MHz. The system has a constant group delay.

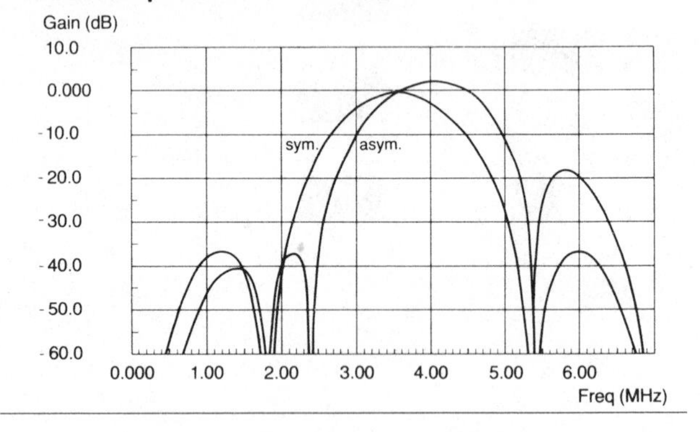

Fig. 5

Fig. 5 shows the characteristics of the chroma band pass filter. Two different responses can be selected by software, one which is assymetrical to the color subcarrier frequency to compensate the IF-response, and the other which is symmetrical to the subcarrier to be used for other signal sources. The filter has a constant group delay for both characteristics. The input word length of 8 bits is continuously increased within the band pass filter to 13 bits thus giving optimum noise performance.

Teletext Processor

The Teletext Processor operates completely digitally and can directly interface with the digital TV chip set.

Teletext application diagram

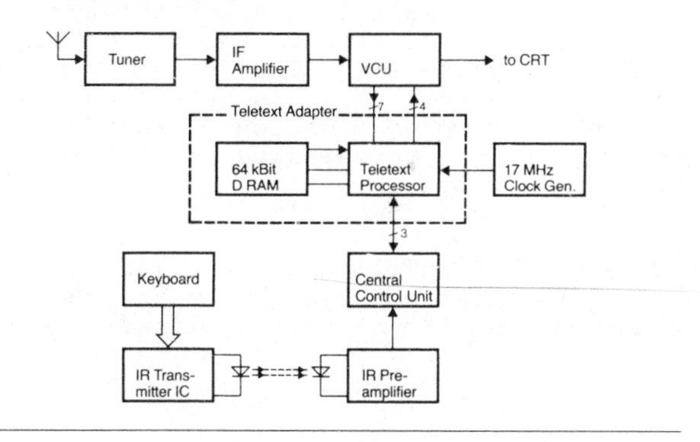

Fig. 6

Fig. 6 shows its implementation together with a standard 64K DRAM as an eight page memory for UK Teletext and German Videotext. As input signals the Teletext Processor receives the digital composite video signal. The output signals are the RGB signals and fast blanking. The four times color subcarrier clock and the H and V synchronisation signals are supplied from the digital TV set. Remote control is possible with the CCU via the IM Bus.

Structure of the Teletext Processor

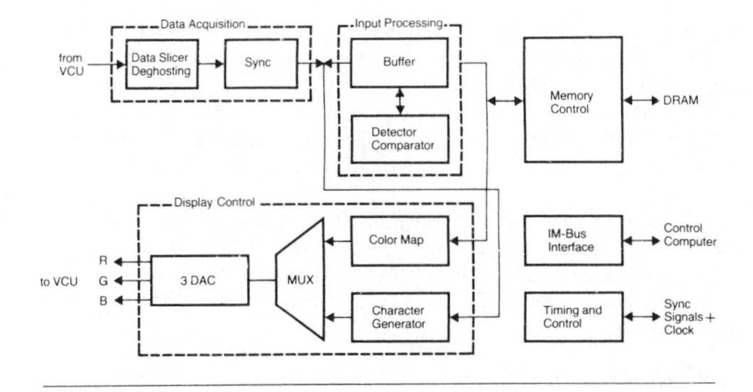

Fig. 7

The main functional blocks of the Teletext processor are shown in Fig. 7. The digital composite video signal is first processed in the data acquisition unit. The signal passes through the data slicer in which the optimum separation level is found. The sliced Teletext signal is then processed in a ghost compensation network. The ghost compensation network consists of a transversal filter with five different coefficients in which ghost signals with a delay time of up to 0,8 µs can be compensated. The calculation of the filter coefficients is done automatically. For this purpose the signal from the output of the ghost filter is correlated in an error detector with a synthetic signal formed out of the Teletext signal at the input of the ghost filter. The results of the correlation controls the filter coefficients.

After the ghost processing the Teletext information is synchronized, identified and buffered in the input processing unit. If a selected page code is identified the memory control unit is activated and the content of the buffer is written into the external DRAM. The maximum size of the external DRAM to be controlled directly from the Teletext processor can be 64 k bit featuring an 8 page memory for UK Teletext and German Videotext. Synchronous to the scanning of the TV set the external DRAM is read by the memory control unit. If the contents of the DRAM represent characters, the input buffer is used to store one complete character line while it is transferred to the character generator of the display control unit to generate the corresponding RGB signals.

If the external DRAM is used as a pixel memory the data are directly transferred to the color map of the display control unit. Mixed operation of character and pixel display is possible; i.e. the Teletext processor with an external 64 K bit DRAM can be used for the Japanese system.

When the DRAM is not accessed from the Teletext processor the memory control unit refreshes the memory content. Via the IM-Bus interface the CCU, or another external processor, can read from and write into all RAM locations.

For the free format systems such as french Antiope and NABTS the teletext processor can be used as an interface circuit between the TV set and the system consisting of microcomputer ROM and RAM. In this application the control commands are handled by the microcomputer, that means, that the control bus has to be connected to the microcomputer system.

Interlace free picture

A digital processing scheme is presented to get a picture display of 525 lines per half frame or 625 lines respectively. This requires a double horizontal scan and double the bandwidth in the video channel compared with a normal TV set.

Comparison Normal Scan / Double Scan

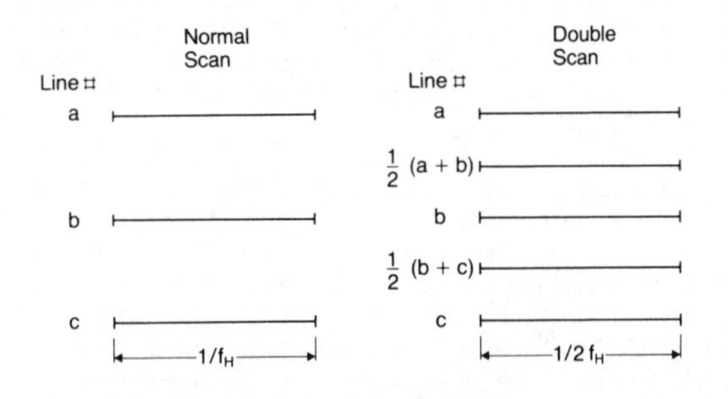

Fig. 8

This diagram shows the organisation of the lines in double scan compared with the conventional TV-display. The broadcasted line number a is stored and then displayed during half the time. During the line with double scan the mean value of the broadcasted line number a and b is displayed, then line number b, then b+c and so on. The internal block diagram of the circuit performing double scan is shown in Fig. 9.

Interlace free system

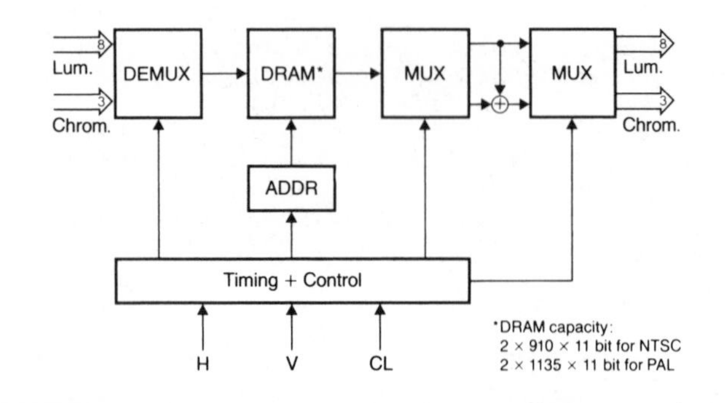

Fig. 9

The 8 bit digital luminance signal and the 3 bit color difference signals sampled at fourtimes color subcarrier are demultiplexed and written into a DRAM which can store the signals for two lines. By reading the signals out with twice the speed of writing them in a time compression of 1 : 2 can be achieved thus generating a video signal to be displayed with double horizontal scan. The two multiplexers and the adder serve to form the mean value of two lines and to put the videosignals in the correct form that they can be routed to the video codec for further processing. Theoretically a one and half line storage would be enough to achieve the time compression but using a DRAM with a storage capacity of two lines the maximum clock frequency can be reduced and the timing becomes simplier.

Two methods of sampling are available to perform the time compression in the DRAM, the normal and the offset sampling.

Double Scan Sampling Principle

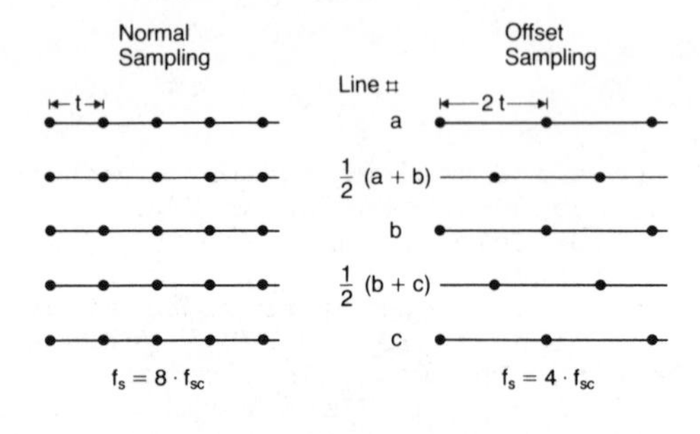

Fig. 10

In both cases the signal is written into the DRAM with a clock of fourtimes color subcarrier. In the normal sampling method the signals are read out with twice the frequency of that with which they were written in, using all stored samples.

In the offset sampling the signals are read out with the same frequency as they are written in using only every second stored sample but from line to line in an offset position.

The advantages and disadvantages of both methods can clearly be seen: The normal sampling requires twice the clock frequency compared with the offset sampling but giving at the same time, twice the signal bandwidth. The normal sampling does not limit the bandwidth of the videosignal because the sampling rate of output signal is eight times color subcarrier frequency resulting in a maximum bandwidth of fourtimes color-subcarrier. This method is therefore well suited for high definition NTSC TV sets using combfilters in the video processing.

The offset sampling is suitable for TV sets in which a color subcarrier trap already limits the luminance bandwidth. The maximum achievable bandwidth is limited to twice the color subcarrier frequency but the offset sampling simulates a better frequency response. This method is well suited for PAL systems in which normally no combfilter is used.

Interlace free System Diagram

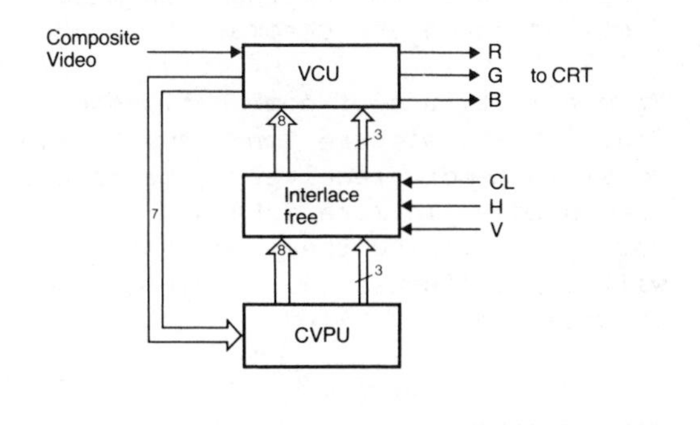

Fig. 11

This figure shows the implementation of the interlace free IC in the digital TV set and its interface.

Further developments are being made to reduce the chip count in digivision and to implement more functions such as video IF and sound IF. New systems will be developed performing better TV pictures such as full frame memory, horizontal and vertical aperture correction , adaptive noise reduction etc. The existing chip set is hereby the start of the digital revolution in TV sets.

(1) Dr. T. Fischer
Electronics Aug. 11, 1981
"Digital VLSI breeds next generation TV-receivers"

(2) Dr. T. Fischer
"Transactions of 1982 IEEE Internationel Conference on Consumer Electronics"
"What is the impact of digital TV"

(3) Fric J. Lerner
Digital TV: Makers bet on VLSI
IEEE Spectrum Febr. 1983

APPLICATIONS OF PICTURE MEMORIES IN TELEVISION RECEIVERS

E. J. Berkhoff, U. E. Kraus, and J. G. Raven
Ned. Philips Bedrijven BV
Eindhoven, The Netherlands

Abstract

Digital serial CCD memory chips with a capacity of 308 Kbit allow to build picture memories with just a small number of packages and few control signals, as the device has inherent addressing.
These picture memories are used for features and picture improvements. From a multitude of possible combinations of features and picture improvements two memory receiver concepts are described.

1. Introduction

For the first time the advance in memory technology makes it possible to introduce picture memories in television receivers on a large scale.
Digital serial CCD memory chips with a capacity of 308 Kbit; a detailed description of which may be found in an accompanying paper(1), allow us to build picture memories with just a small number of packages and few control signals, as the device has inherent addressing.
That the CCD chip is a serial memory is no hindrance in most video processing applications.
The primary motive for using picture memories in television receivers is the fact that only the use of such memories allows a number of significant features and picture improvements.

Features like:

- a Teletext (TXT) background memory (instant response when selecting a page)
- Multi-Picture-in-Picture (MPIP) formed by a sequential scan of the tuner (a real convenience for program selection)
- Still Picture, in order to make a hard copy possible
- Picture-in-Picture, with e.g. the video signal of the Euroconnector as the second video source

and the following picture improvements by reduction of:

- large area flicker; most noticeable in 50Hz TV systems
- line flicker, caused by interlacing
- noise, as present in VCR signals and signals originating in weak reception
- cross-colour and cross-luminance, impairments caused by simultaneous use of one spectrum for the transmission of luminance and chroma.

From a multitude of combinations of features and picture improvements two possible memory receiver concepts have been studied in more detail.
Before describing these concepts we will first discuss the required memory organisation.

2. Memory organisation

The organisation of the field memory has been based on:

- A line locked sampling frequency.
  This choice minimises the complexity of the implementation of most processing algorithms with memories.

- Component video signals instead of composite signal.
  Processing with component video signals also simplifies these processing algorithms.
  Furthermore they simplify compatibility problems with PAL, SECAM and NTSC in 625 lines and 525 lines TV systems.
  With component signals only distiction has to be made between 625 lines/50Hz and 525 lines/60Hz TV systems.

- A 7 bit quantisation of the digitised Y, U and V signals.
  These signals are stored in time multiplexed in the 7 CCD's, which form the 7 bit wide field memory.

The impact of these choices and other aspects like compatibility with external video signals on the receiver architecture are described in (2).

Manuscript received 6/17/83

Reprinted from *IEEE Trans. Consum. Electron.*, vol. CE-29, pp. 251-254, 256-258, Aug. 1983.

3. A receiver concept with one field memory

As far as the use of the field memory is concerned, this receiver has the following specification:

- Noise reduction on the luminance and colour difference signals by means of temporal filtering.

- Cross-colour reduction of the colour-difference signals by a comb filter with a field delay.

- A store-picture facility.

- A TXT background memory for approximately 250 TXT pages.

This concept is intended as a high-performance receiver for TV systems with a 50Hz and in principle as well for TV systems with a 60Hz field repetition rate. The aforecited items are mutually exclusive: either TXT background memory or noise reduction or cross-colour reduction or the store picture facility can be chosen by the user.

3.1. Noise reduction

In principle temporal noise reduction on the digitised Y,U and V signals is done with the well-known scheme of a picture memory forming a first-order recursive filter in time (3), see figure 1. But two deviations exist:

- Although the noise reduction is performed on the complete luminance bandwidth, only the low frequencies are used for movement detection.
  The high frequencies cannot be used for this owing to the presence of the colour subcarrier and its sidebands. This presence would, also in a non-moving scene, force the movement detector to decide on movement.
  The colour difference signals have completely independent movement detectors for their noise reduction.

- With a complete picture memory the time delay is constant, but with the use of 'only' a field memory this delay must be switched every field period from a 312 TV lines delay to a 313 TV lines delay and backwards, in order to let the noise reduction be effective on the same spatial position as indicated in figure 2 (or between 262 and 263 TV lines delay for a 525 lines TV system).

  A sharp horizontal transient in a non-moving picture will now be interpreted by the movement detector as movement and consequently on this transient there is no noise reduction. This effect is hardly disturbing due to the fact that on a transient the visibility of noise is strongly reduced.

Of course the switching in the field delay is done by halting the clock to the memory every second field for one line period more.

3.2. Cross-colour reduction

If the quality of the signal does not make noise reduction necessary the field memory can be used for cross-colour reduction on the colour-difference signals. The CCD's are then applied for a constant delay of 312 and 263 TV lines respectively.
Addition of the direct and delayed colour difference signals now results in a significant reduction of cross-colour.
No movement adaptivity will be used.
The movement effects in the colour difference signals are hardly visible due to the transversal nature of the processing.

3.3. Store picture

With this facility it is possible to store a complete field from a running program and recall it later for viewing or printing.

3.4. TXT background memory

Figure 3 illustrates a normal multipage TXT decoder configuration, consisting of a video input processor (VIP), which slices the TXT data and the actual TXT decoder e.g. the one described in an accompanying paper, the CCT (4).
In Data Entry Window (DEW), being the 16 TV line period during field flyback in which TXT lines can be transmitted, the VIP slices the incoming TXT data to a binary data stream which is applied to the TXT decoder. The multipage facility is realised by adding more display memory capacity to the CCT.

Figure 4 gives a TXT decoder configuration with a background memory, consisting of 7 CCD's.
Coded TXT data are stored in this background memory; in contrary with the former multipage concept in which decoded TXT information is stored.

Between VIP and CCT there is now a decision and control circuitry, called the Background Memory Controller (BMC), with the following functions:

- During DEW, see figure 5, it places all the CCD's in series and, if it detects TXT lines during this period, it stores these data in the CCD's.

- Outside DEW it recirculates the contents of all the CCD's to refresh the memory.

- Recirculation is done in such a way that the contents of the CCD's is put back in the position it had just at the end of DEW.

- Also when the CCD's have been completely filled with TXT data the BMC continues to write in new TXT data, thereby destroying the oldest information. This means that if the memory can contain 250 TXT pages it are always the most recent 250 pages.

- After a request from the CCT for a new page the BMC reacts as follows; during and outside DEW it acts as described before, but is also connects the output of the last CCD to the input of the CCT. After 20ms, one complete refresh period of this CCD, the CCT has seen all the information in this CCD and continues its scan of the background memory with the next CCD, and so on.
  In this way the CCT very rapidly scans all the CCD's, thereby virtually eliminating the normal TXT waiting times, provided of course that the number of pages transmitted does not exceed 250.

- The BMC always completes a full scan and switches the CCT input directly to the VIP output after that scan, in order to cope with subtitle pages, rolling pages and pages not yet stored in the memory.

For the user this concept has the advantage of direct access to all TXT pages. The problem of pre-selecting the contents of the multipage memory of figure 3 is for the TXT decoder with background memory eliminated.

4. A receiver concept with two field memories

This top performance concept is especially intended for 50Hz TV systems, because it features large area flicker reduction by doubling the field rate to 100Hz (5).
As far as the use of field memories is concerned this receiver has the following specification:

- large area flicker reduction
- noise reduction on full bandwidth luminance and colour difference signals.
- TXT background memory
- Multi-Picture-in-Picture (MPIP)
- picture enlargement with a factor 4
- Still Picture

Noise reduction, TXT background memory, MPIP, Still Picture and picture enlargement are mutually exclusive but are always accompanied by large area flicker reduction.

4.1. Large area flicker reduction

The conversion from a 50Hz to a 100Hz field rate is illustrated in figures 6 and 7. The incoming video fields are alternatively written in the two field memories. After an incoming video field has been stored, the appropriate field memory starts the reading and recirculating cycle with twice the write frequency. After a complete recirculation it starts again with a second read cycle at twice the write frequency.
By doing so the incoming video fields are compressed in time with a factor two and twice repeated.
During this action of one field memory the other one stores the next video field and then starts the 100Hz conversion.

4.2. Noise reduction

Figure 8 illustrates that within the concept of the 100Hz converter of figure 7 a picture delay is present which can be used for noise reduction.
During the first read cycle on twice the nominal frequency it is clear that the memory contents must be recirculated because a second read-cycle is necessary.
But now also during this second read-cycle the memory contents are recirculated. If then the next field is stored in that memory, the previous contents, which was stored two fields earlier, is then still available but at normal speed.

The same argument holds for the other field memory and by adding a second 2 to 1 multiplexer after the field memories, a 40ms delayed video signal on the normal 50Hz speed is available. It must be realised that this picture delay can only be used for recursive processing because the memory contents is displayed via the 100Hz output. Only by recursive processing, like noise reduction, the memory contents is a delayed version of the processed signal.
Now noise reduction on the 50Hz video input signal for this 100Hz converter is possible. For this noise reduction we can now use the same circuitry as in the former concept.

4.3. TXT background memory

Due to the fact that no TXT decoder is available with output signals suitable for a 100Hz display, two of the CCD's are always used for a 100Hz conversion of the TXT decoder output data.
The remaining 12 CCD's are used for the TXT background memory and perform this function with the same control circuitry of the former concept.

4.4. Multi-Picture-in-Picture

Due to the fact that two field memories are available we can use one field memory for the display and use the other one as an acquisition memory for collecting e.g. nine small, non-moving pictures by a sequential scan of the tuner. After filling the acquisition memory it copies its contents into the display memory and again starts collecting nine pictures from 9 new channels or a time-update from the former 9 channels.
This feature is intended for easy program selection.

4.5. Picture enlargement by a factor four

This picture processing is also possible with serial field memories by writing in at twice the nominal frequency.
Reading out the contents with the normal read frequency gives an enlarged picture, suitable for display.

4.6. Still Picture

In contrary with the former concept the two field memories are always necessary for the display of video information. This means that now a still picture is possible but not a store picture facility.

References

1. Pelgrom M. et al., "A digital field memory for television receivers", Int. Conf. on Cons. Electronics, June 8-10, 1983, Chicago.

2. Conrads W. "Integrated feature-TV-concept with I^2C-bus control and memories", Int. Conf. on Cons. Electronics, June 8-10, 1983, Chicago.

3. Fanton N.E. "Video digital filter study II: temporal first-order recursive digital filter", BBC Research Department Report No. 1976/26.

4. Crowther G.D. et al., "CCT a VLSI Teletext Decoder Circuit", Int. Conf. on Cons. Electronics, June 8-10, 1983, Chicago

5. Kraus U.E. "Vermeidung des Grossflächenflimmern in Fernseh-Heimempfänger". Rundfunktechnischen Mitteilungen 1981, 6, page 264 to 269.

IEEE Transactions on Consumer Electronics, Vol. CE-29, No. 3, August 1983

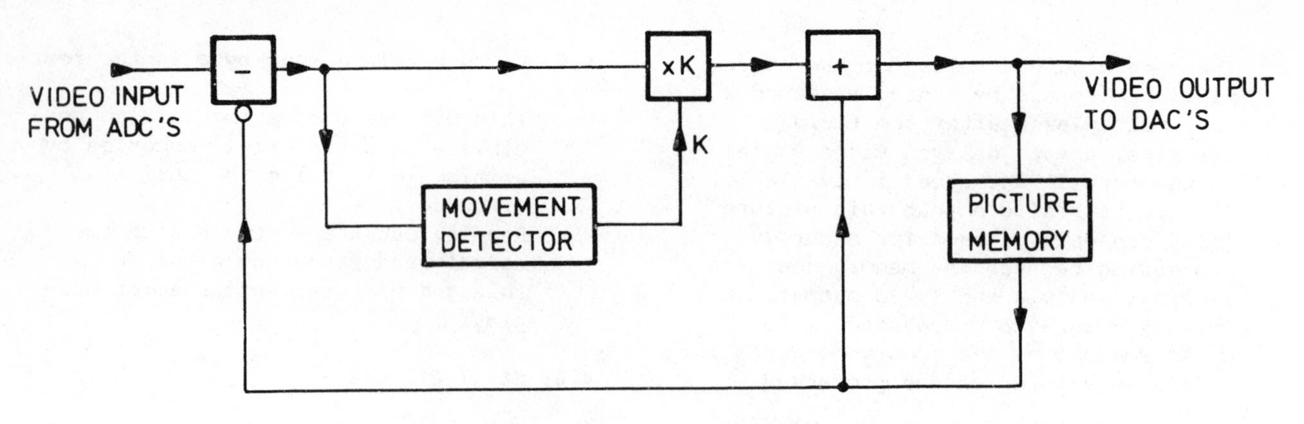

FIGURE 1: NOISE REDUCTION BY TEMPORAL FILTERING

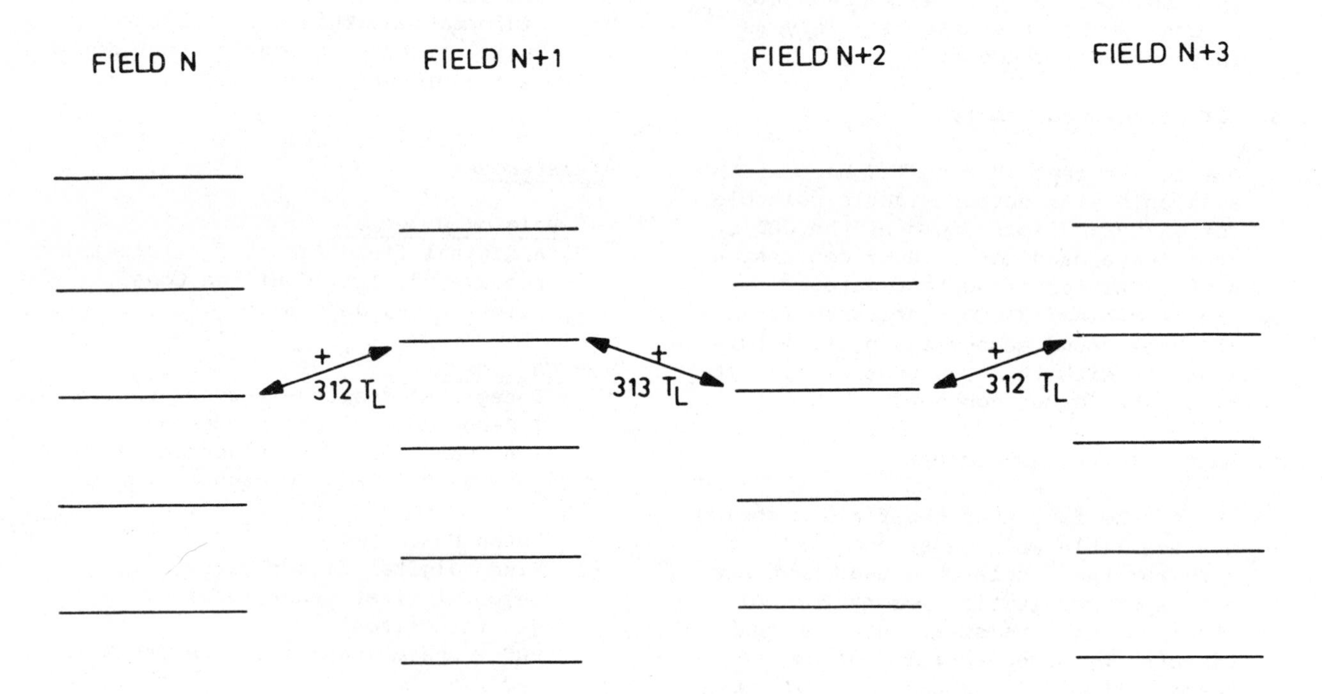

FIGURE 2: ALTERNATING DELAY IN FIELD MEMORY FOR CORRECT SPATIAL FILTERING

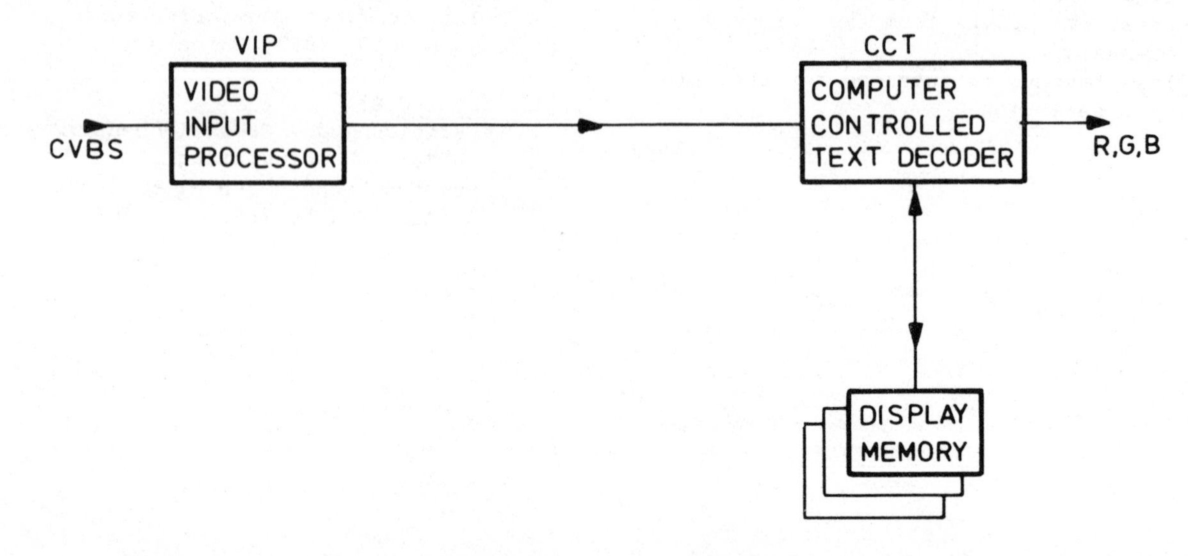

FIGURE 3: NORMAL MULTIPAGE TXT DECODER

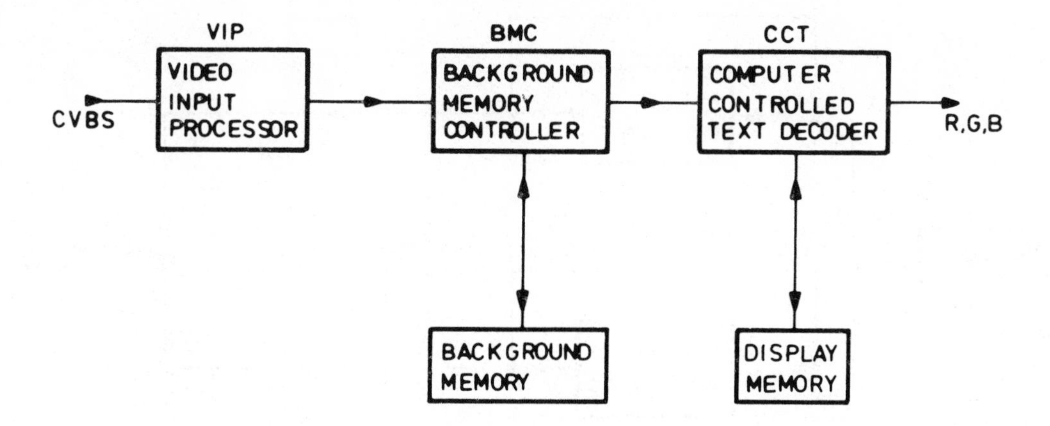

FIGURE 4: TXT DECODER WITH BACKGROUND MEMORY

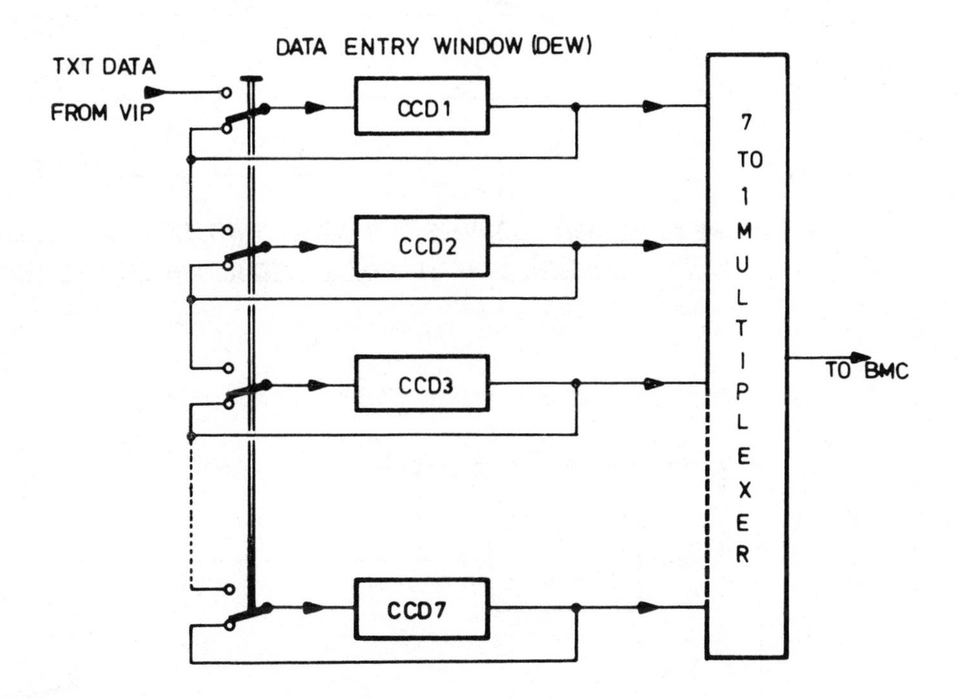

FIGURE 5: CCD MEMORY ORGANISATION FOR TXT BACKGROUND MEMORY

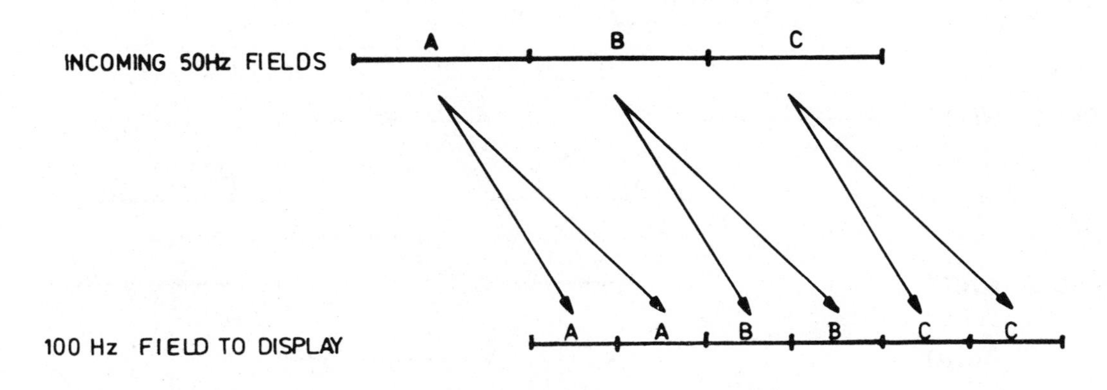

FIGURE 6: PRINCIPLE OF 100 Hz CONVERSION

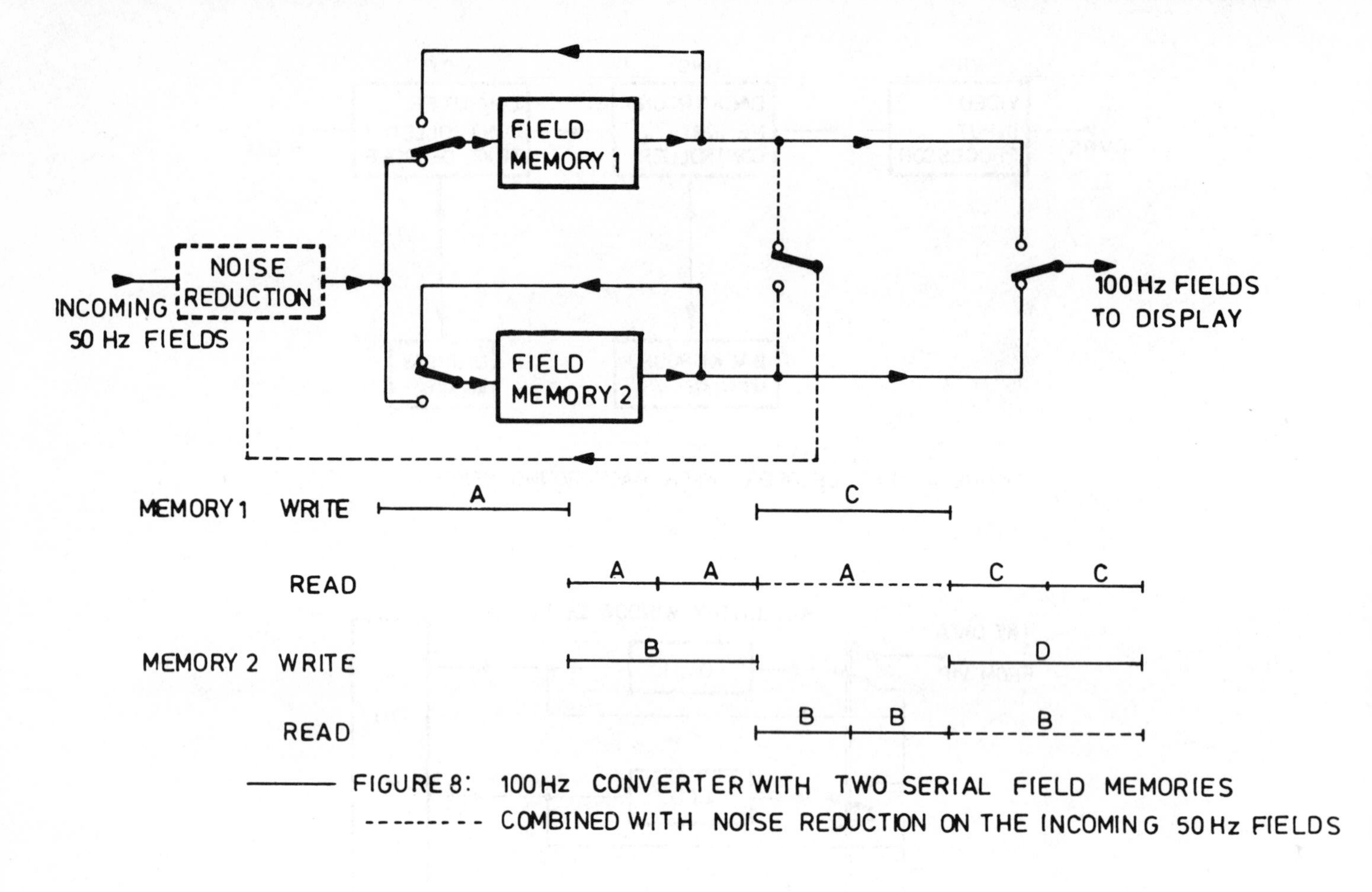

——— FIGURE 8: 100Hz CONVERTER WITH TWO SERIAL FIELD MEMORIES
-------- COMBINED WITH NOISE REDUCTION ON THE INCOMING 50Hz FIELDS

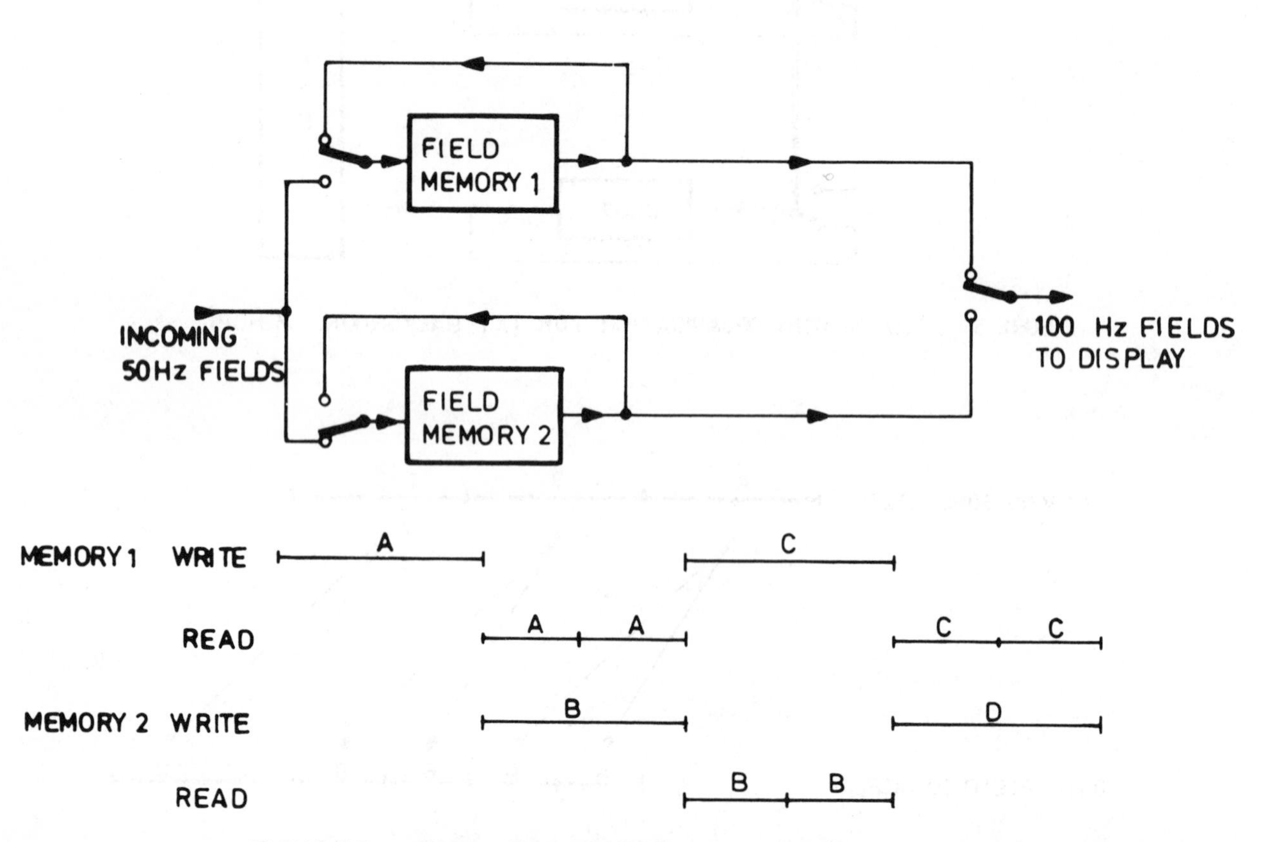

FIGURE 7: 100Hz CONVERTER WITH TWO SERIAL FIELD MEMORIES

Part V
Teletext

TELETEXT is a new television service which has generated a great deal of interest in the last few years. Although a simple teletext system was developed in England in the 1970's, it is still in the development stage in the United States. A number of different teletext systems have been developed: the English system, which has come to be known as the World System, North American Broadcast Teletext Specifications (NABTS), Antiope (French), and Telidon (Canadian). Most of these systems are still being evaluated in the U.S. and only a few viewers have the necessary decoders. However, the World System is further along in introduction and at the present time there are more than two million TV receivers in 14 countries equipped with decoders for this system.

The World System is the simplest system and uses synchronous transmission of data which allows pages to be displayed directly without intervening processing. Antiope uses a nonsynchronous scheme which gives greater flexibility of display, but requires more memory as transmission attributes (DRCS)—character color, shape, or change—are transmitted in parallel rather than serially. Canada's Telidon system uses picture description instructions (PDI) to encode relatively high resolution alpha geometric displays with a minimum number of data bytes. The NABTS system draws upon both the Didon protocol from the Antiope system as well as the display techniques used in the Telidon system. Thus, NABTS is able to display greatly improved graphics at the expense of requiring a more complex decoder with a sophisticated processor and extensive ROM for alpha geometric display.

Deregulation of teletext by the FCC has allowed all of the systems to compete on the marketplace. At this time, it is not possible to predict which system will predominate or how the market will be divided if no clear-cut winner emerges.

This technical part on teletext consists of eight papers with the first three describing United Kingdom teletext, Antiope, and Telidon systems. Unfortunately, there has been very little published at the present time on the NABTS teletext system, except for the specification; thus, no papers on this subject have been included. A general description of television receiver design optimization for teletext reception is covered in the fourth paper. This paper also includes the design of the remote control system for viewer convenience for both teletext and television reception. Next, a paper outlining the requirements for a set top adapter for teletext has been included since it will be a number of years before a significant percentage of TV receivers will have built-in decoders. The sixth paper covers the problem of signal distortions in the transmission path which have substantial effects on teletext reception and may even cause the decoder to be unable to extract the required data. De-ghosting techniques are also covered and how de-ghosting techniques can improve data reliability are shown. The seventh and last paper covers the problem of echo equalization, specifically for teletext, and covers the use of an adaptive transversal filter with automatic coefficient adjustment.

ANTHONY TROIANO
Associate Editor

BIBLIOGRAPHY

[1] Special Issue on Consumer Text Display Systems (Teletext and Viewdata), *IEEE Trans. Consum. Electron.*, vol. CE-25, July 1979.

[2] E. S. Sousa, "Pulse shape design for teletext data transmission," *IEEE Trans. Commun.*, vol. COM-31, July 1983.

[3] T. Rzeszewski, "A new teletext channel," *IEEE Trans. Commun.*, vol. COM-29, Feb. 1981.

[4] J. Lopinto, "The application of DRCS within the North American broadcast teletext specification," *IEEE Trans. Consum. Electron.*, vol. CE-28, Nov. 1982.

[5] M. Obara, "A digital time domain equalizer for teletext," *IEEE Trans. Consum. Electron.*, vol. CE-28, Aug. 1982.

[6] J. P. Chambers, "Enhanced U.K. teletext moves toward still pictures," *IEEE Trans. Consum. Electron.*, vol. CE-26, Aug. 1980.

[7] C. O. Eissler, "Market testing teletext in subscription television medium," *IEEE Trans. Consum. Electron.*, vol. CE-27, Aug. 1981.

[8] S. Lipoff, "Proposed broadcast teletext standard; Field tests, data analysis procedures and sample presentation formats," *IEEE Trans. Consum. Electron.*, vol. CE-26, Aug. 1980.

[9] M. Hirashima, "Sampling clock regeneration in teletext receivers," *IEEE Trans. Consum. Electron.*, vol. CE-26, Aug. 1980.

[10] G. Schober, "Teletext field trials using UHF WETA-TV station," *IEEE Trans. Consum. Electron.*, vol. CE-27, Aug. 1981.

[11] W. Ciciora, "A tutorial on ghost cancelling in television systems," *IEEE Trans. Consum. Electron.*, vol. CE-25, Feb. 1979.

A DESCRIPTION OF THE BROADCAST TELIDON SYSTEM

J. R. Storey, A. Vincent and R. FitzGerald
Federal Department of Communications
Ottawa, Canada

Introduction

Telidon, the Canadian videotex standard, is now a familiar concept, and its operation in the interactive mode via the switched telephone network has been well documented. This paper presents the transmission format used for Telidon data carried as an ancillary signal on broadcast television channels. We shall first review some of the essential characteristics of Telidon, and then show that these are preserved in the broadcast situation.

The fundamental characteristic which distinguishes Telidon from other videotex systems is its ability to construct images from a set of graphic primitives; these are: Point, Line, Rectangle, Polygon, Arc and Text. These are described by codes known as Picture Description Instructions (PDI). It is possible, using PDI's, to encode very complex graphic images in a relatively small number of data bytes. All Telidon terminals, therefore, contain a microprocessor which is capable of interpreting PDI's and drawing the corresponding images.

An important advantage of this "alpha-geometric" coding of graphical information is that the display technology and resolution of Telidon terminals are totally independent of the PDI codes. In other words, manufacturers can produce a wide range of terminals, ranging from low-resolution block-graphics through DRCS to very-high-resolution bit-plane memory designs, all of which interpret the same PDI's.

This terminal-independence of the Telidon PDI's is clearly one characteristic which must be preserved in any transmission scheme, including the broadcast television format discussed in this paper.

SYSTEM REQUIREMENTS

The fundamental requirement of the system is that packets of Telidon data (PDI's) must be inserted into otherwise unused lines of an NTSC video signal at the transmitter such that they can be recognized and interpreted by a terminal at a receiving station. While the first generation of encoding equipment is limited to a one-way i.e., non-interactive service using the Vertical Blanking Interval, these limitations must not be inherent in the transmission format. It must accommodate:

- the use of any or all TV lines. Full channel transmission, where all TV lines are filled with Telidon data, would be implemented primarily on cable television channels,
- one-way non-interactive service, either broadcast or on cable,
- two-way interactive service on cable networks,
- program-related services such as closed-captioning,
- other time-dependent services such as time, news and weather bulletins,
- dynamic TV-line allocation. Since the number of TV lines in the Vertical Blanking Interval is limited, it would be wasteful to permanently allocate lines to specific services such as closed-caption which only use those lines relatively infrequently. The lines used for these ancillary

Reprinted from *IEEE Trans. Consum. Electron.*, vol. CE-26, pp. 578–585, Aug. 1980.

signals should then be dynamically selectable, either at the discretion of the broadcaster or in some programmed manner. Each service category would have an identifying code assigned to it, and several services could share the same set of TV lines.

- error detection and correction: because of the variety of signal impairments which can occur in a television channel, it is felt that some form of error detection and correction is necessary, in order to ensure data integrity and acceptably short waiting times.

Broadcast Telidon Format

The general form of the Telidon packet is shown in Figure 1. Each packet includes an 8-byte header, a one-byte error detection/correction character, and 16, 20 or 24 bytes of data or address information. The three data-rates in use at present are those required to permit one of the three sizes of packet to be transmitted during the active portion of a TV line. These are exactly 251, 291 and 330 times the horizontal line frequency, or approximately 3.95, 4.58 and 5.19 Mbit/sec.

The 8 byte header is shown in Figure 2. It begins with two bytes of alternating 0's and 1's, for data-clock regeneration. This is followed by a one-byte frame-synchronization character which specifies the locations of byte boundaries in the packet.

The synchronization sequence is followed by three bytes for Source Identification (SI). The first SI byte specifies the "service" or "channel" to which the packet belongs. The second and third bytes are either derived from the page number if applicable, or serve to further specify the service. The table of Figure 3 shows a typical list of services available on a

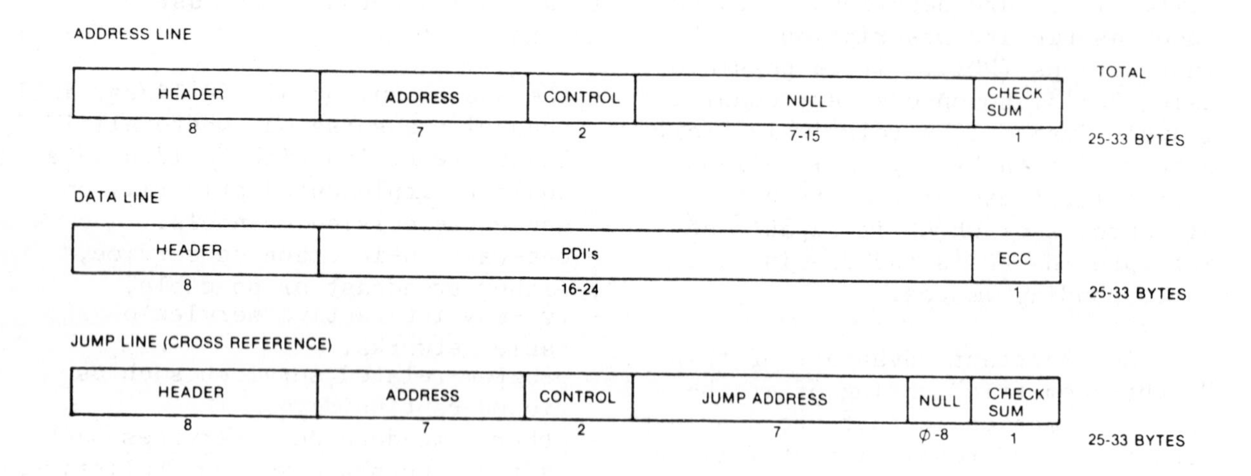

Figure 1. Broadcast Telidon line format

C	C	F	SI_1	SI_2	SI_3	DI_1	DI_2

C = BIT SYNCHRONIZATION BYTE
F = FRAMING BYTE (BYTE SYNCHRONIZATION)
SI_1 - SI_3 = SOURCE IDENTIFICATION
DI_1 - DI_2 = DATA IDENTIFICATION

Figure 2. Header format

SI1	Service	SI2, SI3
1	Program-related	Specify service e.g. closed captioning time, news, sports bulletins
2	one-way Telidon (French)	two most-significant non-zero digits of page number
3	one-way Telidon (English)	
4	two-way interactive	
5	updates	
6	polling	
7	games	specify type: interactive one-way down-line loaded
8	other	

Figure 3. Typical assignment of source identification codes

network and the significance of the SI bytes in each case. It should be noted that this is intended only as an illustration of the concept; individual broadcasters may choose a considerably different structure. The final two bytes of the header are referred to as the data-identification (DI). These two bytes have three principal functions:

- if DI1=0, DI2 specifies an addressing mode
- if 10≤DI≤DF (hexadecimal), the DI provides a continuity counter for successive lines of a page.
- if DI1=F, DI2 specifies a "termination" mode for the page.

There are thus 16 possible addressing modes, as well as 16 termination modes. The present allocations are as follows:

DI=00 : identifies the first line of a page which does not have a page number. This occurs in program-related services such as closed captioning. The data field of this line contains PDI's.

DI=01 : identifies the first line of a numbered page. The data field contains a seven-digit page number XXXXX:YY, followed by two control bytes (described below). The remainder of the data field contains hexadecimal 0's.

DI=02 : specifies a Cross-Reference (jump). The data field contains a seven-digit page number, two control bytes, and a seven-digit target page number. The effect is to cause the target page to be captured instead of the requested page.

DI=F0 : identifies the last line of a page which is the last or only page of a document or index.

DI=F1 : identifies the last line of a page which is not the last page of a document. These two termination modes permit viewing of successive pages of a document by use of a "next-page" command (⟶) at the terminal.

DI=F2 : identifies the last line of a page to which one or more other pages are to be concatenated. This permits transmission of very long pages. The first seven bytes of the data field contain the number of the page which is to be concatenated to the requested page. The remainder of the line is filled with hexadecimal 0's.

The control bytes referred to above contain eight bits, four of which have been assigned:

C1 distinguishes between index and document pages

C2 is a broadcast marker which is set on the first occurrence of any value of SI in the cycle. This can be used by the decoder to detect that a requested page is not in the cycle

C3 is used to inhibit capture of a page

C4 inhibits the display of the page number by the decoder.

Page organization

The DI codes discussed in the previous section yield two distinct page organizations, which reflect the two types of pages which may be present in the transmission at any time:

Numbered pages, which are normally transmitted sequentially in an endlessly-repeating cycle, consist of an address line with DI=01, followed by a variable number of data lines in which DI assumes the values from 10 through to a maximum of DF, and a termination line with DI=F0, F1 or F2.

Un-numbered pages, which occur in program-related services such as closed captioning, are in general time-dependent in that they are transmitted a limited number of times at a specific instant in time. Their organization is somewhat different. As well there is no address line, but rather the first line containing PDI's has DI=00. On subsequent data lines, DI varies from 11 through to a maximum of DF. There is also a termination line, in which DI is normally F0.

Decoding strategies

In the interactive mode via the switched telephone network, the decoder (terminal) sends page requests generated by the viewer to a computer which contains the desired data base. Thereafter, the decoder accepts whatever data are sent to it.

In the interactive mode via a cable television network, the operation of the decoder is similar, except that it must examine each packet of information which it receives, to determine whether it belongs to the requested page, and is addressed to that user.

In the one-way mode, the decoder is constantly receiving packets of information. Upon receipt of a page request from the viewer, it must begin examining each packet to determine whether it belongs to the requested page. It must accept all of the desired packets until it detects a "last-line" code, at which time it discontinues its search. (Note that the words "packet" and "line" are used interchangeably throughout this discussion).

This last mode is discussed in greater detail here, with reference to the Source Identification and Data Identification fields of the broadcast Telidon format. As discussed, the SI bytes specify a service category, and either a service or a page number. For example, using the service listing of Figure 3, a line belonging to page 2794 of service 3, would have SI=327. A line belonging to service 21 of the program-related services category (#1) would have SI=121. The decoder can therefore examine the SI field of all incoming lines of data to determine whether they *might* belong to the desired page. When it detects a match, it examines the DI field until it finds an address line, which is identified by DI=0x. The decoder can then proceed to decode the page number contained in the address line, in order to determine whether it *does* in fact identify the desired page. Once the address line of the desired page has been detected, the decoder then assumes that all subsequent lines bearing the same SI belong to that page and contain PDI's, provided that their DI fields are consistent with the continuity count. The decoder continues to capture all such lines until it encounters a "termination" line (identified by DI=Fx).

In order to minimize hardware costs and to take advantage of the presence in the decoder of a microprocessor, it is desirable to perform as much as possible of the address recognition, DI verification, data transfer and error detection and correction in software. However, low-cost microprocessors cannot possibly perform the required number of calculations in one TV line-time (63 microseconds). For this reason, it is necessary that successive lines from any one page occur at intervals sufficiently long to ensure that each line has been fully processed before the following line is received. It is for this reason that pages are *interleaved* in the broadcast Telidon data stream.

Interleaving

In order to ensure that the microprocessor has sufficient time to process a line of data before receiving another one, it has been specified that at least four milliseconds must separate any two lines of a page.

In the case of Vertical Blanking Interval transmission, this implies that the minimum time separating two successive lines of a page is one TV field, or 16.7 msec.

This separation in time is achieved by interleaving lines from several pages with each other. These lines remain readily identifiable and distinct from one another, *provided that* no two pages having the same SI are interleaved with each other. For example, if page 3456 having SI=234 is in the cycle, no other pages having the same SI may be transmitted until after the last line of page 3456.

Error Detection and Correction Codes

Studies in various countries have produced error statistics for broadcast videotex, which indicate that the use of error-correcting codes would yield shorter waiting times and better data integrity than simple error detection alone. A recent field trial in the Rhone-Alpes region of France has found that 83% of all errors occurred singly. This implies that a code capable of correcting single errors could yield a considerable improvement in service.(4)

There are two codes which have been selected for use in broadcast Telidon. The first, which is applied to address and jump lines (DI=0x or F2), consists of an (8,4) Hamming code on each byte, including the SI & DI fields. The final byte is the twos-complement of the modulo-16 sum of the DI and data bytes. This checksum is also Hamming-encoded.

The other code, which is applied to data lines, is known as a product code. It consists of odd parity on each data byte and a longitudinal odd parity byte in the final position. This code was studied by M. Sablatash and J. R. Storey of the Communications Research Centre, and was selected because it corrects all single errors, detects all double errors, and detects many other patterns of multiple errors. Figures 4 and 5, taken from an unpublished paper by Sablatash and Storey, present a comparison of the product code with simple parity-checking.

Figure 4 shows the improvement due to the product code as a function of the bit error rate. The definition of improvement in this case is the ratio of the probability of decoding failure in a 500-byte page using parity only, to the probability of decoding failure for the same page using the product code. Decoding failure is defined as the reception of an uncorrectable error.

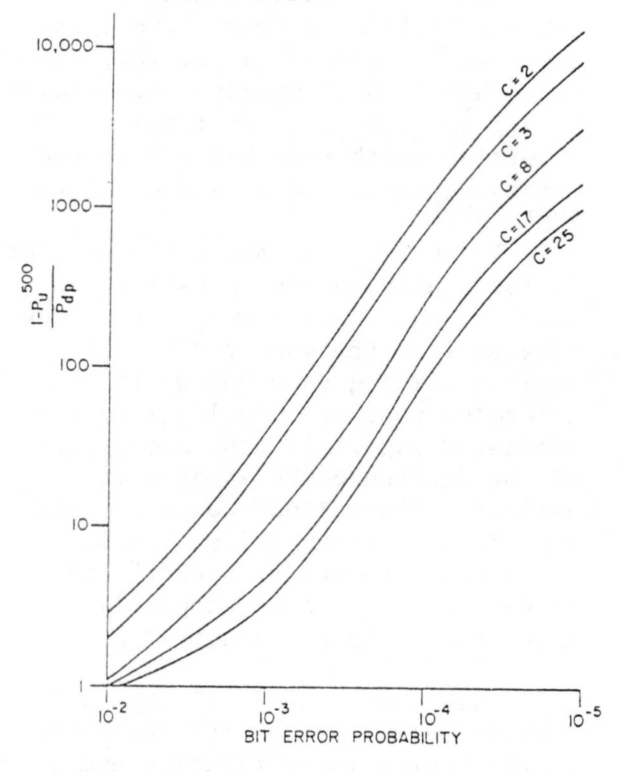

Figure 4. Improvement due to product code

It can be seen that with a block size of 17 bytes (C=17), the improvement factor varies from approximately 10 at a bit error rate of 10^{-3} to over 1000 at 10^{-5}.

Figure 5 shows the transmission efficiency for the product code and simple parity as a function of block size. Efficiency is defined as the product of the number of *information* bits in a page of 500 bytes and the probability that the page is received without decoding failures, divided by the *total* number of bits in the page. We can see, for example, that for a block size of 17 bytes and a bit-error rate of 10^{-4} the product code is approximately 25% more efficient than simple parity.

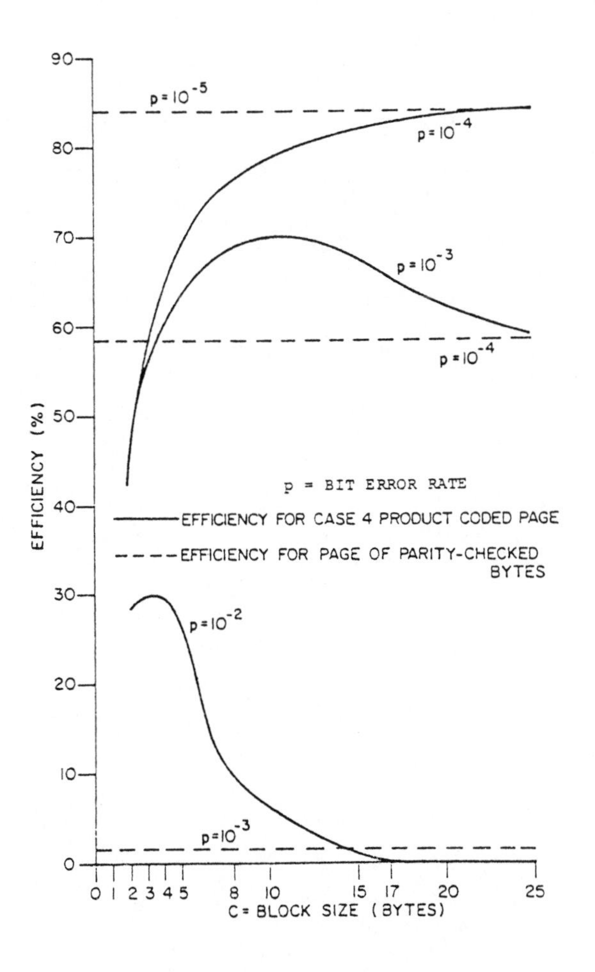

Figure 5. Transmission efficiency

Field trials are presently underway on the television network of the Ontario Educational Communications Authority (OECA), using the codes described in this section. Transmission tests being conducted by the Communications Research Centre over OECA and Canadian Broadcasting Corporation (CBC) channels, as well as cable distribution systems are expected to provide valuable information concerning error statistics, and the performance of the selected codes in an actual operating environment.

Conclusion

The data format described in this paper meets all the requirements for transmission of Telidon data, and provides a flexible basis for time-sharing of the vertical blanking interval TV lines by services requiring access to this resource. The system is totally transparent to the transmitted data, and even the largest PDI files can be accommodated by means of the concatenation feature described in the text, thereby preserving the important feature of terminal-independence which is central to the Telidon concept.

Finally, the use of forward error correction ensures the reception of relatively error-free data with considerably shorter waiting times than would be possible with error detection alone.

REFERENCES

1. Bown, H. G., C. D. O'Brien, W. Sawchuk and J. R. Storey, "Picture Description Instructions for the Telidon Videotex System", CRC Technical Note No. 699-E, Department of Communications, Canada, Nov. 1979.

2. Bown, H. G., C. D. O'Brien, W. Sawchuk and J. R. Storey, "A General Description of Telidon: A Canadian Proposal for Videotex Systems", CRC Technical Note No. 697-E, Department of Communications, Canada, Dec. 1978.

3. Bown H. G., C. D. O'Brien, W. Sawchuk and J. R. Storey, "Telidon: A New Approach to Videotex System Design", IEEE Transactions on Consumer Electronics, Vol. CE-25, No. 3, July 1979, pp. 256-268.

4. Dublet, Guy, "Methodes utilisees et principaux resultats obtenus lors d'une campagne de mesure 'Didon' dans la région centre-est", Revue de Radiodiffusion Télévision, No. 54, 1978.

ANTIOPE AND D. R. C. S.

O. Lambert, R. Brusq, B. Marti and A. Poignet
C. C. E. T. T., France

ABSTRACT

This paper describes a new technique called «D.R.C.S.» which is extremely helpful in text oriented communication systems : it provides through a very low cost user's terminal nearly unlimited repertoire of characters to be displayed for alphabetic applications and also fine graphics where they are needed. The integration of the feature to the French Antiope (1) Teletext or Teletel Videotex services is then discussed.

1 INTRODUCTION

Videotex and Teletext are the family-names given to new communication services ; the former transports the information through interactive networks (packet - switching, circuit switching or P S T N), the later uses, at least in France, the concept of data broadcasting network (2), a new data transmission system, which is implemented on the basis of the existing Television network, wether on «cable» or «on the air». For both services the typical display is the common T V receiver, colour or black and white, although developping technologies such as plasma panels are not precluded in some applications.

The messages delivered consist essentially in textual information, displayed in a 24 rows of 40 characters format for 625 lines T V standard, 20 rows of 40 characters for 525 lines T V standard. Besides, a low resolution graphic facility is provided in some systems, called alpha-mosaïc ; basically, the mosaïc pictorial resolution is 80 (horizontal) by 72 (vertical), each character position being considered as a six dot - pattern where each dot can be switched on or off. Although very limited as far as resolution is concerned, mosaics offer some possibilities for caricatural drawings appropriate to intertainment purpose. Nevertheless they are commonly considered as insufficient in many regards ; some countries, such as Canada, have proposed Videotex systems which, from the start offer fine graphics ; the associated terminal requires a «bit map memory» where each dot on the screen is represented by o pixel of a certain depth (usually 3 or 4 bits) in the memory ; other countries support an intermediate (and not exclusive) system called D.R.C.S. which appears as an extension of character oriented systems. Tne D.R.C.S. (stands for dynamically redefinable character set) allow to define a set of characters whose shapes are dedicated to a page or a set of pages. The communication process requires a preliminary step which is the transfer of the patterns (often called «down loading» step) ; the receiver, after identification of the annoncement sequence, stores the patterns in a RAM character generator ; later on this RAM is used along with the basic ROM character generator ; the character repertoire available is thus dynamically enlarged ; the potential applications of D.R.C.S. will be discussed in the next section.

We claim that it covers an important percentage of the need for fine graphics while not increasing significantly the cost of the basic alpha-mosaïc terminal ; further more it offers a pretty good downward compatibility through adequate care at coding level; this point will be discussed in section 3 along with other technical questions. In the last section we shall describe how the integration of D.R.C.S. is possible in the ANTIOPE or TELETEL services, taking advantage of the modularity of the system, without specific redesign of the LSI chips.

2 POTENTIAL APPLICATIONS OR D.R.C.S.

Up to now three main areas of application have been identified :

2.1 ALPHABETICS :

Written languages can be classified as character - oriented languages such as Latin, Cyrillic, Greek or Hebraïc and as ideogram-oriented such as Japanese or Chinese. If we consider only one subcategory, the latin - based languages, I S O has found that more than 300 different characters are currently used all over the world ; in fact, inside one country, a 96 character-set may be sufficient (English language), or more often 128 (French, Spanish, German) ; consequently many terminals will be implemented with only a national repertoire of characters, for savings on the R O M generator-size. Then we have to face an interworking problem when a subscriber of one country tries to retrieve pages from a data-base using characters not part of his terminal's repertoire. D.R.C.S. solve that problem in an easy way : the international gateway down loads the required repertoire before sending the codes describing the pages themelvesr ; mixed latin and now latin messages can also be displayed in the same way (see photo 1). D.R.C.S. may turn out a very efficient way for ideogram-based communication ; we know the histogram representing the frequency of use of Chinese ideograms ; the most frequently used ones, should be stored in the terminal in a memory ROMS, while the complementary unusual ideograms would be down loaded as D.R.C.S. according to the needs of each page of information (see photo 2).

Figure 1

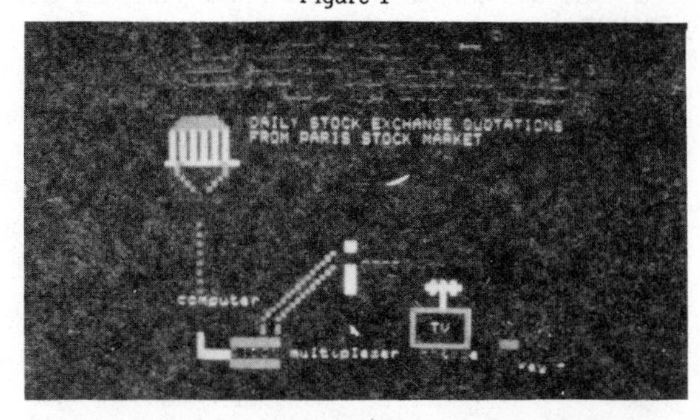

Figure 2

Reprinted from *IEEE Trans. Consum. Electron.*, vol. CE-26, pp. 600–604, Aug. 1980.

2.2 CHARTS, MAPS, SCHEMATICS

A famous quotation says in substance that a good drawing is more efficient than a long talk ; this is especially meaningful in Videotex or Teletext where the written «talk» itself has to be short. Let us review some significant examples :

- a geography training course is hardly acceptable without a map presentation facility ; mosaïcs (fig. 3) provide a distorted presentation of the original picture which allows identification by one who already knows the original but it is not valuable for the student who has no previous knowledge of the topic. Fig. 4 shows a map obtained with D.R.C.S. technique.

- a training course in electronics definitly needs schematics. Using D.R.C.S. to describe common components (transistors, diodes, resistors, capacitors, various logic gates) in their standardized symbolism is excellent regarding both picture quality and coding efficency. The only constraint lies in the fact that the symbols for components have to be justified according to character-boarders ; in fact, this rule, somewhat constraining at editing level may bring out a very neat lay out of the schematics.

- time-tables for air or train transportation are usually made synthetic and attractive through appropriate dedicated symbols indicating for instance wether a meal is served or not during a fly, wether the jet is a 747 or a L 1101 and so on.

. similar applications can be found in various areas such as hotel information (see fig. 5) or games (chess, cards).

- an other important domain where local fine graphics are needed is commercial information delivery - Companies appreciate that their characteristic logo is displayed along with the information itself for clear and fast identification as well as for psychological impact (see fig. 6). Financial and economical results of companies, evolution of incomes are also efficently presented through curves and histograms that can be suitably described by D.R.C.S.

All these examples show that D.R.C.S. fulfill the graphic requirements of many applications : there is a common feature to these applications : only a few characters are needed :

- in some cases because the fine graphical part is only needed in limited areas (logos, map-contours).

- in other cases the same patterns are used several times on the same page (electronic schematics, symbols for transportation time table).

Figure 3

Figure 4

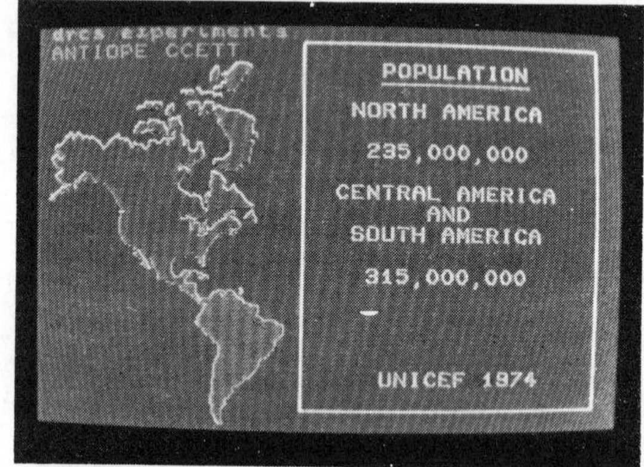

Figure 5

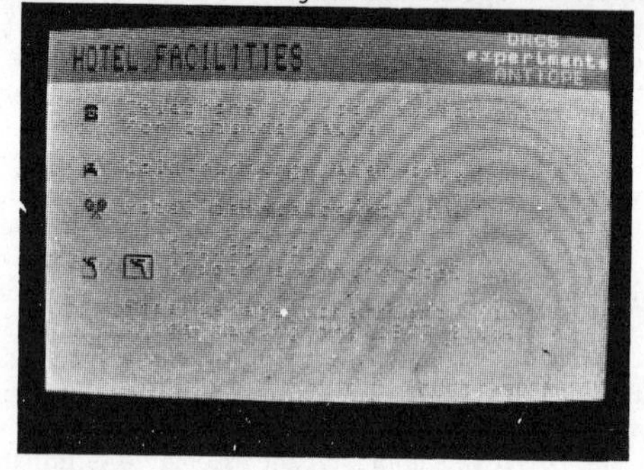

Figure 6

3 TERMINAL PROBLEMS RELATED TO D.R.C.S.

3.1 INTRODUCTION TO THE VIDEOTEX OR TELETEXT DECODER

To ease understanding of the technical problems we shall shortly describe the structure of the Antiope (or Teletel) terminal, focussing on the character generation part : The bloc diagram of the terminal (fig. 7) shows 5 fundamental blocks organized in a standard bus oriented architecture :

Figure 7

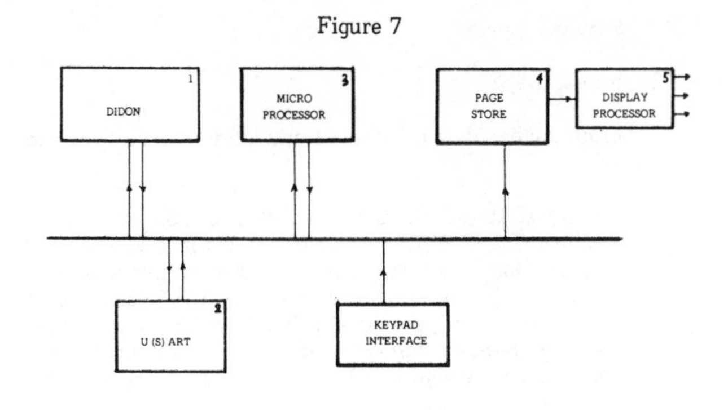

Figure 8

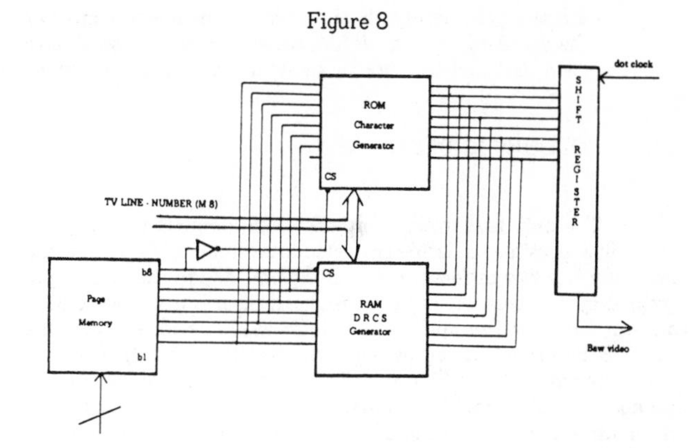

Bloc 1 : is specific to Teletext : its function is to extract data from the video signal transporting data according to DIDON standard. The process is divided into three steps :

demodulation of data (a NRZ modulation is used)

demultiplexing of data : basically DIDON transports packets of data belonging to several data-links ; each packet has a header identifying the data-link to which it belongs (usually a 3 byte-number) ; the main task of the demultiplexer consists in sorting the packets so as to send forward only those packets that belong to the data-link selected by the user.

Buffering : a buffer is needed because of the slowness of the microprocessor that handles the data extracted.

Bloc 2 : is the equivalent of bloc 1 for interactive networks ; in this case the buffer is not necessary.

Bloc 3 : the microprocessor : it controls the various components of the system ; besides, it handles the protocol-aspect of the application (level 5 of I S O - O S A model) and performs interpretation and translation into interval coding of the description of message-presentation (level 6 of I S O model).

Bloc 4 : display store or page memory : the memory (typically 16kbits of RAM) stores in an internal fixed format code the description of the page currently displayed.

Bloc 5 : display-control : it the generates the R, G, B and synchronisation signals to drive the C R T.

Now let us focus on the synthetic video signal generation (fig. 8) a timing chain generates a 10bit-address which selects in the page memory the location where description of the screen character-position to be currently displayed is stored. The output of the page memory in then essentially the seven bit code of the character to be displayed plus one (b_8) acting as a chip-select between standard ROM generator and the RAM generator. The timing chain also provides the generators with a T V line number (modulo 8) to designate the horizontal slice of the matrix to be currently displayed ; the 8bit-output of the selected generator is then serially shifted at dot clock frequency to generate the basic black and white video signal.

It can be easily understood that the introduction of D.R.C.S. in this process is extremely simple as far as all the timing chain of the existing decoders can be kept unchanged.

3.2 DOT MATRIX SIZE

Until the new pressure about D.R.C.S. manufacturers used to feel free about the particular matrix-size implemented in the decoder ; typically 6 (horizontal) by 10 (vertical) formats were proposed as well as 8 by 10 or 10 by 10. Considerations regarding both ROM cost and the actual resolution capability of T V sets (especially small size PIL sets) have led to abandon the 10 by 10 format. 6 by 10 and 8 by 10 are still competiting and D.R.C.S. make then incompatible : in the broadcasting situation, the down loading process has to be independant of such or such type of decoders (double down loading in 6 by 10 and 8 by 10 is prohibited because of transmission ressource wasting).

To solve this problem a high level geometric type description of the patterns has been envisaged backing on the general philosophy of the Telidon PDIs ; unfortunately this technique is not applicable for D.R.C.S. down loading for two main reasons :

1) decoding of such geometric coding schemes may imply an amount of software which seems excessive as compared to the primary goal of D.R.C.S. : upgrading basic typographic terminals through little additional cost.

2) the sampling effect when translating geometric codes into pixels can bring disastrous results in some cases : for instance let us suppose the codes describe a character formed with 4 distinct vertical stockes ; a 6 dot terminal would be totally unable to give a correct display of this information.

At present time no agreement has been reached in the international standardization bodies ; the main support to the 6 dot-option comes from companies already manufacturing decoders using this matrix-size. On the other hand, the eight dot option offers significant advantages :

- it results in «square dots» which is fundamental for upward compatibility with full graphic systems and for hardcopy purposes.

- a 8 is a gold number in the digital domain where most of the standard circuits (RAM, ROM, microprocessors) have been designed to work in an eight bit data bus structure.

3.3 COLOUR PROBLEM

The only real trouble with D.R.C.S. comes from colours : while it is possible to define shapes in total freedom in the matrix, one only has two different colours to paint, due to the basic typographic system feature. In many applications it is desirable to have at least three or four different colours available for instance the the T D F (French broadcasting authority) logo (fig. 9) wants the «t» to be red, the «D» blue and the «F» green, on a white or black background. This could be achieved by very careful application of the matrix grid on the picture so that only two colours are needed inside one basic matrix.

Figure 9

Practically the method results in a huge logo occupying half part of the screen where one wanted a small logo in a corner of the page. The same kind of problem appears when trying to display a map where three countries have common boarders, or for the London underground map at junctions of several lines, each of which being strongly identified by its colour.

A solution to this problem could consist in exchanging «resolution» and «colours» ; in other words exchange a thin brush with a two colour-box of paints and a larger one with a four colour-box of paints. This question is now under study ; the point is that it has deeper incidence on the decoder hardware than it was first anticipated for D.R.C.S. considered as only additional RAM and software.

3.4 CODING SCHEMES AND PROTOCOL ASPECTS

3.4.1 Down loading step

No specific problem has been identified at this level ; the various proposals are very similar : the matrices are transmitted line by line as a sequence of «0» and «1» where «0» indicates a dot belonging to the background and «1» a dot belonging to the foreground. Some compression-techniques have been experimented with but up to now the efficency is not so obvious ; one can observe that when down loading alphabetic characters, the strong natural spatial correlation in the vertical axis is an incitation to transmit column by column rather than line by line. In this case runlenght codes could be applied with benefit.

3.4.2 Protocol aspects

3.4.2.1 Interactive

In interactive data retrieval several kinds of access must be considered :

- A tree structure offers an interesting opportunity for attaching D.R.C.S. to the various levels in the tree. In a given branch, where pages have logical communality, a general D.R.C.S. will cover most of the needs , going down the branch, new local D.R.C.S. will be down loaded that the terminal will either accumulate with higher level ones already stored if it has sufficient memory ressource or simply exchange in the other case.

- If keyword access is used, we propose to attach a subset of D.R.C.S. to each keyword (let us call it S d r c s) ; for instance $S\,d\,r\,c\,s_i$ is attached to $keyword_i$; then an access to a page identified by the compound keyword Key_1 - Key_2 - Key_3 - would automatically down load $S\,d\,r\,c\,s_1$ U $S\,d\,r\,c\,s_2$ U $S\,d\,r\,c\,s_3$.

3.4.2.2 Broadcast

The Antiope coding frame is widely opened to extensions so that the introduction of D.R.C.S. causes no problem. An Antiope message is coded and transmitted as one or several «articles» ; each article is delimited by a start of article (sequence of I S O 646 codes : SOH and RS) ; then come three Hamming protected bytes : values ranging from 000 ro 999 are dedicated to articles corresponding to visual messages ; the rest of the addressing ressource is available for non visual messages ; down loading articles belong to this category. Several D.R.C.S. identified by different numbers can be cyclicly down loaded along with visual articles but this process may result in a double access-time for the user ; in the worst case, this access time is a two cycle period : one cycle to get the page itself and another one to get the needed D.R.C.S.. Nevertheless this situation can be improved espacially if the terminal has a multi D.R.C.S. store.

3.4.2.3 Comptability with mosaïcs :

In graphical applications a good downward compatibility with basic mosaïc terminals can be achieved by a simple rule at coding level : a D.R.C.S. character must be given the same code as the mosaïc character which has roughly the same shape (with less resolution). When two D.R.C.S. characters happen to have the same equivalent in the mosaïc set the two characters must belong to two different sets while keeping the same code inside the set ; in other words, the characters are made distinct through the character set-number rather than the code position in the table ; this distinction is ignored by the mosaïc terminal which displays in both cases the mosaïc character identified by the particular character-code.

4 INTEGRATION OF D.R.C.S. IN ANTIOPE OR TELETEL TERMINALS

In France three manufacturers are designing LSIs for low cost Antiope terminals : EFCIS - THOMSON, PHILIPS - RTC and TEXAS - FRANCE.

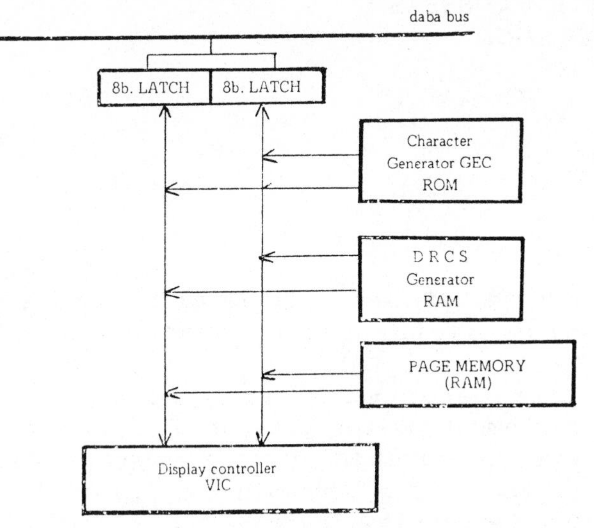

Figure 10 : THOMSON EFCIS display process

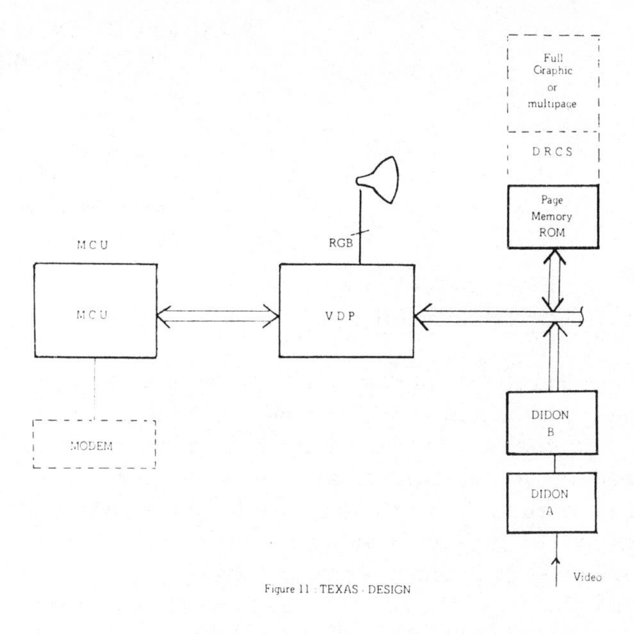

Figure 11 : TEXAS - DESIGN

4.1 PHILIPS - RTC

The first design does not support D.R.C.S. because the display chip (AROM) has the ROM inside it so that no interface is available on the chip to connect an alternative RAM character generator ; this problem will be solved in the second generation of chips (1981).

4.2 THOMSON - EFCIS

Here D.R.C.S. is on Option but the architecture (figure 10) provides with the proper interface ; some SSI standard circuits will be needed to interface between the internal bus and the standard RAM chip so that the RAM generator is seen by the display process exactly as the basic GEC (ROM character generator).

4.3 TEXAS - FRANCE (figure 11)

Here D.R.C.S. is basic along with full graphic capability. At display level the whole process is controled by one single chip (VDP). According to the amount of RAM connected to the VDP, one can build a wide range of terminals : typographic only, multipage store typographic, D.R.C.S. or full bit mapping graphic.

The modularity of the Antiope system has enabled us to introduce D.R.C.S. in the best conditions :

- the transparency of the transport function allows to choice the coding schemes without constraint,

- the microprocesseur already in the decoder handles the new data through very little additional software.

ADAPTATION OF U.K. TELETEXT SYSTEM FOR 525/60 OPERATION

G. O. Crowther
Mullard Limited
New Road, Mitcham, Surrey, U.K.

1. INTRODUCTION

The United Kingdom (UK) Teletext System has now been in service in the UK for four years. It is also being transmitted on a regular basis in more than six European countries and Australia. Experience has confirmed the soundness of the original system design concept which has lead to television sets equipped with decoders from a number of sources. These combine economy, ease of use, and rugged system performance. The small added cost for installing the teletext decoder has been one of the most important factors in establishing the service in Britain.

A paper (Ref. 1), presented to the Spring Conference last year, examined the basic features of the present UK teletext system and showed how these could be extended to meet the requirements of the 525 line NTSC television system without loss of the inherent features of the 625 line system.

This paper discusses the more definitive proposal made by a joint committee consisting of BREMA*, BBC**, and IBA*** for a 525 line system based on the extensive experience in the UK and the future enhancement techniques developed for Europe.

When introducing a teletext system with perhaps only one or two lines in the field flyback period, it is essential to have a simple system to launch the service. This simple system should not carry the overheads of a more sophisticated system designed for the future, either in terms of transmission time overheads or decoder costs.

This paper outlines a hierarchy of systems from the present UK system to a system capable of displaying still pictures with a resolution limited only by the display device.

2. CHARACTERISTICS OF THE BASIC 525 LINE PROPOSAL

In the initial phase of the introduction of a teletext system, it is anticipated that the service will be primarily employed for text transmission with limited picture graphics capability to enhance the text.

The detailed structure of the teletext system suggested for 525 lines NTSC is shown in Figure 1. It has been designed to have an identical structure with the 625 line system so that common integrated circuit (IC) designs can be employed. A number of new concepts have been incorporated in the proposal made a year ago. These are a direct result of the system requirements in the USA which became apparent during recent discussions. These are highlighted in the next sections.

2.1 Display Format

The teletext service in Europe is based on a format of 40 characters per row and 24 rows. In the paper presented last year (Ref. 1) it was shown that 40 characters per row could be accommodated for both the built in decoder system as well as the set top adaptor in the 525 line system.

At that time it was felt that only 20 rows could be accommodated. It has now been demonstrated that by employing

---

* British Radio Equipment Manufacturers Assoc.
** British Broadcasting Corp.
*** Independent Broadcasting Authority.

Reprinted from *IEEE Trans. Consum. Electron.*, vol. CE-26, pp. 587-599, Aug. 1980.

AMERICAN TELETEXT SPECIFICATION PAGE HEADER AND ROW CODING STRUCTURE

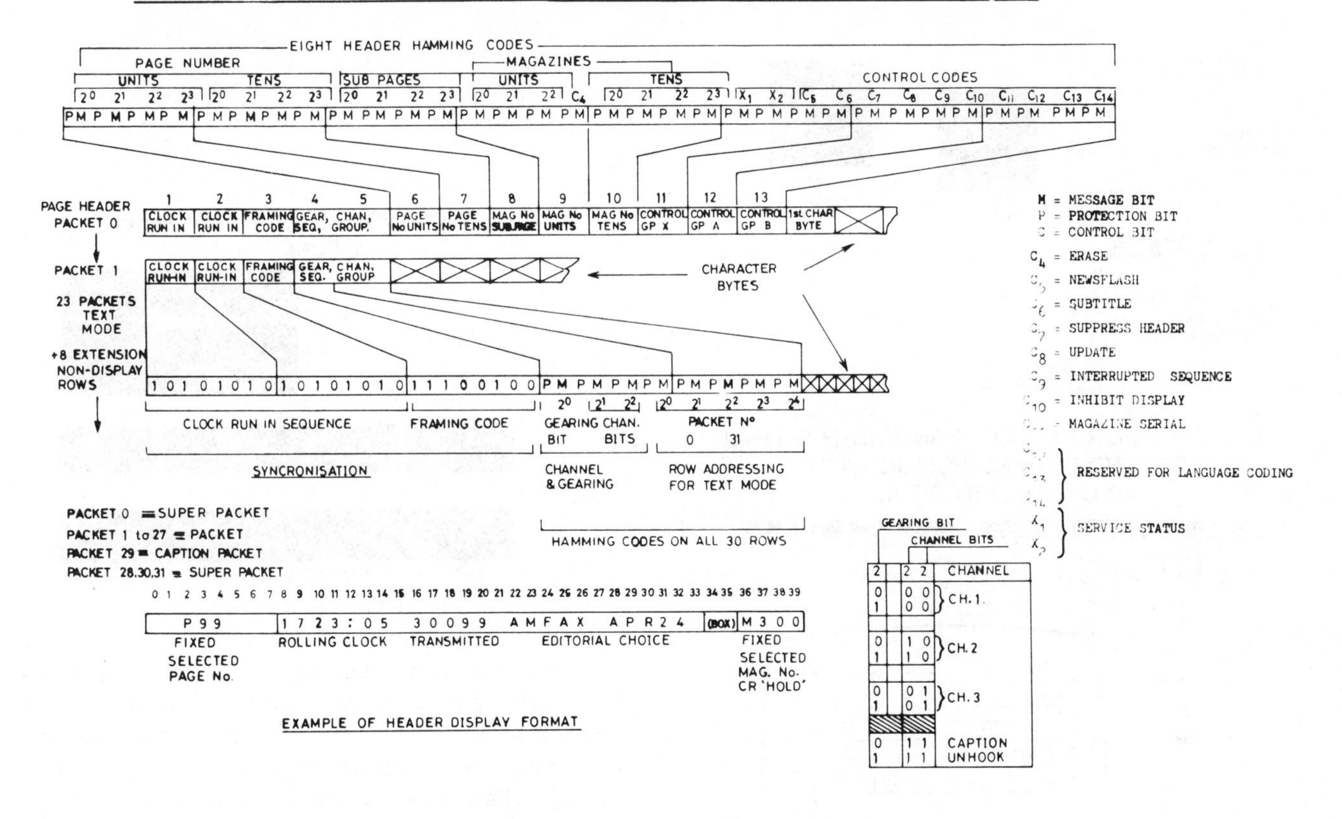

Figure 1.

compression techniques, 24 rows can be readily accommodated by the 525 line system, thus giving the possibility of format compatibility between Europe and the USA.

2.2 Relationship between Display and Transmission Format

The format of the teletext signal on a transmitted TV line consists of three portions:

- Sychronization group
- Address
- Data

It has been agreed to call this a packet. It will be seen that the address is made up of two parts:

- A function part (3 bits).
- A packet number (5 bits, allowing 32 packets).

In the UK system, a decoder has been designed with a simple relationship between the packet numbers, the memory location, and the final display location as illustrated in Figure 2. It is this principle that gives the combination of a low cost decoder and system ruggedness.

In the 625 line system bandwidth, it is possible to transmit one row of data on a TV line. For the 525 line system a slight modification of the strategy is required, and this is achieved with the so called "gearing" concept as shown in Figure 3. Normally a fraction of the display row is transmitted on a TV line, and this would be displayed on the left hand side of the screen. The right-hand part of the screen is filled in by blocks, the size of which depend on the number of data bytes on a TV line.

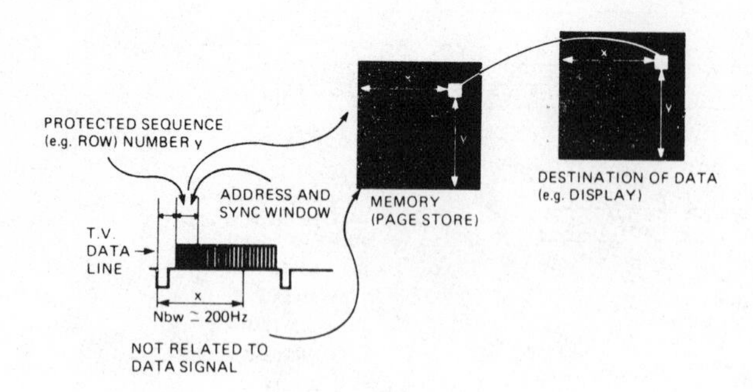

Figure 2.

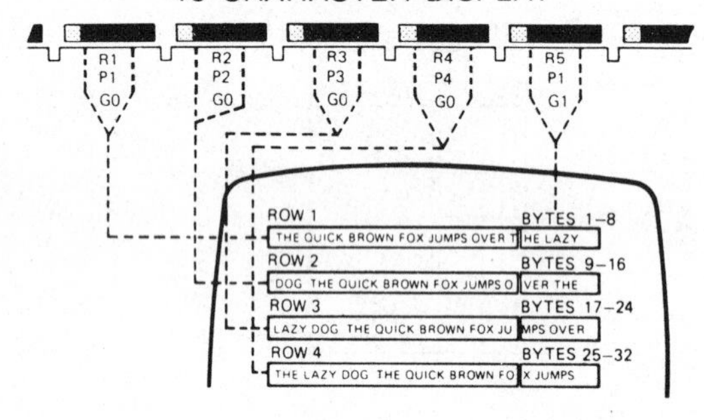

Figure 3.

The location of data on the display is determined by the gearing bit and the packet number (Fig. 4). If the gearing is 0, the data is placed on the left-hand side of the screen. If the gearing bit is 1, then the block on the right-hand side of the screen is filled from the row designated by the packet number. It is not necessary to send a packet which contains only spaces.

This concept is completely independent of the number of data bytes transmitted on the line. Further, it is possible to design the data acquisition circuits to be adaptive to the number of data bytes on the line. Clearly the clock recovery circuit would require adjustment for each bit rate.

2.3 Format of Transmission

As in the UK system, the packet 0 is defined as a super packet and contains more addressing information than the normal data packets 1 to 23. A page

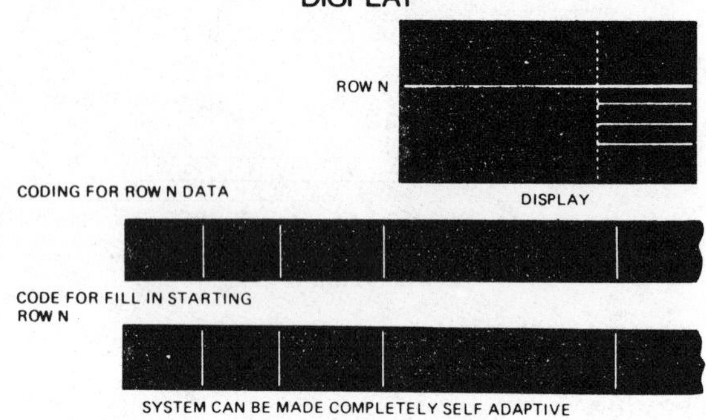

Figure 4.

is started and terminated by a packet 0. The packets of data with the same channel number which are transmitted between two packet 0 all belong to one page. Packets of other channel numbers can however be interleaved as will be discussed later. Basic decoders will only respond to packets 0 to 23.

2.4 Page Addressing Capability

In the early stages of teletext, it is anticipated that data will be transmitted on one or two lines in the field flyback period as is the case in Europe and Australia. Figure 5 shows

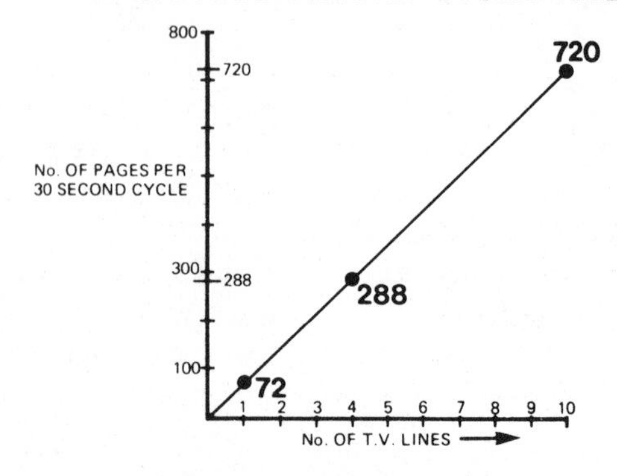

Figure 5.

the relationship between the number of typical pages that can be transmitted within a 30 sec cycle time and the number of TV lines employed for data transmission. Clearly more pages can be transmitted with a proportional increase in the transmission time. Experience suggests that a cycle time of the order of 30 sec is the optimum compromise between the number of pages transmitted and waiting time.

It has also been shown that considerable user frustration can occur if the number of pages that normally can be selected exceeds that transmitted. For these reasons, there is a pressure to minimize the page addessing capability in the early stages. This conflicts with the requirement that decoders should not be made obsolete by enhancements to the system. The most obvious enhancement is the increase in the number of data lines up to the full frame. To overcome this conflict, two system status bits (Fig. 1) have been included in the page header which instructs the decoder on the number of pages available in the transmission in four stages from 300 to 300,000. Extension techniques are also available to further increase this value as discussed later.

3. EXTENSION TECHNIQUES

Figure 6 lists the criteria for extension on the basic system. A number of mechanisms have been incorporated within the basic system. A number have already been tentatively reserved for known enhancements. The final use of others has still to be defined.

3.1 Packet 24 to 31

The use of these has been tentatively defined for Europe, and their use is described in the associated paper by the BBC (Ref. 2). The proposal for the USA allows the same uses to be allocated to these packets. Of particular importance is the packet for linked pages and packet 26 for page associated data. The use of the latter

CRITERIA FOR EXTENSIONS OF TELETEXT SYSTEMS

1. MINIMUM FORM OF LIFE SHOULD NOT BE BURDENED WITH ADDRESSING OVERHEADS ASSOCIATED WITH MORE SOPHISTICATED SYSTEMS.
   a. IN TERMS OF TRANSMISSION TIME
   b. DECODER COSTS
2. EXCESSIVE SPARE CODING CAN LEAD TO USER FRUSTRATION AND APPARENT EQUIPMENT OR SERVICE FAILURE.
3. IT IS DIFFICULT TO PREDICT PRECISELY FUTURE EXTENSION REQUIREMENTS.
4. FUTURE EXTENSIONS SHOULD BE COMPATIBLE OR BE ABLE TO CO-EXIST WITH MINIMUM FORM OF LIFE.

Figure 6.

packet will be described in a later section.

3.2 Unhook Concepts

Figure 7 states the principles of the unhook concept. Two major unhook mechanisms have been defined in the proposal for the USA and are shown in Figures 8 and 9. The packet unhook code shown in Figure 8 allows the whole system to be redefined for the reception of, for instance, Facsimile data. Present day decoders would never be able to address these pages and therefore would not be affected by these page transmissions. The packet unhook comes early in the transmission and therefore neither present or future transmission will have a heavy overhead.

Figure 10 sums up how the addressing can be extended in the future. At each stage of extension one of the new codes will be reserved for further unhook. The extension concept has a close similarity to telephone numbering techniques, where the short number codes are reserved for the most often used local calls while longer numbers are employed for long distance calls.

The unhook concept shown in Figure 9 allows the redefinition of the whole of

CONCEPT OF UNHOOKING FOR THE EXTENSION OF TELETEXT SYSTEMS

IN PRINCIPLE A SINGLE CODE ONLY IN EACH CODE DIMENSION NEEDS TO BE RESERVED FOR FUTURE EXTENSIONS.

1. THESE CODES SHOULD BE IGNORED BY THE MINIMUM FORM OF LIFE.
2. THE TRANSMISSION OF THESE CODES ALLOWS THE CODING TO BE CHANGED IN AN OPTIMUM WAY FOR THE SPECIFIC FUTURE NEEDS.
3. WITHIN ANY EXTENSION FURTHER UNHOOK OR ESCAPE CODES SHOULD BE DEFINED.

THE CONCEPT THEREFORE ALLOWS THE SYSTEM TO GROW OPTIMALLY AS TECHNOLOGY AND T.V. LINES BECOME AVAILABLE.

Figure 7.

UNHOOK MECHANISMS IN MAGAZINE CODE

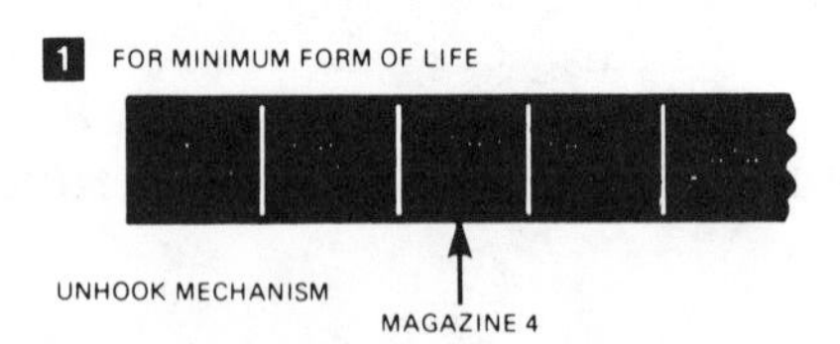

Figure 8.

UNHOOK MECHANISMS
Super Packet Header Formerly Page Header

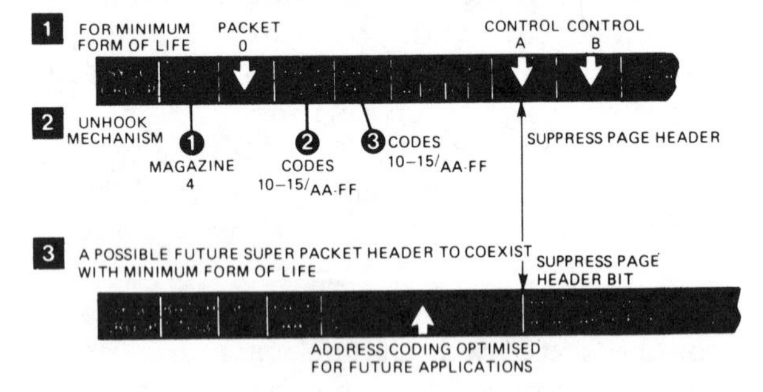

Figure 9.

the page header and all the page associated packets. This unhook mechanism employs the fact that page numbers are sent in BCD code to simplify decoder design. The codes 10-15 can not be requested by the basic decoder and are therefore ignored. The principle is already employed in the present UK transmission. These page numbers, the so called pseudo pages can be employed in two ways:

ADDRESS EXTENSION

1 2 3 4 5 6	MAGAZINE 1 MAGAZINE 2 MAGAZINE 3			
7	CAPTIONS AT HALF RATE			
8	1 2 3	PRIVATE TEXT FURTHER USES		
8	32	1 2 3 255		
8	32	256	1 2 3	

Figure 10.

As an unhook mechanism.
For transmission of nondisplay data such as DRCS and Telesoftware.

Their use in nondisplay transmission is discussed in Ref. 2. For packet 0 unhooking, specific pages will need to be allocated for the application desired. All bytes after the pseudo page number can then be redefined as required by the as-of-now unknown future service.

4. SECOND GENERATION TELETEXT SYSTEM - POLYGLOT C

In Europe the first discussions are underway (see Ref. 1) to enhance the basic system. The prime aim is to give enhanced display techniques within the constraints of a seven or eight bit code scheme which forms the basis of most text and data coding schemes, including teletext. In an interactive wired system, it is normal to employ code extension techniques based on ISO 2022. However, in the noninteractive broadcast situation the error pattern and operating conditions are fundamentally different. A slightly different strategy has been chosen for Polyglot C which is still based, and has a simple one-to-one relationship with the interactive coding scheme. To appreciate the new coding scheme, it is useful to examine the relationships between Teletext and the ISO standards.

4.1 Relationship between UK Teletext and ISO Standards

4.1.1 Background

The UK system proposal has retained a simple relationship between itself and the two ISO code standards. Deviations have been made to achieve ruggedness in reception at minimum decoder cost and data transmission time. ISO 646 and ISO 2022 in their normal formats are based on sevenbit code (128 combinations) of which 32 combinations are reserved for a mixture of formatting, escape routines, machine control, and transmission control commands. Ninty-four codes signify characters, and the remaining two codes are reserved for a printing space and delete.

The positioning of the display on the screen is achieved by counting from a reference point. For example, the following transmission sequence in Figure 11 would be displayed on the top row of the page format at the location indicated.

R is a character in column 4 or 5 which would normally be displayed as the character appropriate to that code.

There is now a discontinuity between transmission sequence and display location. Thus, if for any reason the ESC code can not be recognized, elaborate procedures are necessary in the receiver to prevent the display format from being disrupted. The same arguments apply to the use of codes CR and LF to terminate one row and start the next.

For these reasons, UK Teletext does not employ the codes in column 0 and 1 of the ISO table but achieves their functions in other ways. There are thus no discontinuities in the transmission, and there is a direct relationship between transmission order and display format.

The following sections describe how the control characters from column 0 and 1 of the ISO table are effected in system B.

Location on Row	1	2	3	4	5	6	7	8	9	10	11	12	13	14	-	-
Transmission FF	Sp	Sp	T	H	E	Sp	D	O	G	Sp	S	A	T	Sp	-	-
Display			T	H	E		D	O	G		S	A	T			

Figure 11. Display Positioning.

At this stage there is a strict relationship between the display location and the order of arrival of the data bytes.

If it is desired to give attributes to the text display such as color, further coding is required; this is achieved by transmitting the control code ESC (column 1) and following this with a code in column 4 and 5 which is now interpreted as a display command rather than its normal character representation. Thus, if the word DOG was required to be displayed in RED in the earlier example, the transmission sequence would be as shown in Figure 12.

4.1.2 Implementation of Control Codes of ISO 646

For teletext, not all the control codes from column 0 and 1 are required. Those which are employed are divided into three groups.

- Display format FF, CR. LF
- Display enhancement ESC
- Designation of other character sets S1, SO, SS2, SS3

A common format has been implemented to cover the display formatting for the MFL and the enhanced system. For escape routines, a simple system has

Location on Row	5	6			7	8	9	---
Transmission	E	Sp	Esc	R	D	O	G	---
Display					D	O	G	

Figure 12. Use of Escape Code.

been implemented for the minium form of life. A more general implementation of ESC has been incorporated for the enhanced system and has been linked with the need for extended character sets.

(1) Display Format

The Hamming protected sequence coding at the start of each packet gives precise instructions for the display locations of the data bytes which follow. There is thus no ambiguity of display format, and experience has shown that the byte synchronization mechanism is very reliable.

(2) Escape Routines for Minimum Form of Life (MFL)

In the minimum form of life, a single C_1 set is defined which contains all the picture attributes (color, etc.). According to ISO 2022, these would have a code in columns 4 and 5 and would follow an escape character in the normal ISO data stream. Since in the normal teletext data stream no code in column 0 and 1 are transmitted, there are 32 codes not employed in the transmission of data. These codes are employed to transmit the C1 set. There is thus a very simple translation of the ISO code to teletext code. Again, since the C1 set is now coded by a unique code, there is no ambiguity if they are not proceeded by an escape.

Finally, by ensuring that all attribute codes (serial attributes) are allocated a display location and displayed as a space, they can be inserted into the data stream without fear of a reception error of an attribute code disrupting the display format. The translation from the ISO code is thus well defined.

(3) Control Characters in Higher Form of Life

In the higher forms of life, the codes for ESC, S1, SO, SS1 and SS2 have been implemented in a highly compatible way with the minimum form of life.

In the proposal, the display address at which these discontinuities occur is transmitted with Hamming protection and located in a transmission packet in such a way that the decoder can unambiguously locate the Hamming data in the stream.

The control data is transmitted in a packet marked by the Packet No. 26. This packet has the format shown in Figure 13. It will be seen that the packet is made up of a series of address data pairs consisting of three bytes. The first two bytes are Hamming protected. The third byte is parity protected and contains the associated data. The 16 bits of the first two groups are allocated as follows:

Address	6 bits
Control, etc.	5 bits
Hamming Protection	5 bits

TECHNIQUE FOR RELIABLE "ESCAPE" AND "SHIFT" FUNCTIONS IN TELETEXT
Packet 26 Format

CR CLOCK RUN IN
FC FRAMING CODE
P26 PACKET 26 CODE
A ADDRESS (HAMMING PROTECTED)
D DATA 7/8 BITS

Figure 13.

Table 1 suggests an allocation of the codes in the control section and in the associated parity-protected byte. In its simplest form, the coding could have been implemented in such a way that the control codes had a one-to-one relationship to the ISO control codes and the following parity byte would contain the instruction.

The simple relationship has not been implemented for four reasons:

- Certain instructions can disrupt the display format and are therefore Hamming protected (e.g., double HT).
- Compatibility with minimum form of life.
- Reduced transmission time.
- Simplify the task of the receiver.

As an example, the ISO sequence ESC double height ESC foreground color escape background color ESC flash ESC underline would be sent as a single triplet with control sequence 00101. Three bytes instead of a minimum of ten. Furthermore, the coding concept gives high compatibility between the MFL and the higher forms of life.

From Table 1 there is a simple relationship between ISO and the teletext code. It should be noted that the processing required for the conversion is identical to that which would be required in a receiver to convert an incoming ISO transmission into a suitable format for display purposes. In the enhanced receiver, coding has been provided for the designation of:

- 3 x C_1 attribute control set.
- 3 x G_1 additional sets of 94 graphic characters.
- 1 x G_3 single shift character set.

In addition allowance has been made for implementation of the ISO extended Latin alphabet set.

4.2 Compatability Between the Minimum and High Forms of Life

In the scheme proposed above, the minimum form of life will not respond to data on packet 26. An editor can thus determine the fall back position of any given page. In the case of picture attributes, the higher form of life will continue to take note of the serial attributes but allow these to be overwritten by data on line 26. Thus for example, a minimum form of life decoder could display a word in a single color while the higher form of life would identify a single character by flashing in a second color.

In the case of alphabets, the same also applies, for example ü could be displayed in the higher form of life and be defaulted down to u or ue in the minimum form of life under editor control. The extra space would be taken up at the end of the word.

5. MORE ADVANCED SYSTEMS

Under this heading a number of new display features and services are being explored and field tested in the UK and elsewhere. They include:

- Dynamically redefinable character sets (DRCS).
- Remote trained systems telesoftware.
- Advanced graphic displays.
- Full color picture in text.
- Transparent data.

The precise aims and nature of each of these extensions are covered elsewhere (Refs. 3 to 6). In this section it is proposed to describe how these would be handled within the teletext structure. The difficulty with all the above systems is that the actual transmission is not coded data for display but data which needs further processing with the possible exception of DRCS.

To avoid affecting any of the lower forms of life the data will be transmitted as an allocated group of pseudo

TABLE 1: Allocation of Ancillary Data Codes in Triplets in Packet Number 26.

1. Column Address, Codes 0 thru 39 (Rows 0 - 23)

1.1 Allocation of Associated Mode Bits

CODE		FUNCTION	DATA BYTE (7 BIT + PARITY)
0 0 0 0 0	SI	Normal	Format 'A'
0 0 0 0 1	+	Size 1	"
0 0 0 1 0	ESC	Size 2	"
0 0 0 1 1		Size 3	"
0 0 1 0 0	SI + ESC	Invokes Format B (Link)	Format 'B'
0 0 1 0 1			
0 0 1 1 0	SOA	Block Graphics, Normal	Format 'A'
0 0 1 1 1	SOB	Block Graphics, Separated	Format 'A'
0 1 0 0 0			Format 'C'
0 1 0 0 1	ESC	Two Groups of 16	"
0 1 0 1 0		pastel colors	"
0 1 0 1 1			"
0 1 1 0 0	SOC	Designates DRCS 1	Format 'A'
0 1 1 0 1	SOD	Designates DRCS 2	"
0 1 1 1 0	SS3	Designates G3 Set	Character in G3
0 1 1 1 1	SS2	Designates character in ISO ext'd Latin alphabet*	
1 0 0 0 0		Accent from	
	SS2	Col 5 of ISO	One character
		ext'd Latin	from Go set.
1 1 1 1 1		alphabet	

1.2 Allocation of Codes in Data Byte

1.2.1 Format 'A'

Bits 1,2,3	one of 8 foreground colors.
Bit 4	flashing
Bits 5,6,7	one of 8 background colors.

1.2.2 Format 'B' (extension BYTE via link code)

Bit 1	underline
Bit 2	boxing
Bit 3	reveal/conceal
Bit 4 & 5	flash rates
Bit 6	flashing underline
Bit 7	reserved

1.2.3 Format 'C'

Bits 1,2, & 3	Color Bits 1,2, & 3, foreground color
Bit 4	Flashing
Bits 5,6, & 7	Color Bits 1,2, & 3, foreground color

Fourth color bit for both foreground and background color is incorporated in the mode code. The 4 bits select 1 of 16 pre-selected colors from either a 64 or 4096 color table. The colors are predefined in auxiliary packet number 25.

(*) A character in column 2,3,6 & 7 of ISO extended Latin alphabet.

pages which cannot be requested by lower forms of life. In these page transmissions display formatting is irrelevant. The gearing bit can therefore be employed to increase the number of potential packets in a page. This is illustrated in Figure 14. It shows two blocks being fed sequentially into memory. The size of the block of data is not fixed and is commenced and terminated in the normal way by a Packet 0.

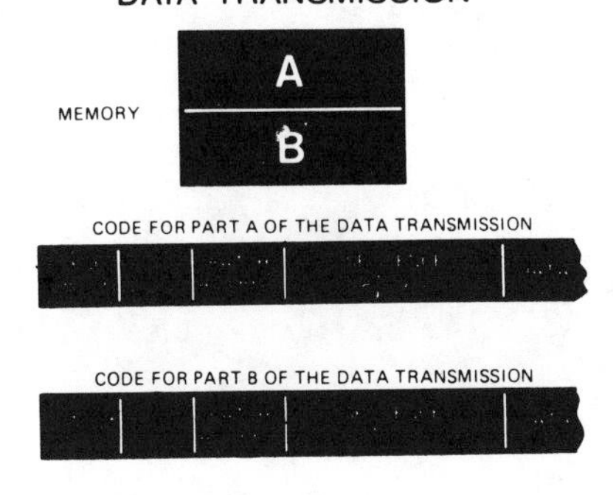

Figure 14.

Again every byte of data has a fixed location on the data stream and is allocated a specific location in memory. It is therefore possible to employ the identical error strategies (see Ref. 1) applicable to normal display data. In some cases further checks may be desirable. These can in fact be incorporated in the data stream. Alternately, a packet has been reserved which contains a Hamming protected CRC check word, to check the whole page. It should be noted that the check is performed on the data stored in memory so that normal integation against errors has already taken place.

Finally this type of data may also be transmitted on channels employing the packet unhook mechanism shown in Figure 8.

6. CAPTIONING

Captioning can be treated as a normal data page with subtitle control bit set to 1 . It is probable that this will be the mode adopted for the minimum form of life. However, if captioning is to be completely effective some more flexible captioning technique is required.

Figure 15 shows a more sophisticated captioning strategy based on packet 29. Packet 29 is ignored by the simple decoders, and in the higher level of decoder it is given a special status. It can be inserted at any time during the transmission without interrupting the normal flow of packets between two packet 0. It will always be a single row of data although two or more may follow one another.

The structure of the packet is slightly different in that the first data bytes are Hamming protected address information specifying the precise point where the caption should be placed on the screen. This allows a caption with a single transmission to bridge the gearing point.

It was felt important that when a specific TV program is recorded on VCR or VTR that appropriate captioning should be retained. However, the normal bandwidth of a recorder is not adequate for teletext to be recovered during playback. For these reasons, it is proposed to operate packet 29 at a half or even one quarter of the normal clock rate. Packet 29 was chosen since it was very close to the "all ones" condition and therefore although at this point the clock would be normal, the bandwidth of the teletext signal was low. Active trials are underway to establish this concept.

7. MULTIPLEXING

It is recognized that teletext data may come from a number of independent sources, and that these will need to be combined at the local transmitter to form the total transmission. Two

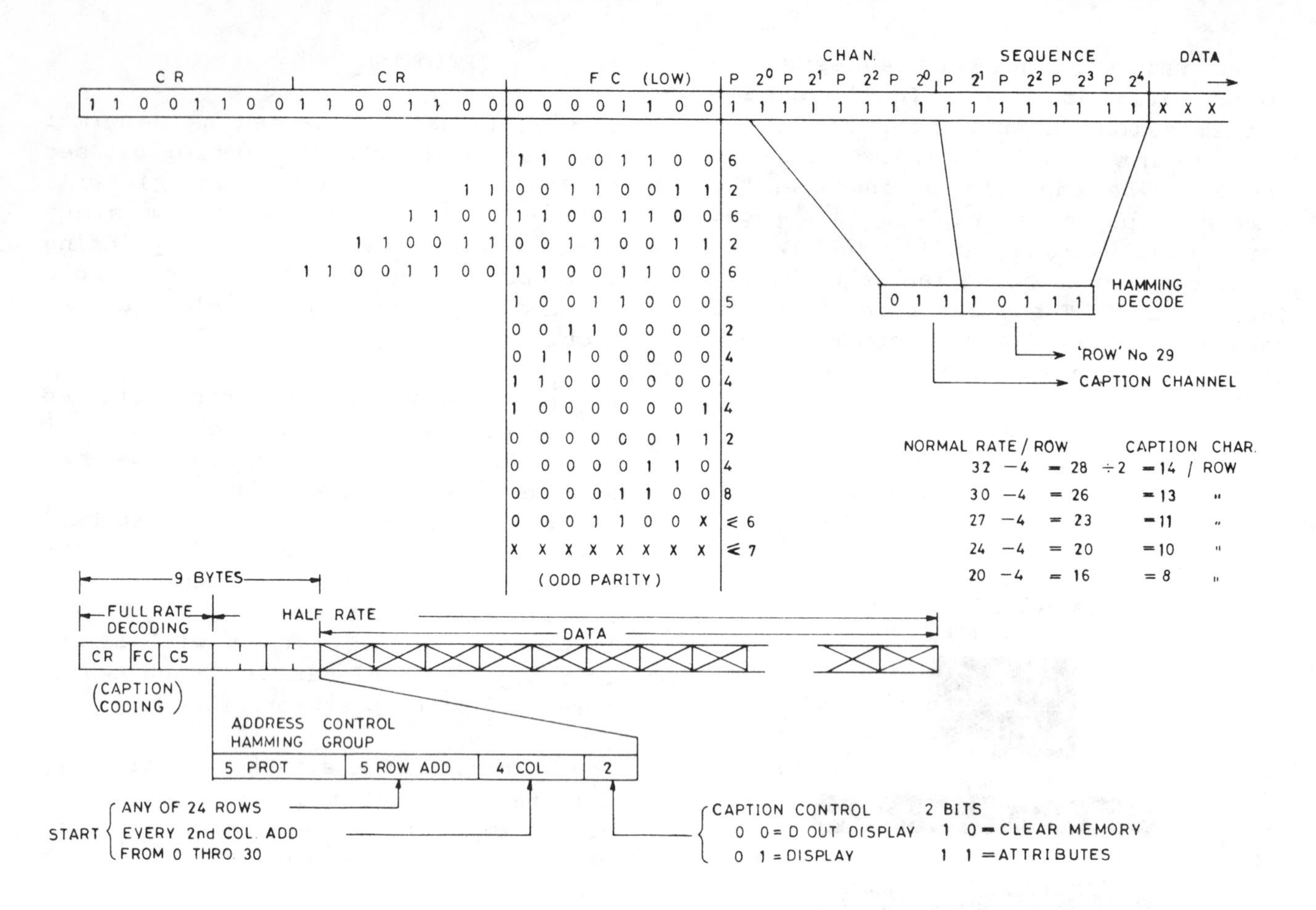

CAPTION CODING

Figure 15.

methods of multiplexing data have been incorporated in the proposal, a purely serial mode and a parallel or interleaved mode. These two concepts are illustrated in Figures 16 and 17, respectively. An alternative representation is shown in Figure 18. An advantage of the parallel system is that time critical data such as captions can be inserted into the data stream without the need to interfere with the main or current page transmission. This method however imposes a considerable address overhead since each packet must now be coded with the channel number. For these reasons, the number of parallel channels in the present proposal has been constrained to four, one of which has been reserved for captions. Further parallel channels can be released by use of the unhook code.

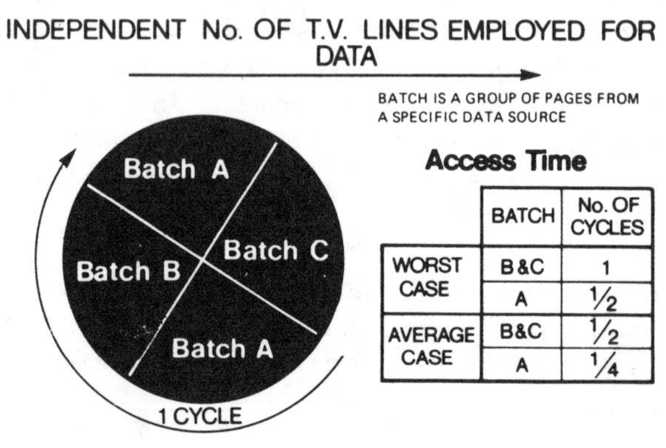

		BATCH	No. OF CYCLES
WORST CASE		B&C	1
		A	½
AVERAGE CASE		B&C	½
		A	¼

Figure 16.

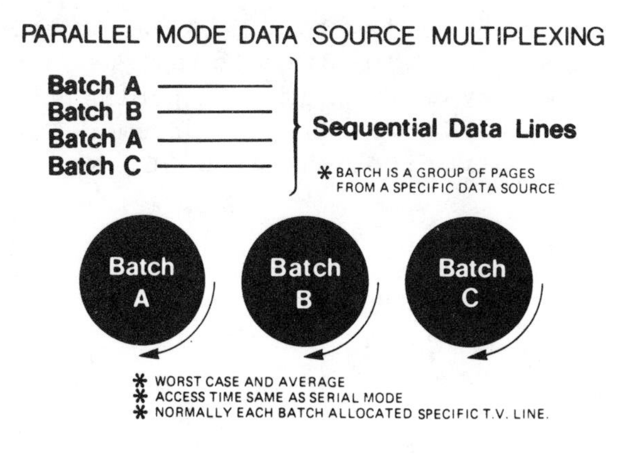

Figure 17.

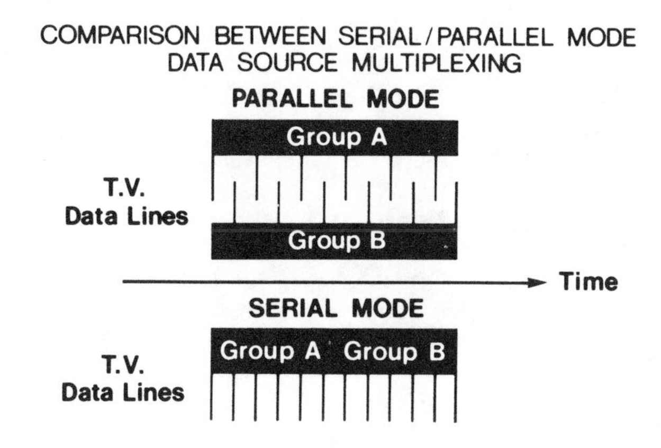

Figure 18.

The more simple multiplexing method is the serial mode where a time allocation is given to each data source probably specified as a number of field periods. The multiplexing in this mode is entirely performed in the header packet (packet 0). In this mode, the addressing overhead is negligible and up to 100 different magazines have been allowed for. The major advantage of the serial mode is that a user can readily determine if a particular magazine is being transmitted by the use of the rolling page numbers. User feedback in the parallel mode is more difficult if a magazine which is not being transmitted is requested there is no data to give any user interaction which leads to user frustration.

It has been demonstrated how to adapt the basic UK teletext text service to the USA 525 line TV system with only minor changes to the original coding structure. The design allows a common chip set to be implemented for both Europe and the USA. Further, based on experience in Europe, it has been shown how the system can be enhanced in stages up to text incorporating high resolution still pictures. It is anticipated that the first enhancements in Europe will include parallel attributes, high definition graphics using DRCS, and a full Latin alphabet consisting of over 300 characters. The coding structure has been so designed that extensions can take place in an optimum time and manner while retaining complete compatibility with early decoders. The proposal discussed in this report has been presented to the EIA Teletext subcommittee.

REFERENCES

1. G.O. Crowther, "Teletext and Viewdata Systems and Their Possible Extension to Europe and U.S.A.", IEEE Trans. on Consumer Electronics, Vol. CE-25, No. 3, July 1979.

2. J.P. Chambers, "Enhanced UK Teletext Moves Towards Still Pictures". (to be published in IEEE Trans. on Consumer Electronics Vol. CE-26, No. 3).

3. G.O. Growther, "Dynamically Redefinable Character Sets in United Kingdom Teletext". (to be published in IEEE Trans. on Consumer Electronics, Vol. CE-26, No. 3).

4. K. Clarke, "Application of Picture Coding Techniques to Viewdata". (to be published in IEEE Trans. on Consumer Electronics, Vol. CE-26, No. 3).

5. J. Hedger, "Telesoftware". (to be published in IEEE Trans. on Consumer Electronics, Vol. CE-26, No. 3).

6. J.R. Storey, A. Vincent and R. Fitzgerald, "A Description of the Broadcast Telidon System". (to be published in IEEE Trans. on Consumer Electronics, Vol. CE-26, No. 3).

TELEVISION RECEIVER DESIGN ASPECTS FOR EMPLOYING TELETEXT LSI

Tadahiko Suzuki, Yasuto Okada, Hajime Yoneta and Richard Jones
Consumer TV Division
Sony Corporation
Shinagawa-ku, Tokyo 141, Japan

1. Introduction

Television multiplex broadcast has caught world-wide interest as a new information media.

In Japan, multiplex sound broadcasting started last September and was an immediate sensation. Preparations for introducing TELETEXT broadcast are now under-way.

In the United Kingdom, TELETEXT broadcasting to an unified specification has already begun service. We are now providing to the U.K. market a 22" compact color television receiver which includes a build-in TELETEXT decoder without changing the appearance of the set at all.

This paper describes the TELETEXT receiver design.

Figure 1 shows the TELETEXT receiver block diagram. Above the dotted line are all the normal blocks found in a conventional TV receiver.

For TELETEXT reception, an existing receiver requires the TELETEXT controls, a TELETEXT decoder, data display circuits, and an additional power supply. The cost of adding a TELETEXT capability to an existing receiver, therefore, depends upon the channel selection circuit system, analog control, video output and power supply of the existing set, and, of course, the control system of the TELETEXT decoder itself.

Table 1.
Number of semiconductors used.

	Present Receiver	Receiver with TELETEXT
IC	10	25*
Transistor	60	65
Diode	65	84
FET	2	2
GCS	1	1

*10 ICs are used for TELETEXT decoder circuit.

Table 1 compares the number of semiconductors in our present 22" remote controlled receiver with that of the receiver with the TELETEXT added.

All the TV and TELETEXT functions can be remote controlled and the TV functions can also be controlled manually.

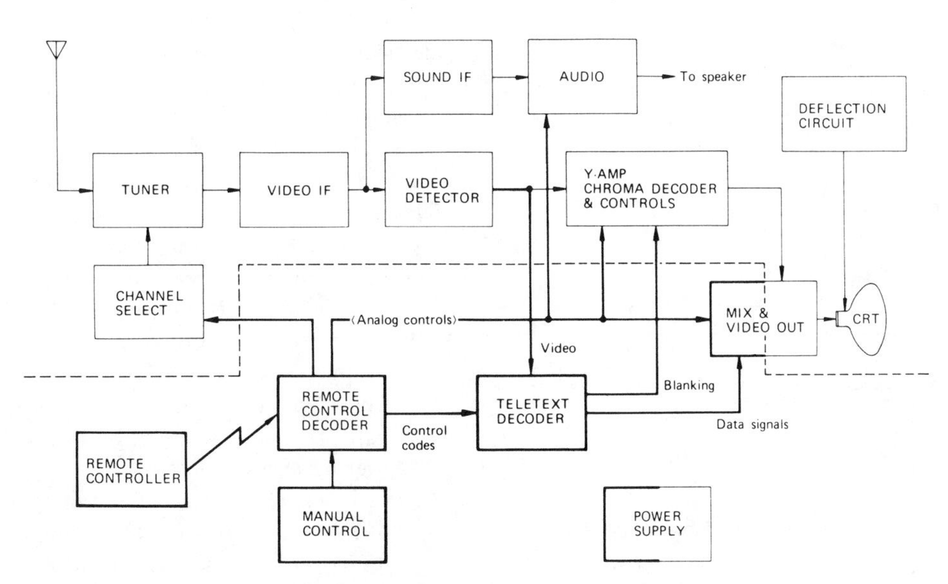

Fig. 1. Block diagram of the TELETEXT receiver.

Reprinted from *IEEE Trans. Consum. Electron.*, vol. CE-25, pp. 400–405, July 1979.

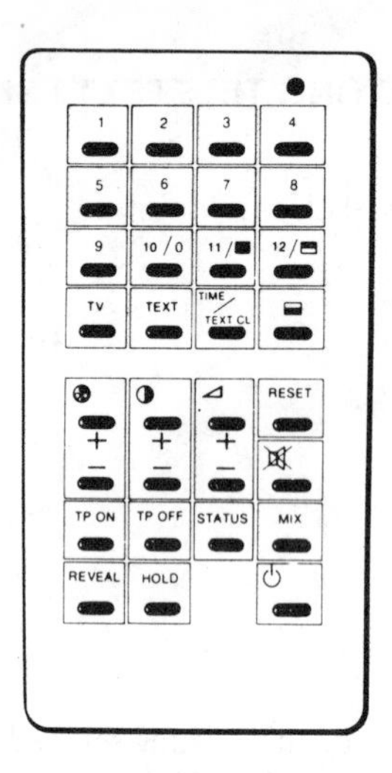

Fig. 2. TELETEXT remote controller.

Figure 2 shows the TELETEXT remote controller. Some of the functions of the remote controller are:

(1) The status key superimposes channel identification on the picture. The time key superimposes the time.
(2) The text key causes the index page (page 100) to appear. A single key press is all that is required.
(3) The character height keys cause the text to be displayed in double height characters. The top and bottom halves of the page are displayed separately.
(4) The hold key stops the reception of TELETEXT data so that updating does not take place.
(5) The TP/ON (Timed Page On) key can select a particular page to be captured at a particular time.
(6) The mix key allows the TELETEXT display to be superimposed on the normal TV picture.
(7) The picture key controls the brightness of the TELETEXT display.

These are the main functions made possible by the use of the TELETEXT decoder combined with the TV receiver and the remote controller.

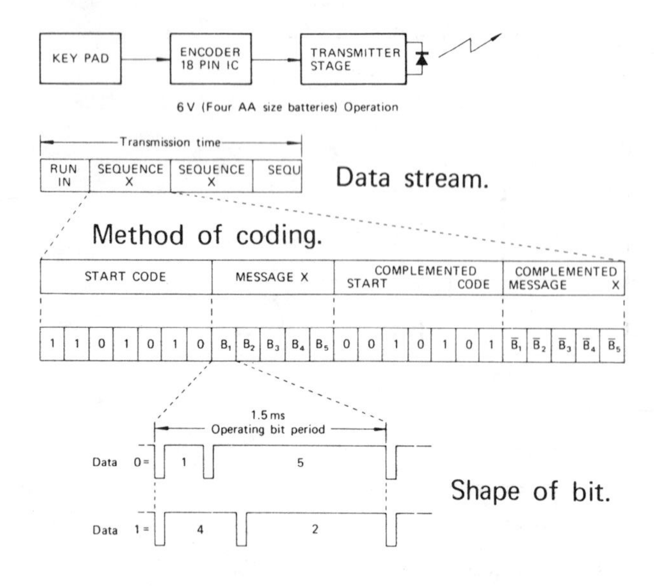

Fig. 3.

2. Design Aspects

2.1 Remote Control

The random access remote controller handles 31 commands with a system using infrared rays.

Figure 3 illustrates the system of transmission.

When a command is entered on the key pad of the remote controller, a short pseudo-random sequence and a 24 bit data sequence are transmitted. The 24 bit data stream consists of a 7 bit start code and a 5 bit message and their binary complements for a total of 24 bits.

The transmission of the binary complements of the start code and message allows checking for false responses caused by adverse transmission path conditions. The five message bits give a maximum of 32 possible commands.

The infrared remote control system is desirable because of its stability. We believe that a code system for remote control commands may be necessary and desirable because of the large number of commands possible. The present model uses the pulse system for transmitting data. But, because a long coded data sequence is susceptible to noise interference, we feel that the carrier system may be more effective than the pulse system. We will consider using the carrier system in future models.

A remote control system with 31 different commands may at first appear complicated. Careful design of the key layout of the controller and the use of a sound "bleep" to enable the user to confirm every command will faciliate use.

Normally, the viewers read TELETEXT information at a comparatively shorter distance than they would watch an ordinary TV program, so there is no need to devise remote controls that must work at a long distance. About 5 meters from the receiver seems to be the required range for a 22" receiver.

Figure 4 shows block diagram of the remote control decoder.

The infrared signal is detected by a photodiode, and is amplified by the amplifier. The manual control generates the same data codes the remote controller, but the function is limited to the TV analog controls. This data is fed to the decoder input.

The remote control decoder is a 24 pin IC. It possesses three functions: analog control, channel selection and TELETEXT control.

(1) Analog control consists of color, picture, and volume, along with stand-by and muting control. Color, picture and volume outputs produce a variable mark space digital waveform adjustable over 62 levels.
(2) Channel selection generates an appropriate number of stepping pulses for the tuning system and LED display for channel identification.
(3) TELETEXT control provides 7 bit serial data, 5 bits of which are identical to the input command message code. The other two bits control TV and

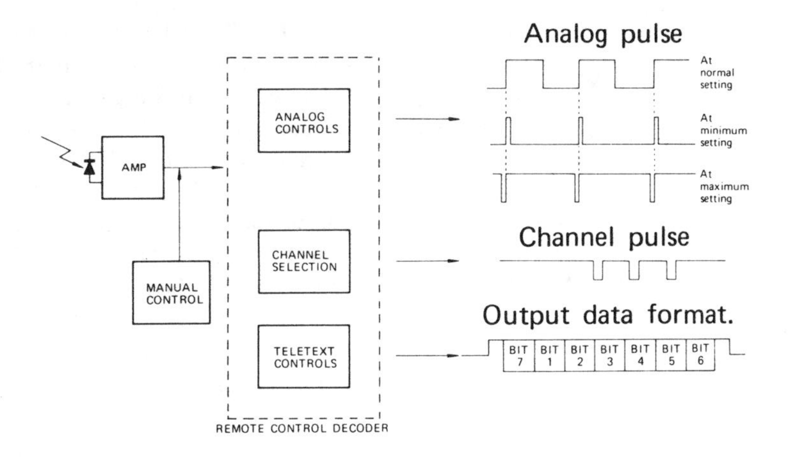

Fig. 4. Remote control decoder.

TELETEXT modes.

It is a great advantage to be able to connect the code directly to the TELETEXT decoder.

2.2 Video IF

The quality of NRZ coded data signals within the field blanking interval of television signals greatly depends upon the characteristics of the transmission path. The IF characteristics are almost the same as that of a conventional TV receiver without pulse equalization circuits.

(1) Circuits

The video IF detection method uses a synchronous detector. The peak detected AGC circuit has good frequency response against flutter.

We have decided to use a LC filter. Certainly the SAW filter has a good reputation for group delay response, but we were afraid that we would not be able to maintain the required accuracy in mass-production at that time.

(2) Characteristics

The theoretical spectrum of data pulses extends to about 7 MHz as a sequence of bits. But they are mostly distributed in the region up to 3.5 MHz (half data rate). In our receiver, group delay response is within 50ns peak-to-peak up to 3.5 MHz and the amplitude response is flat up to 3.5 MHz. (see Fig. 5)

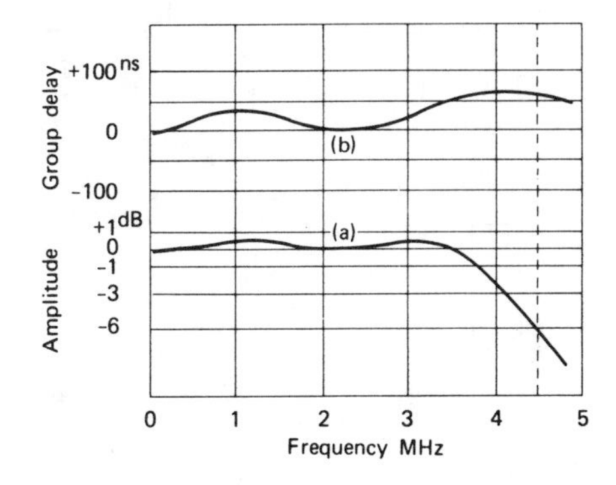

Fig. 5. Video amplitude (a) and Group delay response (b).

Sixty-five to 70% of the eye height can be obtained by using a synchronous detector. This is a better performance than that obtained by envelope detector (see Fig. 6a,b), and it provides an almost constant eye height against variation in the video data amplitude.

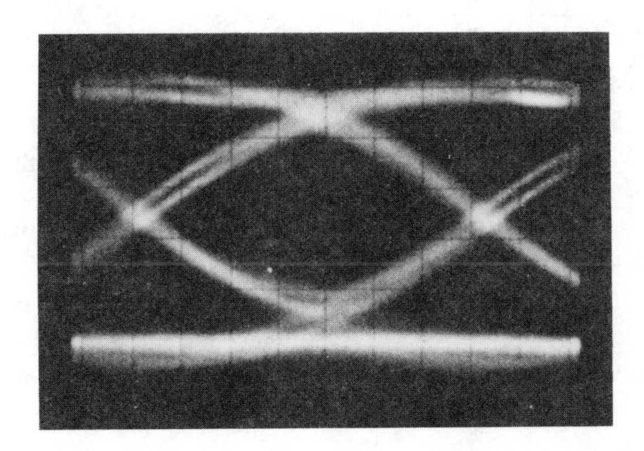

(a) Output of synchronous detector.

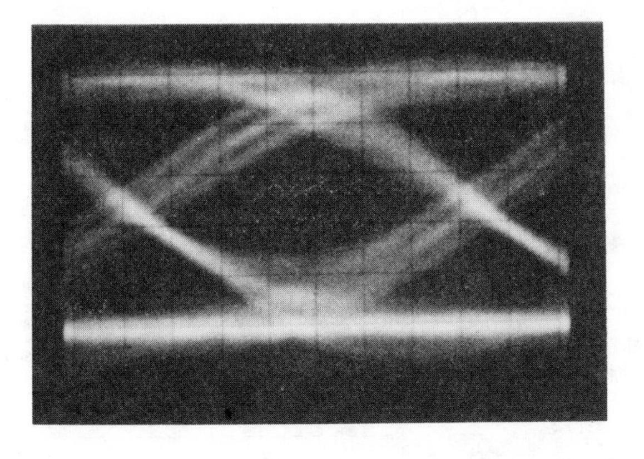

(b) Output of envelope detector.

Fig. 6. Eye height.

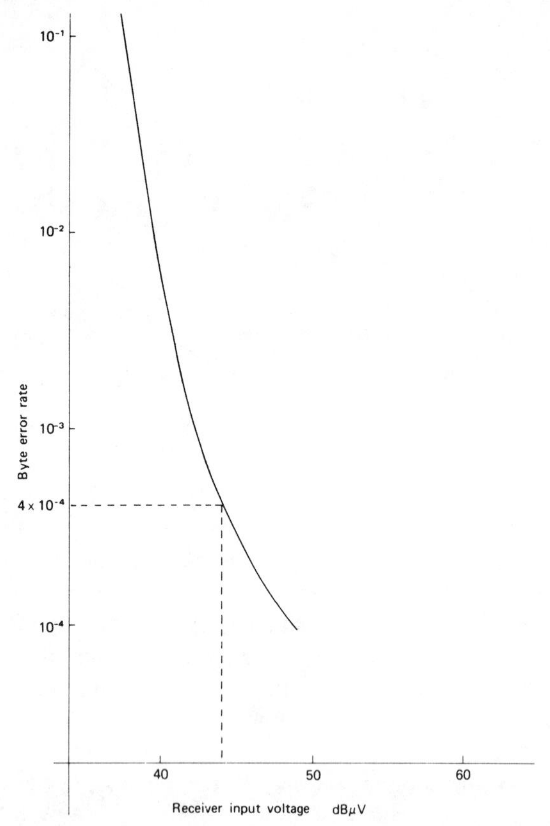

Fig. 7. Byte error rate.

The byte error rate is 4×10^{-4} at the field strength of 44 dBμV (S/N 30 dB), which can be converted to a bit error rate of less than 1×10^{-4}, and this is satisfactory data reception. (see Fig. 7)

Table 2 shows the minimum signal strength for error free TELETEXT reception using an index page and the same receiver at the end of 3 relay stations. These figures come from a field test held in the U.K.

Table 2

Channel Received	At the end of 3 relay stations
41	48 dBμV
44	48 dBμV

2.3 TELETEXT Decoder

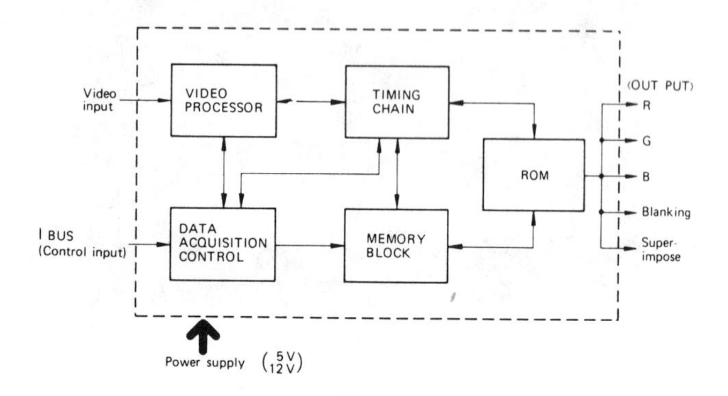

Fig. 8. Inputs and outputs of the TELETEXT decoder.

The TELETEXT decoder is a 160 mm x 120 mm circuit board containing four LSI chips and two 4k bit RAMs for storing a page of data. When the TELETEXT decoder receives the control signals (I bus) from the remote control decoder, it extracts the TELETEXT data signal from the video detector output and obtains red, green and blue data signals which are synchronized with the TV picture. The TELETEXT decoder also generates the necessary blanking signal. Figure 9 shows how the above signals relate to each other. The end result is that the cathode voltage contains the TV and subcaptioning display signals.

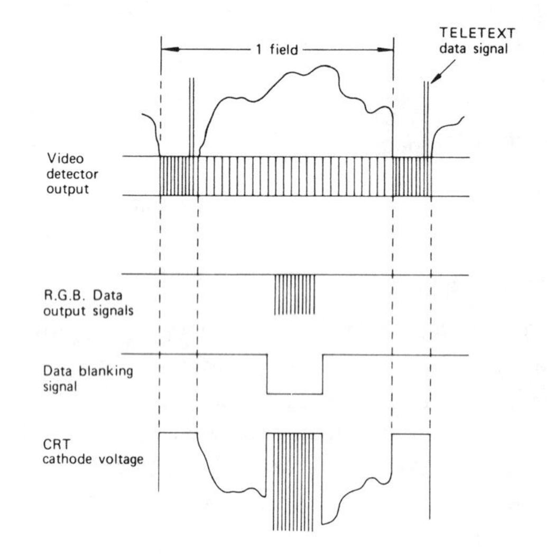

Fig. 9. How the signals combine.

Because the decoder has an output only in the TELETEXT mode, there is no need to interpose switching circuits of R,G,B color signals interfacing the receiver.

Power consumption of the TELETEXT decoder is 5W or less. As a circuit with relatively low voltage and high current will be used, we should consider the power supply to be used and the circuits' ripple immunity.

2.4 Blanking and Display

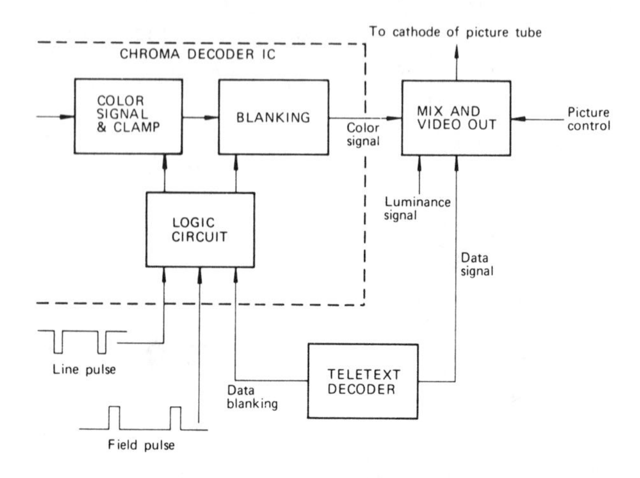

Fig. 10. Blanking and Mix circuit.

The TV mode operates only when TV video signals are supplied at the Mix and Video output stages.

In the TELETEXT mode, a blanking signal from the TELETEXT decoder blanks the TV color signals. The data blanking signal is used for complete TV picture blanking and for caption box blanking as well. The TELETEXT signal obtained simultaneously drives the cathode at the Mix and Video output stages.

In the MIX mode there is no blanking output from the TELETEXT decoder, so that the TV picture and the TELETEXT display appear on the screen at the same time.

The data signal from the TELETEXT decoder consists of R, G, B color signals. It is desirable that the receivers use the same color decode system so that only simple interface circuits between TV and data signals are needed. It is also desirable that both TV and data be driven by a single video output circuit to minimize spurious radiation.

The brightness of the data display is adjustable by the TV receiver's picture control, which controls the Mix and Video output stages.

The luminance of the data display has been set to 110 nits at the standard control condition and provides a good display image on the screen. It is possible to get even clearer displays by using a lower light transmission CRT panel.

Some suggestions have been made that the number of characters in a row be increased by developing a high resolution picture tube, but in order to take full advantage of a high resolution tube, we feel a smaller screen must be used.

We know that our current 13 inch picture tube can give us acceptable data displays on the screen, but better displays are possible by making a finer aperture grill. Table 3 compares the AG pitches and number of slits of a conventional 13 inch CRT with those of a high resolution CRT of the same size.

Table 3

	AG pitch	No. of slits
Conventional CRT	0.6 mm	386
High resolution CRT	0.4 mm	579

The geometrical distortions are not only the deflection distortions, but distortion also occurs when the dynamic regulation of the high voltage circuit is poor, especially when data is displayed in the flash and mix modes.

3. Specifications

Television System:	British color standard
Color System:	PAL
Picture Tube:	56 cm, 22" 114° deflection
Semiconductors:	65 transistors, 2 FETs, 25 ICs, 84 Diodes and 1 Gate Controlled Switch
Aerials:	UHF: 75 unbalanced (coaxial aerial socket)
Channel Coverage:	UHF channels: 21–68 (12 program selection)
IF:	Picture i-f carrier: 39.5 MHz Color subcarrier: 35.07 MHz Sound i-f carrier: 33.5 MHz
Sound System:	6 MHz intercarrier Output power: 2W
Video System:	R,G,B cathode drive
TELETEXT Selection:	Compatible with broadcast Teletext specification, September, 1976
Power Requirements:	200/240V ac, 50 Hz
Power Consumption:	130W 15W ac (in stand-by condition)
Dimensions:	Approximately 669 (w) x 459 (h) x 413 (d) mm 26 3/8 (w) x 18 1/8 (h) x 16 1/4 (d) inches
Net Weight:	Approximately 34 kg

4. Conclusion

(1) A television receiver with a remote controller is essential for the convenient use of the TELETEXT receiver.

(2) An advantage of the TELETEXT decoder is that it can be combined with a remote control system to make a TELETEXT control system, which means that the interface circuitry will be simpler.

(3) The input video amplitude of the TELETEXT decoder varies with the receiver, the transmission quality and other factors. The error immunity and sync stability of the TELETEXT decoder depends on the quality of the input video signal. Therefore, it is desirable that the TELETEXT decoder have a more relaxed tolerance for the amplitude of the signal.

(4) For public approval of TELETEXT, it is important to minimize necessary additional modifications and improvements to existing television receivers.

5. Acknowledgement

This work was performed by the cooperation of the SONY UK Bridgend Plant and SONY Tokyo Consumer TV Division.

The authors would like to express their sincere gratitude to Mr. M. Morita, Managing Director, TV and Consumer Video Group for his encouragement.

They also express their appreciation to Mr. T. Shioda, General Manager, Image Display Division and O. Shibata, General Manager, Consumer TV Division, for their useful advice.

Mr. T. Tokita, General Manager of the SONY UK Bridgend plant, has been especially helpful in giving us generous support and guidance.

SET TOP ADAPTER CONSIDERATIONS FOR TELETEXT

Howard F. Prosser
Oak Communications Inc.
CATV Division
Crystal Lake, Illinois 60014

INTRODUCTION

As soon as broadcast standards are adopted for TELETEXT in the United States, the need for set top adapters will be an immediate reality. Long design time cycles in the television manufacturing industry coupled with the long life of modern solid state receivers will spur the consumer demand for TELETEXT set top adapters for use with any color television receiver. This paper discusses design considerations for such set top adapters.

A set top adapter can be considered as a self-powered stand alone device which is connected between a source of TELETEXT bearing television signals and the antenna terminals of a color television receiver. The TELETEXT set top adapters shown in Figure 1 receive RF signals from varied sources. They differ only in the way that the incoming RF signal is processed. A broadcast adapter must be capable of all channel tuning while cable, MDS (Multipoint Distribution System), and ITFS (Instructional Television Fixed Service), adapters may be fixed tuned to the VHF outputs of their respective converters resulting in lower cost units.

Except for treatment of the RF input signal, the design objectives for any

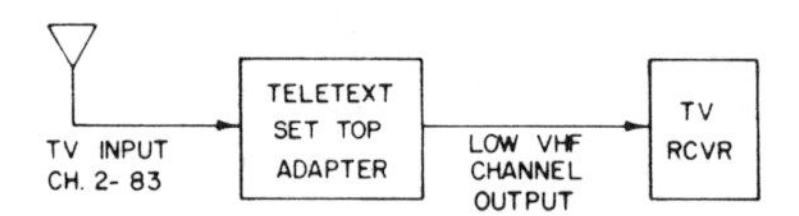

(a.) BROADCAST ADAPTER

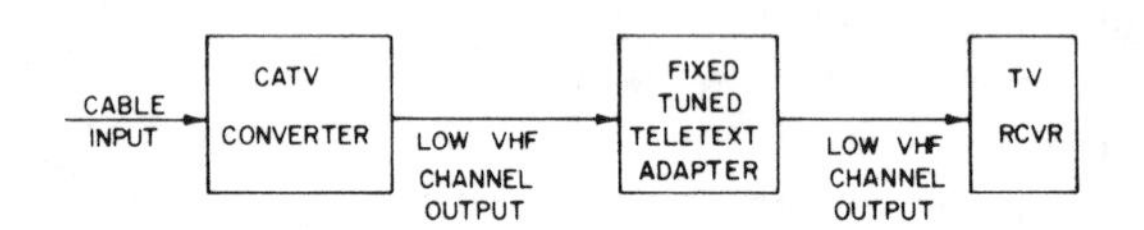

(b.) CABLE ADAPTER

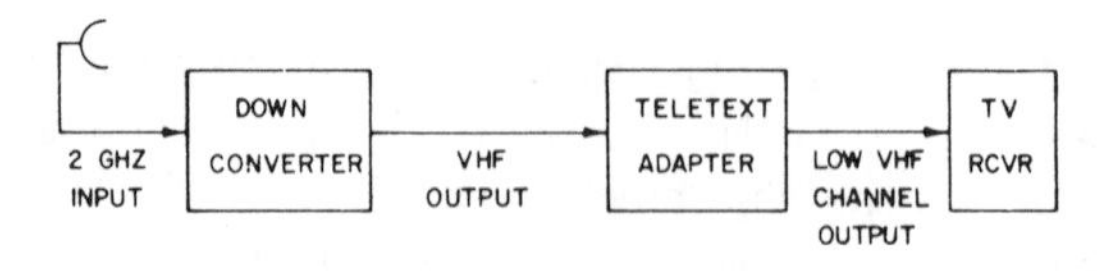

(c.) MDS OR ITFS ADAPTER

FIGURE I. TELETEXT SET TOP ADAPTERS

of these three adapters are shown in TABLE 1. The adapter should have a VHF output signal, switchable between channels 3 and 4, for use in any location. Seven color operation in red, green,

TABLE I

DESIGN OBJECTIVES

1. RF SIGNAL INPUT OUTPUT.
2. MODULATED TEXT SIGNAL OUTPUT ON LOW VHF CHANNEL.
3. SEVEN COLOR OUTPUT SIGNAL (R,G,B,Y,CY,MAG,WHITE).
4. MINIMIZATION OF CROSS LUMINANCE EFFECTS.
5. WELL MATCHED LUMINANCE-CHROMINANCE DELAYS.
6. LUMINANCE PRESHOOT AND OVERSHOOT COMPENSATION.

Reprinted from *IEEE Trans. Consum. Electron.*, vol. CE-25, pp. 393-398, July 1979.

blue, yellow, cyan, magenta, and white should be incorporated in the adapter to comply with the usual TELETEXT standards. Proper band-limiting of the chroma signals should be employed to minimize ragged edges on the characters caused by cross luminance effects (chroma into luminance). Proper registration of the luminance and chrominance in small characters requires that the luminance delay match that introduced by the band limiting chroma filters. Excessive overshoots in the sharply defined luminance signals should be avoided or minimized in the adapter. It must be remembered that an adapter is just that and subject to the limitations of the receiver. TELETEXT decoding circuitry installed in a receiver and connected after all receiver band limiting will by this very fact always produce a sharper display than will the adapter. Nevertheless, observance of these few design objectives in an adapter produces a quite acceptable display on a well adjusted and properly converged color television receiver.

ADAPTER SYSTEM CONSIDERATIONS

A block diagram illustrating broadcast TELETEXT adapter subsystems is shown in Figure 2. The principal subsystems in this adapter consist of UHF and VHF tuners, IF amplifier, video detector, sync processor, digital decoder, color video encoder, video switch, modulator, vestigial sideband filter, and power supply.

The UHF and VHF tuners used should have specifications consistent with those used in high quality color television receivers so as not to limit the bandwidth of the incoming signal excessively. Automatic frequency control should be employed to keep the IF response centered in its bandpass. Delayed AGC

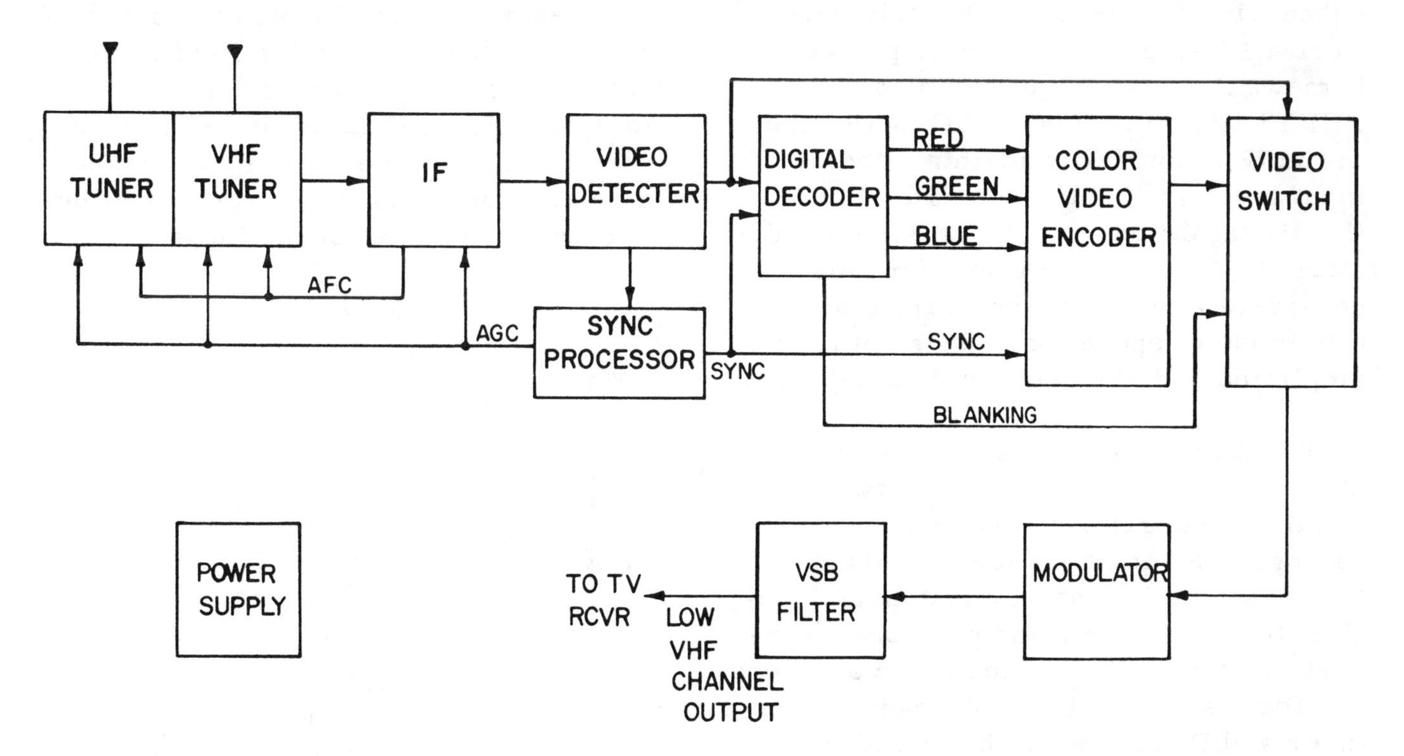

FIGURE 2. BROADCAST TELETEXT ADAPTER SUBSYSTEMS

is required in the tuners for noise and overload considerations.

Guidelines for the IF amplifier group delay and amplitude response are clearly spelled out in the BBC publication, CEEFAX (1). Essentially these requirements dictate that the IF amplifier be specifically designed with the TELETEXT data rates in mind. IF amplitude response requirements are such to make the baseband video response flat up to one-half the bit rate. The group delay should also be flat up to one-half the bit rate. These requirements suggest that a SAW IF filter be employed. Satisfactory operation of a TELETEXT adapter has been obtained with a color receiver IF containing a SAW filter which produces the baseband video amplitude response shown in Figure 3. In this adapter, the video response is down approximately 2 dB at a frequency equal to one-half the bit rate, and down 6 dB at the chroma subcarrier frequency. For this particular IF amplifier, the data pulses are somewhat distorted but clean enough for TELETEXT decoding. AGC is required in the IF amplifier to maintain the data input level constant to the data slicer in the digital decoder. AGC systems which alter the IF bandpass shape for noise considerations should be avoided since this mode of operation affects both the amplitude and shape of the data pulses.

TELETEXT adapters for CATV, MDS, and ITFS applications shown in Figure 1 can realize a significant cost savings over adapters for broadcast use through the elimination of the UHF tuner, VHF tuning system, and possibly a separate IF amplifier. Since these adapters are for a single VHF input channel, the tuner and IF can be combined into a fixed tuned RF amplifier which serves as both tuner and IF. Appropriate selectivity can be provided by SAW filters tuned to the low VHF channels and available for output filtering for Class I RF devices (primarily for the video games market). AGC requirements on these RF amplifiers are much less stringent than for broadcast applications since the input signal level variation to these adapters is in the order of 20 to 25 dB.

The input video processing circuitry of the TELETEXT digital decoders, commonly known as data slicers, convert the data input pulses to logic levels by means of a threshold comparator circuit. Distortions to this train of data pulses can cause erroneous outputs in the data slicer. Video envelope detectors because of quadrature distortion cause non linear distortion in the recovered data pulses which can severly affect the operation of the data slicers. Extensive BBC (1) experience has shown that synchronous detectors reduce the likelihood of data slicer errors compared to envelope detectors. The most satisfactory results are obtained with synchronous detectors using carriers recovered by narrow band phase locked loops.

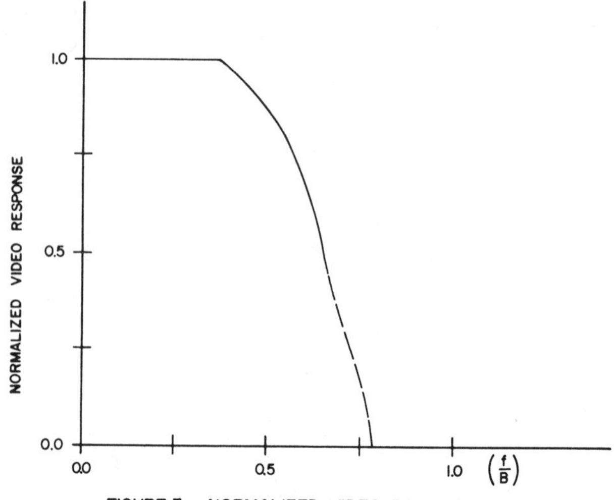

FIGURE 3. NORMALIZED VIDEO BANDWIDTH

When TELETEXT digital decoding circuitry is incorporated as part of a television receiver, it makes use of many receiver subsystems such as sync stripper, horizontal APC, and video output drive circuitry. In a TELETEXT set top adapter, these same receiver subsystems must be duplicated. Reference to Figure 2 shows how this necessary circuitry works in conjunction with the digital decoder to produce an RF signal for use with any color television receiver.

Sync processing circuitry consisting of a sync stripper, and horizontal APC loop is necessary for operation of keyed AGC in the adapter front end. One manufacturer (2) of digital decoding circuitry relies on horizontal flyback pulses for timing information and does not provide a composite sync signal (for after hours operation) which is necessary for subsequent video encoding in the adapter. Other manufacturers (3, 4) provide this composite sync signal needed by the adapter.

The digital decoding circuitry requires a data bearing composite video input signal and in some cases horizontal flyback pulses as discussed above. This decoding circuitry processes the data signal, stores at least one page of text data in a RAM, and then produces the properly timed TTL level output signals corresponding to red, green, blue, and blanking. In a TELETEXT receiver, these output signals are interfaced directly to the video output circuitry. However, in a TELETEXT set top adapter, these properly timed logic level outputs must be re-encoded into a composite video signal for subsequent remodulation for use by a color television receiver.

Figure 4 illustrates the video signal processing in the set top adapter. A video digital to analog converter accepts the red, green, and blue output signals from the digital decoder in addition to sync, horizontal blanking, and burst gate signals. It performs the NTSC color matrixing to obtain the luminance signal, Y (consisting of $0.3\ R + 0.59\ G + 0.11\ B$) and the color difference signals R-Y and B-Y (with a burst gate on the B-Y signal). Composite sync is obtained from the sync processing circuitry while horizontal blanking and burst gate signals are obtained from horizontal flyback pulses by conventional digital techniques. To prevent the extremely fast transitions in the color difference signals from causing ragged edges on the text characters, identical band limiting filters are inserted in the color difference signal lines. These filters are chosen for minimum overshoots and symmetrical impulse response. To compensate for the delay introduced by the chroma band limiting filters, delay is introduced into the luminance signal path. This results in proper registration of the luminance and chroma components of the characters that are subsequently displayed by the receiver used with the adapter.

The composite video encoder shown in Figure 4 contains a chroma subcarrier oscillator which requires a 3.58 MHz crystal. It contains the two balanced modulators for the color difference signals and has provision for an external phase shift network to achieve the necessary quadrature relationship between these color difference signals. This encoder combines the chroma subchannel signal with the luminance signal (containing horizontal

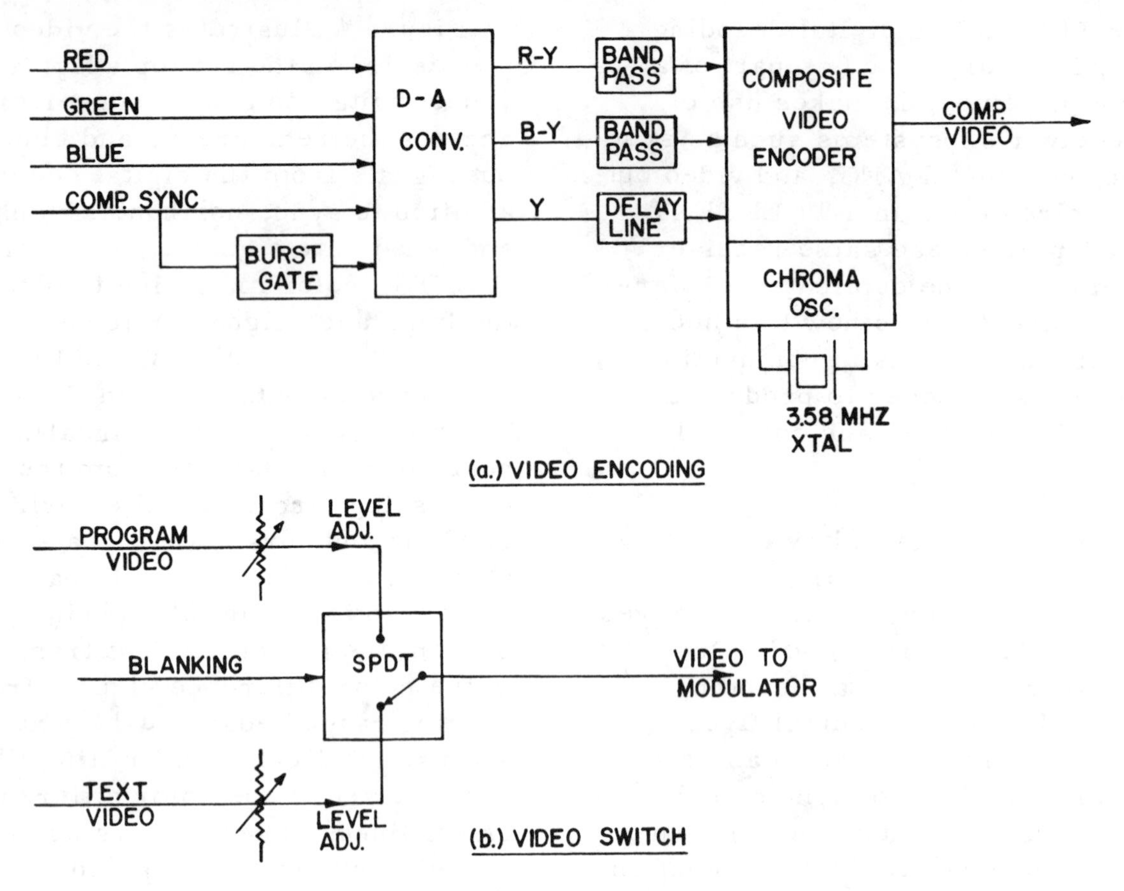

FIGURE 4. VIDEO SIGNAL PROCESSING

sync) to produce a composite color video signal complete with color burst.

Operation of the set top adapter in either the text or picture mode is accomplished with a single pole double throw video switch activated by the blanking signal output of the TELETEXT digital decoding circuitry. Although this video switch can be realized with various circuit implementations, satisfactory operation has been achieved by the use of CMOS bilateral transmission gates.

The output signal of the video switch containing either the program video or the text video is fed to the modulator shown in Figure 2. This modulator produces an output on either channel 3 or 4 depending on the locale in which the adapter is used. The output of the modulator is filtered to comply with appropriate FCC regulations (5). A 75 ohm coaxial output is provided to connect the adapter to the viewer's receiver.

VIEWER INTERFACE

Convenient operation of the TELETEXT set top adapter requires that the viewer be able to operate the adapter from a remote location. The form of remote control is dictated by the digital decoding circuitry employed and cost considerations. A simple form of hard wired remote control requires a 4 by 5 key pad (2) and only nine wires. Another manufacturer (3) provides for cordless remote control using infra-red or ultrasonic transmission. Remote channel switching may be added at extra cost. In either case, set top TELETEXT adapters will most likely be provided with a user operated RF switch on the unit itself to bypass the unit when it is not in use.

EXPERIMENTAL RESULTS

The TELETEXT set top adapter described in this paper has been tested in the laboratory on VHF channels and over the air on a UHF station. Operation was satisfactory in both cases with a data rate of 5.538 M/bits per second (352 H). It should be pointed out, however, that in both cases, tuning of the adapter is very critical. Tuning for best picture operation does not produce best text decoding in all cases. However tuning for best text display usually produces an acceptable picture.

ACKNOWLEDGEMENT

The author wishes to express his thanks to H. Jirka for his helpful suggestions for chroma signal processing and to J. Message for building the experimental models. Both of these men are with the engineering department of OAK Communications, Inc., CATV Division.

REFERENCES

(1) CEEFAX, Its History and the Record of Its Development by the BBC Research Department. BBC Engineering, LONDON, 1978

(2) The TIFAX XMll Teletext Decoder, Application Report B 183. Texas Instruments Ltd., Bedford, England, 1976.

(3) MULTITEXT, Technical Information 050. Signetics. The Yale Press Ltd., England 1978.

(4) TELEVIEW SYSTEM, General Instrument Microelectronics Ltd., Glenrothes Scotland, 1978.

(5) FCC Rules and Regulations, Par. 15.4 Subpart H.

RECEPTION OF TELETEXT UNDER MULTIPATH CONDITIONS

Shri K. Goyal and Stephen C. Armfield
GTE Laboratories Incorporated
40 Sylvan Road
Waltham, Massachusetts 02154

Digital Teletext data is transmitted on a standard television system designed for analog video transmission. Since this system is not ideally suited for data transmission, special attention must be given to reducing the error rate in the received signal. Conditions that cause signal degradation in television systems include system amplitude and group delay distortions, noise and multipath transmission. This paper examines the problems of decoding Teletext data that have been subjected to multipath distortion, and demonstrates how deghosting techniques can improve data reliability. A transversal filter has been considered for this purpose in which standard signals such as truncated sinX/X and raised cosine pulses have been used as training waveforms. The corresponding improvements on displayed video are also discussed.

INTRODUCTION

Information for data broadcasting services such as Teletext is transmitted as digital signals inserted in the vertical interval of the television video waveform. The specifications for these signals are drawn very carefully to ensure compatibility with the existing television system. This paper examines some of these specifications such as data rate and waveform shape, which must be selected to maintain the spectral energy within the television bandwidth. Using computer simulation, the transmitted signal and encoded teletext data (as defined in Broadcast Teletext Specifications[1]) will be examined for different multipath conditions. For all these experiments, echoes have been generated and evaluated at the baseband, ignoring system amplitude and group delay characteristics. These, in general, will have additional influence on decoding reliability.

Adaptive equalization will be seen as a possible method of improving the data reliability when the received signal is distorted. This technique, however, requires a priori knowledge of a "training" waveform which must be an integral part of the transmitted signal. For this purpose, four training waveforms, including truncated sin(X)/X and raised cosine pulse, are examined for their merits through comparative performance evaluations. As one measure of performance, computer generated eye diagrams will be used. Also, briefly documented, are the hardware requirements for these approaches.

Finally, it should be mentioned that, although most data used in the experiments correspond to the British Teletext system, general comments would apply to other systems such as Antiope as well.

TELETEXT DATA FORMAT

Specifications for British Teletext are drawn up by careful considerations of systems requirements. For CCIR standards, the binary data rate has been defined to be 6.9375 Mb/s, the data being inserted in two lines in each field. This permits transmission of four pages per second, each page comprised of 24 rows of 40 characters.

The CCIR standard allows a video bandwidth of 5.0 MHz or greater. The corresponding bandwidth for an NTSC system is 4.5 MHz, which requires a reduced data rate. KSL-TV in Salt Lake City and KMOX-TV in St. Louis have been transmitting experimental Teletext signals using NTSC standards in a 31 character-per-row format. In addition, five bytes per line are used for transmitting auxiliary data. These include two bytes for clock run-in, one byte for framing code and two additional bytes for magazine and row address group, thus bringing the total number of data bits per line to 288. The corresponding clock frequency, which is an integral multiple of the horizontal line frequency, is 5.5384 MHz. Experiments[2] at several clock frequencies have been conducted in Europe to study the reliability of data transmission. Table 1 lists these experimental clock frequencies for CCIR systems and the derived frequencies corresponding to NTSC systems.

Reprinted from *IEEE Trans. Consum. Electron.*, vol. CE-25, pp. 378-391, July 1979.

TABLE 1

Experimental Teletext Transmission Rates for CCIR and NTSC Systems

CCIR		NTSC	
Transmission Rate (Mb/s)	$\times f_H$	Transmission Rate (Mb/s)	$\times f_H$
4.296875	275	3.430069	218
6.203125	397	4.940558	314
6.937500	444	5.538461	352

The experimental Teletext systems in the U.S. are using a two-level nonreturnto-zero (NRZ) signal with special pulse shaping. The binary "0" level is defined at 0 IRE (±2%) and binary "1" at 66 IRE (±6%), where 0 IRE is defined as the black level and 100 IRE as the peak white level. Considerable work has been reported on shaping filters for the data pulses.[3] Pulses with 70% cosine roll-off spectra are generated (See Appendix A) and used in the experiments described in this paper. This pulse spectrum provides reduced intersymbol interference and thus higher decoding reliability and better performance in presence of echoes. This data pulse is generated by passing a 180.56 ns gate pulse through the shaping filter.

An alternative pulse shape often considered is generated by making use of the truncation characteristics of a nominal transmitter. Figure 1 shows the filter shaping characteristics and the corresponding data pulses in the two cases. The two pulse shapes show a certain degree of preshoot and overshoot. In the filtered data train of 288 bits generated for one horizontal line, overshoot and undershoot of 9.1 IRE units were observed.

SIMULATION EXPERIMENT

The simulation facility consisted of the in-house IBM 370/168 computer with on-line plotting and hard-copy capability. A digital frame store system with 8-bit resolution and a sampling rate of four times the subcarrier frequency was also available for subjective evaluation of pictures.

Software was designed to evaluate the system performance under multipath or echo conditions. The input data of 288 bits was inserted in a horizontal line that, in a practical system, occurs during the vertical interval of a field. The data rate was chosen at 5.54 Mb/s. The data consisted of two bytes of clock run-in of alternating 1's and 0's, one byte of framing code, two bytes representing magazine and row address group codes and 31 ASCII coded characters representing A through Z and 0 through 4. The magazine and row address groups of data were Hamming code protected and the ASCII characters were given odd parity. These data bits were shaped by passing them through the shaping filter. In the discrete time format, a full data line of 63.56 μs is represented by 4096 points, providing a time resolution of 15.52 ns per point. Although this greatly exceeds the Nyquist sampling rate, this resolution is essential in providing adequate reproduction capability on the plotter for the data pulse shape of 180.56 ns and subcarrier frequency of 3.58 MHz.

The deghosting systems employed transversal filters with variable tap weights, and used one of the four training waveforms described in a later section. After the tap weights were adjusted using the training waveform, the ghosted data was passed through the filter, producing the deghosted data. Special eye diagrams were generated for both the ghosted and deghosted data and were used to analyze the comparative performance of the deghosting systems. Finally, decoding programs were developed for decoding the video line containing data.

METHODS OF PERFORMANCE EVALUATION AND DATA DECODING

Eye Diagrams

Eye patterns are generally used in monitoring the quality of systems for Teletext data, since they directly relate to the amount of destructive interference which the data can withstand before errors arise in the decoding process. Eye diagrams are generated by plotting the signal amplitude on the vertical axis while varying the horizontal axis as a function of time. The horizontal axis is changed at a quarter of the data rate, i.e., at 1.38 MHz. Even though both plots lead to the same conclusions, two different eye diagrams were generated: one with a linear time axis and the other with a sinusoidal time axis. The linear time axis eye diagram reproduces a data pulse on the plotter as it would appear on a TV waveform monitor and is shown in Figure 2(a). Figure 2(b) shows an eye diagram with a sinusiodal time axis. The eye diagram of the latter type has generally been used in the analysis of Teletext data in Europe.[3,4] Once the eye diagrams have been generated, eye height is calculated by

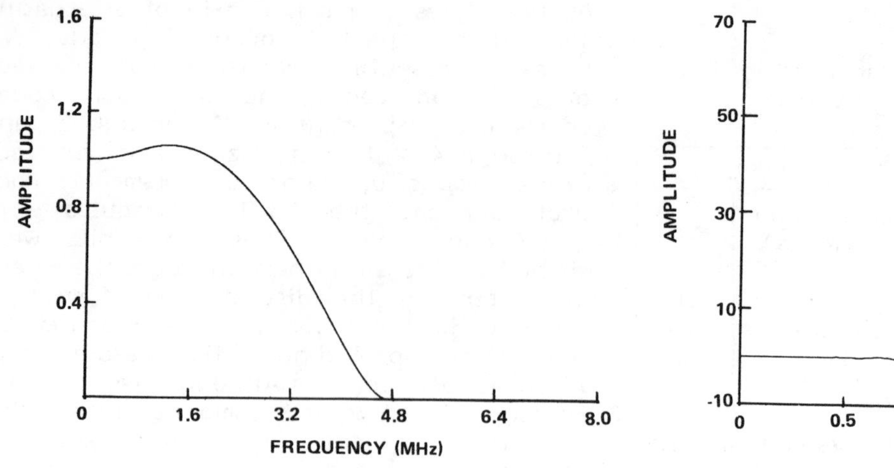

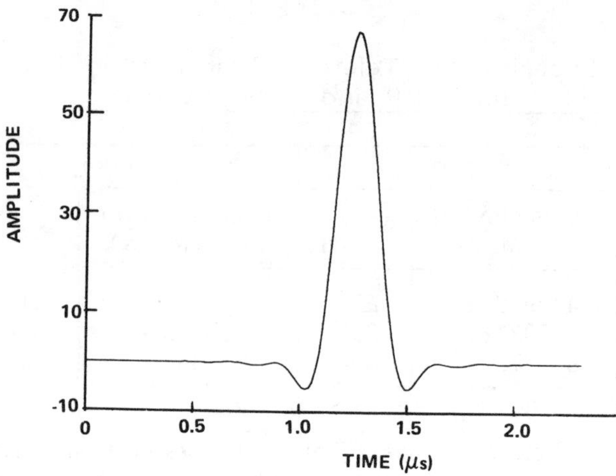

(a) 70% Cosine Roll-Off Shaping Filter and Output Pulse

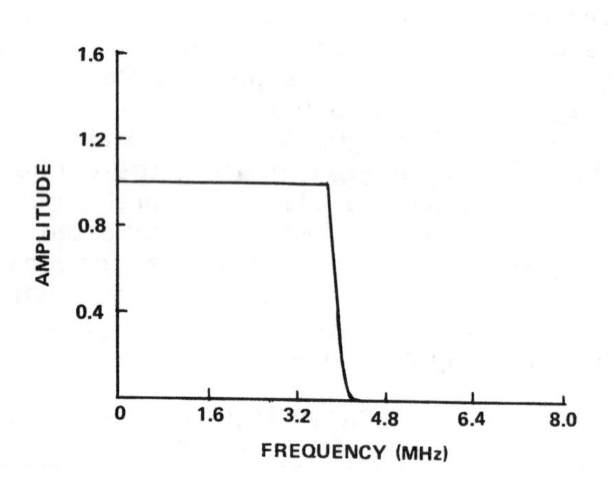

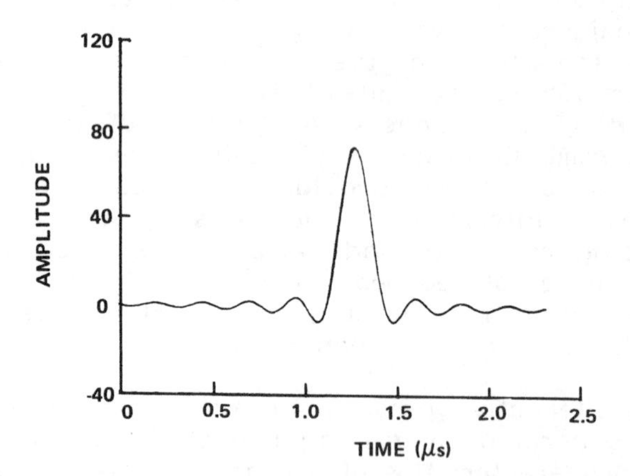

(b) Truncation Filter and Output Pulse

Figure 1. Shaping Filter Characteristics and Data Pulse Shapes

examining the maximum value of 0 and minimum value of 1 and using the relation:

$$\text{Eye Height} = \frac{\text{Minimum 1 - Maximum 0}}{\text{Level 1 - Level 0}} \times 100$$

The decoding threshold of a receiver is defined as the eye height value of the decoder at which significant (1 in 10^3 bit) decoding errors occur. Most receivers used in various field trials in Europe were found to have a decoder threshold[5] in the range of 35 to 60%. The decoder thresholds of 25, 35 and 50% have been used in our experiments for determining the number of errors in a 288-bit data stream. For data analysis, the Teletext data is decoded by setting a fixed threshold to determine the number of decoding errors and eye diagrams are generated for assessing the quality of data signal.

Data Decoding

In the simulated system, data decoding takes place by first establishing a data timing reference point and then decoding the data based on the prescribed threshold levels. In our system, the data timing reference is established by locating the peak of the first clock run-in pulse. Since data is decoded at a time corresponding to the peak of the data pulse, decoding is performed by examining the data amplitude at integral multiples of the bit period (i.e., 180.56 ns) from the data time reference. At each decoding point, by taking

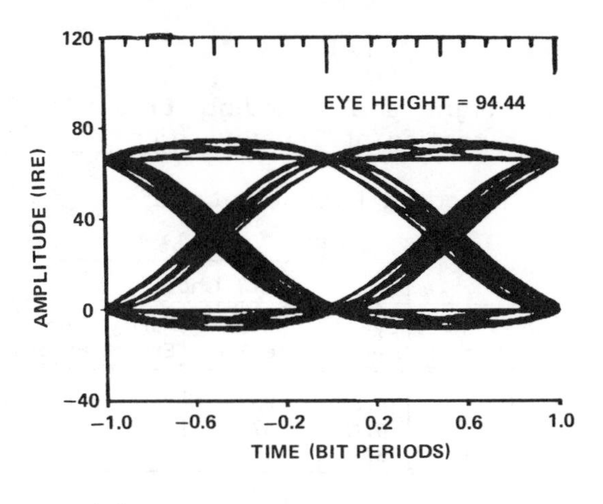

(a) Eye Diagram, Linear Axis

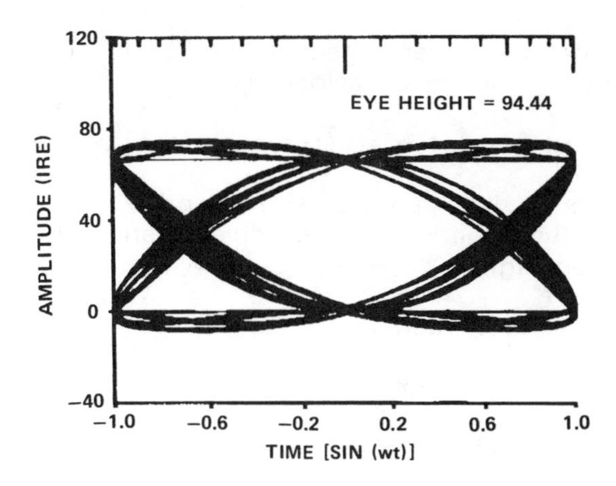

(b) Eye Diagram, Sinusoidal Axis

0 BIT ERRORS
0 INDETERMINATE BITS
0 PARITY ERRORS

MAXIMUM ONE = 67.73
MINIMUM ZERO = −1.79

MINIMUM ONE = 64.12
MAXIMUM ZERO = 1.79

DECODING THRESHOLD = 25% EYE HEIGHT

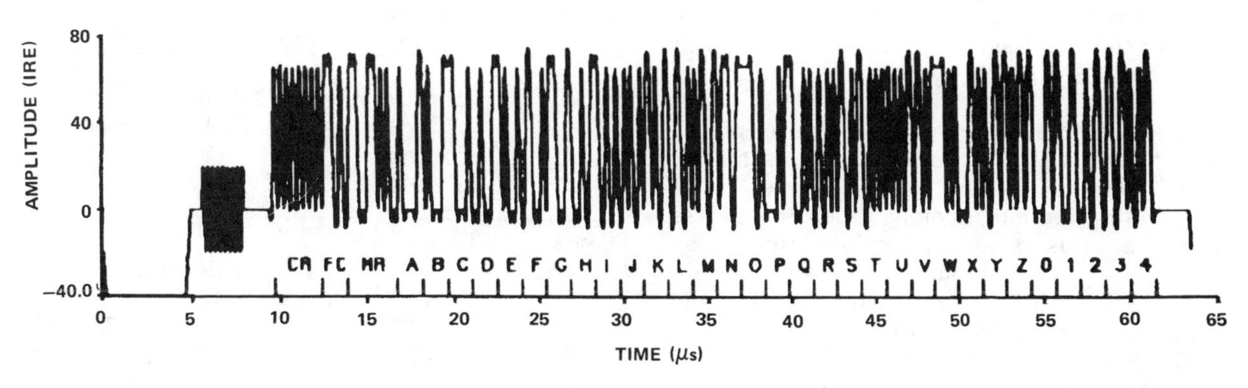

(c) Data Waveform

Figure 2. Teletext Data Decoding and Eye Diagrams

the decoding threshold into account, a decision is made if the data is a "0" or a "1." Decoding is done on a byte-by-byte basis by using an ASCII look-up table. The decoding program also checks for the parity of each data byte and outputs any detected parity error. The decoding eye height is provided to the program which establishes the decoding threshold level for the detection of a "1" and a "0." The decoder program outputs the number of bit errors, the parity errors and the number of decoded levels falling within the indeterminate range (eye opening) of the decoder. Figure 2(c) shows the full line of data (with sync and color burst) which was decoded with a 25% eye opening. The decoded characters are listed at the appropriate byte locations along the time axis.

MULTIPATH SIGNAL RECEPTION AND GHOSTS

In a normal TV reception condition, ghosts are caused by a reflected VHF (or UHF) signal adding to the direct signal. These signals can add in any relative carrier phase. The receiving equipment in such a case would require two synchronous detectors: one for demodulating the in-phase component and the other for demodulating the quadrature component. Such requirements have been examined by several authors,[6,7] and are not dealt with here. By placing constraints on the delay (ghost locations), this analysis can be simplified such that the ghost generation and cancellation can be considered at the baseband and the problem analyzed using computer simulation methods.

A single ghost generated at the RF of frequency ω_c can be described as:

$$Y(t) = x(t)\cos\omega_c t + a_1 x(t+\tau)\cos\omega_c(t+\tau) \quad (1)$$

where x(t) is the video signal and τ is the delay time between the direct and the reflected signal.

By choosing τ such that:

$$\omega_c \tau = 2n\pi \quad (2)$$

Eq. (1) simplifies to:

$$Y(t) = [x(t) + a_1 x(t + \tau)] \cos\omega_c t \quad (3)$$

which represents a single ghost of magnitude a_1 generated at the baseband. This argument can be extended to include multiple ghost conditions as well. The constraint of Eq. (2) implies that the reflected and the direct signals are received at the same carrier phase at the receiving antenna.

Subjective effects of echoes for the NTSC system have been studied by Lessman.[8] Close-in echoes cause considerable subjective distortion in color due to phase disturbance caused at the color subcarrier frequencies. The distortion caused to digital data is of a different nature. The impairment due to echoes, in general, has a more detrimental effect on data reliability than on picture degradation. At the present data rate, pulses occur every 180.56 ns. Echoes of this order of delay will cause impairment to data. For the purpose of simulation, echoes equivalent to 122, 125 and 128 sample intervals (sample interval = 15.516 ns) were chosen to generate 10, 25, 40 and 50% ghosts. Delays were selected such that the ghost adds to the main signal at half, three-quarters and full bit periods of data pulses so that the relative impairment can be evaluated as a function of relative position of the ghost. Table 2 lists the performance at these delays for a 50% ghost. The bit errors listed were counted in a bit stream of 288 bits. Table 3 lists the performance at different ghost intensities.

The ghosted data of different intensities was decoded using decoders with 25, 35 and 50% eye openings, which represent an ensemble of practical receivers.[5] The number of errors in each case for the data stream of 288 bits is shown in Table 4.

TABLE 2

Eye Height and Decoding Errors as a Function of Ghost Position

Decoding Eye Height = 25%

		50% Ghost			
Ghost Delay	Delay in Bit Periods	Eye Height	Indeterminate Bits	Parity Errors	Bit Errors
1.8930	10.48	14.82	26	8	14
1.9395	10.74	4.42	37	11	21
1.9861	10.99	18.77	15	7	8

TABLE 3

Eye Height and Decoding Errors as a Function of Ghost Intensity

Ghost Delay = 1.9395 μs = (10.74 bit periods)
Decoding Eye Height = 25%

Ghost Intensity (%)	Eye Height (%)	Bit Errors	Indeterminate Bits	Parity Errors
50	4.42	21	37	11
40	24.95	0	1	0
25	42.70	0	0	0
10	76.85	0	0	0
0	94.44	0	0	0

TABLE 4

Decoding Errors for Different Decoder Eye Heights

Ghost Delay = 1.9395 μs (10.74 bit periods)

	50% Ghost			40% Ghost		
Decoder Eye Height (%)	Bit Errors	Indeterminate Bits	Parity Errors	Bit Errors	Indeterminate Bits	Parity Errors
25	21	37	11	0	1	0
35	40	79	18	14	26	10
50	58	113	18	40	79	18

ADAPTIVE DEGHOSTING OF TELETEXT DATA

The subject of adaptive deghosting for television has been investigated by several authors.[7,9,10] Ciciora,[6] in his tutorial paper on the subject, examines various deghosting configurations and discusses their practical and theoretical constraints. Despite its limitations, the most acceptable deghosting method to date is a configuration using a feed-forward transversal filter which operates on a baseband signal. Various considerations in this general approach will be discussed in this section and their performances compared for decoding a ghosted Teletext signal.

A general block diagram for an adaptive deghoster (equalizer) is shown in Figure 3. By adjusting the taps appropriately for a particular ghost situation, the transfer function of the transversal filter can be made to equal the inverse of the transfer function of the system that created the ghost. The tap weight adjustment is achieved by comparing a portion of received test signal with the ideal version of this signal, which has been regenerated or stored in the receiving equipment. The difference, or the error signal "ε" shown in Figure 3, is used to determine and adjust tap weights. This method requires that a reference signal "training waveform" be included in the signal format so that it can be identified and compared with the locally generated ideal signal deterministically.

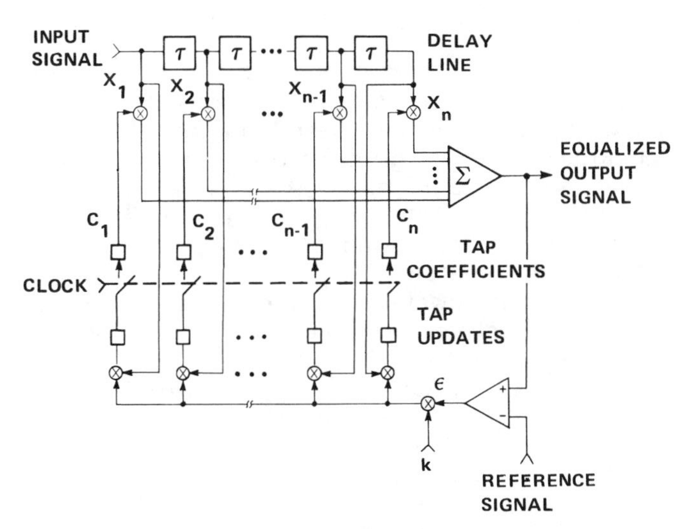

Figure 3. Block Diagram of Adaptive Equalizer

The requirements for the transversal filter are for it to have a flat spectral distribution across the TV band, a period free of signal variation before and after the waveform so that ghosts can be isolated and a shape such that its time location can be determined relatively accurately. Several combinations of signals can be considered as candidates for this purpose. Considering the deghosting hardware, the training waveforms and their usage can be divided into two categories. In the first category, a local regeneration of the ideal training waveform is not required in the deghosting subsystem. Some specific characteristics of the received signal in the existing format can be used to identify the presence and characteristics of ghosts. In the second category, training waveforms have well-defined shapes and these need to be regenerated locally in the receiving equipment as a reference signal (see Figure 3). This makes the hardware system more complex. Also, waveforms from this category will most likely have to be legislated for insertion in the vertical interval at specific locations. The Broadcast Television Systems (BTS) committee of EIA is looking into possible test signals for ghost cancelling.

Training Waveforms

Training waveforms are used in a deghosting system to adjust tap weights for a transversal filter. Since these are part of an actual signal, they carry "signatures" of ghosts distorting the signal. Once the system is trained using these waveforms and the tap weights are adjusted for a particular ghost situation, the signal can be passed through the filter for deghosting.

Considering the first category of training waveforms, the initial transition in the vertical serrations after the equalizing pulses (corresponding to line three in field two) was used as the training waveform for equalization. The system considers any transitions other than the one described as the ghosts and tries to cancel them. Figure 4 shows the training waveform and its spectral distribution. The waveform contains insufficient energy at the higher end of the TV spectrum but satisfies the other training waveform requirements.

Candidates for the second category of test signal for training waveform include: (a) windowed or truncated sinX/X pulse, and (b) 2T raised cosine pulse. The existing sync and burst waveform can also be used for this purpose and its ideal version can be regenerated at the receiving end.

Figure 5 shows a sinX/X waveform which has been truncated to include seven cycles (corresponding to about 25 sample points)

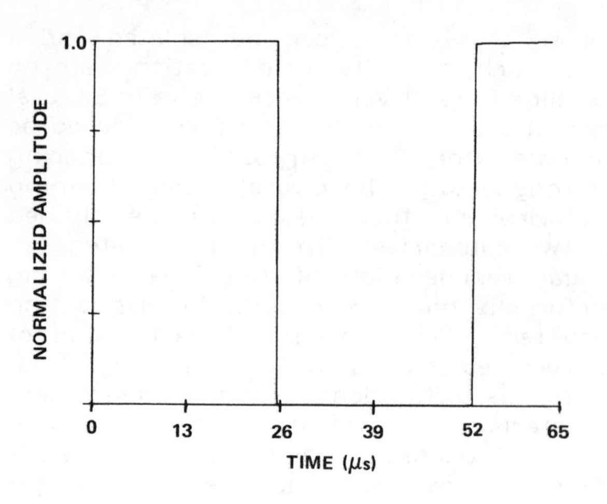

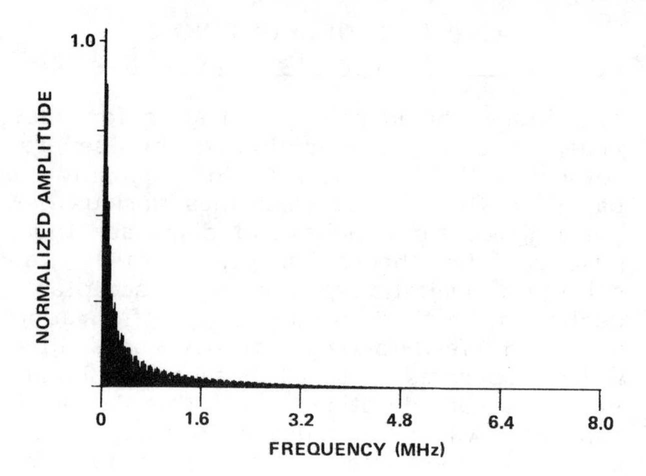

Figure 4. Vertical Serration Transition of Line Three and Its Spectral Distribution

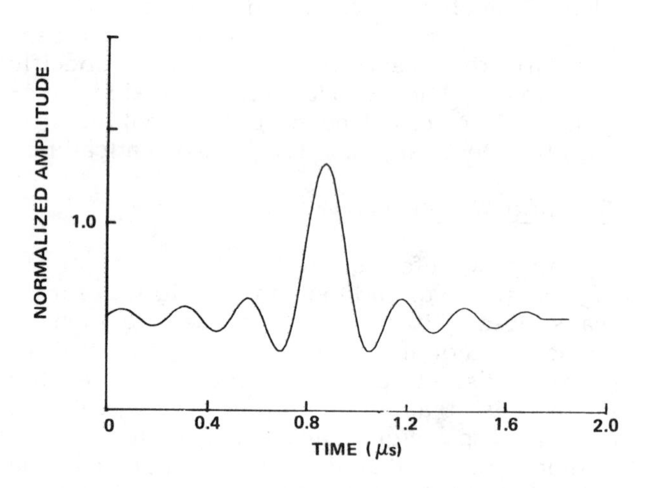

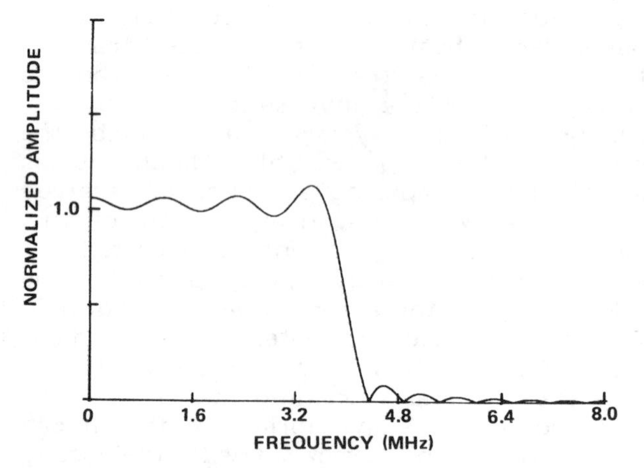

Figure 5. Truncated sinX/X Waveform and Its Spectral Distribution

of the waveform sampled at four times the subcarrier frequency. The normalized spectral distribution of this waveform shows fairly uniform distribution of energy in the full television band of 4.2 MHz. Such a waveform could easily be generated in the receiving equipment to provide an ideal reference signal.

The other candidate waveform, a 2T raised cosine pulse and its spectrum, is shown in Figure 6. This pulse has relatively little energy at the higher end of the TV spectrum and negligible energy above 4.0 MHz. Figure 7 shows the sync and burst signal and its spectral distribution. This also shows little energy in the higher end of the spectrum. The advantage this waveform offers is that it is available for usage at this time and no special TV line need be dedicated for its use.

The adaptive deghosting system of Figure 3 was simulated using truncated sinX/X, 2T raised cosine, and sync and burst training waveforms. The simulated system used a 128-tap transversal filter, the input signal being one of the ghosted training waveforms sampled at four times the subcarrier frequency. The tap weights were adjusted by comparing the equalized output of the transversal filter with the ideal version of the training waveform (reference signal). The respective tap coefficients, C_i's, were changed in every field in proportion to the product of error signal ε and the signal component X_i at the tap location (i). Thus, the updated tap weight, in every field, is given by:

$$C_{i,(t+1)} = C_{i,t} + k\varepsilon X_{i,t} \qquad i = 1, n \qquad (4)$$

where n is the number of taps, (t+1) refers to the time after the update, and t refers to the time before the update.

After the tap weights have been adjusted by running the training waveform for many

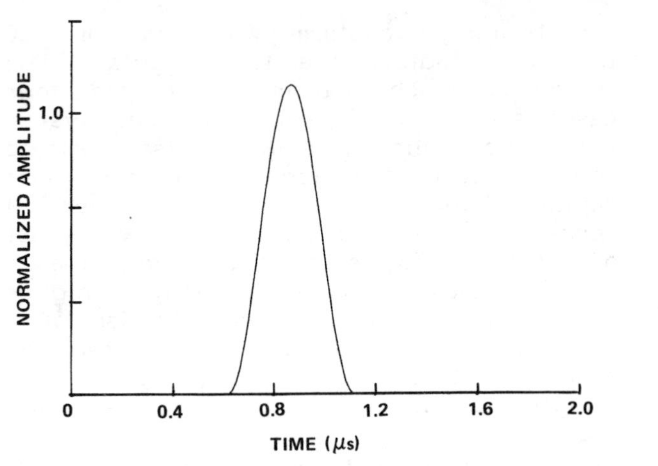

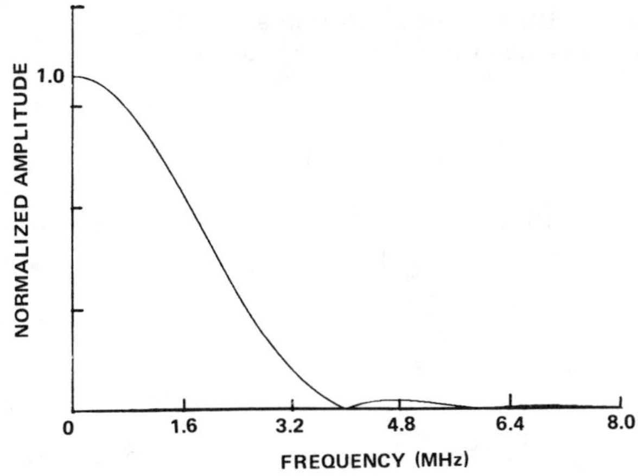

Figure 6. 2T Raised Cosine Pulse and Its Spectral Distribution

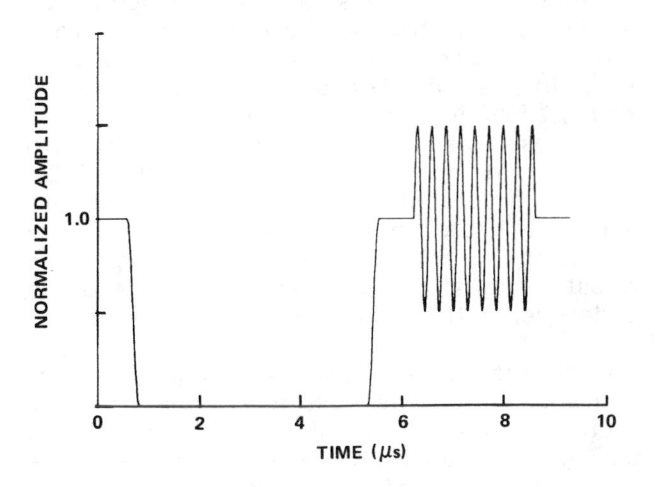

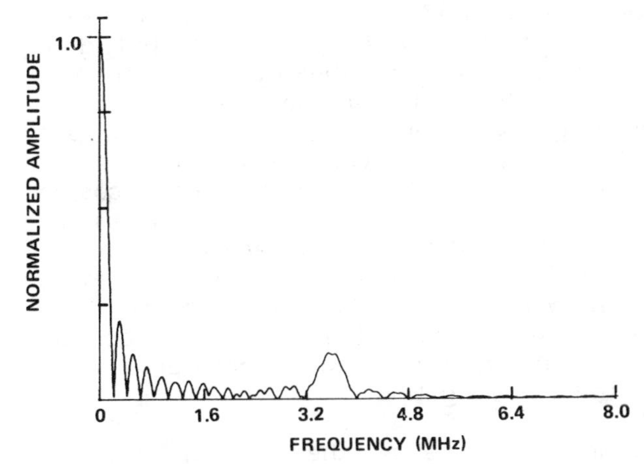

Figure 7. Sync and Color Burst Waveform and Its Spectral Distribution

passes, the Teletext signal was passed through the system and the deghosted output was decoded and the data analyzed.

The deghosting system using the vertical serration transition as the training waveform was similar in concept. In this case the reference signal (which corresponds to a dc level) is compared with the differentiated version of the equalized signal, and in every field, the tap coefficients are incremented or decremented by a fixed value. The updated tap weights in this case are given by:

$$C_{i,(t+1)} = C_{i,t} \pm \Delta \qquad (5)$$

where Δ is a constant value for the tap weight update.

The relative performance of the deghosting systems was compared by calculating the signal-to-noise (S/N) ratio of the output data and plotting eye diagrams for each case. Before this data was generated, system parameters were approximately adjusted in each case to get the maximum value of output S/N ratio. Table 5 shows that the deghosting systems using truncated sinX/X and vertical serration training waveforms improve the data eye by nearly the same degree. The eye heights for the deghosted data using the 2T raised cosine pulse and the sync and color burst were somewhat less than the eye height obtained with the sinX/X pulse. The table also lists the S/N ratio improvement observed in each case by running the respective training waveforms through the deghosting system.

The output signal-to-noise ratio for the data was calculated by computing:

$$(S/N) = 10 \log \frac{\sum_{i=1}^{k} \left[(S_{ref})_i \right]^2}{\sum_{i=1}^{k} \left[(S_{rec})_i - (S_{ref})_i \right]^2} \qquad (6)$$

where

$(S_{ref})_i$ is the ith sample of reference data

$(S_{rec})_i$ is the ith sample of received data

and k is the total number of data samples

and is also listed in Table 5. A close examination of the eye height and output data S/N ratio shows that the two values are not very closely related in the case of different training waveforms. Looking at the output S/N ratio performance, one would expect a better eye height with a 2T raised cosine pulse than was obtained. A close examination of the output data showed larger overshoots in this waveform than were present in the waveforms deghosted using the truncated sinX/X training pulse. The correlation between these two parameters needs further investigation.

In an actual system, the training waveform will continue to adjust the transversal filter taps in every field until these show no further change. In our experiment, the the training waveform was run for 500 passes to adjust the tap weights. The convergence characteristics of the four cases are shown in Figure 8. The system using the truncated sinX/X pulse seems to converge to its final tap settings much more rapidly than the system using the vertical serration transition. Figure 9 shows a plot of the final tap weight distribution in each of the cases. The generation of higher order ghosts, which is a limitation of a feed-forward transversal filter, can be easily identified in these tap weight distribution plots.

To demonstrate the improvement caused by deghosting systems, Figures 10, 11 and 12 show eye diagrams and data signal plots of a 50% ghosted signal, a deghosted signal using the truncated sinX/X and a deghosted signal using the vertical serration transition, respectively. The eye height of 4.4% in the ghosted signal becomes 73.14% and 74.59% in the two cases. The data in each case was decoded using a decoder with a 50% eye opening. The number of bit errors, indeterminate bits and parity errors are also listed in these figures. It is interesting to note that in each case the deghosted signal was decoded error-free using a decoder with a 50% eye height.

For completeness, Figure 13 shows the performance of a deghosting system for the picture signal. In this case, the vertical serration transition was used as the training waveform for an 80% ghosted signal. The deghosted output in this case showed very little residual ghost.

TABLE 5

Comparison of Deghosting System Performance for Different Training Waveforms

Ghost = 1.9395 ns 50%
Ghosted S/N Ratio = 9.37 dB
Ghosted Signal Eye Heights = 4.4%

Training Waveform	Number of Passes for Convergence	Eye Height After Deghosting	Output Data S/N Ratio (dB)	S/N Ratio Improvement of the Training Waveform (dB)
Truncated sinX/X	90	73.14	20.68	24.78
2TRaised Cosine	200	64.83	21.41	24.16
Sync and Color Burst	500	68.90	17.04	39.61
Vertical Serration Transition	160	74.59	21.42	32.10

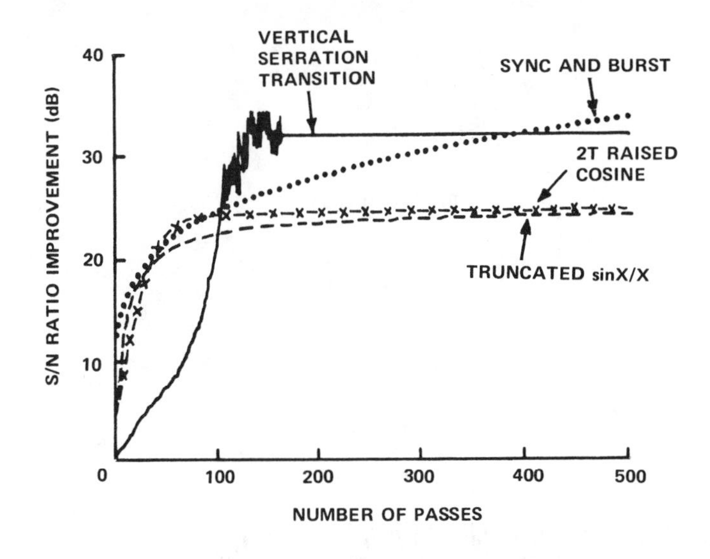

Figure 8. Convergence of Deghosting Algorithms (S/N ratio is computed using the training waveform as the data)

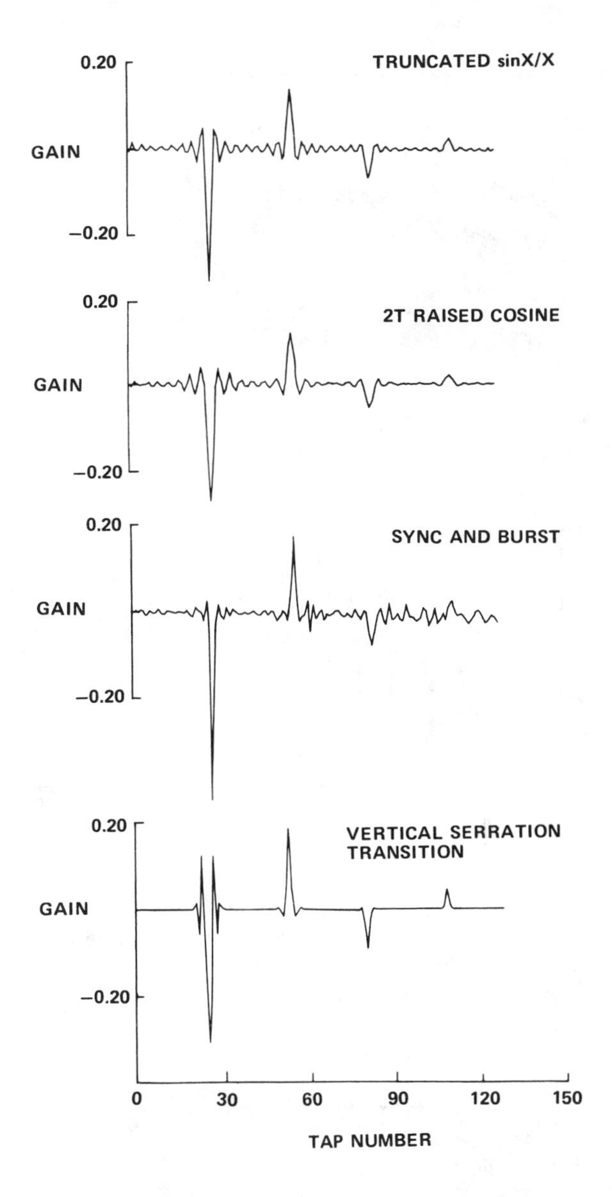

Figure 9. Tap Weight Distribution

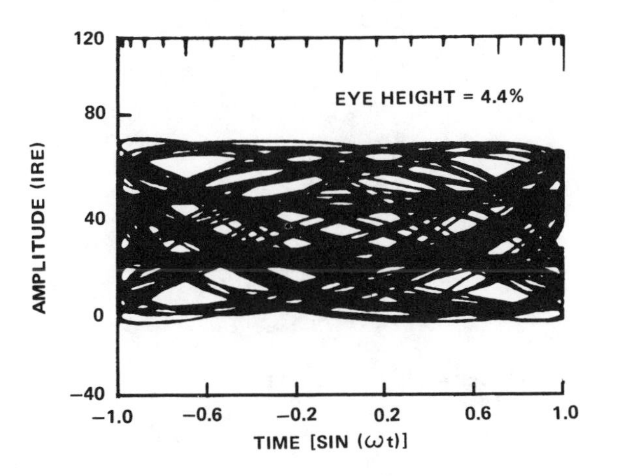

58 BIT ERRORS
113 INDETERMINATE BITS
18 PARITY ERRORS

DECODING THRESHOLD = 50% EYE HEIGHT

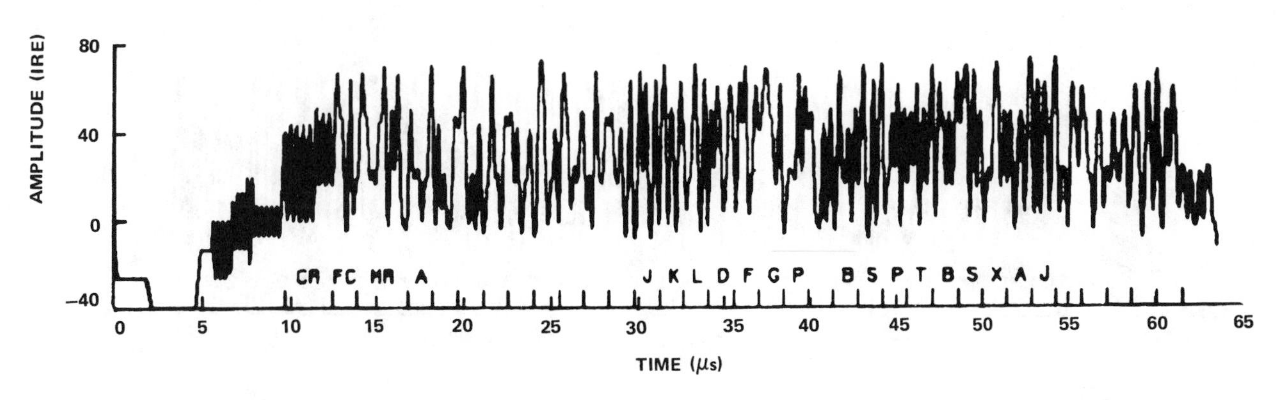

Figure 10. Eye Diagram and Data Waveform for a 50% Ghosted Signal

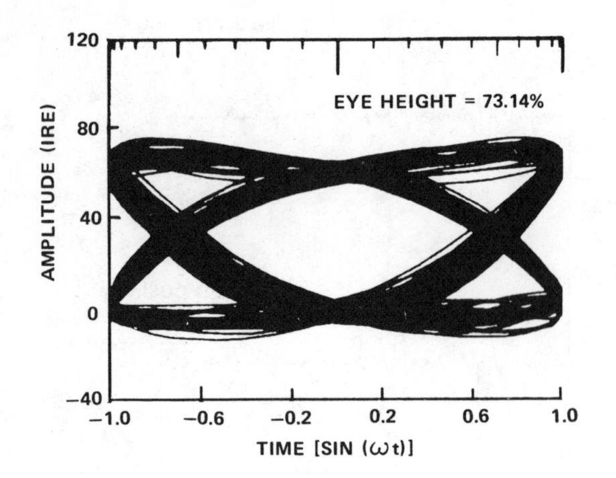

DECODING THRESHOLD = 50% EYE HEIGHT

0 BIT ERRORS
0 INDETERMINATE BITS
0 PARITY ERRORS

Figure 11. Eye Diagram and Data Output of Deghosting System Using Truncated sinX/X Training Waveform

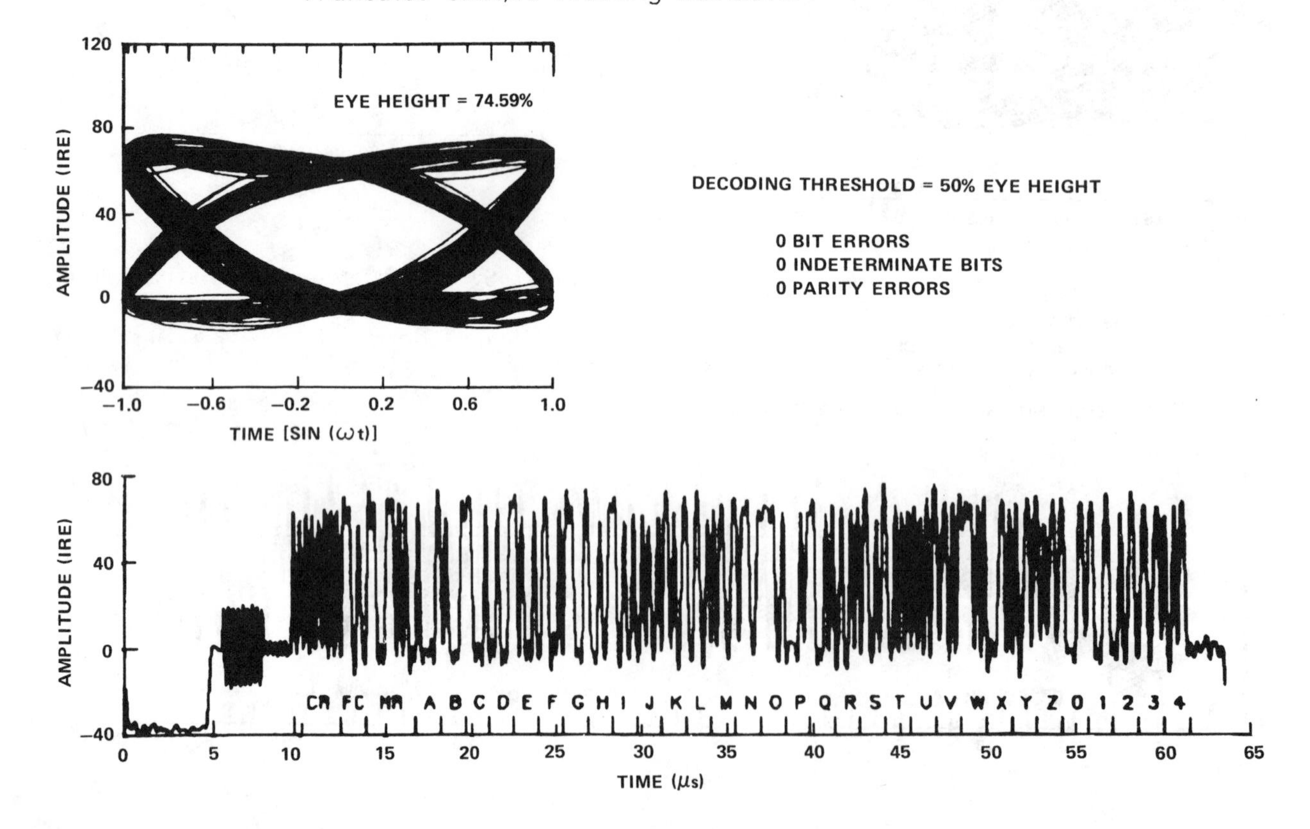

Figure 12. Eye Diagram and Data Output of Deghosting System Using Vertical Serration Transition

(a) Ghosted

(b) Deghosted

Figure 13. SMPTE Slide with 80% Ghost at 2.5 μs and its Deghosted Reproduction

DISCUSSION

Ghosts cause distortion to Teletext data as well as to picture video. It would be desirable to design a deghosting system in a TV receiver which improves the performance of both signals, even though the level of improvement may be different in the two cases. This requires that the training waveform legislated for the purpose is suitable for both of the signals.

All four of the adaptive methods of equalization discussed in this paper demonstrate potential as practical systems. It must be noted, however, that constraints layed out earlier in this paper represent serious impediments to the practical realization of a general-purpose deghoster for video or data. Namely, the prerequisite for synchronous demodulation to adequately define the nature of the ghost(s) in the baseband, and the necessary legislation for inserting a standard waveform are both problems that must be given attention.

Assuming these difficulties have been overcome, what single method and training waveform might produce the best performance? Our experiments indicate that the technique using the vertical serration transition gives the best eye height and has good convergence (see Table 5). In addition, this signal is currently available in the NTSC signal format and would require no special legislation. However, the spectrum of this training waveform [Figure 4(b)] shows the lack of high frequency information, which could limit the ability of this method to deghost a Teletext signal. Even though no such limitation was observed, it is felt that it would provide less than adequate improvement for a picture containing highly saturated colors.

The two training waveforms whose spectra are more nearly flat over the video bandwidth are the truncated sinX/X and 2T raised cosine pulses. Table 5 shows that there is no clear agreement concerning which of these waveforms offers better performance. Although the truncated sinX/X shows significantly better eye height (73.14% vs. 64.83%), the 2T raised cosine gives a marginally better signal-to-noise figure (21.41 dB vs. 20.69 dB). This conflict can be partially explained by the fact that the eye height calculation considers values of the signal only at bit cell boundaries, whereas the signal-to-noise calculation is made over the entire video line. Signal excursions within the bit cell are thus ignored by the decoder.

There has been considerable interest in optimizing the window characteristics for the sinX/X waveform. Its performance for different windowing characteristics need be evaluated before final conclusions are drawn. This subject needs further study.

The sync and burst training waveform is available at the present time for the purpose and it may not be difficult to regenerate it in the receiver once its format is more rigidly defined. Eye height improvement obtained with this training waveform was adequate (4.4% to 68.9%), and with it one would expect a better performance for pictures with saturated colors than obtainable with the vertical serration transition approach. On the other hand, there is reason to believe that the broader spectral content of a sinX/X or 2T raised cosine pulse will ultimately enable better ghost reduction for the picture video. Our limited experiments also indicate that the systems using these waveforms show greater insensitivity to loop parameters and thus may offer higher system stability.

SUMMARY

The problem of multipath distortion in video signals is one that affects both picture information and, in the case of data broadcasting services, digital information. This paper has presented experimental results pertaining to this problem. Through computer simulation it has been shown that moderately ghosted Teletext data cannot be received and decoded reliably. However, by applying many common methods of baseband equalization, even grossly distorted signals can be decoded error-free. The advantages of deghosting are not limited solely to highly ghosted or distorted signals. In a modest ghosting situation, it would offer higher quality pictures and enable improved noise margin for decoders. For these and other reasons, this type of equalization demonstrates much promise for improving the reception of Teletext under multipath conditions.

APPENDIX A - DATA GENERATION

All data used in the simulation experiments were generated by the scheme outlined in Figure A-1. First, the binary bit stream was created in accordance with the Teletext Specifications.[1] The specifications call for two bytes of clock run-in, followed by a one-byte framing code and two bytes of Hamming-code protected addresses for magazine and row numbers. The remaining 31 bytes were available for the ASCII characters comprising the message for that line.

The bit stream is next converted to a nonreturn-to zero (NRZ) signal with 15 ns rise and fall times. This signal is then passed through a special shaping filter for obtaining the pulse shape which minimizes intersymbol interference. The spectrum $P(\omega)$ of such a data pulse is found to have 70% cosine roll-off characteristics.[3]

The shaping filter characteristics, $S(\omega)$, are derived by dividing the spectrum of the gate function, $G(\omega)$, into $P(\omega)$, that is:

$$S(\omega) = \frac{P(\omega)}{G(\omega)} \tag{A-1}$$

The spectrum of the gate pulse, data pulse and shaping filter characteristics are shown in Figures A-2(a), (b) and (c), respectively.

Finally, the shaped data signal is added to a blank video line. The timing is such that the seventh peak of the clock run-in sequence arrives 12 μs after the falling edge of the horizontal sync pulse. A video line produced by this method is shown in Figure 2 in the main body of this paper.

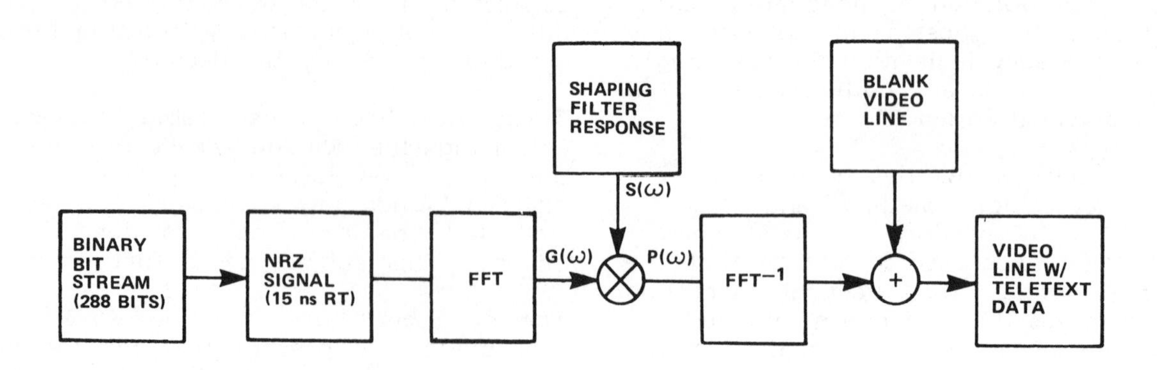

Figure A-1. Flow Diagram for Generation of Teletext Data

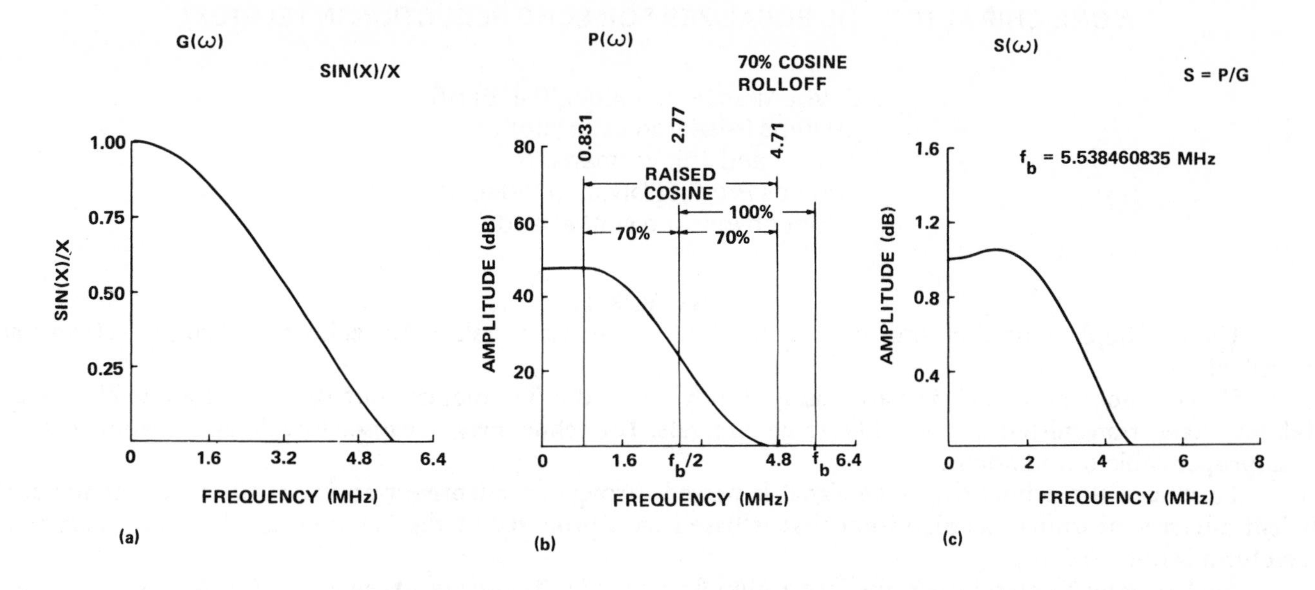

Figure A-2. (a) Spectrum of Desired Pulse Shape; (b) Spectrum of Raw Data (Gate Function); and (c) Response of Shaping Filter

ACKNOWLEDGMENT

The authors acknowledge the contribution of Stuart McGraw in generating and running several computer programs reported in this paper. Helpful discussions with John Blank on the subject and the permission of the management of the General Telephone and Electronics Corporation to publish this work are also gratefully acknowledged.

REFERENCES

1. Broadcast Teletext Specifications (September 1976), published by BBC, IBA and BREMA.

2. Swiss PTT Research Department Report GTI 094 (Sept. 1978), "Data Transmision Within the Television Channels."

3. M.J. Kallaway and W.A. Mahadeva, "Optimum Transmitted Pulse-Shape," BBC Research Report RD 1977/15.

4. Teletext Field Tests in Bavaria (August 1976), published jointly by BBC, IBA and Institute for Rundfunktechnik.

5. L.A. Sherry, "A Summary of ORACLE Field Trials," Teletext Transmission Working Group, Note 51, Independent Television Authority.

6. W. Ciciora, et al., "A Tutorial on Ghost Cancelling in Television Systems," IEEE Trans. on Consumer Electronics Vol. CE25, No. 1, pp. 9-44 (February 1979).

7. H. Thedick, "Adaptive Multipath Equalization for TV Broadcasting," IEEE Trans. on Consumer Electronics Vol. CE23, No. 2, pp. 175-181 (May 1977).

8. A. Lessman, "The Subjective Effects of Echoes in 525 Line Monochrome and NTSC Color Television and Resulting Echo Time Weighting," Journal of SMPTE Vol. 81, pp. 907-916 (December 1972).

9. E. Arnon, "An Adaptive Equalizor for Television Channels, IEEE Trans. on Comm. Tech. Vol. Com. - 17, pp. 726-734 (December 1969).

10. K. Yamamoto, N. Yamaguchi and M. Miyata, "Ghost Reduction Systems for Television Receivers," IEEE Trans. on Consumer Electronics Vol. CE23, pp. 327-344 (August 1977).

A ONE-CHIP AUTOMATIC EQUALIZER FOR ECHO REDUCTION IN TELETEXT

J.O. Voorman, P.J. Snijder, P.J. Barth
Philips Research Laboratories
and J.S. Vromans
Philips Product Division Video
Eindhoven, The Netherlands

ABSTRACT

Theory, implementation and test results of an automatic equalizer for echo reduction in Teletext are described.

Short echoes need not be particularly annoying in the TV picture, but they can completely disturb Teletext data, transmitted in the field retrace intervals. The echoes may be caused by reflections from obstacles or improper cable terminations.

To reduce the echoes the video signal is passed through an adaptive transversal filter. Automatic coefficient adjustment utilizes an algorithm that is based on a property of the Teletext signal. A specific training waveform is not needed.

Analog implementation permits integration on one chip. Therefore, a Laguerre filter (cascade of tapped first-order phase-shifters), analog correlators and tapweight multipliers have been developed. Coefficient values are stored as charges on integrated dielectric capacitors.

1. INTRODUCTION

The introduction of new services such as Teletext and Viewdata requires the implementation of many new functions. This stimulates process development, circuit design as well as systems research to solve the accompanying problems. One of these problems concerns the improvement of the reception of Teletext data.

Teletext data are transmitted in certain lines of the video field retrace intervals. There are many variations in system and data format. Although the problem of improving the reception and its solution are common to all of them, we shall fix our attention on the U.K. Teletext standard in order to simplify the explanations given below. For clarity some details of the transmission format are set out in figs. 1 and 2.

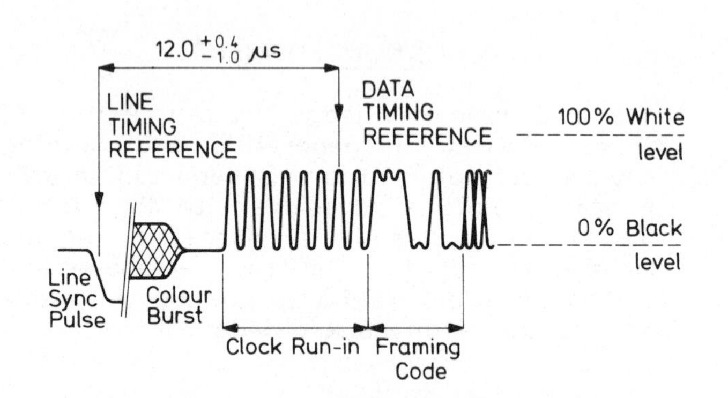

Fig. 2.
Each Teletext line starts with two clock run-in bytes (10101010 – 10101010) and a byte for word synchronization (framing code – 11100100).

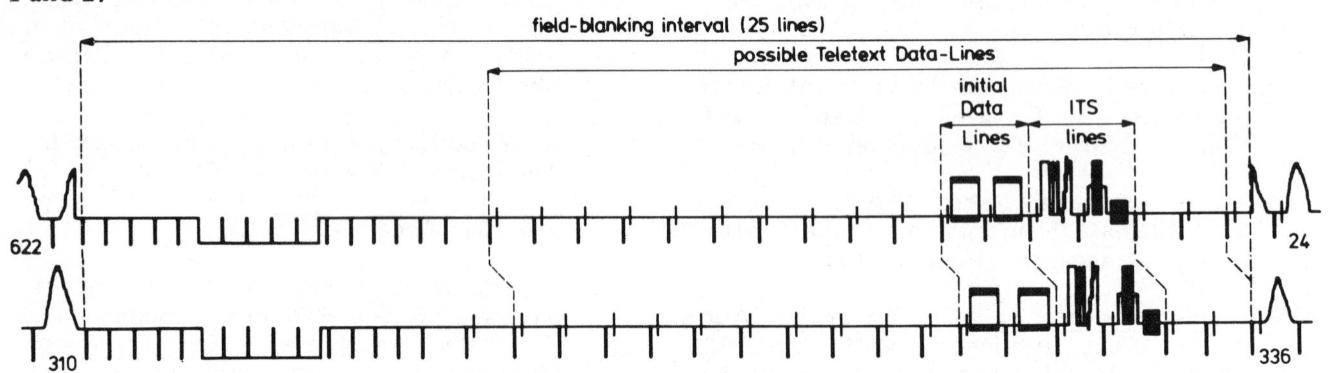

Fig. 1.
The field blanking intervals with frame synchronization and equalizing pulses on the left. The free lines (on the right) are used for test signals and Teletext data. Each line with Teletext comprises 42 eight-bit bytes preceded by clock run-in and framing code bytes (see also fig. 2). The bit length is 144 ns.

Reprinted from *IEEE Trans. Consum. Electron.*, vol. CE-27, pp. 512-529, Aug. 1981.

It is a reasonable precondition to prescribe for the introduction of a new service, such as Teletext, that it should work satisfactorily in all cases where there is a proper reception of the TV picture. By checking the above requirement against the various impairments that can influence the TV picture and the Teletext data, it is found that almost all impairments first degrade the TV picture to an unacceptable level before Teletext breaks down. This applies to noise (fig. 3), co-channel interference and almost all echoes.

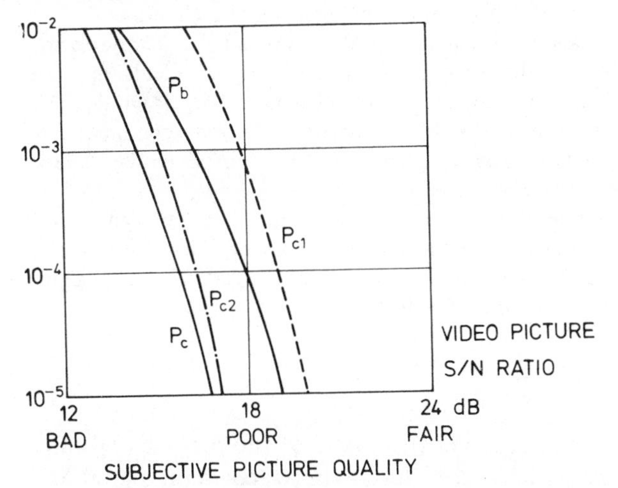

Fig. 3.
Error rates and (white Gaussian) noise.
Pb = bit error rate, Pc1 = character (byte) error rate at first reception, Pc2 after a second reception (correction of bytes with an incorrect parity), Pc3 = Pc4 = . . . = Pc.
The corresponding subjective picture quality has been taken from Brand and Hügli (1972).
The figure illustrates that, with increasing noise, first the TV picture degrades to poor, before the Teletext breaks down.

Only short echoes can seriously disturb the Teletext before they are found to be annoying in the TV picture. Short echoes are subjectively masked by the main signal (fig. 4). Many people even appreciate some occurence of short echoes. Short echoes can accentuate contours in the picture.

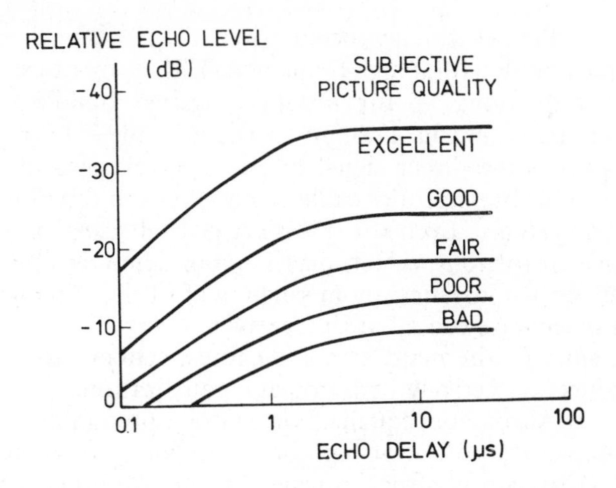

Fig. 4.
The subjective influence on the TV picture of a single (in-phase) echo as a function of relative level and the distance from the main signal. Short echoes are significantly less annoying.

On the other hand, relative echo levels of the order of magnitude of −6 dB completely close the eye pattern of the Teletext data, almost independently of the echo delay. Combinations of short echoes – for instance, a positive and a negative echo close to each other – can be even less visible in the TV picture and more effective in destroying the Teletext data.

This leads us to the conclusion that echoes with a delay of up to 0.5 – 1 μs can cause erroneous Teletext reception, whereas they need not be very annoying in the TV picture.
This region of echoes needs echo reduction if Teletext is to be satisfactorily introduced.

More details on impairments and their influence on the reception of Teletext can be found in the recent literature (Klingler 1980, Ishigaki et al. 1980). A weighting of (multiple) echoes for estimating the degree of visibility (annoyance) in the TV picture has been proposed by Yamazaki and Endo (1980).

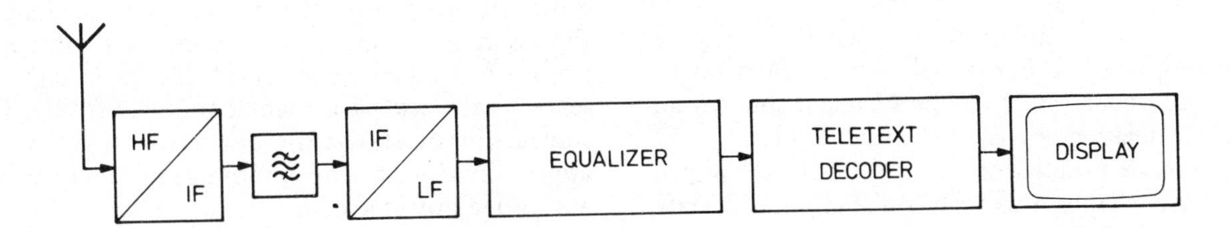

Fig. 5.
Equalizer arrangement.
In view of the implementation, equalization is done preferably at baseband (video), that is behind the (quasi-synchronous) demodulator at the input of the Teletext decoder.

Echoes usually occur when the TV signal is still modulated (at high frequencies). They are caused by reflections – for instance, against buildings, mountains and in a valley to the ground –, by reception of a direct signal and a repeater signal together or by improper cable terminations in distribution systems. Even the TV receiver itself can introduce distortions, which may be considered as short echoes, for instance, by misalignment of the antenna or inaccurate tuning of the receiver.
In spite of the many possible causes, echoes can be reduced effectively by automatic equalization.

Automatic equalization is done preferably at baseband (and not at high or intermediate frequencies) to simplify the implementation. We prefer to have the arrangement in fig. 5.

An automatic equalizer consists of a variable filter, the coefficients of which are controlled according to some criterion.

At the time that we started with the equalizer design several questions had still to be answered:
- which criterion could be used for automatic coefficient control?
- did we need a special training waveform?
- which implementation was most cost-effective?

We were helped by the knowledge already available on automatic equalization in data transmission (for a review, see: Proakis 1975) and some early studies on equalization for TV signals.
More recently Ciciora et al. (1979) have presented a tutorial overview of problems and possibilities of removing echoes from TV pictures.
Computer simulations on the influence of amplitude, group delay and multipath distortions and the remedy via automatic equalization using several types of training waveform have been reported for Teletext (Goyal and Armfield 1979) and for TV pictures (Goyal et al. 1980).
Various implementations for equalizers for the TV picture are being studied (Makino et al. 1980, Ishigaki et al. 1980).

2. ECHOES BEFORE AND AFTER DEMODULATION

Before demodulation any echo is just a delayed copy of the original signal. Distributed echoes may occur. They can be seen as a sum (or an integral) of delayed copies of the main signal.

After demodulation of the VSB AM signal echoes have shapes that can substantially deviate from the original. Two effects occur.
This is illustrated in fig. 6. Fig. 6a shows an echo-free pulse. In fig. 6b we see the same pulse with an echo (video signal).

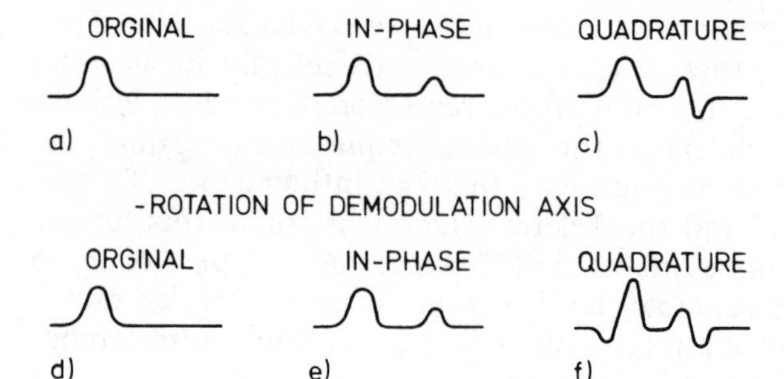

Fig. 6.
Hf echoes: their appearance at baseband.
TV signals are broadcast as VSB AM signals. Demodulation relative to the carrier of the main signal yields in-phase echoes in cases where the echoes have delays of whole numbers of wavelengths (b). A quarter of a wavelength extra or less yields quadrature types of echo (c). Of course, mixed echoes can occur.
In practice the carrier is derived from the TV signal itself (with echoes). With quadrature types of echo the phase of the (regenerated) carrier wave is influenced. The main signal shows a distortion as well (f).

The echo preserves the shape of the main signal only in cases where the delay (between main signal and echo) is equal to a whole number of hf carrier wavelengths (we have an in-phase echo).
An extra half wavelength gives a negative echo of the same shape (180 degrees echo). In the cases of extra 1/4 or 3/4 wavelengths quadrature types of echo result (fig. 6c; the echo is demodulated by the carrier of the main signal, i.e. with a 90 or 270 degrees wrong phase for the echo).

Apart from the different appearance of the demodulated echoes, the phase of the regenerated carrier wave is influenced by quadrature echoes. This results in a distortion of the main signal too (fig. 6f).
The two effects are of the same order of magnitude.

An equalizer has to be able to remedy the above effects.
Although the echoes may have shapes which are quite different from the main signal, we shall nevertheless treat them as being composed of combinations of delayed versions of the main signal (this saves us the use of a twin (complex) filter). For instance, fig. 6c shows the quadrature echo, to a first approximation, as the combination of a positive and a negative in-phase echo.
In fig. 6f we see a distortion of the front part of the main signal. This implies that the equalizer should be able to handle this type of very short 'pre-echoes'.

Referring to the above we should keep in mind that an echo equalizer for Teletext does not need to be as accurate as an equalizer for the TV picture. The eye pattern of the data has to be opened. Residual distortions do not bother us too much.
The lower accuracy requirements also permit us to make the adaptation of the equalizer very fast and to cope with time-variable distortions.
The latter may come from the demodulator.
Also moving echo patterns can be tracked. Non--stationary echo patterns have been found in cases where an antenna was moving (in the wind).
Continuously in-phase echoes became quadrature type echoes, etc. Fig. 7 shows that the speed of the reflection point does not need to be very high for the above effects.

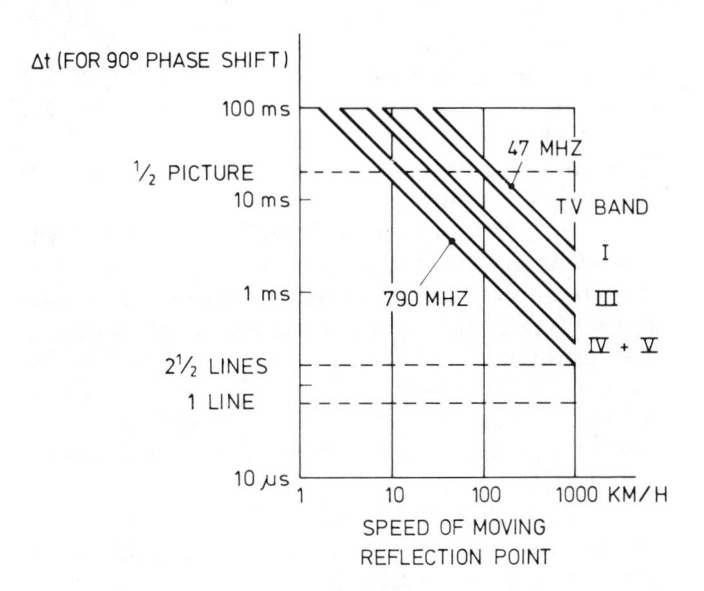

Fig. 7.
The influence of a moving reflection point.
Let the velocity of the reflection point be v. In Δt the maximum variation of the echo delay is: 2 v Δt (source, reflection point and receiver are at a straight line). A phase variation of 90 degrees corresponds with a variation of the echo delay of 1/4 wavelength (= c/4f, c = velocity of light). In the figure this Δt has been depicted as a function of the normal speed of the moving reflection point. The frequency f is parameter. A speed of the reflection point of 10 km/h can already cause strong variations in the echo patterns, even within one frame interval.

3. THE USE OF A TRANSVERSAL AND/OR A RECURSIVE VARIABLE FILTER

Having seen the video signal as consisting of a main signal, possibly disturbed by some leading and lagging echoes, which are copies of the main signal, we shall have to remove the echoes.

An optimum filter for the complete removal of lagging echoes is a recursive filter. This is indicated for one echo in fig. 8.

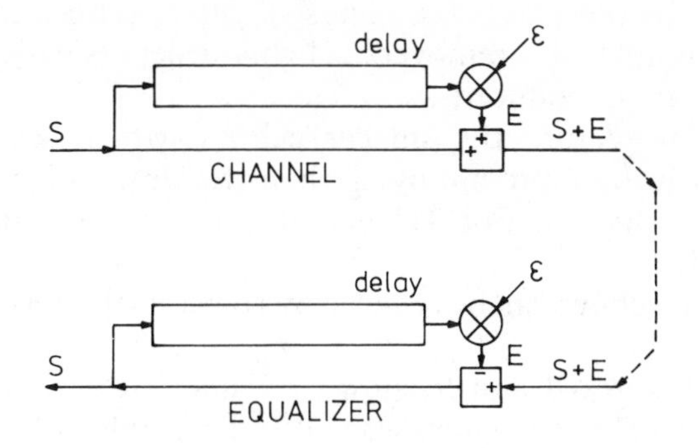

Fig. 8.
Principle of echo removal by recursive filtering.
The transmitted signal s, the channel with main signal path and echo path giving the combined signal s+e and the arrangement for echo removal (below) are given. The latter is a recursive filter.
The (lagging) echo is removed completely.

The recursive equalizer filter is the inverse of the channel filter. Lagging echoes can be removed ideally.
Observe that we define the main signal as the largest 'echo'. This also defines which echoes are lagging and which are leading. It guarantees that the loop in the equalizer is stable (echo e is smaller than the main signal s, by definition).

Instead of a recursive filter it is possible to use a transversal filter. The principle of echo reduction with transversal filtering is depicted in fig. 9.

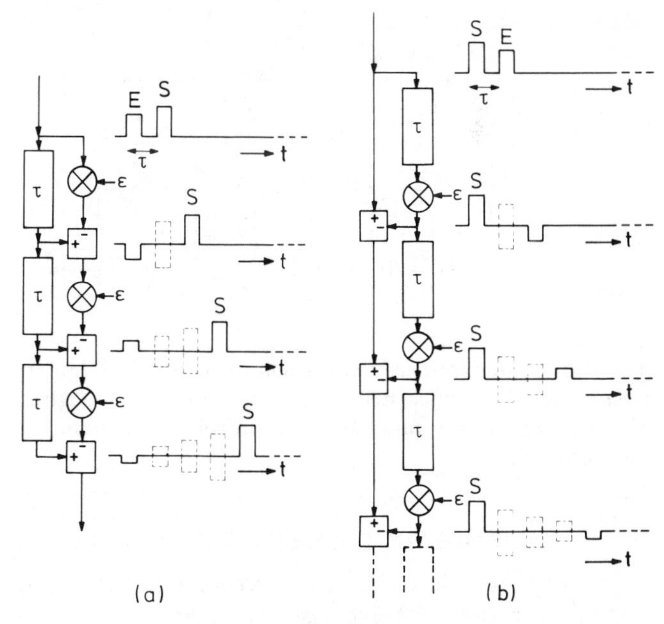

Fig. 9.
Principle of reduction of a leading echo (a) and a lagging echo (b) by transversal filtering. The echoes are removed but second-order (smaller) echoes are introduced, and these, in turn can be removed, etc. A higher-order echo always appears. Which one it is, depends on the filter length.

Unlike in the case of a recursive filter, echoes are not completely removed. Higher-order (smaller) echoes are introduced.
In a TV picture the smaller echoes with a larger delay can be more annoying than the larger original one (see fig. 4). For Teletext the situation has improved.
Leading echoes can be reduced by transversal filtering (fig. 9a).

Figures 8 and 9 show principles of echo reduction. The principles can also be employed together, for instance: transversal filtering for leading echoes and recursive filtering for lagging echoes.

For echo reduction with Teletext we do not need very high accuracy. We have chosen therefore what leads to the simplest arrangement: a transversal filter equalizer.

In practice, however, we do not know the delay of the echo(es) in advance (as is assumed in fig. 9). We should split up, the delay line in as many pieces as we expect to have echo delays. This would require an infinite number of taps. However, equalization of a bandlimited signal only requires a bandlimited impulse response of the filter. In the case of pure delay sections the impulse response is equal to a series of signal samples. The samples do not need to be finer spaced than is prescribed by the Nyquist criterion (e.g. 0 – 5 MHz: tap spacing ⩾ 100 ns). Interpolation yields the overall response. We arrive at a transversal filter arrangement of the type as in fig. 10.

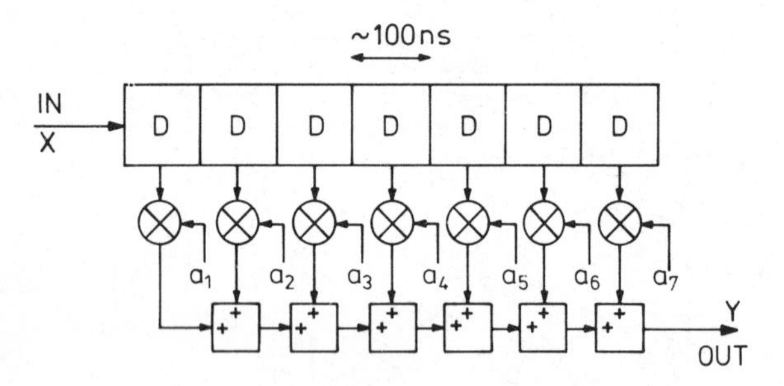

Fig. 10.
Transversal filter arrangement for echo reduction in Teletext. It consists of a tapped delay line with variable tapweights. The tapweights are automatically controlled according to some criterion.

4. AUTOMATIC TAPWEIGHT CONTROL

Let us describe the automatic tapweight control for the simplest case: we ignore noise and assume to have ideal elements.

Although one is used to a delay line with pure delays the description below is generalized to the use of any type of 'delay' section. This is done in view of the implementation with first-order phase-shift sections (Laguerre filter).

Let the response at tap k of the filter to a unit impulse at the input be $h_k(t)$. The response at tap k to an input signal x(t) is $x(t)*h_k(t)$, where the $*$ sign denotes a convolution. The output signal of the filter is:

$$a_1x(t)*h_1(t) + a_2x(t)*h_2(t) + a_3x(t)*h_3(t) + \ldots,$$

where a_k $(k = 1, 2, \ldots, n)$ are the variable tapweights.
For our convenience we add a term a_0 (some DC) to the above in order that the data levels of the Teletext become symmetrical with respect to zero.
Hence,

$$y(t) = a_0 + a_1x(t)*h_1(t) + a_2x(t)*h_2(t) + \ldots,$$

$$= \underline{a}^T\underline{x}, \qquad (4.1)$$

where $\underline{a}$ and $\underline{x}$ are column vectors with components $a_0, a_1, a_2, \ldots, a_n$ and $1, x(t)*h_1(t), x(t)*h_2(t), \ldots, x(t)*h_n(t)$ respectively.
T denotes a transposition, which is needed to form the inner vector product.

For the automatic coefficient control the echoes have to be measured in one way or another. It is common practice to apply an echo-free reference signal $y_r(t)$ to the equalizer during a start-up period. The difference $y-y_r$ (that is: the echoes and the influence of the equalizer filter together) is reduced to zero by a proper control of the coefficients.
Let us describe the convergence during the start-up phase.

We shall assume that – by a correct choice of the number of coefficients and of the impulse responses $h_k(t)$ – the echo-free reference signal can be written as:

$$y_r(t) = \underline{a}_r{}^T\underline{x}, \qquad (4.2)$$

We wish to know whether $\underline{a}$ converges to $\underline{a}_r$, or more precisely, what is the type of control law for the filter coefficients that should be chosen so as to obtain convergence?

The squared error $(y-y_r)^2$ can be minimized. The squared error is a time-variable quadratic function of the coefficients a_k $(k = 0, 1, \ldots, n)$. It has a steady zero minimum for $\underline{a} = \underline{a}_r$.
Minimization by way of gradient methods (steepest descent) controls the coefficients according to:

$$\frac{d}{dt}a_k = -f\frac{\partial}{\partial a_k}(y-y_r)^2 = -2fx_k(y-y_r), \qquad (4.3)$$

where f is a positive loop gain.

In vector notation:

$$\frac{d}{dt}\underline{a} = -2f\,\underline{x}(y-y_r). \qquad (4.4)$$

Let us see what happens with the distance $|\underline{a} - \underline{a}_r|$ as a function of time.

More specifically we consider:

$$\frac{d}{dt}|\underline{a} - \underline{a}_r|^2 = 2(\underline{a} - \underline{a}_r)^T \frac{d}{dt}\underline{a}. \qquad (4.5)$$

N.B. $|\underline{a} - \underline{a}_r|^2 = (\underline{a} - \underline{a}_r)^T(\underline{a} - \underline{a}_r)$.
With the coefficient control (4.4) and expressions (4.1) and (4.2) is obtained:

$$\frac{d}{dt}|\underline{a} - \underline{a}_r|^2 = -4f(y-y_r)^2. \qquad (4.6)$$

This means that – with this particular control law (stochastic iteration algorithm) – the coefficient vector $\underline{a}$ converges to the coefficient vector $\underline{a}_r$ of the reference signal as long as the output signal of the filter y differs from the echo-free reference signal y_r.

Observe that, instead of minimizing the squared error, we could have minimized the absolute value of the error signal: $|y-y_r|$ equally well.
This would have lead to a coefficient control:

$$\frac{d}{dt}\underline{a} = -f\underline{x}\,\mathrm{sign}\,(y-y_r). \qquad (4.7)$$

The latter control (sign algorithm) leads to:

$$\frac{d}{dt}|\underline{a} - \underline{a}_r|^2 = -2f\,|y-y_r|, \qquad (4.8)$$

with similar convergence properties.
Different procedures for minimization can be chosen.

The above description is the description of non-autonomous equalization (during start-up), i.e. of an equalizer to which, in addition to the input signal x(t), a reference signal $y_r(t)$ is fed. After convergence, the reference signal might be derived from the (echo-free) output signal of the equalizer (tracking mode). The control has become autonomous.

External supply of a reference signal, however, presumes the existence of some known training waveform. This can be a pulse in a free line, a step, a truncated sin(x)/x waveform, a pseudo-random data line, etc. The clock run-in and framing code bytes might also be used for the purpose. In all cases the reference waveform has to be synchronized with the actual waveform.

When we found out that the automatic tap-weight control (during tracking and during start-up) could be performed uniquely on a property of the Teletext data itself, the latter method was chosen. Apart from the advantages that we need no space and no standardization for an extra training waveform and that synchronization is almost automatical, there is the added advantage of continuous tracking (during the data lines).

Variations in the echo pattern are corrected for continuously.

The criterion is based on the following.

- Echo-free Teletext is essentially a two-level (binary) signal.
- With echoes there will be more than two levels (provided that the data are sufficiently random)!

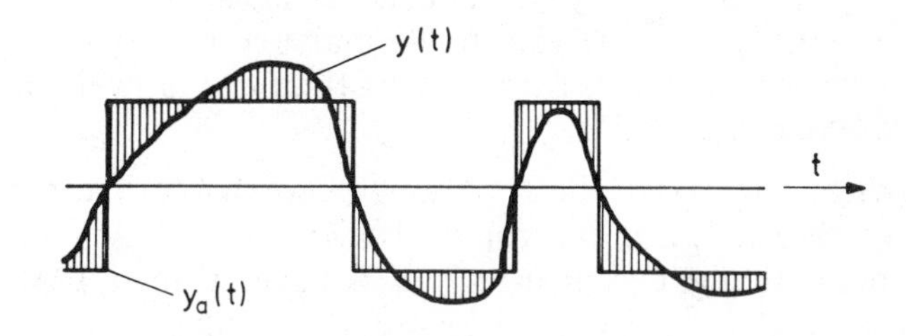

Fig. 11.
A filter output waveform y(t).
The auxiliary waveform $y_a(t)$ is more 'data-like'. It is used as a reference signal. The two signals should converge to the echo-free two-level data signal.

A filter output signal waveform y(t) is depicted in fig. 11. On the basis of the output signal we shall define an auxiliary signal $y_a(t)$, which is more 'data-like':

$$y_a(t) = a\,\mathrm{sign}(y). \qquad (4.9)$$

The auxiliary signal is a binary signal but its amplitude 'a' – an extra coefficient which has to be included in the automatic control – and its zero-crossings will initially be different from those of the echo-free signal. The auxiliary signal will be employed as a reference signal.
We have an autonomous control during tracking and start-up.

The above description of convergence (by way of equations (4.2)-(4.8)) cannot be followed for the autonomous equalizer behaviour.

Equation (4.3) modifies to:

$$\frac{d}{dt}a_k = -f\frac{\partial}{\partial a_k}(y-y_a)^2 = -2fx_k(y-y_a), \qquad (4.10)$$

where we have ignored the dependence of y_a on a_k.

It can be shown that this is permitted by the argument that $(y-y_a)^2$ is continuous over the zero-crossings of y.
Equation (4.4) modifies to:

$$\frac{d}{dt}\underline{a} = -2f\underline{x}(y-y_a). \qquad (4.11)$$

In equation (4.5): $\frac{d}{dt}\underline{a}_a = \underline{0}$ is not correct.

Instead, with (4.11) and (4.1), we arrive at:

$$\frac{d}{dt}|\underline{a}|^2 = -4f\,y(y-y_a) = -4f|y|(|y|-a). \qquad (4.12)$$

There are three equilibrium states: $y = 0$, $y = +a$ and $y = -a$.

In the state $y = 0$ all coefficients a_k $(k = 0, 1, \ldots, n)$ vanish. To make certain that we never arrive in this state, one of the filter tapweights is fixed: $a_m = 1$.
With one fixed filter tapweight we have that the automatic gain control coefficient $a > 0$. At moments that y(t) is near to zero we have (from (4.12)): $\frac{d}{dt}|\underline{a}|^2 = +4f|y|a \geqslant 0$. This equilibrium point has become unstable.

Two stable equilibrium states are left: $y = +a$, $y = -a$. Three types of behaviour may occur: $y = +a$, $y = -a$ or y(t) is alternately $+a$ and $-a$. The two first cases are latch-up states. The third is the required mode.

To force the equalizer adaptation into the third mode the zero-crossings of y(t) are monitored. When, during a prescribed period of time (e.g. during half a data line), there are no zero-crossings, the coefficient 'a' is set at zero. The DC correction a_0 is controlled according to (eq. (4.10)):
$\frac{d}{dt}a_0 = -2fy(t)$, which means that y(t) becomes DC-free (symmetry restoration). There are zero-crossings again and the echoes are reduced in the correct mode.

In the case that the loop gain is too high limit-cycles may occur ($|y|$ oscillates around 'a'), but by a decrease of the loop gain they are damped. The filter output signal converges to the auxiliary signal. The filter output signal becomes binary. It will be echo-free (provided that the input signal – and therewith y(t) – has sufficiently random zero-crossings).

Let us introduce a short-hand notation for the error signal:

$$e(t) = y(t) - y_a(t). \qquad (4.13)$$

Expression (4.10) of the tapweight control can be written for the DC correction (a_0) and for the filter tapweights (a_k) separately. There may be different values for the loop gains. Minimization of the squared error also yields an expression for the control of the AGC coefficient (a):

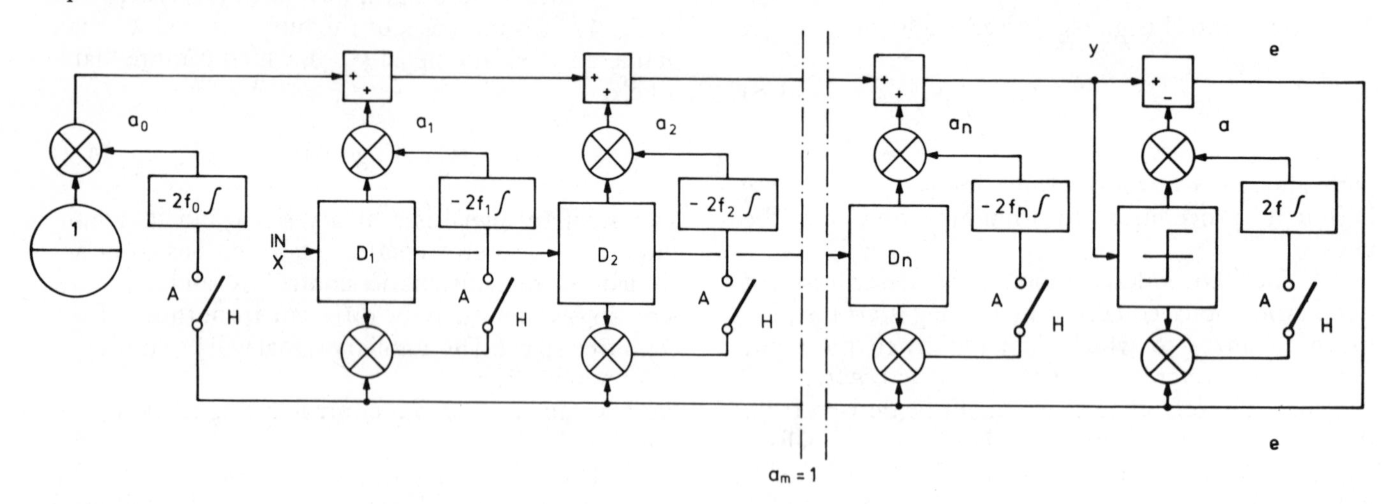

Fig. 12.
Arrangement of the automatic equalizer.
We have a cascade of delay sections. Tap signals are provided on the upper side as well as on the lower side of the delay sections. The tapweights of the filter ($a_1, a_2, \ldots, a_n$), the level corrector (a_0) and the automatic gain control (a) are automatically controlled (with the exception of filter tapweight a_m, which has a fixed value). The filter output signal (y) is sliced. The error signal is used for control of the coefficients according to expressions (4.14). Adapt/hold switches are closed during the adaptation of the coefficients (on the Teletext data). They are opened to hold the coefficient values fixed during the remaining parts of the TV frames.

$da_0/dt = -2f_0 e(t),$

$da_k/dt = -2f_k e(t) x_k(t), \; k = 1, 2, \ldots, n, \; k \neq m,$

$(a_m = 1),$

$$da/dt = +2f\, e(t)\, \text{sign}\,(y), \qquad (4.14)$$

The expressions above have been implemented in the equalizer arrangement in fig. 12.
One of the filter tapweights is fixed. All others are adapted when the adapt/hold switches are closed (during the Teletext data). The coefficients are kept constant when the switches are open (during the main part of the TV frames).

In the case where in 4.14 the error signal e(t) is replaced by a sign (e) we arrive at a sign algorithm. Different (and intermediate) solutions are conceivable.

In the case of additive noise the coefficient control continuously tries to obtain a two-valuedness of the output signal of the filter, and an optimum open data eye is thus strived for. The criterion takes into account the combination of noise and echoes.

Non-linear saturation characteristics can be included in the above description. Proper saturation characteristics will not cause overflow latch-up or limit cycles.
An example of saturation is where the values of all filter tapweights are limited to being smaller than 1 (a_m = 1). This forces the main signal to pass via tap m. The echoes are reduced via the other taps. This is consistent with the definition of the main signal (as being the largest 'echo'), removes ambiguity and significantly improves the speed of adaptation of the equalizer.

In conclusion, an automatic coefficient control has been designed that is free from unwanted states or limit cycles. It removes the echoes from the Teletext signal, provided that the data is sufficiently random. An auxiliary signal is employed as a reference signal.

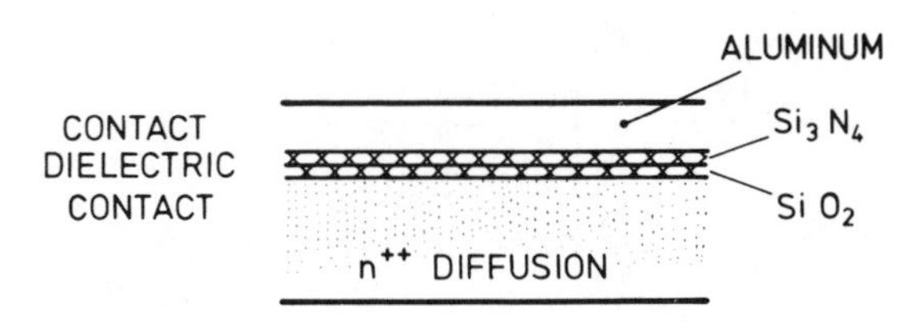

Fig. 13.
Cross-section of an integrated dielectric capacitor.
The capacitors are made on the epitaxial layer with diffusions of the bipolar process. A thin silicon oxide/nitride sandwich is the dielectric. The top electrode is of the first interconnection layer.

The method can be seen to be related to decision feedback, but with natural sampling (asynchronous). The arrangement is a non-linear analog filter.

An analysis of the autonomous control problem for the case of all sampled signals has recently been published by Mazo (1980).

5. IMPLEMENTATION PRINCIPLES

Although the advantages of digital circuits (e.g. accuracy, stability and reproducibility) are well recognized, chip area and supply power consumption still prevent their use for a competitive solution in this case. The high speed, the many multiplications and the three to four orders of magnitude difference between the smallest and the largest time constants in the equalizer (time constants of the delay per section and of the control loops, respectively) pose serious problems with regard to the simplicity of digital implementation.
Analog circuitry on the other hand, is more difficult to design, but we have obtained a one-chip solution with an acceptable supply power consumption (< 500 mW).

We started from a common bipolar process with two layers of interconnection. It was extended with integrated dielectric capacitors. High quality capacitors are needed for the integrators in the control loops (low leakage; also in view of the hold function).
The construction of the capacitors has been sketched in fig. 13. Capacitor values of the order of magnitude of 1 nF/mm^2 are made with a thin silicon oxide/nitride sandwich as a dielectric. The common diffusions form the lower electrode, and the first layer of aluminum the upper plate. Second-layer interconnections may cross the capacitors.

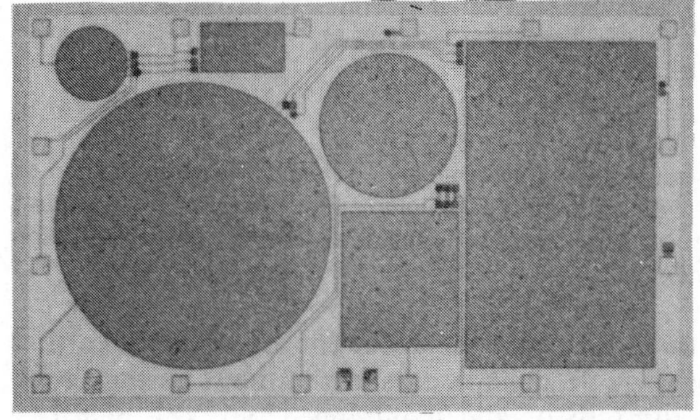

Fig. 14.
Photograph of a test chip with integrated dielectric capacitors.

Fig. 14 shows a photograph of a test chip with capacitors.

Once we have high quality integrated dielectric capacitors, they can be used for the implementation of the delay elements as well. An analog 'delay' line has been made which consists of a cascade of first-order phase-shifters. More precisely, the transfer function to tap k is:

$$H_k(p) = \frac{1}{1+pRC}\left(\frac{1-pRC}{1+pRC}\right)^{k-1}, \quad k = 1,2,..., n. \qquad (5.1)$$

Each transfer function has the same first-order lowpass character, but they have different numbers of allpass sections.

The lowpass character is not only of advantage in the implementation. The transfer functions are Laplace transforms of Laguerre functions (Laguerre filter; see also: Lee 1960). Any causal impulse response with a finite energy can be expanded in a series of Laguerre functions (which are orthogonal on the half axis $t > 0$).

The corresponding electronic allpass type circuits have parallel parts (see, for instance: Holt and Gray 1967, Deliyannis 1969). The parts should be properly matched in order to obtain a frequency-independent allpass gain. The bandwidth should be of the order of magnitude of 5 MHz. Parasitic effects of integrated transistors cannot be ignored in this wide frequency band.
It should be possible to compensate for the above imperfections.
In this respect the circuit principle in fig. 15 is favourable.

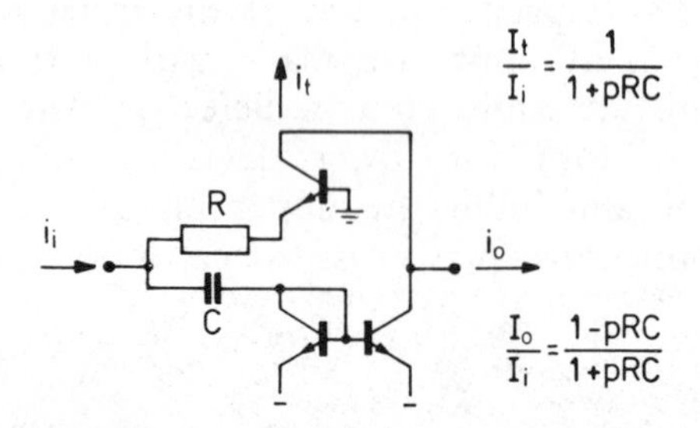

Fig. 15.
Principle of a first-order phase shift cell with indication of a tap for an integrated Laguerre filter. Parasitic effects can be greatly eliminated by adjustment of the gain of the current mirror.

The resistor and the capacitor are virtually earthed. The low frequency part of the input current passes via the resistor and it is tapped (i_t). The high frequency part goes via the capacitor and an inverting current mirror. By adjusting (increasing) the gain of the current mirror a (first-order) compensation can be obtained for various parasitic effects (the gain can be made frequency-independent up to high frequencies).
The principle permits the integration of a delay line with up to some twenty sections.

In addition to the integration (and hold) function and the Laguerre filter we shall consider the multiplication. Multiplication is based on the logarithmic relationship between base-emitter voltage and collector current of bipolar transistors. Addition/subtraction of base-emitter voltages corresponds to multiplication/division of collector currents.
From fig. 16 we have (in the case of identical transistors):

$$(I+i)/_{(J+j)} \cdot (J-j)/_{(I-i)} = 1 \quad :: \quad J = \frac{J}{I}\,i. \qquad (5.2)$$

(I and J are supply currents, i and j are signal currents). The result can be made independent of gradients in transistor properties (on the chip) by proper positioning of the transistors (see, for instance: Gray and Meyer 1977).

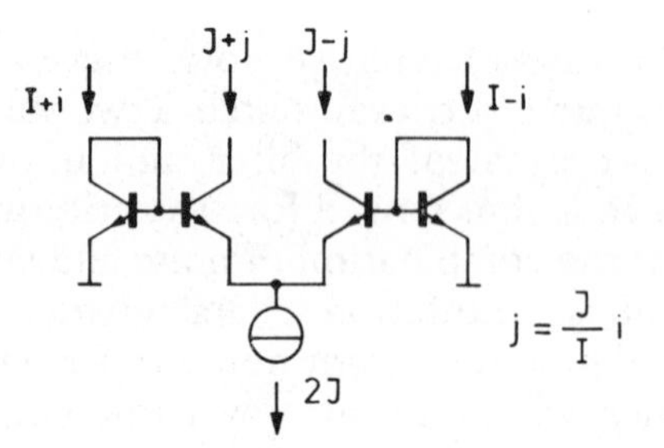

Fig. 16.
Multiplication.
Addition/subtraction of base-emitter voltages of bipolar transistors corresponds to multiplication/division of collector currents, respectively.

6. CIRCUIT DESIGN

Below we shall describe in greater detail the electronic circuit of the adaptive Laguerre filter cell (delay section, tapweight multiplier and control, see fig. 17). The set-up of further circuits is sketched

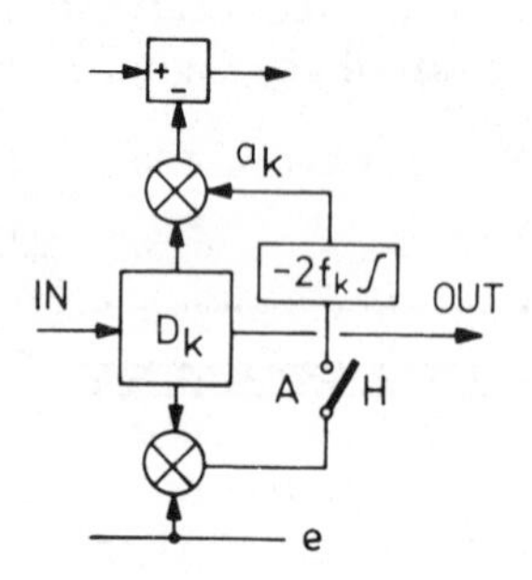

Fig. 17.
Typical adaptive delay cell.
Input and output of the delay element are indicated. The error signal e, multiplied by a tap signal, is integrated when the adapt/hold switch is closed. A second filter tap signal is multiplied by the tapweight. The result passes to the summation line.

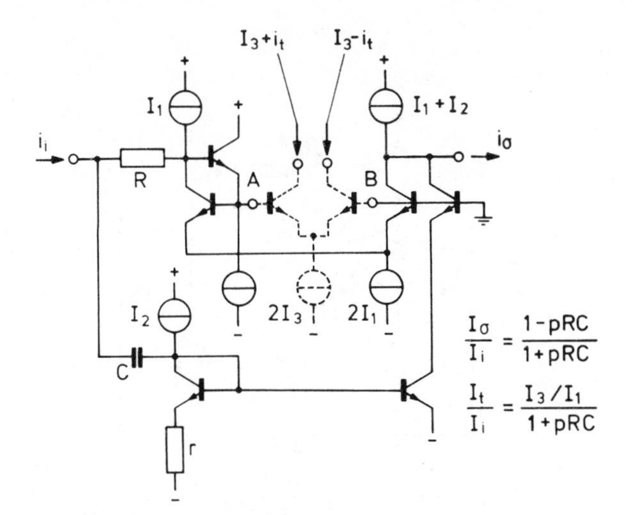

Fig. 18
The Laguerre delay section in greater detail.
A tap has been inserted in the resistive path. Variation of supply current 2 influences the gain of the current mirror in the capacitive path (adjustment for flat gain).

in rough. Let us start from the Laguerre delay section principle in fig. 15. It is further completed in fig. 18. Taps can be connected between points A and B. The final delay cell, as part of the adaptive filter cell, is shown in fig. 19 (left part).

Fig. 19 shows a drawing of the Laguerre delay cell and, in the centre, a tapweight multiplier, which is controlled by the voltage on the (upper) capacitor. On the right is the correlator multiplier. A second (unweighted) tap signal of the filter is multiplied by the error signal (see also the simplified diagram in fig. 20).

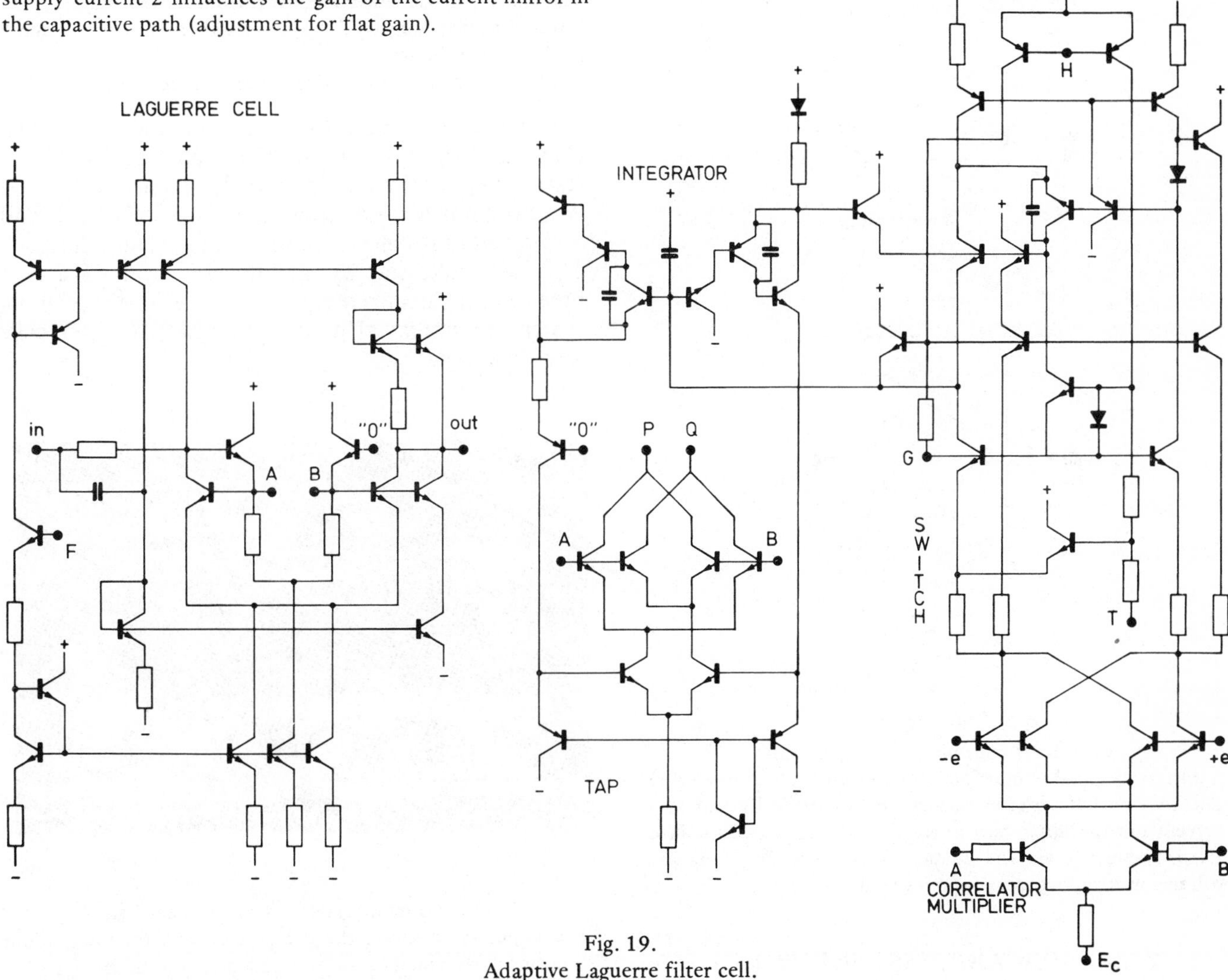

Fig. 19.
Adaptive Laguerre filter cell.
It consists of a Laguerre delay section which is tapped between A and B.
The tapweight multiplier is shown in the centre. P and Q are + and − current summation points for all taps. The multiplier is controlled by a capacitor voltage, which is sensed with the aid of complementary super emitter followers.
The capacitor (integrator) is charged by the correlator multiplier on the right. The adapt/hold switch is controlled by the voltage at point T.

The correlator multiplier and the adapt/hold switch in particular have been designed with care to avoid DC offsets, have a sufficiently wide bandwidth in the multiplier and decrease any influence of the switch on the integrator charge.

Of the other circuits, cell m (with a fixed tap) does not need the righthand part of the adaptive delay cell in fig. 19, whereas the DC corrector does not need the lefthand part.

The slicer should be a very fast one. Its principle is shown in fig. 21. It combines a sensitivity of the order of 1 mV with a high speed.
It has a variable tap for obtaining the auxiliary (reference) signal.

An amplifier gives the signal y/a at the output. It acts as an accurate automatic gain control (AGC). In addition the sliced signal sign(y) can be taken from the equalizer. The latter signal is well suited for feeding to the Teletext decoder.

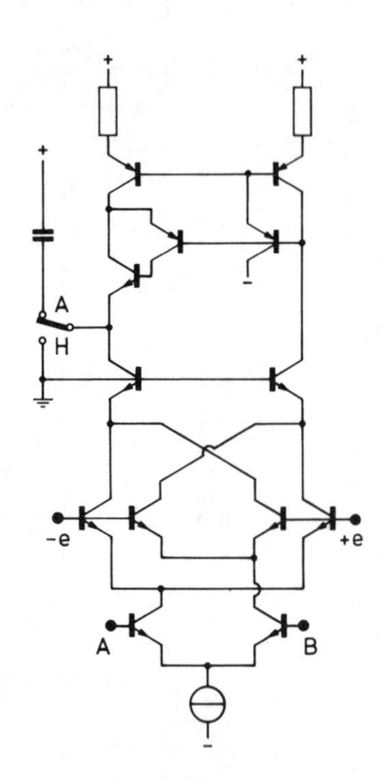

Fig. 20.
Correlator multiplier circuit principle.
A filter tap signal is multiplied by the error signal (− e, + e). The two output currents, which are combined by way of a current mirror (upper part of the figure), can charge the integrator capacitor via the adapt/hold switch. The capacitor voltage, in turn, controls the tapweight.

7. BREADBOARD, TEST SET-UP AND FIELD TRIALS

After the echo removal problem for Teletext had been set out and analyzed some years ago, and equalizer convergence had been examined for various

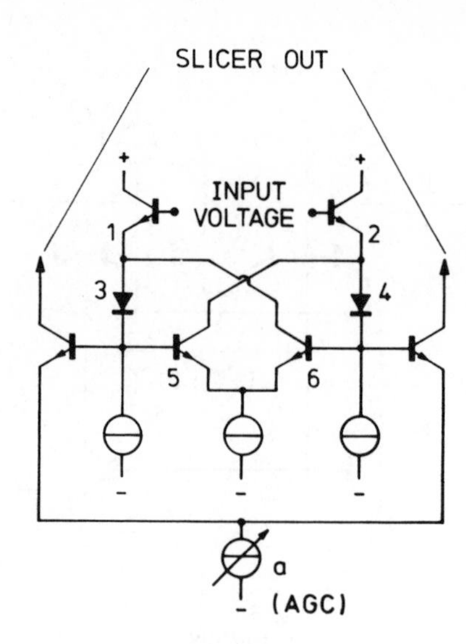

Fig. 21.
Slicer principle.
The emitter diode resistances of the differential stage 1,2 are lowered by the 'negative' diode resistances 5 and 6. 3 and 4 are level-shift diodes.
A variable tap gives the auxiliary signal: a sign (y).

test signals and criteria by way of computer simulations, and the final set-up had been established, we started preparations for the implementation.
A configuration was chosen for the delay cell and optimized in rough with the use of CAD methods.
This was followed by the development of a first breadboard arrangement, which served to show system feasibility. Up to eight adaptive filter cells were inserted.

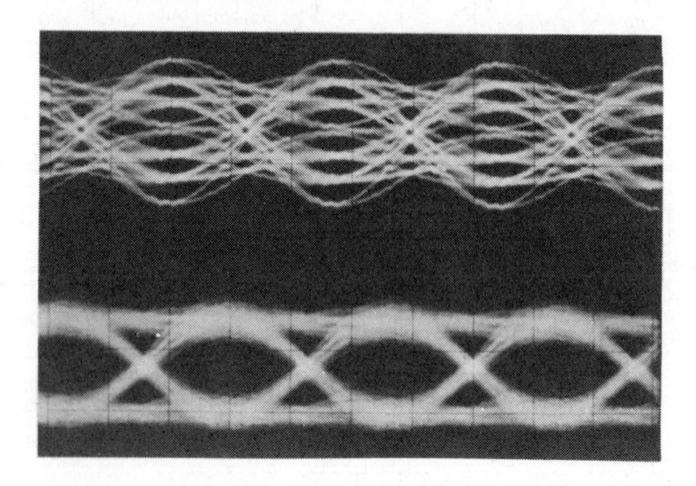

Fig. 22.
Pseudo-random data signal with echoes.
The upper trace shows the eye pattern of a pseudo-random data signal, to which echoes are added (in this case at 144 ns: − 30%, 2x144 ns: + 30% and 3x144 ns: − 30%).
The signal with echoes is passed through a video lowpass filter and fed to the equalizer.
The lower trace shows the eye pattern of the output signal of the equalizer.

During the preparatory study and the computer simulations some test equipment was developed.

Firstly a pseudo-random data generator was made with provisions for echo addition. Positive as well as negative echoes and combinations of echoes can be added at delays of multiples of half the clock period (if running at 7 Mb/s, up to a delay of 2 μs). Results like those shown in fig. 22 are typical.

Secondly a more realistic (but less versatile) test set-up was built (fig. 23).

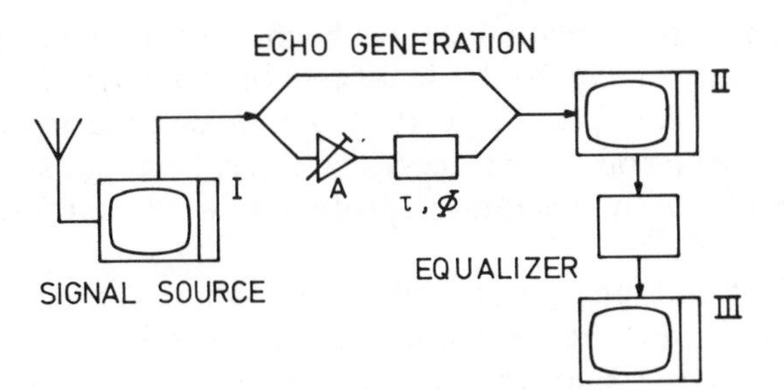

Fig. 23.
Principle of test set-up.
A first TV set is employed to receive TV signals with Teletext. It is used as a signal source for the test set-up.
An echo is introduced (at i.f.) via a piece of coaxial cable (delay: 270 ns). 1/4-wavelength pieces of cable permit the echo to appear as an in-phase, 90, 180 or 270 degrees echo. The relative amplitude is adjustable.
The signal (with echo and modulated to h.f.) is fed to receiver II. It displays results of signals with echoes. Its video signal is passed via the equalizer to the third receiver.
Receiver III can show the improvement. Receivers II and III have Teletext decoders. For a comparison of the performance of receivers II and III the equalizer can be left out of the signal path.

There are three TV receivers.
The first receives TV signals (with Teletext) and is used as a signal source for the test set-up. Signals from Dutch and German broadcast stations have been used. An example of a signal from receiver I is given in fig. 24. The upper trace shows a part of a field retrace interval with some empty lines, two lines with Teletext data and some test signals. The lower trace shows an adapt/hold command. It indicates the adaptation intervals of the equalizer. N.B. the duration of the adaptation intervals can be reduced to one short interval before the equalizer performance breaks down.
The adapt/hold command is to be generated outside the equalizer.

The output signal of receiver I is modulated to i.f. One echo with a delay of 270 ns can be added by way of a piece of coaxial cable. It can be chosen in one of four phases: 0,90, 180 or 270 degrees.
The echo amplitude is adjustable.

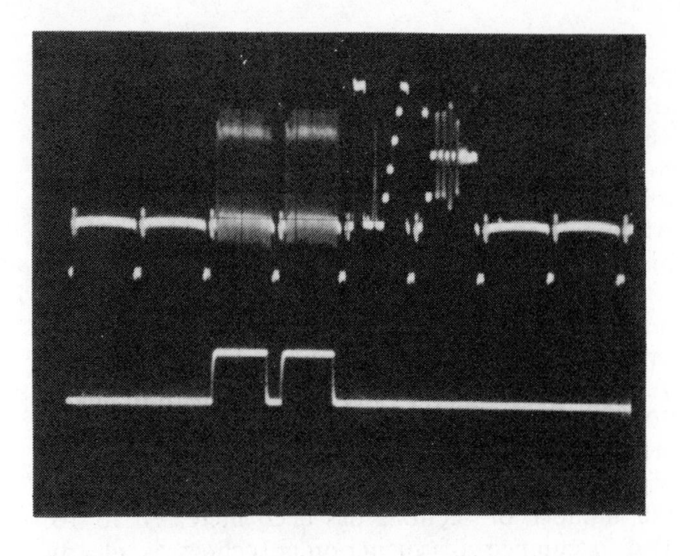

Fig. 24.
Part of a field flyback interval with: some empty lines, two Teletext data lines and some test signals (as received by TV receiver I in the test set-up in fig. 23).
The lower trace shows an adapt/hold command.

The resulting signal (with echo) is modulated to h.f. and fed to the antenna input of receiver II (fig. 23).
Receiver II displays results of a signal with an echo. The influence of the cho on TV pictures (see, for instance, the example of a quadrature echo in fig. 25) and on Teletext can be seen.

Fig. 25.
Part of a TV test pattern, which is corrupted by a quadrature echo (270 degrees, 270 ns, 30%). In-phase echoes and echoes of 90 or 180 degrees are even less visible. They can seriously affect the Teletext reception (see fig. 26).

Fig. 26 gives a photograph of the righthand part of the test set-up in fig. 23. Receivers II and III and our first equalizer breadboard are shown. The same echo that disturbs the picture in fig. 25 disturbs the Teletext which is displayed by the lefthand receiver (II).
After the signal has passed via the equalizer, the picture of receiver III is error-free.

Of the many results obtained from the test set-up, some are presented in figures 27 and 28.
The behaviour of the equalizer on an echo-free Teletext signal and on Teletext signals with in-phase, 90, 180 and 270 degrees echoes is shown. All results are from the equalizer with 7 sections (m = 1, n = 7), with the exception of the results in fig. 28f, where the number of sections has been increased to 10. The occurence of higher-order echoes is clearly visible.
Almost all echoes have a relative amplitude of 45%. In fig. 27f the echo level has been increased to 70%. Echoes as high as this one can still be handled. Signals from the output amplifier as well as sliced output signals are shown. The correction of the frequency spectrum (multiburst) is clearly visible in figures 28c and 28e.

Although the equalizer has been designed for echo reduction in Teletext data, it can also handle the TV picture.
The effect is not a direct improvement.
The time constants in the control loops for the coefficients are of the order of magnitude of one line time. The filter coefficients jitter depending on the data that have passed just before each frame. The first-order echo has largely gone, but time-varying higher-order echoes are seen instead.
The equalizer has not been designed for echo reduction in the TV picture. It has been optimized for Teletext data.

A second equalizer breadboard has been designed with further optimized electronic circuitry and with the possibility of inserting up to 16 adaptive delay cells.
The latter breadboard has been used for field trials, mainly in The Netherlands and in Switzerland.

Reception directly from an antenna as well as distribution via cable systems has been considered. Sites with reception problems were particularly visited.
Not one place has been found where the equalizer decreases the performance of reception. In most cases the performance was significantly improved. At locations with too low a signal-to-noise ratio the equalizer was (of course) hardly of any help. Reception improvements observed went from erroneous Teletext (or even no Teletext at all) to error-free Teletext or reception was found no longer to be as critically dependent on, for instance, antenna alignment.

Fig. 26.
Part of the equalizer test set-up.
The left hand receiver (II) displays erroneous Teletext of an input signal with an echo. After the video signal has passed through the breadboard arrangement of the automatic equalizer (in the front), the right hand receiver (III) displays the Teletext error-free. Oscilloscopes show signals and eye patterns of the data before and after equalization. The line selector defines the adapt/hold command.

Fig. 27.
Results of the equalizer in the test set-up.
The upper traces are input signals. Lower traces are output signals.

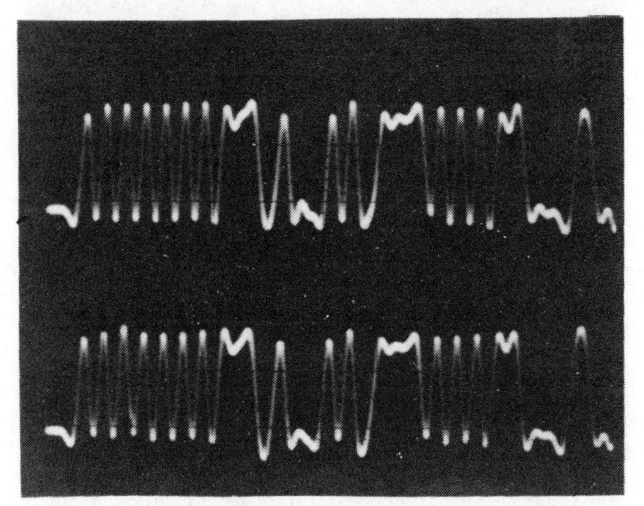

(a) in : echo-free Teletext signal.
out : the equalizer introduces hardly any distortion.

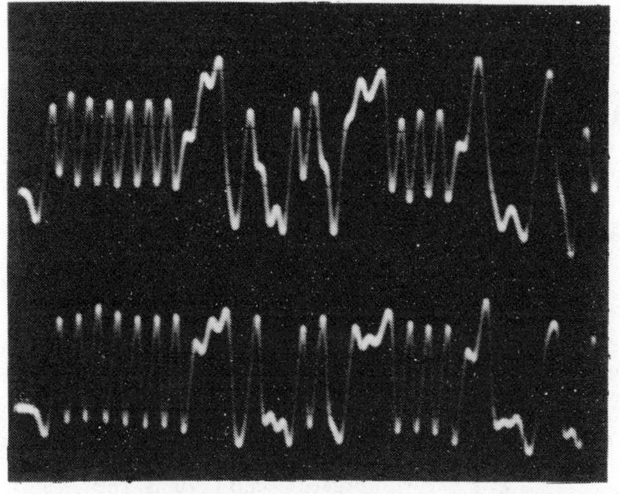

(d) in : Teletext with an echo of 90 degrees (270 ns, 45%),
out : y(t)/a.

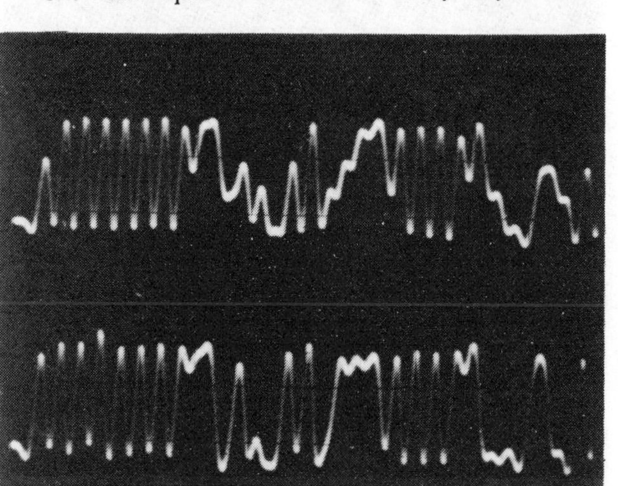

(b) in : Teletext with an in-phase echo (270 ns, 45%),
out : equalizer output signal y(t)/a.

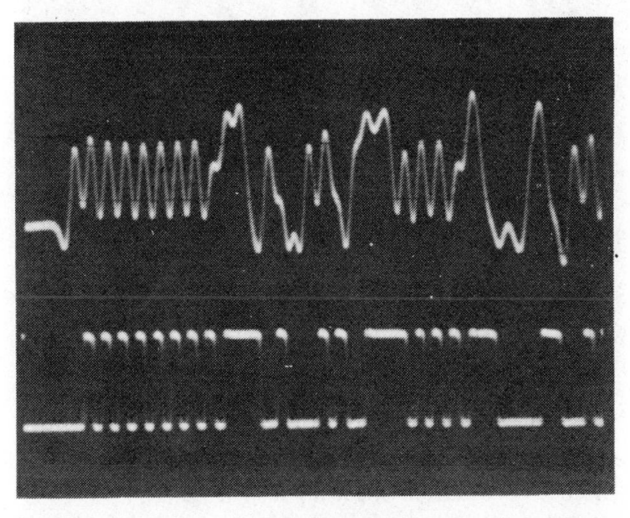

(e) in : as in (d), out: the sliced signal sign (y); the latter signal is suited for feeding to the Teletext decoder.

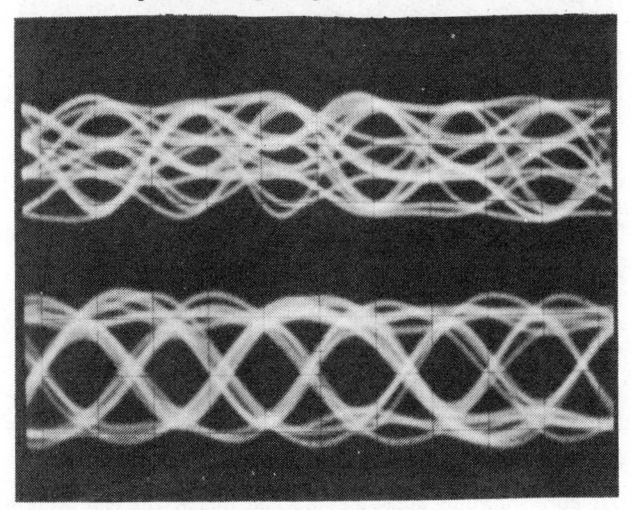

(c) data eye patterns of the signals in (b).

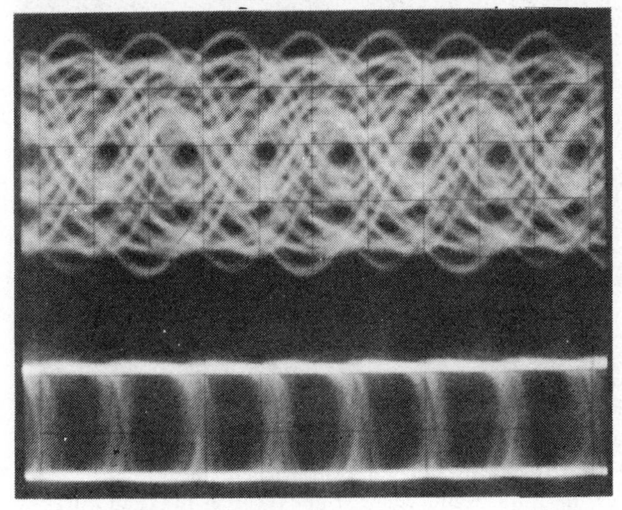

(f) eye patterns of signals in (e) with larger echo (70%).

Fig. 28.
Results of the equalizer in the test set-up (continued).

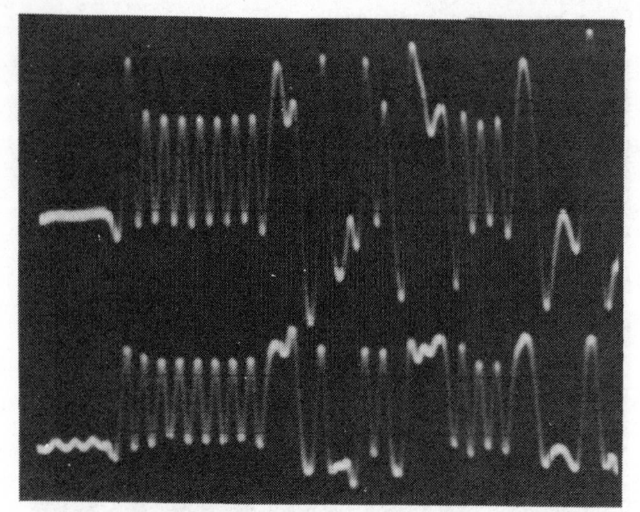

(a) in : Teletext with negative echo (270 ns, 180°, 45%),
out : the equalizer output signal y(t)/a.

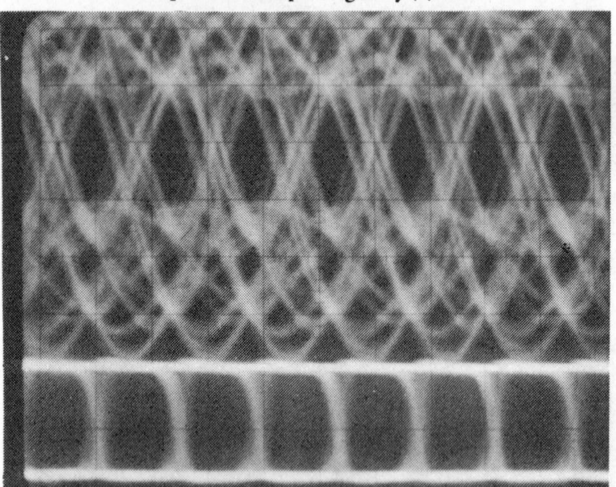

(b) in : eye pattern of the input signal in (a),
out : eye pattern of the sliced output signal sign (y).

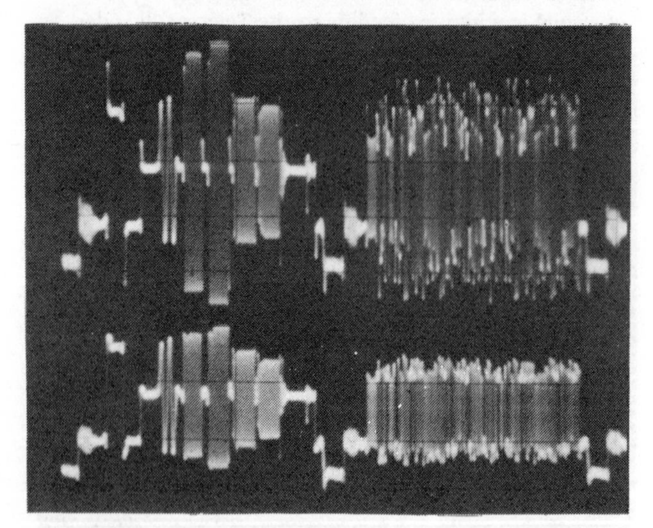

(c) multiburst and Teletext line (echo as in (a) and (b)); data signal undershoots come below top sync level.

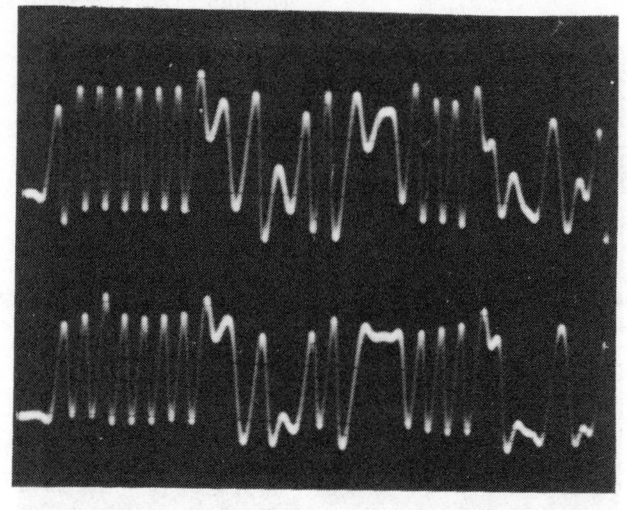

(d) Teletext with an echo of 270 degrees (270 ns, 45%).

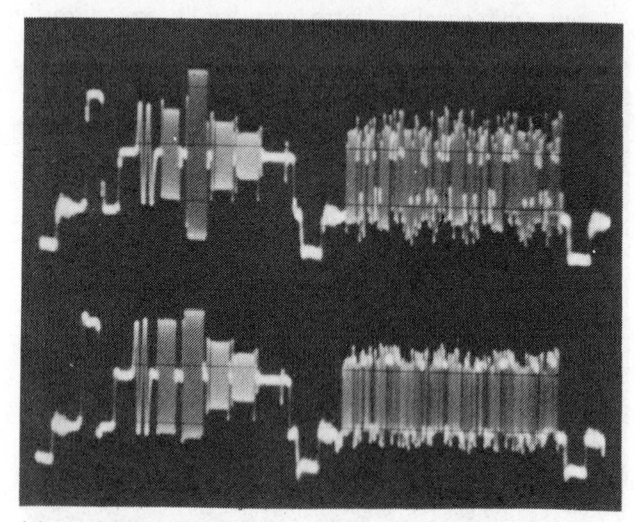

(e) multiburst and Teletext line, corresponding to (d).

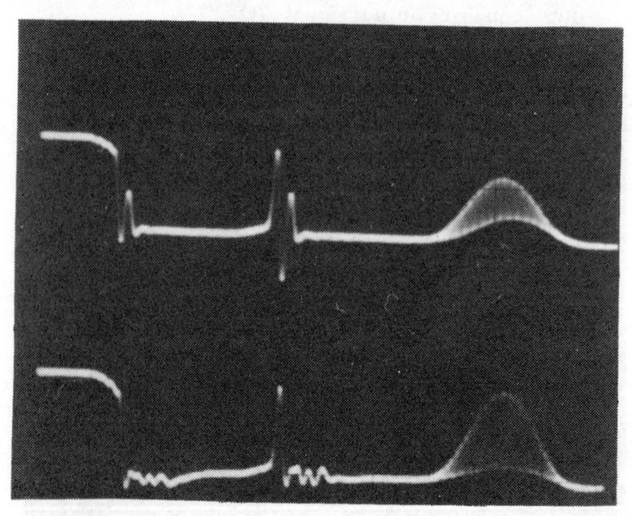

(f) corresponding step, 2T and 20T pulses; the echo is reduced, the error is spread out; n has been increased to 10.

The field trial in Switzerland, where we were guests of the Swiss PTT, was carried out in attendance of the Italian RAI. Eleven different locations, which earlier field trials had shown to be critical were visited.
Apart from the types of result as we have found in the test set-up, it was confirmed that the equalizer can handle multiple echoes (one of the more spectacular examples is shown in fig. 29) and fluctuating echo conditions (see also: Klingler 1980).

Another result is shown in fig. 30. It concerns the reduction of a short leading echo.
Although we have found (experimentally) that for our purpose, the optimum choice for the position of the fixed tap is at the first delay element (m = 1), it is remarkable to see that very short leading echoes can still be reduced to some extent.

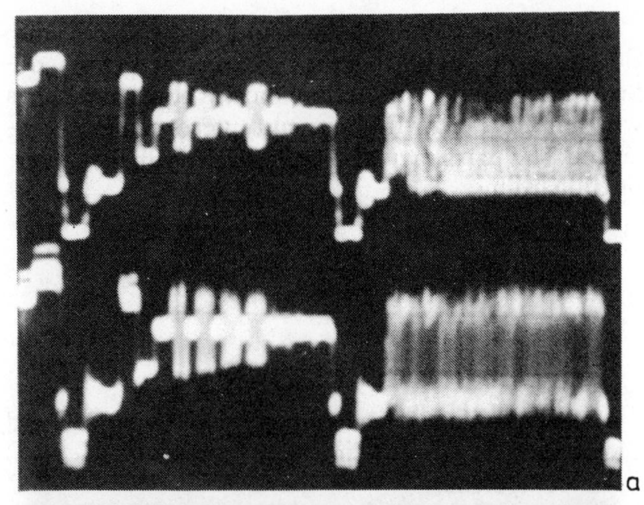

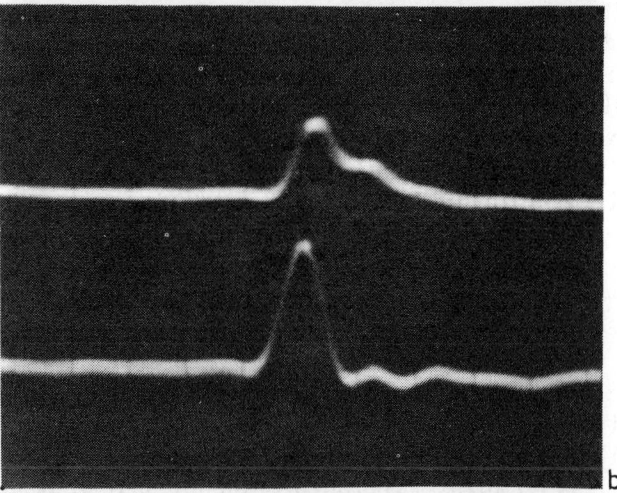

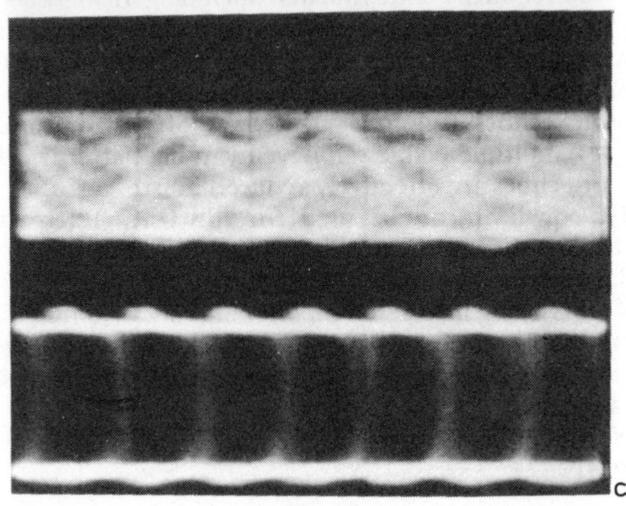

Fig. 29.
Result of field trial in Switzerland.
Figures (a), (b) and (c) refer to the same situation with a very bad multiburst (a), due to a multiple echo (see 2T pulse in (b)). The lower traces show the improvements due to the equalizer. The eye pattern of the sliced output signal (c) results in an error-free Teletext picture.

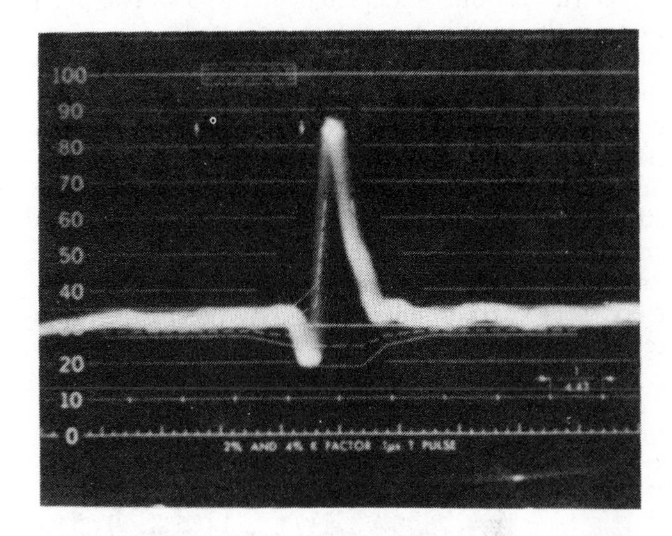

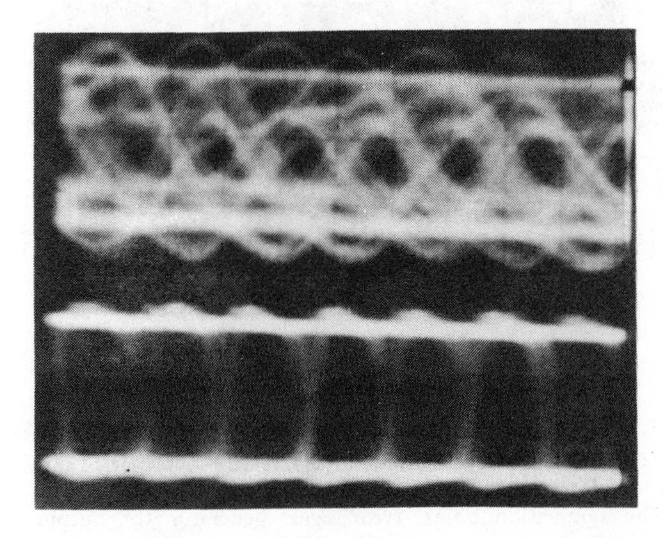

Fig. 30.
A short leading echo observed in the field trial in Switzerland. The eye pattern was significantly improved.

Experiments have been carried out with different numbers of delay cells (up to 12). From the field trials we got the impression that an equalizer with one fixed (m = 1) and six variable cells might be a good compromise between performance and costs.

8. FINAL BREADBOARD, LAY-OUT AND INTEGRATION

A third (and final) equalizer breadboard has been optimized based on the above number of cells (m = 1, n = 7). A saving in supply power consumption has been obtained (to < 500 mW at the nominal supply voltage of 12 V). The circuitry has been stabilized against temperature and supply voltage variations (e.g. a supply voltage of 8 to 16 V is permitted).

The lay-out is depicted in fig. 31. The positions of the equalizer parts are indicated.

Two of the 16 bonding pads are used for the supply (+ and −), two for input and output of the delay line, two for the equalizer output signals (from the output amplifier and for the sliced signal), one is for the adapt/hold command and five for adjustments (to obtain a flat gain of the delay line and to fix various loop gains).
Provisions have been added for cascading of equalizers. Some extra bonding pads have been added for testing (on the wafer).

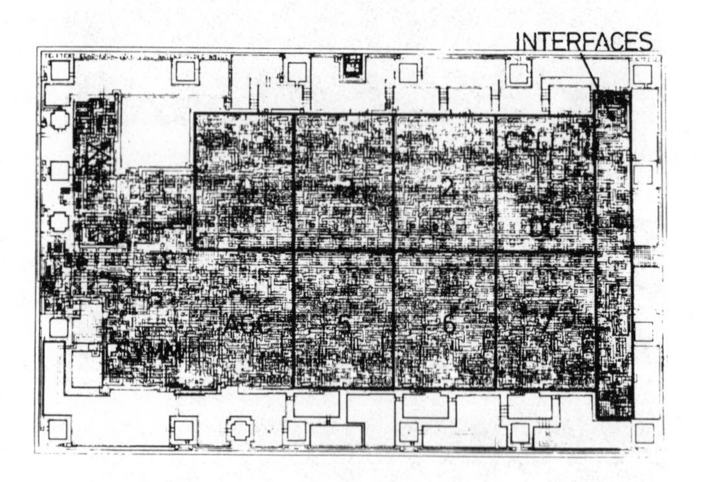

Fig. 31.
Equalizer lay-out.
Fixed cell 1 and the adaptive delay cells 2, 3 . . ., 7 are indicated; DC = DC level corrector (a_0); AGC = automatic gain control (a).
The summation point, error signal generator (e), output amplifier, symmetry restorer (SYMM) and some interfaces are also indicated.
Around the circuitry we have the integrated capacitors (with crossing second-layer interconnections) and bonding pads.
The chip size is: 2.9x4.6 mm^2. It fits into a DIL 16 package.

The complete equalizer and at the same time the equalizer parts (together in a twin pattern) are being integrated. The action on the equalizer parts will simplify error-finding and performance evaluation. The parts may also be used for a higher-order breadboarding.

9. CONCLUSION

A non-linear analog filter has been devised that is capable of reducing short echoes (0.5 - 1 μs) in Teletext signals.
It can extend the area served by Teletext to at least all places where an acceptable TV picture is obtained.

The arrangement consists of a transversal filter (Laguerre filter: cascade of tapped first-order phase--shifters) with variable tapweights.
The tapweights are automatically controlled.
The control uses a property of the Teletext signal:
- echo-free Teletext is (essentially) a two-level signal,
- with echoes there are more than two levels (provided that the data is sufficiently random).

Convergence is shown to be guaranteed on common Teletext.
There is no need for a special training waveform.

After many experiments (including field trials) with breadboard arrangements, the final equalizer was defined as having seven filter sections (a compromise between performance and costs).
Single echoes (in-phase as well as of quadrature type), multiple echoes and even varying patterns of short echoes are effectively reduced in size.
The equalizer can be used for the U.K. Teletext standard, but also for Antiope and different binary data formats (and bit rates).

It is being integrated on one chip (2.9x4.6 mm^2) in a bipolar process, which is extended with a possibility for integration of high quality dielectric capacitors.
We expect to have the first equalizer chips in June 1981.

ACKNOWLEDGMENTS

We have pleasure in thanking Messrs.
J.G. de Groot and L.A. Daverveld of our laboratories for their decisive technological work on the integration of dielectric capacitors.

We would also like to express our gratitude to the many other people who have actively stimulated, and contributed to the accomplishment of the equalizer integrated circuit.

REFERENCES

H. Brand und H. Hügli (1972)
Fernseh-Empfangstechnik I
Hallwag Verlag, Bern.

W. Ciciora, G. Sgrignoli and W. Thomas (1979)
A tutorial on ghost cancelling in television sytems
IEEE Transactions on Consumer Electronics CE-25, 9-44.

T. Deliyannis (1969)
RC active allpass sections
Electronics Letters 5, 59 - 60.

S.K. Goyal and S.C. Armfield (1979)
Reception of Teletext under multipath conditions
IEEE Transactions on Consumer Electronics CE-25, 378-392.

S. Goyal, S. Armfield, W. Geller and J. Blank (1980)
Performance evaluations of selected automatic deghosting systems for television
IEEE Transactions on Consumer Electronics CE-26, 100-119.

P.R. Gray and R.G. Meyer (1977)
Analysis and design of analog integrated circuits
Wiley, New York.

A.G.J. Holt and J.P. Gray (1967)
Active allpass sections
Proceedings of the IEE 114, 1871-1872.

Y. Ishigaki, Y. Okada, T. Hashimoto and T. Ishikawa (1980)
Television design aspects for better Teletext reception
IEEE Transactions on Consumer Electronics CE-26, 622-628.

Y. Ishigaki, K. Oouchi, K. Utsunomiya, H. Yamada, C. Kuriki, and J. Saitoh (1980)
An automatic TV signal equalizer 'Picture Clinic'
IEEE Transactions on Consumer Electronics CE-26, 638-656.

R. Klingler (1980)
Influence of receivers, decoders and transmission impairments on Teletext signals (U.K. Teletext and French Antiope)
Paper contributed to the International Conference on New Systems and Services in Telecommunications, Liege, Belgium, Editor: G. Cantraine.

Y.W. Lee (1960)
Statistical theory of communication
Wiley, New York.

S. Makino, J. Murakami, M. Sakurai, S. Ohnishi and M. Obara (1980)
A novel automatic ghost canceller
IEEE Transactions on Consumer Electronics CE-26, 629-637.

J.E. Mazo (1980)
Analysis of decision-directed equalizer convergence
The Bell System Technical Journal, 1857-1876.

J.G. Proakis (1975)
Advances in equalization for intersymbol interference
in: Advances in Communication Theory and Applications IV, 123-198, Editors: A.V. Balakrishnan and A.J. Viterbi, Academic Press.

S. Yamazaki and Y. Endo (1980)
A quantitative measurement of subjective effects of TV multiple-ghost images by 'perceived DU ratio'
IEEE Transactions on Broadcasting BC-26, no. 3, 62-69

Part VI
Multichannel Television Sound

TELEVISION sound has been receiving much greater emphasis in the past few years due to the interest in stereo sound as well as the possibility of adding an additional language or program with the use of an added channel. In December 1983, the Electronic Industries Association chose the BTSC system, and submitted the system to the Federal Communications Commission in January 1984. The BTSC multichannel sound system, which is fully compatible with the TV receivers now in use, employs subcarrier techniques for the stereo information (or additional audio channel) and DBX companding of the audio signal for noise and interference reduction. Two papers on the BTSC system, by Eilers and Tyler *et al.*, have been included in this part with a number of additional papers included in the bibliography. Eilers' paper, the third in this part, describes the BTSC and transmission standards; the next paper by Tyler *et al.* describes the audio signal companding. Additional papers on this subject are listed in the bibliography ([8], [9], and [12]) and give more information on some of the requirements for the encoding, reception, decoding, and DBX noise reduction of multichannel sound.

A relatively recent development, a technique which has a number of advantages over the traditional methods, especially in multichannel or stereo sound, is the use of SAW filters for TV sound separation. The use of SAW filters for the separation of sound signals from picture signals is described in the first paper of this part, by Yamada and Uematsu. The next paper, by Fockens and Eilers, is on intercarrier buzz, which has always caused difficulties and is even more critical now with the emphasis of high quality and multichannel sound. A number of additional papers dealing with the design of IF amplifiers and filters with the required response, as well as the design of the FM sound detector and detector IC's are included in the bibliography.

ANTHONY TROIANO
Associate Editor

BIBLIOGRAPHY

[1] D. Holmes, "Stability considerations in transistor IF amplifiers," *Transistors 1*, RCA Laboratories, Mar. 1956.

[2] J. Avins, "It's a television first-receiver with integrated circuits," *Electronics*, Mar. 21, 1966.

[3] J. P. GrosJean, "FM IF filter design with group delay and amplitude response considerations," *IEEE Trans. Broadcast Telev. Receivers*, vol. BTR-16, Nov. 1970.

[4] W. Luplow, "Skewing and AC parameter tolerances in small-signal single tuned amplifiers," *IEEE Trans. Broadcast Telev. Receivers*, July 1968.

[5] J. N. Denenberg, "The power mean frequency estimator: Another approach to the FM detector," *IEEE Trans. Broadcast Telev. Receivers*, vol. BTR-20, Aug. 1974.

[6] P. Menniti, "An integrated-circuit sound for television receivers," *IEEE Trans. Consum. Electron.*, vol. CE-21, Feb. 1975.

[7] C. Eilers, "Television multichannel sound broadcasting—A proposal," *IEEE Trans. Consum. Electron.*, vol. CE-27, Aug. 1981.

[8] V. Mycynek, "TV multichannel sound: Reception and decoding," *IEEE Trans. Consum. Electron.*, vol. CE-30, Aug. 1984.

[9] R. Lee, "TV multichannel sound: Encoding and transmission," *IEEE Trans. Consum. Electron.*, vol. CE-30, Aug. 1984.

[10] B. Cocke, "A 2.5-watt monolithic TV sound system with refined DC control," *IEEE Trans. Broadcast Telev. Receivers*, vol. BTR-2, Aug. 1984.

[11] J. Weigand, "Second-generation sound channel: Bipolar IC," *IEEE Trans. Consum. Electron.*, vol. CE-25, Aug. 1979.

[12] K. Shinohara, "The development of bipolar IC's for dbx noise reduction system," *IEEE Trans. Consum. Electron.*, vol. CE-28, Nov. 1982.

[13] L. Blaser, "A comparison of integrated television sound systems," *IEEE Trans. Broadcast Telev. Receivers*, vol. BTR-17, Feb. 1971.

NEW COLOR TV RECEIVER WITH COMPOSITE SAW IF FILTER SEPARATING THE SOUND AND PICTURE SIGNALS

J. Yamada and M. Uematsu

Consumer Products Research Center and Yokohama Works,
Hitachi Ltd., 292 Yoshida-machi, Totsuka-ku, Yokohama 244, Japan

1. Introduction

Numerous researches and developments of Surface Acoustic Wave (SAW) TV IF filter have been performed for cost reduction and improvement of reliability of TV circuit. Under these circumstances we reported in 1978[1] that we had established the design and the fabrication techniques of the SAW IF filter[2]-low noise amplifier system in massproduction size, and put it into the market. This year we have completed a high performance TV IF circuit for sound multiplexing broadcasting, which consists of a more advanced SAW filter. The signal detections are individually performed by the intercarrier signal in IF without any external signal. This paper describes the design techniques of newly developed composite SAW IF filter, which is the key device to realize the new circuit and the sound-picture-color crossbeat interferences to picture and sound quality in color TV sets.

2. System conception

We consider the sound-picture-color crossbeat interferences and the causes of the generation. At first the picture quality is investigated as follows. (1) sound-picture crossbeat to color signal (sound cross color) : the difference frequency 3.58MHz intermixes to color signal as a result of the interference between the sound and the 920kHz response of the picture signal in IF frequency, and appears undesirable color response in the picture. (2) color-sound crossbeat to picture signal (920kHz beat) : the sound interferences the color signal, and in consequence the difference frequency 920kHz intermixes to the picture signal, then many strips appear in the picture. Next the sound quality is investigated. (3) sound-picture crossbeat (buzz) : the sound signal is phase-modulated by the picture signal when the detector has a large differential phase angle, and appears the buzz noise in speakers. (4) sound-picture crossbeat (buzz beat) : the difference frequency appears as the beat noise in a subchannel speaker in consequence of the interference between the subcarrier response in sound multiplexed signal and the harmonics of horizontal synchronized signal. (5) sound-picture crossbeat (boom) : the harmonics of picture signal, especially 2.25MHz response interfere the sound IF signal 4.5MHz, and then discomfort boom noise appears.

It is well known since old time that the separate carrier system reduces the above mentioned interferences in color TV sets. However it is compelled to increase electronic parts for stability IF frequency.

Fig. 1 shows new system configuration to settle the difficulties. The answers are : (1) concentrate each frequency characteristics requirement for the sound-picture separation system to the composite SAW filter. (2) the sound and picture signals can get the most suitable detector individually, which are multiplied by the intercarrier signal. (3) the picture filter in the SAW filter has a deep sound trap for the suppression of sound interference to the picture quality. (4) the sound filter has twin peaks of the sound signal and picture carrier for the reduction of picture interference to the sound quality.

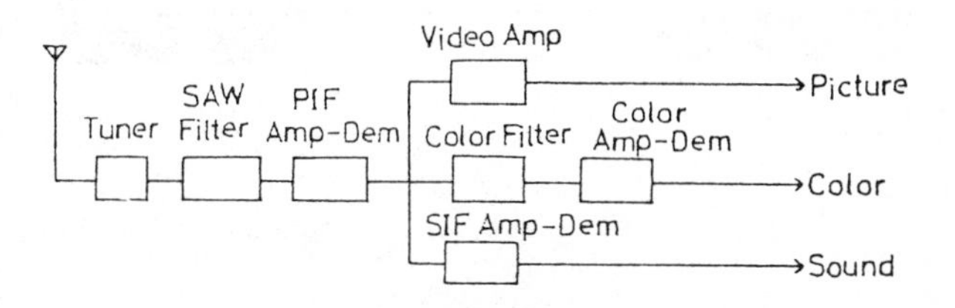

(a) Conventional System

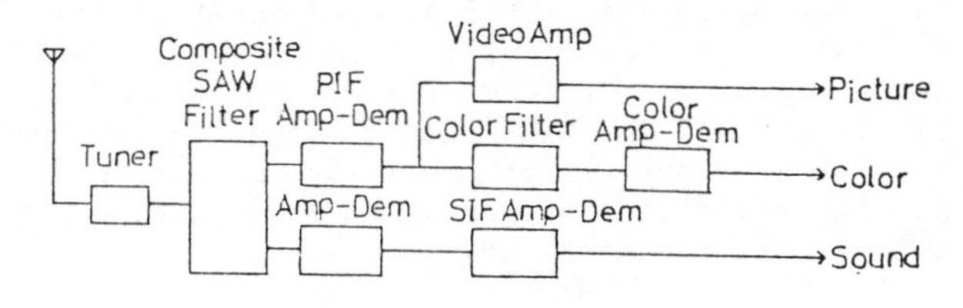

(b) New System

Figure 1. System configuration

3. Design of SAW filter

Impulse model[3] is used for fundamental design of the composite SAW filter having asymmetric amplitude characteristics. Input transducer is 15 pair unapodized IDT for reservation of adjacent TV channel traps and two output transducers are 60 (picture filter) and 70 (sound filter) pair apodized varing pitch IDT. This paragraph is described on keeping a deep sound trap in the picture filter particularly. It is well known that surface wave diffraction degrades adjacent sound trap. We get deep sound trap in band by the control the distance of two transducers and their apertures. The SAW filter response is written as follows.

$$H(k_1) = \int_{-\infty}^{\infty} \frac{\sin \frac{b \cdot k_1}{2}}{k_1^2} \left\{ \sum_{n=1}^{N} \sin \frac{a_n k_1}{2} \exp i(-k_1 h_n + k_3 d_n) \right\} \times \left\{ \sum_{m=1}^{M} \exp i(-k_1 l_m + k_3 g_m) \right\} dk_1 \quad \text{------- (1)}$$

Manuscript Received 6/11/82

Reprinted from *IEEE Trans. Consum. Electron.*, vol. CE-28, pp. 192–194, Aug. 1982.

where b,a_n are apertures of unapodized and apodized IDTs,and (l_m,g_m),(h_n,d_n) are strength and locations of their impulses. Wave numbers are identified as follows.

$$k = k(\theta)\sin(\theta) \quad \text{------------(2)}$$
$$k = k(\theta)\cos(\theta) \quad \text{------------(3)}$$
$$k(\theta) = w/v(\theta) \quad \text{------------(4)}$$

Velocity anisotropy v(θ) around X-axis in 128° rotated Y cut $LiNbO_3$ is calculated.[4] Using this techniques we get the IDT pattern shown in Fig.2 for Japanese NTSC system.

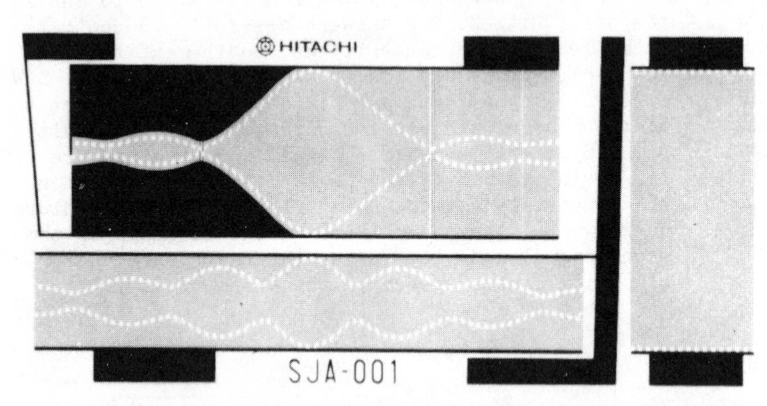

Figure 2. IDT pattern of composite SAW filter

4. Total performance

The solid lines in Fig.3 and in Fig.4 show the frequency response of picture and sound filter respectively under 50Ω terminations. The traps are sufficiently deep and wide enough where the temperature deviation of the filter substrate is considered. And the deviations of group delay are little within ±50 nanosec in pass band.

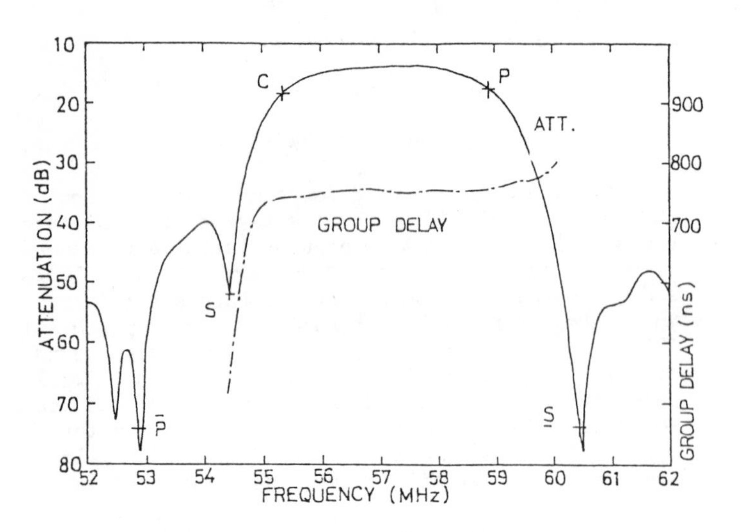

Figure 3. Frequency response of picture filter

Fig.5 shows the schematic diagram of new IF circuit in which IC201 is a picture detector and IC202 is a sound detector. Passive parts placed before and behinde the SAW filter are selected for the triple transit suppression. Individual carrier filters which extract reference carrier from the IF signal are adjusted in different selectivity. The carrier filter in sound detection is regulated for minimizing differential phase angle in order to suppress the buzz beat interference. In the new system we get greatly improvement on crossbeat interferences in Table 1 ,which is shown against our conventional system.

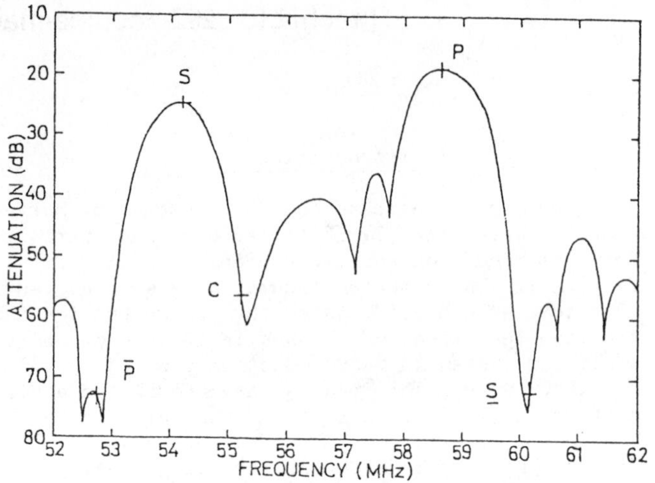

Figure 4. Frequency response of sound filter

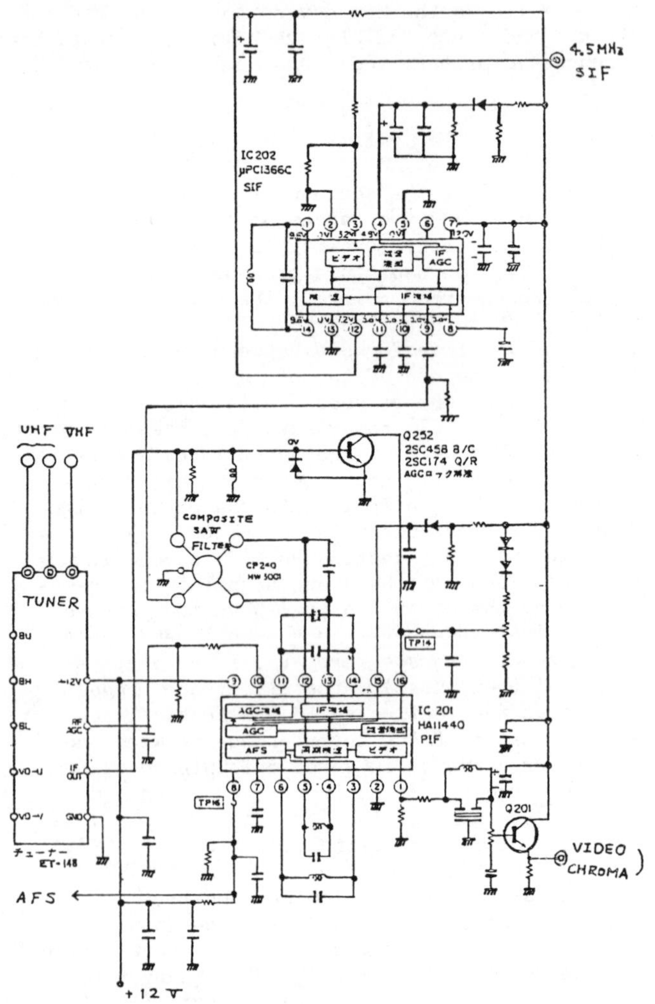

Figure 5. Schematic diagram of new IF circuit

The crossbeat interferences are measured in accordance with Japanese testing method for color TV receivers (JIS-C6101). (1) sound cross color : S/N ratio in detected color signal when P/S=0 dB signal applied. (2) 920kHz beat : 920kHz/100kHz ratio in

Table 1. Reduction of crossbeat interference

	improvement
sound cross color	13 dB
920 kHz beat	14 dB
buzz beat	3 dB
buzz	4 dB
boom	20 dB

detected picture signal when P/C=12 dB and P/S=0 dB signal applied. (3) buzz beat : buzz beat/desired sound ratio when sound signal is 30% modulation (modulated frequency 200Hz) and picture signal is fully modulated to white level. (4) buzz : buzz/desired sound ratio when sound modulated frequency is 1kHz. (5) boom : audible boom noise threshold level when P/S ratio varies.

In the result,the improvement of sound cross color and 920kHz beat interference depends on deep sound trap in the picture filter. Also the high sound quality is obtained by reasons of high sound level,low differential phase angle in detector and the suppression of the band between sound and picture carrier.

5. Acknowledgement

The authors would like to express their sincere appreciation to Dr.K.Shibayama and Dr.T.Moriizumi for valuable advices, and Dr.F.Mari,Mr.M.Kanamori, Mr.T.Toyama,Mr.Y.Noro and Dr.K.Hazama for helpful encouragement.

References

1)H.Nabeyama et al.,IEEE Trans.,CE-25,1,50 (1979)
2)K.Hazama et al.,IEEE Ultrason.Symp.Proc.,504 (1978)
3)C.S.Hartmann et al.,IEEE Trans.,MTT-21,162 (1973)
4)J.J.Campbell et al.,IEEE Trans.,SU-15,209 (1968)

INTERCARRIER BUZZ PHENOMENA ANALYSIS AND CURES

Pieter Fockens
Carl G. Eilers
Zenith Radio Corporation

Abstract

This partly tutorial paper defines buzz as the result of video related incidental phase modulation (IPM) of the intercarrier. The main sources of IPM are 1. parametric and transit time effects, both non-linear in character and 2. AM to PM conversion of a strictly linear character. 1. is the main source of transmitter visual IPM but can also take place in the receiver. Linear AM to PM is prominent in the receiver as Nyquist IPM.

Visual IPM is transferred to the intercarrier in the familiar kinds of second detector. Visual IPM cannot transfer to the aural carrier in the transmitter but it can in receiver tuner and IF. Co-processing of visual and aural carriers can result in IPM of the aural carrier through a parametric nonlinearity.

Calculation results in the form of signal-to-buzz ratios are given for pure sine wave video signals and for composite blank, with a linearly varying parametric nonlinearity and with a linear Nyquist slope.

Buzz sensitivity increases linearly with increasing audio frequencies just as FM thermal noise. Thus subcarriers cause more buzz.

Transmitter IPM monitoring is briefly touched upon.

Several modified receiver circuit arrangements are analyzed as to buzz performance to show typical interactions and tradeoffs.

When operating with audio subcarriers significant buzz improvement is to be expected from 1. elimination of the Nyquist-slope-caused IPM, 2. elimination of transmitter IPM, 3. elimination of parametric nonlinearity in the receiver.

I. Introduction

Buzz results when video related, phase modulated components of the visual carrier (= Incidental Phase Modulation (IPM)) are transferred to the channel that carries the sound. The blank and sync components provide the characteristic buzz sound. In monophonic sound, especially when reproduced with limited low and high frequencies, the buzz is, most of the time, barely noticeable and certainly acceptable. But when multichannel sound is attempted with subcarriers it is found that these elevated base-band frequencies are much more vulnerable to IPM. In fact the property of FM by which thermal noise increases 6 dB for an octave frequency increase is also active on buzz interference (and is of course caused by the fact that the same peak angle deviation at twice the frequency has twice the time rate of change of the angle, to which change the FM detector output is

Reprinted from *IEEE Trans. Consum. Electron.*, vol. CE-27, pp. 381-396, Aug. 1981.

proportional). It thus becomes important to have preferably quantitative data on the various causes of IPM. There is a rather extensive literature available on IPM as relating to chroma [1],[2], [3],[4],[5],but less is available concerning sound reproduction [6],[7].

It is the intention of this paper to set the causes of IPM in perspective, give quantitative results, to analyze current transmitter and receiver design practices and to indicate ways to improvement. Suggestions are made for transmitter monitoring.

II. Analysis of Intercarrier Buzz

A. The generation of buzz is principally a 2-stage process, first of IPM of the visual carrier and secondly - by a distinct process - transfer of this IPM to the intercarrier. (In this paper "Intercarrier" is to be understood as merely a 4.5 MHz FM modulated intermediate frequency signal and not necessarily the result of an "intercarrier detection process.")

Here is a list of IPM sources:

1. Level-dependent phase shift
   a. Parametric effects
   b. Transit time effects
2. AM-PM Conversion
   a. Unequal sideband amplitude ("Nyquist IPM")
   b. Unsymmetric sideband phase
   c. Straight FM (AFC circuits)
3. Miscellaneous
   a. Overmodulation
   b. Reverse mixer feedthrough
   c. 3-Tone intermodulation (?)
   d. CATV equipment

B. Transfer of IPM to the Intercarrier; Detection

The simplest type of second detector which is still extensively used is the envelope detector. The visual carrier acts as local oscillator for the suitably attenuated aural carrier and this produces the Intercarrier as "beat" product. If the visual carrier is amplitude modulated only and if this amplitude modulation is transferred to the intercarrier it will be eliminated by the limiter. Any angle modulation which is only on the visual carrier, however, is fully transferred to the intercarrier. (See Appendix I for an analytical demonstration.) The FM sound detector detects these video signal related angle modulations. In case they are in the audio frequency range or in the composite baseband range of frequencies for multichannel sound they become audible. The sync and blank components cause the characteristic buzz sound.

When a synchronous second detector is used IPM transfers by a different process, although the result may be the same. Assume that the Reference Channel (into which the IF signal is branched in order to derive a local carrier) has a bandwidth which is wider than twice the composite baseband width of the multichannel audio. As a result the locally generated carrier still contains the IPM. Appendix I also includes a demonstration of how IPM on a carrier transfers to the output in a linear multiplier. Thus IPM is transferred to the intercarrier.

Next assume a synchronous detector with a reference channel only a few kHz wide. All IPM faster than a few kHz is now eliminated from the local carrier and thus will not transfer to the intercarrier.

There is yet another process of interest in the second detector. Assume that both visual and aural carriers have acquired equal amounts of IPM. This could, for example, happen in a tuner that has insufficiently bypassed AFC. Consequently video products are applied to the frequency control in the tuner l.o. This is a straight FM modulation process and it acts on both visual and aural carriers. The interesting result in an envelope type second detector is that the IPM is eliminated from the intercarrier since it is the result of differencing of visual and aural carriers. A synchronous detector with wide band reference channel acts similarly on the intercarrier. When the reference channel is narrow, however, the local oscillator has no significant IPM while the aural carrier does. The result is an intercarrier with IPM. Note that the envelope detector and the synchronous detector with sufficiently wide reference channel process IPM in like manner which is opposite from the way a synchronous detector with narrow reference channel processes IPM.

In the early days of television receivers had separate processing of visual and aural carriers. This required extra vacuum tubes and it made tuning much more critical, especially on UHF, so intercarrier detection was an improvement. But the split processing had no buzz!

They had no buzz because the aural carrier was received without IPM, the tuner did not introduce any (no AFC) and after the tuner both carriers were separated. The aural carrier had its own bandpass filter and no Nyquist slope introduced IPM (see section 2. below). Thus with present-day sophisticated tuning systems and with integrated circuits split-sound processing is becoming attractive again provided that no tuner IPM is introduced. This is taken up in section III.

1. Level Dependent Phase Shift; Transmitters

When in some stage or device the phase shift of a small signal varies with bias level (= luminance level) the small signal experiences differential phase distortion while the luminance modulated carrier experiences IPM. This is a nonlinear effect but not in the familiar sense of the curved (static) transfer characteristic of some active device. Such a nonlinearity causes harmonic distortion, intermodulation and differential gain distortion but not, per se, differential phase distortion. This distortion can best be explained by parametric effects and/or by transmit time effects. The former include dependence of junction capacities, dynamic resistances, etc. on current and/or voltage. Thus reactive properties of such a circuit become dependent on bias level. Transit time variations, as take place, for example, in Klystron amplifiers cause similar phase effects. The consequences for the chroma signal are described by Behrend [1] Blair [2] and Kuroki [4]; especially the transmitter is the source of the IPM in these references. In U.S. Television broadcasting the aural carrier is processed separate from the visual carrier until the passive diplexer. Thus no direct transfer of IPM to the aural carrier takes place in the transmitter, although similar parametric effects may take place in receiver tuner and IF. Mostly, though, the IPM reaches the intercarrier via the second detector as described above.

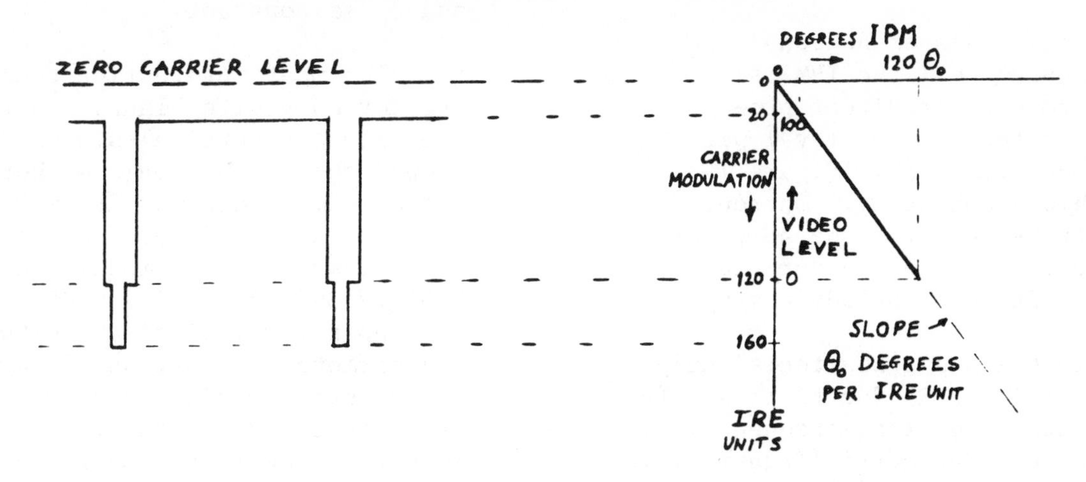
IPM DUE TO PARAMETRIC NONLINEARITY
ZERO CARRIER LEVEL
DEGREES IPM
120 θ_0
0
20
100
CARRIER MODULATION
VIDEO LEVEL
120
0
160
SLOPE
θ_0 DEGREES PER IRE UNIT
IRE UNITS

FIG. 1

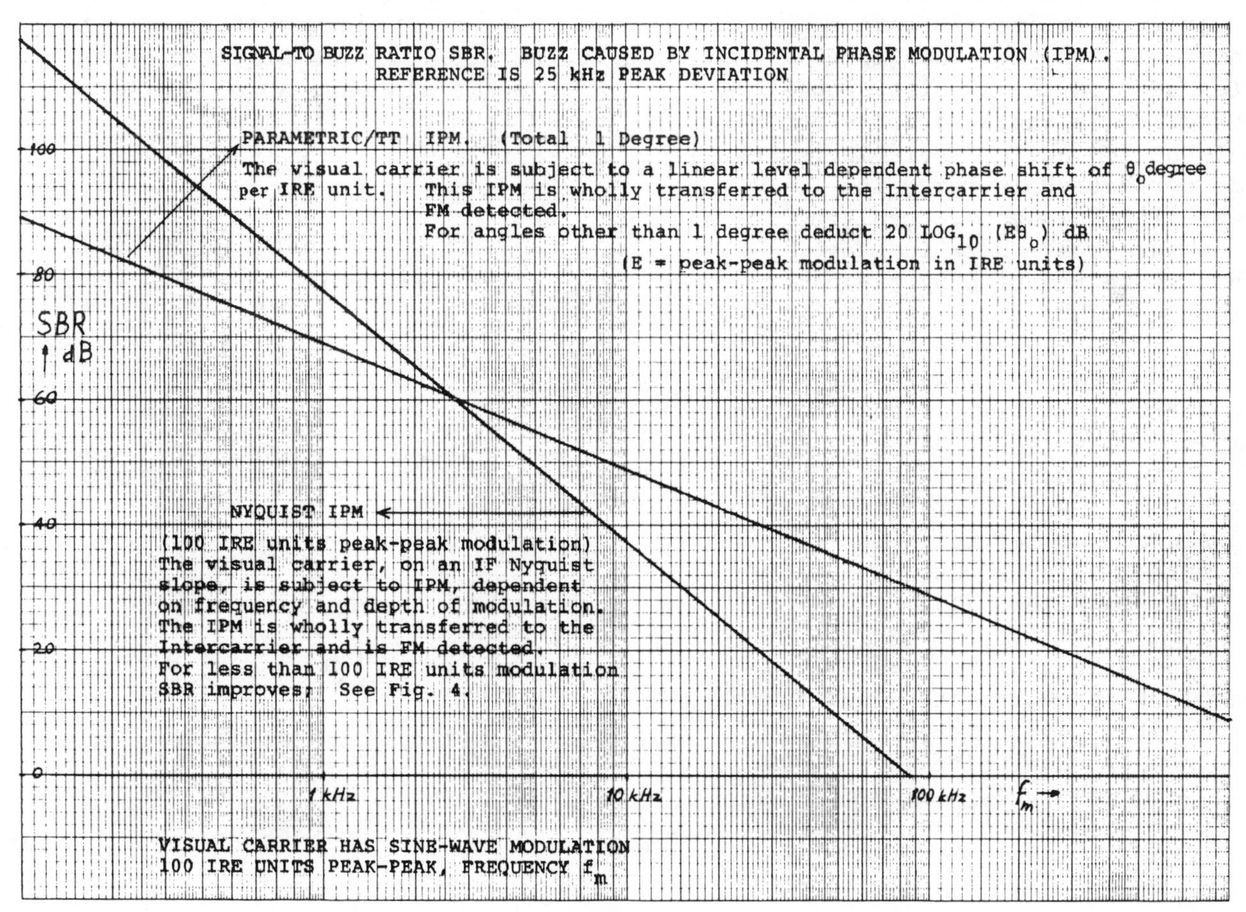
SIGNAL-TO BUZZ RATIO SBR. BUZZ CAUSED BY INCIDENTAL PHASE MODULATION (IPM).
REFERENCE IS 25 kHz PEAK DEVIATION
PARAMETRIC/TT IPM. (Total 1 Degree)
The visual carrier is subject to a linear level dependent phase shift of θ_0 degree per IRE unit. This IPM is wholly transferred to the Intercarrier and FM detected.
For angles other than 1 degree deduct 20 LOG_{10} ($E\theta_0$) dB
(E = peak-peak modulation in IRE units)
SBR dB
NYQUIST IPM
(100 IRE units peak-peak modulation)
The visual carrier, on an IF Nyquist slope, is subject to IPM, dependent on frequency and depth of modulation.
The IPM is wholly transferred to the Intercarrier and is FM detected.
For less than 100 IRE units modulation SBR improves; See Fig. 4.
100
80
60
40
20
0
1 kHz
10 kHz
100 kHz
f_m
VISUAL CARRIER HAS SINE-WAVE MODULATION
100 IRE UNITS PEAK-PEAK, FREQUENCY f_m

FIG. 2

Quantitative Analysis

On the basis of Behrend's measurements [1] the IPM is assumed to vary in direct proportion to the carrier level with θ_o degrees per IRE unit. Figure 1 shows this graphically. Current transmitters appear to be able to hold the IPM within 1° over the 120 IRE units not including sync.

Figure 2 shows how the signal-to-buzz ratio (SBR) due to this effect varies with modulating frequency. The curve marked "Parametric/Transit Time IPM" is for a sine wave of 100 IRE units peak-peak and .01 degrees/IRE unit IPM. Curves for other levels are parallel and can be found in the way indicated in Figure 2. Note that the SBR decreases 6 dB per octave. This is the result of the frequency independence of θ_o (which is justified for the range of a few dozen kHz around 45 MHz) combined with the triangular interference sensitivity of FM. When in a later section the Nyquist slope caused IPM is analyzed it is found that the IPM angle itself increases 6 dB per octave increase in modulating frequency, resulting in 12 dB per octave buzz increase. (This is represented by the other curve in Figure 2.)

Sine wave modulation is easily treated but composite blank and sync combined with a constant grey level (including black and white) is more realistic. SBR numbers for these signals are collected in Table I. The first entry is also found at 31.5 kHz in Figure 2.

The other entries are mainly for blank and composite blank and all are for constant white level. For a level other than white level deduct 20 log ($E\,\theta_o$) dB (representing an increase in SBR when E decreases and θ_o is constant).

The most interesting numbers are for composite blank. In the one case the vertical sidebands around the 2nd harmonic of horizontal are AM demodulated by $2f_h$ = 31.468 kHz to the audio band (52.8 dB) and the other case is for demodulation by $2.5f_h$ = 39.335 kHz (57.6 dB) of the upper vertical sidebands of the 2nd harmonic of horizontals and the lower vertical sidebands of the 3rd harmonic of horizontals.
Note that the last number includes de-emphasis. The vertical sideband terms are RMS added and the numbers are again to be corrected for a total IPM angle different from 1 degree.

2. AM-PM Conversion; Nyquist Slope

a. Under consideration here are strictly linear effects. When a double sideband AM modulated signal

IPM DUE TO UNEQUAL SIDEBAND AMPLITUDE (a) OR PHASE (b)

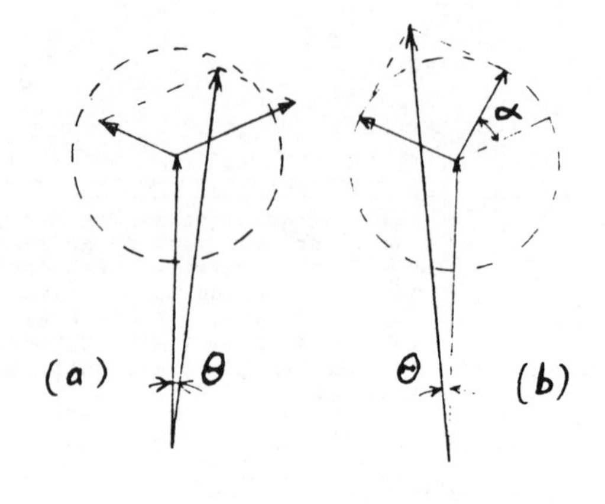

Figure 3

TABLE I

PARAMETRIC/TRANSIT TIME IPM

TOTAL IPM ANGLE ASSUMED = 1 DEGREE

MODULATING SIGNAL (No De-emphasis applied, except as noted)	SIGNAL-TO-BUZZ RATIO SBR * dB
SINE WAVE frequency = 31.468 kHz	39.2
VERTICAL BLANK (RMS sum of harmonics)	72.6
HORIZONTAL BLANK (15.734 kHz component)	35.7
HORIZONTAL BLANK (31.468 kHz component No demodulation)	44.9
HORIZONTAL BLANK (47.202 kHz component No demodulation)	43.8
COMPOSITE BLANK (RMS sum of vertical sidebands of 31.468 kHz = 2nd harmonic of horizontal; AM demodulated by 31.468 to baseband)	52.8
COMPOSITE SYNC (RMS sum of vertical sidebands of 31.468 kHz = 2nd harmonic of horizontal; AM demodulated by $2f_h$ = 31.468 kHz to baseband)	71.9
COMPOSITE BLANK (As above; AM demodulated by $2\frac{1}{2} f_h$; De-emphasis included)	57.6

* The numbers in the Table are for a total of 1 degree IPM. For different angles deduct $20 \text{ LOG}_{10} (E\ \theta_o)$ dB.

E is the peak-to-peak modulation level in IRE units and θ_o is the IPM phase slope in degrees/IRE unit (assumed constant).

Example: If E = 50 IRE units and θ_o = .01 degree/IRE unit total IPM angle 0.5 degree) 20 LOG(0.5) = -6 dB has to be deducted from (or 6 dB to be added to) the number in the Table.

is passed through a passive network that results in unequal sideband amplitude or unsymmetrical sideband phase shift or both, the result is phase modulation. This is demonstrated by the well-known vector diagram of Figure 3 where the resultant vector "wobbles" around the unmodulated position at the modulation rate.

Deutsch [8] has given expressions for $\theta = \theta(t)$ as well as for its time derivative, to which the FM detector output is proportional.

Quantitative Analysis

Assuming a linearly decreasing Nyquist slope from 750 kHz below

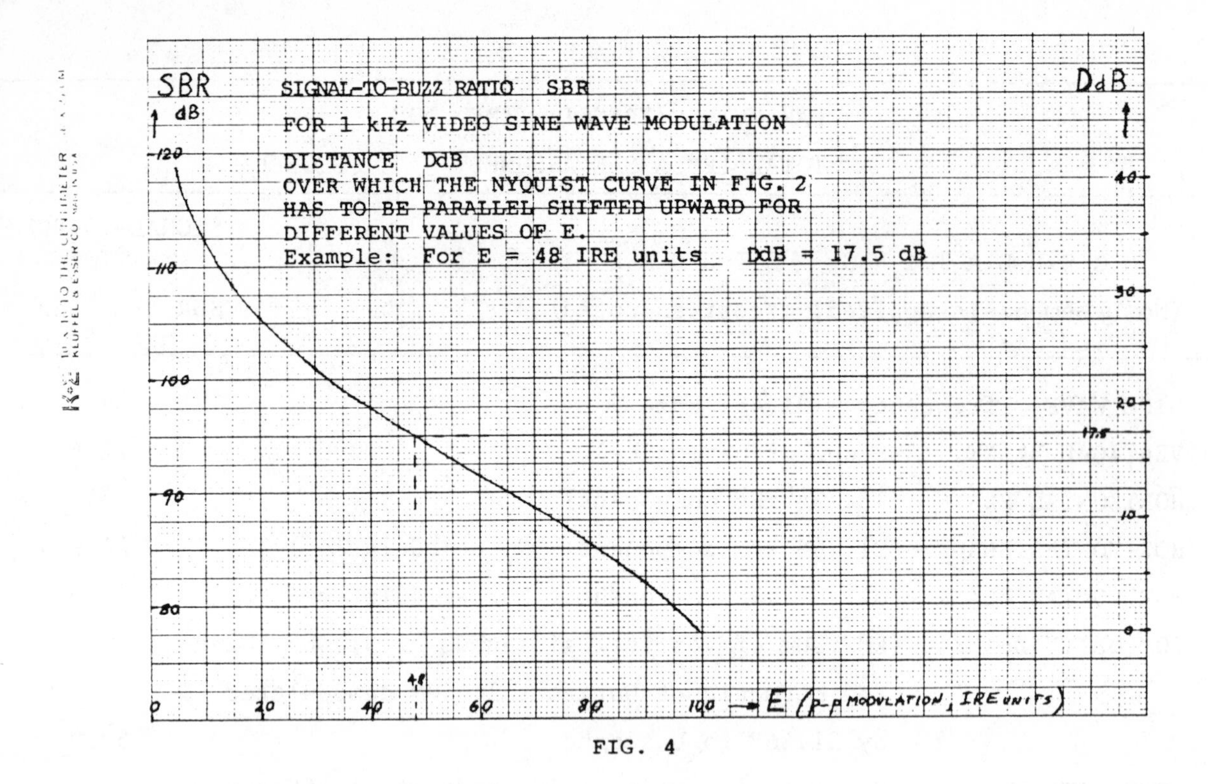
SBR
dB
SIGNAL-TO-BUZZ RATIO SBR
FOR 1 kHz VIDEO SINE WAVE MODULATION
DISTANCE DdB
OVER WHICH THE NYQUIST CURVE IN FIG. 2
HAS TO BE PARALLEL SHIFTED UPWARD FOR
DIFFERENT VALUES OF E.
Example: For E = 48 IRE units DdB = 17.5 dB
DdB
120
110
100
90
80
40
30
20
17.5
10
0
0
20
40
48
60
80
100
E (p-p MODULATION, IRE UNITS)

FIG. 4

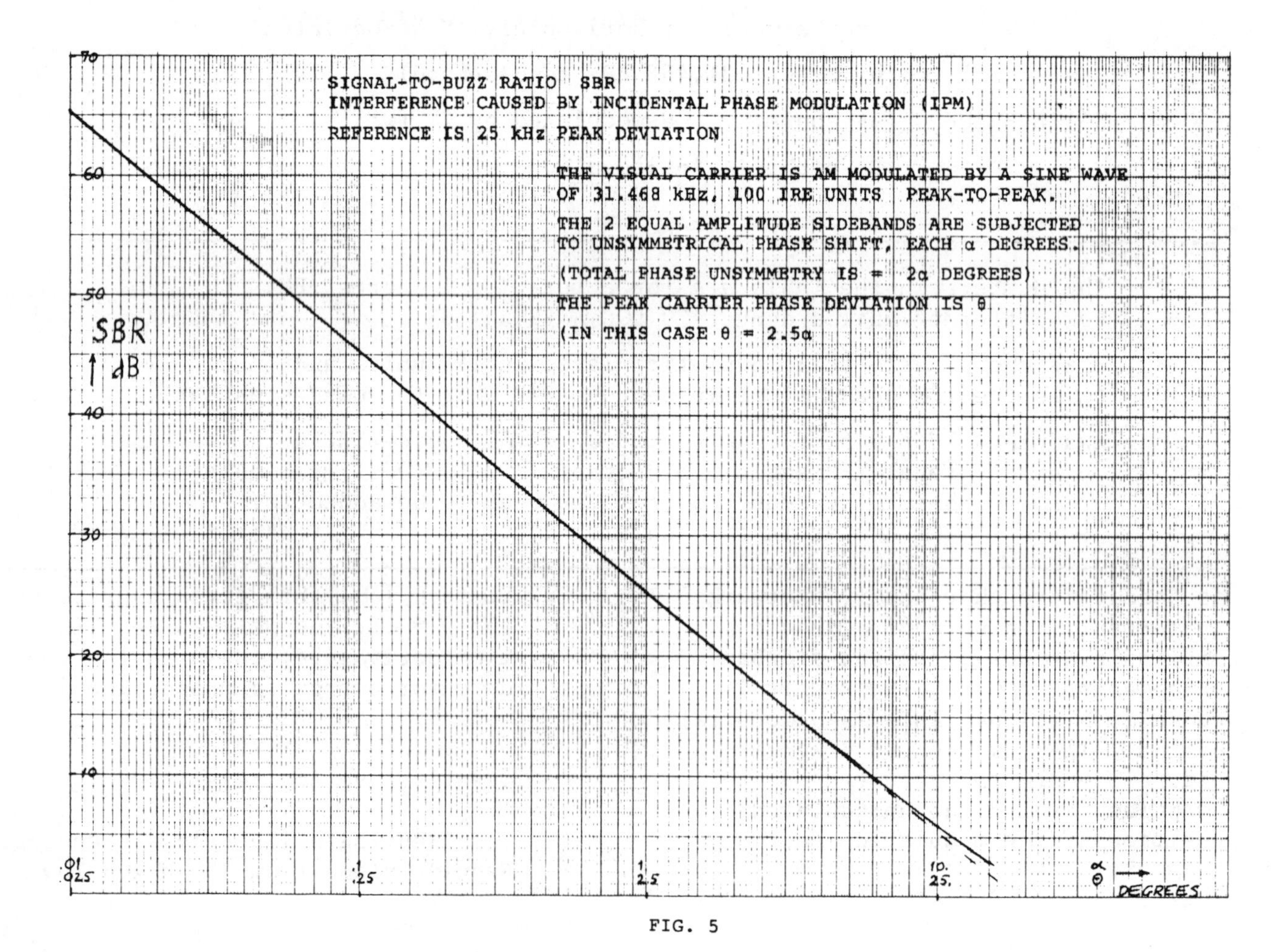
SIGNAL-TO-BUZZ RATIO SBR
INTERFERENCE CAUSED BY INCIDENTAL PHASE MODULATION (IPM)
REFERENCE IS 25 kHz PEAK DEVIATION
THE VISUAL CARRIER IS AM MODULATED BY A SINE WAVE
OF 31.468 kHz, 100 IRE UNITS PEAK-TO-PEAK.
THE 2 EQUAL AMPLITUDE SIDEBANDS ARE SUBJECTED
TO UNSYMMETRICAL PHASE SHIFT, EACH α DEGREES.
(TOTAL PHASE UNSYMMETRY IS = 2α DEGREES)
THE PEAK CARRIER PHASE DEVIATION IS θ
(IN THIS CASE θ = 2.5α
SBR
dB
70
60
50
40
30
20
10
.01
.025
.1
.25
1
2.5
10.
25.
α
θ
DEGREES

FIG. 5

the carrier to 750 kHz above the carrier the IPM is completely determined for given modulating frequency and depth of modulation. For 100 IRE units peak-to-peak sine wave the resulting signal-to-buzz ratio is graphed in Figure 2. For less than 100 IRE units modulation the graphs are parallel to the given one and shifted upward by an amount of DdB which is found in Figure 4.

b. The case of unsymmetrical sideband phase shift and sine wave modulation of 100 IRE units peak-to-peak is plotted in Figure 5. The SBR is plotted as a function of the angle of unsymmetry for a frequency of 31.5 kHz (actually the result of this IPM is not buzz but a"tone"of 31.5 kHz). Note that a phase unsymmetry of 0.2 degrees causes 45.3 dB signal-to-interference ratio. The sideband amplitudes are assumed equal. Calculations for a practical case of a minimum phase IF and combining amplitude and phase unsymmetry are currently in the planning stage.

c. The results of some calculations for blanks are collected in Table II. Unlike for the parametric IPM case with linear phase slope the SBR is here not proportional to the peak-to-peak modulation. Calculated SBR values for 4 different depths of modulation are listed.

An important comparison for the viability of multichannel sound is between the composite blank data of Table I and Table II, both demodulated to audio band. Table I: 52.4 dB (1 degree total IPM) and Table II: 29.6 dB (100 IRE units peak-to-peak modulation).

The conclusion to be drawn from these numbers is that for a successful multichannel sound system it is a first necessity to eliminate the Nyquist IPM. Once this is accomplished the transmitter IPM should be limited to 1 degree or preferably less.

d. Another source of IPM is the tuner AFC to which reference was made in section II. B above. One of the objectives of TV receiver design is to have a fast acting AFC. This is

TABLE II

AM-TO-PM CONVERSION BY LINEAR NYQUIST SLOPE

MODULATING SIGNAL	SIGNAL-TO-BUZZ RATIO SBR (db) *			
IRE Units →	100	80	60	40
31.5 kHz SINE WAVE	17.6	25.5	31.6	35.1
VERTICAL BLANK - BASEBAND	57.0	63.3	68.8	74.4
HORIZONTAL BLANK - FUNDAMENTAL	39.8	47.5	53.4	59.4
- 2nd HARMONIC (NO DEMOD.)	15.8	28.4	35.6	42.2
- 3rd HARMONIC (NO DEMOD.)	15.9	25.2	31.6	37.8
COMPOSITE BLANK (RMS SUM OF VERTICAL SIDEBANDS OF $2f_h$ = 31.468 kHz, AM DEMODULATED BY $2f_h$	29.6	34.9	39.8	45.2

* NO DE-EMPHASIS INCLUDED

desirable to allow a quick scan of the TV channels without having to noticeably wait for the channels to lock in. Faster action requires more bandwidth and this allows more video components to reach the tuner VCO. This causes direct FM modulation of the l.o. with subsequent transfer to visual and aural carrier alike. A steeper roloff of the AFC filter is constrained by the stability requirement for the AFC feedback loop. Fortunately the described IPM need not cause buzz. In a previous section was described how IPM will be eliminated if the 2 signals that "beat" against each other have the same IPM in the envelope detector and in the synchronous detector with a wide enough reference channel. The split sound or parallel sound receiver type will have to contend with this IPM as will have the receiver with a synchronous detector with narrow reference channel. Microphonics and l.o. phase noise cause similar problems.

3. Miscellaneous IPM Sources

a. Overmodulation

Video overmodulation generally causes severe buzz. When the visual carrier comes close to zero the 4.5 MHz "beat" from an envelope detector disappears. The FM detector loses capture and noise comes up. The carrier always returns at blanks which gives the noise the buzz character. In fact, it would be better described by "rattle" than by buzz.

The synchronous detector with wideband reference channel, on visual overmodulation, loses reference and could thus cause the same effect in the sound as the envelope detector, depending on the fly-wheel action of the 45.75 MHz oscillator.

In the case of separate processing of visual and aural carriers overmodulation has no effect on the sound.

b. Reverse Mixer Feedthrough

With insufficient balancing and buffering the RF signal may reach the tuner local oscillator through the mixer. The AM may cause the instantaneous phase of the l.o. to vary. This constitutes IPM and is, really, of the parametric kind. The oscillator Q is, generally, not high enough to eliminate any of this IPM, just as in the case of AFC - introduced IPM.

c. Video Intermodulation Products

The following process has been suggested as a possible buzz source. Two horizontal harmonics, $2f_h$ apart and in the passband of the 4.5 MHz band pass filter, arrive at the limiter input in combination with the 4.5 MHz intercarrier. Third order intermodulation produces $2f_h$ sidebands of the 4.5 (A+B-C and A-B+C) the vertical sidebands of which produce buzz after stereo demodulation. The present estimate is that this process will not significantly contribute to the buzz but no measurement results are in yet.

d. CATV Processing

CATV systems universally use a 15-20 dB visual-to-aural carrier ratio rather than the 7-10 dB specified by the FCC for over-the-air broadcasting. This means that in a CATV system there is 8-10 dB increased susceptibility to buzz. This is important input to the receiver designer.

A new source of IPM could be in CATV heterodyne processors. The IF type need not pose any problem but when demodulation to baseband is included many of the IPM sources mentioned above for receivers are potentially harmful.

C. Summary

Buzz is caused by video related incidental phase modulation of the intercarrier. This can be caused by transfer of IPM from the visual carrier in a detector sensitive to this transfer. IPM of the visual carrier can be caused by parametric or transit time effects in the transmitter or by unsymmetric sideband treatment anywhere but especially on the Nyquist slope. When aural and visual carriers are coprocessed parametric effects can cause direct IPM of the aural carrier. If identical IPM is present on aural and visual carrier the conventional types of 2nd detector will eliminate this IPM from the intercarrier.When a synchronous second detector has a narrow band reference channel (few kHz) the significant IPM is removed from the regenerated carrier and this will produce an unadulterated intercarrier if no IPM of the aural carrier is caused in circuits preceding the second detector.

D. Monitoring

After receivers have been cleared of Nyquist and other kinds of IPM it becomes important for high quality practicing of multichannel sound to eliminate significant amounts of transmitter IPM. This requires monitoring.

There is currently a high quality demodulator on the market for which an IPM measurement procedure has been worked out, using a waveform monitor [9],[10]. The demodulator generates the in-phase and the quadrature output video components which are subsequently applied to horizontal and vertical deflection plates of the waveform monitor, respectively. The modulating signal is unmodulated stairstep which produces a series of dots on the screen. If these dots line up vertically the quadrature component and thus the IPM are zero. Horizontal deflection indicates IPM. The angle can be read on a radial scale centered at zero carrier; maximum resolution is approximately 1 degree.

It is desirable to obtain greater resolution and accuracy.

One possible method would be to equip the demodulator with a split or quasi-split aural channel (see section III below). The FM sound detector would be followed by a subchannel decoder for the system chosen as the standard. The subchannel output without modulation, measured RMS in a 15 kHz band, would be a sensitive indication of buzz caused by transmitter IPM.

III. CURES FOR INTERCARRIER BUZZ

A. General Considerations

Until recently buzz has not been a significant problem. Add to this the complexity of the problem and it will not be a surprise that there is no good and economically attractive solution at this time.

Lately, however, the pressure for such solutions has been rising. There is increased interest in good quality monophonic sound; multichannel sound with its increased baseband width requires solutions; some Pay TV systems use subcarriers which should not result in decreased sound quality.

Generally speaking, there are several conceivable philosophies for the elimination of IPM:

Correction at the transmitter not only of transmitter caused IPM but receiver precorrection as well.

Elimination of transmitter visual IPM vs imparting the same IPM to the aural carrier.

The first philosophy is attractive from an economics point of view. But technically severe limitations would be imposed on the receiver designer provided it would work at all, which is not sure. The second decision would also have to be carefully weighed.

After a few short paragraphs on transmitter cures for buzz will follow descriptions of a number of alternate TV receiver circuit arrangements from a buzz point of view. The purpose is to provide insight into the interactions and tradeoffs involved. It could serve as the basis for further needed work.

B. Transmitter Cures for Buzz

Aural transmitters in the U.S. are separate from visual transmitters and thus are free of IPM.

Visual transmitters require, in general, different correction methods for the high efficiency, high level circuits than for the low level circuits. For example, in a low level modulator the IPM can often be cured by optimum adjustment of injection levels and/or by feeding small amounts of quadrature carrier around the modulator.

High level circuits require precorrection. For example, IPM could be detected from a demodulator free from IPM of its own. This is negatively fed back to the input of a phase modulator which could be included in the carrier path to some IF modulator.

C. Receiver Cures for Buzz

1. The Separate Sound Receiver

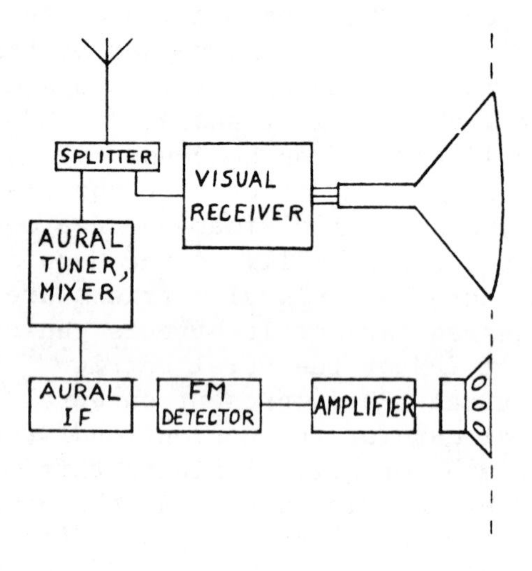

Figure 6

The aural carrier is transmitted without IPM. After the receiving antenna visual and aural carrier are split and separately processed. No IPM can transfer from the visual to the aural or to the intercarrier in this way. Such a circuit arrangement should be free of all buzz. The only way in which buzz can enter is through CATV or rebroadcast equipment, if improperly designed from a buzz point of view. Since this source exists for all systems to be discussed below it will not be further mentioned.

A separate sound receiver as part of a standard television receiver is obviously a very expensive addition. Only the circuits from the 4.5 MHz bandpass filter, or equivalent, downstream are needed in any case; tuner, mixer, l.o., IF, AGC, AFC, and some of these possibly twice for double conversion, are all extra.

2. The Split Sound Receiver

Also named: "Parallel Sound Receiver" or "Split-Carrier Sound Receiver"

QUASI-SPLIT SOUND RECEIVER

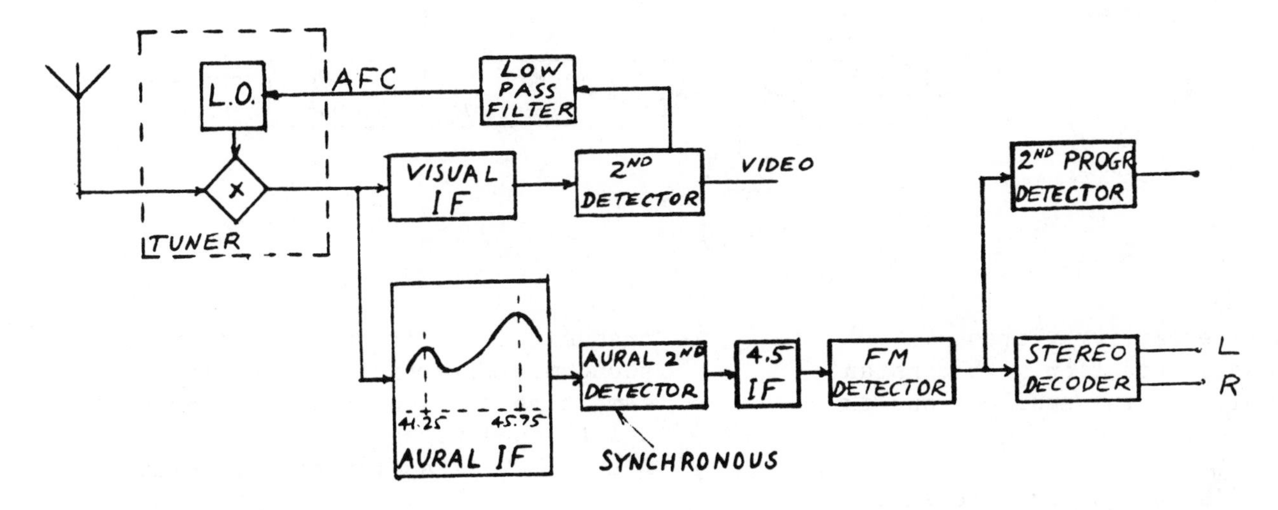

Figure 7

This method of sound processing was used in the early days of television before intercarrier detection was introduced. It was discussed above in the last paragraphs of section I B. Only tuner IPM can cause buzz because only in the tuner are visual and aural carriers co-processed. Video derived AFC can also cause IPM.

Compared to the Separate Sound Receiver as to complexity,only one tuner is needed but special care has to be taken to keep IPM out of it.

3. The Quasi-Split Sound Receiver

Also named "Quasi Parallel Sound Receiver"

A more descriptive name would be "Split-Intercarrier Sound Receiver"

This system also uses a separate sound channel but with intercarrier sound detection and is thus insensitive to IPM caused in the tuner. The Nyquist IPM is eliminated by a specially designed IF filter that has symmetrical response around the visual carrier. The aural 2nd detector will not eliminate transmitter IPM if it is of the envelope type or of the synchronous type with wide reference channel.

The latter type has a thermal noise advantage over the former. The visual carrier is contaminated by, among other, quadrature noise. The smaller the carrier (white signal, overmodulation) the larger the resulting phase angle. Thus in envelope detection the largest noise angle can be transferred

SPLIT SOUND RECEIVER

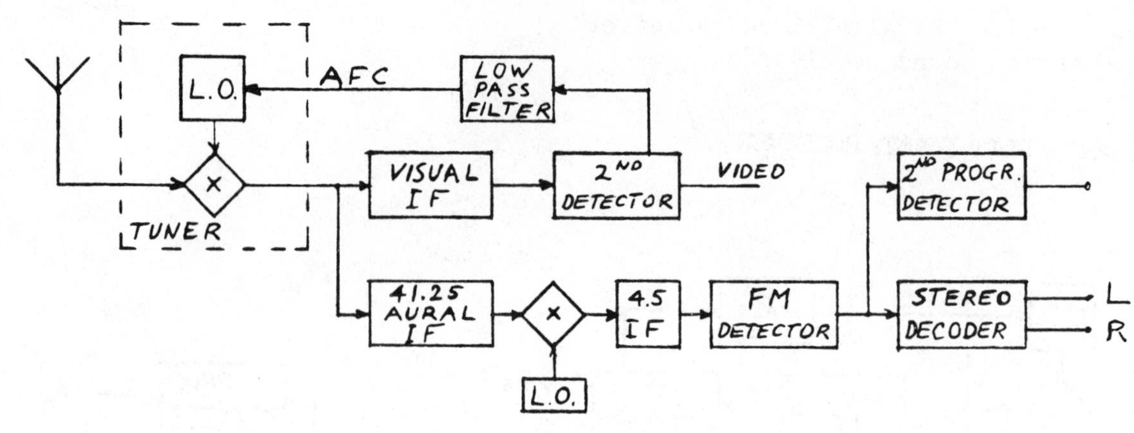

Figure 8

to the intercarrier whereas in synchronous detection with constant l.o. amplitude the smallest one is transferred.

If a narrowband reference channel were used for the synchronous detector l.o. then transmitter IPM would also be eliminated, to the degree that it falls outside of the reference channel. An additional advantage would be reduced phase noise transfer to the intercarrier especially at the higher audio baseband frequencies where it counts most. Unfortunately the narrowband requires extra tuning accuracy.

There have been reports of dual-output-SAW devices that can provide the conventional Nyquist-sloped output as well as the one indicated in Figure 8.

The circuits needed are not simpler than for the Split Sound Receiver but several advantages are gained. Insensitivity to tuner IPM; the advantage of normal tuner tolerance as compared to the much lower tolerance needed for the split receiver tuning. If transmitters that convert to stereo sound minimize their IPM at the same time the sensitivity of quasi-split sound processing to transmitter IPM is no longer a problem.

4. Additional Improvement

Another possibility for buzz improvement exists and will be discussed here even though it is more system than receiver based.

Noise Reduction Systems

At the present time there are available several companding type of noise reduction systems and more work is being done in this field. Though not specifically developed for buzz reduction, any system that features low level signal enhancement would also reduce the subjective effect of buzz. Since a new service is involved a unique chance exists to include such a system from the start. If it is not done from the start compatibility requirements greatly reduce the possible benefits. At this time comparative lab tests of several systems are being conducted under auspices of the EIA.

IV. Conclusions

Intercarrier buzz is a well-known phenomenon that receiver designers have contended with for years. Recent interest in improved (monophonic) television sound, in stereo and multichannel sound and in other subcarrier uses has stimulated work in decreasing buzz. This paper attempts to analyze the causes, partly quantitatively, and points out the road to cures with the compromises and tradeoffs involved.

First the receiver Nyquist slope effect has to be eliminated, next the transmitter IPM must be minimized and finally any combined visual aural processing must be done in circuits with sufficiently low parametric nonlinearity.

One final conclusion.

Buzz is a systems problem. It will take cooperation between transmitter manufacturers and operators, cable equipment manufacturers and cable system operations and receiver manufacturers to bring it down to acceptable levels - if not eliminate it.

REFERENCES

[1] W. L. Behrend, "Effects of Incidental Phase Modulation of TV Transmitters, or Other Circuits, on TV Signals", IEEE Trans. on Broadcasting, VOL. BC-19, #3, Sept. 1973, pp 53-63.

[2] R. H. Blair, "Comments on [1]", IEEE Trans. on Broadcasting, VOL. BC-21, #2, June 1975, pp 38-39.

[3] S. Kuroki, "Comments on [2]", IEEE Trans. on Broadcasting, VOL. BC-22, #1, March 1976, pp 18-19.

[4] S. Kuroki, "Effects of AM-to-PM Conversion of TV Transmitters on Video Signals", IEEE Trans. on Broadcasting, VOL. BC-22 #3, Sept. 1976, pp 57-67.

[5] T. Rzeszewski, "A System Approach to Synchronous Detection", IEEE Trans. on Consumer Electronics, VOL. CE-22, #2, May 1976, pp 186-193.

[6] F. Hilbert and N. Parker, "Technical Considerations for Stereophonic Broadcasting on TV", IEEE Trans. on Broadcast & TV Rec., VOL. BTR-11, #2, July 1965, pp 82-89.

[7] H. O'Neill, "Incidental Phase Modulation and Sound Channel Noise in Intercarrier Systems", E&D Report 106/72, Engineering Division IBA (Great Britain).

[8] S. Deutsch, Theory and Design of TV Receivers. New York, McGraw Hill, 1951, pp 465-467.

[9] S. A. Roth and C. W. Rhodes, "A New Precision TV Measurement Demodulator", IEEE Trans. on Broadcasting, VOL. BC 23, #4, Dec. '77, pp 109-112.

[10] C. W. Rhodes, Testing and Using Synchronous Demodulators, Tektronix TV Products Application 28, 1979.

[11] C. Eilers and P. Fockens, "Television Multichannel Sound Broadcasting - A Proposal", IEEE Chicago Spring Conference on Consumer Electronics, June 1981.

APPENDIX I

TRANSFER OF IPM FROM VISUAL TO AURAL CARRIER

INTERCARRIER ENVELOPE DETECTION

Assume an unmodulated picture carrier $f_1(t)$ with incidental phase angle $\theta = \theta(t)$

$$f_1(t) = A\cos(\omega_c t + \theta)$$

The aural carrier $f_2(t) = B\cos[\omega_c t - (\omega_s t + \phi(t))]$

where

$\omega_s = 2\pi \times 4.5 \times 10^6$

$\phi(t)$ = Desired angle modulation

The envelope E of $f_1(t) + f_2(t)$ is represented by

$$E = A\sqrt{1 + \left(\frac{B}{A}\right)^2 + \frac{2B}{A}\cos(\omega_s t + \phi + \theta)}\ .$$

When $B/A \ll 1$

$$E \approx A + B\cos(\omega_s t + \phi + \theta)$$

The incidental phase angle θ is transferred to the Intercarrier.

SYNCHRONOUS DETECTION

Assume a local carrier $f_3(t)$

$$f_3(t) = \cos(\omega_c t + \alpha)$$

Detector output = Baseband of $[f_3(t) \times (f_1(t) + f_2(t))]$

$$= \frac{1}{2} A \cos(\theta - \alpha) + \frac{1}{2} B\cos(\omega_s(t) + \phi + \alpha)$$

The Intercarrier is the second term.

If $f_3(t)$ is derived from $f_1(t)$ in a wideband reference channel

$$\alpha = \theta = \theta(t)$$

and the IPM is present on the Intercarrier.

As the reference channel is narrowed (to say f_o) $\alpha = \alpha(t)$ will have a spectrum limited to f_o.

TV MULTICHANNEL SOUND—THE BTSC SYSTEM

Carl G. Eilers
Zenith Electronics Corporation
Glenview, IL 60025

Introduction

In late 1978, the Broadcast Television System Committee of the Electronic Industries Association, on behalf of the television industry, formed a subcommittee for the purpose of formulating standards for the broadcasting and reception of multichannel television sound which was to include stereophonic as well as second program (second language, for example) enhancements of the main audio program. On December 22, 1983, the industry chose the Zenith transmission system coupled with the dbx noise reduction system and submitted the combined system, the BTSC system, to the Federal Communications Commission on January 30, 1984. On April 23, 1984, the FCC released a Report and Order in Docket No. 21323 wherein the choice of technical standards was left to the marketplace, but protection was afforded to the pilot frequency of the BTSC system. In addition, if a pilot is transmitted at 15.734 kHz, the BTSC system must be employed as described in the FCC office of Science and Technology Bulletin No. 60.

A Brief Description of the BTSC System

The transmission standards are illustrated in Figure 1 and summarized in Table I.

The main channel modulation consists of an (L+R) audio signal. The pre-emphasis is 75 microseconds. The L-R audio signal is subjected to level encoding according to Figure 2 which is part of the dbx Companding System that includes complementary decoding (expansion) in the receiver. The encoded L-R signal causes double sideband, suppressed carrier amplitude modulation of a subcarrier at $2f_H$. The audio bandwidth of pre-emphasized L+R and of encoded L-R is 15 kHz.

The main channel peak deviation is 25kHz. With level encoding temporarily replaced by 75 microseconds pre-emphasis the subchannel peak deviation is 50 kHz. When L and R are statistically independent, the peak deviation of the main channel and the stereophonic subchannel combined is also 50 kHz due to the interleaving property. When L and R signals are not statistically independent or when (L+R) and (L-R) signals do not have matching pre-emphasis characteristics (as is the case when (L-R) is encoded), the combined deviation of main channel and stereo-phonic subchannel is constrained to 50kHz and the separate components assume their respective natural levels dictated by the acoustic scene.

A CW pilot subcarrier signal of frequency f_H is transmitted with a main carrier deviation of 5 kHz.

The subcarrier for the SAP channel has a frequency of $5f_H$ (78.670 kHz) and is frequency locked to $5f_H$ in the absence of modulation. The SAP audio signal is subjected to level encoding identical to that of the L-R signal. The resulting SAP modulating signal is bandlimited to 10 kHz and frequency modulates the SAP subcarrier to a peak deviation of 10kHz. The main carrier deviation by this subcarrier is 15kHz.

The professional subchannel has a subcarrier located at approximately 6.5 f_H and modulates the main carrier by 3 kHz peak deviation.

Other Baseband Configurations

The foregoing description illustrated the case for a fully loaded multi-channel sound base band.

Of course, some transmissions will consist of monophonic audio with second audio program (SAP) with or without non-public subcarrier(s). This is illustrated in Figure 2.

Another base band configuration might consist of main (L+R) channel, pilot and stereophonic (L-R) subchannel with or without non-public subcarrier(s). This is illustrated in Fugure 3.

The last base band configuration, consisting of monophonic audio with or without non-public subcarrier(s) is shown in Figure 4.

Broadcast Standards

The radiated signal parameters and accompanying performance standards are summarized in Chart I.

Some Comparisons with FM Stereophonic Broadcasting

In this section, some comparisons will be made between the television multichannel sound transmission system which includes the dbx encoding process and the familiar Broadcast FM stereo system.

Both Broadcast FM Stereo and Television Stereo Sound use the (L+R) signal on the main channel; they also both use an AM, double-sideband, suppressed carrier modulated subcarrier and a pilot at half the subcarrier frequency.

The new AM subcarrier has a frequency of $2f_H$ (f_H = 15.734 kHz, the picture horizontal scanning frequency). Another difference is the modulated stereophonic subcarrier injection which varies in accordance with the dbx algorithm. The L-R signal is encoded (compressed) while the L+R signal is not. This requires a different arrangement of pre-emphasis networks, lowpass filters and matrix circuits. The pilot

Manuscript received June 11, 1984.

Reprinted from *IEEE Trans. Consum. Electron.*, vol. CE-30, pp. 236–240, Aug. 1984.

subcarrier has increased injection. Broadcast FM stereo generator lowpass filters typically have loss poles at 19 kHz. For television stereo audio, these filters have a losspole at f_H = 15.734 kHz while maintaining 15 kHz audio bandwidth. Second Audio Program (SAP) signals are similar to SCA (Subsidiary Communications Authority) signals except for injection levels and the intended audience. SAP is a public service whereas SCA is a non-public service.

The SAP subcarrier is frequency-locked to a reference (5 times horizontal scan rate) in the absence of modulation, while SCA subcarriers need no special frequency stability.

The pilot zero crossings in relation to the stereophonic subcarrier zero crossings have a tolerance of 3 degrees for the television case while for FM broadcast standards, no tolerance is given in the FCC rules.

In the FM broadcast case both the (L+R) and (L-R) audio modulation signals are subjected to 75 microsecond pre-emphasis, whereas in the television case, only the (L+R) audio modulation signal is subject to 75 microsecond pre-emphasis while the (L-R) audio modulation is subject to variable pre-emphasis.

Audio Signal Processing and Companding

Audio signal processing is not new to the broadcaster and can be used optionally in conjunction with television multichannel sound source material. The dbx Companding System is, however, the mandatory noise and interference reduction companion to the chosen transmission system and is based on complementary audio processing at the transmitter and at the receiver.

In order to obtain respectable stereo signal-to-termal noise ratios at the Grade B contour, it was clear at an early stage in the system development that noise reduction would be a desirable feature. In order to maintain monophonic (L+R) compatibility, it was decided not to compand the main channel but to compand only (L-R). Noise reduction was accomplished because most of the noise is introduced in the subchannel. The dbx System includes a Stereo difference compressor and a separate SAP compressor on the transmitting end and a single expander for both Stereo and SAP in the receiver. Stereo separation will be influenced by the degree to which compressor and expander processing are complementary.

Professional Channel

As noted before, the BTSC system can accommodate a subchannel of modest bandwidth at a center frequency of approximately 6.5 f_H. If used for audio, the quality will be similar to voice grade with approximately 3-4 kHz of modulation capability. The subcarrier deviation should be held to approximately 3 kHz. The BTSC standards call for a 3 kHz deviation of the aural carrier by the subcarrier. However, the FCC in the recent Report and Order allows a total aural carrier deviation of 75 kHz, thus, perhaps, allowing the broadcaster to inject the non-public professional channel at a 5 kHz aural carrier deviation level. Of course, the broadcaster must show no degradation to the program services when so doing.

Hardware for the Total System

The hardware configuration for a typical television transmitter and a typical television receiver for the entire Zenith/dbx system is shown in Figure 5. Greater detail is presented in the companion papers to this one -- "TV Multichannel Sound-Encoding and Transmission" and "TV Multichannel Sound-Reception and Decoding."

Television Transmitter

For the television transmitter, a source of left (L) and right (R) stereo audio as well as a second audio program (SAP) source is required. Additionally, horizontal synchronizing pulses are derived from the incoming program video to lock the pilot and SAP subcarriers. Sum (L+R) and difference (L-R) signals are derived from the incoming L and R signals and processed with 75 sec pre-emphasis and dbx compression, respectively, and, in turn, are processed into a stereo composite signal. The SAP audio is submitted to dbx compression and applied to the SAP FM subcarrier modulator, the output of which is combined with the stereo composite signal (including pilot) and an optional professional channel. The entire composite signal is impressed on the frequency modulated aural exciter/transmitter which, in turn, is combined with the visual transmitter output and fed to the transmitting antenna.

Television Receiver

In the television receiver, the aural channel baseband signal is provided from the received television broadcast by an appropriate FM detector in an intercarrier mode or split-sound mode with due attention paid to sound buzz and delivered to the stereo decoder and SAP FM detector. The stereo decoder provides sum (L+R) and difference (L-R) signals to a 75 sec de-emphasis circuit and a dbx expander circuit, respectively, which, in turn, are matrixed to make left (L) and right (R) audio and further power amplified to drive left and right loudspeakers. The SAP FM detector provides a signal to a dbx expander (maybe the same expander, above) which, in turn, feeds a power amplifier (perhaps the above power amplifier in a monophonic mode) and loud speaker arrangement.

Conclusion

The BTSC system delivers a stereophonic program audio quality essentially limited by the main channel (monophonic) quality, and a SAP audio program of somewhat lesser quality.

References

[1] "Intercarrier buzz phenomina analysis and cures" IEEE Transactions on Consumer Electronics, Vol. CE-27, No. 3, August, 1981, pp 398-409.

[2] "Compandor Complexity Analysis," EIA, December 12, 1983, pp 22-48.

[3] "Intercarrier Buzz in Television Receivers," Proceedings, 37th NAB Engineering Conference, pp 251-266.

[4] "The Zenith Multichannel TV Sound System," Proceedings, 38th NAB Engineering Conference, pp 352-368.

[5] "TV Multichannel Sound-Encoding and Transmission," R. Lee, R. Granstrom, D. DeWeger (Companion Paper).

[6] "TV Multichannel Sound-Reception and Decoding," V. Mycynek and P. Fockens (Companion Paper).

[7] FCC Report and Order, Docket 21323, "In the Matter of the Use of Subcarrier Frequencies in the Aural Baseband of Television Transmitters."

[8] FCC Office of Science and Technology Bulletin No. 60, "Multichannel Television Sound Transmission and Audio Processing Requirements for the BTSC System."

CHART I

Radiated Signal Parameters and Tolerances

Main Channel

1. Modulating Signal		L+R
2. Frequency Range		50 Hz - 15 kHz
3. Pre-emphasis		75 sec
4. Aural Carrier Deviation by Main Channel	max.	25 kHz

Pilot Subcarrier

1. Frequency (color program)		f_H = 15.734 kHz
2. Frequency (black-white program)(no burst)		15,734 ± 2 Hz
3. Aural Carrier Deviation by pilot subcarrier		5 kHz
4. Pilot-to-Interference Ratio (1,000 Hz band) (Reference 5 kHz deviation)	min.	40 dB

Stereophonic Subchannel

1. Modulating Signal		L-R
2. Frequency Range	(dbx encoded)	50 Hz - 15 kHz
3. Subcarrier Frequency		2 f_H = 31.468 kHz
4. Subcarrier Modulation Method		AM-DSB-SC
5. Aural Carrier Deviation by modulated stereophonic subcarrier	max.	50 kHz
6. Aural Carrier Deviation by main channel signal plus modulated stereophonic subcarrier	max.	50 kHz
7. Stereo Subcarrier suppressed to Aural Carrier Deviation of	max.	0.25 kHz
8. The Stereophonic subcarrier shall cross the time axis with a positive slope simultaneously with each time axis crossing of the pilot subcarrier		
9. The difference between time axis crossings of the pilot subcarrier and of the stereophonic subcarrier (in pilot frequency degrees) is	max.	3 degrees

Composite Stereophonic Modulation

1. Stereophonic Separation (50 Hz - 15 kHz) (no dbx Encoding)			40 dB
2. Equivalent Input Separation (@ 10%, 75 microsec equivalent modulation):	50 - 100 Hz	min.	26 dB
	100 - 8,000 Hz	min.	30 dB
	8,000 - 15,000 Hz	min.	20 dB
3. Total Distortion (30 kHz band; includes dbx Encoding):	50 - 100 Hz	max.	3.5%
	100 - 7,500 Hz	max.	2.5%
	7,500 - 15,000 Hz	max.	3%
4. Crosstalk of stereo subchannel signal into main channel (reference 25 kHz deviation)		max.	-40 dB
Crosstalk into main channel (reference 25 kHz deviation)		max.	-60 dB
Crosstalk of main channel signal into stereophonic subchannel (reference 50 kHz deviation)		max.	-40 dB
Crosstalk of all multiplex signals into the stereophonic subchannel (reference 50 kHz deviation)		max.	-60 dB
5. FM Noise on aural carrier in main channel range (reference 25 kHz deviation)		max.	-58 dB
FM Noise on aural carrier in stereophonic subchannel range (reference 50 kHz deviation)		max.	-55 dB
AM Noise (50 - 15,000 Hz band; (reference 100% AM)		max.	-50 dB

CHART I (continuation)

Second Audio Program Subchannel

1.	Modulating Signal	(dbx Encoded)	SAP Signal
2.	Frequency Range		50 Hz - 10 kHz
3.	Subcarrier Frequency		$5f_H$ = 78.67 kHz
4.	Subcarrier Frequency Tolerance	max.	±500 Hz
5.	Subcarrier Modulation Method		FM
6.	Subcarrier Deviation	max.	10 kHz
7.	Aural Carrier Deviation by SAP Subcarrier	max.	15 kHz
8.	Total Distortion (20 kHz band; (includes dbx Encoding) 50 - 100 Hz	max.	3.5%
	100 - 7,500 Hz	max.	7%
	7,500 - 10,000 Hz	max.	3%
9.	Crosstalk from stereo into SAP (reference 10 kHz deviation)	max.	-50 dB
10.	FM Noise on aural carrier in SAP Channel Range (reference 10 kHz deviation)	max.	-50 dB

dbx Encoding

1.	The gain through the dbx Encoder for a: 300 Hz tone, causing 14.1% modulation, equals		0 dB
	8,000 Hz tone, causing 32% modulation, equals		18.4 dB
2.	The equivalent input tracking (50 Hz - 10 kHz or 15 kHz band; including equivalent input noise) is	max.	0.3 dB
3.	The equivalent Encoder input noise level (10 or 15 kHz band; reference 100 Hz, 100% 75 microsec. equivalent modulation) is	max.	-70 dB

Aural Transmitter

1.	Deviation range	min.	73 kHz
		recommended	100 kHz
2.	Modulation bandwidth	recommended	120 kHz

Visual Transmitter

1.	The amplitude of a 4.5 MHz sine wave video signal in the radiated signal referenced to the amplitude of a 200 kHz sine wave video signal in the radiated signal shall be	max.	-30 dB
2.	Incidental Phase Modulation of the visual carrier (1 - 94 kHz band) for a: carrier amplitude between white and blanking level is	max.	3 degrees
	carrier amplitude between blanking level and sync tip is	max.	5 degrees

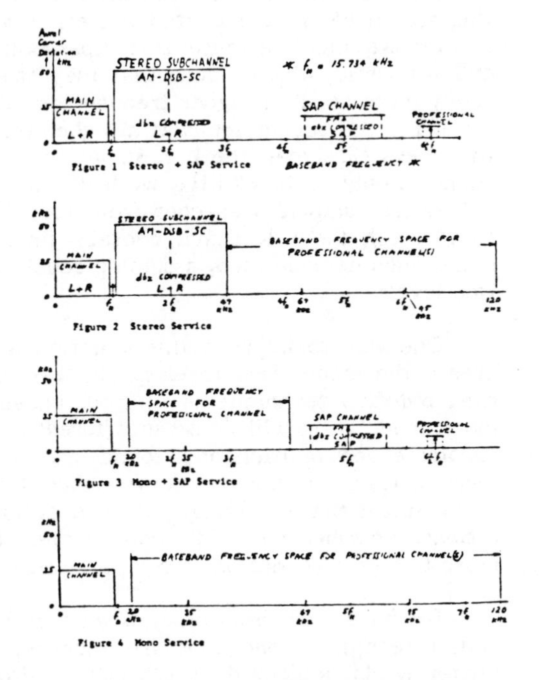

Figure 1 Stereo + SAP Service

Figure 2 Stereo Service

Figure 3 Mono + SAP Service

Figure 4 Mono Service

TABLE I

AURAL CARRIER MODULATION STANDARDS							
Service or Signal	Modulating Signal	Modulating Frequency Range kHz	Audio Processing or Pre-Emphasis	Subcarrier Frequency *	Subcarrier Modulation Type	Subcarrier Deviation kHz	Aural Carrier Peak Deviation kHz
Monophonic	L+R	.05-15	75 µsec				25 †
Pilot				f_H			5
Stereophonic	L-R	.05-15	dbx Compression	$2f_H$	AM-DSB SC		50 †
Second Program		.05-10	dbx Compression	$5f_H$	FM	10	15
Professional Channel	Voice or Data	.3-3.4 0-1.5	150 µsec •	$6\frac{1}{2}f_H$	FM FSK	3	3
						Total	73

* f_H = 15.734 kHz

† Sum does not exceed 50 kHz

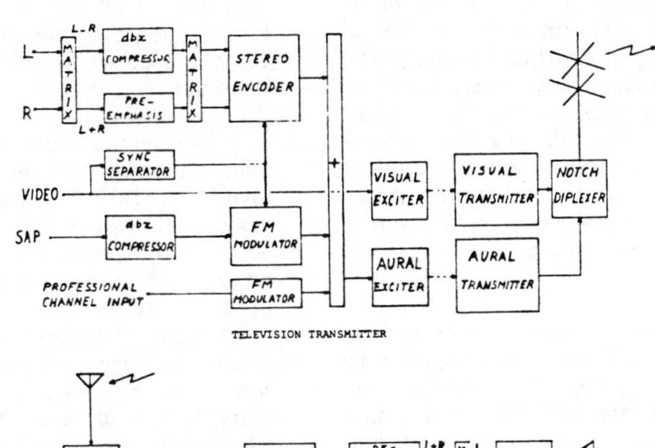

TELEVISION TRANSMITTER

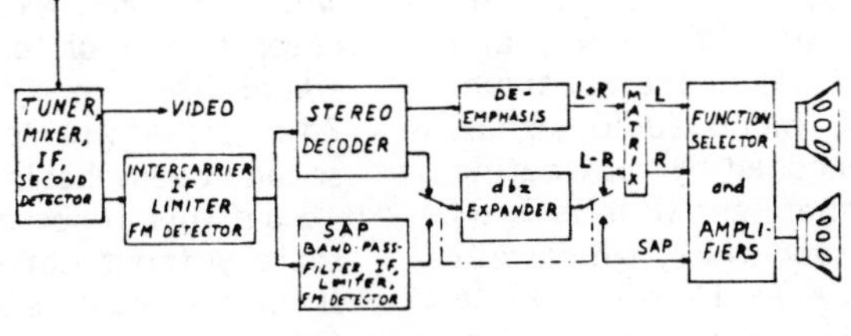

TELEVISION RECEIVER

Fig. 5 The BTSC Multichannel TV Sound System

A COMPANDING SYSTEM FOR MULTICHANNEL TV SOUND

Leslie B. Tyler,
Mark F. Davis,
William A. Allen
Engineering Department
dbx Inc.
Newton, MA USA

SUMMARY

The US Federal Communications Commission recently authorized the broadcast of multichannel sound for television. A five-year industrywide effort sponsored by the Electronics Industries Association resulted in a new standard for transmission of stereo information with the simultaneous opportunity for broadcasting a separate Second Audio Program (SAP) channel. This standard includes a unique three-part companded noise-reduction system intended to allow broadcasters to increase their programing options without reducing present coverage area.

This paper discusses the design goals for the noise-reduction system, a mathematical model of the system, and its performance.

INTRODUCTION

Without companding, the BTSC* multichannel-TV-sound-transmission system is capable of very high-quality transmission and reception of stereo audio. However, its full potential will never be realized without a good companding system. The reasons for this lie in the nature of the stereo television signal itself and in the practical limits to the quality (signal level, interference level, etc.) of the received signal. In order to understand these limits and how the recommended companding system can circumvent them, this paper will first discuss the basic characteristics of the sound-transmission system and then explain how the compander works to improve it.

The millions of mono television sets now in use make compatibility of the stereo signal with them of prime importance. Accordingly, the BTSC system transmits the sum of the left and right stereo audio signals (L+R) in the spectrum space now occupied by the mono audio signal. The stereo information is encoded by subtracting the left audio signal from the right (this is L-R) and transmitting it over a subcarrier, which will be received only by new, stereo TV sets. While an existing, mono set will ignore this subcarrier, a stereo set will use the L-R signal to reconstruct the original left and right audio signals by adding, subtracting, and scaling the L+R and L-R signals.

*The EIA's Broadcast Television Systems Committee system combines a proposal for subcarrier allocation and modulation from Zenith with a proposal for companding from dbx. The BTSC recommended to the FCC that this combination be the standard for US multichannel television.

Manuscript received September 28, 1984.

STEREO-NOISE LEVELS[1]

The L-R signal is transmitted via an AM subcarrier located at 31.468 kHz (twice the video horizontal-scanning frequency of 15.734 kHz), superimposed on the present FM audio carrier. Except for the subcarrier frequency, the system is essentially the same as that used for transmitting stereo FM in the United States. FM transmission systems like these have a parabolic noise characteristic, which means that they have relatively more noise at higher frequencies. Because the L-R's AM subcarrier is at a higher frequency than the L+R signal (the L+R signal contains frequencies only up to 15 kHz, while the modulated subcarrier bandwidth extends from about 17 kHz to 46 kHz), the L-R signal contains much more noise when demodulated, assuming equal modulation levels.

One way to improve this situation is to increase the modulation level of the L-R subcarrier, but this technique is limited, because too much modulation will cause interference. In the recommended multichannel-sound system, 6 dB more L-R modulation is allowed than for L+R. The result is that even under ideal reception conditions, the subcarrier adds approximately 15 dB of noise to stereo reception as compared with mono.

To make matters worse, when transmission and/or reception conditions are impaired (transmitter ICPM, multipath, etc.), buzz and/or hum may be introduced into the audio. This would further degrade the stereo signal-to-noise ratio as compared with mono if no companding system were used.

The significance of this added noise can be appreciated by calculating the expected S/N ratio under typical, less-than-ideal reception conditions. Grade-B reception describes the condition wherein the picture is somewhat snowy but still acceptable to most viewers. The mono S/N ratio in this case is about 65 dB, which means that most mono listeners are not bothered by the noise. In stereo, however, the situation is 15 dB worse, for an S/N ratio of 50 dB. These figures refer to the difference between peak-sinewave-signal levels and unweighted noise floors after 75-microsecond deemphasis. (Without 75-microsecond deemphasis, the S/N ratio is about 43 dB.) To put this in perspective, a Philips compact cassette without noise reduction provides 55-60 dB from the peak signal to the unweighted noise floor. Without companding, grade-B audio reception is not even this good -- certainly unacceptable for high fidelity, and obtrusively noisy for uncritical listening as well.

Perhaps the most important consequence of this noise is its effect on coverage area. A TV viewer who becomes excited about stereo reception and runs out to purchase a stereo-equipped set will be very disappointed if he lives in a grade-B reception area. Imagine bringing the new set home, hooking everything up, turning on a favorite

Reprinted from *IEEE Trans. Consum. Electron.*, vol. CE-30, pp. 633-640, Nov. 1984.

channel (one that has been touting the new stereo service) and hearing hiss that was never before audible! Worse, the hiss disappears when he switches to mono! Will the customer blame the set manufacturer, the TV station, or both? The only parties likely to escape blame are the system inventors.

SAP-NOISE LEVELS

The situation is worse for the Second Audio Program (SAP) channel. Because the SAP subcarrier is at 78.67 kHz, even more noise is introduced due to the FM parabolic noise curve. Owing to the potential for interference at this frequency, the SAP signal-modulation level may not be allowed to be as large as that of the L+R signal. Furthermore, this is an FM subcarrier, which makes it additionally subject to buzz beat, an intermodulation of the picture with the audio that causes a particularly obnoxious type of non-harmonically related distortion.

The SAP S/N ratio with 75-microsecond deemphasis is about 33 dB in grade-B reception and reaches only about 43 dB in grade A (without 75-microsecond deemphasis, these S/N ratios are 26 and 36 dB). Again, these figures compare poorly even to a cassette without noise reduction.

AM-INTERLEAVE EFFECTS

Another factor that must be considered in understanding the limits to performance of the sound-transmission system is that the modulation level of the composite signal (L+R audio plus L-R modulated subcarrier) is dependent on the sum of the L+R and L-R signals. Not only must the individual signal levels be limited in order to prevent overmodulation, but the sum of the two must also be limited. In the recommended system, if full modulation of the L-R subcarrier is achieved, the L+R signal cannot simultaneously reach its full modulation level. If the noise-reduction system used with the sound-transmission system reduces the possibility of the L-R subcarrier's reaching full modulation, the L+R signal need not be constrained as much.

Fortunately, early on in the multichannel-TV-sound evaluation program, the EIA recognized the likely limits to performance of all the proposed transmission systems, and sought out companding proposals that would reduce or eliminate these problems.

DESIGN GOALS

The recommended noise-reduction system was designed to aid the multichannel-sound-transmission system in delivering a clean, noise-free audio signal into the home. It is capable of providing consistently high-quality audio in the face of the variety of possible channel degradations. Specifically, the system was to:

1) Provide significant amounts of noise reduction even in poor-reception areas (grade-B or worse).
2) Preserve input-signal dynamic range without headroom loss or other anomalies.
3) Prevent the stereo subcarrier from interfering with overall transmitted power levels (AM-interleave effects).
4) Ensure reliable, effective performance even in the face of severe manmade-noise and transmission/reception-system impairments.
5) Provide this noise reduction at reasonable cost and simplicity.

WHERE TO USE NOISE REDUCTION

It was clear that the noise in the sound-transmission system without companding would come from the subcarriers and not the main (L+R) audio channel. Therefore, the system was designed to work only on the L-R and SAP channels and to transmit the L+R signal without companding. This decision also addressed compatibility of the compander with the TV sets now in use: if no changes were made to the mono signal before transmission and the new subcarriers would not interfere with present receivers, then compatibility would be guaranteed.

Another decision that was made early on in the compander design program was to use the same compander for both the L-R and SAP channels if at all possible. This would allow TV-receiver manufacturers to produce sets at the lowest possible cost, because only one noise-reduction decoder need be included per set, switchable between the L-R and SAP channel.

A block diagram of a system embodying this philosophy is shown as Fig. 1.

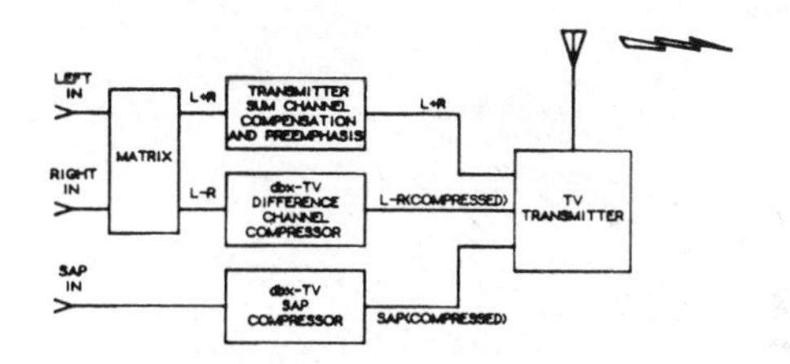

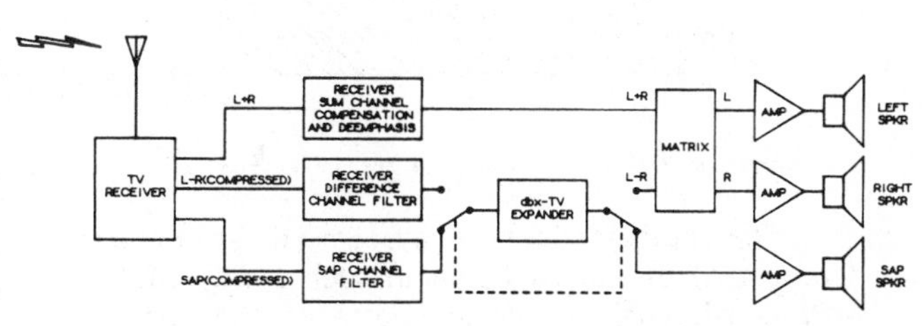

Figure 1: Block Diagram, Complete Audio System

In view of the limited dynamic range in the impaired channels and the performance goals listed above, it was necessary to design a new compander specifically for this task. What follows is a brief discussion of the psychoacoustics of noise reduction and a description of the system.

MASKING

All audio-noise-reduction systems work on the principle of masking. This principle can be stated simply:

If a desired program signal (music or speech) is loud enough and broad enough in its spectral content, then the ear's attention will be captured by this signal rather than by the noise of the transmission medium.

For example, if the program consists of a single low-frequency sinewave, it must be transmitted at a very high level relative to the background noise of the stereo-subcarrier channel in order for the ear's attention to be captured by the low-frequency note and for the listener to be unaware of the background noise (Fig. 2). On the other hand, if the music is a raucous electric guitar, its spectrum is so broad that it does not need to be much higher in level than the background noise for the noise to fade below the listener's threshold of perception (Fig. 3).[2]

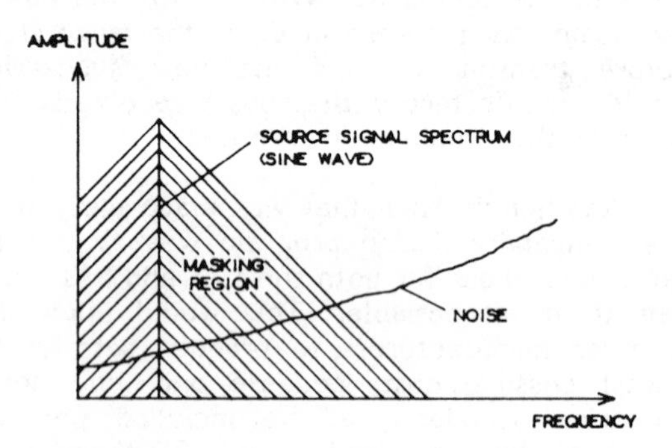

Figure 2: Masking of Noise by a Low-Frequency Tone

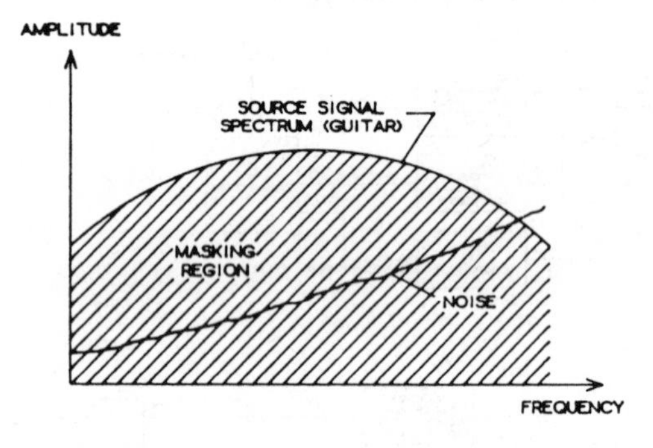

Figure 3: Masking of Noise by a Broad-Spectrum Signal

The noise-reduction system must encode (compress) the audio signal in such a way that it will consistently mask the noise of the channel during transmission. Then it must decode (expand) the received signal to recover the original audio. Of course, there should be no distortion or other degradation of the audio itself in passing through the encode/decode (companding) cycle. And in the decoding process, all the audible noise should be eliminated. As noted above, this requires two conditions:

1) The level of the transmitted audio must be high relative to the background noise.
2) The spectrum must be conducive to masking.

The background-noise spectrum of the recommended transmission system's stereo subcarrier is white, having a 3-dB/octave rising characteristic, while the SAP subcarrier has a 9-dB/octave rising characteristic. Masking of this noise in the presence of signal will therefore take place only if the transmitted-signal spectrum contains substantial high-frequency energy, especially in the case of the SAP channel. If the program material itself could be relied upon to have sufficient high-frequency content, then the compander would only have to keep amplitude levels high through the transmission channel. However, most program material has its dominant energy at low frequencies.

PREEMPHASIS AND DEEMPHASIS

The simplest way to transform such program material into a signal conducive to masking is to apply fixed preemphasis to the signal before transmission. This is done in present mono TV audio and in FM broadcasting by using 75-microsecond preemphasis. This preemphasis changes the spectrum of the average program material (containing mostly low frequencies) to be more evenly balanced between highs and lows. In the TV or FM receiver, the signal is deemphasized, which restores correct tonal balance and at the same time reduces the audibility of the noise.

Fixed pre- and deemphasis is used in the companding system, too, for the same purpose. In it, however, two preemphasis networks are used. One is essentially the same as 75 microseconds (actually 72.7). The other is 390 microseconds, but the rising frequency response created by this preemphasis is curtailed at 30 microseconds. The transfer function F(f) describing this preemphasis is shown below.

$$F(f) = \frac{(jf/408\text{Hz})+1}{(jf/5.23\text{kHz})+1} \cdot \frac{(jf/2.19\text{kHz})+1}{(jf/62.5\text{kHz})+1}$$

The frequency-response curve of the complete preemphasis (Fig. 4) has a steep section between about 2 and 5.5 kHz, which helps the

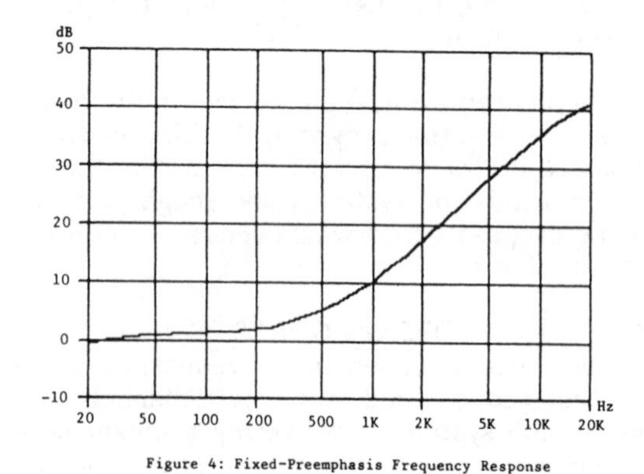

Figure 4: Fixed-Preemphasis Frequency Response

companding system overcome the large amounts of high-frequency noise present in grade-B reception. In the TV receiver, a corresponding deemphasis restores tonal balance to the program material and reduces the audibility of hiss picked up in transmission. The transfer function $F^{-1}(f)$ describing this deemphasis is shown below.

$$F^{-1}(f) = \frac{(jf/5.23kHz)+1}{(jf/408Hz)+1} \cdot \frac{1}{(jf/2.19kHz)+1}$$

Unfortunately, audio program material is inconsistent in its spectral balance and its level. With fixed preemphasis alone, two problems would remain. First, some audio signals contain predominantly high frequencies. Strong preemphasis would boost them too much, causing overmodulation. This is called lack of headroom -- insufficient room for the peaks of the program to be transmitted cleanly. Second, some audio signals would be too low in level and too lacking in high frequencies to mask the channel noise properly even with preemphasis.

SPECTRAL COMPANDING

In order to make the audio more consistent in its spectral balance before transmission, the companding system uses a second stage, in which the frequency response of the system varies to suit the signal. We call this spectral companding. The spectral compressor (in the encoder) monitors the spectral balance of the input signal ("How much high-frequency material is there?") and varies the high-frequency preemphasis accordingly. When very little high-frequency information is present (and masking is least likely), the spectral compressor provides large high-frequency preemphasis. When strong high frequencies are present (and overload is most likely), the spectral compressor actually provides deemphasis, thereby reducing the potential for high-frequency overload. The system frequency response is dynamically adjusted so that the resulting encoded signal consistently contains a substantial proportion of high frequencies before transmission. This increases the likelihood of masking the channel noise.

HOW SPECTRAL COMPRESSION WORKS

The spectral compressor consists of a variable pre/deemphasis stage and a band-limited rms-level detector (Fig 5). The filtering that precedes the detector restricts its sensing range

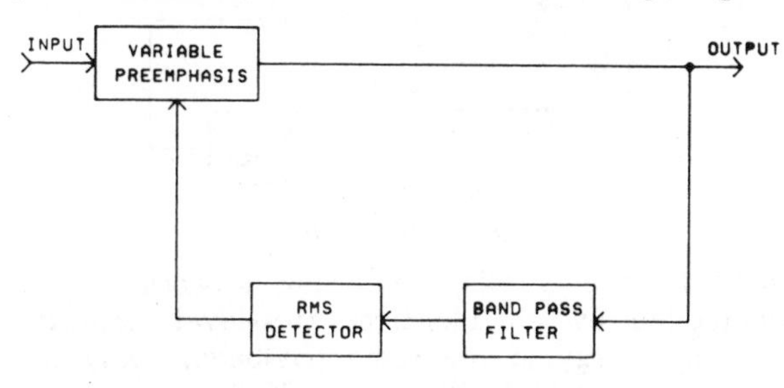

Figure 5: Spectral Compressor

to high audio frequencies. The detector output is actually a voltage proportional to the log of the sensed energy (therefore proportional to the "decibel-rms" value of the high-frequency energy). This voltage is used to control the pre/deemphasis characteristic.

The variable-preemphasis/deemphasis stage works by varying the gain of a voltage-controlled amplifier (VCA) embedded in a frequency-selective network (Fig. 6). When the gain is low, no

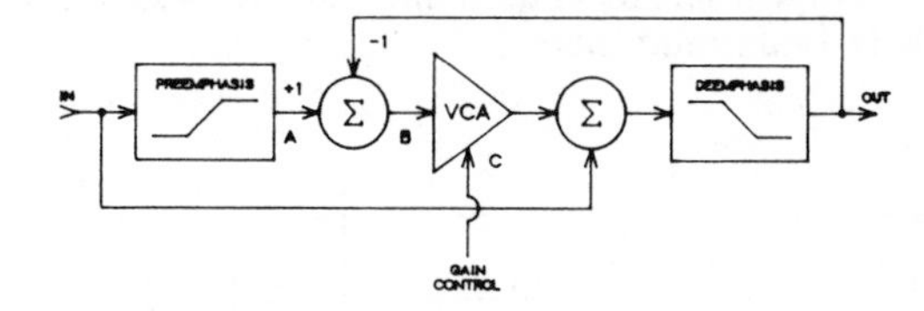

Figure 6: Variable-Preemphasis/Deemphasis Network

signal appears at point B. The only transfer from the input to the output must go from point C to the output, which passes through the deemphasis network. This attenuates high frequencies.

Alternatively, when the VCA gain is high, the response between point A and the output is essentially flat (this is analogous to an op amp with a closed feedback loop). The signal at point C is small compared with the signal at point B (because of the high VCA gain), so the output signal is essentially the same as that at point A, which is the input signal after preemphasis. This provides gain to high frequencies.

At intermediate VCA-gain settings, the frequency-amplitude response varies smoothly between these two extremes, with flat occurring at unity (0-dB) VCA gain.

The range of variation in the frequency-response curves that the spectral compressor can

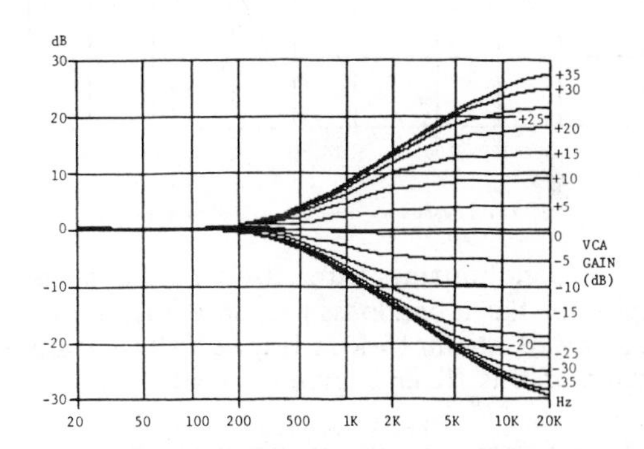

Figure 7: Spectral Compressor, Frequency-Response Range

produce is substantial: from +27 to -27 dB at 15 kHz (Fig. 7). The transfer function S(f,b) describing this response is shown below.

$$S(f,b) = \frac{1 + (jf/F)(b+51)/(b+1)}{1 + (jf/F)(1+51b)/(b+1)}$$

where F = 20.1 kHz, $b = 10^{(d/20)}$, and d is the decibel-rms value of the signal sensed by the bandlimited rms detector. Since the VCA used is exponential-responding, the conversion from the variable d (representing the log-scaled control voltage) to the variable b (representing the linear voltage gain of the VCA) is easily accomplished.

Because the spectral compressor and the fixed preemphasis are in series in the signal path, the range of possible responses from the system as a whole (Fig. 8) varies from nearly flat (spectral compressor at maximum cut) to +55 dB at 15 kHz (maximum boost).

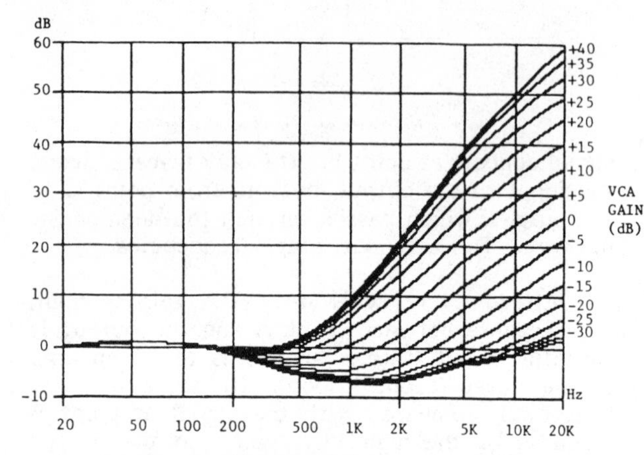

Figure 8: Frequency-Response Range of Spectral Compressor and Fixed Preemphasis

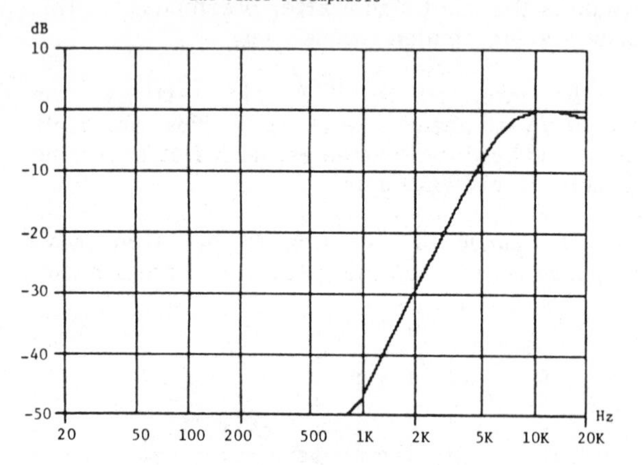

Figure 9: Spectral Compander, Frequency Response of Rms Detector

The bandlimiting filter used in the spectral compressor has a bandpass response with a center frequency of about 10 kHz (Fig. 9). The transfer function of this filter is given below.

$$Q(f) = \frac{(jf/7.66\text{kHz})^2}{[(jf/7.66\text{kHz})^2 + (jf/7.31\text{kHz}) + 1]} \cdot \frac{1}{[(jf/26.9\text{kHz}) + 1]} \cdot \frac{(jf/3.92\text{kHz})}{[(jf/3.92\text{kHz}) + 1]}$$

Because of this bandpass filter, the spectral-compressor detector senses the level of high frequencies in the compressor output. When high levels of high-frequency energy are detected, the VCA gain is increased. This results in deemphasis, reducing high-frequency gain. When low levels of high-frequency energy are detected, the VCA gain is decreased. This results in preemphasis, increasing high-frequency gain. In effect, the range of spectral variation at the output is reduced, "compressing" the spectral dynamic range.

The detector output corresponding to 0 dB (d = 0 dB, the point at which the variable-preemphasis network produces a flat response) is an important parameter in determining the characteristics of the system. We have chosen this to occur when an 8-kHz sinewave is at 5.16% modulation at the output of the system. (Note that this refers to the output-signal level, not the level at the input to the rms detector itself.) The practical consequence of this alignment is to make the compressor reduce the amplitude of strong high frequencies so that they will tend not to cause overmodulation of the channel.

An rms detector is used because of its unusual combination of properties, including relative insensitivity to phase shifts in transmission and a unique blend of signal-dependent time constants governing its acquisition and release behavior. A complete description of the rms-level detector is beyond the scope of this paper.

SPECTRAL EXPANSION

During reception, the spectral expander (in the decoder) will restore high frequencies to their proper (original) amplitude. If the original input signal contained predominantly low frequencies, the decoder will attenuate the high-frequency background noise, leaving the low-frequency signal and only the low-frequency background noise, which will be masked by the signal. If the original input signal contained predominantly high frequencies, the decoder will not need to attenuate the high frequencies to restore correct frequency response; instead, high-frequency gain may be necessary. However, the signal itself will provide masking for the noise in this case.

The spectral expander (Fig. 10) is a mirror image of the spectral compressor. The same

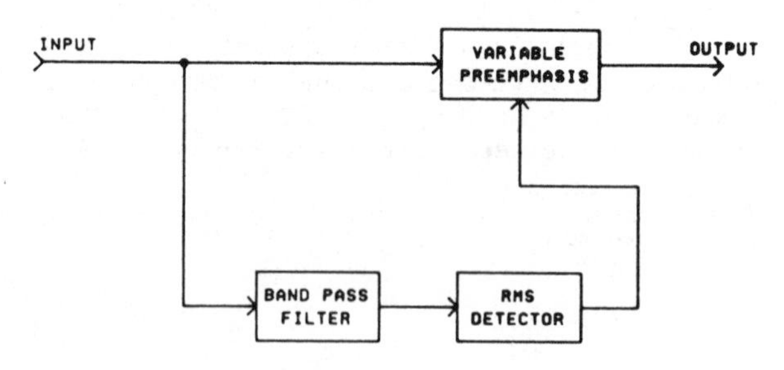

Figure 10: Spectral Expander

variable preemphasis/deemphasis network is used, controlled by an identical rms-level detector sensing energy in the same passband. Only the control polarity is changed. This changes the transfer function of the network to $S^{-1}(f,b)$, as shown below.

$$S^{-1}(f,b) = \frac{1 + (jf/F)(1+51b)/(b+1)}{1 + (jf/F)(b+51)/(b+1)}$$

where F = 20.1 kHz, $b = 10^{(d/20)}$, and d is the decibel-rms value of the signal sensed by the bandlimited rms detector. As in the spectral compressor, the VCA used in the spectral expander is exponential-responding. Therefore, the conversion from the variable d (representing the log-scaled control voltage) to the variable b (representing the linear voltage gain of the VCA) is easily accomplished.

Also as in the spectral compressor, the detector output corresponding to 0 dB (d = 0 dB, the point at which the variable-preemphasis network produces a flat response) occurs when an 8-kHz sinewave is at 5.16% modulation at the input of the system. (Note that this refers to the signal level at the input to the expander, not the level at the input to the rms detector itself.)

Since the input signal to the expander is identical to the output signal from the compressor (except for noise), the rms-level detector monitors the same signal in each case. This ensures that the expander mirrors the action of the compressor, thereby maintaining audio transparency.

By using the spectral compressor, two simultaneous requirements are met:

1) The system is extremely forgiving of high-background-noise environments, because the spectral shaping of the input signal is adjusted according to the needs of the input signal to provide high masking at all times.
2) Headroom is maintained throughout the frequency range because extreme preemphasis is used only when it is really needed. (The complete compander's high-frequency-overload characteristic is quite close to the 75-microsecond preemphasis characteristic used in the mono -- L+R -- channel.)

WIDEBAND AMPLITUDE COMPANDING

Neither the spectral compander nor the fixed pre-/deemphasis will help reduce noise when the signal is very low in level, especially if the signal has little high-frequency content. This is where the third stage of the system design becomes important: the wideband compander. This element is responsible for adjusting the level of all frequencies simultaneously to keep the signal level in the transmission channel high at all times.

The wideband compressor reduces the dynamic range of input signals by a factor of 1/2 in decibels. Not only are small signals raised in level but large signals are reduced (Fig. 11). The output level tends to hover around 14% modulation (an "unaffected-level" point of -17 dB), which has three benefits:

1) The signal is consistently above the noise floor.
2) The signal is consistently below 100% modulation, reducing AM-interleaving effects.
3) Headroom is maintained for transient peaks to overshoot the nominal level at the compressor output without overmodulation.

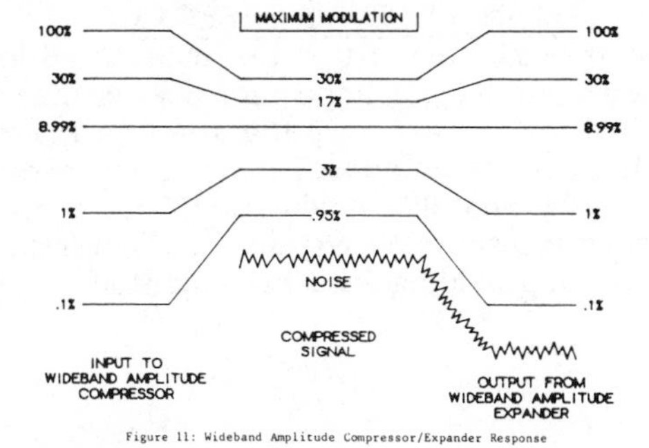

Figure 11: Wideband Amplitude Compressor/Expander Response

HOW WIDEBAND AMPLITUDE COMPRESSION WORKS

The wideband amplitude compressor works by controlling the gain of a VCA in response to an rms-level detector that senses the low- and mid-frequency-energy level at the output of the compressor (Fig. 12). The VCA acts to reduce gain when signal amplitudes are high and boost gain when they are low. Since all frequencies in the input signal pass through this VCA, the action is the same for all frequencies. The transfer function W(a), below, describes this action.

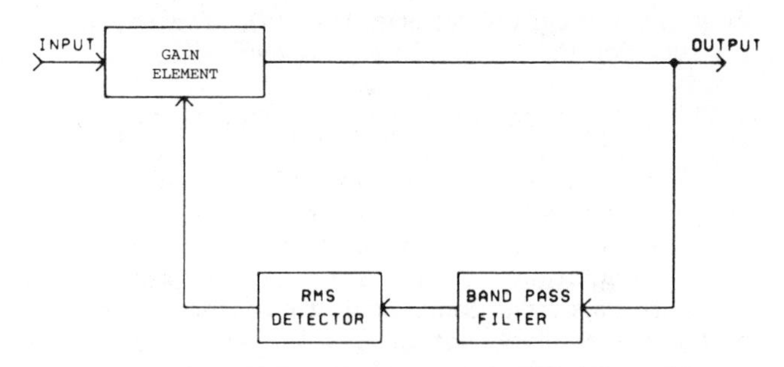

Figure 12: Wideband Amplitude Compressor

$$W(a) = \frac{1}{a}$$

where $a = 10^{(c/20)}$, and c is the decibel-rms value of the signal sensed by the bandlimited rms detector. Since the VCA used is exponential-responding, the conversion from the variable c (representing the log-scaled control voltage) to the variable a (representing the linear voltage gain of the VCA) is easily accomplished. This combination provides precisely "decilinear" response (Fig. 13).

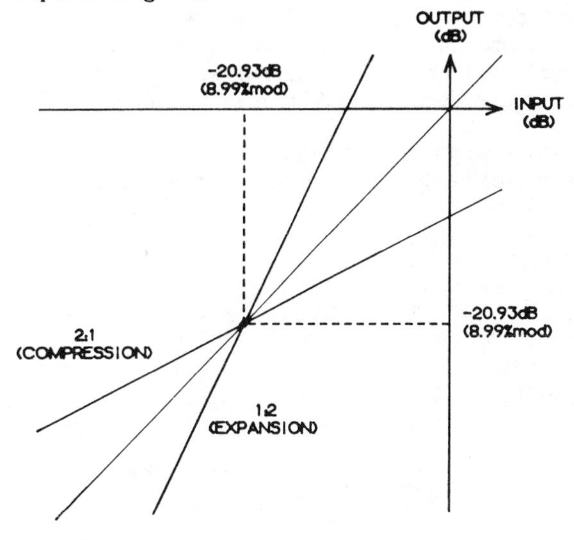

Figure 13: Wideband Amplitude Compander, Response Characteristics

The filter restricting the rms-level sensor rolls off at about 35 Hz and 2.1 kHz, thereby limiting the sensed energy to the dominant ener-

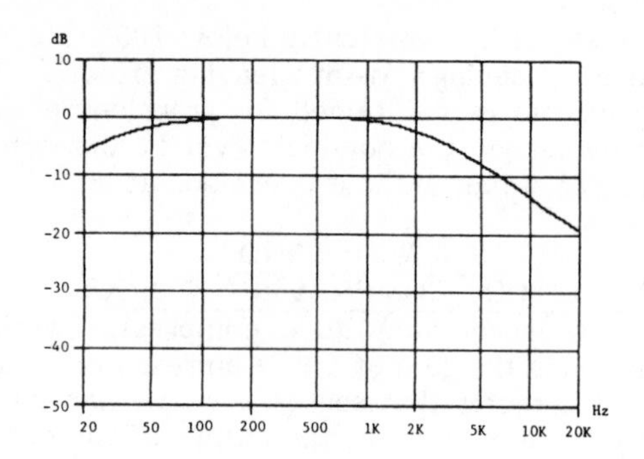

Figure 14: Wideband Compander, Frequency Response of Rms-Detector Filter

gy in most program material (Fig 14). The transfer function P(f) of this filter is shown below.

$$P(f) = \frac{(jf/35.4Hz)}{[(jf/35.4Hz)+1]\ [(jf/2.09kHz)+1]}$$

The detector output corresponding to 0 dB (d = 0 dB) was chosen to occur when a 300-Hz sinewave is at 8.99% modulation (-20.9 dB re 100% modulation) at the output of the system. (As in the spectral compander, this refers to the output-signal level, not the level at the input to the rms detector itself.) This alignment causes the compressor to allow room for overshoots at the system output without overmodulation.

An rms-level detector, albeit with slower time constants, is used here for the same reasons as in the spectral compressor.

WIDEBAND AMPLITUDE EXPANSION

During reception, the wideband amplitude expander (in the decoder) must restore the signal levels to their proper, original amplitude. If the signal in the transmission channel is high in level, the decoder will increase the level further, restoring the signal to its original amplitude; if the signal in the transmission channel is low in level, the decoder will attenuate it further, restoring it to its original amplitude and pushing channel noise down toward inaudibility. When there is only noise in the transmission channel, it is sensed as a small signal and is strongly attenuated by the decoder, making it inaudible.

The wideband amplitude expander (Fig. 15) is a mirror image of the wideband amplitude

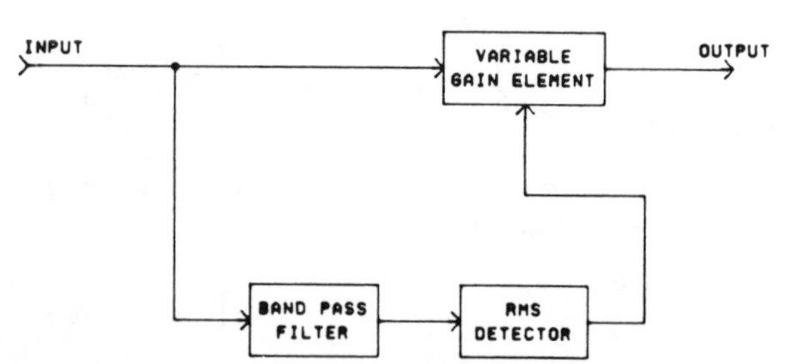

Figure 15: Wideband Amplitude Expander

compressor. The same VCA is used, controlled by an identical rms-level detector sensing energy in the same passband. As in the spectral compander, only the control polarity is changed. This changes the transfer function to $W^{-1}(a)$, as given below.

$$W^{-1}(a) = a$$

where a = $10^{(c/20)}$ again and c is the decibel-rms value of the signal sensed by the bandlimited rms detector. As before, since the VCA used is exponential-responding, the conversion from the variable c (representing the log-scaled control voltage) to the variable a (representing the linear voltage gain of the VCA) is easily accomplished.

As in the spectral compander, the rms-level detector in the wideband expander monitors the same signal as in the wideband compressor. This ensures that the expander mirrors the action of the compressor, maintaining audio transparency.

TRANSIENT PROTECTION

Another important aspect of the compander design was to protect against very large transients' causing excessive modulation of the transmission system. In a peak-limited medium, such as the chosen television-transmission system, the peak excursion of the compressor output is an important parameter to control. This requirement can be most easily met by a clipper/limiter incorporated into the noise-reduction compressor and set to operate at 100% modulation.

Clippers are relatively inaudible in operation if the amount of clipping that takes place is small. That is, the circuit must clip only transients that last under a few milliseconds. In order to accomplish this:

1) The unaffected level point of the wideband compressor is set to be substantially (approximately 17 dB) below the point that would cause 100% modulation in the transmisson channel.
2) The wideband and spectral compressors, which precede the clipper, are allowed to operate quickly enough to let only brief overloads reach the clipper.
3) The static preemphasis precedes the clipper, further reducing the duration of transient overloads.

By including a clipper of known characteristics within the noise-reduction circuitry, compliance with FCC rules is assured while transparent compander operation is maintained.

AM-INTERLEAVE EFFECTS

Another side benefit of this unaffected level alignment, but a most important one, is that the signal level broadcast over the stereo subcarrier tends to have its amplitude distribution average between 10% and 30% modulation. For the recommended transmission system, this is extremely important in providing for large amplitudes in the mono carrier without exceeding the allowable modulation limits of mono-plus-stereo carriers. In this case, since the stereo subcarrier tends to average below 30% modulation, the mono carrier will be allowed to stay around 70% modulation, which is not too different from current practice.

DIFFERENCE-CHANNEL FILTERING

In order to protect the 15.734-kHz pilot, it is necessary to prevent the audio channel from containing any information above about 15 kHz. This is accomplished with a lowpass filter in the compressor. This filter is included within the feedback loop of the compressor (Fig. 16), so that

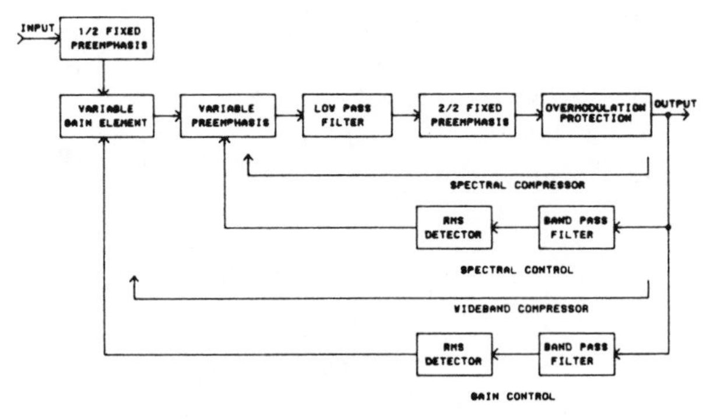

Figure 16: Compressor (Encoder)

the rms-level detectors in both the compressor (encoder) and expander (decoder) sense the same bandlimited signal. An 11-pole Cauer filter, catalog no. C11-20-73, has been tested for this application, although the exact choice of characteristics may vary. The phase shift introduced by this filtering must be compensated by an identical phase shift in the sum (L+R) channel, or stereo separation will suffer. In practice, this is accomplished by inserting an identical filter in the sum channel, since both channels must be bandlimited for proper operation.

In order to provide for future improvements to the system as a whole, any filtering required at the transmission end must be compensated fully at the transmission end of the system only. Similarly, any filtering required at the receiver end will be compensated only in the receiver. In this way, hardware limitations to the phase linearity of filters, ac coupling required to eliminate dc offsets, and other such limitations of the present state of the art will not be perpetuated by the system implementation.

The received, demodulated difference (L-R) signal will have high-frequency components that could interfere with proper decoding if they are not attenuated before reaching the expander detectors (particularly the spectral-expander rms detector). Filtering is therefore necessary to eliminate mistracking. The design of this filter, too, is left up to receiver manufacturers, to allow optimization of the system to suit the specific situation. It may be placed in either the L-R signal path or the expander-control path. If it's to be in the signal path, a compensating filter with the same phase and amplitude characteristics must be placed in the sum-channel path or separation will suffer. If it's in the control path, only one filter is needed (Fig. 17).

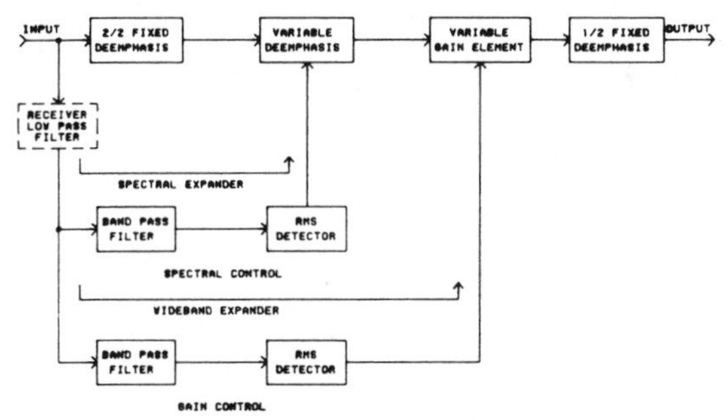

Figure 17: Expander (Decoder)

CONCLUSION

The BTSC system provides true high-fidelity stereo performance for all viewers within the present range of TV signals. When switching into stereo, there will be none of the hiss added that accompanies the switch into stereo with FM. Neither the fidelity nor the coverage area need be compromised.

For the SAP channel, BTSC provides adequate performance to be achieved out to the grade-B contour. City-grade (grade-A) performance will be quite good, consistent with a 10-kHz-bandwidth channel. The uncompanded-channel S/N ratio is so low that some compander artifacts are inevitable in grade-B reception, but service quality remains acceptable.

Full compatibility with TV sets now in use is maintained, while new receivers may include stereo and SAP reception at reasonable cost. Add-on receivers that allow stereo or SAP reception with a mono TV set also are possible.

ACKNOWLEDGMENT

The system presented here represents the work of many individuals other than the authors; particular note should be taken of the diverse contributions to the EIA committee effort, frequently crossing company lines. The result reflects contributions from all. The authors would like to express appreciation to the EIA and to the individuals and companies making up the EIA Multichannel Television Sound Subcommittee, particularly those in the Companding Working Group. Specific thanks go to Jim Gibson, RCA, and Ron Lee, Zenith, who provided some of the mathematics herein.

REFERENCES

1. See the appropriate sections of the EIA BTSC Report on Multichannel Television Sound, Vols. 1-A (pub. NAB) and 2-A (pub. EIA), and the Supplement thereto.
2. I. M. Young and C. H. Wenner, "Masking of White Noise by Pure Tone, Frequency-Modulated Tone, and Narrow-Band Noise," J. Acoust. Soc. Am. 41, 700-705, 1967.

Part VII
Projection Television

PROJECTION systems are not new, having been demonstrated as far back as the late thirties. However, none of these early approaches resulted in a viable consumer product, because they were either much too costly and complex or because they suffered from inadequate brightness and contrast.

Over the next 35 years or so, the television industry was busy with the transitions from monochrome to color, from vacuum tubes to transistors and integrated circuits, and from 15 in (12 V) round, metal-funnel, delta-gun, shadow-mask kinescopes to rectangular, all glass, in-line CRT's in a variety of sizes that now range from 5 to 30 in. These new direct-view picture tubes provided higher brightness, greater contrast, better stability, and easier setup and, for the most part, seemed to satisfy the public's taste for an improved display.

Henry Kloss of Advent Corporation is generally credited with reviving interest in projection TV with the introduction, in 1973, of a three-tube system using specially designed CRT's containing Schmidt optics within the bulbs. At about the same time, Sony announced its projector which utilized a 13 in Trinitron CRT as the picture source.

Today, all major television receiver manufacturers have projection sets in their product lines and the market has grown considerably from 20 000 units in 1977 to an estimated 175 000 for 1984. While this is just a little over 1 percent of the market in terms of units, because of the premium prices commanded by projection receivers, the dollar volume is probably in the 5 percent range. The thrust of today's activity centers on increasing brightness, viewing angles and contrast ratio, improving resolution, reducing cabinet size, and lowering cost.

The first two papers in this part, by A. Robertson, provide an overview of video projectors that have made it to the marketplace as either consumer or industrial products. Included are a variety of light-valve models, others containing reflective (Schmidt) optics, and still others using a refractive system of lenses and mirrors to project scanned rasters onto a large screen. Most use front projection.

Currently, rear-projection sets have achieved the greatest popularity in the American market as cleverly folded optical paths result in more compact designs. For example, 37 in diagonal screens are now available in cabinets $38\frac{1}{2}$ in high by $34\frac{7}{8}$ in wide by $23\frac{1}{2}$ in deep. In comparison, a typical 25 in diagonal screen direct-view console (100° deflection angle) is $32\frac{1}{2}$ in high by $38\frac{1}{4}$ in wide by $19\frac{1}{2}$ in deep with a $3\frac{1}{8}$ in cup extending beyond the cabinet back.

The next three papers deal with key components in rear-projection systems: CRT's, lenses, and screens. M. Kikuchi *et al.* report on a liquid-cooled $5\frac{1}{2}$ in projection CRT which permits operation at higher current densities with a resultant increase in brightness. Additional benefits accrue in the form of extended cathode life, less tendency for glass solarization, and a thermal environment suitable for plastic lenses.

In the fourth paper, R. L. Howe and B. H. Welham review the important systems considerations in the design of refractive projectors and discuss plastic lenses which have been developed for projection applications.

Next, Tominaga *et al.* describe a new rear-projection screen consisting of a uniform pattern of "unit lenses." These lenses, having two spherical surfaces on a common axis, provide improvement in luminance distribution, color-shift, and contrast compared to the conventional Fresnel-diffuser-lenticular type.

Finally, the last paper, by Grigor'ev *et al.*, which appeared in the *Soviet Journal of Quantum Electronics*, considers the use of electron-beam-scanned semiconductor lasers for large-screen projection.

In reading Part III on advanced television systems, one should appreciate that enhanced definition and high definition techniques will be most effective in improving the appearance of large-screen displays which may well engender additional consumer interest in projection.

HAROLD J. BENZULY
Associate Editor

BIBLIOGRAPHY

[1] E. Bauman, "The Fisher large screen projection system," *SMPTE J.*, vol. 60, p. 344, 1953.

[2] E. Betensky, U. S. Patents 4 300 817 and 4 348 081.

[3] O. Bogdankevich, "The use of electron-beam pumped semiconductor lasers in projection television," *IEEE J. Quantum Electron.*, vol. QE-14, Feb. 1978.

[4] R. J. Doyle and W. E. Glenn, "Lumatron: A high resolution storage and projection display device," *IEEE Trans. Electron Devices*, vol. ED-18, p. 739, 1971.

[5] J. R. Fendley, Jr., "Resolution of projection TV lenses," *SID Int. Symp. Dig.*, 1982.

[6] W. E. Glenn, "Large screen displays," presented at IEEE Conf. Electron Device Techniques, 1970.

[7] W. E. Glenn, "Principles of simultaneous-color projection television using fluid deformation," *SMPTE J.*, vol. 79, p. 788, 1970.

[8] W. E. Good, "A new approach to color television display and color selection using a sealed light valve," *IEEE Trans. Broadcast Telev. Receivers*, vol. BTR-15, no. 1, p. 21, 1969.

[9] W. E. Good, "Projection television," *IEEE Trans. Consum. Electron.*, vol. CE-21, p. 206, Aug. 1975.

[10] H. Hagiwara *et al.*, "45 inch projection receivers," *Toshiba Rev.*, p. 37, Mar.–Apr. 1981.

[11] R. Hockenbrock and W. Rowe, "A self-converging three-tube projection system," *SID Int. Symp. Dig.*, 1982.

[12] H. Howden, "Production of optical correction plates for projection television," *Philips Tech. Rev.*, vol. 39, no. 1, p. 15, 1980.

[13] R. Howe, "Big optics for big screen television," *Opt. Spectra*, Mar. 1978.

[14] K. Kubota *et al.*, "An 8′ X 8′ display using a laser-addressed white-on-black mode liquid crystal light valve," *SID Int. Symp. Dig.*, p. 44, 1983.

[15] V. Kuklev, "Sealed scanning semiconductor laser with transverse electron-beam pumping," *Sov. J. Quantum Electron.*, vol. 9, p. 206, Feb. 1979.

[16] B. M. Lavrushin and E. S. Shemchuk, "Prospects for the use of

electron-beam-pumped semiconductor lasers in projection television," *Sov. J. Quantum Electron.*, vol. 6, p. 1434, Dec. 1976.

[17] G. Marie, "Projection d'images de television sur grand Écran" (in French), *ACTA Electron.*, vol. 18, no. 3, p. 221, 1975.

[18] A. S. Nasilov and E. S. Shemchuck, "Use of laser electron-beam tubes in projection television," *Sov. J. Quantum Electron.*, vol. 8, p. 1082, Sept. 1978.

[19] K. Schiecke, "Projection television: Correcting distortions," *IEEE Spectrum*, p. 40, Nov. 1981.

[20] T. Schmidt, "A new monochrome data/graphics video projector," presented at IEEE Chicago Spring Conf. Consumer Electron., June 19, 1980.

[21] "Radiometry and photometry for the electronics engineer," Tektronix Application Note, Analytical Instruments no. 4.

[22] W. Thust, "Laser beam deflection in large scale display writing," *ACTA Electron.*, vol. 18, no. 3, p. 233, 1975.

[23] T. T. True, "Color television light-valve projection systems," presented at IEEE Int. Conv., Session 26, Mar. 24, 1973.

[24] J. A. van Raalte, "A new Schlieren light valve for television projection," *Proc. SID*, vol. 12/2, Second Quarter, 1971.

[25] B. H. Welham, "Optical design and system considerations for video projection," in *SID Int. Symp. Dig.*, vol. XIV, 1983, p. 38.

[26] M. A. Zaha, "Shedding some needed light on optical measurements," *Electronics*, p. 91, Nov. 6, 1972.

Projection television

A review of current practice in large-screen projectors

by Angus Robertson

Although projection television has been around since television was originally developed, it is only in recent years that considerable research has been directed towards developing new techniques for producing large-screen television projectors. Several TV projectors have been produced specifically for the consumer market, although at present their prices are far higher than direct-viewing receivers.

In the early days of television, it was not easy to manufacture cathode-ray tubes larger than 30cm diameter. To obtain larger pictures, manufacturers used a lens in front of a small, high-intensity c.r.t. and projected the raster onto a screen contained within the cabinet. Mirrors were usually used to fold the light path and enable smaller cabinets to be used. During the fifties, larger and brighter c.r.ts were manufactured and projection TV faded out. Although larger tubes have been produced for special applications, sizes have levelled off with diagonals of 66cm: larger pictures require special techniques.

The first large-screen television projector was invented by Professor Fischer at the Swiss Federal Institute of Technology in 1939. At that time, Prof. Fischer thought that the growth of television would come from the development of networks of neighbourhood "television theatres" and he invented the Eidophor with the capability of projecting TV pictures onto cinema-sized screens. The earliest Eidophors were cumbersone machines, which could project only black and white pictures in a darkened or semi-darkened room. They were not the most reliable of machines and for a number of years the Eidophor system was little known or used.

Later the American space programme called for a reliable, high-performance, large-screen projection system capable of working for long periods of time, to provide data displays in NASA flight control centres. Gretag AG, Zurich, a subsidiary of Ciba Geigy and patent holders and manufacturers of the Eidophor, successfully developed the projector's capability to meet NASA specifications. The latest Eidophors are able to project full-colour television pictures onto screens 18m wide. The Eidophor is still the only commercially-available projector able to project cinema-sized pictures, but colour versions cost over £100,000. Cheaper techniques have therefore been developed to provide projectors for industry, education and the home.

Three basic projection techniques are used. Eidophor, General Electric, Hughes, Westinghouse, IBM and Titus (Philips) use light-valve projectors, in which varied techniques are employed to modulate a light source, which is then projected onto the screen. The second method is to use Schmidt optics (like the telescopes) to magnify and project the image from a small, high-intensity c.r.t. Advent, Pye (Mullard/Philips), Image Magnification, Ikegami, Kalart Victor and Pro-

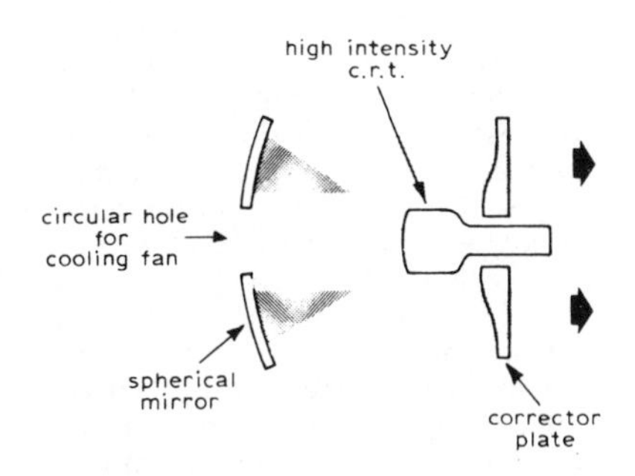

Fig. 1. Light path in Schmidt optical system.

Fig. 2. Magna Image 111H colour Schmidt projector.

Reprinted with permission from *Wireless World*, vol. 82, pp. 47-52, Sept. 1976.
Copyright © 1976, I.P.C. Business Press Ltd.

Fig. 3. CV3 Superscreen Schmidt projector.

jection Systems all use this technique. Finally, the refractive technique is the cheapest method, in which a glass or acrylic lens is placed in front of an ordinary c.r.t., usually a 33cm colour Trinitron, and the picture is projected onto a high-gain screen. We can expect to see lenses available separately soon to enable the handyman to modify his own TV set for projection.

Screens

Most video projectors, except possibly Eidophor, have low light outputs when compared with cine projectors. For instance, refractive projectors have an output well below 50 lumens, Schmidt projectors emit between 200 and 500 lumens and the high power Eidophor produces 7,000 lumens. An ordinary 16mm ciné projector, with a 250W, 24V lamp, provides around 650 lumens while in motion and an overhead projector is usually rated at between 1,500 and 2,500 lumens.

To increase the apparent brightness of the projected image, it is common practice to use special screens which have "gain". Since no more light may be reflected from the screen than falls on it, the screen is made directional so that available light is concentrated into a small angle.

Projectors such as the Advent use an Ektalite-foil screen material, developed by Kodak. The material provides an on-axis gain of between eight and ten, with a 40° viewing angle but, to eliminate hot spots, the screen must be compound curved (both horizontally and vertically) and this necessitates a solid screen frame, usually fibreglass. Zygma Electronics use a specially-developed, solid, compound-curved screen that uses a highly-reflective spray-on paint instead of the more usual foil. Gain is 4.7 times with a horizontal viewing angle of 90°. While these are fine for permanent installations, they usually need a van for transportation and may only be concealed with difficulty, unlike a roll-up screen. A matt screen surface has a unity gain, while a beaded screen usually has a gain of 1.8. Projection Systems Inc. market a silver lenticule screen with a 2.5 gain and the exceptional viewing angle of 170°. This screen material, which is rollable, also has excellent ambient light rejection characteristics, this light being reflected sideways away from the viewers — ideal for hire work where the locations used rarely seem to have adequate black-out.

Mechanische Weberei, in West Germany, manufacture a projection screen with a gain of 3.5. This is also roll-up but it does not have the same ambient characteristics as the previous screen. It is however perfect for use where adequate black-out is provided. Ideal Image Inc. has produced a plastic lenticular screen with a gain of five but the lenticule (lens) size is such that minimum viewing distance is 6m and it must be compound curved. The company is presently developing a new plastic material similar to the Ektalite foil but which requires only horizontal curving rather than the compound curve. Single curves are used for large cinema screens and only require a curved frame lattice rather than the solid frame required for Ektalite. Although the screens being presently manufactured are 25m by 7.5m for wide-screen (and why not?) television projection, smaller sizes are available to order.

Fig. 4. Videobeam 1000A from Advent.

Rear projection. Although front projection is common there are certain applications where rear projection is preferred. For instance the Eidophors installed at the NASA flight control centre are rear projected for convenience; it would be impractical in that particular location to project the image onto the front of the screen. Screens used for rear projection are usually of high density and low gain in order to minimize "hot spots". To obtain the projected image the correct way round current to the horizontal scanning coils is reversed. On some projectors scan reversal is achieved by a switch, on others, it is a case of resoldering two wires on each scanning yoke.

Although not yet commercially available, a rear projection screen has been developed in the USA with a gain of about eight. No further details are known nor do I know of any other rear projection screens with a gain greater than unity.

Screen brightness. Projector light output is usually quoted in lumens, which have the advantage of being identical in imperial and metric units. The illumination falling on the screen is measured in lumens/m^2 (lux) but since most screens provide some gain, luminance in candelas/m^2 (nits) is the unit used (except by the Americans who still use ft lamberts). **Example:** A projector light output is 500 lumens. Screen size is 3m x 4m = 12m . Screen illumination is

$$\frac{500}{12} = 41.7 \text{ lux.}$$

If the screen gain is two, screen luminance is

$$\frac{41.7 \times 2}{\pi} = 26.53 \text{ cd/m}^2.$$

Resolution. A 625-line, 50-field television signal has a bandwidth of 5.5MHz and a horizontal resolution of 570 lines. The resolution of colour c.r.ts is limited by the number of phosphor dots or stripes; therefore, the larger the screen, the higher the possible resolution. Since most colour projectors project each colour separately, resolution is usually only limited by the electronics associated with each channel. For the display of computer-generated data and command applications, higher definition is often required and the use of 1,029 lines per frame and video bandwidths of up to 40MHz enables a horizontal resolution of over 1,000 lines to be achieved. Digital techniques are sometimes used to obtain accuracy in the corners of the picture. When displaying characters which combine more than one colour, registration accuracy is critical; otherwise double characters will be displayed.

Schmidt projectors

Fig. 1 shows the principle of the external Schmidt c.r.t. projector, in which tube diameter can vary between 75mm and

150mm for different projectors. The reflector needs to be two or three times the raster size, which precludes the use of large colour tubes on purely physical grounds. The centre of the mirror is removed to prevent light being reflected directly back towards the c.r.t. faceplate and a fan is inserted in the hole to cool the high-intensity faceplate.

Light output from Schmidt projectors depends upon tube and reflector size. RCA quote a light output of 450 lumens from a 125mm c.r.t. when operating with 45kV and 500μA. Tubes operating with more than about 30kV produce X-rays and care is required in the construction of such projectors to provide adequate shielding. Some projectors include interlock circuits which remove e.h.t. when the protection covers are removed for maintenance.

For colour projection, three optical systems are used with green, red and blue c.r.ts. Usually, these are mounted in-line, but one manufacturer uses a triangle formation. Complex analogue circuits are included to enable registration of the three images in much the same way as a shadow mask tube – a process which is made easier by the in-line layout. Projectors usually have a built-in cross-hatch generator and basic registration controls are often mounted on a separate control unit for convenience.

Another facility usually included is keystone correction. When a projector is mounted on the ceiling (out of the way), the projected image is angled down at the screen giving a picture wider at the bottom than the top. Keystone correction enables the verticals to be electronically corrected, usually to correct for angles of ±15°. An optical keystone corrector may be provided which adjusts the c.r.t. position to provide equal focus over the screen. The tube is usually mounted on a carriage which enables it to be moved in relation to the reflector to provide optical focus for differing projection distances. The corrector plate is designed to compensate for deficiencies in the optical system and is optimized for a particular projection throw. Although most Schmidt projectors allow focusing at variable distances, often the corrector plates must also be changed if a wide range of projection distances is to be accommodated.

Image Magnification Inc. The Magna Image I is a monochrome projector which uses a 125mm c.r.t. Picture width is variable between 1.2m and 6m with a resolution of 600 lines in the centre. Price: £3,500. The Magna Image IIIH, shown in Fig. 2, is a colour projector using three heads in-line. Picture widths between 2.4m and 6m with a resolution of 500 lines. Price: £12,750.

Kalart Victor Corp. The Telebeam II projects a monochrome picture between 1.8m and 3.6m wide (specified when ordered) with a resolution of 550 lines. Light output is 384 to 576 lumens from a 125mm tube. Price: £3,810.

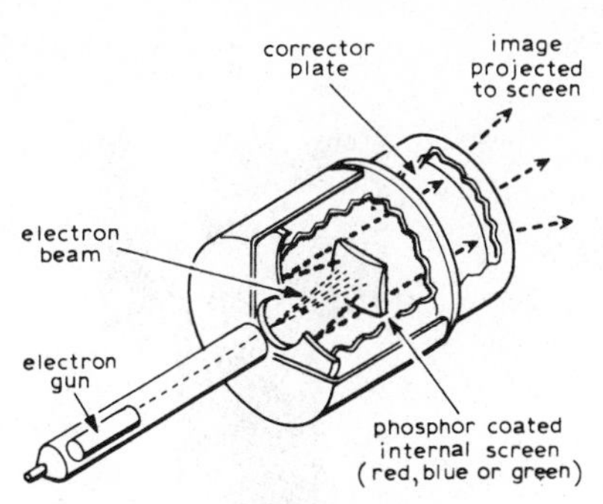

Fig. 5. Advent Lightguide tube, with complete Schmidt system inside envelope.

Fig. 6. Zygma Teleprojector.

Projection Systems Inc. The CV3 Superscreen, Fig. 3, projects a colour picture with widths between 1.8m and 2.4m from 75mm tubes. Light output is about 200 lumens. Price: £4,490. Model 270A is a monochrome projector with a light output of 800 lumens for screen widths between 1.8m and 6m. Resolution is 1,000 lines in the centre. Price is $11,800. Model 560 is a colour projector with a light output of 600 lumens for a screen width of 3.6m. Price: $14,750. Amphicolor 1000 uses three 150mm tubes for colour projection with 4800 lumens. Screen widths of either 2.4m or 4.2m. Price: $29,500.

Pye TVT. The Mammoth is a colour projector with a claimed light output of 800 lumens onto a 4m wide screen. Centre resolution is 600 lines and price is on application.

Ikegami. The TPP-2C is a colour projector with a light output of 360 lumens, intended for a maximum 4m by 3m screen. Price: £16,000.

Advent Corp. The Videobeam 1000A in Fig. 4 uses the Schmidt technique, but instead of having separate tube, reflector and corrector plate, all are vacuum sealed in the same envelope, as seen in Fig. 5. The electron beam scans a 75mm phosphor coated target (red, green or blue). Emitted light is then reflected onto the spherical mirror and back through the corrector plate to the screen. Although this approach has manufacturing advantages, the projection distance is fixed at 2.54m exactly, and light output is low since all parts are sealed within the tube and it is not possible to cool the target. Although not specified by Advent, working backwards from the screen brightness gives about 60 lumens light output. Thus a bulky, high gain screen is required to obtain an adequate screen brightness. Screen dimensions are 1.32m by 1.75m. Advent has however, recently announced a set of lenses which may be attached externally to the sealed tubes enabling a 2.4m x 1.8m picture on a flat screen. The screen brightness thus obtained would necessitate a well darkened room.

Advent are intending to introduce a new, cheaper projector on the American market this summer. No further details are available nor has a British launch date been announced.

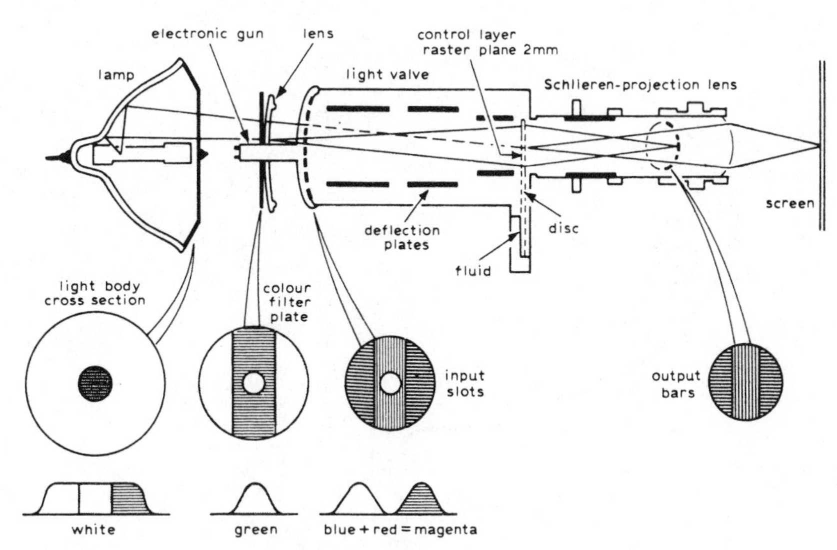

Fig. 7. General Electric light-valve projector.

Zygma Electronics manufacture a projector which has characteristics which are very similar to the Advent Videobeam. The Type 2001 Teleprojector shown in Fig. 6, uses internal Schmidt optics tubes with a fixed projection distance of 2.54m onto a high gain screen 1.75 x 1.32m. Screen brightness is 140cd/m and the price is £4,950.

Light valves

General Electric use transmission light valves in monochrome and colour projectors. Colour pictures are produced from a single projection tube using a diffraction grating to separate the colours. The projector, seen in Fig. 7, uses a separate xenon light source, a fluid control layer in the light valve, and a projection lens. Optically it is similar to a slide or ciné projector.

Miniature grooves are created on the deformable surface of the fluid control layer by electrostatic forces from the charge deposited by the electron beam, which is modulated with video information. These groove patterns are made visible by use of a "dark field" or schlieren optical system consisting of a set of input slots and output bars. The resulting television picture is imaged on the screen by the projection lens.

Cross sections of the light body, colour filters and input and output slots are shown below the light valve in Fig. 7. Green light is passed through the horizontal slots and is controlled by modulating the width of the raster lines themselves, by means of a high-frequency carrier applied to the vertical deflection plates and modulated by the green video signal. Magenta (red and blue) light is passed through the vertical slots and is modulated by the diffraction

Fig. 8. General Electric PJ6000.

gratings created at right angles to the raster lines by velocity modulating the electron spot in the horizontal direction. This is done by applying a 16MHz (12MHz for blue) signal to the horizontal deflection plates and modulating it with the red signal. The grooves created have the proper spacing to diffract the red portion of the spectrum through the output slots while the blue portion is blocked. For the 12MHz carrier the blue light is passed and the red blocked. Thus, simultaneous and superimposed primary colour pictures are written with the same electron beam and projected to the screen as a completely-registered full-colour picture.

Because of problems of heat dissipation, and the avoidance of frequent repairs (the light valve costs about $12,000), the xenon lamp is limited in power to 650W. The PJ7000 monochrome projector has a light output of 750 lumens and is suitable for picture widths between 0.75m and 3.6m, with a typical horizontal resolution of 1,000 lines. Three lenses are available to accommodate different throw/width distances. The PJ6000, Fig. 8, and PJ5000 colour projectors have a light output of about 280 lumens, a resolution of 600 lines and the same focusing ability, although a 2.4m screen width is optimum for colour. Light output of the colour projectors is less than the monochrome projectors since light is lost in the diffraction process. Life of the light valve is usually over 3,000 hours but 7,000 hours has been achieved in the laboratory. Price of the PJ7000 is $46,000, and the PJ6000 and PJ5000 cost $52,500.

Titus light valve. Developed by the Laboratoires d'Electronique et de Physique Appliquée (LEP), part of the Philips organisation, the Titus projector uses the Pockels effect in a refractive light valve. This works on the principle that certain crystals, in this case potassium di-hydrogen phosphate, rotate the plane of polarization of a beam of incident light through an angle proportional to modulation by the accelerating voltage of a constant-current electron beam. Fig. 9 shows a tube using these principles in a monochrome projector. A peltier cooler is required to keep the temperature of the plate just above its Curie temperature of about —50°C.

The target is bombarded by an electron beam whose accelerating voltage lies between 500 and 1,000V, a grid

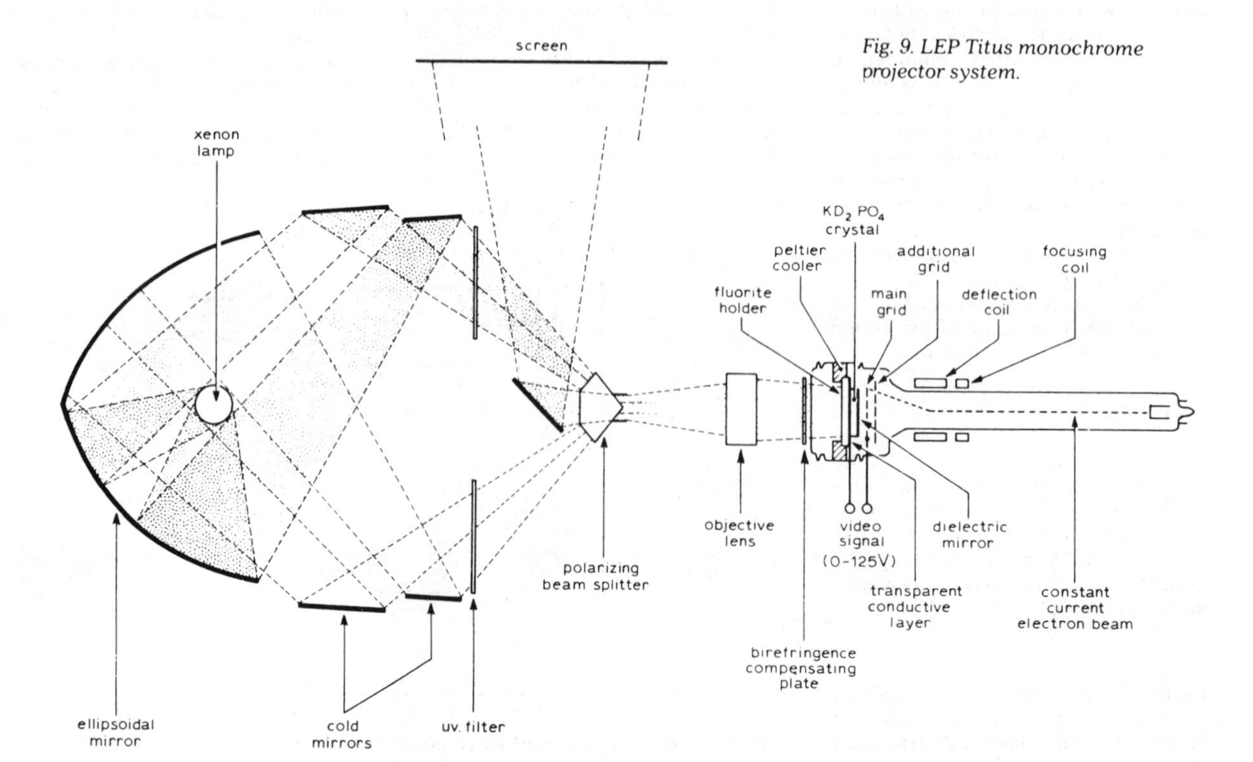

Fig. 9. LEP Titus monochrome projector system.

being placed in front of the target at a distance of about 40µm. The electron beam, of constant intensity, functions as a flying-spot short circuit between this main grid and the point of impact of the target, which thus reaches a potential close to that of the grid. The video signal is applied between the transparent conductive layer and the grid, to ensure that the various points of the target are charged to the corresponding video voltage when they are hit by the electron beam, irrespective of their previous potential. Erasure and writing are therefore simultaneous and this, coupled with the long discharge time-constant of the target, results in flicker-free operation. In addition, since the voltage pattern stored on the target does not depend on the intensity of the electron beam, it is found that no line structure is apparent on the picture. The absence of line structure, however, is not accompanied by loss of vertical resolution.

Twin ellipsoidal mirrors are used to provide high collection efficiency from the 2.5kW xenon lamp. A calcite polarizing beam splitter which transmits only light whose electric vector is parallel to the plane of Fig. 9 is used to transmit light to the Titus tube. The projection lens is placed between this polarizer and the tube and acts as a collimating lens so that the luminous beam incident on the plate has a mean directional normal to the latter. When the light beam is reflected at the dielectric mirror and passes through the lens and beam splitter again, only the light component with its electrical vector perpendicular to the plane of Fig. 9 is transmitted to the screen. In practice, light output from the monochrome projector is about 2,500 lumens, with a horizontal resolution reaching 750 lines. A 4kV xenon lamp may be used to increase this output.

A colour projector using Titus tubes is shown in Fig 10, in which two dichroic mirrors are used to split away blue and red beams. Ellipsoidal mirrors similar to those of the monochrome projectors are used but not shown here. Using a 4kW xenon lamp, 3,200 lumens output have been obtained from an experimental prototype.

The Titus is the only television projector that has an output capability comparable to the Eidophor. Efficiency is about half that of the Eidophor since half the light is lost in the original polarization.

Hughes liquid crystal. This projector uses a liquid-crystal reflective light valve which is addressed by a c.r.t. Fig. 11 shows the various layers which make up the liquid-crystal light valve. In operation the cadmium sulphide photoconductor acts as a high-resolution, light-controlled voltage gate for the liquid-crystal layer. The dielectric layer serves to reflect the projection light while the cadmium telluride light-blocking layer prevents residual pro-

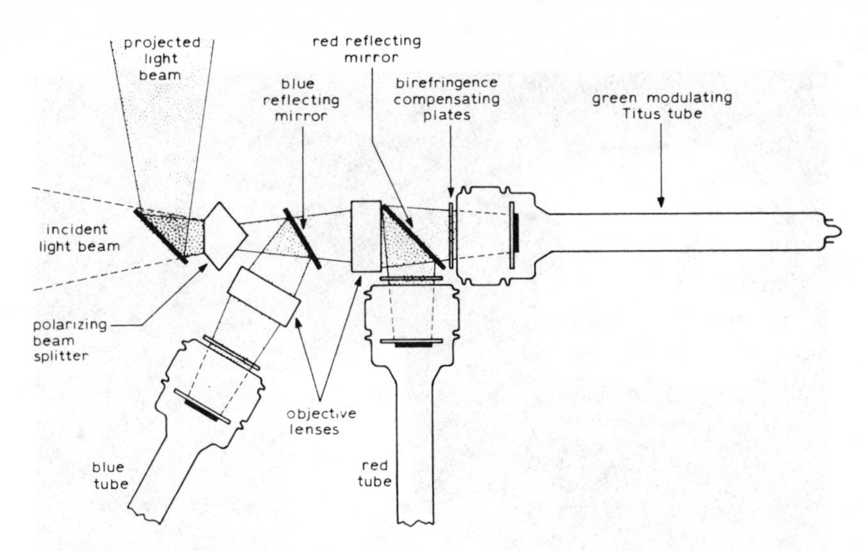

Fig. 10. Three Titus tubes used for colour projection.

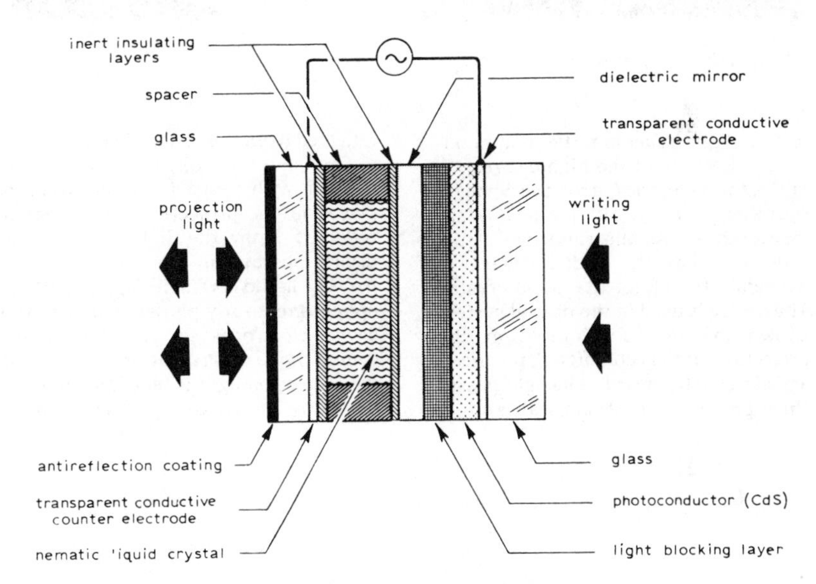

Fig. 11. Sectional view of a liquid-crystal light valve.

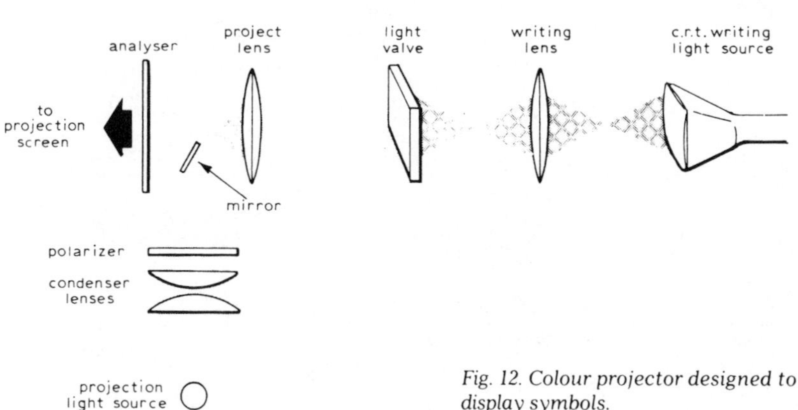

Fig. 12. Colour projector designed to display symbols.

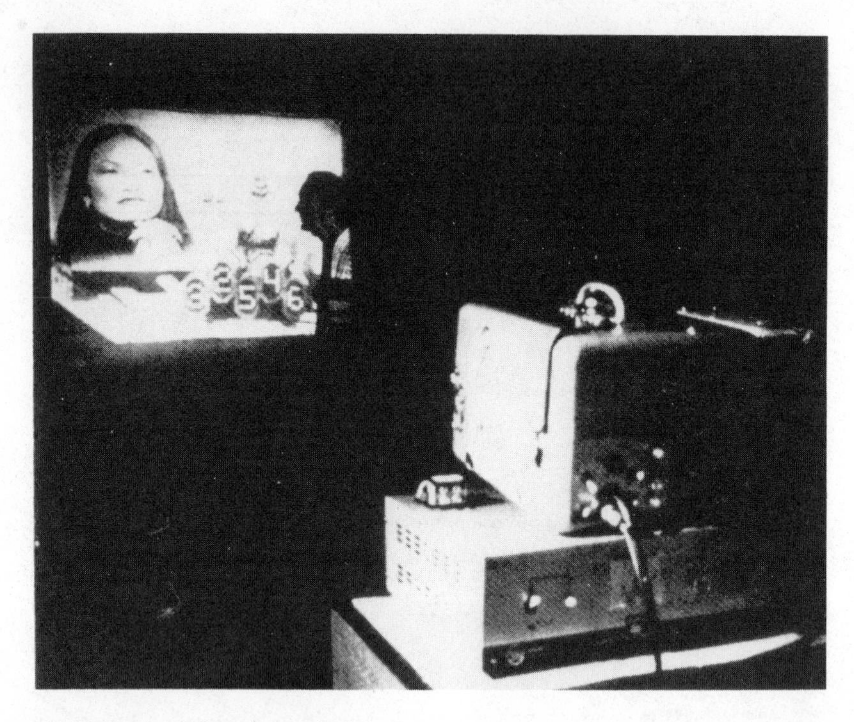

Fig. 13. Hughes light valve projector.

jection light reaching the photoconductor. Because of the high d.c. resistivity of the dielectric mirror, the device is operated with an alternating voltage impressed across the sandwich structure. This has the added benefit of extending the life of the liquid crystal. The device is used in the optical system shown in Fig. 12. Light from the projection lamp is collimated, polarized and directed to the cell. The light passes through the liquid crystal and is reflected from the dielectric mirror.

The second polarizer, which is crossed with respect to the first, is placed in the projection beam that is reflected from the light valve. The display operates in the following manner. The liquid crystal is aligned with its optical axis nearly perpendicular to the device electrodes so that, in the off state, no phase retardation occurs and the projection light is blocked from the screen by the crossed polarizers. With imaging light incident on the photoconductor, a voltage above the field effect threshold is switched onto the liquid crystal layer. The liquid crystal has negative dielectric anistropy, so that the molecules tend to align normally to an applied electric field. Thus the applied voltage rotates the molecules from their initial state parallel to the field and introduces a phase retardation between ordinary and extraordinary rays in proportion to the spatial intensity variation of the imaging light. This phase retardation changes the polarization of the projection light and, due to the dispersion effect, allows selected colours to pass the crossed polarizers.

In the display of symbols, the colour of a selected character in the projected image may be selected by the level of intensity of the input imaging light. Low imaging light intensity (low voltage switched to the liquid crystal) leads to a white (all colours present) on black image, while higher imaging intensity increases the alternating voltage on the liquid crystal and leads to the selection of certain colours; blue, green, yellow and magenta. These colours have only one luminance and are thus suitable only for displaying coloured characters or symbols superimposed upon a black and white grey-scale picture

So far, images of 350 lumens have been projected, using a 150W lamp, by the equipment shown in Fig. 13. This projector was developed by Hughes in co-operation with the US Navy, and in its present form is intended for symbol displays.

To be continued next month with a look at refractive projectors

Projection television—2

Refractive projectors

by Angus Robertson

The simplest form of television projection is to project the raster of an ordinary c.r.t. onto a screen using a lens. Light output depends upon the accelerating voltage and the efficiency and aperture of the projection lens; wide-aperture glass lenses are extremely expensive. There are two groups of refractive projectors, which are dealt with separately.

Trinitron projectors

Sony. The principle of the Sony VPP-2000, seen in Fig. 14, is shown diagrammatically in Fig. 15. A special 33cm Trinitron colour c.r.t. is used as the source, the light from which is directed, via a mirror, through the projection lens, with six elements, a focal length of 29cm, a diameter of 12.5cm and aperture f/2. A high-gain, solid screen, 1.02m by 0.76m, is used to obtain a quoted 34 nits (cd/m^2) luminance and, working backwards from this, the calculated light output is 10 lumens. Price is £1,500. Sony also produce a self-contained projection system, the KP-4000, whose principle is illustrated in Fig. 16. A Trinitron tube projects onto an internal 81cm by 61cm high-gain screen with a brightness of about 34 nits. Although not presently available in the UK, price in the States is similar to the VPP-2000. The VPK-1200E uses three 33m single colour tubes to project a 1.8m by 2.4m picture onto a high gain screen. Light output is about 200 lumens. Only two lenses are used; the outputs of the red and blue tubes are combined by a dichroic mirror. Lenses are f/2 with 30cm focal lengths. Price is expected to be around £16,000.

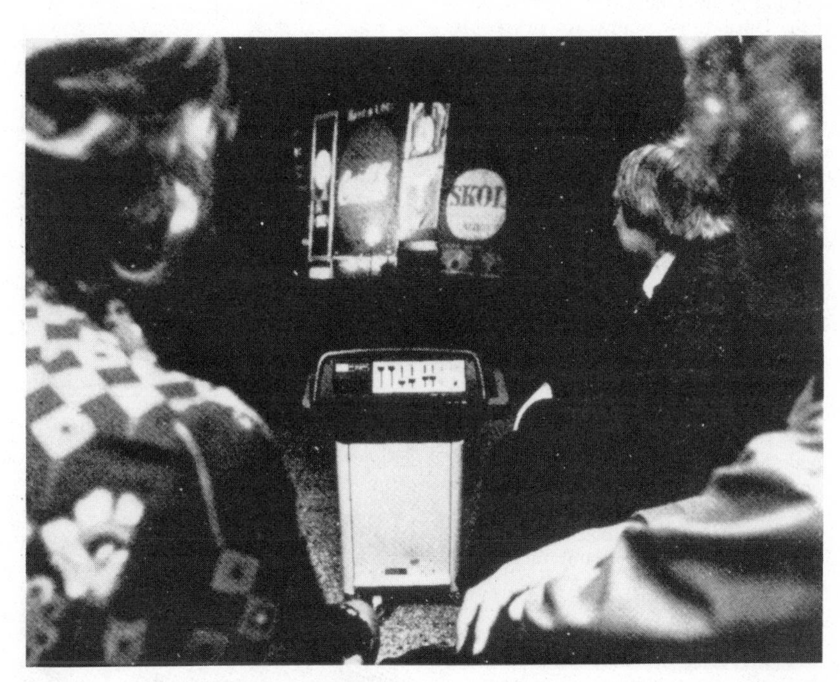

Fig. 14. Sony VPP-2000 uses a Trinitron source

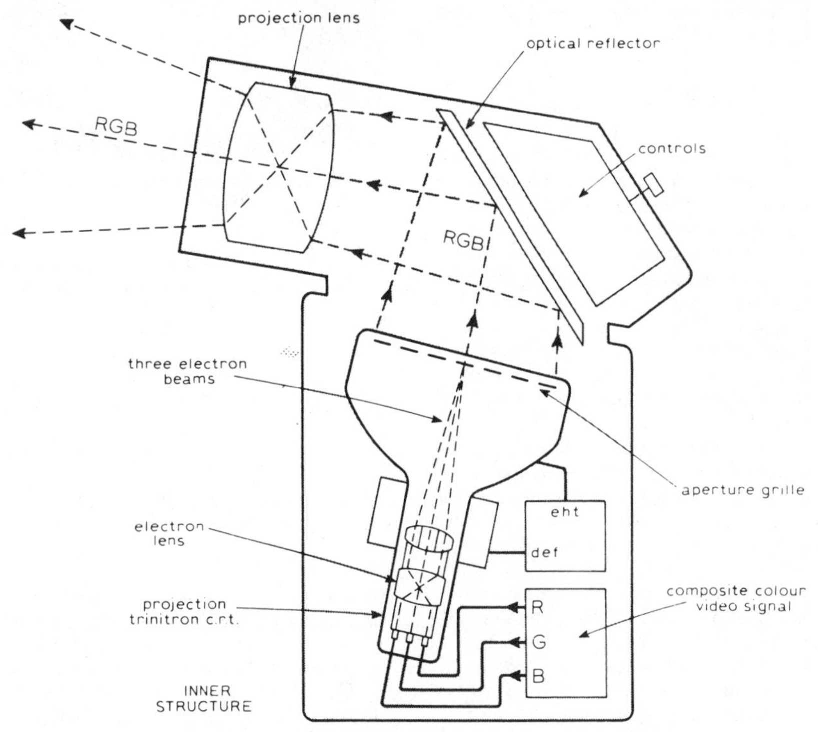

Fig. 15. Principle of the Sony VPP-2000

Muntz/Markoff Theatre Vision Inc. and **Tele-Theatre** (Fig. 17) both manufacture refractive projectors using a Trinitron colour tube and a separate 1.02m by 0.76m high gain screen. Tele-Theatre price is £1,000.

Shannon Communications Inc. has developed a 30cm diameter lens system moulded from acrylics. It fits in front of a Trinitron tube and unlike the previous systems is focusable. Unfortunately little experience exists in the moulding of lenses of this diameter and the British company who were originally going to

Reprinted with permission from *Wireless World*, vol. 82, pp. 67-72, Oct. 1976.
Copyright © 1976, I.P.C. Business Press Ltd.

manufacture the acrylic lenses now feel unable to guarantee production. Whether the Shannon projector will be marketed in the near future now remains to be seen.

The problem with the manufacture of acrylic lenses is still to be solved. Lenses may be ground down from a solid blank which is an expensive process. Alternatively they may be hot moulded from acrylics. However when using this process to mould large diameter lenses (30cm), problems with expansion and cooling make accuracy extremely difficult. Until these snags are overcome, Mullard are going to try polishing lenses hot moulded to obtain greater accuracy.

The basic difficulty encountered with the previous refractive projectors is that the smallest colour tube available is the 33cm Trinitron. This uses an aperture grill and vertical stripes of coloured phosphor on the screen. About 400 groups of stripes are deposited across the screen and as mentioned earlier, this limits the resolution to 280 lines. It is not presently feasible to manufacture a colour tube of smaller dimensions because of the difficulties of maintaining sufficient resolution. Thus projector manufacturers are stuck with the problems of large diameter, wide-aparture lenses. There difficulties are not so prevalent when separate high-intensity, single-colour tubes are used. Shadow mask tubes may of course be used instead of the Trinitron tube.

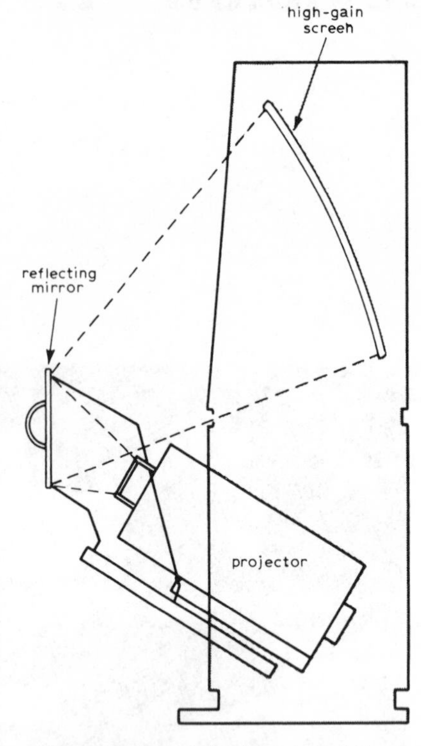

Fig. 16. Self-contained version of the Sony VPP-2000, the KP4000

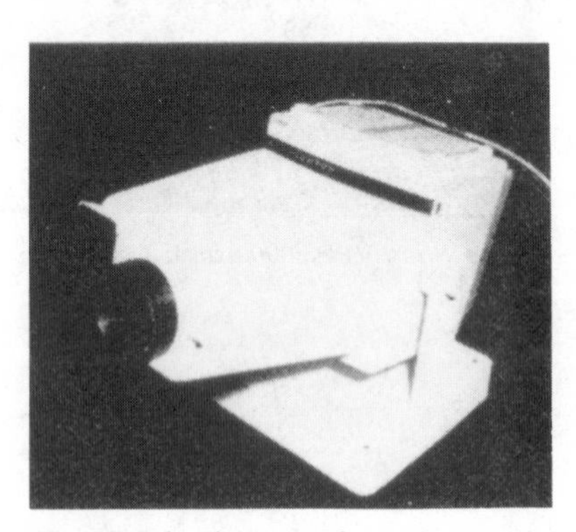

Fig. 17. Tele-Theatre refracting projector

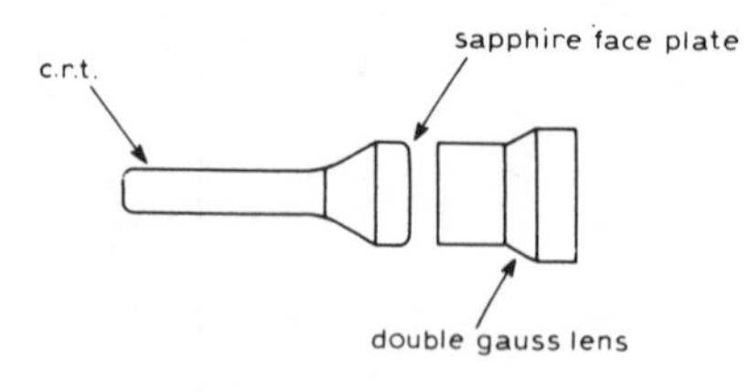

Fig. 18. Refractive projector from Aeronutronic Ford

Aeronutronic Ford

The refractive principle used by these projectors is perhaps the simplest to explain, but requires highly specialised techniques to obtain sufficiently high light outputs.

A 175mm c.r.t. is used as a basis for the projectors, shown in Fig. 18. The standard tube uses a glass faceplate with an active area of 10cm by 12.5cm; accelerating voltage used is 60kV with a very high beam current. To obtain a higher light output a sapphire faceplace may be used, which has a much higher thermal conductivity than glass, allowing power dissipation of up to 40W (six to seven times that of glass) from the faceplate. Such a large plate is bonded to the tube by an exclusive process that compensates for varying thermal stresses between the glass tube and sapphire faceplate.

The projectors are fully refractive using nine element lenses with a speed of f/0.87. Analogue circuits are used for coarse colour registration with a digital correction system using semiconductor storage to obtain an accuracy of half a picture element over the *entire* screen, not just the centre. To combine the outputs of the three colour tubes, two of which are not normal to the screen, the Scheimphlug condition is used where the plane of the screen, plane of the c.r.t. and plane perpendicular to the axis of the lens, intersect at a common line to preserve focus at expense of distortion (which can be corrected at the same time as the convergence errors).

Surprising compactness is obtained with these projectors; a single c.r.t. unit is 76cm high, 40cm wide and 106cm deep. Three units mounted vertically are used for colour and two colour units mounted side by side for high-power colour projection.

The ATP-1000 is a monochrome projector with a light output of 280 lumens using a glass-faceplate c.r.t. Resolution is 1,000 lines and contrast ratio 1:8. Price is $35,000 to $70,000. The ATP-3000 projects a 1,000 lumen colour picture, which is sufficient for a 2.7m by 3.6m picture. Price is $200,000 to $400,000. Finally the ATP-6000, shown in Fig. 19, is basically two ATP-3000 units providing 2,000 lumens with a 1,000 line horizontal resolution. Price is $400,000 to $500,000.

Advent has just introduced a refractive projector in the USA, fig 19a, which is effectively a baby version of the Aeronutronic Ford projectors. The Videobeam 750 uses three 12.5cm diameter c.r.t.s focused onto a 1.55m x 1.15m high gain screen using three 12.5cm refractive acrylic lenses. Price is presently $2,495 and introduction is expected in the UK during 1977.

Eidophor light valve

The Eidophor is a projector which, by means of a high-intensity light source and an oil layer influenced by the video signal, can project a black-and-white or coloured image by way of light valves.

The principle of the Eidophor is sketched in Fig. 20. The light source is a high-pressure xenon lamp which uniformly illuminates an aperture and is projected onto mirror bars with the aid of a condenser lens. The aperture image is reflected from the bars onto the concave mirror within the tube envelope. This arrangement of mirror bars is called a "dark field" projection system and ensures that no light can fall on the picture screen in spite of the concave mirror being intensely illuminated by the xenon lamp.

An oil layer 0.1mm thick is applied to the concave mirror and as long as its surface is completely smooth, the reflected light is not deflected and the picture screen remains dark.

However if this oil layer is deformed, part of the light is slightly deflected from its normal path and will pass between the mirror bars. This deflected light is focused by the projection lens onto the screen, the brightness increasing with increasing deformation of the oil layer.

The deformations are caused by an electron beam which exerts electrostatic forces on the oil and causes deformation. The electron spot size on the oil is chosen so that the resultant lines touch each other, distributing the charge evenly over the entire scanned area (72mm by 54mm): consequently the layer remains smooth. However, if the electron spot size is reduced, the lines no longer touch and the charge distribution assumes a line structure because the interline spacing carries no charge. This causes deformation of the surface and the smaller the spot, the larger the deformation. Thus, light reflected from the oil layer on the concave mirror is deflected past the

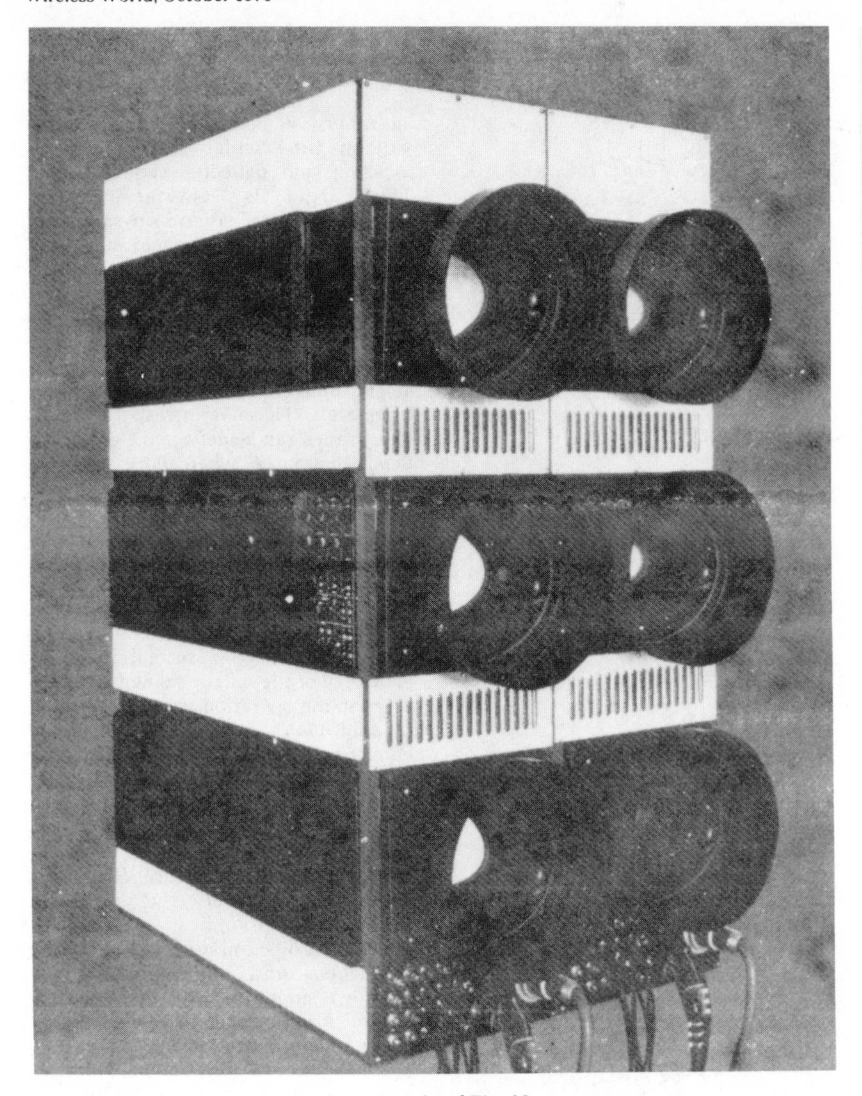

Fig. 19. Colour projector using the principle of Fig. 18

Fig. 19(a). Advent Videobeam colour projector – essentially a smaller version of half the Fig. 19 type

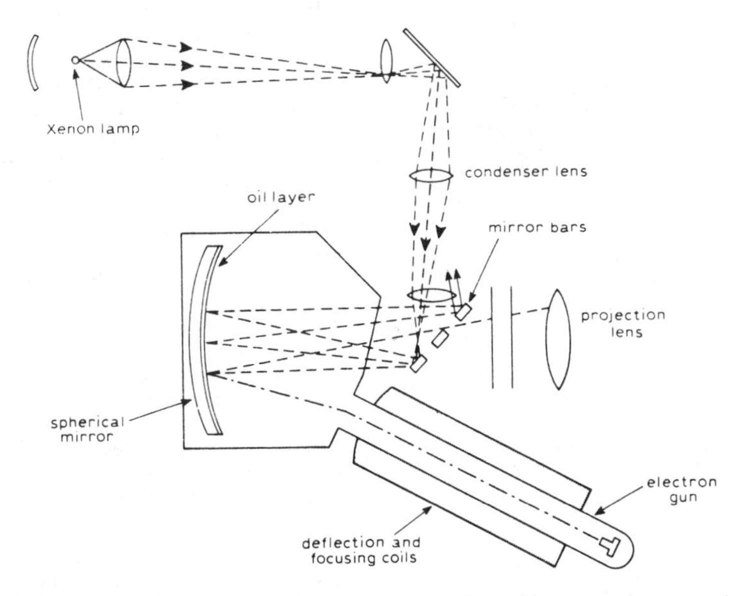

Fig. 20. Principle of the Eidophor. The lower ray is reflected between the mirror bars by the deformed oil layer

mirror bars and onto the screen. By continuously varying the spot size, a full brightness range may be obtained.

The nonconductive oil is made partially conductive and this, combined with the surface tension of the oil, provides for the oil to be smoothed after each field in readiness for new deformations. The great advantage of this technique is that the light source intensity is independent of the electrical power of the electron beam. Another feature inherent in the dark field technique is the contrast ratio of 1:100.

A colour Eidophor is arranged as shown in Fig. 21. The light from a single xenon source is split, using dichroic mirrors, into red, green and blue light, each beam being separately treated in a similar tube to the monochrome projector.

The EP8 monochrome Eidophor projector has an output of 4,000 lumens from a 2.5kW lamp. Maximum picture size is about 12m by 9m. A wide range of lenses are available for varying projection distances. Price is 250,000 Swiss Francs. A new range of monochrome projectors will soon be available. The 5170 in Fig. 22 is a colour projector with 3,000 lumen output. Resolution is 800 lines in the picture centre, registration accuracy being 0.1% in a circle 80% of picture height. Price is 600,000 Swiss Francs. Finally the 5171 is a high intensity colour Eidophor which has a 7,000 lumen output with a 4.8kW lamp. Maximum picture size is 18m by 13.5m. Seven different lenses are available. This unit, the highest-powered TV projector available, costs 690,000 Swiss Francs.

Westinghouse mirror matrix

This is a reflective light valve which uses a matrix of mirrors built into a vidicon camera tube.

The light-reflecting Schlieren system employed is shown in Fig. 23, where the light valve is seen to be the faceplate in

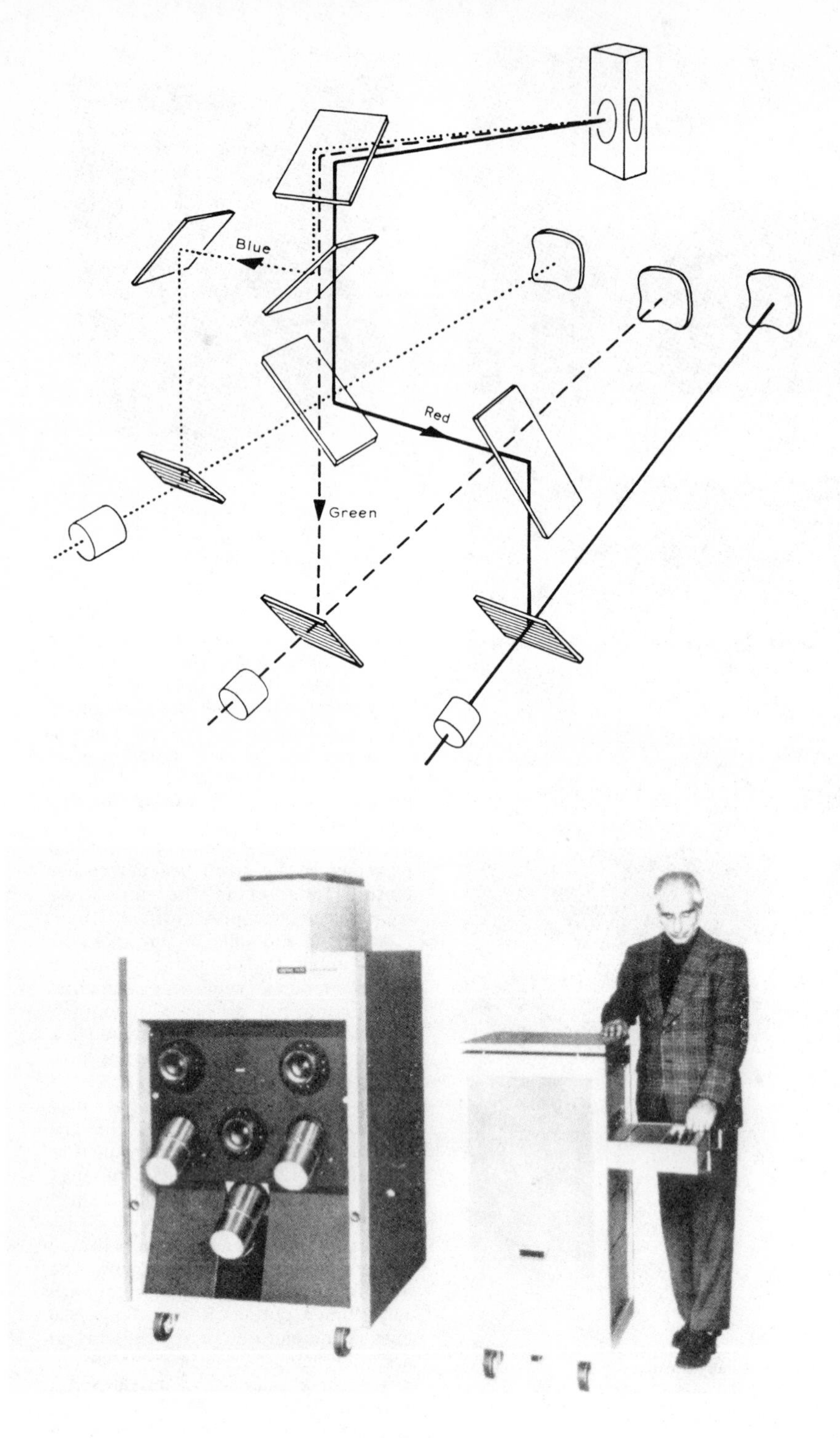

Fig. 21. Optical system of the colour Eidophor

Fig. 22. Gretag 5170 colour Eidophor. Control circuitry is in separate console on right

an otherwise conventional sealed-off vidicon tube which uses standard focusing and deflection components. The target is fabricated from monocrystalline silicon-on-sapphire substrates, using high yield semiconductor techniques. It is composed of a dense matrix (25.5 elements/mm) of aluminised silicon dioxide membranes (about 3,000Å thickness) supported centrally on small silicon posts 4-5μm in height above the transport sapphire faceplate. These flat, stress-free oxide membranes can be deflected electrostatically by up to 4° when addressed with the electron beam. Thus, because light scattered by activated mirror elements is directed around the central stop in Fig. 23, an intensity-modulated display of the deposited charge pattern on the mirror matrix is produced on the screen.

Mechanical and optical considerations have led to a special four-leaf geometry of the mirror elements in Fig. 24, enabling operation at a voltage level of only 175V) and an optical gating efficiency of about 50% to be achieved. The latter stems from the fact that light from activated mirror elements is spatially separated from the fixed diffraction background produced by the segmented target structure. Since the modulated light is effectively directed away from the optical axis of the Schlieren projection system, high screen brightness and high contrast are provided simultaneously by use of a central, cross-shaped Schlieren stop.

The mirror matrix is fabricated using chemically-inert, low vapour-pressure materials, so its inclusion within the sealed-off vidicon envelope shows no detrimental effects on tube life. In addition, the electrical insulation properties of the mirror matrix structure give long storage times for the charge pattern and its low thermal impedance suggests its suitability for high light-level flux handling capabilities.

At present, the write and erase time of 1/30s is such that real-time video cannot be projected, but the storage time inherent in the tube (many hours) makes the projector very suitable for single frame display such as might be used for computer displays, Travel indicators and such applications. A 1.3m × 1m screen was used with the prototype projector, which exhibited a 400-line resolution and 15:1 contrast ratio, with full grey scale. Total gated light output was about 90 lumens using a 150W xenon lamp and f/3.5 lens. Substantially higher luminous flux outputs can be expected with a larger light source and improved optics. A limiting resolution of 600 lines has been

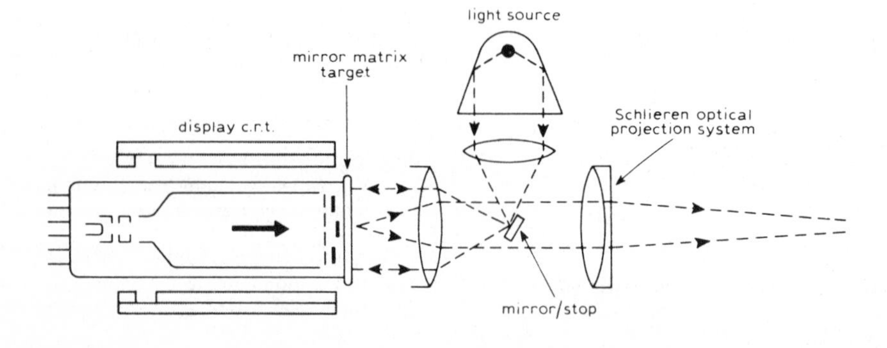

Fig. 23. Reflective type of light valve used in the Westinghouse mirror-matrix system

achieved using a 750,000 element array with fewer than 30 defective elements in a 50mm sealed tube. Price is expected to be between $2,500 and $3,500.

General Dynamics

A projection system similar to that of the Schmidt type is used by General Dynamics. The techniques uses a correction lens mounted between the c.r.t. and spherical mirror. Although providing better optical correction than the standard Schmidt, the large physical size and cost of the lens have prevented its wide use.

Lasers

Various attempts have been made to construct a projector using lasers as the light source. Although it is reasonably simple to modulate the laser light, scanning must be mechanical, using mirrors or prisms rotating at high speeds. However since the light path is very complex, efficiency is very low and this coupled with mechanical problems has frustrated development.

IBM deformographic

This uses a light valve containing a deformable target. Fig. 25 shows the principle of operation of the deformable storage display tube (d.s.d.t.) which uses a Schlieren optical system to convert the deformations of an optical surface into visual imaging points.

The heart of the d.s.d.t. is a dielectric membrane (target) which consists of an electronically-controllable storage substrate, a deformable material layer and a reflective layer. The target is mounted in the tube envelope so that the storage substrate faces the electron gun chamber of the tube. Deformations, created in the deformable material as a result of negative electrostatic charges deposited by the write gun of the d.s.d.t., are converted into a visual image by the off-axis Schlieren optical system. Since the substrate is a good insulator, it provides long-term image storage. Also, because of its secondary emission characteristic, the effective polarity of the deposited charge may be varied as a function of electron beam energy. Thus a deposited charge can be written and erased in a controlled manner by appropriately directed electron beams of selected energies.

Since the deformable material is isolated from the electron gun chamber, cathode poisoning is eliminated. By employing an elastomer as the deformable material, a simple mechanical restoring force provides the actual erase function once the deforming charge is removed. An additional advantage is the placing of the reflective layer on the deformable material, which allows an efficient reflective optical system to be employed instead of a complex transmissive system. Two electron guns are used mounted in the tube's rear. The write gun provides a magnetically deflected pencil beam while the erase gun is designed to cover the entire substrate with its electron beam cloud. Electronic control of these guns with time sequencing provides for such facilities as storage, variable persistance, selective erase, coloured images and optical processing.

Storage is achieved by sequentially writing and erasing. A single writing operation places information on the target where it is stored until neutralized by the erase gun up to several minutes later. Variable persistance is arranged by simultaneous operation of write and erase guns, the degree of persistence being controllable by varying the erase current. Selective erase may be arranged by altering the potential of the writing gun to produce a directional erase beam. Although theoretically possible, deflection problems with two different beam potentials could cause problems.

Coloured data may be displayed by exploiting the Schlieren plane effect. Fig. 26 shows two methods of producing coloured symbolic data. The top portion shows two E's generated from horizontal and vertical strokes. Because of the optical pattern created as a result of the stroke patterns, and crossed configuration separates the two characters into different colours at the image plane. The lower characters show colour generation by controlling depth of deformation. The lightly drawn character limits the reflective pattern to the inner annular filter ring while a heavily written character causes a large share of the reflected light to fall on the outer filter. The relative sizes of the characters shown in Fig. 26 are significant since the coloured characters made from directional strokes must be larger than black and white characters. Various other filter arrangements may be

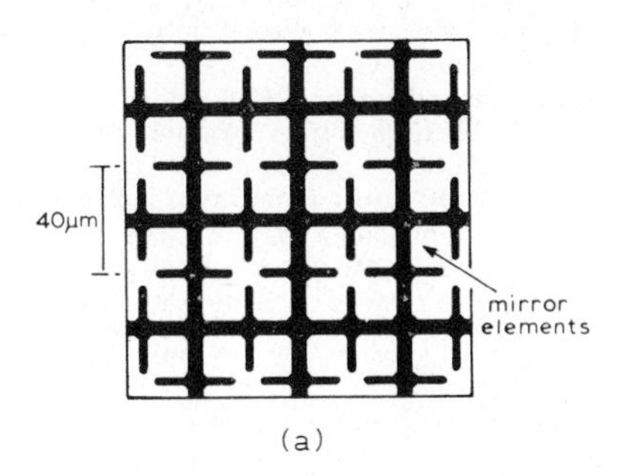

Fig. 24. Mirror-matrix light valve. Cross-section of one element is shown at (b)

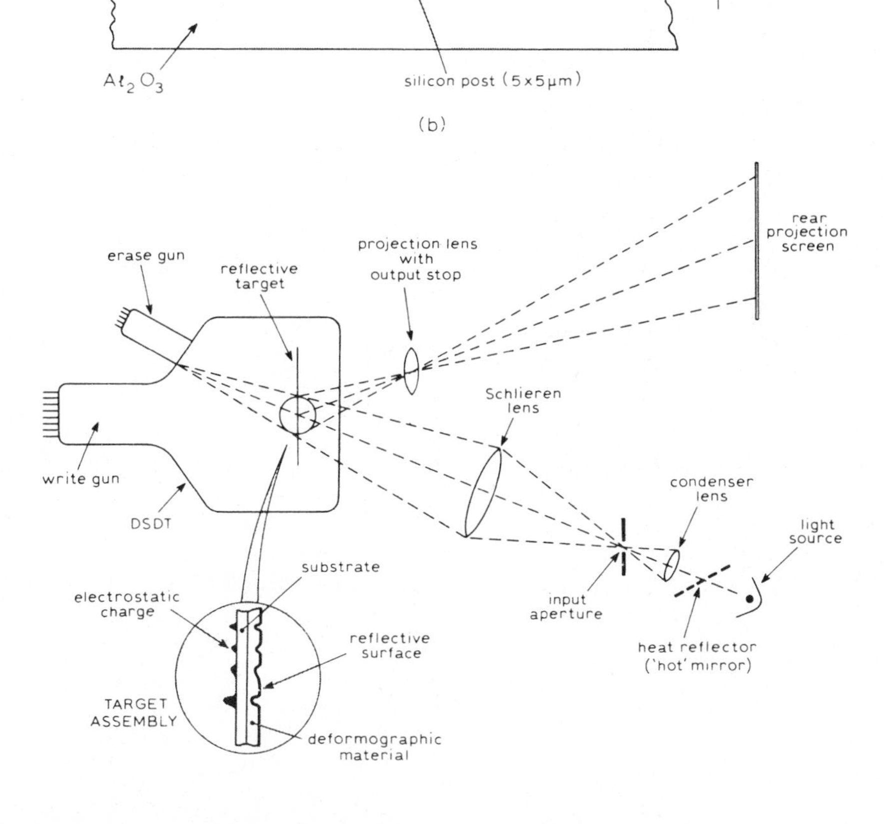

Fig. 25. System by IBM using a reflective, deforming surface – similar in some respects to that of Fig. 23

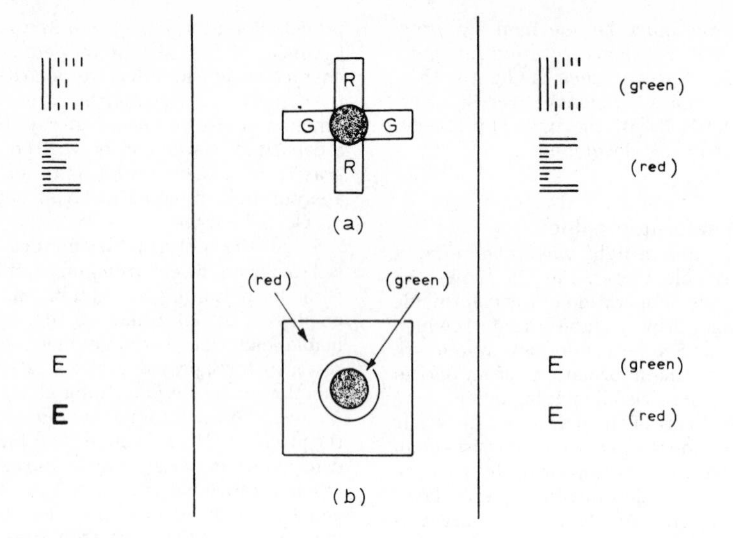

Fig. 26. Production of coloured symbols by crossed filters and symbols made up of vertical or horizontal lines is at (a), while (b) shows the method using differing degrees of deformation to include varying amounts of dissimilar filters

used in the Schlieren plant to enhance or interrogate the displayed tube.

Presently, a 150W xenon lamp is used as the light source and this provides a light output of between 300 and 500 lumens. Up to four different colours may be displayed, with a contrast ratio of 40:1. Rear-projected display systems suitable for computer generated alphanumerics and graphics projected on a 1.5m square screen are now available for purchase from IBM. Projectors are constructed individually by hand and cost around $200,000 but this will be reduced to about $28,000 when production starts.

Prices quoted in sterling indicate that the equipment is imported into the UK while those quoted in dollars would have to be purchased from the USA with all the additional costs that are incurred (freight, taxes etc).

I should like to thank the following people from both Europe and the USA who have provided the information necessary to write this article. Dr G. Baenziger, Gretag AG; Jacques Donjon, LEP; Vincent Donohoe, Speywood Communications Ltd; Patrick Gamuti, Projection Systems Inc; William Good, General Electric; Tom Holzel, Ideal Image Inc; John Huggett, Crown Cassette Communications Ltd; Alexander Jacobson, Hughes Research Laboratories; Harvey Nathanson, Westinghouse Research Laboratories; Michael Spooner, Redifon Flight Simulation; Jerrett Stafford, Aeronutronic Ford; Ronald Freeman, IBM.

Makers and importers

Advent: Crown Cassette Communications Ltd, 3 Soho Street, London, W1.
Aeronutronic Ford, 3939 Fabian Way, Palo Alto, California, USA.
Eidophor: Televictor Ltd, Channing House, Wargrave, Berks.
General Electric Co, Building 6, Electronics Park, Syracuse, New York.
IBM Corp, Department 102B, Building L91, Awego, State of New York 13827.
Ikegami: Dixon's Technical Ltd, 3 Soho Square, London, W1.
Image Magnification: REW Audio Visual Ltd, 10-12 High Street, Colliers Wood, London SW19.
Kalart Victor: British Films Ltd, 260 Balham High Street, London SW17.
Projection Systems: Speywood Communications Ltd, Northfield Industrial Estate, Beresford Avenue, Wembley, Middlesex.
Pye TVT Ltd, PO Box 41. Coldhams Lane, Cambridge.
Sony (UK) Ltd, 134 Regent Street, London W1.
Tele Theatre: Speywood Communications Ltd, Northfield Industrial Estate, Beresford Avenue, Wembley, Middlesex.
Westinghouse Electric Corp, R. and D. Centre, Beulah Road, Pittsburgh, Pennsylvania 15355.

A NEW COOLANT-SEALED CRT FOR PROJECTION COLOR TV

Masahiro Kikuchi, Katsumi Kobayashi*
Tomosuke Chiba and Yasumoto Fujii

Picture Tube Division
*Image Display Division
TV and Consumer Video Group
Sony Corporation
Ohsaki, Shinagawa
Tokyo 141, Japan

INTRODUCTION

Standard-sized color TV receivers are now commonplace. A new attraction, many viewers are discovering, is the viewing of TV programs on a very large screen. Considerable work has recently been done on the cathode ray tubes, circuits, projection lenses and screen necessary to realize a projection TV system for consumer use.[1-3]

The brightness in a white field produced by the three CRTs of a typical projection system can reach 10,000 fL. To obtain a picture this bright, the CRTs in the projection system must be operated at a higher anode voltage and a higher current density than conventional CRTs. These operating conditions generate considerable heat at the glass panel and this can cause a number of problems, such as thermal quenching of the brightness of the phosphor screen. Generally, the problems caused by high temperatures at the glass panel have been eliminated by an air-cooling system, but the use of a fan introduces new problems with the noise and magnetic interference from the fan's motor. We have addressed the heat problem in another way by developing a new coolant-sealed CRT for projection TV.

THE STRUCTURE OF THE NEW PROJECTION CRT

Figure 1 is a photograph of the new projection CRT and Fig. 2 shows the structure and the dimensions of the projection CRT, together with its typical operating parameters. Note that the diameter of the tube neck is a relatively narrow 30.6 mm and that the CRT employs a

Fig. 1. Photograph of a new coolant-sealed CRT for the newly-developed color projection system.

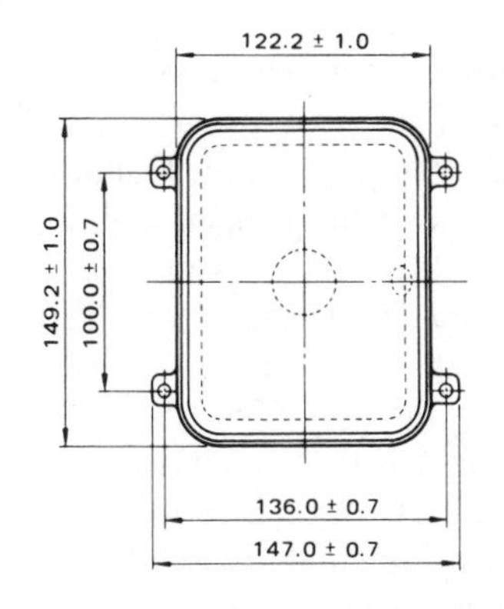

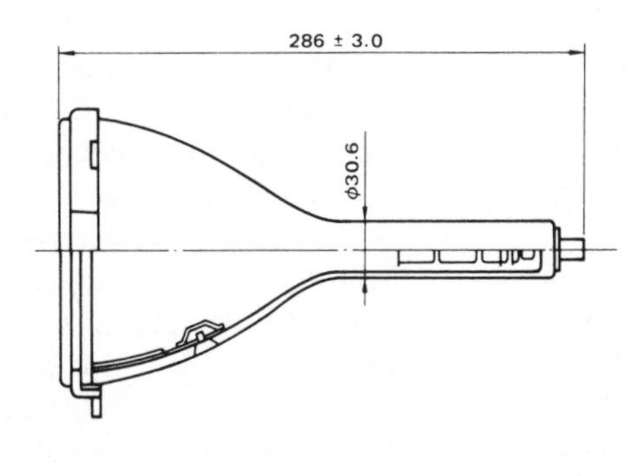

Effective screen size	5.5V" diagonal
Aspect ratio (H : V)	4 : 3
Deflection angle	55°
Focus of electron gun	Hi-Unipotential
Typical operating parameters	
Anode voltage	26 kV
Focusing voltage	6.8 kV
Average cathode current	320 μA

Fig. 2. Structure, dimensions and typical operating parameters of the projection CRT.

Reprinted from *IEEE Trans. Consum. Electron.*, vol. CE-27, pp. 478–484, Aug. 1981.

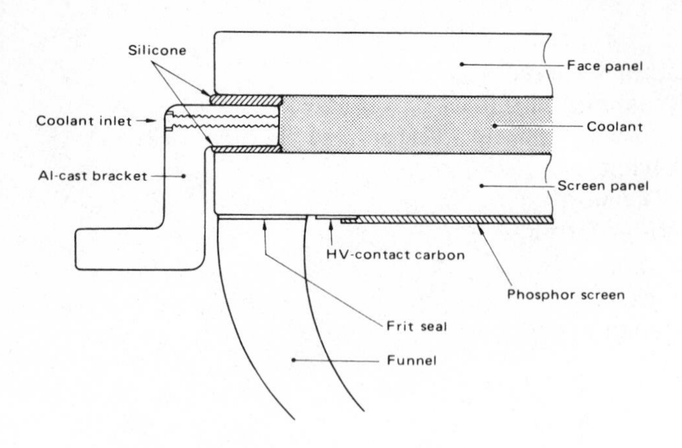

Fig. 3. Detailed structure of the sealed cell.

unipotential gun. Figure 3 is a detailed structural drawing of the arrangement of the new CRT's front panel, which is composed of two flat glass plates, one for the face panel and one for the screen panel, between which a low-viscosity mixture of ethylene glycol and water is sealed. A flat glass plate screen panel has been used in the past only in CRTs designed for institutional use. The use of a flat phosphor screen in a projection CRT is essential if the field curvature aberration of the projection lens is to be minimized. The face panel and the screen panel are held in place by an aluminum-cast bracket. Temperature-vulcanized silicone is used as a sealant to firmly join the panels and the bracket. Silicone after vulcanization becomes plastic: its hardness is 30 shore A and its expansion is 400%. Its adhesive strength is over 30 kg/cm^2.

The CRT's screen panel is made of strontium glass which is not subject to solarization. The maximum amount of stress on the screen panel and funnel glass is less than 1,000 psi, the same as in a conventional color picture tube.

The distance between the screen and the outside surface of the glass plate is 16.5 mm, which takes into account the design of the optical lens and heat radiation. A total thickness of 11.5 mm is required if the glass plates are to be an effective X-ray radiation shield. The coolant, which is 5 mm deep, is free to circulate by natural convection. After the silicone is thermally vulcanized the coolant is injected through a screw hole in the bracket.

The coolant expands as its temperature rises. See Table I for its average expansion coefficient. An increase or a decrease in pressure inside the cell is absorbed by the elasticity of the silicone sealant. The slight increase in pressure inside the cell which develops when the CRT is operated continuously causes the coolant vapor to permeate the silicone, but the thickness of the silicone is such that the loss is insignificant.

COOLANT REQUIREMENTS

The coolant must be transparent and thermally conductive. Anticipating the conditions under which a projection TV may be operated, the coolant should also have the following qualities:

(1) The melting point should be below -40°C and the boiling point above 100°C. The vapor pressure of the coolant must be low and stable over the range of temperatures encountered during CRT operation.

(2) The viscosity of the coolant should be low enough to enable it to circulate by convection.

(3) The coolant should be electrically conductive to the extent that is effective in preventing the accumulation of static electricity on the front panel, but it should induce no current leakage to the circuit.

(4) The coolant must not be expensive, poisonous or inflammable.

(5) The coolant should have a refractive index close to that of the glass plate.

Table I. Physical properties of the 80 wt.% ethylene glycol aqueous solution.

Boiling point	126°C	Electrical conductivity	$10^5 - 10^6$ Ω·cm
Freezing point	-45°C	Natural convection ratio	0.21 (water = 1.0)
Density	1.10 (g/cm^3)	Thermal expansion coefficient	5.1% (25°C – 100°C)
Refractive index (25°C)	1.41	Vapor pressure	(atm)
Specific heat	0.64	80°C	0.22
Viscosity (25°C)	11.4 (cm·poise)	100°C	0.47
Thermal conductivity	0.73 x 10^{-3}	120°C	0.94

An aqueous solution of 80% ethylene glycol satisfies the above requirements. Table I indicates the physical properties of this liquid. As the table shows, the boiling point of the solution is 126°C and the freezing point -45°C. In temperatures below -45°C, the coolant takes the form of slush and cases no damage to the CRT. Even at temperatures as low as -10°C or as high as +50°C, the CRT operates without difficulty. The coolant, which is transparent in the visible wavelength region, has a refractive index of 1.41, similar to that of glass, so contrast and brightness, which are affected by multiple reflections, can be maintained. Because the liquid has an adequate electric conductivity of 10^5 to 10^6 Ω·cm, no dust particles are attracted to the surface of the face panel. Experimental results indicate that it is free from discoloration and remains stable during changes in temperature, even after aging.

An additional advantage of the coolant-sealed arrangement is that the coolant can be dyed to match the spectral energy distribution of the phosphors in order to enhance the contrast of the projected image. This method of enhancing the contrast is more economical than using pigmented phosphors or a colored front panel and it does not sacrifice brightness.

A NEW PHOSPHOR-SCREENING PROCESS

We have developed a new phosphor screening process in which the substrate film and HV contact carbon as well as the phosphor film are screened by stainless steel mesh screens. This method is made possible by the flatness of the screen panel. A flow chart of the process is shown in Fig. 4.

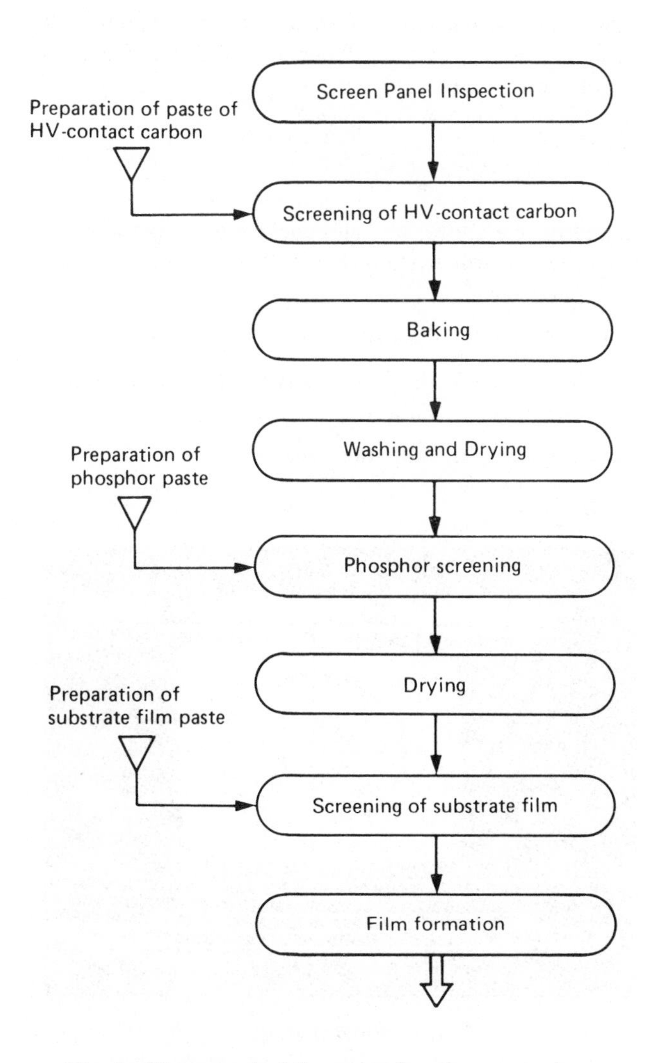

Fig. 4. Flow chart of the new phosphor-screening process.

The HV contact carbon paste is a mixture of carbon, aluminum phosphate, and a thixotropic agent. After screening, the panel glass is baked. The resistivity of the screened film is 150 Ω·cm and the thickness is about 15 μm. Then follows the washing and drying of the panel. The phosphor paste consists of a mixture of phosphor powder, ethylcellulose (as a binder), and a solvent with a high boiling point of 250°C (to make the paste viscous). Particle size and specific gravity are factors to be considered in deciding on the percent of phosphor powder in the paste, which ranges from 70% to 90%, depending on the color. The following are the phosphors employed in the CRT and their weight when coated on the panel.

Red	$Y_2O_3:Eu^{3+}$	7.5 mg/cm^2
Green	$Gd_2O_2S:Tb^{3+}$	5.7 mg/cm^2
Blue	ZnS:Ag	5.7 mg/cm^2

After coating, the panel is dried. This new method produces a dense phosphor film containing fewer pinholes than usual film, which helps to inhibit solarization of the glass.

The substrate film paste consists of an aqueous emulsion of acrylic polymer, a thickening agent, a solvent to make the paste viscous and a PH-control agent.

Fig. 5. Microphotograph of the phosphor and substrate film screened by the new process.

Figure 5 is a microphotograph of the phosphor and the substrate film. The thickening agent is identical with the emulsion in its chemical structure, but the average molecular weight of the thickening agent is 400,000 and that of the emulsion 300,000. This paste can be made into a film without dissolving the screened phosphor layer. The film was dried with use of a far infrared heater, after which the film weighed about 0.5 mg/cm^2.

The most noteable characteristic of this process is the use of a silk-screening technique to form the substrate film. This screening technique, which is easily automated, can reduce the manufacturing time to two thirds of that presently needed. In addition, the amount of expensive rare-earth phosphors required can be reduced to one third of that previously necessary. Since this process has no influence on arcing in the CRT, it permits a gun with a 30.6 mm diameter neck to be employed so that the deflection power can be reduced. The new screening process is still in the developmental stage and we expect further research to improve it.

COMPARISON OF THE COOLANT-SEALED CRT WITH A CONVENTIONAL PROJECTION CRT

We will now describe the mechanism of heat radiation from a conventional projection CRT. Most of the energy generated by electrons on a phosphor screen takes the form of thermal energy; less than 10% of the power is transformed into light energy. A large portion of the thermal energy thus obtained is radiated through the screen panel in all directions but only a small fraction of the energy is conducted to the funnel. The temperature at the screen panel then rises and natural convection releases the heat from the screen panel into the air. The temperature distribution and gradient on the surface of the panel are shown in Fig. 6 (a). The distribution was measured by a thermal imager, model TIP 1400HS manufactured by Nippon Aviotronics Co. The thermal conductivity of the panel glass is as small as 0.65 kcal/m•hr•°C. The temperature gradient is proportional to the input power. In the conditions under which conventional projection CRTs operate, heat radiation from the screen panel induces the vertical convection of air. The difference in temperature between the screen panel and the atmosphere $\triangle Q$ is given by[4]:

$$\triangle Q = \frac{1}{Num^4} \cdot \frac{(0.811+Pr)}{0.638^4 Pr^2} \cdot \frac{r^2}{h^3 g \beta} k,$$

where

Num : Nusselt number
Pr : Prandtl number
r : density of air (kg/m^3)
h : screen panel height
g : gravitational constant
β : thermal expansion coefficient of air
k : a proportional constant.

The thermal resistivity of the radiation induced by natural convection is so high that a large temperature gradient results. The temperature difference $\triangle T$ at the center of the screen is given by:

$$\triangle T \propto P^{0.8} = VI_k,$$

where

P : input power (W)
V : anode voltage (kV)
I_k : cathode current (mA).

The experimental data agrees with the results obtained from the above equation. When a CRT is operated at an anode voltage of 26 kV, the temperature difference at the center of the panel is given by:

$$\triangle T = 90.4 I_k^{0.8} \text{ (mA)}.$$

Based on the above, we can conclude that the following measures are effective in lowering the temperature of the phosphor screen:

(a) Reducing the thermal resistivity of the panel
(b) Reducing the thermal resistivity of air by means of forced air-circulation
(c) Expanding the radiation area.

Normal operation

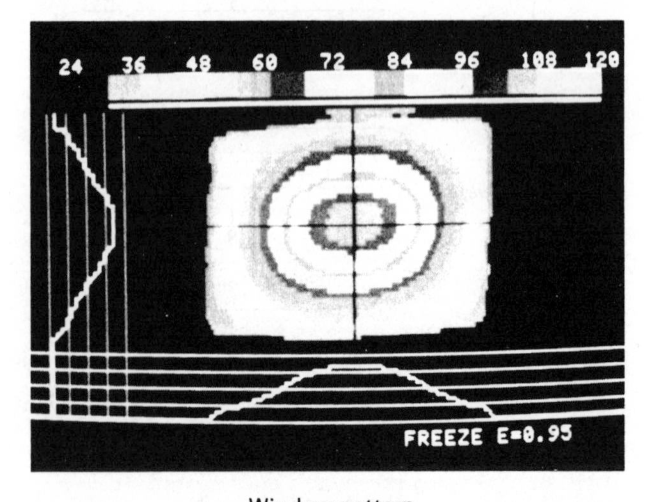

Window pattern

Fig. 6 (a). Temperature distribution in a conventional tube.

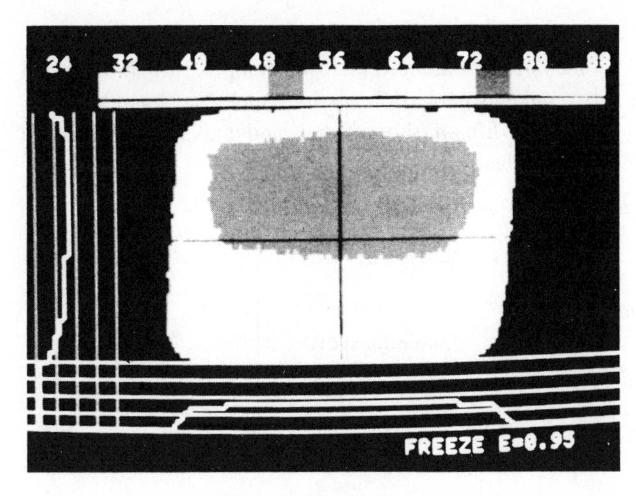

Normal operation

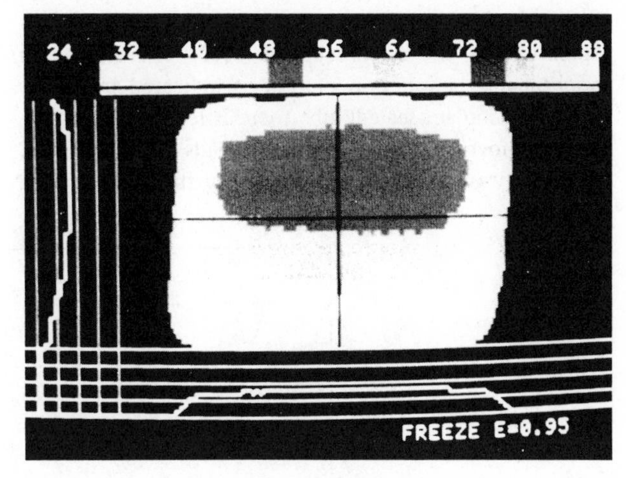

Window pattern

Fig. 6 (b). Temperature distribution in the coolant-sealed CRT.

The new CRT, which was developed based on measures (a) and (c) above, has the following advantages over a conventional CRT:

(1) The temperature difference between the phosphor screen and the surface of the face panel is reduced and temperature variations on the phosphor screen and the panel surface are reduced.

(2) Fluid convection in the coolant effectively enlarges the radiation area and facilitates heat radiation[5]. The heat transmitted to the highly-conductive Al bracket and the face panel radiates from their surface into the air, so that thermal conductivity is improved and the temperature of the phosphor screen is significantly reduced. Figure 6 (b) shows the pattern of temperature distribution on the surface of the face panel of the coolant-sealed CRT.

Table II compares the new coolant-sealed CRT and a conventional CRT. The temperature in the center of the surface of the face panel of the coolant-sealed tube is reduced by as much as 20°C. Moreover, this new tube significantly reduces thermal quenching of the brightness of the P43 phosphor[3]: we have achieved a more than 15% increase in brightness under normal operating conditions and a more than 50% increase with a window pattern signal. The temperature at the panel of a conventional CRT and of the coolant-sealed CRT are plotted as a function of time in Fig. 7. The figure shows clearly that in the coolant-sealed CRT the normal-operation and window-pattern temperature curves are similar, while those of a conventional CRT are very different. We assume that it is the natural convection of the coolant that holds the increase in temperature to the same level dur-

Table II. Comparison between a conventional CRT and the coolant-sealed CRT.

	Raster size	Input power	Temperature rise on the phosphor screen from room temperature	Average temperature gradient	Relative brightness
Conventional CRT	Normal operation	0.17 (w/cm²)	+68°C	32°C	0.68 (0.81)
	Window pattern	0.67	+120°	45°	0.27 (0.32)
Coolant-sealed CRT	Normal operation	0.17	+46°	15°	0.835 (1.0)
	Window pattern	0.67	+78°	20°	0.645 (0.77)

*Normal operation: 96 mm x 71 mm
*Window pattern 48 mm x 35.5 mm

*HV = 26 kV
I_K = 430 μA
Green CRT

ing normal operation and during the window-pattern operation in the coolant-sealed CRT. The change in brightness as a function of time is indicated in Fig. 8. With the coolant-sealed tube there is less change in the color temperature of white, which leads to an enhanced uniformity of colors on the screen. Furthermore, distortion caused by thermal stress is reduced and the CRT becomes more reliable. The lower temperature at the panel also makes solarization of the panel glass less likely, extends the life of the cathode and makes it possible to adopt a plastic lens.

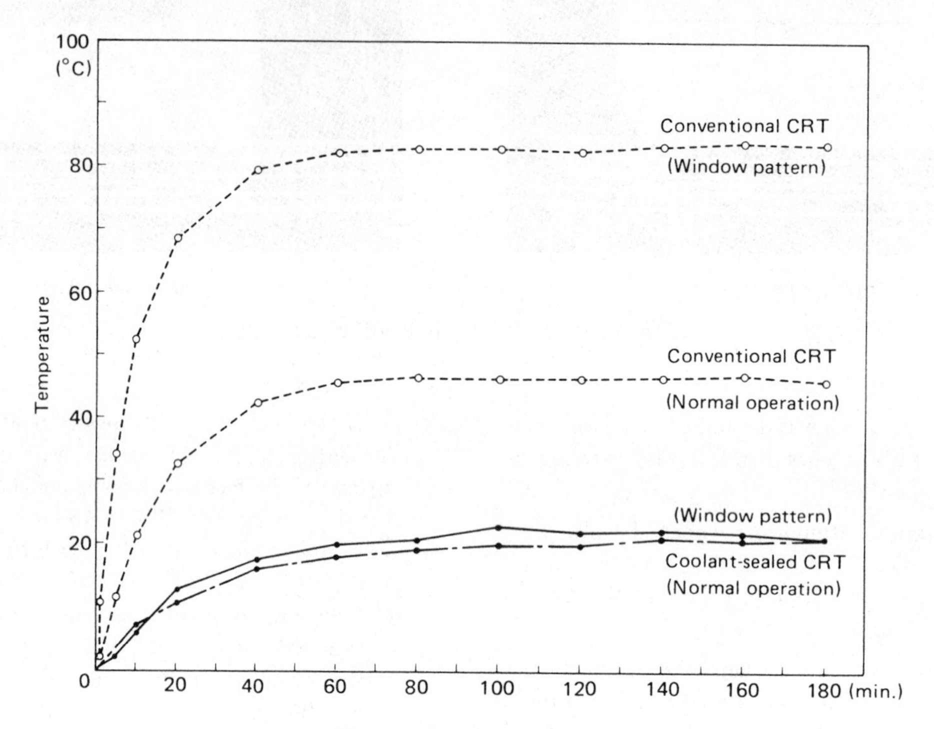

Fig. 7. Temperature change in a conventional tube and the coolant-sealed CRT as a function of time. (Operating conditions as in Table II.)

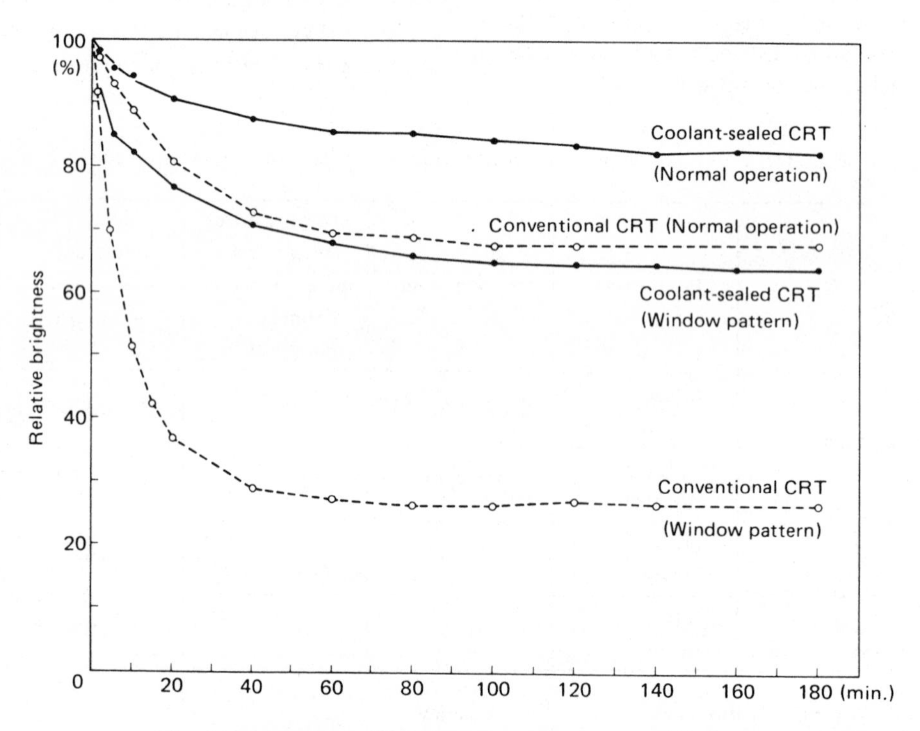

Fig. 8. Thermal quenching of P43 phosphor brightness. (Operating conditions as in Table II.)

SUMMARY

Our newly-developed projection CRT is equipped with a sealed cell containing coolant in front of the screen. The coolant has the same function as the fan which is used in conventional projection CRTs to keep the temperature at the face panel low. The coolant reduces the temperature at the face panel by 20°C, with a resultant improvement in brightness and reliability. Further reduction of the temperature is possible with an expanded radiation area.

This coolant-sealed CRT, combined with a computer-designed three-element plastic lens, has been employed in a coffee table-type video projection system consisting of a projection unit and a screen. This system provides a clear picture of 120 fL for a 50" screen and 60 fL for a 72" screen. This same tube has been adopted for airborne use.

ACKNOWLEDGEMENTS

The authors wish to thank Dr. Akio Ohkoshi, General Manager, Electronic Devices Development Division for his technical suggestions. The authors are indebted to the members of the engineering staff assigned to the development of this CRT for their contributions, especially to Mr. Koichi Momoi, General Manager, Picture Tube Division, Sony Corporation. The constant encouragement of Mr. Masaaki Morita, Senior Managing Director of this corporation, is also greatly appreciated.

REFERENCES

1) S. Shimada, "Large Screen Video-Display", J. Inst. of Television Engineers of Japan, Vol. 27, No. 5, May 1973, p.53.
2) M. Chivsans, "Three Little Tubes Make a Big TV Picture", Design News, OEM 6–4, 1973, p.42.
3) I. Itoh et al., "New Color Video Projection System with Glass Optics and Three Primary Color Picture Tubes for Consumer Use", IEEE Chicago Spring Conference on CE, June 1979.
4) J. P. Holman, Heat Transfer, McGraw-Hill, 1963.
5) T. Schmidt, "A New Monochrome Data/Graphics Video Projector", IEEE Trans. on CE, Vol. CE-26, No. 3, August 1980, p.414.

DEVELOPMENTS IN PLASTIC OPTICS FOR PROJECTION TELEVISION SYSTEMS

Roger L. Howe and Brian H. Welham
3997 McMann Road
Cincinnati, Ohio 45245

Introduction

In 1973, Advent Corp. introduced the first high-brightness, color projection-television system designed specifically for the consumer market. Although General Electric Co. was making excellent light-valve color systems, and Sony Corp. had produced single-tube systems, the Advent system, utilizing Schmidt-type reflective optics, is thought by many to be the most instrumental in stirring general public interest in projection television. This was followed by a less expensive unit introduced in 1976 using aspheric plastic refractive optics. These earlier developments have led to the current intense activity to produce improved systems that will win consumer acceptance. It is no longer a question of when projection television will happen, but rather in what form, at what quality level, and at what price.

Virtually all knowledgeable observers forecast a substantial demand for high-quality, economical and unobtrusive projection television. Precisely accurate market information is not readily available. However, the composite estimate of industry observers throughout the world sees the U.S. market for projection television growing from 20,000 in 1977 to 60,000 by the end of 1979. And in 1983, the U.S. market is expected to be in the 350,000 to 500,000 range. Some manufacturers believe that by 1981 projection television sets will be sold in Europe at an annual rate exceeding 60,000 units and growing rapidly. Opinions on other foreign markets are more varied, making speculation difficult.

Viewed from another aspect, the current U.S. market for conventional color-television sets is about 10-million units. If the 1983 market forecast is valid, then in that year projection television could comprise 5% of the entire present market. However, these figures are only speculation. Anyone familiar with consumer electronics knows that market forecasts for relatively new products seldom prove to be accurate.

These market projections stem from the assumption that the availability of one-piece systems aesthetically compatible with a variety of home environments will create a substantial market. This assumption indicates the need for a variety of sizes and configurations much like those available in small conventional television systems. It is further assumed that these sets will have good small-area contrast, a wide viewing angle, and sufficient brightness for average room lighting in which nonviewers can read comfortably. Lastly, price is assumed to range from slightly below $2,000 for simple, front-view, one-piece systems up to $3,000 for rear-screen systems with a complement of special features.

Principal systems limitations, to meet cost and performance levels demanded by the consumer market, have been cathode ray tubes (CRT's), lenses and screens. The increasing market potential is rapidly spurring tubemakers to develop improved CRT designs specifically for projection applications. Previously available optical system designs limited image brightness and cabinet configuration. It is in this area that recent developments in plastic lens design and manufacturing technology provide greater freedom for adjusting system parameters to get optimum performance without major compromise. More specifically, the use of low cost plastic as an optical raw material coupled with high volume aspheric manufacturing techniques has allowed the development of new high-speed lens systems ideally suited for projection television. Screen developments for cleanable front and rear systems with controlled viewing angles are lagging CRT and lens development, but the intensity of effort is increasing and much progress is anticipated.

System Considerations

The prime objectives of any projection-television system design are high brightness, wide viewing angle, sharp contrast/resolution, low cost and attractive configuration. Brightness and resolution are optimized by design compromises and tradeoffs between all system elements, including the human eye, Fig. 1. These elements are interactive;

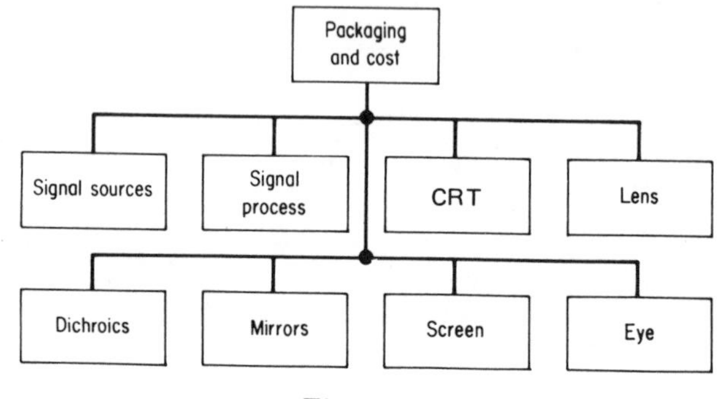

Figure 1

Received January 28, 1980.

Reprinted from *IEEE Trans. Consum. Electron.*, vol. CE-26, pp. 44–53, Feb. 1980.

changing the characteristics of one affects the performance of all others and, thus, overall system performance. Since lens characteristics have the greatest influence on brightness and resolution, the lens becomes the key factor about which all other elements are selected to get adequate brightness, screen viewing angle, and picture clarity. First, however, it is necessary to examine dominant system variables to determine their effect on optical performance. The following discussion is limited to 3-tube refractive designs, which out-perform comparable single-tube projector designs from the standpoints of cost, brightness, design flexibility, and compactness.

Although reflective Schmidt-type optical systems are in limited use, because this paper deals with plastic refractive optics, they will not be discussed other than to say both optical systems can produce acceptable image quality. Cost, angular coverage and flexibility considerations open questions that must be considered by the system designer.

System component evaluation is based primarily on required screen brightness, which is calculated from: $Bs = \frac{Bcrt\ T\ G\ R^N\ D}{4f^2\ Gw\ (1+m)^2}$ *. Equation 1.
Where Bs = projection-screen brightness (white light), FL (Foot Lamberts); Bcrt = green CRT brightness, FL; T = lens transmittance, %; G = screen gain; f = f/# of lens; R = mirror reflectivity; N = number of mirrors; D = dichroic efficiency for green light; Gw = green component in desired white light, %; and m = magnification.

Expressing brightness in another way,

$L = \frac{Bs\ As}{G}$; As = Acrt m^2, where L = flux, lumens; As = screen area, sq. ft.; and Acrt = CRT image area, sq. ft. In terms of foot-candles,

$Es = \frac{Bs}{G}$, where Es = screen illuminance, foot-candles.

The graph (Fig. 2) displays the effect of lens f/# on system brightness relative to the CRT diagonal selected. The CRT brightness, lens transmission, mirror reflectance, screen gain, and CRT green component are held constant.

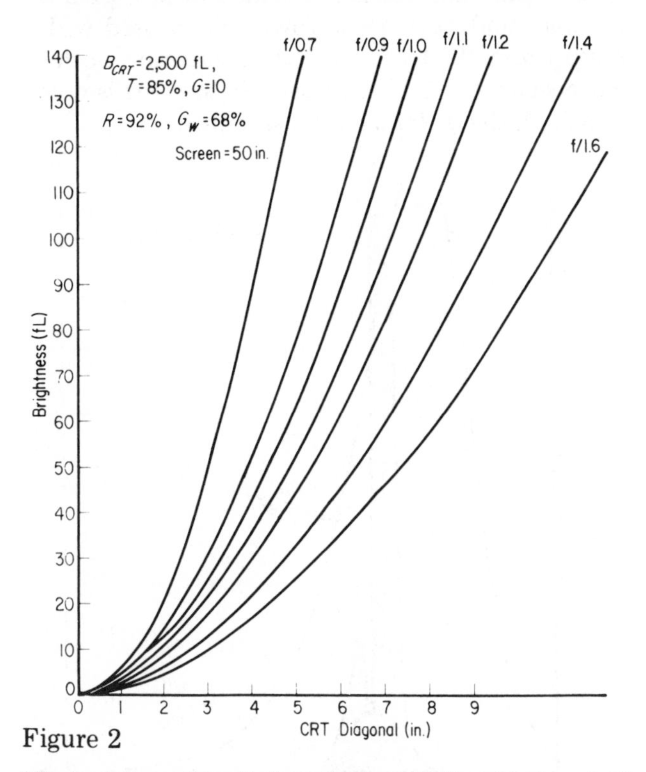

Figure 2

*Derived from "Applied Optics and Optical Engineering", edited by Rudolph Kingslake, Academic Press, 1965, New York and London.

Three-tube system configurations are restricted to variations of three basic concepts. In an in-line or triangular configuration, Fig. 3, the images in the viewing plane are tipped and must be held within the system depth of focus to provide acceptable image quality. This problem can be relieved partially by applying the Scheimpflug* condition, commonly used in view cameras and photogrammetry. Since the image is tipped, "keystoning" is evident in the images of the outer two CRT's. This condition must be corrected electronically for good registration. Also, because the two outer lenses are off-axis and their axial bundles converge on the screen at different angles, an image color shift is evident when viewed from left or right. This effect can be minimized by using a lower-gain screen and masking or apodising the peripheral bundle on the appropriate side of the outer lens.

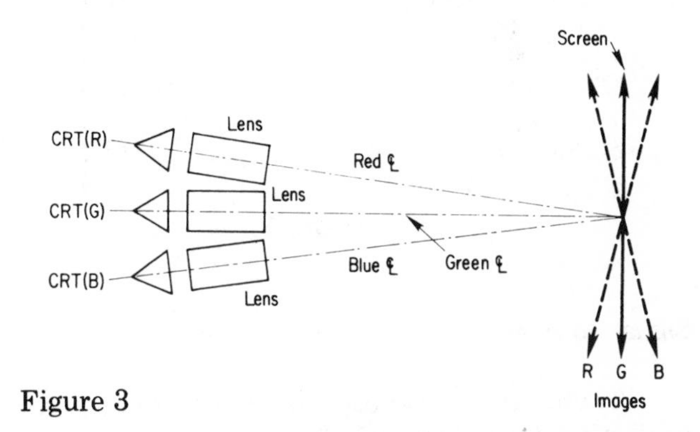

Figure 3

The second system, Fig. 4, is a variation of the first. Here, some corrections for keystoning performed electronically in the in-line system are traded for added field of view in the lens. The lens axes are all parallel and perpendicular to the screen, but the outer CRT's are displaced laterally outward so that the outer optical axes converge at

*See Scheimpflug; derived from "Applied Optics and Optical Engineering", edited by Rudolph Kingslake, Academic Press, 1965, New York and London.

the screen. The disadvantage is that the field coverage of the lens is increased, which generally means a longer focal length lens and thus increased object-to-image distance.

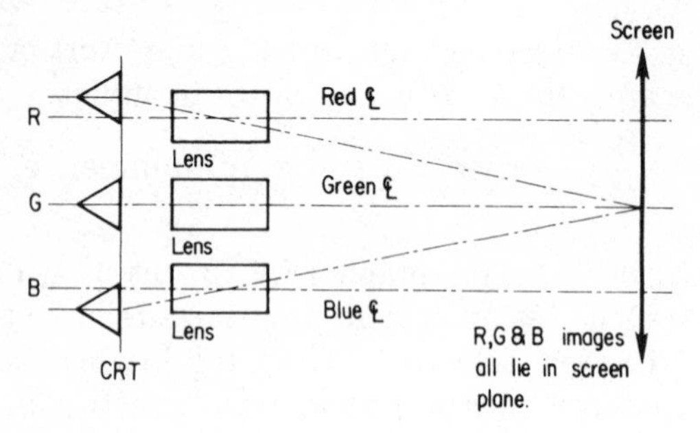

Figure 4

However, this arrangement greatly simplifies the mounting-bulkhead design. Also, if the distortion is corrected optically and matched for the three lenses, keystoning is eliminated. Finally, the offset CRT's introduce off-axis optical vignetting within the lens. However, color shift is still evident when viewing the screen off-axis.

The third system combines the separate optical axes prior to reaching the screen, Fig. 5. This system eliminates both optical keystoning and screen color shift caused by three converging axes. However, the dichroic box introduces significant light loss and has a color shift different from that of the other 3-tube systems. The dichroic mirrors do not necessarily need to be crossed; they can be located within, behind, or in front of the lens.

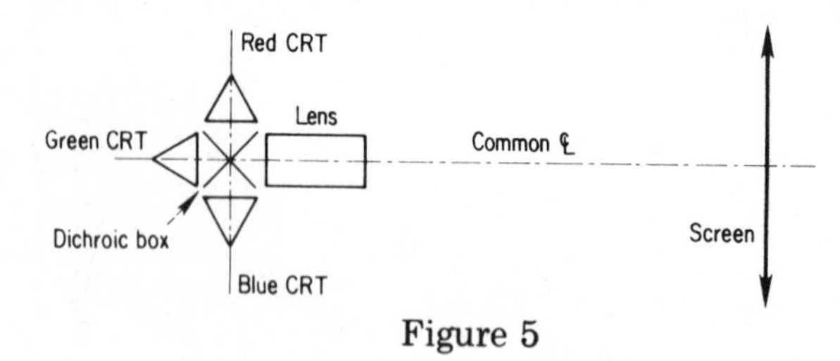

Figure 5

Signal Sources

There are a variety of signal sources for projection television including tuners, tape decks, video discs and video cameras. Since NTSC signals are standardized, at present the signal cannot be considered as a controlling factor since no significant options or variations are available.

Signal Processing

Because the final output image from a projection television is magnified, signal noise, resolution, and transient response all require special attention. Additional signal filtering should be considered since normal signal irregularities will be magnified on the viewing screen.

CRT Design

This discussion concentrates on the characteristics of the CRT as they interface with the optics.

Output: The ultimate objective of the optics system is to produce imagery at various wavelengths which, when viewed under normal conditions, are compatible with the color perception of the human eye. The normal method of producing the visual spectrum is to combine the three primary colors — red, green, and blue — in the ratios required to produce white light. These ratios (for example, 10% blue, 70% green, and 20% red) are determined by the peak wavelengths produced by various phosphors responding to excitation. In all cases, brightness of the resultant combination is inversely proportional to the percentage of green present. For example, Equation 1 indicates that $\underline{G_W}$ component can dramatically change screen brightness with merely an increase from 0.6 to 0.7. Therefore, it is desirable to consider minimizing the green phosphor component in the resultant mix.

A second consideration for selecting the type of phosphor is its emitted spectral bandwidth. For example, Fig. 6 shows the radiant energy output of the two green-producing phosphors. In Case 1, the spectral bandwidth is broad, and in Case 2 the spectral bandwidth is narrow. When used with a monochromatic lens, Case 2 is superior both in image sharpness and contrast. If the lens is color-corrected, the difference is less important.

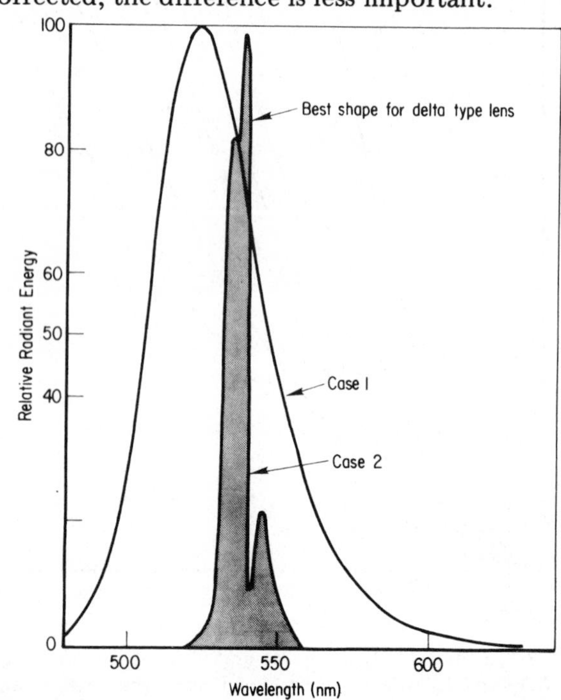

Figure 6

X-Rays: In addition to producing visible wavelengths, a CRT also generates heat and X-rays. To meet Federal requirements, X-ray emission can be reduced either by inserting an X-ray-absorbing glass plate between the lens and the CRT faceplate, or by using the CRT faceplate made from a material with sufficient thickness to absorb the rays before they reach the lens. Using a thicker CRT faceplate is more desirable optically because an extra plate introduces back reflections, surface losses, and transmission losses — all of which degrade overall optical performance.

Heat: The faceplate also generates heat. If plastic optics are used, excessive temperature variations will cause focus shift because the thermal expansion of plastic is high (approximately 3.6×10^{-5} in./in./^{o}F). The maximum continuous operating temperature of the plastic used in plastic lenses can vary up to 170°F to 200°F without performance degradation. If the anticipated faceplate temperature is higher, then the heat must be dissipated by a bulkhead heat sink, forced air, or circulating fluid.

Size: Projection television systems have used CRT sizes ranging from 3-in. diagonal to 9-in. diagonal. Optimum size selection is often based on cost of the CRT-lens combination for a given power-brightness ratio. Thus, the optical parameters of the screen image must be considered. Table 1 shows that viewing-screen resolution is the same for a range of CRT sizes. The problem, then, is selecting a lens that can relay an image from the CRT to the screen with sufficient brightness and contrast to form an acceptable display. Consideration should also be given to registration, which becomes more demanding on the CRT faceplate for high magnifications since the relay lens magnifies any registration error at the CRT faceplate.

Phosphors: In a direct-view system, the viewer looks directly at the scanning spot on the CRT faceplate. Consequently, spot size and clarity are a function of viewing distance. In projection television application, the spot is magnified and imaged onto a screen where it appears larger, and edge clarity is generally degraded. Therefore, it is advantageous to reduce spot size and make spot brightness a sharp step function from one edge to another without blooming.

Center-to-Edge Brightness: Every lens has a center-to-edge fall-off which, at best, follows a $\cos^4$ function of the angular coverage. When determining CRT center-to-edge brightness variation, optical vignetting characteristics must be considered along with the CRT to effectively evaluate relative illumination of the projected image.

Faceplate Curvature: Fig. 7 shows that the natural field curvature of a projection lens curves toward the lens. Deviations from this curvature require extra work within the lens. Since the CRT has a natural tendency to curve in the opposite direction, a compromise must be made which is compatible with both lens and CRT characteristics. An acceptable solution is a flat phosphor image plane.

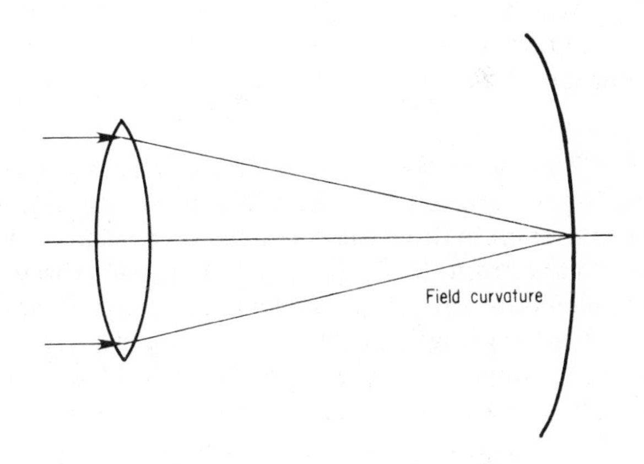

Figure 7

Thickness and Material: High-speed lenses are used to maximize brightness at the screen. The higher the lens speed, the greater the acceptance

Table 1 — Viewing-Screen Resolution vs CRT Size

CRT Diagonal (in.)	Height (in) (3 x 4 aspect ratio)	Resolution at 525 Scan Lines (line pairs/mm)	Magnification	Resolution at Screen for 50-in. Diagonal (line pairs/mm)
3	1.8	5.75	16.66	0.34
4	2.4	4.30	12.50	0.34
5	3.0	3.40	10.00	0.34
6	3.6	2.80	8.33	0.34
7	4.2	2.46	7.14	0.34
8	4.8	2.15	6.25	0.34
9	5.4	1.91	5.55	0.34

angle at the phosphor. The greater the acceptance angle, the more sensitive the light rays become to bending within the CRT faceplate. Therefore, the CRT faceplate is a critical component of the optical system. Thickness, index of refraction at the three primary wavelengths, and dispersion should be specified. Also, high-transmittance and nonbrowning materials must be used.

Liquid Cooling: If cooling liquid contacts the CRT faceplate, the liquid becomes an optical component and must have good optical transmission characteristics.

Internal Reflections: Internal reflections within the CRT faceplate detract from system contrast. The three reflective sources are the phosphor, the faceplate, and the glass-air interface. When light emitted from the phosphor reaches the glass-air interface, up to 4% is reflected back toward the phosphor. This 4% is then either absorbed or reflected by the phosphor. Selected phosphors should be examined for reflectivity, and the use of a coating should be considered for the external glass surface. Nonbrowning glass should have maximum transmittance and minimum internal scattering.

Lenses:

Projection television systems presently use a variety of lens types, including glass only, glass-plastic combinations, and plastic only. The f-numbers vary from f/1.0 to f/1.8 using from three to six elements to obtain the required magnification and field coverage. Total conjugate length TCL is a more important consideration than lens focal length because package size varies directly with TCL. Magnification, focal length fl, and TCL are related by:

$$TCL = \frac{fl\,(1+m)^2}{m} + NS.$$

Where NS = lens nodal separation. This dimension can vary from +8 in. to -4 in. for a given lens design.

For a given magnification (which is fixed by the screen size and CRT size selected), TCL is influenced mostly by focal length. For a given CRT size, as focal length decreases the half field angle coverage of the lens increases, Fig. 8. Half field angles of up to 25° are readily available, but greater angles require more lens elements, thereby increasing complexity and cost.

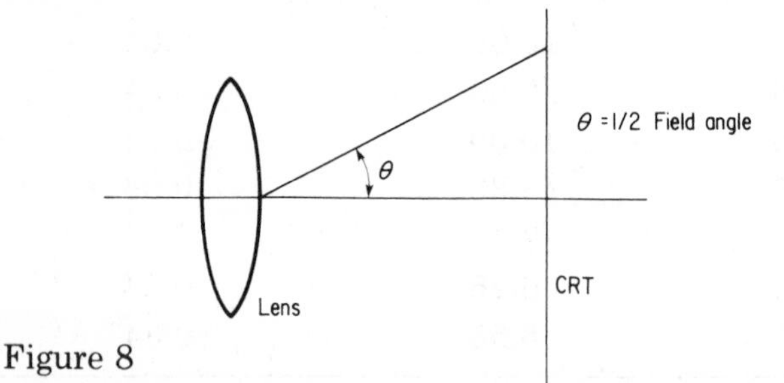

Figure 8

Color Correction: When a photographic lens produces an image on film having a broad spectral response, color correction is needed to form red, green, and blue images in the same image plane and at the same magnification. In a projection television system using three independent color sources, color correction is not essential. Each lens can be focused independently for its own color, and small magnification errors can be corrected electronically.

Distortion: Distortion is the percent deviation between the theoretical and actual length of a radial vector in the image plane. In Fig. 9, for example, Point A is the theoretical location of the imaged point, and Point B is the actual location. The error is expressed as a percentage of the theoretical position and is plotted as a function of field position. Fig. 10 shows a radial distortion plot using a flat CRT and a flat screen. Distortion, expressed as $100\left(\frac{\Delta x}{x}\right)$, can be controlled to within ±2%.

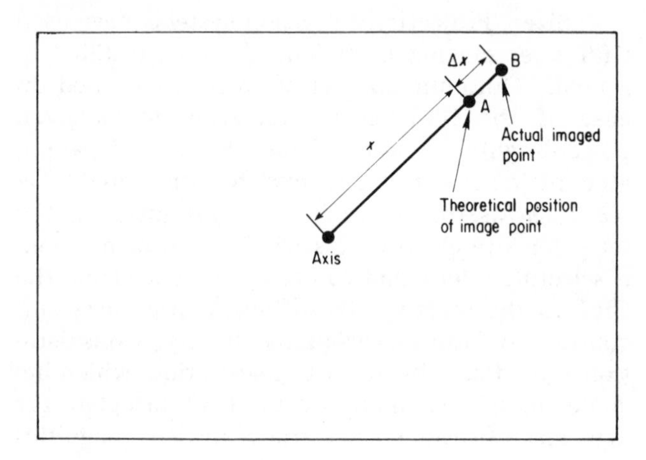

Figure 9

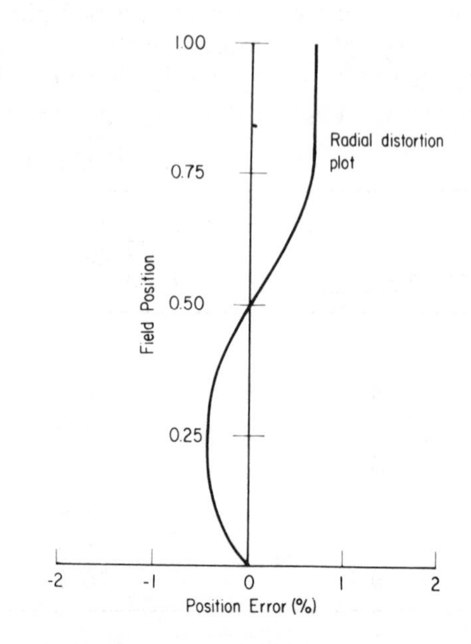

Figure 10

Resolution and Contrast: The maximum theoretical lens resolution is approximately $\frac{1600}{\text{f-number}}$ in line pairs/mm. However, television CRT rasters in the USA contain up to 525 scan lines. Therefore, lens resolution should be limited to $\frac{525}{2H}$, where H = screen height, mm, and resolution is in line pairs/mm. The design should be optimized for maximum modulation transfer function (MTF) at this and lower frequencies.

Depth of Focus: The equation for depth of focus DOF for a diffraction-limited (perfect) lens is: DOF = $4 \lambda (f/\#)^2$. Since the lenses used in this application are not diffraction-limited, it is more useful to plot the usable depth of focus at the design frequencies. Longitudinal magnification is a square function. Therefore, to determine the depth of focus at the screen, the CRT image position depth of focus must be multiplied by the square of magnification.

Transmittance: The current material used for the plastic lens manufactured for projection television is acrylic (methyl methacrylate), which has a transmittance of over 99%. However, each uncoated air-plastic interface scatters 3.9% of the light. This loss is the same as glass of the same index of refraction. Applying a single quarter wavelength of magnesium fluoride coating reduces this surface loss to approximately 1.6%. The transmittance of a three-element lens of the DELTA type described later is a minimum of 74% uncoated and 82% coated.

Relative Illumination: Relative illumination (RI) is corner illumination as measured relative to the axis illumination. Three factors determine RI: optical vignetting, mechanical vignetting, and the $\cos^4$ characteristic. The $\cos^4$ contribution is a direct function of the angular coverage. Mechanical and optical vignetting are controlled by adjusting the physical and performance characteristics of the design.

Plastic Designs: To meet high-performance and maximum-brightness demands at the lowest possible cost, a new form of plastic lens designated the DELTA* type has been designed. This lens uses low-cost acrylic material for the optical elements and, through innovative use of aspheres, has optimized optical performance with a minimum number of elements. Fig. 11 shows a plastic lens designed specifically for projection television. Designated the DELTA IV, the lens has a 114mm focal length and an f-number of f/1.0. The lens accommodates flat CRT faceplates measuring 5.1-in. diagonal and can be adjusted for varying faceplate

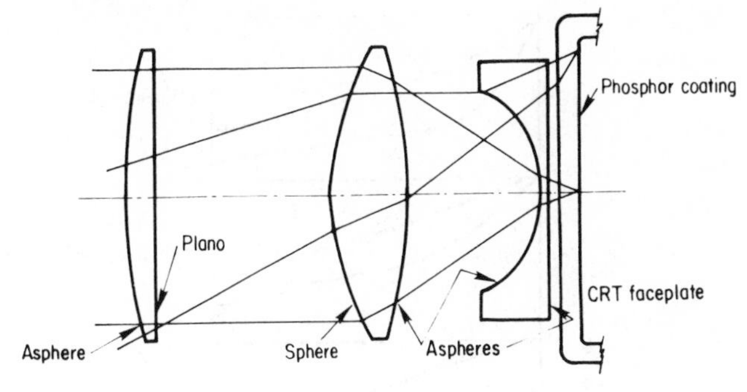

Figure 11

thicknesses. Screen sizes can range from 45 to 55 in., and relative illumination is 35% in the corners. Depth of focus and resolution are shown in Fig. 12 and Fig. 13 respectively.

Distortion is shown in Fig. 10. Other DELTA type lenses are available for 5-in. CRT's to facilitate different cabinet configurations and screen sizes. The DELTA design can be 'scaled' to accommodate larger or smaller CRT sizes, and the f-number varied for different applications.

Dichroic Combinations:

Red, green, and blue picture components can be combined prior to reaching the screen with dichroic filters. The advantage of this arrangement is that it forms a single exit pupil which eliminates screen color shift and, when combined behind the

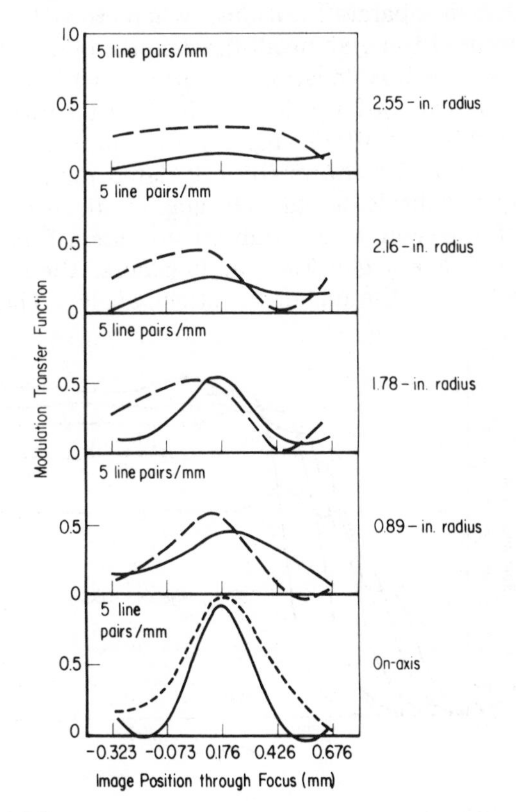

Figure 12

*Patent Pending

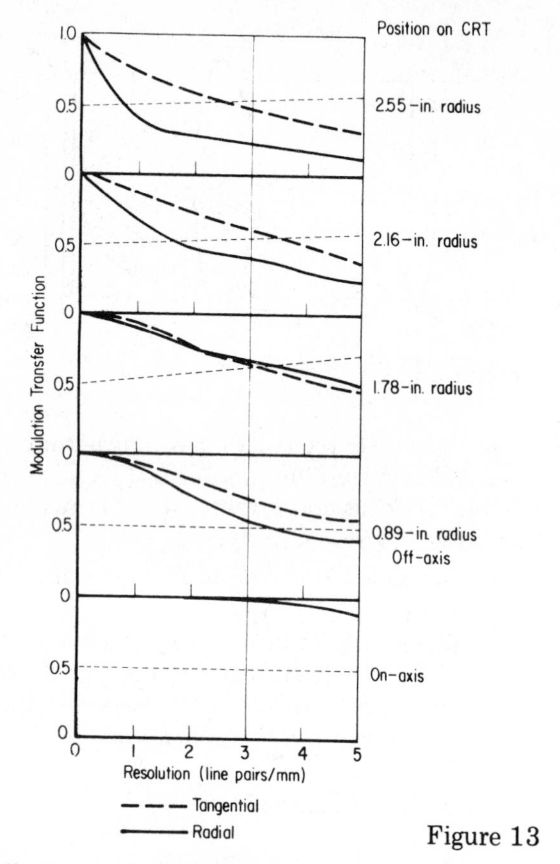

Figure 13

lens, allows easy refocusing for a variety of screen sizes. No convergence correction is required. See Fig. 5.

Dichroics can be located in front of, within, or behind the lens. One or two dichroics can be used in separate locations. When crossed dichroics are used to combine all three colors in one dichroic box, the box location can also be within, behind, or in front of the lens. The best location is dictated by the attack angles of the rim rays.

Fig. 14 shows the calculated data for a 15-layer dichroic design. The graph indicates that cutoff wavelength and transmission are a function of the attack angle, Fig. 15. In general, the faster the lens speed the higher the attack angles. Therefore,

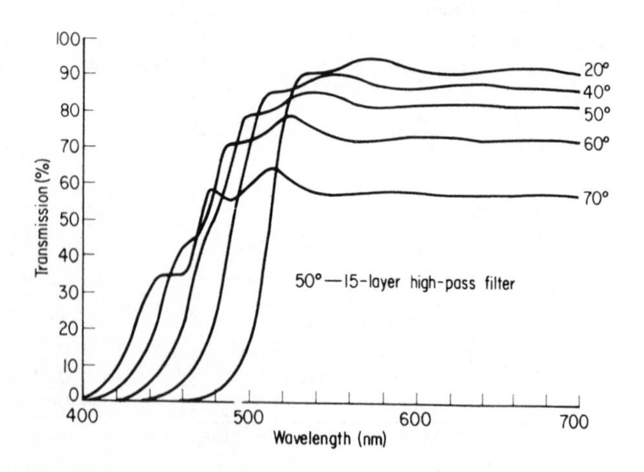

Figure 14

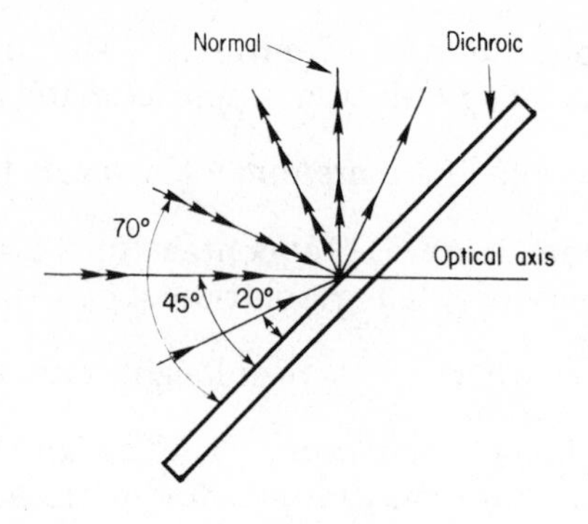

Figure 15

Fig. 14 suggests two items for consideration. First, a bundle of light rays passing through a dichroic filter suffers a light loss up to 30% when the wavefront is integrated over the entire field. The faster the lens, the greater the loss in intensity. Second, the dichroic produces a color shift which is different from that of a screen. If the dichroics are located within a bundle of converging rays instead of a bundle of parallel rays, the color shift is different.

The color spread relative to attack angle can be located between the primary color spikes if the phosphor spectral response is sharp (For example, Case 2 in Fig. 6). The resultant dichroic effect is then more light loss as a function of attack angle, which would appear visually as normal vignetting.

Since the dichroic is located within the lens optical path, dichroic thickness and quality must be evaluated using the same criteria used for mirror evaluation. Surface quality must be good and back reflections must be eliminated or minimized.

Mirrors

Compact projection-television packages often result from clever use of mirrors.

Front vs. Rear Surface: Fig. 16 shows how light reflects from a front-surface and a rear-surface mirror. On front-surface mirrors the only light loss is through absorption or scattering by the coating. Rear-surface mirrors are used occasionally because the mirror is exposed and requires frequent cleaning. However, on rear-surface mirrors each back reflection of up to 4% of the light results in additional light loss and, consequently, contrast reduction.

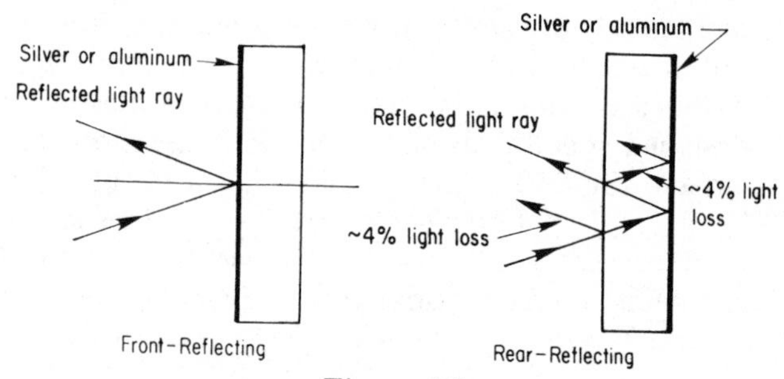

Figure 16

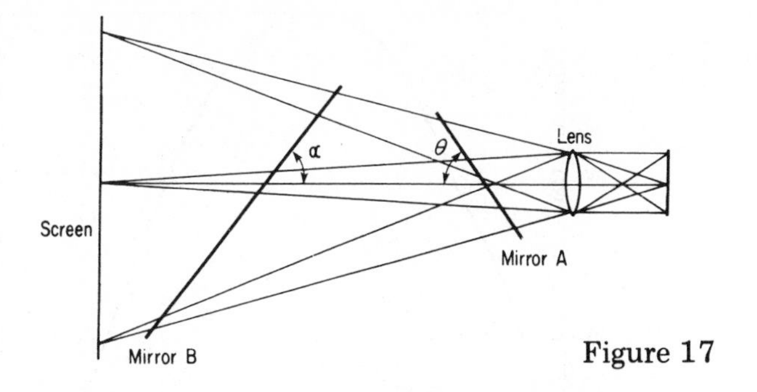

Figure 17

Size, Thickness, and Weight: Required mirror size is determined by drawing tunnel diagrams in two planes, one of which is shown in Fig. 17. Once mirror location and angle are fixed by package geometry, minimum size can be determined. To minimize weight, thickness should be minimized without compromising mechanical strength required for strain-free mounting and adjustment.

Surface Quality: Mirror surface quality is determined by required optical performance, location within the system, and the angles relative to the optical axis. Normal mirror specifications include reflectivity, power, irregularity, and "orange peel". These specifications apply only to unmounted mirrors, so good strain-free mounting techniques should be used to maintain specified quality.

Reflectivity: The number of mirrors in the system and reflectivity both affect system brightness (See Equation 1). Reflectivities from 80% to 95% are available. Consequently, reflectivity should be balanced against surface endurance and cost. Occasionally, a reflecting surface produces polarization, which could cause light loss when the surface is used with a second mirror under the wrong conditions.

Screens

The screen width, based on viewing distance that a viewer can observe comfortably, is large and is dependent on how dominant within a room the observer wishes to make his television. Fig. 18.

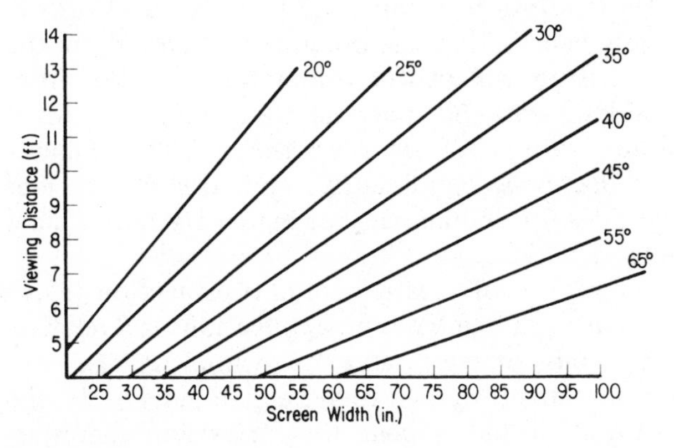

Figure 18

Front or Rear: Curved front view screens are used for one and two-piece systems, and thus far have been much easier to manufacture than rear screens. They are curved to reject ambient lighting and control the reflected image. Rear transmitting screens are flat. Transmitted images are controlled by coating, embossing and molding techniques.

Horizontal vs. Vertical Viewing Angle: Optimum horizontal and vertical viewing angle is easily established. See Fig. 19. However, screen gain requires special consideration. Equation 1 indicates that screen gain affects brightness significantly. The lower the screen gain, the wider the viewing angle. Also, lower screen gains minimize color shift.

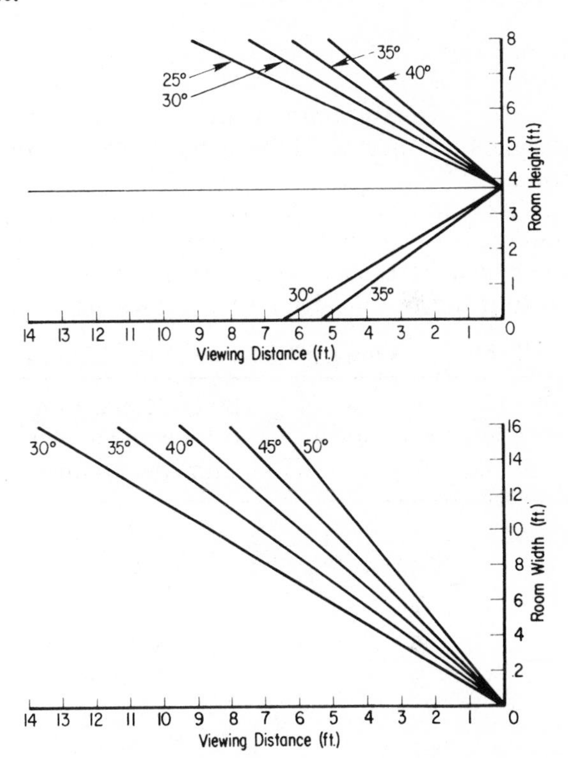

Figure 19

Color Shift: Color shift, producing a different dominant color when viewed on either side of the screen axis, is evident when three separate

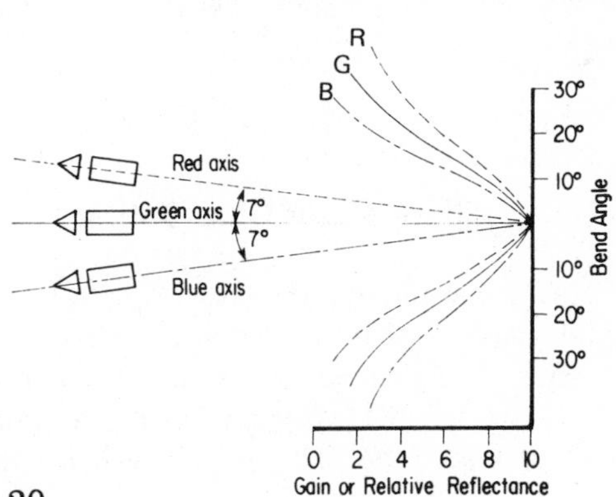

Figure 20

optical paths for red, green, and blue images are combined at the viewing surface. Color shift is proportional to the angular separation of the lenses, Fig. 20. The effect of color shift, due to lens separation, is less apparent with lower gain screens. The reflected light distribution shown in Fig. 20 is typical for a reflecting screen with a gain of approximately 10.

Contrast: Imagery falling on a front-reflecting screen with high surface scattering loses contrast. On a rear screen, surface reflections, material transmittance, and internal scattering cause loss of contrast. Other considerations include ambient-light rejection and maintainability (for example, cleaning, dusting, etc.).

Human Eye

To correctly present the information on a screen, the human sight mechanism must be understood.

Resolution: Resolution of the human eye, assuming 20-20 vision, is approximately 2 min of arc, which can be translated to line pairs/mm on the screen if the viewing distance is known, Table 2.

Table 2 — Viewing Distance vs Resolution

Viewing Distance (ft.)	20-20 Resolution at Viewing Screen (line pairs/mm)
5	2.25
6	1.88
7	1.60
8	1.41
9	1.25
10	1.13
11	1.02
12	0.93

Color Sensitivity: The color sensitivity of the eye is well known. Fig. 21. Consideration must be given to color balance and mixing. Each manufacturer must select the balance best suiting his needs.

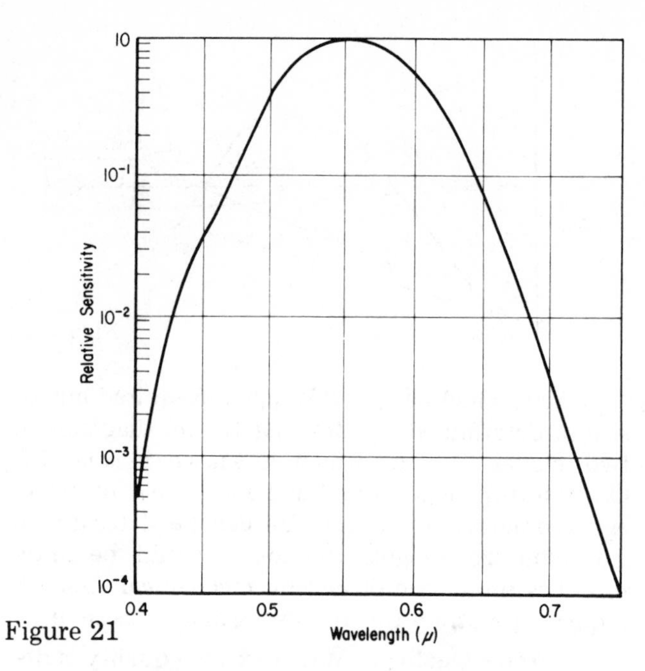

Figure 21

Plastic versus Glass Optics

The development of plastic optics for projection television has been encouraged for a variety of reasons, the primary ones being low cost and superior design performance using a minimum number of elements. This is achieved through the use of aspheric surfaces.

Glass lenses cannot be economically produced with aspheric surfaces, so special designs with low f/#'s, wide angular coverage and flat fields are difficult to configure without using numerous spherical elements. Although weight is not a critical factor, many feel it is significant. Plastic lenses with similar performance are much lighter than their glass counterparts not only because of the optical material but also due to lighter mounting systems.

Until recently, glass-projection television lenses, compared to plastic, demonstrated considerably superior on-axis contrast and resolution, but the edge performance was inferior because of field curvature. Now, through new design techniques and manufacturing improvements, plastic lenses have significantly narrowed the on-axis quality gap without compromising edge quality.

The future of plastic optics for projection television applications appears strong, but it is difficult to make definite predictions. Much will depend on new requirements imposed on the optics by developments in other systems components. Also, other technologies to produce projection television optical systems will play a role. Most observers believe that the predominant activity in the near future will center round f/1.0 plastic lenses combined with 5" diagonal CRT images. This is because with CRT sizes less that 5", the available light decreases significantly; and over 5" the lens cost goes up dramatically or higher f/# lenses must be used.

The use of plastic aspheric projection television lenses in high-volume applications is a relatively new phenomenon, and the technology of design and manufacture is progressing rapidly. In the future it would appear that television manufacturers will have new flexibility as progress is made.

Conclusion

This paper has touched on the many variables a projection television system designer must take into consideration, with emphasis on the inter-active role of refractive plastic optics. Because there are so many variables with complex inter-relationships, theoretical development of an optimum, cost-effective system is difficult, if not impossible, to precisely develop. However, by understanding the variables, the designer can come close and then, through trial and error, "fine tune" to get the desired system. The knowledgeable system designer makes his final decision when he sees a high quality image produced in what he believes to be an acceptable cabinet configuration at an acceptable cost.

A NEW PROJECTION SCREEN CONSISTING OF MINUTE LENSES

M. Tominaga*, T. Aoba** and L. Mori*
*Research and Development Center
**Consumer Products Materials Technology Appliance Center
Toshiba Corporation
Komukai, Saiwai-ku, Kawasaki-City, 210, Japan

1. Introduction:

The following two types of projection screens have been applied to the rear projection television system:
(1) Fresnel-diffuser type.
(2) Fresnel-diffuser-lenticular type.

Requirements for the screens [1),2)] are as follows:
(1) a high screen gain as well as a wide viewer area.
(2) a uniformity of luminance and color, independent of positions on the screen.
(3) no color shift, independent of viewing direction.

An example of the three-tube, in-line color projection system is shown in Fig. 1. For this system, the luminance and the chromaticity of a picture element are determined by the optical superposition of the light beams of three primary colors — red, green and blue — on the screen. As Fig. 2 shows, the angle of incidence of the projected beam onto the screen varies with the change of position on the screen. Also the incident angles of the three primary-color beams which are projected onto the same position are different, because of the in-line alignment of the three cathode-ray tubes.

Figure 3 shows the intensity distribution of the diffused light to viewers when a diffuser type of screen is applied. The direction of the peak intensity depends on the incident angle of the projected beam. Its luminance distribution against the viewing angle is shown in Fig. 4. It has a maximum value at the viewing angle of θs, which equals to the beam-incident angle. At the different angles from θs, the luminance drops. This figure shows that the luminance of some point on the screen varies with the change of the viewing position.

When the three primary-color beams are projected onto the diffuser type of screen, their light-intensity distributions are such as shown in Fig. 5. As their angles of incidence are different, the chromaticity of this point on the screen depends on the viewing angle and the viewing position.

Such characteristic causes the non-uniformity of luminance and color of the screen. This is also true of a lenticular type of screen which consists of many cylindrical-lens components. Such undesirable problems remain even if those screens are combined with a Fresnel lens.

2. New Screen Consisting of Minute Unit Lenses:

The newly developed screen is one of the lens-type screens as shown in Fig. 6. It consists of a lot of minute unit lenses. The axes of the unit lenses are parallel with each other. Each unit lens (Fig. 7) has spheroidal surfaces on both sides. The two surfaces

Manuscript Received 6/11/82

Reprinted from *IEEE Trans. Consum. Electron.*, vol. CE-28, pp. 284–289, Aug. 1982.

have a common axis. The focal plane of one surface is near the other surface. The edges of the unit lens adjoining its neighbours form a rectangle. Horizontal and vertical pitches determine the light intensity distributions in the horizontal and vertical direction, respectively.

Figure 8 shows the refracted rays toward viewers which are projected nearly parallel onto the first surface of the unit lens. The incident beam is focused near the second surface and diverges to the viewer area. The axis of the light-intensity distribution of the refracted beam is nearly perpendicular to the screen surface. So a Fresnel lens is not required.

Figure 9 shows the direction of the refracted rays of three primary colors which are projected onto the center of the first surface. They have different incident angles. However, such rays are refracted in parallel toward viewers. So, the chromaticity of the mixed rays toward viewers is the same as that of the mixture of the incident beams. In the picture element, the chromaticity of the diverging beam varies little even if the viewing angle against the screen is changed.

The geometrical parameters which determine the size of viewer area (maximum bend angle) and the luminance (screen gain) are (1) horizontal and vertical pitches p, (2) thickness t, and (3) radius of curvature r. They are shown in Fig. 10. As an example, maximum bend angle against the pitches in case of the lens thickness of 2 mm is shown in Fig. 11. The five curves show the maximum bend angles at the different radiuses of curvature — 0.6, 0.7, 0.8, 0.9, and 1.0 mm. The screen-gain characteristic against the maximum bend angle is shown in Fig. 12. Increasing the maximum bend angle, of course, the screen gain drops. These figures show that the desired combination of the viewer area and the screen gain can be rather freely selected by changing such parameters.

The new screen is made of polymethylmethacrylate material. Figure 13 shows the screen gain against the viewing angle when the light beam is projected perpendicularly to the screen. A gain of 6.8 has been obtained at the angle of 0°. The maximum bend angles are horizontally ±30° and vertically ±15°, respectively. The screen gains at the angles of ±30° in the horizontal direction are greater than 60 percent of that at the angle of 0°.

The uniformity of luminance and color is very important for the projection screen. Figure 14 shows the viewing position where the measuring instrument was placed for the evaluation of the uniformity of the new screen. The luminance and the chromaticity of each point on the screen were measured with a telescope type of luminance- and color-meter. The results are shown in Figs. 15 and 16.

Figure 15 shows the luminance distributions on the new screen and the conventional Fresnel-diffuser-lenticular screen. As this figure shows, in case of the conventional screen, luminance on the lower side of it is much lower than on the upper side. Comparing these two uniformities, the new screen has better performance.

Figure 16 shows the color-difference distributions on those screens. The arrows indicate the the color-shift vectors from the screen center. The origins of the arrows correspond to the measured points on the screen. They are plotted on the u'v' uniform chromaticity scale diagram of CIE 1976. The upward arrows show that the colors of the corresponding points shift toward yellow and the downward arrows show that the colors of the corresponding points shift toward blue. The lengths of the arrows correspond to the magnitudes of color shift. The four arrows shown on the lower side of this figure have the lengths which correspond to the color difference of 0.04 on the u'v' diagram. The difference of 0.04 correspond to about the difference between 6500K and 15000K in color temperature. In the conventional screen, the colors on the right side of it shift toward blue and the colors on the left side turn toward orange. The colors on both sides are very different. That implies that the color of some point on the screen varies with the change of viewing angle. The new screen has better color uniformity.

In addition to the good uniformity of luminance and color, the new screen gives a high picture contrast image due to the refracting optics instead of the diffusing one.

Three beams of primary colors projected onto the screen are focused on the second surface of the unit lens, as mentioned before. At the center of the screen, the three beam spots of primary colors are in line at the center of the second surface, and at the corner, they are at the corresponding corner. Such beam spots occupy a very small part of the total area of the second surface. The extra part of it is not used for light-beam refraction. If the second surface is black-coated excluding such beam-focused area, ambient light is effectively absorbed by it. As a result, a higher picture contrast can be achieved.

3. Conclusion:

The new screen gives a better performance than the conventional screen. It is more advantageous in that a desired combination of screen gain and viewer area can be rather freely selected by changing the geometrical parameters of the unit lenses. Also its uniformity of luminance and color is much improved. The new screen gives a high picture contrast image. If the screen surface is black-coated a higher contrast can be achieved.

References:

1) Raw, R. R. & Maloff, I. G. : " Projection Screens for Home Television Receivers ", J. O. S. A., Vol. 38, No. 6, pp. 497 (1948)

2) Howe, R. L. & Welham, B. H. : " Developments in Plastic Optics for Projection Television System ", IEEE Trans., CE-26, pp. 44 (1980)

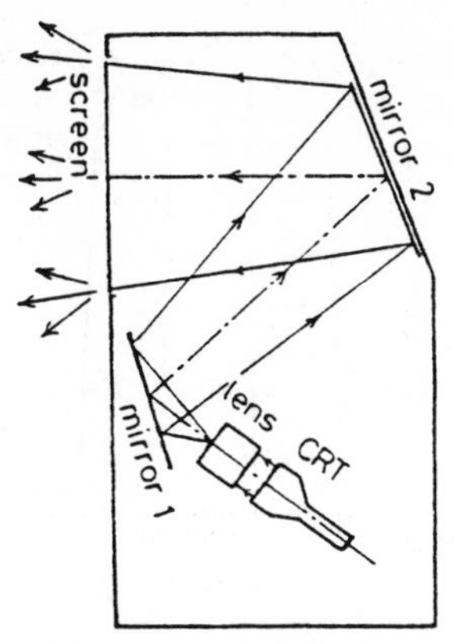

Fig. 1 Color Projection Television System.

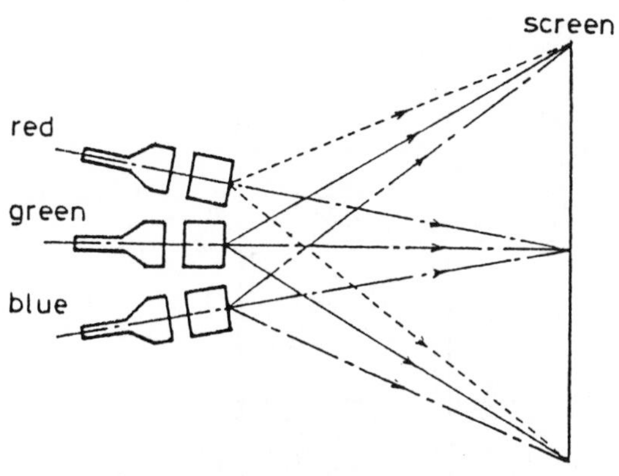

Fig. 2 CRT-Lens-Screen Alignment.

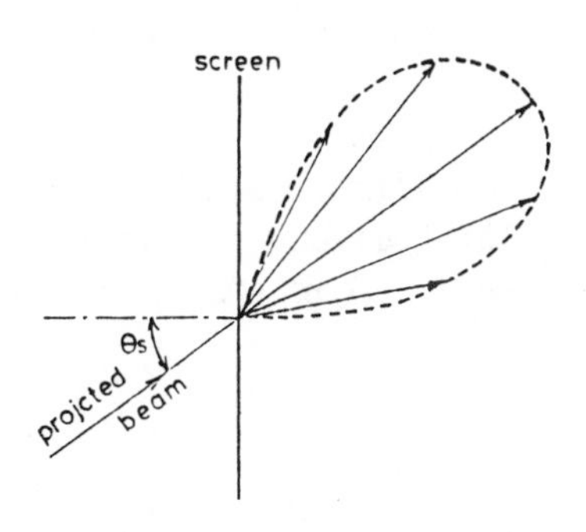

Fig. 3 Light Intensity Distribution of a Diffuser Type of Screen.

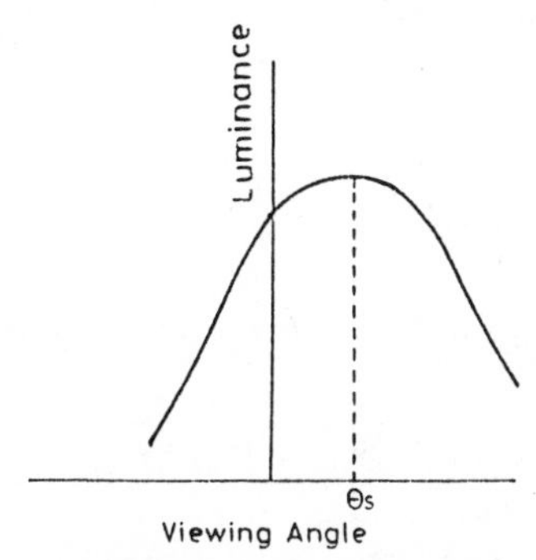

Fig. 4 Luminance Distribution against the Viewing Angle of a Diffuser Type of Screen.

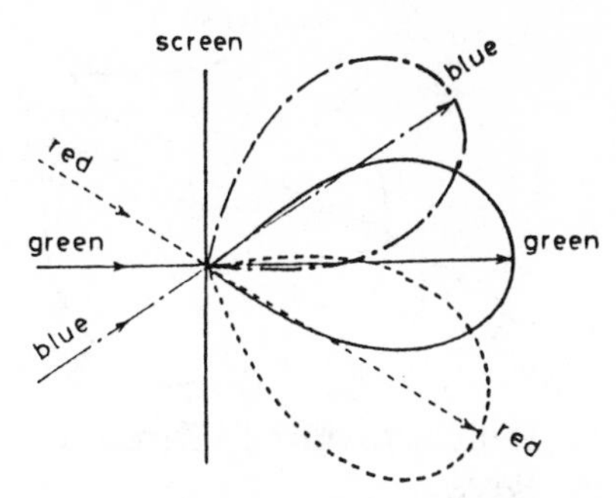

Fig. 5 Light-Intensity Distributions of Three Primary-Color Beams.

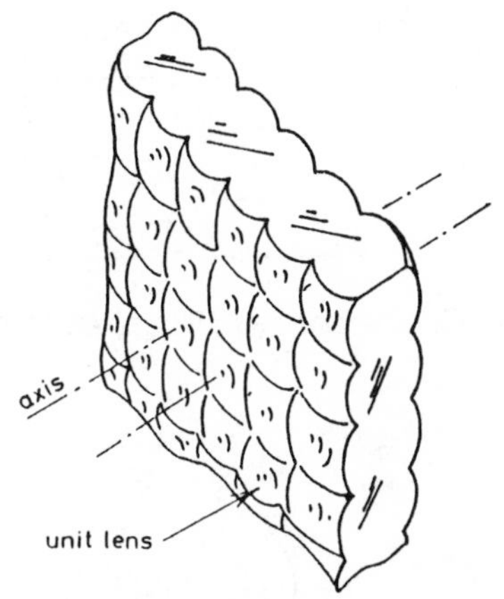

Fig. 6 View of the Newly Developed Screen.

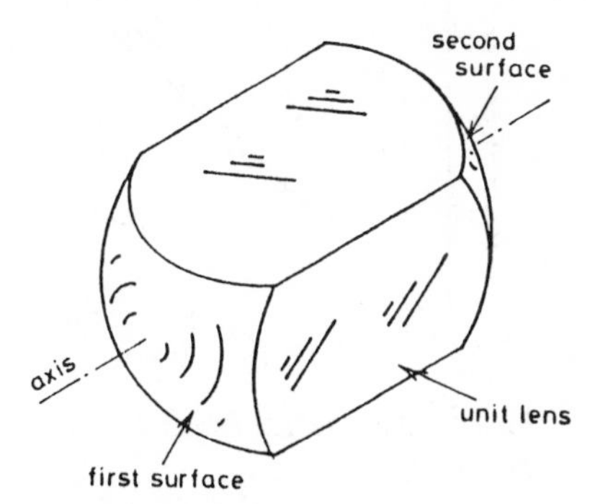

Fig. 7 Unit Lens.

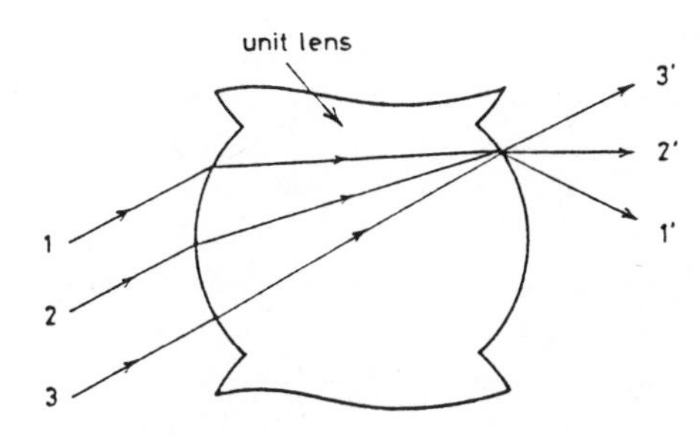

Fig. 8 Refraction of Nearly Parallel Incident Rays.

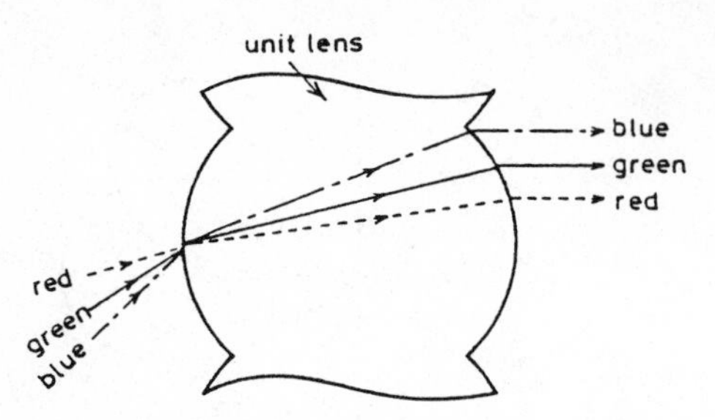

Fig. 9 Refraction of Three Primary-Color Rays.

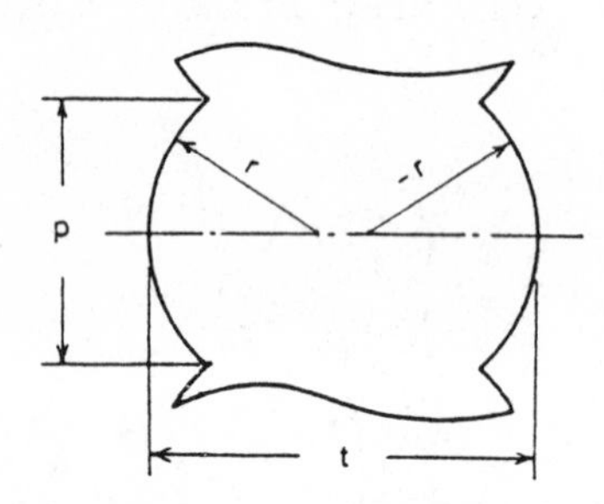

Fig. 10 Geometrical Parameters of Unit Lens.

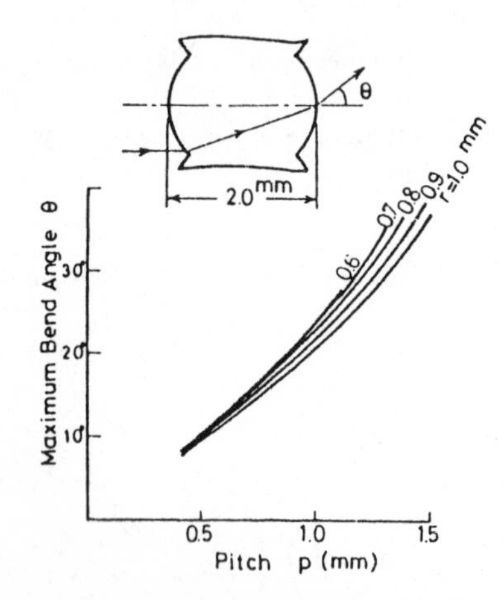

Fig. 11 Maximum Bend Angle against Pitches of Unit Lens.

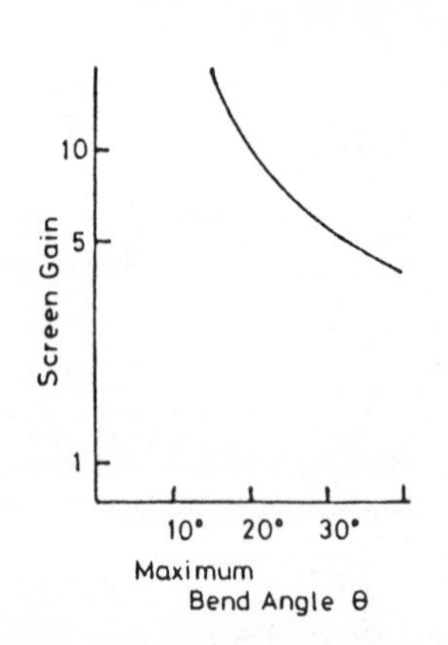

Fig. 12 Screen Gain against Maximum Bend Angle.

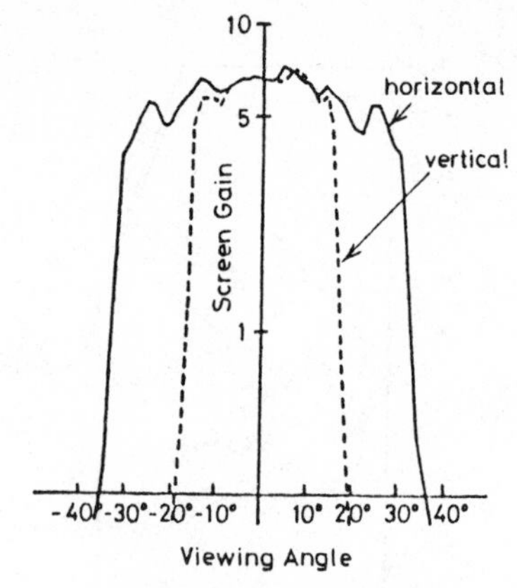

Fig. 13 Screen Gain against Viewing Angle in Horizontal and Vertical Direction of the New Screen.

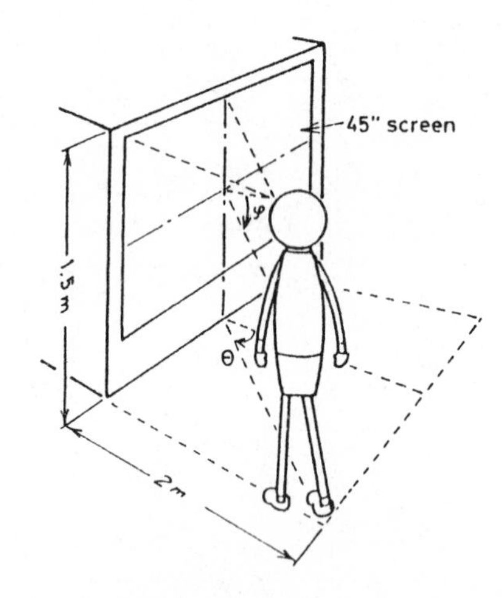

Fig. 14 Viewing Position where Uniformity of Luminance and Color was Evaluated.

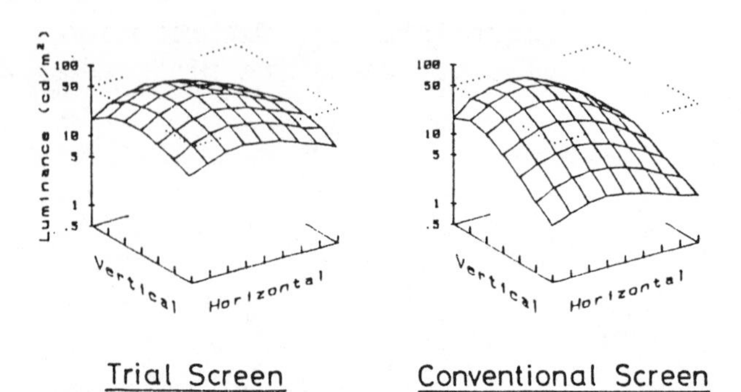

Fig. 15 Luminance Distributions on
(1) the New Screen and
(2) the Fresnel-Diffuser-Lenticular Screen.

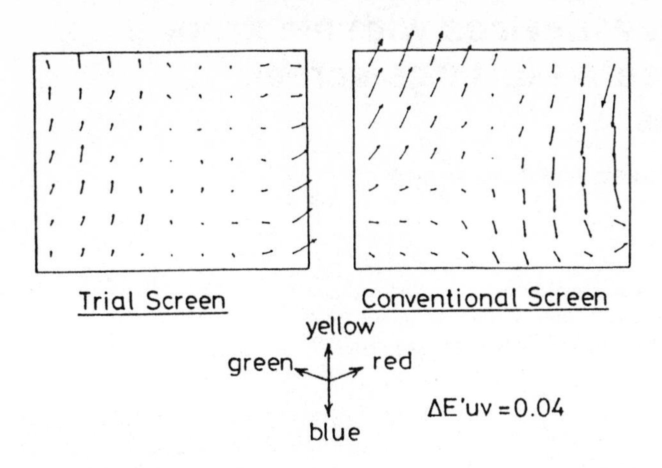

Fig. 16 Color-Difference Distributions
(1) the New Screen
and
(2) the Fresnel-Diffuser-Lenticular Screen.

Possible use of quantoscopes—new devices with electron-beam-scanned semiconductor lasers—in large-screen color projection television systems

V. I. Grigor'ev, V. N. Katsup, V. P. Kuklev, V. E. Sitnikov, and V. N. Ulasyuk

(Submitted June 30, 1979)
Kvantovaya Elektron. (Moscow) 7, 489–494 (March 1980)

A brief description is given of the use of quantoscopes, which are new devices based on electron-beam-scanned semiconductor lasers, in projection of images on a screen of 12 m^2 area. Consideration is given to the possibility of construction of a color television projector with three quantoscopes generating the primary colors.

PACS numbers: 42.55.Px, 85.60.Jb

Several systems, operating on the basis of different principles, are now available for projecting images on screens of up to 10 m^2 area.[1,2] The best parameters are the systems with light-valve relay tubes belonging to the Eidophor family. For example, the Gretag 5170 television projector can be used to produce color images on a directional screen up to 100 m^2 area with a resolution of 800 television lines at the screen center. The output of the projector can reach 3000 lm (Ref. 2). Soviet light-valve monochrome teleprojectors of the Ariston type produce a flux of 2000 lm and an image with a resolution of 500 television lines at the screen center.[3]

However, a wider use of these systems is hindered mainly by their exceptional complexity resulting from the need to pump continuously the vacuum chamber and to replace frequently burnt-out cathodes; other shortcomings are a long warmup time, limited period of continuous operation, need to remove the ozone evolved in the operation of a high-power xenon lamp, high power consumption, large dimensions, and considerable weight.

An analysis of future developments in projection television indicates that the problem of qualitative improvement in the parameters and operation characteristics of projectors for creating images on screens of about 10 m^2 area cannot be solved without new devices based on fundamentally new principles. In this connection there is a considerable interest in electron-beam-scanned semiconductor lasers. The first experiments on the use of such lasers in producing television images on a screen of 6 m^2 area were reported in Ref. 4. A constantly pumped unit, based on an electron microscope,[5] was employed. However, commercial television projectors must be based on sealed electron-beam tubes free of the need for constant pumping and they must have optimal parameters of the exciting electron beam.[6]

The first electron-beam devices of this kind—quantoscopes[7,8]—had a serious shortcoming of a small diameter of the laser target. Moreover, some inconvenience in the operation resulted from the need to apply a high voltage to the laser target, mainly because of the difficulty of wide-band transfer of a video signal to the modulating electrode of the system, which was under a high voltage when the laser target was grounded. This arrangement made it very difficult to bring a quantoscope in contact with cooling devices and gave rise to the problem of screening of the associated soft x rays.

The above disadvantages were overcome by developing a new type of quantoscope with a larger laser target operating at a high cathode voltage. The external appearance of the new quantoscope is shown in Fig. 1.

The laser target of the new device is a semiconductor single-crystal plate with a diameter of about 50 mm and about 20–30 μ thick; the plate is polished by a mechanochemical method and it carries the following reflection coatings: an output multilayer dielectric mirror with a reflection coefficient 0.8–0.9 and a nontransmitting metallic mirror bombarded by an electron beam. The semiconductor plate is bonded (with an optically transparent adhesive) to a leucosapphire exit window, which also acts as the heat sink of the laser target.

When the cathode voltage is from −50 to −75 kV, an electron projector with a principal magnetic lens produces an electron beam whose diameter is 25–35 μ on the laser target and which carries a current up to 2 mA; this beam is spread into a television raster by an electromagnetic deflection system.

Quantoscopes have been built with laser targets generating red and green radiation with the following wavelengths at liquid nitrogen temperature: ~500 nm (CdS); ~520 nm ($CdS_{0.9}Se_{0.1}$); ~610 nm ($CdS_{0.48}Se_{0.52}$); ~695 nm (CdSe). Typical emission spectra of the quantoscopes are shown in Fig. 2.

Some of the main parameters of the quantoscopes are as follows: length about 750 mm; diameter of the "neck" part 36–38 mm; resolution 1000–2000 lines; stopping voltage 100–130 V; video signal amplitude 70–90 V; radiation beam divergence 6–10°; angle of electron-beam deflection less than 15°. The average radiation power emitted by a liquid-nitrogen-cooled target operating according to the first and second television inter-

FIG. 1. External appearance of a quantoscope.

Reprinted with permission from *Sov. J. Quantum Electron.*, vol. 10, pp. 279-282, Mar. 1980.
Copyright © 1980 American Institute of Physics.

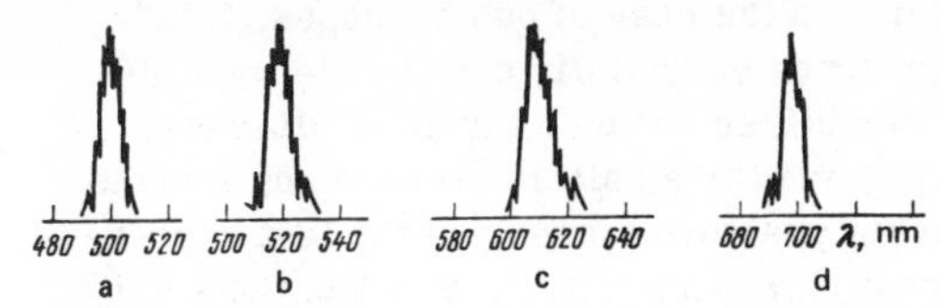

FIG. 2. Emission spectra of quantoscopes with CdS (a), $CdS_{0.9}Se_{0.1}$ (b), $CdS_{0.48}Se_{0.52}$ (c), and CdSe (d) single-crystal targets.

lacing standards is 1.5–4 W and 2–5 W, respectively. The width of the spectrum is then 6–7 and 4–5 nm, respectively. The quantoscope power is higher in the second case because the switching time of an interlacing element is then close to the optimal value (20–30 nsec) and the fall in the radiation power due to the heating of the active region directly under the electron beam is slight.[9]

The dependence of the quantoscope radiation power on the beam current is linear (Fig. 3) and, therefore, the modulation characteristic of quantoscopes is similar to the corresponding characteristics of conventional kinescopes, so that there is no need for additional gamma correction of the video signal in projecting television images with a quantoscope.

The quantoscopes described above are used in a small prototype monochrome television projection system, whose external appearance is shown in Fig. 4 (without the cooling unit). The main components of the projector are shown schematically in Fig. 5.

The dimensions of the high-voltage power supply unit together with a quantoscope are 1400×700×500 mm and those of the video control unit are 700×500×400 mm. The power consumption is about 1 kW.

The quantoscope target is grounded, which simplfies greatly the cooling system and makes it more effective. A high voltage is applied to the quantoscope cathode. A video signal at the carrier frequency is transferred to the "high" side by two coupled circuits which are insulated from one another sufficiently to withstand the total voltage. This signal is next detected, amplified, and applied to the control electrode of the quantoscope. The high cathode voltage is supplied by a voltage-multiplication source which includes an oscillator operating

FIG. 4. External appearance of a quantoscope-based television projection system.

at about 20 kHz. The voltage is stabilized by a feedback circuit with a dc amplifier, whose input receives a signal from a high-voltage divider connected in parallel with the output of the high-voltage source.

This system can be used to project broadcast transmissions or it can be employed in a closed-circuit application together with a camera, which is included in the projector. The arriving information is imaged on an EPAR-3000 screen of 3×4 m dimensions. The image brightness is 30–50 cd/m² and the number of transmitted brightness gradations is 6–8. Part of an image is shown in Fig. 6.

This prototype projector can serve as the basis for industrial development of color quantoscope projectors.

A color image can be obtained by combining, on a common large screen, images from three quantoscopes generating the primary colors. There should be no difficulty in combining the images generated by the quantoscopes because the small size of the laser targets as well as the relatively small electron-beam deflection angle and radiation divergence relax significantly the requirements in respect of the projection optics and electronic devices for raster correction.

A major advantage of quantoscopes is the narrowness of the emission spectrum compared with phosphors used in colored kinescopes. This means that the high purity of the color produced by quantoscopes will make it possible to reproduce a wider range of colors than that handled by television systems in current use.

However, the possibility of a considerable expansion

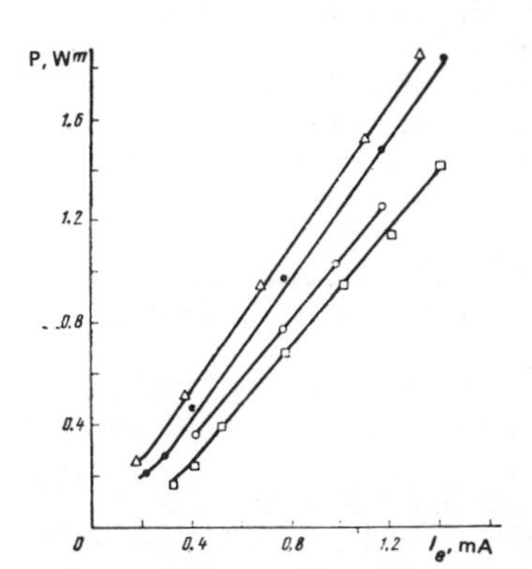

FIG. 3. Dependences of the output radiation power of quantoscopes on the electron beam current incident on laser targets made of CdS (□), $CdS_{0.9}Se_{0.1}$ (○), $CdS_{0.48}Se_{0.52}$ (●), and CdSE (△).

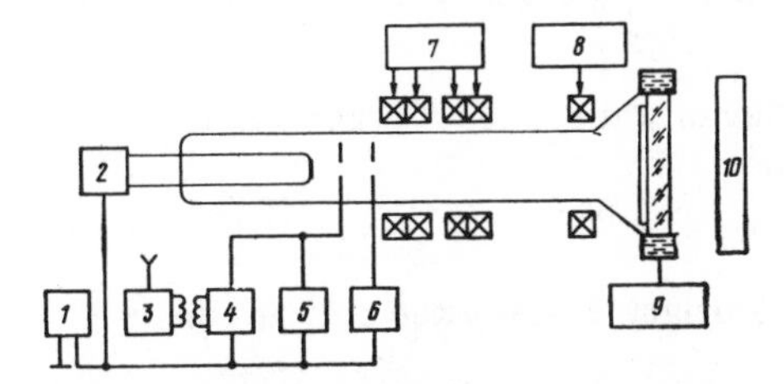

FIG. 5. Schematic diagram of a quantoscope-based television projection systems: 1) high-voltage power supply unit; 2) heating supply unit; 3) television receiver with a unit for transmitting a video signal to the modulation electrode of a quantoscope; 4) video amplifier; 5) bias voltage source; 6) acceleration voltage source; 7) focusing and aligning coils, and a stigmator with a supply unit; 8) deflection coil with a scanning unit; 9) cooling system; 10) projection objective.

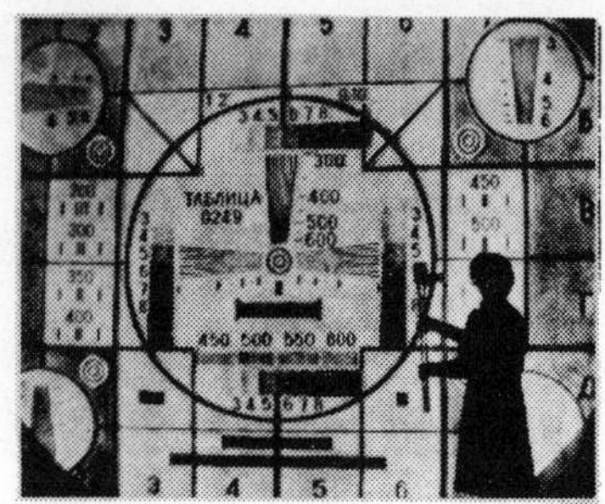

FIG. 6. Image of a test chart displayed on a large screen by a television projection system.

of the primary color triangle with the aid of quantoscopes, compared with the currently available color kinescopes, is limited by the circumstance common to all information imaging systems and related to the need to increase the radiation power at longer (red) and shorter (blue) wavelengths.[10] For example, the output power of a quantoscope has to be increased by more than one order of magnitude as we go over from the red color of kinescopes to the emission of the 695 nm wavelength (typical of CdSe single crystals), if the same image brightness is to be retained. Moreover, when the color triangle differs from that of color kinescopes, one has to introduce an additional matrix circuit with cross links to correct for the noncorrespondence of the quantoscope colors and the main television broadcasting colors.

Naturally, the problem will be simpler when we have quantoscopes generating radiation with the same chromaticity as the primary colors of the conventional kinescopes.

The necessary color triad can be obtained using the following semiconductor single crystals $CdS_{1-x}Se_x$ ($x \approx 0.52$) generating red light of the wavelength $\lambda_R = 612$ nm (one can use also GaSe with $\lambda \approx 605$ nm); $CdS_{1-x}Se_x$ ($x \approx 0.16$) generating green color with $\lambda_G = 535$ nm (one cal also use ZnTe with $\lambda \approx 530$ nm); $Cd_{1-x}Zn_xS$ ($x \approx 0.27$) generating blue color with $\lambda_B = 470$ nm. (The values of x are calculated for liquid nitrogen temperature on the basis of the data given in Ref. 11.)

We can now determine the proportions of the colors emitted by quantoscopes which have to be mixed to obtain white light: $P_R:P_G:P_B = 0.98:1:1.235$, where P_R, P_G, and P_B are the powers of the red ($\lambda_R = 612$ nm), green ($\lambda_G = 535$ nm), and blue ($\lambda_B = 470$ nm) quantoscope radiations.

The total light flux of white color F generated by quantoscopes is then

$$F\,[\mathrm{lm}] = 1020\,P_G\,[\mathrm{W}],$$

and the ratio of the brightnesses of the color fields is

$$L_R : L_G : L_B = 0.46 : 0.915 : 0.112.$$

The maximum image brightness on an external large screen of area S with a coefficient of axial brightness r_0 is then

$$L_W = F r_0 \eta / \pi S,$$

where η is the transmission coefficient of the optical projection system; in the case of quantoscopes, this coefficient is governed essentially only by the light absorption in the objectives because the small diameter of the laser target and the small divergence angle make it possible to avoid aperture losses. In the last expression it is assumed, for simplicity, that η has the same value for all the radiations.

For example, in the case of a standard ÉPAR-3000 screen with $r_0 = 1.6$, $S = 3\times4$ m, and $\eta \approx 0.8$, we obtain $L_W = 70$ cd/m² for $P_G = 2W$. For comparison, we shall mention that the normal brightness of high-quality images in motion-picture theatres should be 35^{+15}_{-10} cd/m².

The greatest problem in the construction of color television projectors based on quantoscopes is still that of growing large perfect single crystals of the compounds $CD_{1-x}Zn_xS$ and $Cd_{1-x}Zn_xSe$ suitable for generating blue radiation with the 470 nm wavelength.

It would be more realistic to expect construction of television projectors in which quantoscopes have ZnSe single crystals emitting $\lambda_{B'} = 450$ nm at liquid nitrogen temperature.

The energy characteristics are then determined by the following expressions:

$$P_R : P_G : P_{B'} = 0{,}78 : 1 : 0.79;$$
$$F\,[\mathrm{lm}] = 900\,P_G\,[\mathrm{W}].$$

Quantoscopes based on ZnSe single crystals 40 mm in diameter (grown from the melt) are now available but television images obtained with their aid are highly nonuniform and the brightness is low because of the high stimulated emission threshold and also because of the presence of twins and blocks in the original single crystals. These results are only preliminary and there are reasons for expecting a considerable improvement in the quality of these single crystals.

Availability of ZnSe single crystals 50 mm in diameter with a perfect structure, generating efficiently coherent blue radiation, will make it possible to use quantoscopes in constructing a high-quality color television projector capable of creating images of 50–100 cd/m² brightness and at least 1200 television lines resolution on a screen of about 10 m² in area.

[1]A. Robertson, Wireless World 82, No. 1489, 47 (1976).
[2]A. Robertson, Wireless World 82, No. 1490, 67 (1976).
[3]I. B. Ivanenko, Tekh. Kino Telev. No. 6, 38 (1975).
[4]N. G. Basov, O. V. Bogdankevich, A. S. Nasibov, V. I. Kozlovskiĭ, V. P. Papusha, and A. N. Pechenov, Kvantovaya Elektron. (Moscow) 1, 2521 (1974) [Sov. J. Quantum Electron. 4, 1408 (1975)].
[5]A. S. Nasibov, V. P. Papusha, and V. I. Kozlovskiĭ, Kvantovaya Elektron. (Moscow) 1, 534 (1974) [Sov. J. Quantum Electron. 4, 296 (1974)].
[6]G. S. Kotovshchikov, G. A. Meerovich, and V. N. Ulasyuk, in: Display Techniques (In Russian), Naukova Dumka, Kiev (1976), p. 190.
[7]G. S. Kotovshchikov, V. P. Kuklev, N. P. Lantsov, G. A. Meerovich, A. G. Negodov, and V. N. Ulasyuk, Kvantovaya Elektron. (Moscow) 1, 428 (1974) [Sov. J. Quantum Electron. 4, 242 (1974)].
[8]T. A. Il'yashenko, V. N. Katsap, E. U. Kornitskiĭ, G. S.

Kotovshchikov, V. P. Kuklev, and V. N. Ulasyuk, Elektron. Promst. No. 1, 44 (1977).
[9]I. N. Andreev, O. V. Bogdankevich, M. V. Gushchin, G. A. Meerovich, and V. N. Ulasyuk, Kvantovaya Elektron. (Moscow) **6**, 789 (1979) [Sov. J. Quantum Electron. **9**, 468 (1979)].
[10]B. M. Lavrushin and E. S. Shemchuk, Kvantovaya Elektron. (Moscow) **3**, 2605 (1976) [Sov. J. Quantum Electron. **6**, 1434 (1976)].
[11]R. S. Feigelson, A. N'Diaye, Shaiw-Yih Yin, and R. H. Bube, J. Appl. Phys. **48**, 3162 (1977).

Translated by A. Tybulewicz

Part VIII
Videotape

THIS collection contains several technical papers relating to the latest developments in consumer video cassette recorders (VCR's).

Since Ampex Corporation was successful in introducing its first practical broadcast-use VTR employing a rotary four-head system in 1956, video technology has shown tremendous progress in various fields, ranging from broadcasting and industrial to home uses. With the transition from open reel videotape to video cassette systems which involve the developments in automatic tape threading made in the early 1970's, the popularity of consumer VCR's has dramatically soared over the years.

During this period, numerous technological development and technical innovations have been made in video recording. Needless to say, video recording and playback systems require comprehensive technology, ranging from the basic formats to signal processing, servo, and system control circuits, together with efficient integration techniques to pack all this circuitry into a compact unit. Video technology also covers magnetic tapes, magnetic heads, and precision mechanical engineering and other arts. The variety and scope of these technological fields are wide and extensive.

Especially for the consumer acceptance of VCR's, new technical innovations were required to increase recording density, improve tolerance against color signal jitters, make trick play, perform special functions (slow motion, picture search, etc.), and to make programmed (timer) recording possible. Higher circuit integration has been pursued to help miniaturization and cost reduction in order to produce smaller and lower priced VCR's. In addition, interface technology is becoming increasingly important for VCR's to be used with TV receivers and tuners. As a matter of fact, most of these requirements have been already fulfilled, and so many technical papers have been published covering this extensive and rapid progress.

There are quite a few interesting and remarkable papers, but it is impossible to introduce all of them within the given space. The aim of this collection is to provide some basic principles on the two or three most popular systems among the current home-use VCR's and to present a number of the latest papers focusing on them. As peripheral material, recent papers on the most advanced software duplication techniques are included, because the ever-increasing popularity of VCR's is expected to generate greater demand for pre-recorded videotapes, resulting in keener attention and interest given to new, economical, high-speed, and mass duplicating techniques for the future.

The editors hope this collection of reprint papers may help video and circuit engineers and graduate students of all backgrounds who are interested in this particular field of electronics engineering.

ACKNOWLEDGMENT

The editors are greatly indebted to Mr. Shigeo Shima, Fellow of the IEEE and Corporate Advisor to the Board of Sony Corporation, who has provided valuable advice and practical suggestions.

AKINAO HORIUCHI
Associate Editor

BIBLIOGRAPHY

Classical (Early Stage Literature and Disclosures of New Systems)

[1] J. T. Mullin, "VTR, a video magnetic tape recorder," in *IRE Conv. Rec.*, pt. 7, 1954.

[2] H. F. Olson, W. D. Houghton, A. R. Morgan, J. Zenel, M. Arzt, J. G. Woodward, and J. T. Fische, "A system for recording and reproducing television signals," *RCA Rev.*, Mar. 1954.

[3] H. F. Olson, W. D. Houghton, A. R. Morgan, M. Arzt, J. A. Zenel, and J. J. Woodward, "A magnetic tape system for recording and reproducing standard FCC color television signals," *RCA Rev.*, pp. 330–392, Sept. 1956.

[4] "Home television magnetic tape player," *RCA Labs. Monthly News Magazine*, p. 4, Oct. 1956.

[5] C. P. Ginsburg, "Comprehensive descriptions of the Ampex video tape recorder," *SMPTE J.*, Apr. 1957.

[6] C. E. Anderson, "The modulation system of the Ampex video tape recorder," *SMPTE J.*, Apr. 1957.

[7] E. Cambi, "Trigonometric components of a frequency-modulated wave," *Proc. IRE*, Jan. 1948.

[8] R. M. Dolby, "Rotary-head switching in the Ampex video tape recorder," *SMPTE J.*, Apr. 1957.

[9] P. E. Axon, "The BBC vision electronic recording apparatus, VERA," *E.P.U. Rev.*, pt. A, May 1958; *J. Telev. Soc.*, Nov. 1958; E. Pawley, "BBC engineering 1922–1972," *BBC*, pp. 492–495, 1972.

[10] E. E. Masterson, "Magnetic recording of high frequency signals," U.S. Patent 2 773 120, Dec. 4, 1956.

[11] E. Schueller, "Magnetischen aufzeichnung und wiedergabe von fernsehbildern," German Patent 927 999, Sept. 23, 1954.

[12] N. Sawazaki, G. Inada, and T. Tamaoki, "A new video-tape recording system," *SMPTE J.*, vol. 69, pp. 868–871, Dec. 1960.

[13] N. Kihara and M. Morizono, "A compact slant track video tape recorder for television signals," presented at Int. Conf. Magnetic Recording, London, July 6, 1964.

[14] K. Suzuki, E. Kimura, and K. Yokoyama, "Magnetic recording apparatus," Japanese Patent 480 366, Apr. 22, 1979.

[15] N. Kihara, "Magnetic recording and reproducing system," U.S. Patent 3 188 385, June 8, 1965; Japanese Patent 450 256.

[16] Y. Suzuki and K. Uchida, "Design concepts in the development of the SONY BVH-1000 one-inch VTR," presented at 10th Int. Telev. Symp., Montreux, Switzerland, June 1977.

[17] N. Kihara, "Colour cassette system for the NTSC and Japanese colour-television standards," *E. B. U. Rev.*, pt. A, no. 125, Feb. 1971.

[18] Y. Shiraishi and A. Hirota, "Magnetic recording at video cassette recorder for home use," *IEEE Trans. Magn.*, vol. MAG.-14, Sept. 1978.

[19] T. Kono, T. Kamai, and S. Kakuyama, "Video recording and playback systems (Beta hi-fi VCR)," *IEEE ICCE Tech. Dig.*, June 1983; also in *IEEE Trans. Consum. Electron.*, Aug. 1983.

[20] K. Mohri, Y. Yumde, M. Umemura, Y. Noro, and S. Watatani, "A new concept of handy video recording camera," *IEEE Trans. Consum. Electron.*, vol. CE-27, Aug. 1981.

[21] N. Takano and I. Segawa, "Betacam—A VTR in camera," presented at 123rd SMPTE Tech. Conf., Oct. 1981; also IBC Tech. Conf., Brighton, England, Sept. 1982.

Review Articles

[22] H. A. Chinn, "Status of video tape in broadcasting," *SMPTE J.*, vol. 66, Aug. 1957.

[23] T. J. Merson and J. Roisen, "Helical scan recording," in *Conf. Rec., 2nd Int. Telev. Symp.*, Montreux, Switzerland, Apr.-May 1962, pp. 198-201.

[24] D. Kirk, Ed., *25 Years of Video Tape Recording*. 3M United Kingdom Ltd., May 1981.

[25] A. E. Alden, "The development of national standardization of the one-inch helical video tape recording systems," *SMPTE J.*, Dec. 1977.

[26] F. Remley, "The new 1-inch VTR formats—A symposium," *Educ. Industr. Telev.*, Jan. 1978.

[27] T. Mehrens, "Innovations in ENG recording equipment," *SMPTE J.*, July 1983.

[28] K. Sadashige, "Overview of time-base correction techniques and their applications," *SMPTE J.*, Oct. 1976.

[29] F. Granum and A. Nishimura, "Modern developments in magnetic tape," in *Proc. Conf. Video Data Rec., IERE Conf. Proc.*, July 1979.

[30] E. Adams, "Recent developments in soft magnetic materials," *J. Appl. Phy.*, Mar. 1962.

[31] H. Sugaya, "Recent advances in video tape recording," *Trans. IEEE Magn.*, vol. MAG-14, p. 636, Sept. 1978.

Introductions of Important New Technologies

[32] S. Okamura, "Recording system with provision for fast or slow reproduction," U.S. Patent 3 170 031, Dec. 27, 1963.

[33] N. Kihara, "Magnetic recording of signals containing synchronizing information," U.S. Patent 3 215 772, Nov. 2, 1965.

[34] S. Okamura, "Magnetic recording processing equipment," Japanese Util. Patent 39-23924, Aug. 18, 1964.

[35] H. Sugaya, F. Kobayashi, and M. Ono, "Rotary-head type magnetic recording/reproducing apparatus," Japanese Util. Patent 54-9346, Mar. 24, 1979.

[36] R. M. Dolby, "Video system with transient and dropout compensation," U.S. Patent 2 996 576, Feb. 20, 1959.

[37] S. Amari, "Magnetic recording and/or reproducing apparatus with chrominance crosstalk elimination," U.S. Patents 4 007 482 and 4 007 484.

[38] Y. Ishigaki, "Magnetic recording and/or reproducing system," U.S. Patent Re.29975, Apr. 24, 1979.

[39] T. Numakura, "Methods for recording video signal," Japanese Patents 45-28613 and 50-61287.

[40] M. Fujita, "Methods for recording video signal," Japanese Patent 59-9928.

[41] "Digital time base corrector for heterodyne VTR's," *Int. Broadcast Eng.*, pp. 24-27, Oct./Nov. 1977.

[42] M. W. Tallent, W. B. Hendershot, A. L. Swain, and R. M. Harrison, "Television signal time base corrector," U.S. Patent 3 900 885, May 23, 1974.

New Developments

[43] Marvin Camras, "Magnetic transducer head," U.S. Patent 3 079 470, Feb. 26, 1963.

[44] M. Mizushima, "Mn-Zn single crystal ferrite as a video-head material," *IEEE Trans. Magn.*, vol. MAG-7, Sept. 1971.

[45] H. Sugaya, "Newly developed hot-pressed ferrite head," *IEEE Trans. Magn.*, vol. MAG-4, pp. 295-301, Sept. 1968.

[46] K. Yamakawa, "System for recording and/or reproducing color television signals," U.S. Patent 3 845 237, Oct. 29, 1974; Japanese Patent 870524, July 13, 1977.

[47] NTSC/PAL/SECAM and NTSC/PAL compatible home VCR examples, Multi-standard models (1979-1982), VHS:NV-8600; NV-7500; NV-390. Beta:SLT-7; SLT-9.

[48] Y. Watanabe, "Helical scan VTR with deflectable head," U.S. Patent 4 203 140, May 13, 1980; Japanese Patent 49-136554.

[49] Y. Machida, M. Morio, and N. Kihara, "Simplified standards converter as a video tape reproducing system," presented at IEEE Chicago Fall Conf. Consum. Electron., Nov. 1978.

[50] R. A. Hathaway and R. Ravizza, "The development and design of the Ampex auto scan tracking (AST) system," *SMPTE J.*, vol. 89, Dec. 1980.

[51] H. Tanimura, Y. Fujiwara, and T. Mehrens, "A second generation 'type C' one-inch VTR," *SMPTE J.*, Feb. 1983; also in *IBC Conf. Rec.*, 1983, pp. 216-220.

Peripheral Technologies

[52] M. Camras and R. Herr, "Duplicating magnetic tape by contact printings," *Electronics*, Dec. 1949.

[53] J. Greiner, E. Eichler, and K. Krones, "Curie point magnetic recording process," U.S. Patent 3 364 496, Jan. 16, 1968; German Patent 432 796, Feb. 15, 1965.

[54] J. R. Morrison and D. E. Speriotis, "The magnetic transfer process," *IEEE Trans. Magn.*, vol. MAG-4, Sept. 1968.

[55] J. Hokkyo and N. Ito, "Theoretical analysis of contact printing of magnetic recording," presented at 3rd Conf. Magn. Rec., Budapest, Hungary, Sept. 1970.

[56] C. W. Crum and H. W. Town, "Recent progress in video tape duplication," *SMPTE J.*, Mar. 1971.

[57] J. E. Dickens and L. K. Jordan, "Thermoremanent duplication of magnetic tape recordings," *SMPTE J.*, Mar. 1971.

[58] W. B. Hendershot, III, "Thermal contact duplication of video tape," *SMPTE J.*, Mar. 1971.

[59] J. A. O'Neil, "Record for reproducing sound tones and action," U.S. Patent 1 653 467, Dec. 20, 1927.

[60] Fritz Pfleumer, "Lautschrifttraeger," German Patent 500.900, Jan. 31, 1929.

[61] M. Camras, "Magnetic impulse record member, magnetic material, and method of making magnetic material," U.S. Patent 2 694 654, July 25, 1947.

[62] R. A. V. Behren, "Magnetic recording tape and method of making same," U.S. Patent 2 711 901, June 28, 1955.

[63] F. J. Darrel, "Magnetization process in small particles of CrO_2," *J. Appl. Phys.*, July 1961.

[64] T. J. Swoboda, P. Arthur, Jr., N. L. Cox, J. N. Ingraham, A. L. Oppegard, and M. S. Sadler, "Synthesis and properties of ferromagnetic chromium oxide," *J. Appl. Phys.*, vol. 32, Mar. 1961.

[65] S. Iwasaki and K. Nagai, "Some considerations on the design of high output magnetic tape for short wave-length recording," Scien. Rept. Inst. Tohoku Univ., Japan, vol. B-15, p. 86, 1963.

[66] M. Kawasaki and S. Higuchi, "Alloy powders for magnetic recording tape," *IEEE Trans. Magn.*, Sept. 1972.

[67] Y. Imaoka, S. Umeki, Y. Kubota, and Y. Tokuoka, "Characteristics of cobalt adsorbed iron oxide tapes," *IEEE Trans. Magn.*, vol. MAG-14, Sept. 1978.

Important Relevant Engineering Standards

[68] "Proposed American standard, dimensions for video, audio and control records on 2-in. video magnetic tape," SMPTE RH 22.120, Feb. 1960.

[69] "Transverse track video recorders," IEC Standard Pub. 347, 1982.

[70] "Type B helical video recorders," IEC Standard Pub. 602, 1980.

[71] "Basic system and transport geometry parameters for 1-in type C helical-scan video tape recording," ANSI C98.18M-1979, Aug. 1979.

[72] "Dimensions and location of records for 1-in type C helical-scan video tape recording," ANSI C98.19M-1979, Aug. 1979.

[73] "Frequency response and reference level of recorders and reproducers for audio recorders for 1-in type C helical-scan video tape recording," ANSI C98.20M-1979, Aug. 1979.

[74] "Tracking control records for 1-in type C helical-scan video tape recording," *SMPTE Recommended Practice*, RP 85-1979, Feb. 1979.

[75] "Video record parameters for 1-in type C helical-scan video tape recording," *SMPTE Recommended Practice*, RP-86-1979, Feb. 1979.

[76] "Type C helical video recorders," IEC Standard, IEC Pub. 558, 1982.

[77] "Helical-scan video tape cassette system using 0.5 in (12.70 mm) magnetic tape (50 Hz-625 lines)," IEC Standard Pub. 511, 1977.

[78] "Helical-scan video tape cassette system using 0.5 in (12.70 mm) magnetic tape (60 HZ-525 lines)," IEC Standard Pub. 511A, 1977.

[79] "Helical-scan video-tape cassette system using 19 mm (3/4 in.) magnetic tape, known as U-format," IEC Standard Pub. 712, 1982.

[80] "Helical-scan video-tape cassette system using 12.70 mm (0.5 in) tape on'Beta format'," IEC Standard Pub. 767, 1983.

[81] "Helical-scan video-tape cassette system using 12.70 mm (0.5 in) tape on 'VHS format'," IEC Standard Pub. 774, 1983.

Emerging New Technologies for the Future

Digital Video Recording:

[82] J. L. E. Baldwin, "Digital television recording," in *IERE Conf. Proc.*, no. 26, 1973, pp. 67–70.

[83] H. Yoshida and T. Eguchi, "Considerations in the choice of a digital VTR format," *SMPTE J.*, no. 7, July 1982.

[84] H. Yoshida and T. Eguchi, "Digital video recording based on the proposed format from SONY," in *Proc. SMPTE Conf.*, Paper no. 124-98, Nov. 11, 1982; also in *SMPTE J.*, no. 5, May 1983.

[85] W. Haberman, "Progress in the development of the future digital video recording format," *EBU Rev.*, no. 198, Apr. 1983.

[86] M. Morizono, H. Yoshida, Y. Hashimoto, and T. Eguchi, "Digital video tape recording with increased packing density—Progress report," in *Conf. Record, SMPTE Winter Conf., Television Technology in the 80's*, San Francisco.

[87] Y. Fujiwara, T. Eguchi, and K. Ike, "Tape selection and mechanical considerations for the 4:2:2 DVTR," *SMPTE 1984 Winter Telev. Conf.*, Montreal, Canada, Feb. 1984.

Perpendicular Magnetic Recording Technologies:

[88] S. Iwasaki and Y. Nakamura, "An analysis for the magnetization mode for high density magnetic recording," *IEEE Trans. Magn.*, vol. MAG-13, Sept. 1977.

[89] S. Iwasaki, "Perpendicular magnetic recording," *IEEE Trans. Magn.*, vol. MAG-16, Jan. 1980.

[90] S. Iwasaki, Y. Nakamura, and K. Ouchi, "Perpendicular magnetic recording with a composite anisotropy film," *IEEE Trans. Magn.*, vol. MAG-15, Nov. 1976.

[91] J. Hokkyo, K. Hayakawa, I. Saito, and K. Shirane, "A new W-shaped single-pole head and a high density flexible disk perpendicular magnetic recording system," *IEEE Trans. Magn.*, vol. MAG-20, Jan. 1984; presented at Int. Conf. Magn. Record. Media, Ferrata, Italy, Sept. 1983.

New Materials for Video Head:

[92] Y. Makino, K. Aso, S. Uedaira, S. Ito, M. Hayakawa, K. Hotai, and Y. Ochiai, "Amorphous alloys for magnetic head," in *Proc. 3rd Int. Conf. Ferrites*, Sept./Oct. 1980, Kyoto, Japan.

[93] ——, "Induced magnetic anisotropy of co-based amorphous alloys," *J. Appl. Phys.*, vol. 52, Mar. 1981.

[94] K. Shiiki, S. Otomo, and M. Kudo, "Magnetic properties, aging effects and application potential for magnetic heads of Co-Fe-Si-B amorphous alloys," *J. Appl. Phys.*, vol. 52, Mar. 1981.

Magneto-Optical Video Recording:

[95] L. Meyer, "Thermomagnetic writing in Mn-Bi films," *J. Appl. Phys.*, vol. 29, 1958.

[96] D. Chen, "Magnetic materials for optical recording," *Appl. Opt.*, vol. 13, Apr. 1974.

[97] K. Sugano, S. Matushita, and Y. Sakurai, "Thermomagnetic writing in Tb-Fe films," *IEEE Trans. Magn.*, vol. MAG-12, no. 6, 1976.

[98] N. Imamura, Y. Mimura, and T. Kobayashi, "Magnetic recording on Gd-Fe amorphous alloy films in contact with some magnetic materials," *Japan J. Appl. Phys.*, vol. 15, no. 4, 1976.

[99] T. Nomura and H. Tokumaru, "Magneto-optical memory experiments on chromium dioxide flexible disk," *SPIE Optic. Data Storage*, vol. 38, pp. 245–249, 1983.

[100] Y. Togami, K. Kobayashi, M. Kajiura, K. Sato, and T. Teranishi, "Amorphous Gd-Co disk for thermomagnetic recording," *Appl. Phys.*, vol. 53, Mar. 1982.

[101] I. Sander and M. Urner-Wille, "Digital magneto-optical recorder," *SPIE Opt. Data Storage*, vol. 382, pp. 240–244, 1983.

[102] Y. Sakurai, K. Onishi, T. Numata, H. Tsujimoto, and K. Saiki, "RE-TM amorphous films for magneto-optical recording," *IEEE Trans. Magn.*, vol. MAG-19, vol. 5, Sept. 1983.

DEVELOPMENT OF A NEW SYSTEM OF CASSETTE TYPE CONSUMER VTR

Nobutoshi Kihara, Fumio Kohno
and Yoshio Ishigaki
SONY Corporation
Shinagawa, Tokyo, Japan

1. INTRODUCTION

Videotape recorders intended for the consumer market began to appear about 1964. Since then, many new developments have appeared in the recording of the chroma signal, in the increase of recording density, and in the mechanism of recorders. These developments have brought about the introduction of cassettes to house the videotapes. Representative of such a trend is the U-matic videocassette, the superior performance of which has opened up many technical applications. Recently, the U-matic system has begun to be applied to broadcast uses, drawing worldwide attention.

The technology of this U-matic system has been further modified through new innovations and inventions to reduce further the size of the videocassette and recorder/player. This new system has been named the "Betamax".

The recording system, recording density and related features of this Betamax system are shown in comparison with other systems in Charts 1-1 and 1-2. These charts show how much the tape area and volume have been reduced for the recording of a unit time of television signals. The Betamax videocassette has thus become 1/2.7 times the volume of the U-matic videocassette. This smaller videocassette promises to become widely applicable for home consumer use. Details of this system are given below.

ITEMS \ SYSTEMS	Betamax	U-matic	EIAJ-1	Units
TAPE WIDTH	12.65	19.00	12.65	mm
TAPE THICKNESS	20	27	30	μm
TAPE SPEED	40	95.3	190.5	mm/sec
DRUM DIAMETER	74.5	110	115.8	mm
RELATIVE SPEED	6.97	10.3	11.1	m/sec
CARRIER FREQUENCY	3.5~4.8	3.8~5.4	3.2~4.6	MHz
MINIMUM RECORDED WAVE LENGTH	1.45	1.9	2.4	μm
VIDEO TRACK WIDTH	58.5	85	110	μm
GUARD BAND WIDTH	0	52	63	μm
AZIMUTH	±7	0	0	Deg

CHART 1-1 COMPARISON WITH OTHER SYSTEMS

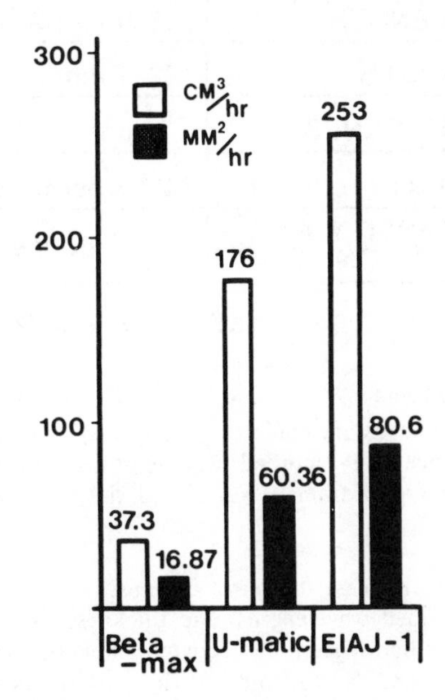

CHART 1-2
COMPARISON OF TAPE CONSUMPTION

2. MAGNETIC RECORDING

2–1 High-density Recording

To increase the recording density of the image signal per unit volume of magnetic medium, the usual methods employed are (1) shortening the recorded wave-length on the medium, (2) using a narrower track head, and (3) reducing the thickness of the magnetic tape. These three methods raise the density in three dimensions. Also in two dimensions, involving the tape area, there is extra space on the tape in which no information is recorded, represented by the guard band. If a recording system is developed that eliminates the need for this guard band, it would be possible to increase substantially the recording density.

The Betamax utilizes all three methods given above and also eliminates the guard band. Such high-density recording was achieved by increasing the performance of the video-tape and video head and also utilizing a slanted azimuth recording pattern.

2–2 The Videotape

The recorded wavelength in the Betamax system is 1.5 μm at white peak of the FM carrier and 1.2 μm at peak point of the pre-emphasized image signal. In order to obtain a playback signal having an acceptable S/N ratio at such short recorded wavelengths, it was necessary to improve the magnetic characteristics of the tape. For example, magnetic material of the U-matic videotape has a coercivity of 500 oersteds. In the Betamax, the coercivity was increased to 600 oersteds.

The surface characteristics of the tape and the dispersion of magnetic particles, which have a direct bearing on the S/N ratio at short wavelengths, were also improved. Also the thickness of the tape was reduced to 20 μm, as compared to 27 μm for the U-matic videotape. This was made feasible through improvements in the tape transporting system.

Received January 5, 1976

Reprinted from *IEEE Trans. Consum. Electron.*, vol. CE-22, pp. 26–35, Feb. 1976.

Usually, a very thin tape having a slick surface does not have good running characteristics, because it is prone to adhere to the head drum. However, in the case of the Betamax tape, this problem was overcome by paying attention to the binder and lubricating material. Chart 2-1 lists the various factors of the Betamax tape.

TAPE LENGTH		150m (60min)
TAPE WIDTH		12.7 mm
TAPE THICKNESS		20μm
COERCIVITY	Hc	600 oersteds
FLUX DENSITY AT SATURATION	Br	1300 gauss

CHART 2-1 Betamax TAPE FACTORS

2–3 Video Head

The video heads are made of single-crystal MnZn ferrite material. Since the recorded wavelength is as short as 1.2 μm, the gap width is 0.6 μm, and the track width is 60 μm.

In order to increase the playback sensitivity of the head at short recorded wavelengths, the thickness of the head piece is more than twice the effective track width, as shown in Figure 2-1, and the thickness of the portion for the coil has been made as large as feasible to raise the efficiency of the magnetic circuit. Also, this shape helps to decrease the wear on the head as well.

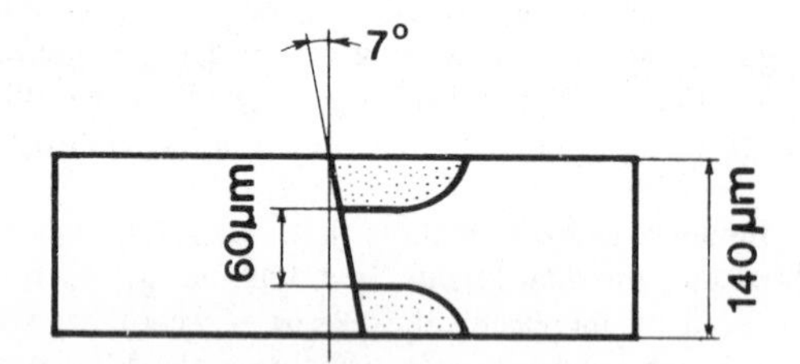

FIG 2-1 SHAPE OF VIDEO HEAD CONTACT SURFACE

In order to record on a slanted azimuth, as described in more detail later, the azimuth of the head gap is ± 7° from perpendicular (making a total of 14° difference between the two heads used on the drum). In principle, the head output at this azimuth is slightly less than the output for an azimuth at perpendicular. In other words, the effective speed between the tape and head is decreased to an amount equivalent to the cosine of 7°. If V_o is the relative speed between the head and tape and the effective speed between the head and tape is V, then

$$V = V_o \cos 7°$$

Since $\cos 7° = 0.99254$

the decrease in head output due to the slanted azimuth is negligible.

2–4 Slanted Azimuth Recording System

In order to increase the recording density on the Betamax tape, the guard band between tracks has been eliminated. This type of recording format, however, raises the problem of crosstalk.

The Betamax recording system uses two rotating heads. By slanting the aximuth in opposite directions on the two video heads, the luminance crosstalk between adjacent tracks during playback can be reduced by utilizing this azimuth loss. This effect becomes greater as the recorded wave-length becomes shorter. The ratio of the crosstalk with respect to the main signal is given by the following equation.

$$C/S = \frac{\sin(\frac{\pi W}{\lambda}\tan\theta)}{\frac{\pi W}{\lambda}\tan\theta} \times \frac{W}{T - W}$$

where C = crosstalk output,
S = main signal output,
W = amount of mistracking,
θ = azimuth angle difference between the two heads,
T = tracking width of video head,
and λ = recorded wavelength.

At long wavelengths, C/S is mainly governed by the term $\frac{W}{T-W}$. However, as θ becomes larger, $|\tan\theta|$ becomes greater, and C/S decreases. When $\theta = 90°$, C/S theoretically becomes zero. At short wavelengths, where $\lambda \leqq \pi W \tan\theta$, the following relationship holds,

$$0 \leqq \left|\sin(\frac{\pi W}{\lambda}\tan\theta)\right| \leqq 1,$$

so that the peak value of the term

$$\frac{\sin(\frac{\pi W}{\lambda}\tan\theta)}{\frac{\pi W}{\lambda}\tan\theta}$$

decreases as λ decreases.

Therefore, as long as the amount of mistracking W is not extremely large, it is possible to reduce C/S to an acceptable range at short wavelengths of recording without making θ too large. (See Figure 2-2.) Moreover, if the

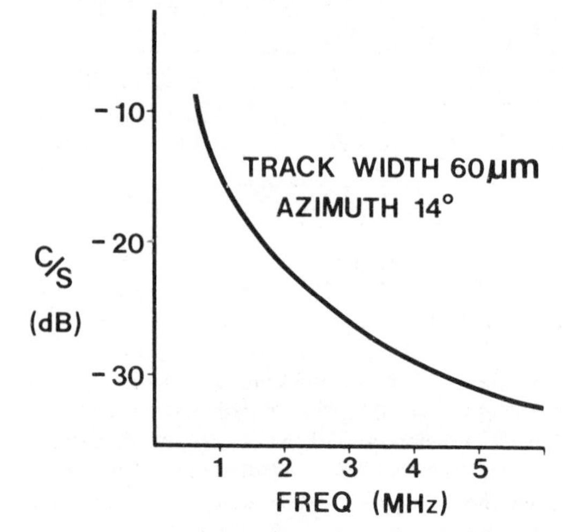

FIG 2-2 FREQUENCY CHARACTERISTIC OF CROSSTALK AT 14° AZIMUTH DIFFERENCE

tracking pattern on the tape has the correct H alignment, the crosstalk is visually not noticeable. In the Betamax, the tape pattern has the correct H alignment.

On the other hand, in a recording system using two rotating heads at different relative azimuth angles, the playback signals from the two heads, when there is mistracking, will not have continuity with respect to time, thereby causing a discontinuity on the time base at the point of switchover from the signal output of one head to the signal output of the other head. On the reproduced picture, this appears as a dihedral error (see Figure 2-3). This time base discontinuity, or "jump", τ_ν can be expressed in the following manner:

$$\tau_\nu = \frac{W \cdot 2 \tan \frac{\theta}{2}}{v}$$

When θ is small, $\tau_\nu = \frac{W \tan \theta}{v}$

Here, v is the relative speed between the tape and head. Therefore, if θ is made too large, then even a slight mistracking will cause a large dihedral error and the reproduced picture will not be acceptable.

In the Betamax, θ was chosen as 14°, which was the result of considering the reduction of crosstalk between adjacent tracks and avoiding unacceptable dihedral error. The azimuth of one video head is set at +7° from perpendicular and that of the other head at -7° from perpendicular.

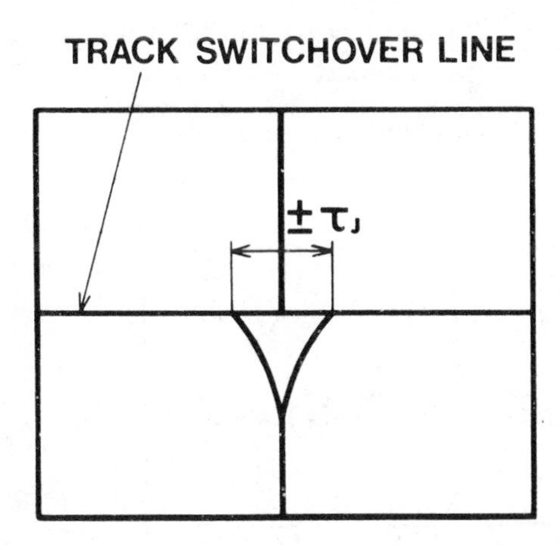

FIG 2-3
SPLIT SCREEN DUE TO MISTRACKING

3. RECORDING SYSTEM FOR THE CHROMA SIGNAL

3–1 Recording Method

In home and industrial video recording systems, the differential gain and differential phase caused by distortions in transmission and the Moire patterns due to mutual interference between video and FM signals are usually avoided by frequency modulating the luminance signal and combining this frequency-modulated signal with the chroma signal that has been frequency converted to a lower band.

When slanted azimuth recording is employed, as described in section 2, C/S is attenuated an acceptable amount at frequencies of 2MHz or more where the transmission band lies for the FM luminance signal, as seen in Figure 2-2. However, at 1.3 MHz or lower, where the transmission band for the chroma signal is located on the spectrum, the attenuation is not adequate. Because of this, when there is mistracking, the crosstalk from the chroma signal of the adjacent track causes problems involving color flicker and color distortion. The slanted azimuth recording method was already known many years ago, but it was not practically used because of color problems.

To solve these problems, the Betamax system makes use of the high vertical correlation of the color signals between adjacent scanning lines. This new method of recording the chroma signal consists of interleaving the frequency spectrum of the chroma signal for one line with the spectrum for the adjacent line, and at playback using a comb filter having a one-H delay line, so that the crosstalk of the chroma signal from the adjacent track is eliminated.

By using a comb filter, the following advantages are obtained in addition to the elimination of crosstalk:

1. Hue distortion due to residual phase errors is improved.
2. Distortion in color saturation due to deviations of head contact with the tape is improved.
3. The cross-color due to the high frequency portion of the luminance signal is improved.

As a result, color pictures of very good quality are obtained.

To obtain the two chroma signals interleaving each other on the frequency spectrum, it would be possible to select frequencies f_1 and f_2 that have the following relationship:

$$|f_1 - f_2| = (2n-1)\frac{1}{2} f_H$$

where n = an integer,
f_H = horizontal scanning frequency.

However, the time required for the phase of the chroma signal output to stabilize at the changeover point of the tracks, as well as the ease of integrating the circuitry, became factors that led to a method where the two frequencies interleaving each other are obtained from one frequency in the Betamax system. This is outlined below.

Let us assume that F_A and F_B are the chroma input signals to be recorded on track A and track B respectively. Then,

$$F_A = \sum_{k=-\infty}^{\infty} V_k \exp\{j(\pm\omega_s \pm K\omega_H)t\}$$

and

$$F_B = \sum_{m=-\infty}^{\infty} V_m \exp\{j(\pm\omega_s \pm m\omega_H)t\}$$

where V_k and V_m are coefficients,
ω_s = angular frequency of the chroma subcarrier.
and ω_H = angular frequency of the horizontal scanning frequency.

If the phase of F_B is inverted at each scanning line, then the inverted signal $G(t)_{rec}$ will be

$$G(t)_{rec} = \sum_{\ell=-\infty}^{\infty} V_{2\ell-1} \exp\{j(2\ell - 1)\frac{\omega_H}{2} t\}$$

and the chroma signal F_B' recorded on track B is

$$F_B' = F_B \cdot G(t)_{rec}$$

$$= \sum_{m=-\infty}^{\infty} \sum_{\ell=-\infty}^{\infty} V_m \cdot V_{2\ell-1} \cdot \exp\left[j[\pm\omega_s + \{(\ell \pm m) - \frac{1}{2}\}\omega_H]t\right]$$

Thus F_A and F_B' will be mutually interleaved.

When track A is played back, the crosstalk cF_B' from track B will be mixed with the main signal F_A ("c" represents the degree of mistracking causing the crosstalk), but by means of a comb filter having null points at $\omega_s \pm (2n-1)\frac{\omega_H}{2}$, it is possible to eliminate cF_B'. Next, when playing back track B, the combination

$$F_B' + cF_A$$

is obtained. When the phase is inverted to the same phase as that at the time of recording,

$$(F_B' + cF_A) \cdot G(t)_{pb} = F_B \cdot G(t)_{rec} \cdot G(t)_{pb} + cF_A \cdot G(t)_{pb}.$$

Since $G(t)_{rec} \cdot G(t)_{pb} = 1$, the above equation becomes

$$(F_B' + cF_A) \cdot G(t)_{pb} = F_B + cF_A \cdot G(t)_{pb}.$$

This means that the crosstalk component has the same frequency content as the spectrum contained in the signal played back from track A. Thus, just as in the process where crosstalk was eliminated in playing back track A, the crosstalk cF_A is removed, so that only the signal F_B is reproduced.

In the actual system, the chroma signal recorded on the tape is at a lower carrier frequency ω_L obtained by means of a conversion carrier ω_C applied to the subcarrier ω_s :

$$\omega_L = \omega_C - \omega_s$$

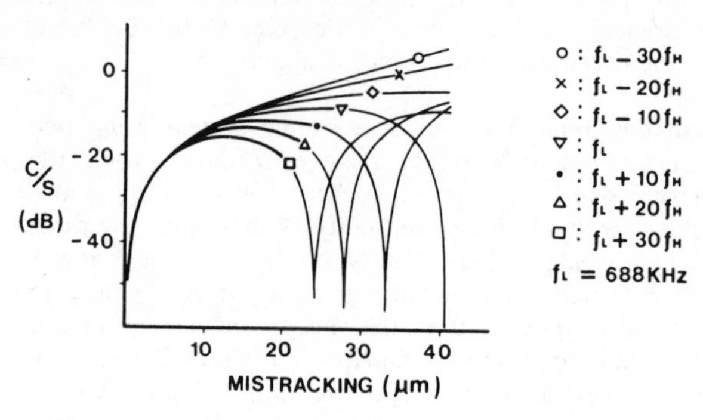

FIG 3-1 CHROMA SIGNAL CROSSTALK DUE TO MISTRACKING

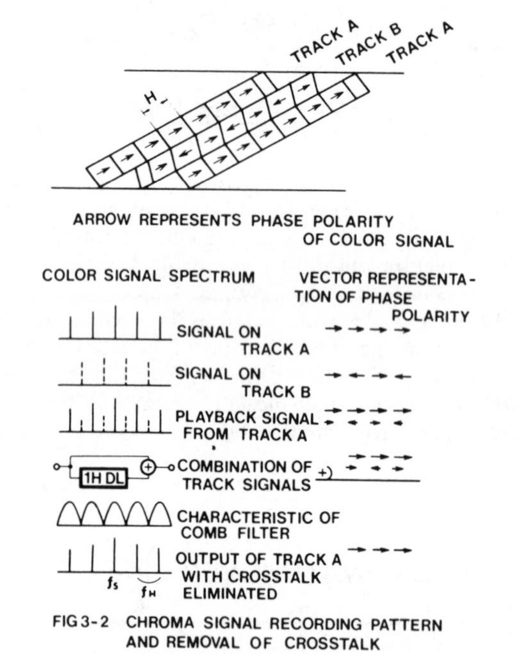

FIG 3-2 CHROMA SIGNAL RECORDING PATTERN AND REMOVAL OF CROSSTALK

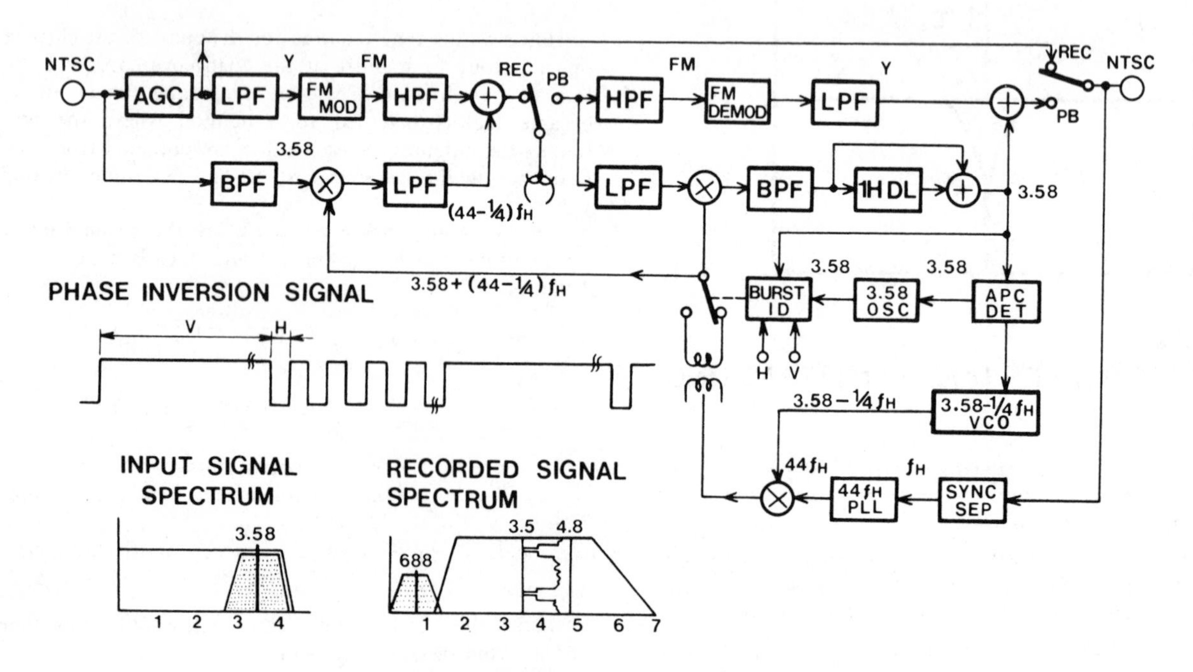

FIG 3-3 BLOCK DIAGRAM SIGNAL PATHS

Moreover, the phase inversion process on each horizontal scanning line at the time of recording and playback is actually obtained by inverting the phase of ω_C to arrive at the equivalent phase inversion of ω_L, in order to prevent undesirable transients and spurious disturbances.

In Figure 3-1, the variation of crosstalk in the chroma signal due to mistracking is shown with respect to parameters of the chroma signal side-band. Figure 3-2 shows the recording pattern of the chroma signal on the tape and how the crosstalk is removed by means of the comb filter. Figure 3-3 is a block diagram of the signal paths.

3–2 Burst Identification

When playing back track B, if the phase inversion taking place at that time is different from the phase inversion that took place at the time of recording, then

$$G(t)_{rec} \cdot G(t)_{pb} = -1,$$

and the signal after phase inversion would become

$$-F_B + cF_A \cdot G(t)_{pb}.$$

This means that the phase of the chroma signal output would be reversed for the entire track (that is, field). Such a situation is shown in Figure 3-4.

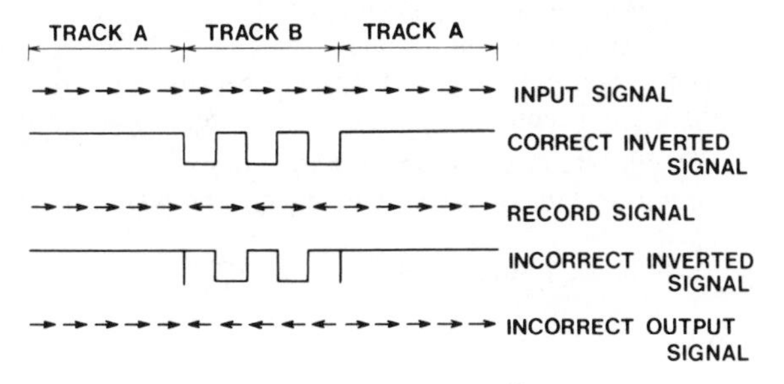

FIG3-4 FUNCTION OF BURST IDENTIFICATION SYSTEM

When this takes place, the automatic phase control of the TV receiver is not able to follow and correct for this adequately. As a result, color distortions appear on the upper part of the picture.

To solve this problem, the Beramax system employs a phase detector that is separate from the one used in the APC for time-base correction, in order to identify separately the burst signal of the chroma playback signal. This instantaneously identifies and corrects the phase of the inverted signal. A block diagram of the burst identification circuit is shown in Figure 3-5.

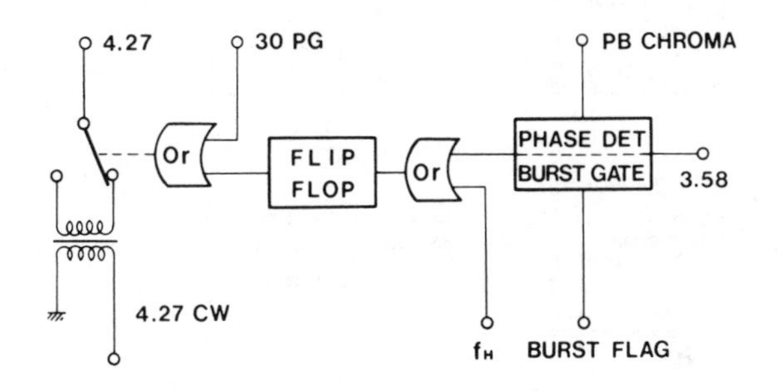

FIG3-5 BLOCKDIAGRAM OF BURST IDENTIFICATION CIRCUIT

Since it is only necessary to differentiate between 0° and 180°, the burst identifying circuit operates at even a higher stability than the identification method of the PAL system. Moreover, even when the phase of the inverted signal is disturbed by noise or dropouts, the burst identifier promptly makes corrections so that the proper phase is restored within two scanning lines, thereby preventing disappearance of color.

3–3 The Comb Filter

In order to remove chroma signal crosstalk from the adjacent track, a comb filter is used, which has null points along the crosstalk spectrum. The one-H glass delay line used for this purpose does not have ideal characteristics, because of (1) delay error, (2) unwanted reflections, and (3) variation of equivalent impedance with frequency.

A typical frequency characteristic curve of the signal obtained from such a filter, taking into consideration these factors, is shown in Figure 3-6. The attenuation curve deteriorates along the frequency axis away in either direction from the central subcarrier frequency. However, this deterioration is practically acceptable for the following reasons:

(1) There is some azimuth loss still remaining.
(2) The deviation in tracking is adequately small enough even when videocassettes are interchanged.
(3) The chroma signal content is mainly concentrated around the central frequency.
(4) The attenuation provided by the comb filter near the central frequency is adequately large.
(5) The crosstalk from side-bands is hardly recognizable on the picture.

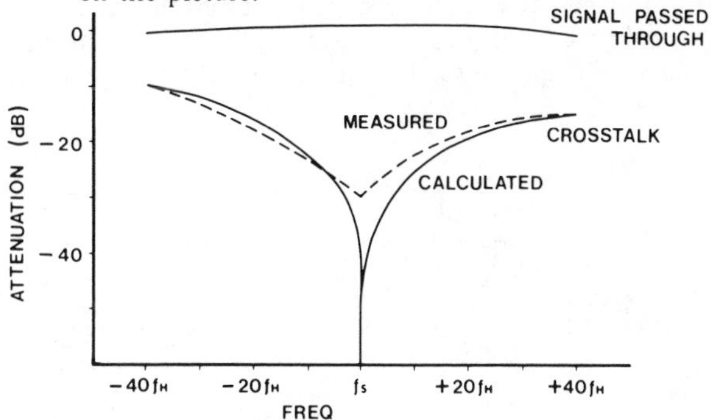

FIG 3-6
FREQUENCY CHARACTERISTIC OF SIGNAL THROUGH COMB FILTER

3–4 Effect of Tape Speed Deviation

Because the tape speed when recording and the tape speed when playing back are slightly different in VTR equipment in general, there is a slight deviation of frequency in the playback signal. Although the central frequency of the chroma signal is held at the base frequency through the AFC circuit, any deviation of the tape speed will cause the band width of the chroma signal to differ slightly between recording and playback. On the other hand, since the one-H delay time of the comb filter is fixed, there will be a slight difference between it and the one-H of the playback signal, causing the spectrum of the color crosstalk to shift off of the null point frequencies of the comb filter, to a greater degree at points further out on the color sidebands. A typical set of frequency characteristic curves of the signal using the comb filter are shown in Figure 3-7 for deviations of tape speed ± 0.5%. As stated in section 3-3, this amount of deterioration does not pose problems from a practical standpoint.

When the speed varies by d%, the time difference between the luminance signal and the chroma signal will be $d/200f_H$(sec). For a deviation of 0.5% in tape speed, the time difference would be 0.16 microsecond, which is not practically a problem.

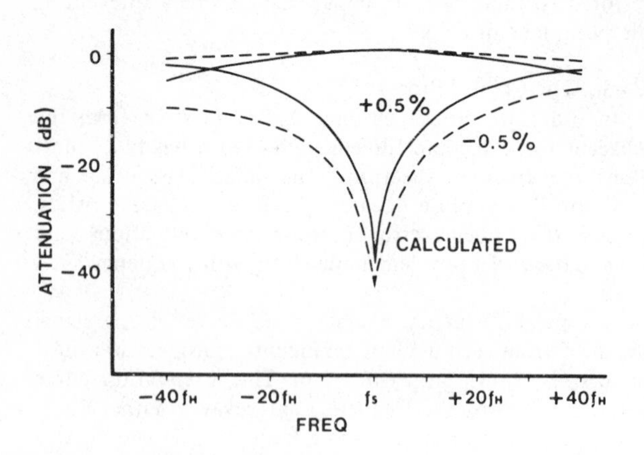

FIG 3-7 EFFECT OF TAPE SPEED VARIATION ON FREQUENCY CHARACTERISTIC OF SIGNAL THROUGH COMB FILTER

4. INTEGRATED CIRCUITS

Basic Policies: To maintain high quality and stable mass productivity of low-cost home videotaping units, the circuitry was integrated to as high a degree as the IC's used in color television sets, which probably have the highest degree of integration among various home electronic products. Compared to discrete circuits that might have been used, the number of parts were reduced to about one-third, which contributed to the conservation of material and production costs, as well as kept down fluctuations of quality.

In integrating the circuitry, the following basic guidelines were used for developing the necessary IC's.

(1) IC's were designed to be optimum for the Betamax system, but at the same time made applicable as widely as possible to other types of VTR units.

(2) The degree of integration was increased to handle a large number of functions. A new process for producing IC's was developed for this purpose. Eleven types of bipolar IC's, each containing an average of 200 elements, were produced.

(3) From the standpoint of IC technology, it is not always rational merely to integrate discrete circuits used in previously known systems. Therefore, signal processing methods and control methods that are most suited to circuit integration were especially developed and designed.

(4) To obviate external adjustments as much as possible, automatic control circuits were incorporated to a very high degree, which include the following.
   - Automatic gain control
   - Automatic color level control
   - Automatic phase control
   - Automatic frequency control
   - Automatic color killer

(5) The power unit was left discrete because of power dissipation requirements. The audio circuitry was also left discrete, because many parts are external to the circuit and the advantages of integration were not so great.

Figure 4-1 shows a block diagram of the IC's.

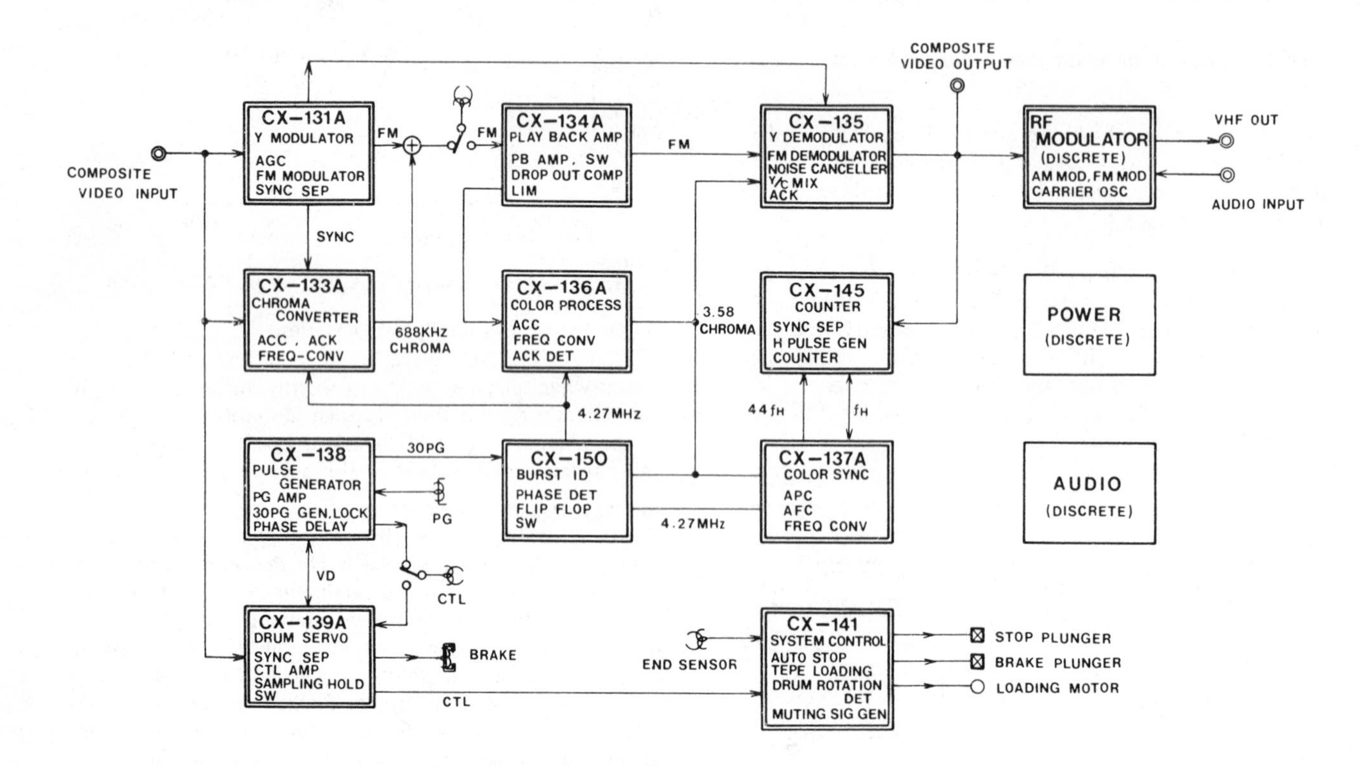

FIG 4-1 BLOCK DIAGRAM OF IC'S

5. TAPE TRANSPORT MECHANISM

5–1 Threading System

Many different methods of threading the videotape could be considered, but in the Betamax, a system using a loading ring was employed. The loading ring method provides high precision and stability. To increase the precision of this loading ring, the ring was made from a precision zinc die cast and then the surface was tooled carefully so that it would spin without fluctuations.

When threaded, the tape is ready in the "forward" position. As soon as the pinch roller makes contact, the tape will quickly pick up speed and start running normally and accurately. The guides have been precisely attached to the die cast ring, so that the tape will run without twisting. The tape is threaded in the order shown in Figure 5-1.

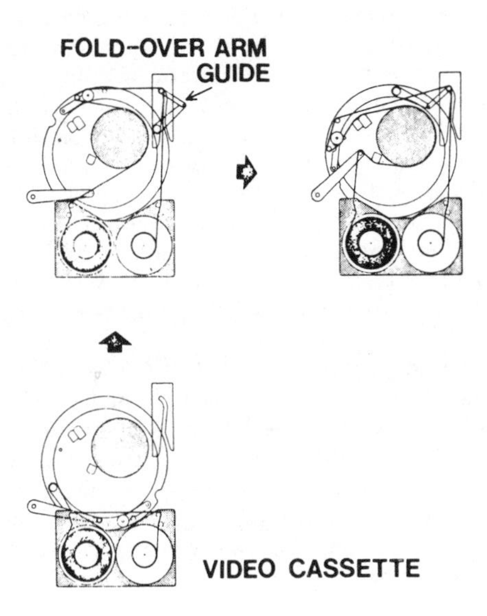

FIG5-1 TAPE LOADING MECHANISM

PHOTOGRAPH 5-1 VIEW OF DRUM EXIT SIDE

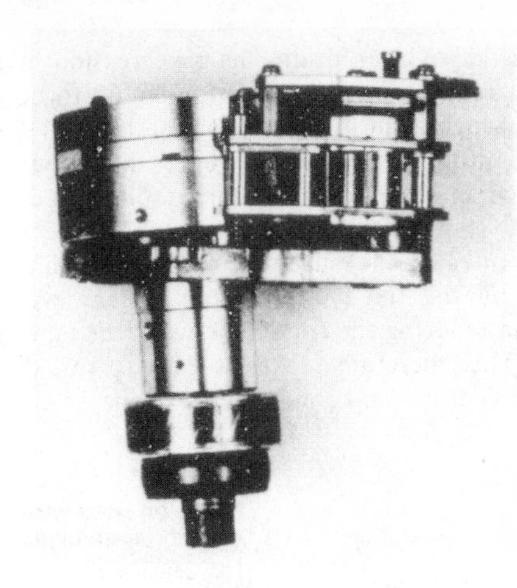

PHOTOGRAPH 5-2 VIEW OF DRUM ENTRY SIDE

5–2 Tape Guiding System

To have successful high-density video recording, it is important for the tape to run very accurately. To achieve this accuracy, the guides must be very precise. Various schemes were therefore used in designing the guides. For the rotating guides, bearings made of resin containing solid lubricating oil were used, which made it possible for the roller to rotate with high precision at 4000 r.p.m. or more without the bearings burning out. Also, to reduce the number of guides required, a fold-over arm guide attached to an elbow is used. This fold-over arm guide pulls the tape around the outside of the loading ring, and this makes it possible to use fewer guides, reducing the chances of damaging the tape edge.

The head drum also has arrangements to stabilize the running of the tape. A spring mechanism lightly presses down on the tape at the midportion of the drum. This gently pushes the lower edge of the tape down into the groove of the lower drum, so that slight irregularities of tape motion will not build up into any wavering motion. Deviations of tape width will not result in the tape lifting out of the groove guide. The tip of the spring is protected with ceramic, which will not damage the tape edge and will not wear out for at least 2000 hours of running time. At the entry and exit of the drum position, an erasing head, audio head, and control head are located. To guide the tape along these fixed heads, long frames keep the tape pressed in position. This keeps the audio, video, and control tracks from deviating and makes it possible, therefore, to use narrow track heads. Photographs 5-1 and 5-2 show this structure.

5–3 Head Drums

The rotating drum contains two heads and is controlled by a brake servo system. The drum assembly consists of three cylindrical layers, the lower fixed drum, the middle rotating drum (a disk holding the two heads), and the upper fixed drum. With the upper and lower drums fixed and the middle drum rotating, the best features of previous systems (where the upper drum rotates in one case and where the upper drum is fixed in another case) are embodied in the Betamax.

In this three-layer drum, the tape tension at entry of 50 grams results in a tension of about 60 to 80 grams at the exit, which is an increase of about 1.2 to 1.7 times. In previous drum arrangements where the upper drum is fixed, such an entry tension would result in an exit tension of two to three times. (See Figure 5-2). The three-layer system possesses the advantage of tape-running stability featured by the fixed upper drum system, while at the same time reducing the friction between the tape and drum surface. This therefore makes it easy to run the tape at forward, fast forward, and rewind while it is still fully threaded.

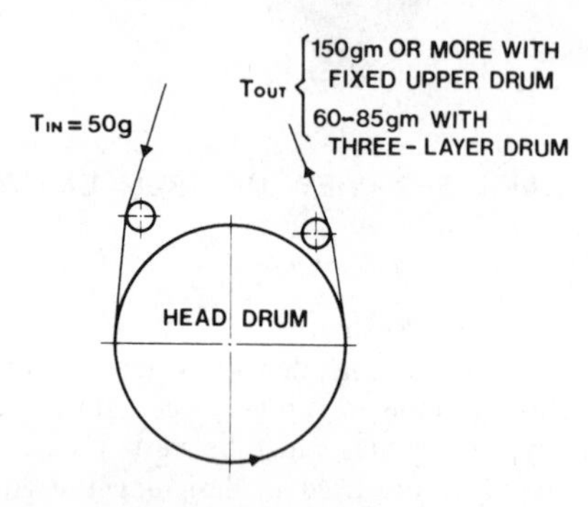

FIG 5-2 TAPE TENSION AT ENTRY AND EXIT POINTS OF DRUM

The starting point on the tape is easy to set accurately. Thus the continuity from one scene to another in video recording can be accurately located, with little deviation. There is another reason for the deviation being small: Because the density of recording is high, the tape speed is slow, allowing for a shorter tape and hence a smaller diameter of reel. This means that the inertia is small and hence the tape attains proper speed very quickly.

Another feature of the three-layer assembly is that about half of the width of the tape is running over either the upper or lower fixed drums, so that the rotating drum will not suddenly jerk the tape and damage it in case of wrinkles or scratches on the tape or adherence due to moisture on the drum or tape. Also the controls for running the tape are coupled with the tape end sensors, so that any operation of the controls will not cause the tape to sag inside the player. If the power source cuts off during any operation, the auto-stop mechanism functions to prevent the tape from sagging or breaking. Also automatic start, recording and stopping will function with a timer turning the external power source on and off.

6. CASSETTE TAPE

6–1 Strength of Tape and Casing

Because of the high density of recording, the cassette tape could be made very small. The outer dimensions of the cassette was 96mm (width) x 156mm (length) x 25mm (thickness). Loaded with tape that can contain 60 minutes of color recording, the entire videocassette weighs only 210 grams. This light weight helps the toughness of the cassette, which can be dropped from the height of a desk onto a wooden floor without breaking.

The casing consists of top and bottom halves and a hinged edge lid. The top half has a window that allows the user to see the amount of tape on the supply reel. The inner side of the top half has a leaf spring that pushes the reels down against the bottom half of the casing. The leaf spring keeps the reels firmly in place, so that the cassette could even be used in a vertical position with the reels rotating stably. The lower half of the casing has brake guides and a locking mechanism for the edge lid. Also the under side of the bottom half contains a tab that can be broken off to prevent accidental erasure of a recording. By covering the hole where the tab was with a piece of plastic adhesive tape, the cassette can be used to record again.

The lid on the edge of the cassette is hinged and can be opened or closed. Guards are attached to the inside of the lid to prevent the tape from being directly exposed, so that the user will not be able to finger the tape or touch it underneath the lid. Another guard fence is located between these guards so that the tape will not sag at the exit of the cassette.

6–2 Two Independent Reels

The cassette has two identical reels having upper and lower flanges, which keep the rolled tape from becoming untidily rolled even when the cassette is dropped on the floor. Even if there is some untidiness in the roll, there will be no change in the required pull-out tension. This prevents any failure of the take-up reel to wind the tape due to the cassette itself. Also, since the two reels are mounted on the same plane, the thickness of the cassette is held to a minimum. The bottom side of the leaf flange has ratchet-like teeth, which are caught by the brake to prevent the reel from running loose at the wrong time.

6–3 Brakes and Lid Lock

The brake mechanism is coupled to the lid movement, so that it releases when the lid opens up to an angle of 75° or more. The reels, therefore, do not rotate until the lid is open. The tape does not sag when the cassette is transported. Also, there is a slight braking effect applied to the tape itself, so that the tape will not sag or become loose at the exit point.

The brakes on the reels operate in such a way that the reels will rotate in the direction tensing the tape but will not rotate in the direction loosening the tape. (See photo 6-1.) With this scheme, it is not necessary to provide a brake release mechanism in the VTR.

The lid lock can be released from three directions. However, it cannot be released by using the fingers. A sharp-pointed instrument such as a lead pencil is needed to release it. The ordinary user, therefore, cannot easily open the lid, and the tape is thus well protected.

6–4 Tape End Sensors

At the beginning and end of the tape, a leader and trailer made of aluminum foil laminated with a polyester base, have been attached. Sensing coils are located in the VTR at the entry and exit points of the tape, so that the start and end of the tape can be detected without the coils touching the tape. System control is therefore very much simplified.

Because the volume of tape used has been reduced radically, the inertia of the reels and tape is very low, allowing for quick start and stop. This therefore made it possible to use extremely short leader and trailer sections of the tape for sensing, so that the video heads will never touch either of these sections of the tape. This simple, non-touching method has made it possible to achieve fool-proof system control of high reliability at low cost.

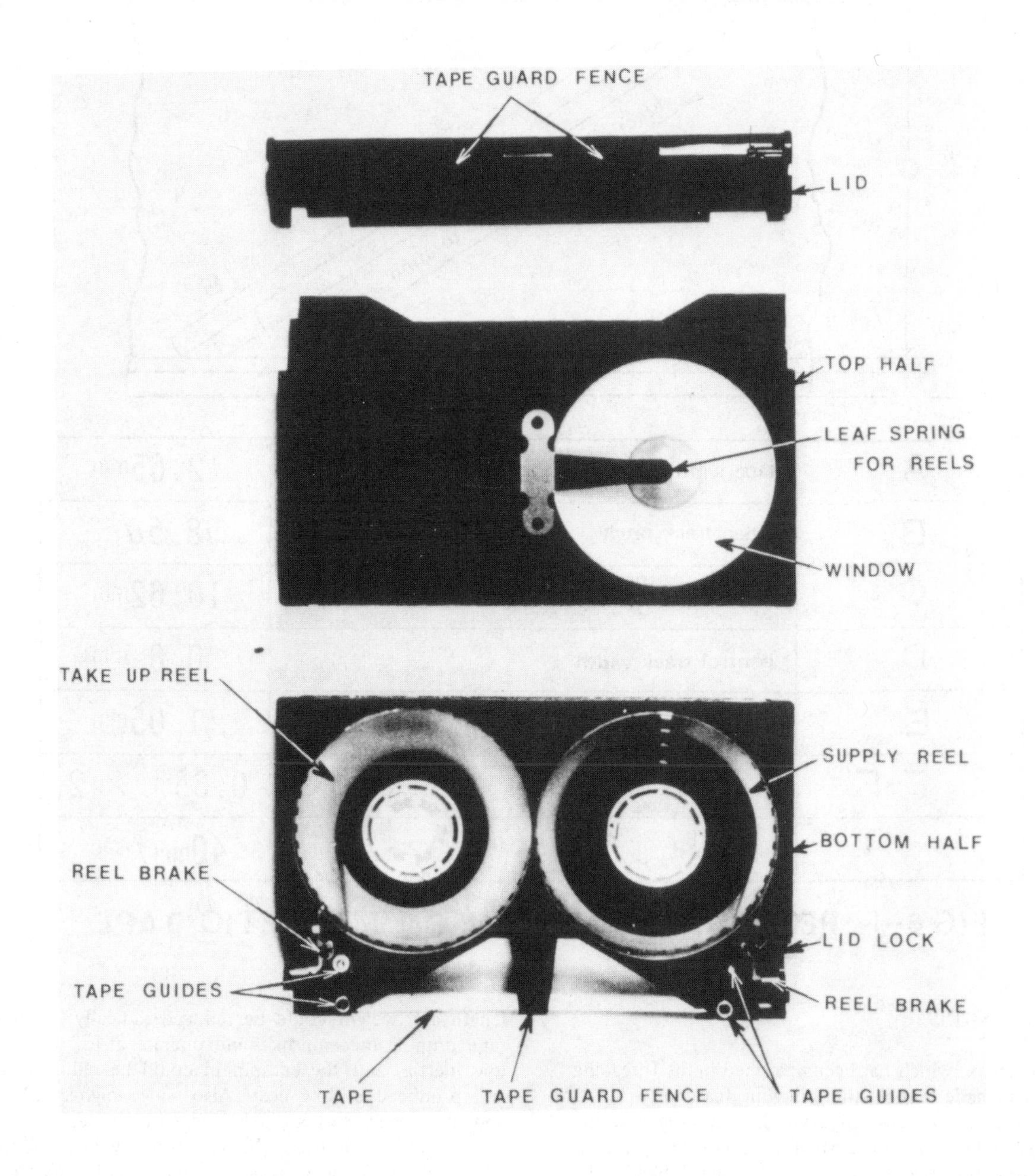

PHOTOGRAPH 6-1 DISASSEMBLED VIDEOCASSETTE CASING

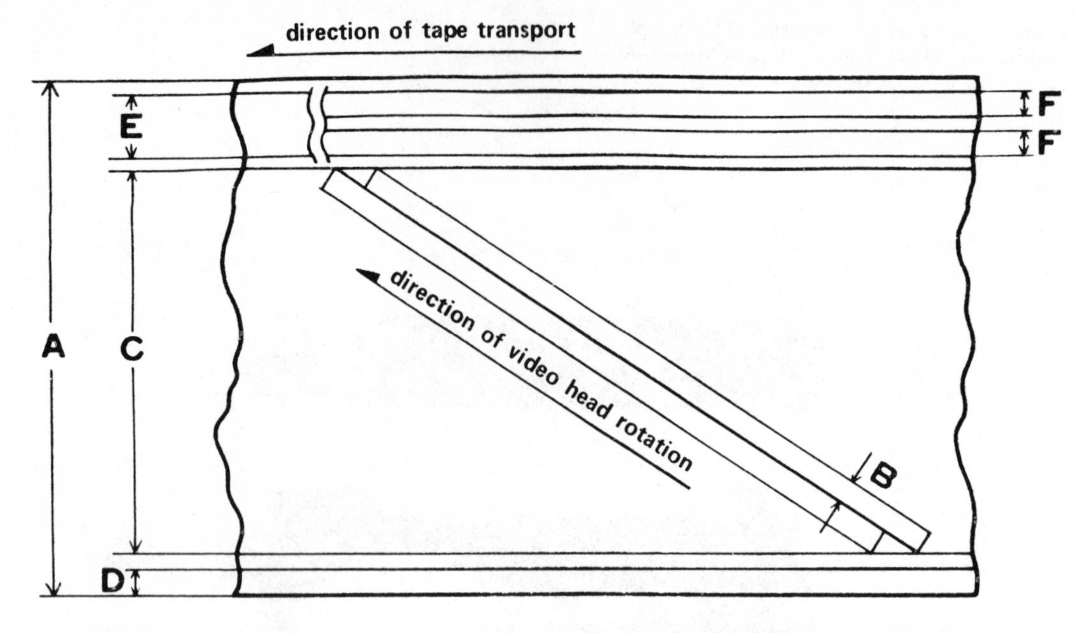

A	tape width	12.65mm
B	video track pitch	58.5μ
C	width of tape used for video track	10.62mm
D	control track width	0.6 mm
E	audio track width (monaural)	1.05mm
F•F′	audio track width (stereo)	0.35mm × 2
	tape speed	40mm / sec

FIG 6-1 RECORDING PATTERN ON MAGNETIC TAPE

7. CONCLUSIONS

The Betamax, which has been described in the foregoing sections, has made feasible the following features:

1. By the use of slanted azimuth recording and the invention of a new system of color recording, it became possible to eliminate the guard band. The development of a new, high-coercivity tape and a new recording head has made feasible recording at even shorter wavelengths than before ($\lambda = 1.2\ \mu m$) and the area of tape used therefore has been extremely reduced.
2. The use of extra-thin tape and the development of a new tape guiding system have aided in the perfection of a stable tape running mechanism that assures good tape interchangeability.
3. The full integration of the circuitry has simplified the physical circuit assembly, and this has contributed to the reduction of cost and size of the recorder and player.

Through this high-density recording technology, which does not sacrifice performance or picture quality, the tape length and weight could be reduced radically, so that the tape running mechanism could operate at low torque and low inertia, and the equipment could be reduced in size and produced at low cost. Also, since high-speed access and high-speed start are achieved with the tape remaining wound around the head, the recorder/player can be operated in the same way as an ordinary audio-cassette tape recorder.

The Betamax not only opens up advantages in the professional field but also promises to expand fields of use by consumers in general.

8. ACKNOWLEDGEMENTS

The authors express their deep appreciation for the advice provided by Masaru Ibuka, Chairman, and Akio Morita, President, the guidance in production technology given by Susumu Yoshida, Senior Managing Director, and the support provided by the Executive Staff, all of Sony Corporation.

VIDEO CASSETTE RECORDER DEVELOPMENT FOR CONSUMERS

Yuma Shiraishi and Akira Hirota
Victor Company of Japan, Limited
Yokohama, Japan

Introduction

The first 2-head, helical scan type color video tape recorder appeared on the scene 18 years ago. Since that time numerous technical innovations have made possible an increase in the recording density, development of new color recording systems, utilization of cassette tapes, and so forth. Based on these achievements, a video cassette recorder designed and developed especially for home usage has recently met with completion and this system has been named "VHS".

Development policy

Progress in VTR technology and engineering during the past 20 years has been quite remarkable. Many VTRs utilizing the latest contemporary techniques have been released and employed for broadcasting and other professional or educational purposes. Meanwhile, an increasing demand for home-oriented VTRs has come to the fore. Home-use appliances must be such that they satisfy the various needs of as many people as possible. Taking this into account, our development of the VHS system initially centered upon a "Total Concept Method" in which various aspects would be harmoniously integrated so as to encompass the requirements of as many people as possible.

Main requirements would be: connectability to a TV receiver; picture quality equivalent to TV broadcast programs; continuous recording time of more than 2 hours; excellent compatibility; application flexibility, such as recording and playback of off-the-air programs, playback of prerecorded cassettes, recording of live scenes using a video camera, slow-motion or speed-up playback, etc.; low cost; easy operation; trouble-free operation and easy servicing; transportability; good productivity; and capability of interchanging common parts and subassemblies with different models.

Technically speaking, our efforts were focused onto the selection of technical items for meeting various requirements in a harmonizing way, and then onto the realizing of high-density recording and finally onto the development of small-sized, lightweight and easy-to-handle mechanisms.

High-density recording

In the VHS system, high-density recording had to be established in order to make long recording time, compact design and low cost physical realities. Reduction of the writing speed was the first subject of our investigations and then reduction of the track pitch.

For overcoming the first obstacle, a magnetic tape of Hc 600 oersteds was selected to improve surface smoothness without deteriorating runnability, thereby enhancing the S/N ratio. Also, a ferrite video head having a gap of 0.3 μm was developed for recording and playback of short wavelengths and two limiters were employed in the FM playback circuit in order to utilize a high-frequency band while reducing over-modulation. These factors contributed to making a writing speed of 5.8 m/sec for NTSC and 4.8 m/sec for PAL a reality. With this writing speed, a minimum recorded wavelength is 1.3 μm for NTSC and 1.0 μm for PAL and SECAM.

The second problem was solved through employment of a slanted azimuth recording system to eliminate crosstalk of FM signals between adjacent tracks and a phase-shift color signal processing system to eliminate crosstalk of color signals between adjacent tracks. Improvement in machining accuracy up to the order of a few microns also contributed to this solution. As a result, the track pitch is 58 μm for NTSC and 49 μm for PAL and SECAM.

Simplified loading system

For the purpose of obtaining an uncomplicated tape transport mechanism, a parallel loading system was conceived and a stable loading mechanism, shown in Fig. 1, was completed after several improvements devised for securely holding the loading poles. The travel of the loading poles is 8 cm. The tape is positioned between the capstan shaft and pinch roller upon inserting the cassette within the recorder. The main base of the loading mechanism is on a horizontal plane while those parts which are tilted in relation to the main base are only the drum and the two guide poles.

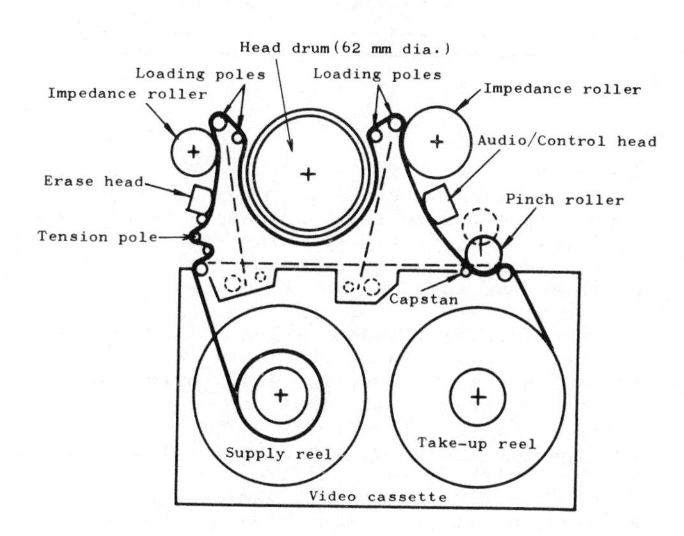

Fig. 1 Tape loading mechanism

Received June 24, 1978

Reprinted from *IEEE Trans. Consum. Electron.*, vol. CE-24, pp. 468-472, Aug. 1978.

Specifications of the VHS system

Table 1 compares the technical data of the VHS system with those of the 3/4" U-format, virtually for professional usages. Fig. 2 shows the recording pattern on the VHS format tape.

	Items \ Systems	Units	U-3/4"	VHS
	Continuous recording time	hr	1	2
A	Tape width	mm	19	12.65
	Tape thickness	µm	27	20
	Tape speed	mm/sec	95.3	33.35
	Tape consumption	m^2/hr	6.52	1.52
	Drum diameter	mm	110	62
	Writing speed	m/sec	10.26	5.8
P	Video track pitch	µm	137	58
	Guard band width	µm	52	0
R	Audio track width (monaural)	mm	-	1.0
D, E	Audio track width (2-track)	mm	0.8	0.35
	Angle of video track	Deg.	5°57'33.2"	5°58'9.9"
	Tape coercivity	Oe	500	600
	FM carrier frequency	MHz	3.8 - 5.4	3.4 - 4.4
	Minimum recorded wavelength	µm	1.9	1.3
	Head gap azimuth	Deg.	0	±6
	Recording Method Luminance Chrominance		FM Converted subcarrier 688 kHz	FM Phase-shift & converted subcarrier 40 fH
	Cassette size	mm	221x140x32	188x104x25

Table 1 VHS specifications compared with U-3/4" format (NTSC)

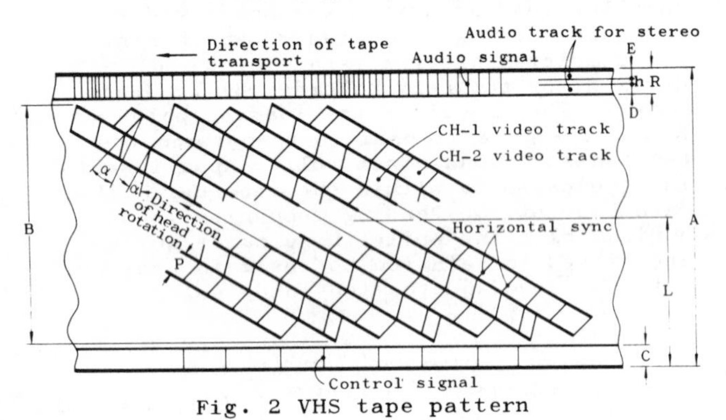

Fig. 2 VHS tape pattern

Cassette

The external view and internal structure of the VHS cassette are shown in Fig. 3 and Fig. 4, respectively. The cassette is so constructed that two independent flanged reels are arranged in the same plane and locked by brakes to prevent slackening of the tape during transportation. These brakes are released only when the cassette is loaded into the recorder. Also, only when in the recorder is the tape exposed, since the edge of the cassette from which the tape is drawn is equipped with a protective cover. At the center of the cassette is provided a hole for light to enter. This light is allowed to pass to either side of the cassette and is employed for detecting the tape ends. The ending and beginning of the tape are detected by a sensor device provided on either supply reel side or take-up reel side, respectively. The tape length is 247 m for 2-hour recording in the NTSC system.

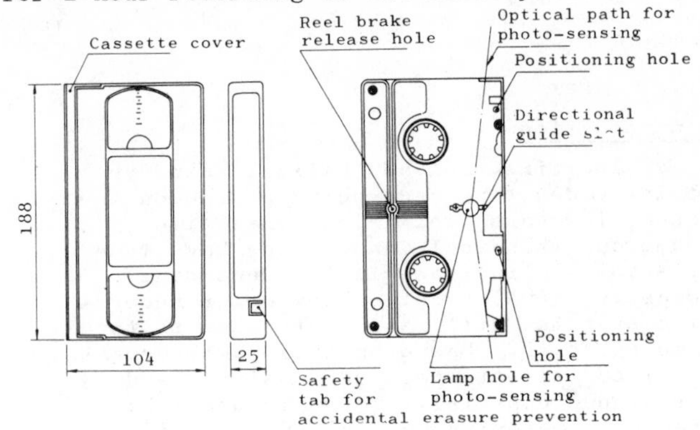

Fig. 3 External view of the cassette

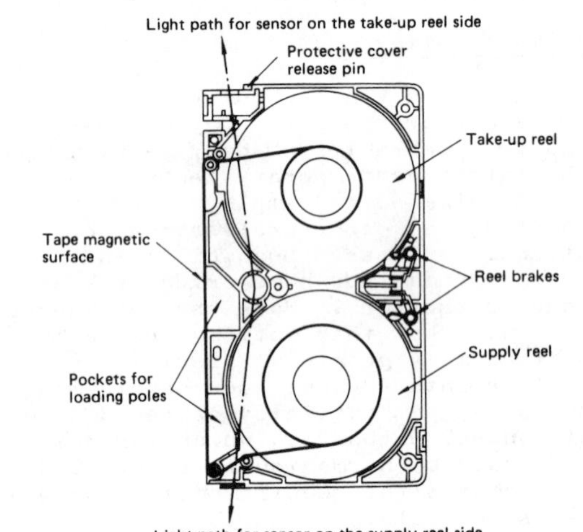

Fig. 4 Internal structure of the cassette

Slanted azimuth recording

In the VHS system the gap azimuth of the two video heads are determined so that they are at ±6° from perpendicular in respect to the direction of head movement. The equation for crosstalk between two adjacent tracks is given below:

$$C = 100\,\frac{t}{T}\;\frac{\left|\sin\left(\frac{2\pi f t}{v}\tan\alpha\right)\right|}{\frac{2\pi f t}{v}\tan\alpha}$$

where c = crosstalk output (%),
t = amount of mistracking,
T = video track width,
v = writing speed,
f = frequency of crosstalk, and
α = azimuth angle.

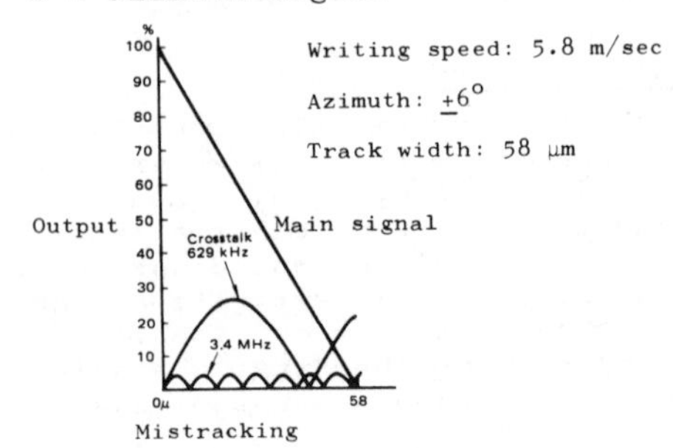

Fig. 5 Crosstalk due to mistracking

Fig. 5 graphs the results of calculations. Though the amount of crosstalk greatly differs depending on that of mistracking, the slantec azimuth recording system is found sufficiently effective in the FM band as judged from the maximum crosstalk value.

The slanted azimuth recording VTR was formerly considered difficult to perform at a variable speed playback since no playback output is available when the track which was recorded by one head is traced by the other head. However, considerations from a viewpoint of frame playback gave rise to the following adventages inherent with the slanted azimuth system:

(1) With regard to the crosstalk signal, mistracking does not result in any critical effect upon adjacent tracks.

(2) No noise band is visible in the picture frame if the following condition is satisfied:

$$V_p = V_r + \frac{2}{n} V_r$$

where V_p = tape speed during noise-free playback,

V_r = tape speed during recording, and

n = a positive or negative integer.

(3) It is possible to broaden the video head width, even to exceed the width of the video track.

Moreover, in order to prevent distorted pictures, continuity of the horizontal sync signals must always be maintained. The VHS system standard tape pattern, shown in Fig. 2, satisfies this requisite.

Video signal circuit

Fig. 6 shows a basic block diagram of the video signal circuit for recording and playback. During playback the FM-converted luminance signal passes through a newly developed circuit called "Double-Limiter circuit", which is shown in Fig. 7. The carrier or high-frequency component of the FM signal is first separated from the side band or low-frequency component and enters the first limiter, so that high frequencies close to the upper limit are fully reproduced. Then the carrier is mixed with the side band and passes through the second limiter.

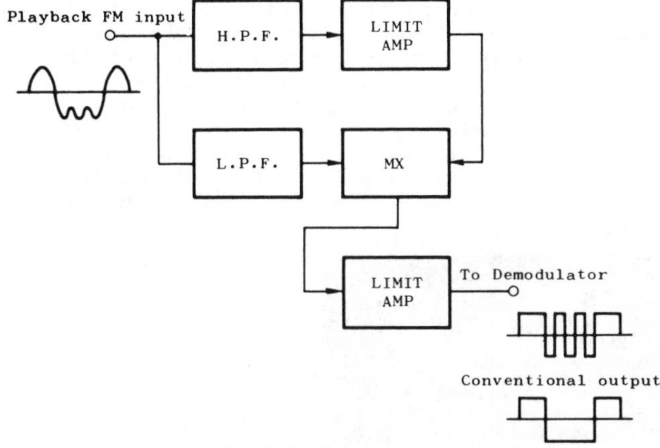

Fig. 7 Double limiter circuit

With the input signal as shown in Fig. 8, ordinary limiter circuits produce an output as shown in Fig. 9, generating over-modulation. The double-limiter circuit promotes obtaining a good output characteristic as shown in Fig. 10.

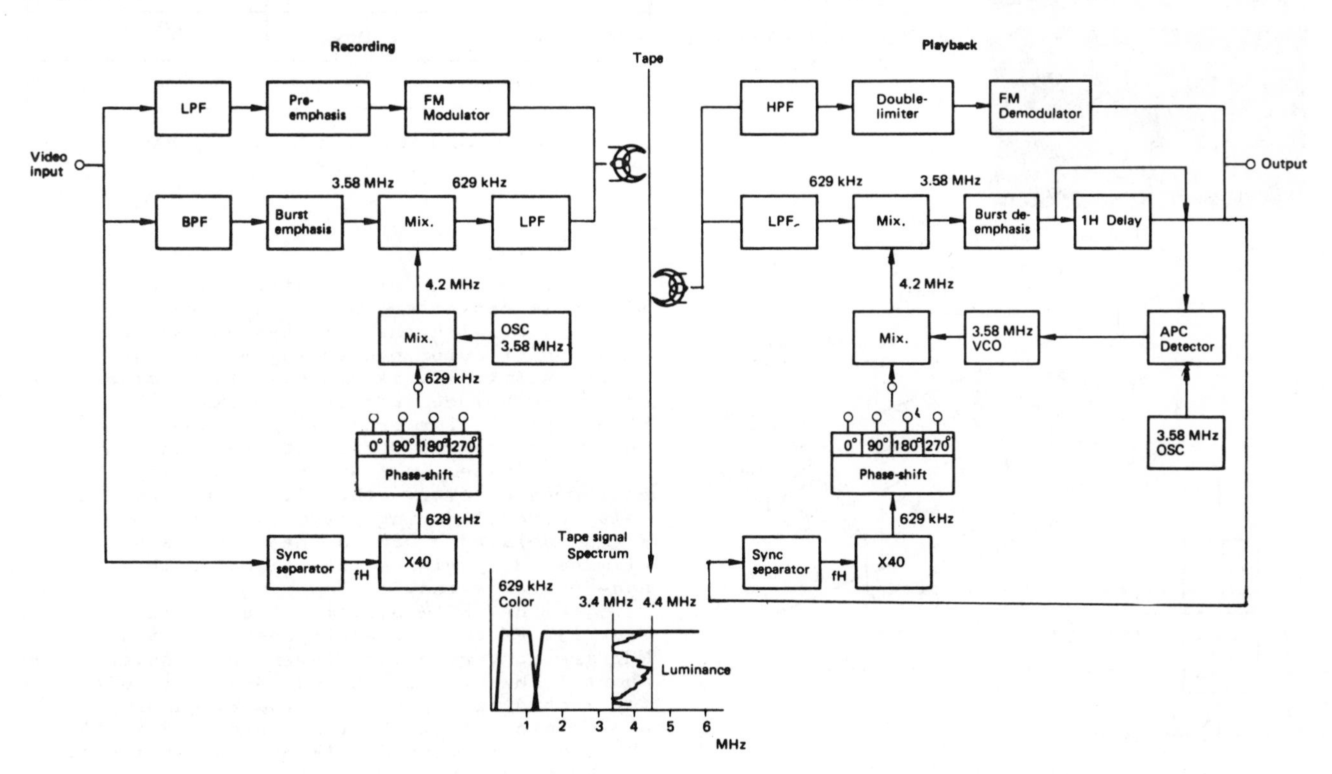

Fig. 6 Block Diagram of Video signal circuit

The center frequency of the low-frequency converted chrominance signal is 629 kHz and the maximum crosstalk in this frequency range amounts to 25 %, with no improvement made by the slanted azimuth recording system. Therefore, another countermeasure had to be devised to cope with this problem.

Fig. 11 shows a color signal processing method for attaining elimination of crosstalk of the chrominance signal. During recording, the chrominance signal is phase-shifted at the point of the horizontal sync signal by 90° forward on one video track and then by 90° backward on the adjacent video track (or field). During playback, reverse phase-shift is applied to restore the original phase relationship and then a signal which is delayed by 1H is added to the phase-shifted chrominance signal, thereby to eliminate crosstalk.

In order to improve color convergence, the burst signal is boosted by 6 dB during recording and lowered by the same 6 dB during playback.

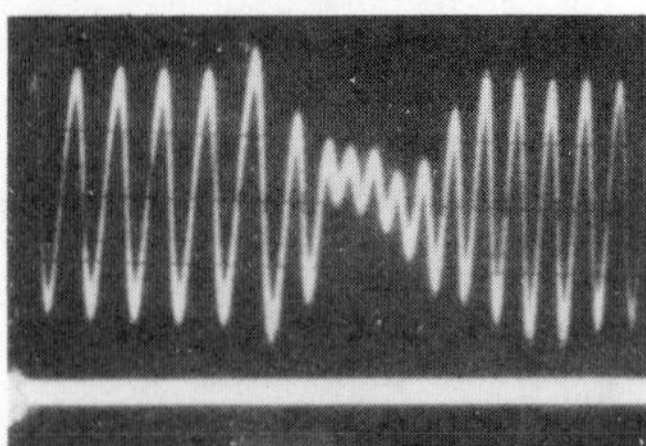

Fig. 8 Input

Fig. 9 Rated limiter output

Fig. 10 Double limiter output

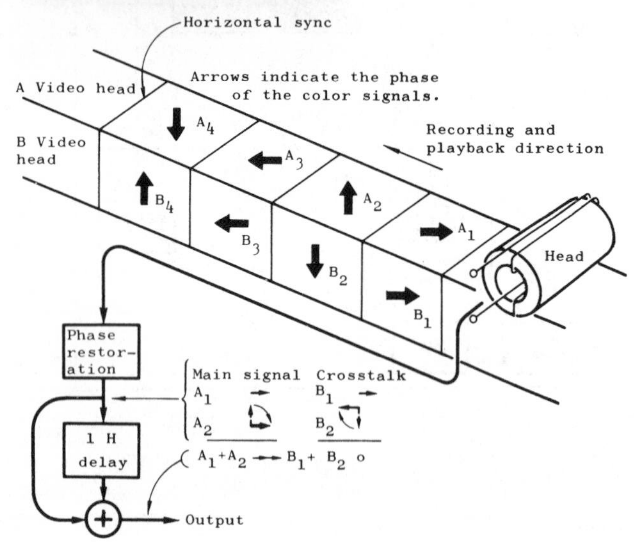

Fig. 11 Color signal processing in the NTSC system

Application to different television systems

In European areas, the number of fields per second is 20 % smaller. Considerations to this fact were given at the very beginning of the VHS development, as attested to through those previously mentioned techniques towards realizing high-density recording. Consequently, it is possible to adapt the number of revolutions of the drum to the number of fields. The resulting writing speed is 4.87 m/sec and the minimum recorded wavelength is 1.0 μm, which have been already applied to practical use. The specifications of the VHS system for the PAL and SECAM standards are shown in Table 2.

Item	Units	Specifications
Continuous recording time	hr	3
Tape width	mm	12.6
Tape thickness	μm	20
Tape speed	mm/sec	23.39
Tape consumption	m^2/hr	1.06
Drum diameter	mm	62
Writing speed	m/sec	4.87
Video track width	μm	49
Tape coercivity	Oe	600
Carrier frequency	MHz	3.4 - 4.8
Minimum recorded wavelength	μm	1.0
Head azimuth	Deg.	$\pm 6^{\circ}$
Cassette size	mm	188 x 104 x 25

Table 2 Details of the VHS PAL and SECAM system

For the PAL color system, the color signal is converted into a low-frequency carrier, as with the NTSC system, and a phase-shift color processing method, as shown in Fig. 12, is employed to eliminate crosstalk between adjacent tracks. The phase-shift color processing method for the PAL system is different from that for the NTSC system in that phase-shifting is performed on every other field and that the delay circuit during playback provides a signal delayed by 2H so that crosstalk is eliminated by mixing this signal with a non-delayed signal.

In the SECAM system, as shown in Fig. 13, FM color signals centered at 4 MHz are converted into lower frequencies of about 1 MHz by the 1/4 count-down circuit for recording. During playback, quadruples are obtained from the 1 MHz band. Fig. 14 shows a maximum crosstalk between adjacent tracks. Crosstalk occurring in the 1 MHz band is less than 15 %. Also, the FM signals are not likely to be subject to interference. These factors enable quality video signals to be obtained.

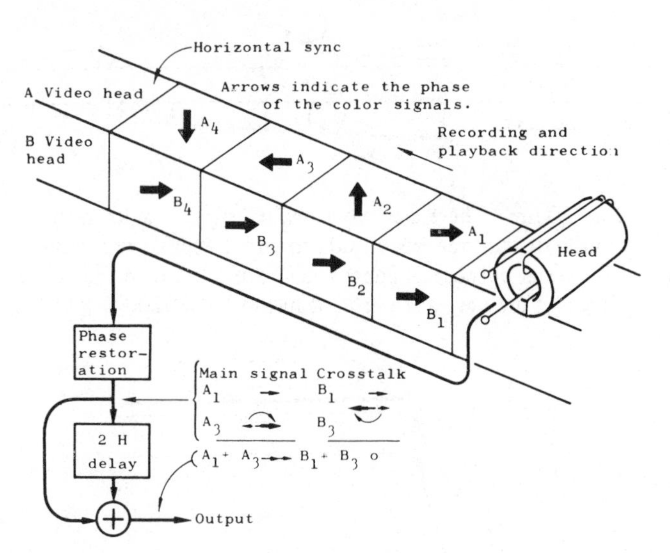

Fig. 12 Color signal processing in the PAL system

The tape pattern is so constructed that a signal continuity is maintained as in the NTSC system, allowing the possibility of variable speed playback. The two video heads are arranged at an interval of 180°. Differences from the NTSC system are found only in the tape speed, the drum speed and the color signal processing circuit; with most of the various components conforming to all standards within the VHS system.

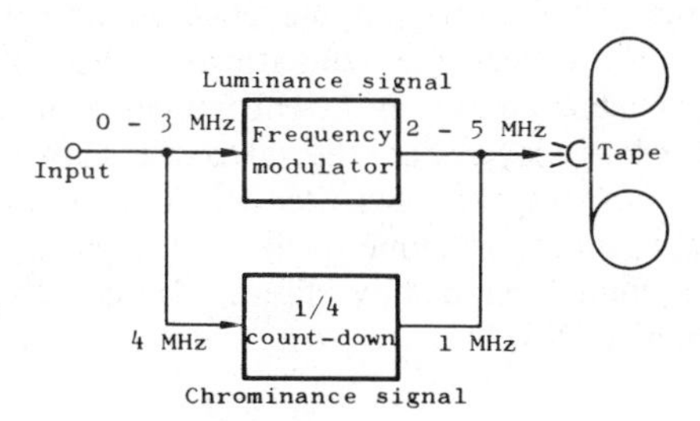

Fig. 13 Color signal processing in the SECAM system

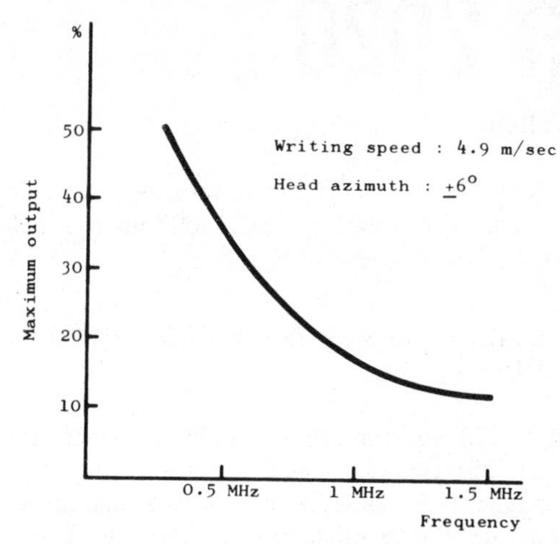

Fig. 14 Crosstalk frequency characteristic

Application expansion potentials

The VHS system is favored with reserves for future expansion. One example is that in the NTSC configuration a 4-hour continuous recording function can be added as an extra feature. Another example is a model which is equipped with still-frame and double-speed playback facilities. Clean, noise-free pictures are obtainable even in the double-speed playback and the sound is sufficiently discernible through employment of the technique of a time base extension.

As a third example, a VHS portable version has already met with completion. Its size is 338 mm wide, 328 mm deep and 132 mm high and its weight is 9.3 kg including battery pack and cassette.

Conclusion

The VHS system evolved through a "Total Concept Method" aiming at meeting various divergent requirements including those from the viewpoint of flexible applications. The minimum recorded wavelength of 1 μm has come to be a reality and techniques for its realization have already been established. The VHS total design was conducted for accommodating a long-time recording of more than 2 hours and its specifications are capable of being adapted to various applications subject to future expansion. A recent analysis has also been made into the feasibility of a variable speed playback employing the slanted azimuth recording system.

The Philips Video Cassette Recorder VR 2020

1. INTRODUCTION

The VR 2020 is the first video cassette recorder to operate on the basis of the new "VIDEO 2000" quarter inch Video Recording System.

2. THE PRINCIPLES OF THE VIDEO AND SOUND RECORDING

The VR 2020 works in accordance with the helical scan principle, with two video heads mounted on a rotating drum. Here the magnetic tape is houses on two adjacent spools in the newly developed video compact cassette. The head drum, which has a diameter of 65 mm, is made up of two parts. The bottom part is incorporated in the tape drive, while the upper part contains the two video heads. The tape, which follows an M-shaped path through the recorder (see Figure 1) is run through the recorder at a speed of 2.44 cm/s. The

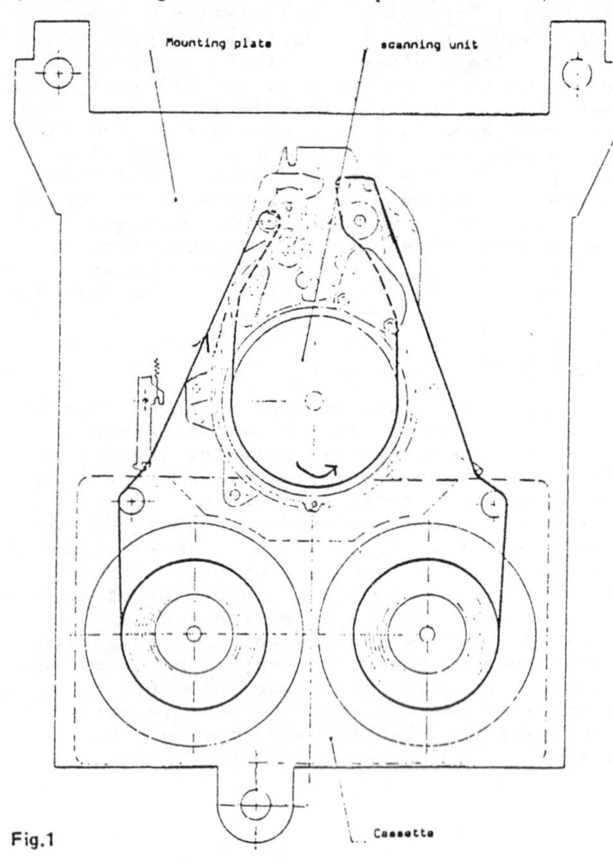

Fig.1

reading or writing speed of the video heads is 5.08 m/s, the direction of rotation of the video heads being the same as the direction in which the tape is transported through the recorder (see Figure 2). The reversibility of the cassette is obtained by using the ½-inch tape in two separate parts along its length (Figure 3). The width of the video tracks, which lie next to each other, is 22.6 μm. The audio tracks are located at the side edges of the tape and each has a width of 650 μm (see Figure 4). The "Video 2000" system offers the possibility of stereophonic sound reproduction (not yet in the VR 2020), the width of the audio tracks in that case being

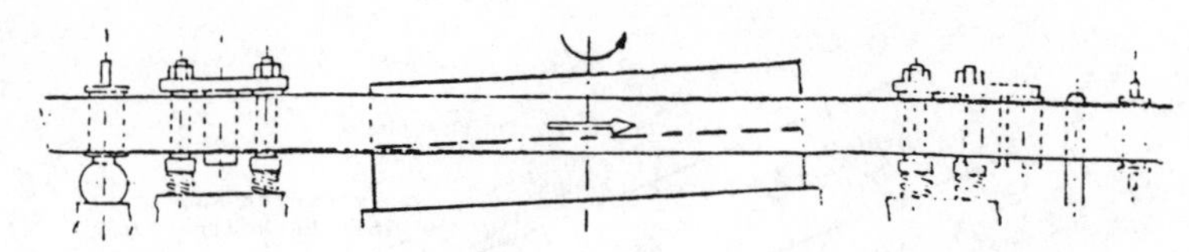

Fig.2

2 x 250 μm. There are two auxiliary signal tracks in the middle of the tape, each 300 μm wide, which are reserved for special purposes. These tracks are present in the system, although they are not yet used in the VR 2020.

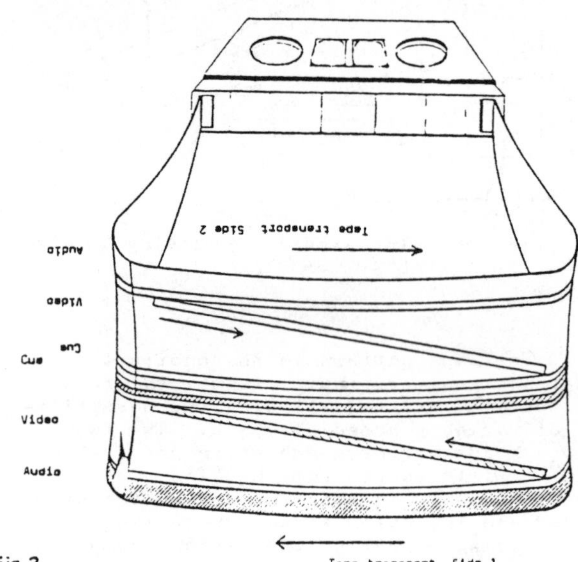

Fig.3

Degradation of picture by crosstalk between adjacent tracks is minimised during playback by, amongst other things, incorporating a special filter consisting of a 128 μs delay line with an adder circuit, in the chrominance circuitry. As in the VCR long play system the video heads are angled at $+15^\circ$ and -15° respectively (tilted gaps). In addition, special attention is paid to the sound quality, through the application of a new system developed by Philips, "Dynamic Noise Suppression" (DNS).

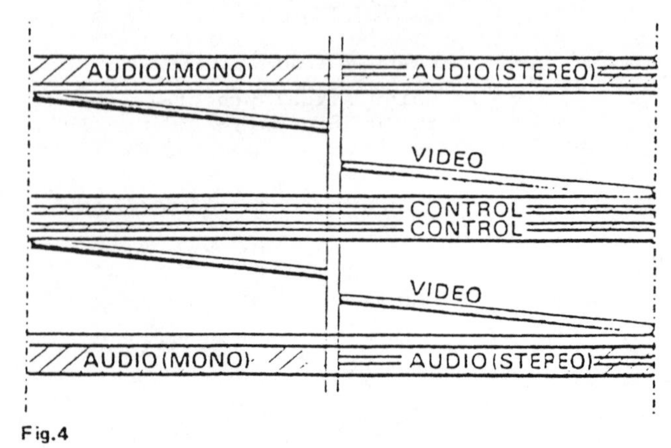

Fig.4

3. SELF-CORRECTING VIDEO HEADS

The video heads are placed on small plates of piezo-ceramic material (PXE) which are attached to the upper drum unit

Reprinted with permission from *Electron. Technol.*, vol. 13, pp. 233-236, Nov./Dec. 1979.
Copyright © 1979, Society of Electronic & Radio Technicians.

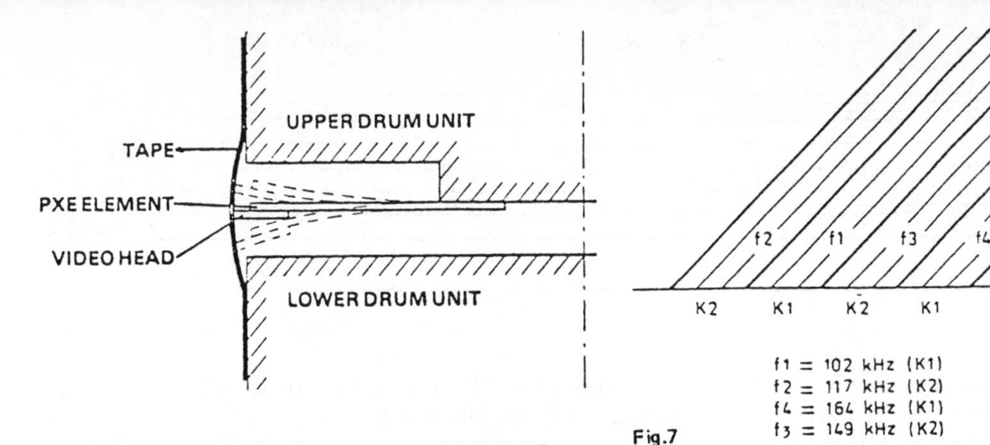

Fig.7

PXE ELEMENT SERVO CONTROLLED

Fig.5

(Figure 5). In this way it is possible to move the video heads vertically by applying a specific voltage to the PXE material. The amount of movement is directly proportional to the voltage level on the plates. This method of control ensures recorded cassettes are interchangeable with other Video 2000 system machines. It also enables slow motion, fast motion, and still frame facilities without interference and degradation of picture quality.

4. THE DYNAMIC TRACK FOLLOWING SYSTEM

The system of self-correcting video heads requires precise electronic control of the video heads on the relevant tracks. This control is achieved by means of a special circuit, indicated by the letters DTF, which is an abbreviation of Dynamic Track Following. This system is used both during recording and playback. The DTF system works in the following manner. In the frequency range below the 625 kHz chrominance carrier (Figure 6), during "recording", supplementary signals with a constant frequency are written by the video heads K_1 and K_2 into the video track on the tape, these signals occuring in the sequence:

K_1 with f_1 = 102 kHz

K_2 with f_2 = 117 kHz

K_1 with f_4 = 164 kHz

K_2 with f_3 = 149 kHz.

During playback, the relevant video head scans the tracks written during recording. Here, apart from the desired video signal and the auxiliary signal present due to crosstalk, signals originating from the auxiliary signals which are written in the

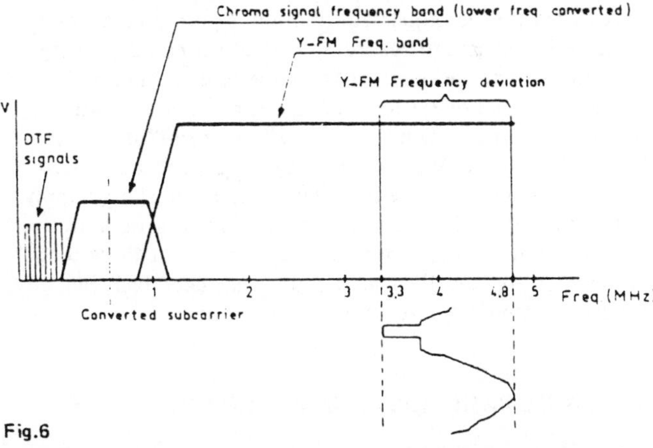

Fig.6

adjacent tracks also arise. By mixing these crosstalks signals are produced which are ultimately converted into a control voltage. If, for example, the video head K_1 is running over the track with f_4 (164 kHz), but deviates a little too far downwards, then this results in a signal with a difference frequency of $\Delta f = f_4 - f_2 = 164 - 117 = 47$ kHz (Figure 7). If on the other hand the head K_1 deviates more in an upward direction, then a signal with a difference frequency of $\Delta f = f_4 - f_3 = 164 - 149$ kHz = 15 kHz, is produced. Thus for a downward deviation, an auxiliary signal with a high difference frequency is produced and for an upward deviation, one with a low difference frequency. For the other video head, an upward deviation results in an auxiliary signal with a high difference frequency and a downward deviation results in an auxiliary signal with a low difference frequency. By now comparing the difference signals with each other, it becomes possible to keep the video heads always in the right position with regard to the written tracks.

During the period in which the video head is not in contact with the tape, the actuator is pulled back to the average starting position during scanning of the preceding track. As soon as the head is back in contact with the tape again, it is back in the correct starting position.

The head drum is heated to minimise changes in head drum diameter, thus ensuring head wear and frictional losses due to tape sticking to tape are reduced to a minimum.

5. AUTOMATIC TRACKING

If, during playback, the two video heads show an equal deviation in the same direction, this can be remedied by moving the tape. For this, the control voltage coming from the DTF discriminator is fed to the tape servo, with the result that the tape is moved relative to the head and correct alignment is restored.

6. THE SELF-CORRECTION OF THE HEADS DURING RECORDING

During recording, the heads must be lined up in exactly the right position with respect to each other, because the video tracks lie directly next to each other without a guard band. However, due to various tolerances, for example the hysteresis of the piezo-ceramic actuator material, no permanent static condition exists between the two video heads during recording, so dynamic control is necessary. Consequently, during recording, one actuator is placed in the nominal position and the other is included in a control loop. The required control voltage is obtained through periodic measurements at the beginning of each track within the vertical blanking interval. The amplitude of a wave train f_5 is measured which is written into the track during 1.5 lines with a frequency of approximately 223 kHz (a–b in Figure 8). Writing is then interrupted during the following 1.5 lines (b–c in Figure 8).

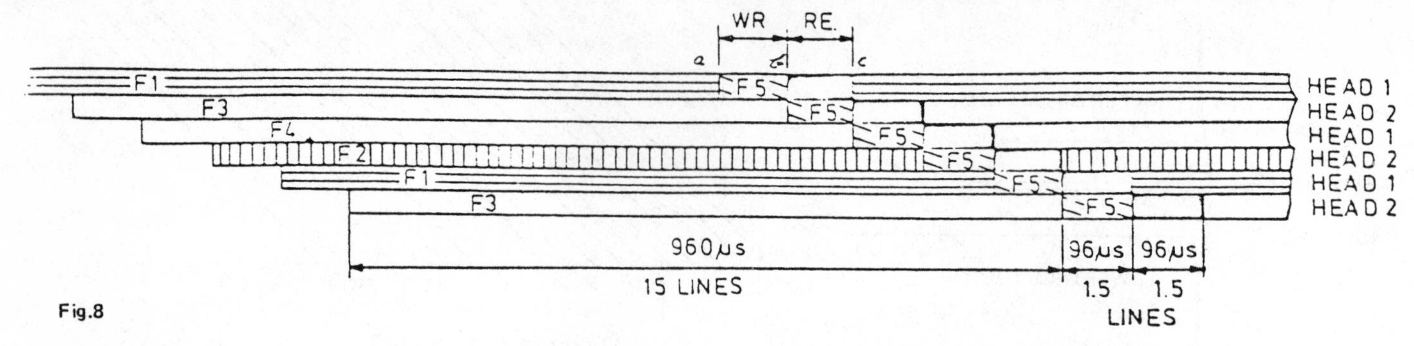

Fig.8

During this time, the video heads are switched over to "read" and register the amplitude of the preceding track with wave train f_5. The control voltage for the actuators is obtained from the result of the measurements from head K_1 and head K_2.

7. TAPE SERVO

The magnetic tape is run through the recorder at a constant speed of 2.44 cm/s. To achieve this the capstan is directly connected to the motor spindle, on which a 216–pole tacho-generator is mounted (direct drive). The speed of the motor is 116.4 rpm. For recording, the required speed is obtained by making a comparison between the phase of the tacho frequency and a fixed reference frequency of 419.376 Hz produced by a quartz-controlled oscillator. For playback, instead of the phase comparison, the control voltage of the automatic tracking system is used to move the tape into the required phase position.

8. HEAD SERVO

The top part of the drum (head disc) is also driven directly, the motor being coupled to a 125–pole tacho-generator. The speed of the head disc is 1500 rpm. For comparing the phases, a phototransistor coupled to the head disc emits one pulse per revolution, or 25 pulses per second. These pulses are compared to a reference frequency of 25 Hz. During recording, the 25 Hz reference frequency is obtained by dividing the vertical frequency; during playback it is produced by a quartz-controlled oscillator.

9. SEARCH TUNING

Once the "search" button has been pressed, the recorder automatically scans all TV bands. When a station is found, accurate tuning takes place automatically. The tuning obtained can then be stored in the memory by means of the "store" command, followed by a channel number to be chosen by the user. A maximum of 26 programmes can be stored in the memory in this way.

10. PROGRAMMABILITY

The VR 2020 has five memory blocks for storing programmed recordings. Each block can contain: the starting time, the stopping time, the day of recording and the relevant channel number. Programming can be done up to a maximum of 16 days in advance. A warning that the recorder has been programmed in this way, is given by the indication "attention timer". In addition, the cassette lift is opened if there is no cassette or a protective cassette in the machine, to draw the user's attention to the fact that he cannot record under these circumstances.

11. PROTECTING THE CASSETTE AGAINST UNINTENTIONAL ERASURE

The cassette has a separate switch for each side which can be set in one of two positions. In the first position recording is possible. In the other position, marked by a red indicator, recording and erasure is not possible.

12. AUTOMATIC REWINDING

The recorder automatically winds back to the beginning of the tape if the "auto rewind" button is pressed at the end of the pre-programmed recordings.

13. ELECTRONIC TAPE COUNTER

The VR 2020 is equipped with a four-digit electronic tape counter. To return to a certain position on the tape quickly, the "go-to" button is pressed and the required counter number is keyed in on the keyboard. The recorder then automatically winds on or back to the requested position.

At the beginning of the tape, the counter is set automatically to zero. If required the counter can also be set manually to zero.

14. REMOTE CONTROL

The recorder can be operated remotely for channel selection and all tape drive functions by means of a detachable infrared transmitter and receiver system.

15. MICROPROCESSOR SYSTEM

The VR 2020's simplicity of operation is attributable to its use of an advanced microprocessor. If any key is pressed at random, a check is first made by the microprocessor program to see whether the requested function is permissible, for instance in the case of recording with a protected cassette. This means that the recorder is completely safeguarded against incorrect operation. Thus it is possible, to switch over directly from "wind" to "play", or from "off" to "record". All the necessary intermediate steps are carried out automatically by the recorder. If the electricity supply is interrupted, all the transmitter tunings and programme orders stored in the memory are kept for about three months. Moreover, after the electricity supply has been restored, any function which was interrupted is continued.

16. WIDE-BAND ANTENNA AMPLIFIER

The recorder contains a wide-band antenna amplifier, which feeds off air signals to the TV receiver without losses, together with the recorder signal.

17. SCANNING UNIT

This unit has been designed in such a way that the precise positioning of video head, audio heads and tape guides can be automatically pre-adjusted in the factory.

This ensures that service replacement as an assembly can easily be carried out. The unit is made up of a number of sub assemblies which can also easily be replaced.

The scanning unit consists of: –

(a) The micro-chassis: this is a precision formed pressure die-cast component, to which are fixed the tape guiding elements, audio head, erase head and the end positions of the threading-in elements.

The micro-chassis also has a reference plane so that the head disc motor can be accurately positioned should replacement be necessary.

(b) The head disc motor. This head disc motor is integrated with the stationary tape drum, on which is mounted the extremely precise straight edge.

(c) The head disc. This highly accurate unit comprises the video heads and the head actuators. The special feature of the head disc is a clamping construction with which it can be accurately assembled on the motor spindle.

18. CONCLUSION

With the unwrapping of the V.C.C. (Video Compact Cassette) by Philips and also Grundig, a united European front has been presented to Japanese competition. Only time will decide the outcome of the ensuing battle. The V.C.C. format has much on its side, low tape speed with good video and sound reproduction, and a format that is upwards compatable. That is recordings made on this generation of V.C.C.'s will still be playable on more elaborate machines from the V.C.C. stable in the future.

While this is yet another standard on the market this is inevitable if development is not to be frozen. The tape economy resulting from the V.C.C. system will keep the cassette competitive with video disc and this must be a good thing for the consumer.

19. ACKNOWLEDGMENT

This paper is based largely on material supplied by Philips and permission to reprint is gratefully acknowledged.

DEVELOPMENT OF AN EXTREMELY SMALL VIDEO TAPERECORDER

M. Morio, Y. Matsumoto, Y. Machida,
Y. Kubota and N. Kihara
VTR Engineering Development Division
Sony Corporation
Shinagawa-ku, Tokyo 141, Japan

A New Compact Video Movie Unit

1. Introduction

We have developed a video recording system for consumer use in which a color video camera and a video recorder are integrated into one handy portable unit. The unit comprises a solid-state micro CCD camera[1-3] which employs a single image-sensor chip and a small videocassette recorder which can record up to twenty minutes on a video microcassette employing newly-developed high-energy tape. Figure 1 is a photograph of the new unit, which measures 191 × 171 × 60 mm and weighs two kg, including an F:1.6, 3 power zoom lens and a rechargeable battery pack which can power the unit for forty minutes.

The unit is one part of a Video Movie system, the other part of which is a Home Editor unit. When the Video Movie unit is connected to the Home Editor unit the audio and video can be played back on any TV set and can be edited on any standard videocassette recorder.

Because it is inevitable that the picture quality suffers to some extent when the tape is edited from the Video Movie unit to a video taperecorder, the quality of the playback picture is higher than that of a conventional VTR to make up for this loss.

We decided to employ a two-rotating head helical-scanning drum configuration because, in addition to allowing high-density recording, a two-rotating head drum system has these advantages over a longitudinal recording system: separately-recorded scenes can be played back without a break, the circuitry is relatively simple and the power consumption is relatively low. A two-rotating head system has these advantages over a single-rotating head system: both the cassette and the tape-threading mechanism are relatively simple.

Fig. 1 The Video Movie unit and the micro videocassette.

As our primary goal was a highly compact unit, it was necessary to develop a more compact tape-transport system, a micro videocassette and a technique for packaging electronic components with very high density, and to reduce the size of the drum. If the size of the drum was to be reduced, a short-wavelength recording technique would have to be developed.

2. The Short-Wavelength Recording Technique

First, we set a target of reducing the drum diameter to a half of the diameter of the Betamax drum, which is equivalent to reducing the head-to-tape speed to 3.5 m/sec and the recording wavelength to $\lambda = 0.74\,\mu m$ at the white peak of the FM carrier. Figure 2 compares the frequency response of the RF signal at $V_h = 3.5$ m/sec and that of the Beta-III format using CrO_2 tape.

From the experimental formula reported by Muramatsu,[4] the reproduced head output level E_h is given by

$$E_h = a \cdot W \cdot V_h \cdot B_r \cdot L\,(B_r/H_c,\ s/\lambda,\ g), \quad (1)$$

where W: track width

V_h: head-to-tape speed s: head-to-tape spacing
B_r: remanance g: head gap length
H_c: coercivity a: constant.

The loss function $L\,(B_r/H_c,\ s/\lambda,\ g)$ can be expressed as

$$L\,(B_r/H_c,\ s/\lambda,\ g) = L_d \cdot L_s \cdot L_g, \quad (2)$$

where L_d is the self-demagnetization loss, L_s the spacing loss and L_g the gap loss. Here L_d is the idealized self-demagnetization loss at $g = 0$ and $s = 0$. The loss at $g \neq 0$ and $s \neq 0$ is included in the spacing loss L_s.

From Fig. 2 it is evident that at 1 MHz the ratio of the output level between the two speeds is −6 dB, which is directly dependent on the speed ratio. This is because the loss function is negligible in the longer wavelength region. In the shorter wavelength region, on the other hand, the ratio of the output levels at 5 MHz is −13.1 dB. The additional loss −7.1 dB from 1 MHz can be attributed to the increase in the spacing loss, which is equivalent to 0.11 μm spacing. The gap length of a reproducing head at $V_h = 3.5$ m/sec is 0.25 μm, which is half the Betamax gap length. Both speeds produce the same gap loss.

Reprinted from *IEEE Trans. Consum. Electron.*, vol. CE-27, pp. 331-339, Aug. 1981.

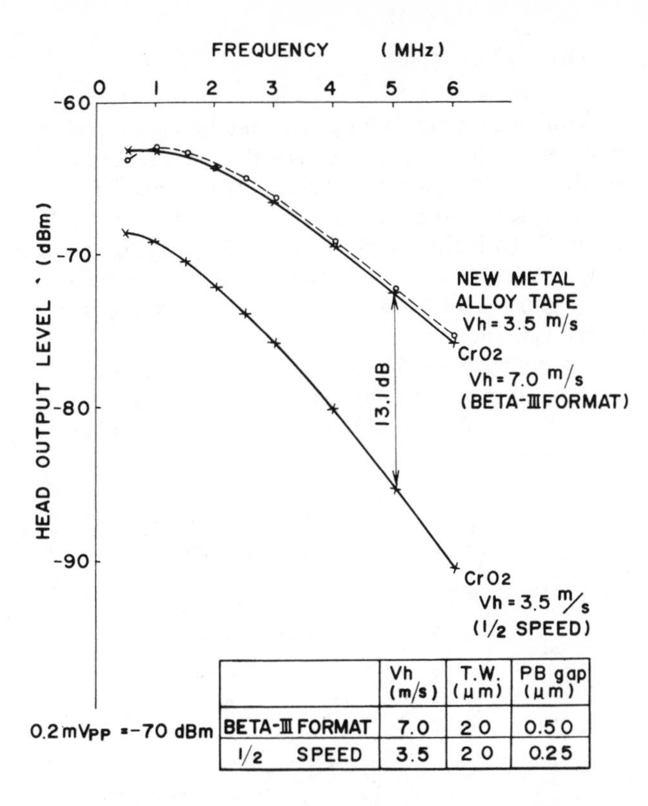

Fig. 2 Frequency response of CrO_2 ($V_h = 7.0$ m/sec), CrO_2 ($V_h = 3.5$ m/sec) and new metal alloy tape ($V_h = 3.5$ m/sec).

This simple experiment makes clear the difficulties of short wavelength recording: the decrease in the differential gain of flux changes caused by the reduced head-to-tape speed and the increase of spacing loss caused by the shortened wavelength. Shortening the wavelength also affects the signal-to-noise ratio, since the number of magnetic particles in one wavelength is decreased. Required characteristics of the tape to be used in a micro videocassette recorder are:

1. High B_r and high output.
   (H_c in proportion to B_r.)
2. Smooth surface.
3. High S/N with fine magnetic particles.

2-1. The New High-Energy Video Tape

We have employed a metal alloy tape with high remanance B_r and high coercivity H_c for the first time in a video application.

Let us discuss the head output of this tape. The ferromagnetic metal powder possesses magnetic characteristics of $B_r = 2000$ to 3000 G and $H_c = 1000$ to 2000 Oe, values which are two to three times larger than those of oxide powders. An increase in B_r brings about higher output and an increase in H_c reduces the self-demagnetization loss and the spacing loss[5]. The ratio B_r/H_c should be determined to minimize the loss function.

The dependence of L_d and L_s on B_r/H_c was calculated for $s = 0.06\,\mu m$ and $0.11\,\mu m$ and $\lambda = 0.7\,\mu m$ using the experimental formula of Muramatsu.[4] (See Fig. 3.) It was found that the dependence becomes steep when $B_r/H_c > 2$. Using Eq. (1), the output contour map for B_r and H_c was calculated as in Fig. 4. It is clear that if the head output is to be increased, H_c should relate to B_r as $1 < B_r/H_c < 2$. Because of the head saturation during recording, however, a higher H_c is undesirable. We determined the characteristics of the newly-developed metal alloy tape to be $B_r = 2400$ G, $H_c = 1200$ Oe. ($B_r/H_c = 2.0$ is similar to that in conventional oxide tape.) In the new tape, the output at $\lambda = 0.7\,\mu m$ was 6.8 dB higher than in CrO_2 tape when both had the same spacing loss.

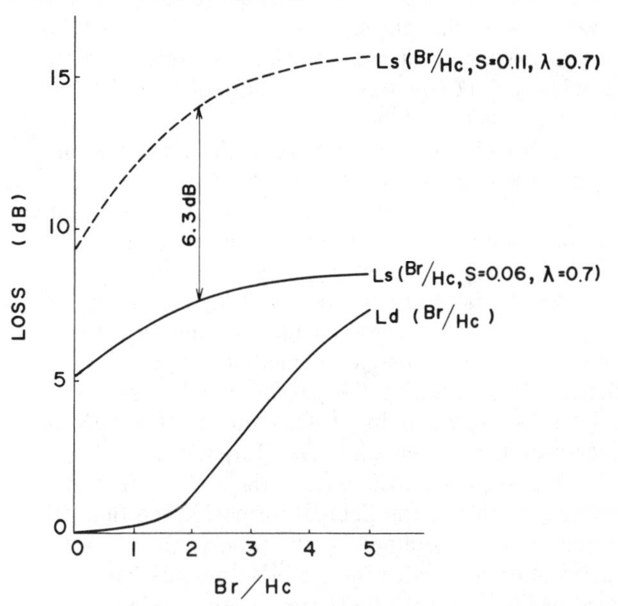

Fig. 3 B_r/H_c dependence of Self-demagnetization loss L_d and spacing loss L_s.

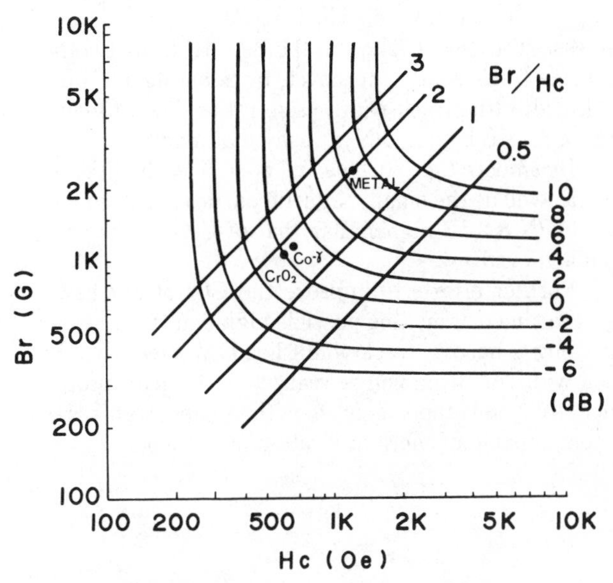

Fig. 4 Output contour map. ($s = 0.06\,\mu m$, $\lambda = 0.7\,\mu m$)

Moreover, as is shown in Fig. 3, we reduced the head-to-tape spacing s from 0.11 μm to 0.06 μm, resulting in a 6.3 dB reduction in the spacing loss.

The broken line in Fig. 2 shows the frequency response of the newly-developed metal alloy tape, showing the same output as the Beta-III format. The output is 13.1 dB higher at 5 MHz compared with the output of CrO_2 tape because of the higher B_r and H_c and the smoother tape surface.

After increasing the head output level over the total noise (consisting of the amplifier, erase and rubbing noise), we must then increase the signal-to-noise ratio of the tape itself. The tape modulation noise near the carrier frequency is closely related to the smoothness of the tape surface; further away from the carrier frequency the tape modulation noise is more closely related to the size of the magnetic particles, the packing density and dispersion.

C/N is defined by the ratio of the carrier output level to the noise level and is given by[6]

$$C/N = (k \cdot n \cdot W \cdot f_c)^{\frac{1}{2}} \lambda, \qquad (3)$$

where n: number of particles per unit volume

f_c: carrier frequency k: constant.

As the depth of the recorded layer in a tape is proportional to the wavelength λ, the effectively recorded volume of one wavelength in a tape is proportional to λ^2. Assuming the particle noise is gaussian, C/N is proportional to λ. This means that C/N is decreased 6 dB when λ is reduced by a half.

We set a goal of making the C/N in this unit better than that of the Beta-III format. In the Beta-III format, $C/N_{mod.}$, defined by the ratio of the 4.5 MHz carrier output level to the 3.5 MHz modulation noise level, is 50 dB at the 10 kHz resolution bandwidth.

By using small magnetic particles, which are one fourth of the volume of CrO_2 tape, the newly-developed metal alloy tape attains 6.2 dB higher $C/N_{mod.}$ than CrO_2 tape, which means the new tape possesses the same $C/N_{mod.}$ as the Beta-III format with the same track-width. Moreover, by adopting a 37 μm track-width (the track-width used in the Video Movie unit), a 2.7 dB higher $C/N_{mod.}$ can be attained.

By smooth tape surface and good particle dispersion, as well as the small magnetic particles, a chroma signal C/N 8.2 dB higher than that of CrO_2 tapes has been achieved.

Further efforts to improve the C/N of our new tape are aimed at making possible high-density recording using a narrow track-width head. We feel that a track-width of 10 μm will be realized in the near future when this new tape is used with a tape transport system capable of more accurate tape transport.

2-2. The Video Head

Available tape coercivity is limited by the saturation of the video heads during recording. To record a tape, the recording magnetic field must exceed the saturation field of the tape H_s, which is normally two or three times higher than H_c. A narrow gap length reduces the recording loss but the magnetic flux density at the gap cannot exceed the saturating flux density B_s of the head material.

Figure 5 shows the calculated saturation limit[7] for the metal alloy tape with $H_c = 1200$ Oe. It is clear that in order to adopt a narrower gap length, a higher B_s is required.

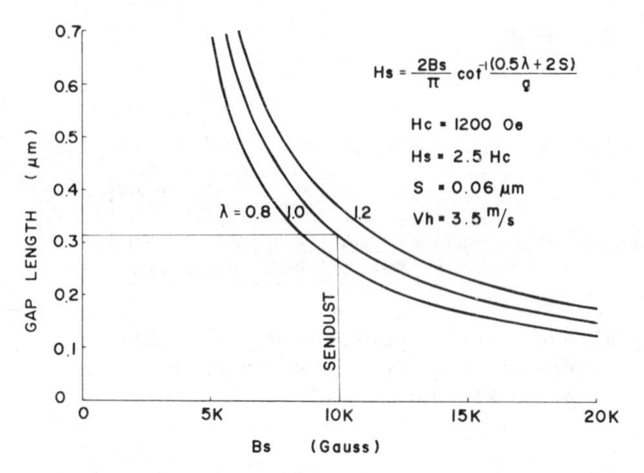

Fig. 5 Head saturation effect of recording head.

For a system in which the chroma signal is converted to a lower band and recorded with a frequency-modulated luminance signal, head saturation increases the third order distortion so that the recording current and the output of the chroma signal are restricted.

In the Video Movie unit, by utilizing high-B_s sendust[8] heads for recording only and by adopting a 0.35 μm gap length, the head saturation effect is eliminated.

The gap length for a reproducing head must satisfy opposing requirements for high reproducing efficiency and low gap loss. Figure 6 shows the gap length dependence of the product of the reproducing efficiency and the gap loss. From Fig. 6 it is clear that a gap length of between 0.22 and 0.25 μm is appropriate for maximizing the FM carrier output at 5 MHz, a range which yields the same gap loss as a Betamax unit.

A two-head system using two recording/reproducing heads is now under development. Because of their high B_s and high permeability μ, the use of amorphous heads in this system is highly likely.

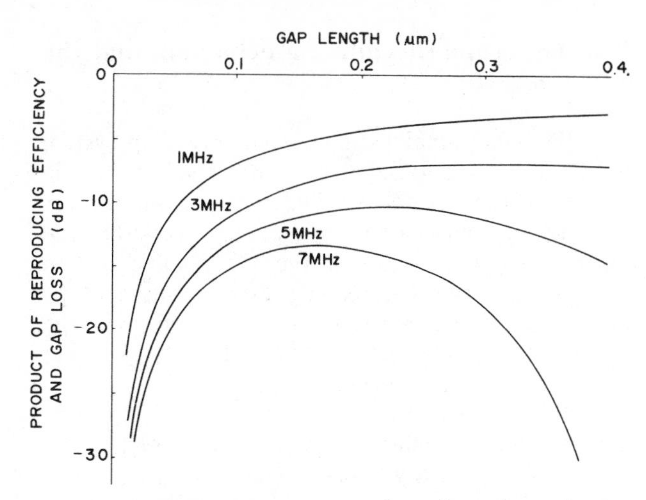

Fig. 6 **Relation between gap length and product of reproducing efficiency and gap loss (ferrite head).**

3. The Tape-transport Mechanism

3-1. The Drum System

In short-wavelength recording, the drum system must insure a small and stable head-to-tape spacing.

The drum systems employed in conventional video taperecorders are:

1. Upper Drum Rotating System (UD)
2. Middle Drum Rotating System (MD)
3. Head-on-Propeller Rotating System (HOP)

We modified the MD and HOP systems to be applicable for this unit. The modified system is named the Head-on-Disc Rotating System (HOD). Figure 7 illustrates these four drum systems.

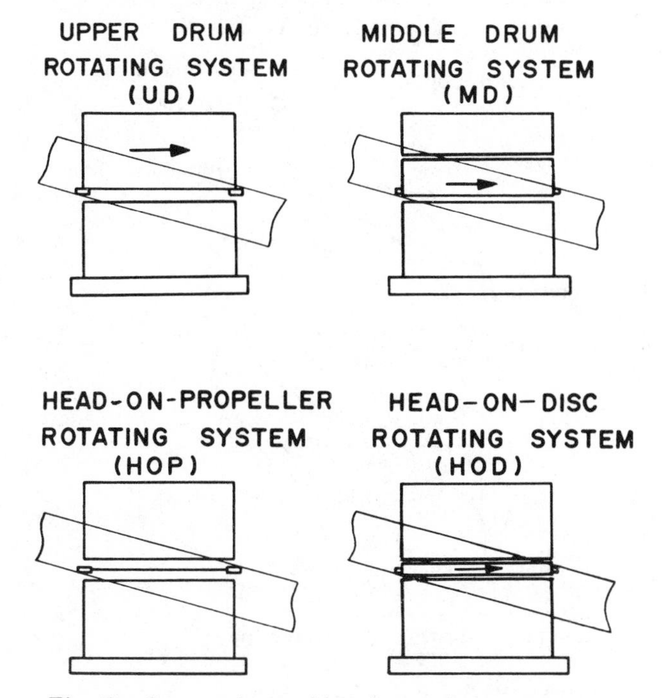

Fig. 7 **Comparison of head rotating systems.**

In the UD and MD systems the thickness of the dynamic air film is not uniform. It is known that the thick air film at the point where the tape first contacts the drum decreases exponentially as the tape goes around the drum.

Conventional video taperecorders have solved this problem by making the video heads protrude and by using high tape tension. This solution has never been totally acceptable because high tape tension creates tape stick slip, and it is becoming increasingly unacceptable as the trend toward even thinner tape makes low tape tension a necessity.

The new HOD system was designed to assure a uniform air film around the drum even though tape tension is low. To investigate the degree of uniformity of the air film we measured the output level at $\lambda = 0.6\,\mu m$ when the head contacts the tape and calculated the ratio of 80% output time to full time. Figure 8 compares the contour map of this ratio for the variables of tape tension and head protrusion in the HOD system with that in the UD system. It is clear that the HOD system has a wider stable area, which makes this system more applicable at a low tape tension.

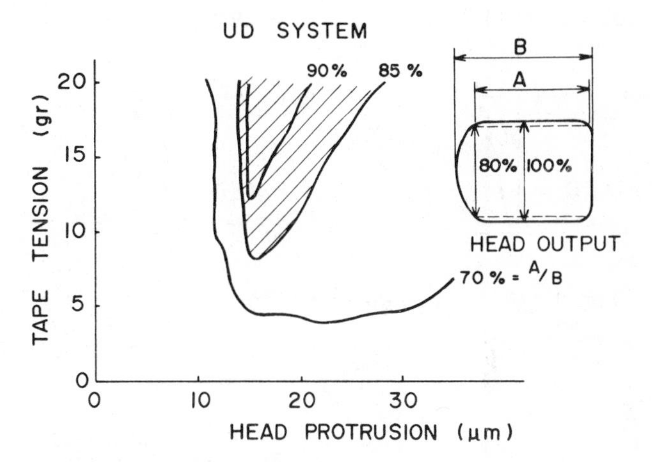

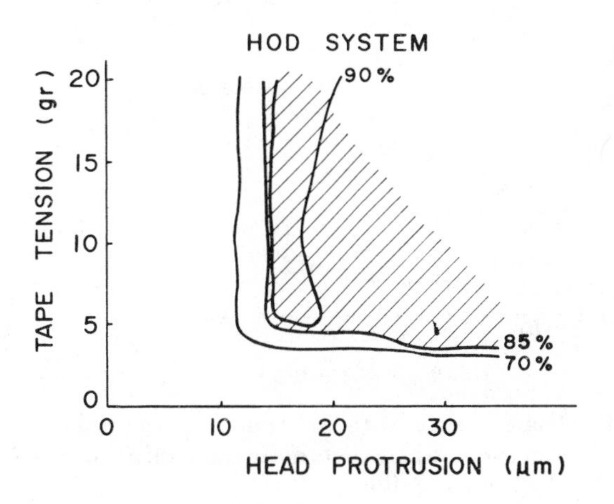

Fig. 8 **Contour map of 80% head output time relative to full time.**

It should be noted that, in the HOD system, instead of a dynamic air film, a static air film is used to reduce the friction between tape and drum. Each time the heads lift the tape, air accumulates in the space delineated by the tape, the upper and lower fixed drums, and the rotating disc. The static air film was measured by the optical method and was found to be thin (2 to 3 μm) but uniform in both the transverse and longitudinal tape directions.

Figure 9 shows our measurements of the air film. In analyzing the dimensions of the drum system, we found that the thickness of the static air film as well as the increase in tape tension around the drum is heavily dependent on the radius of the rotating disc. Figure 10 indicates that the smaller the space between the tape and the disc, the easier the air film can be formed.

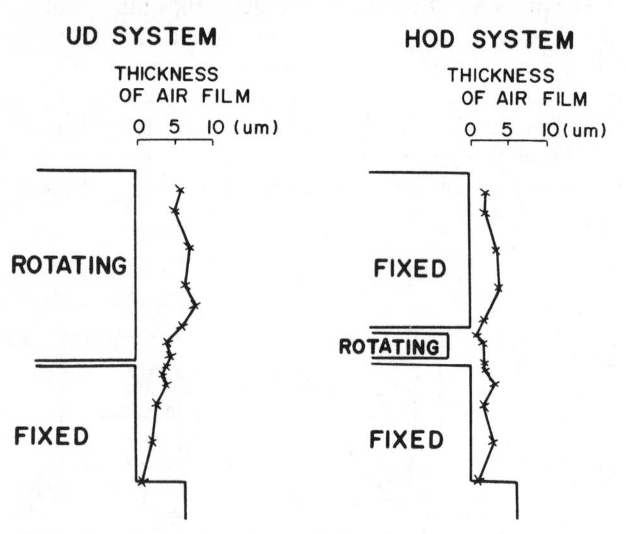

Fig. 9 Cross-sections of air film

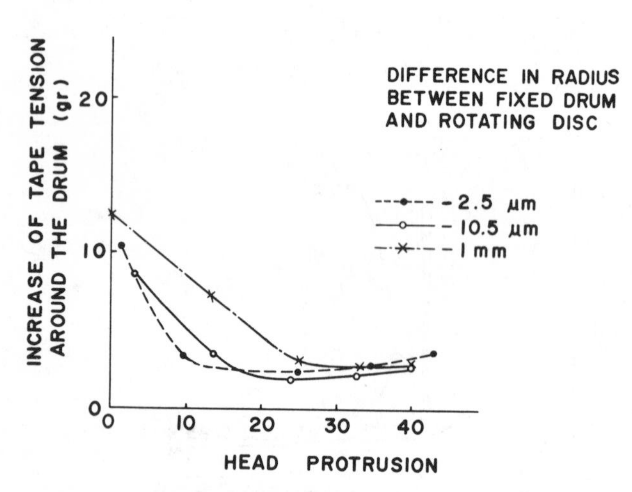

Fig. 10 Dependence of the increase of tape tension around the drum on the radius of the rotating disc.

3-2. The Tape-threading Mechanism and the Cassette

This unit employs the U-type threading system used in U-matic and Betamax VTRs. The U-threading system was chosen because its fewer wrap angles subject the very thin, narrow tape to less stress than the more larger wrap angles in the M-threading system and because the U-threading system's fixed tape guides can be more accurately adjusted than the movable tape guides in the M-threading system so that the tape tracks more accurately.

With compactness in mind, we made the diameter of the threading ring as small as possible, but we found it necessary to place three tilted tape guides (see Fig. 11) on a threading ring this small to guide the tape from the capstan to the take-up reel in the fast wind/rewind modes as accurately as in the forward mode.

The cassette measures 56 × 35 × 13 mm, slightly larger than an audio microcassette. As can be seen from Fig. 12, the diameter of the drum of a U-threading unit essentially dictates how small the cassette can be. A further reduction in the size of the cassette would not yield a corresponding reduction in the overall dimensions of the unit because the characteristics of the new tape govern how small the drum can be.

Because the unit is portable, it must operate well in all positions. With this requirement in mind, we chose a cassette configuration very similar to that of an existing 1/2-inch videocassette, with double-flange reels. In order to take up slack so that transportation would be stable in all positions, we applied a magnetic chuck to the reel table, as is shown in Fig. 13.

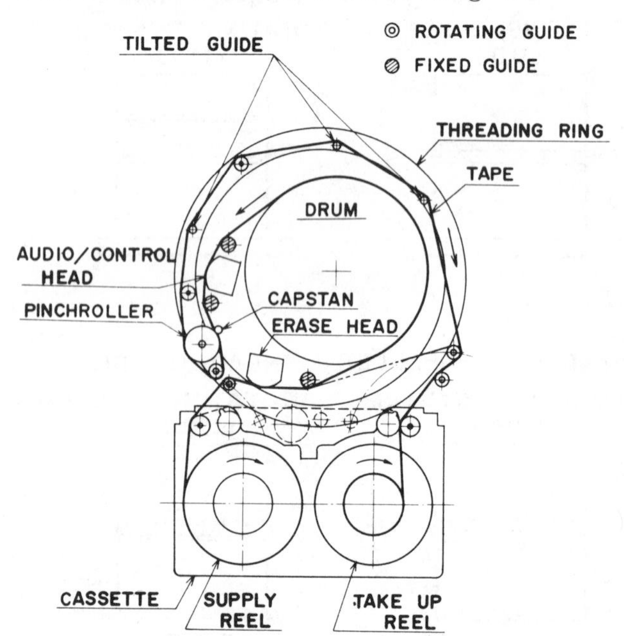

Fig. 11 Tape-threading mechanism.

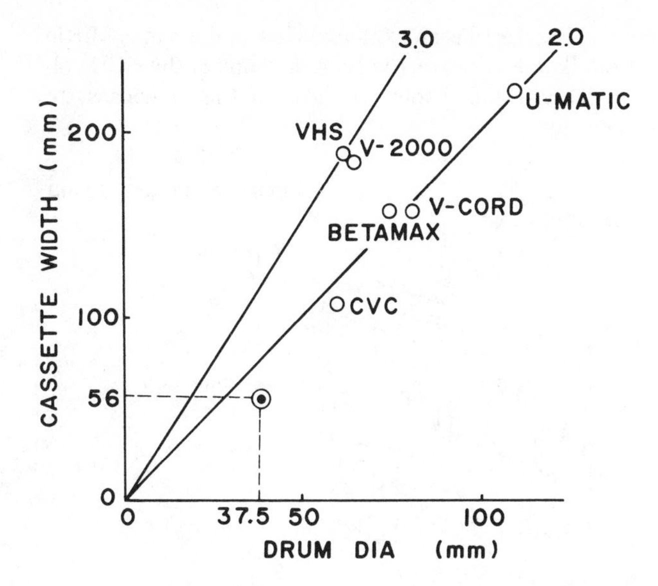

Fig. 12 Drum diameter vs. cassette width in various systems.

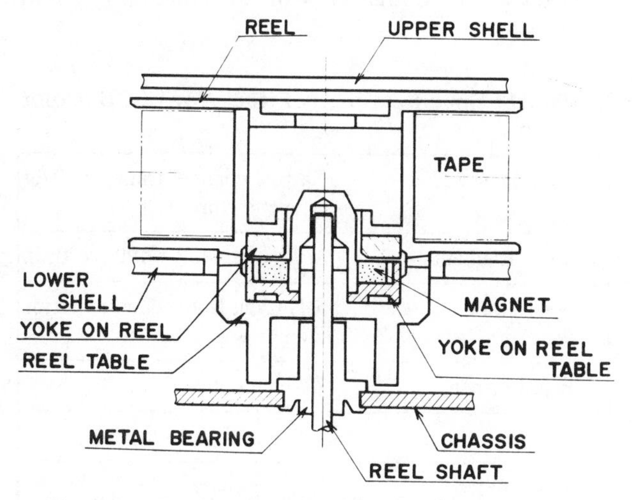

Fig. 13 Reel and reel table.

3-3. The Drive System

Figure 14 shows the new unit's drive system. The drum shaft is driven by one small coreless motor and a second coreless motor drives the capstan shaft and reel tables by belts. The small size of the unit, the new drum, the low tension on the tape and low friction allow these small coreless motors to be used, with the result that power consumption is low, an essential requirement for any battery-driven unit. Table I shows the measured power consumption.

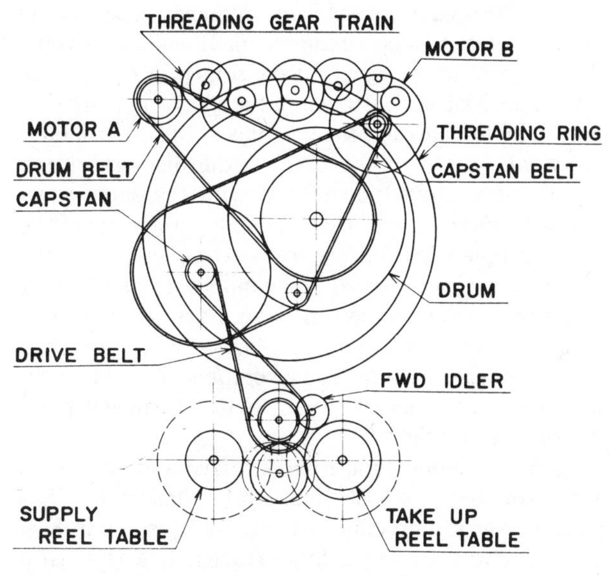

Fig. 14 Drive system.

Table I. Measured Power Consumption.

	Drive For	Driven By	Power Consumption
Motor A	Drum	Belt	0.16 W
Motor B	Capstan	Belt	0.06 W
	Reel	Belt and Idler	
Total Power Consumption in FWD Mode			0.22 W

4. High-Density Packaging of the Electronic Circuits

The VTR block has circuits similar to those of a conventional VTR except for certain color reproduction circuits (which are built into the Home Editor unit) and for the circuits which make possible special functions such as picture search and slow-motion. The circuits are integrated into six large-scale hydrid modules and inter-connected by flexible sheet wiring.

Power consumption in the VTR block of less than three watts is made possible by using micro switching regulators and by reducing the supply voltage for the video circuits from 12 volts to 9 volts. Ten percent of the power is consumed in mechanical parts and 90% is consumed in the electronic circuits.

The total area of the modules is 133 cm^2 which is about one tenth of the area of the discrete circuit boards of a conventional VTR with the same functions. To increase the packing density (the ratio of the area occupied by the electronic components to the area of the ceramic base) of the modules, the resistors and two conductive-pattern layers are printed on a ceramic base. A large number of the resistors are positioned under other electronic components. The packing density is from 0.9 to 1.2 (when sufficient components are stacked.)

All components are miniaturized and LSI chips are mounted on 1.27 mm-pitch dual-inline flat packages or square chip carriers. As is shown in Fig. 15, capacitors having a large capacitance (1 to 100 microfarads) are presently much larger than micro electronic components. To minimize the size of the module, it was necessary to design the circuit so that these large capacitors could be replaced by equivalent circuits or smaller capacitors.

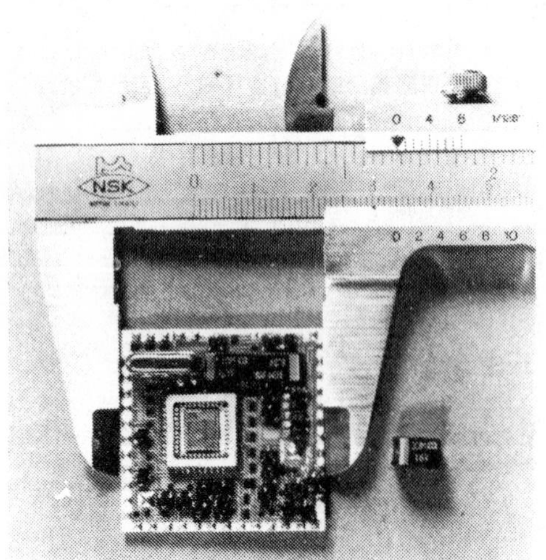

Fig. 15 Large-scale hybrid module compared with capacitor.

Figure 16 is an exploded view of the Video Movie unit. Table II shows the specifications of the CCD color camera and Table III those of the videocassette recorder.

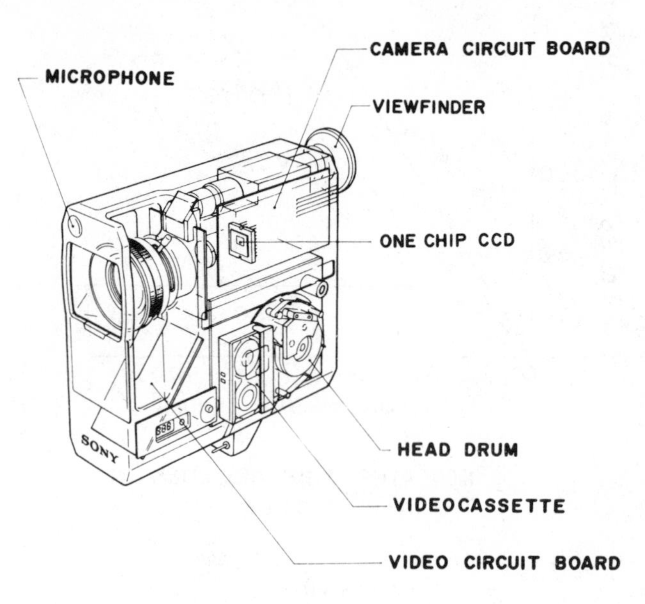

Fig. 16 Exploded view of the Video Movie unit.

Table II. Specifications of the New CCD Color Camera

Image Sensor	Narrow channel frame transfer type one chip CCD
Picture Elements	570 (horizontal) × 490 (vertical)
Color Coding	R-G-Cyan Dot-sequential stripe filter
Image Format	2/3-inch optical
Chip Size	10.1 mm × 12.1 mm
Unit Cell Size	16 μm × 14 μm
Lens	3 power zoom (14 to 42 mm) F:1.6 with macro capability, auto iris, optical viewfinder
Horizontal resolution	250 TV lines
Signal to Noise Ratio	Over 45 dB
Minimum Scene Illumination	70 lux (F:1.6 S/N 35 dB)
Power Consumption	4 watts
Weight	600 grams with lens and view finder

Table III. Specifications of the Videocassette Recorder

Recording System	Rotating heads helical scanning, FM modulation, chroma under azimuth recording
Horizontal Resolution	250 TV lines
Signal-to-Noise Ratio	Over 45 dB
Drum Diameter	37.5 mm
Track Pitch	36.7 μm (H alignment: 1.5H)
Audio Recording System	Stationary head, monaural
Audio Frequency Response	100 Hz to 10 KHz
Weight	600 grams
Battery	Silver-zinc, 9 v, 0.5 Ah, rechargeable

5. Future Work

Work continues to reduce still further the size and the weight of the Video Movie unit and to increase its reliability. Clearly, electronic circuits which consume less power are desirable so that the unit could be operated on dry batteries. To reduce the power consumption, construction must be simplified, video LSIs capable of operating on 5 volts must be developed and more sophisticated devices, such as motors with driving circuits and erasing heads with oscillators, must be made available.

A longer recording time is also desirable. To achieve this, the tape speed will have to be brought down to as low as 7 mm/sec. At a tape speed this low, however, as it will be very hard to record acceptable sound with a stationary head, a rotating head will be required. Various approaches to the problem of sound recording have been proposed, such as FM multiplx recording on the video track and PCM recording on a track after the video signal. Considering the sound-editing problems in FM recording and the rapid development of PCM recording techniques, we feel the solution is most likely to lie with PCM sound recording.

A high-density recording technique using a track-width less than 10 μm is in the final development stage. If the present very smooth-surfaced tape is to run stably at a very low tape speed, however, more work must be done on the tape-transport mechanism.

We continue our work on high-output, low-noise tape for high-density recording as well as work on a recording/reproducing head with high B_s and high μ.

6. Conclusion

A very compact Video Movie unit combining a color video camera and a micro videocassette recorder has been developed. The unit, when combined with a Home Editor unit, makes a complete Video Movie system which can playback the recorded tape on any TV set and edit it on any standard videocassette recorder.

To realize this new unit, it has been necessary to develop a new high-energy tape, a narrow gap length head, a short-wavelength recording technique, a compact tape-transport mechanism, a micro videocassette and a method of packaging components with very high density.

We feel the compact video camera/recorder unit will make conventional 8 mm-film home movie systems obsolete.

Acknowledgment

The authors wish to thank Sony Magnetic Products Incorporated who developed the tape and video head, and the Semiconductor Division of Sony Corporation who developed the CCD camera.

REFERENCES

(1) Y. Hirata et al., "2/3-in. Narrow Channel CCD Imager," in Tech. Papers, Symp. TEBS, Inst. TV Eng. of Japan, No. TEBS69-3, Feb. (1981).

(2) M Shimada et al., "Single Chip Color Camera using Narrow Channel Frame Transfer CCD Imager," in Tech. Papers, Symp. TEBS, Inst. TV Eng. of Japan, No. TEBS70-4, Mar. (1981).

(3) Y. Takemura et al. "Present and Recent Movement in Solid-State Color Television Camera," J. Inst. TV Eng. of Japan, vol. 35, Mar. (1981).

(4) S. Muramatsu, "An Analysis of the Recording Characteristics in Short Wavelength," presented at the Tech. Group Meeting of IECEJ, MR75-28 (1976).

(5) J. Hokkyo, "Theoretical Analyses on Output and Recording Limit of Magnetic Recording," presented at the Tech. Group Meeting of IECEJ, MR-65-11, (1965).

(6) S. Nakagawa, "Optimization of Track Width and Relative Speed for Broadcasting Use VTR Determined by SN Ratio," presented at the Tech. Group Meeting of TV Eng. of Japan, 8-7 (1974).

(7) S. Muramatsu, "Limit of Video Tape Coercivity by Head Saturation," presented at the 7th Conf. of TV Eng. of Japan, 9-4 (1971).

(8) M. Tada, Tech. Rept. No. 1 Sony Corp. (1961). E. Adams, "Recent Developments in Soft Magnetic Alloys," J. App. Phy. Supp. vol. 33, No. 3, Mar. (1962).

RECORDING VIDEO CAMERA IN THE BETA FORMAT

Seiji Sato, Koichi Takeuchi and Masanobu Yoshida
Consumer Video Division
Sony Corporation
7-4, Kohnan 1-chome, Minato-ku,
Tokyo, 108 Japan

ABSTRACT

In this paper we shall introduce the small diameter head drum VCR System that uses exactly the same recording format as the original large diameter head drum system. Having identified the different factors of the tape pattern with the mechanical parts, we were able to develop a new technique of recording the same information as is done in the original format. As this is specialized for only recording, we were able to drastically reduce the number of both the mechanical parts and circuits in our compact all-in-one camera/VCR.

I. INTRODUCTION

Recently, there has been remarkable penetration of VCRs. They are being used not only passively for recording TV programs, but also actively, as the trend of making original productions with video cameras continues to grow. Manufacturers, in trying to meet these increasing demands, have put much effort into making compact, lightweight portables. However, they have almost reached the physical limitations which are brought about by the systems original standards. Even though there have been great advancements in density technology for VCR hardware, it is very difficult to change the software (cassette) once the standard has been set. When the BETA Format was developed, the first thing that was decided was the size of the small, compact cassette. Now, making the most of this small cassette, we developed a revolutionary system that is completely compatible with the large diameter drum format. A cassette tape recorded by this all-in-one VCR/Camera system can be playedback on any BETA Format VCR reproducing the same high performance and high picture quality, as the large diameter drum.

Manuscript received 6/17/83

II. SMALLER-DIAMETER GUIDE DRUM

1. Weight and Volume

In the present VCR models, the tape of the VCR was wrapped about 180° around the drum, where it met two video heads (one on each side), and recorded or played back material. The drum diameter and the tape wrap angle are almost inversely proportionate to the video track length. In the new system, the tape wrap angle is about 300° and one double-azimuth head is used for recording. $300\,D^* = 180\,D_0$ (D_0: normal drum diameter; D^*: new drum diameter), therefore, $D^* = 180/300\,D_0$ or the new drum diameter is $3/5^{ths}$ that of the normal Beta drum diameter. In actual numerical values, the original drum diameter is 74.487 mm and the new one is 44.671 mm, the volume of the new drum is about $(3/5)^3$ that of a normal drum-500 g (as used in the latest Beta format portable model) compared to 90 g.

Reprinted from *IEEE Trans. Consum. Electron.*, vol. CE-29, pp. 365–374, Aug. 1983.

In addition to the drum itself, the weight and volume of the drum mount, the chassis, the loading mechanism, etc. have also been greatly reduced. Reducing the cassette size, itself, only makes that part smaller in a VCR, but reducing the size of the drum diameter, makes it possible to reduce the size, weight, and volume of the whole unit. When the VCR part has been drastically reduced in size, it is possible to add the camera part, and make an all-in-one model that is remarkably compact and lightweight. Photograph 1 is a picture for comparing the size of the normal Beta drum and this new system's drum.

Photo. 1. The Normal Beta Drum & The New System's Drum

2. Double-Azimuth Head

In this system, due to the direction of the head gap, it is ideal to use a VCR that has a Guardband for one direction (for example the U-Matic System). However, in order to achieve high density recording, there has been a trend in the penetration of models using a guardband-less Azimuth recording. As the BETA Format uses two $\pm 7^\circ$ Azimuth Heads for recording, this system in keeping with the format, has a head gap of 667μ (1.5H on the track), and is equipped with one Double-Azimuth head for recording. Since this is only one chip, when it is put on the drum, only the height from the base, and the point where the head sticks out of the drum has to be adjusted. Previous models with two heads 180° opposite of each other (which include almost all popular models) are much more complicated to adjust, because all adjustments of one head have to be made in relation to the head opposite it. Photograph 2 shows the upper drum of this system. And Fig. 1 is one double-azimuth head.

Photo. 2. The Upper Drum

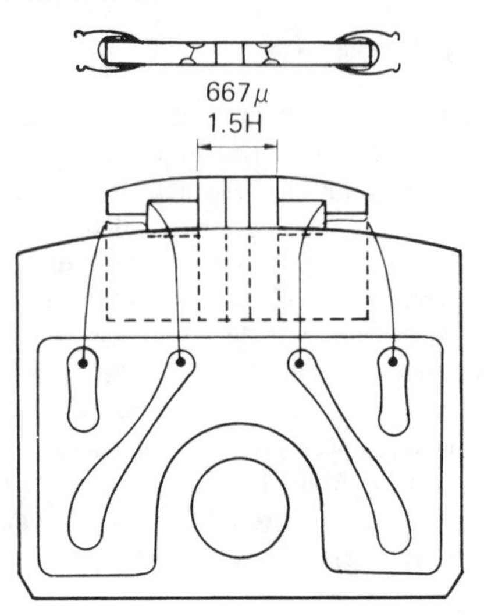

Fig. 1. One Double-Azimuth Head

III. FUNDAMENTAL PRINCIPLES

In order to achieve the same tape pattern on the track length, the tape wrapping angle had to be increased. (W = the wrapping angle) If W is 360° the diameter of the drum is exactly half of the wrapping angle, but taking into consideration the mechanical system (Input/Output Guide Posts, etc.) W becomes about 300°. This means that a gap is produced where the head does not have contact with the tape. When the head is in contact with the tape, the processing of the signal from the pick-up tube results in one field of information recorded. Namely, the horizontal frequency of the pick-up tube is 360/W, & at the gap of the angle W, 262.5H of information are recorded. In the area where the tape and the head drum are not in contact, the pick-up tube overscans. In order for an ordinary monitor TV to playback the recorded information, the field frequency has to be the same as the original frequency 60 Hz-NTSC, 50 Hz-CCIR. The drum's rotation is synchronized to this value. It rotates 360° every $1/60^{th}$ of a second, but with the wrapping angle W it is shortened to $1/60 \times W/360$. The track length is the same, in that the relative speed between the tape and the head is increased by 360/W. 1H of the track length, also becomes the same, as the recorded signal also is increased by 360/W. However, the angle of the recorded tape pattern is determined by the lead angle to the drum, the tape speed, and the relative speed of the tape to the head. In attaining the original tape speed, just the relative speed has to be modified, and the original lead angle has to be corrected. In studying the above, if the right values are chosen for the wrapping angle, the drum diameter, and the lead angle, a system with a small diameter drum, can produce exactly the same recorded tape pattern as a large diameter drum system. Figure 2 shows the relationship between the pick-up tube, the tape, and the drum.

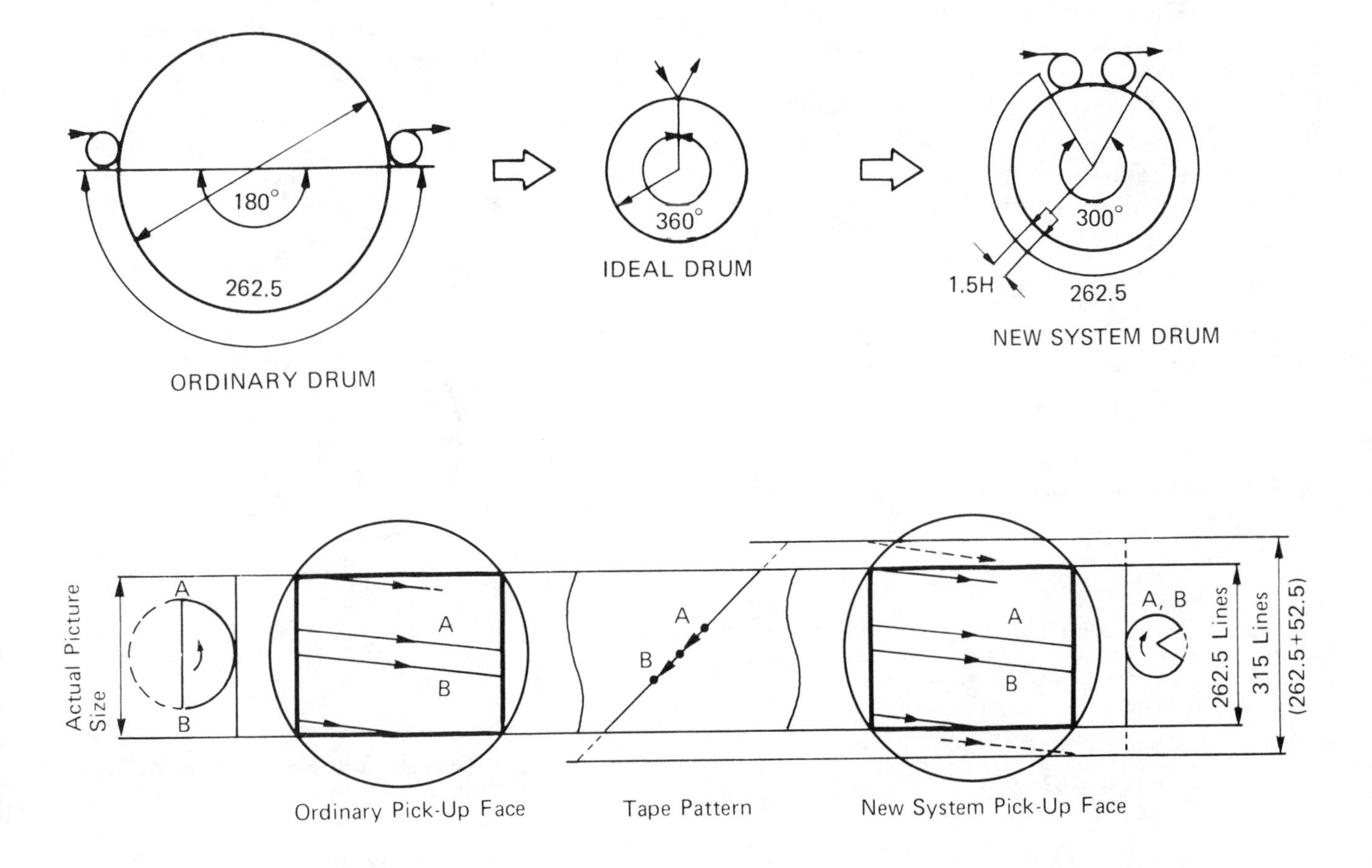

Fig. 2. The Relationship between the Pick-Up Tube, Tape, and the Drum

IV. RECORDING SYSTEM FOR IDENTICAL PATTERNS

When increasing the tape angle and reducing the diameter of the drum, the recording tape pattern must have complete compatibility with the existing Beta Format's recording tape pattern. Several techniques are used to assure this compatibility.

1. Mechanical Identification

Before processing the signals, first of all, the mechanical factors, involved in achieving compatibility with the original recording pattern had to be uniform. Thus the video track length, angle and pitch all had to be equivalent to those of the present format. In calculating these factors, first the wrapping angle (W) was determined, then the other factors were figured accordingly. It is desirable that the drum and other parts be manufactured so that they can be used for both the NTSC and CCIR Systems. Therefore, a wrapping angle appropriate for both systems was derived. The interval where information is recorded for this W° is (525/2)H (H: one horizontal scanning interval) in the NTSC System, and (625/2)H in the CCIR System. When the tape is wrapped 360° around the drum, this interval should have the recording ability of (525/2 + X)H in the NTSC System and (625/2 + Y)H in the CCIR System. We can derive the following equation from Fig. 3.

$$360:W = (\frac{525}{2} + X):\frac{525}{2} = (\frac{625}{2} + Y):\frac{625}{2}$$

From the right side of the equation we see that;

$$(525 + 2X):525 = (625 + 2Y):625 \text{ or}$$

$$25 \times (2X) = 21 \times (2Y)$$

Since 25 and 21 are prime numbers relative to each other, we get;

$$2X = 21m, \ 2Y = 25m$$

$$360:W = (525 + 21m):525 = (625 + 25m):625$$

and it follows that;

$$W = \frac{525 \times 360}{525 + 21m} = \frac{625 \times 360}{625 + 25m} = \frac{25 \times 360}{25 + m}$$

As stated above in part III, we have the relationship of the camera's horizontal frequency in this system at $f_H^* = (360/W)f_H$ (where f_H is the horizontal frequency in the existing system). If we substitute this W value in the previous equation we get;

$$f_H^* = \frac{25 + m}{25} f_H$$

In the original system, the f_H was a constant times f_V.

$$f_H = 25K \, f_V$$

(where K is a constant integer)

and it follows that;

$$f_H^* = K(25 + m) f_V$$

Therefore, if m is a whole number, then both f_H^* and f_V are increased by a whole number making it easy to process the signals.

Thus when m = 4, W = 310.345°
and when m = 5, W = 300°

As an overlap of about 10° is necessary, and in order to control the entering and exiting of the tape a guide is also necessary. Thus 300° for the wrapping angle is appropriate for manufacturing. The following is a continuation of the explanation of when W = 300°.

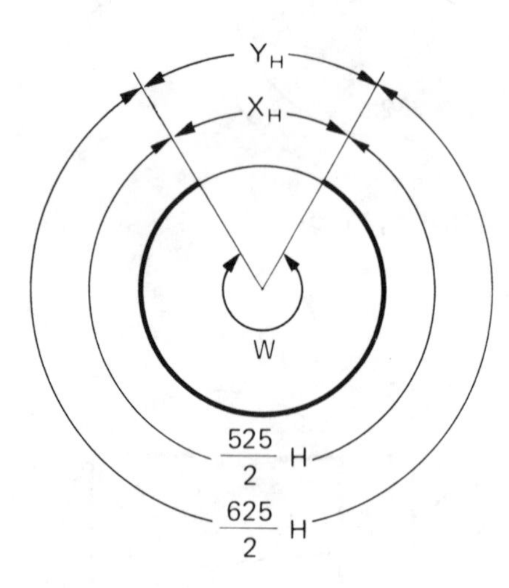

Fig. 3. The Relationship between the Wrapping Angle, the NTSC & CCIR Systems

If the wrapping angle is 300°, we can determine the diameter "D*" of the new drum. If we apply the cosine theorem to the three points A, B, and C of the triangle shown in Fig. 4, we get the following relationships;

$$(\overline{AC})^2 = (\overline{AB})^2 + (\overline{BC})^2 - 2\overline{AB}\cdot\overline{BC}\cos(180 - b)$$

$$\overline{AC} = \pi D^*,\ \overline{AB} = (360/300)L,\ \overline{BC} = V_t/f_V$$

$$D^* = \frac{1}{\pi}\sqrt{\left(\frac{360}{300}L\right)^2 + \left(\frac{V_t}{f_V}\right)^2 + 2\times\frac{360}{300}L\times\frac{V_t}{f_V}\cos b}$$

where D^* : new drum diameter
L : track length
V_t : tape speed
f_V : field frequency
b : recording track angle

Since the L, V_t, f_V, and b values are pre-determined for the Beta format, the small drum diameter can be derived. Substituting these values, we find that D^* = 44.671 mm which is about 60% of the normal 74.487 mm size drum.

Next we'll calculate how much the lead angle θ must be corrected;

As Fig. 4 shows; $\overline{FG} = \frac{300}{360}\pi D^*$, $\overline{DE} = \frac{1}{2}\pi D_o$

where D_o : the normal drum diameter

Here the sine components of FG and DE are equivalent, thus;

$$\frac{300}{360}\pi D^* \sin\theta = \frac{1}{2}\pi D_o \sin a$$

$$\theta = \arcsin\left(\frac{180 D_o}{300 D^*}\sin a\right)$$

where θ : new drum lead angle
a : the normal drum lead angle (5°)

When we fill in the figures with D_o = 74.487, D^* = 44.671, and a = 5°, we get θ = 5°00′09″. This angle has to be corrected by 9″ in comparison with the normal Beta drum lead angle. This is equivalent to about 5μ on both lead edges on the drum's circumference. The above is how we determined the drum's diameter and the lead angle, that allows the mechanical factors of the recording tape pattern's, tape speed, field frequencies, etc. to be completely the same as the existing Beta format's.

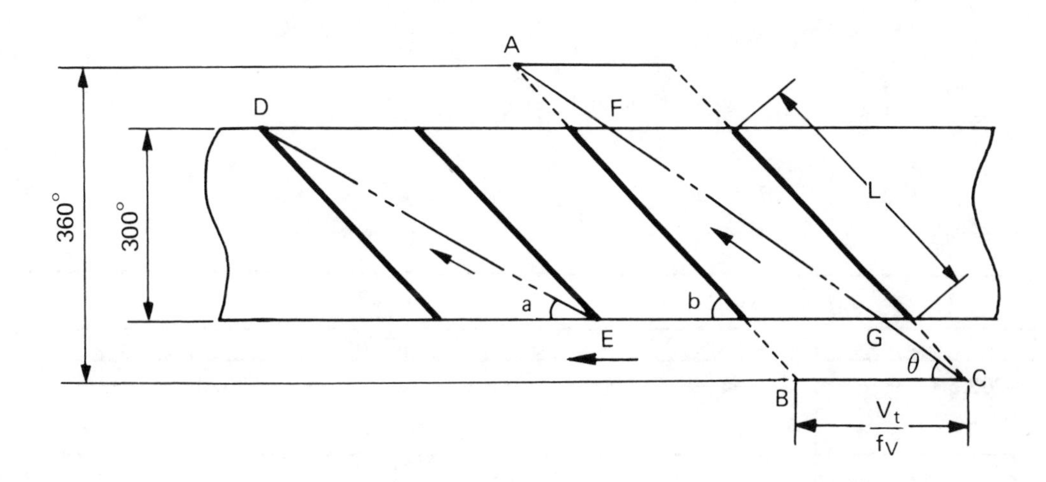

Fig. 4. The Video Track Pattern

Table 1. Beta Format (βII)

L	Track Length	116.672 mm
V_t	Tape Speed	20 mm/sec.
f_V	Field Frequency	59.94 Hz
a	Normal Beta Drum Lead Angle	5°00′00″
b	Recording Video Track Angle	5°00′51″
θ	Normal Drum Diameter	74.487 mm

Table 2. The Determined Figures

W	Wrapping Angle	300°
D^*	New Drum Diameter	44.671 mm
θ	New Drum Lead Angle	5°00′09″

The following is an explanation of the processing of the signal systems.

2. Electrical Identification

The tape speed and track pitch have already been determined; the drum rotation is synchronized to the same field frequency of 59.94 Hz as the original system. In order for the 300° – wrapped tape to record 262.5 lines of information, the camera's horizontal frequency is multiplied by 1.2; (360/300). One field then is $262.5 \times 1.2 = 315$ scanning lines, however 52.5 lines (315 – 262.5) are overscanned outside of the actual picture size. However since the head is not in contact with the tape at this time, nothing is recorded. When using a Double-Azimuth head there is a vertical deflection space between the head, of 1.5H, therefore, every other field, is shifted. The reason for the 1.5H head gap, is because when the tape is wrapped at a 300° angle, the horizontal scanning line is 315 or a whole number, thus, the horizontal scanning can not be interleaved. In the following explanation $f_H^* = 1.2f_H$. (* shows that the frequencies have been re-calculated to match popular frequencies.) As Fig. 5 shows if different Azimuth heads are used for recording video tracks T_1 and T_2, f_H^*'s horizontal scanning signal along with the standard field frequency f_V of the vertical synchronizing signal are supplied from the reference signal generator. f_H^* can enter the camera's horizontal deflection circuit just as it is, but every other field in f_V has to be switched by a 1.5H delay before it can enter the vertical deflection circuit. This specialized field frequency becomes S_V (special vertical synchronized signal). Head H_A begins recording 262.5H as soon as it reaches the f_V. Head H_B begins recording only after a delay of 1.5H from the f_V. Thus, track 1 records a signal that has been interleaved with track 2. With this system, there is no common standard Azimuth jitter and jumping in the playback picture, as found in most VCRs. Also, the drum's rotation can be synchronized to a vertical synchronized signal that has not gone through the delay circuit. Figure 6 shows the synchronizing signal and the time chart of the tracks.

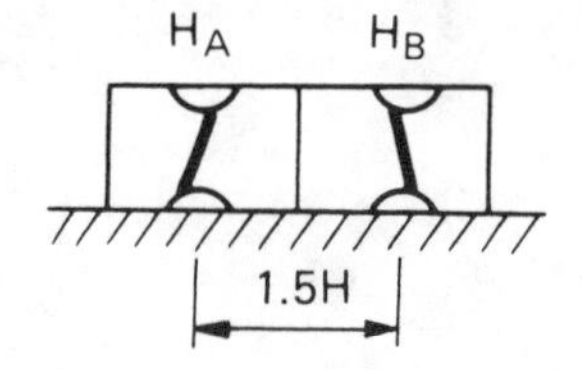

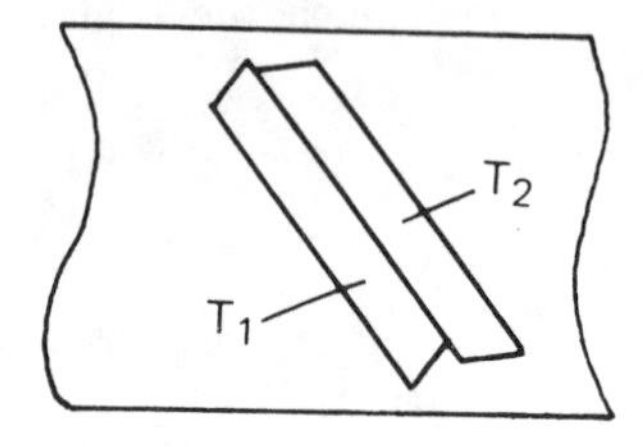

Fig. 5. Head and Track

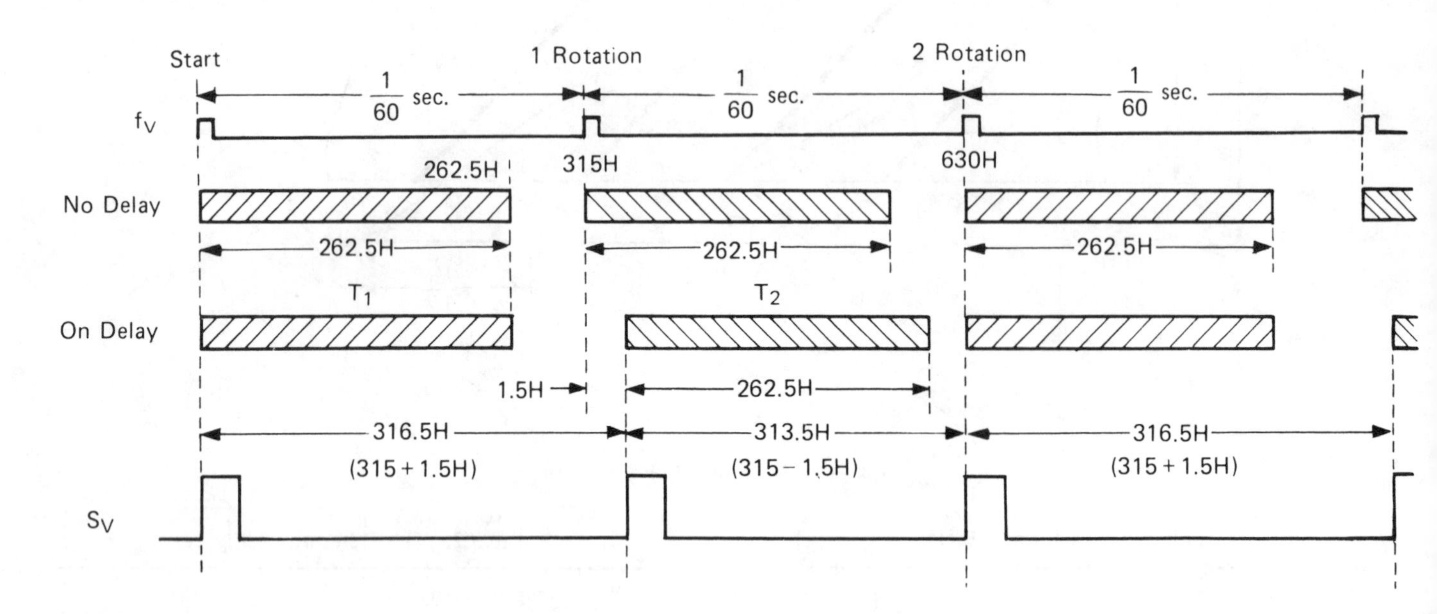

Fig. 6. Recording System

V. TAPE TRANSPORT MECHANISM

There is no threading mechanism before or after the tape is automatically wrapped 300° around the drum. The main reason we incorporated the U-loading system, is because it has a smooth tape path that does not twist the tape. When the tape comes off the supply reel, the first inclined guide takes it gently through the tension regulator arm in an upward direction, so that by the time it reaches the second inclined guide, it is almost horizontal when it is wrapped around the head drum. Where the tape comes out, the drum lead is positioned so that it is parallel to the chassis, when it approaches the drum. After the tape leaves the tape exit guide it runs parallel to the chassis. It then goes by the audio CTL head, and a pinch roller stabilizes the speed to 20 mm/sec, before the take-up reel receives it (See Fig. 7). This main system is all run by one motor. As this is only a recording system, the reduction of the rotation speed of the capstan is simply carried out by means of a belt from the drum. The take-up reel's speed is then slowed down by another belt from the capstan axis to the reel (See Fig. 8).

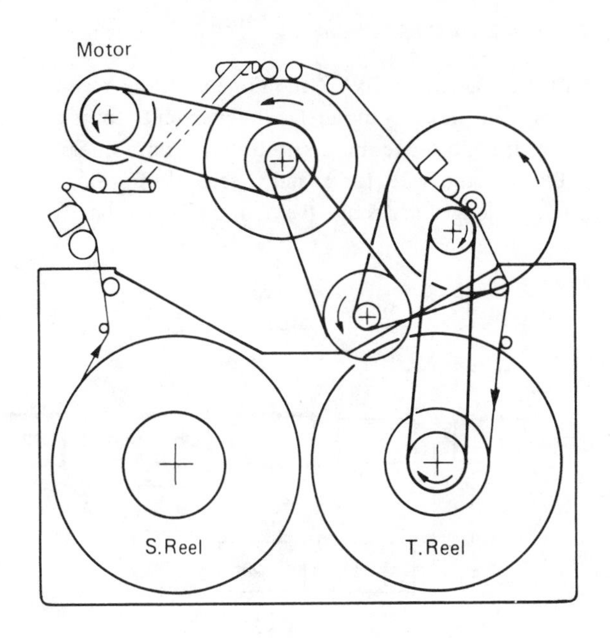

Fig. 8. The Belt Drive System

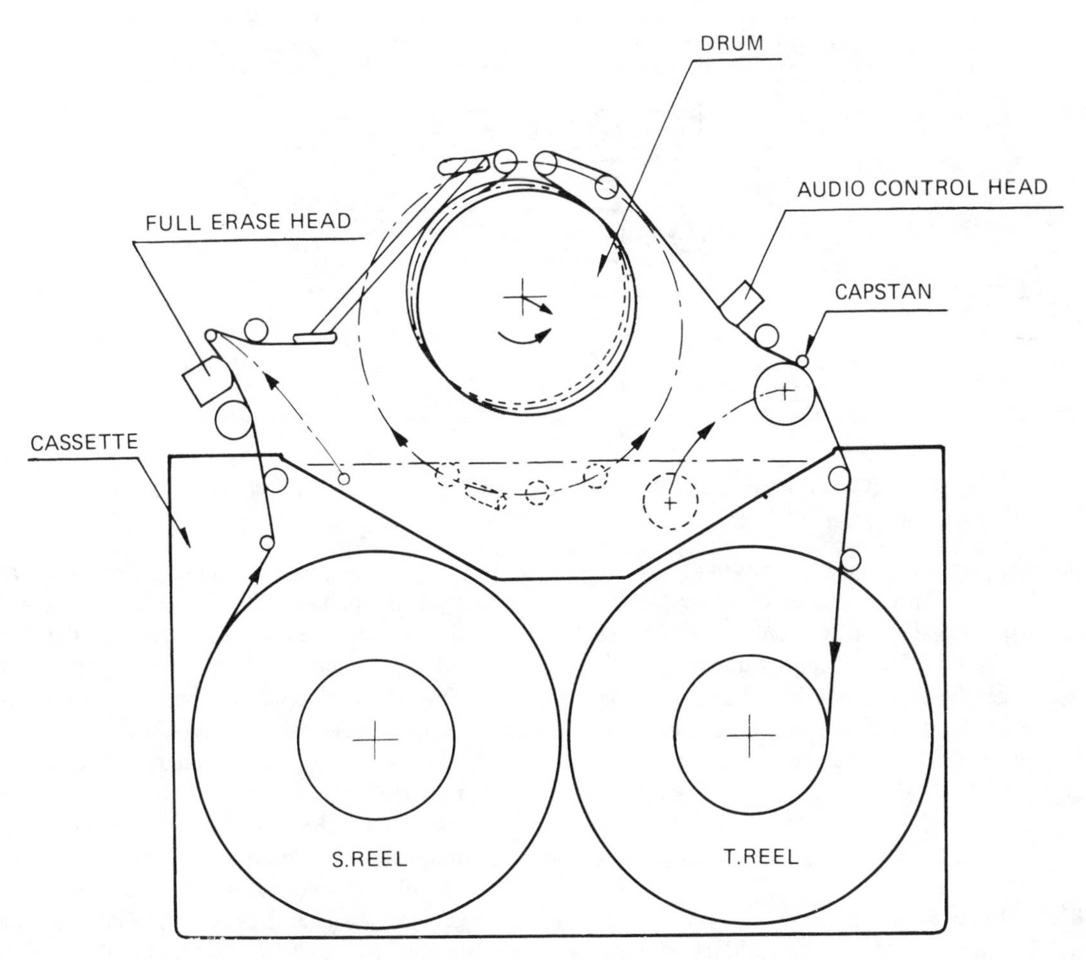

Fig. 7. The Tape Transport Mechanism

VI. CAMERA PARTS

With the camera and VCR all in one unit, the Y and C signals can have completely independent routes. Therefore, it is no longer necessary to have comb filters, LPFs, BPFs in the VCR for Y and C separation. Since these are no longer necessary the Y and C signal bands do not have to be stuffed in limited bands, but can use much wider ones. This results in improved resolution and better color reproduction. The dark out-lined blocks of the block diagram part A (camera) and part B (VCR) of Fig. 9, have been simplified in part C, (all-in-one system).

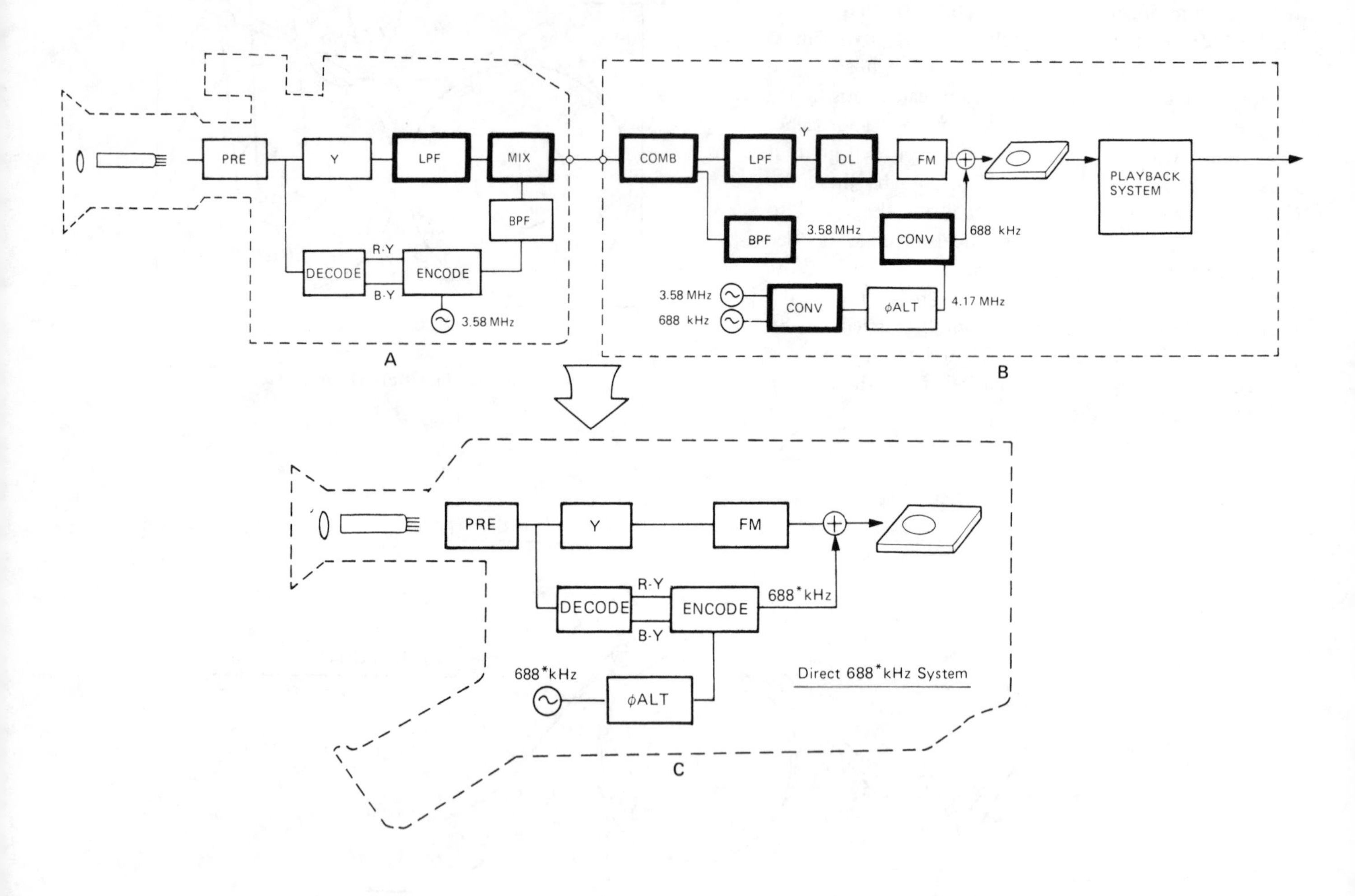

Fig. 9. A (Camera) and B (VCR) Simplified and Combined in C (All-In-One Technique)

The basic frequency for this system is $(44 - 1/4)f_H^* \times 16 = 700f_H^* = 840f_H$. This is counted down by $1/16^{th}$, and $(44 - 1/4)f_H^* = 688\,kHz \times 1.2$ is used as the chroma sub-carrier. Also, the 140^*f_H is counted down by $1/5^{th}$ and clocked to get the Sync, HD, VD, BLK, etc. As mentioned in Part IV, if "m" is a whole number, then the necessary pulse signals all are counted down by the whole number.

Direct 688*kHz Encoder

Present VCR models modulate the 3.58MHz chroma signal of the NTSC System, down to $(44 - 1/4)f_H = 688\,kHz$ before recording it. However, in an all-in-one system, the chroma component can be directly encoded from the pick-up tube at 688*kHz. In the Beta Format, the B track is encoded as is, and on the A track in every other 1H the 688*kHz polarity is inverted by 0, π. This inverted 688*kHz and the normal 688*kHz are then encoded alternately, every 1H. This results in being able to get the chroma signal recorded directly on the tape. Thanks to this method of being able to directly encode at 688*kHz there is no longer a need for a frequency modulation generator and the frequency modulator circuitry, which means that the circuitry, in general, has been greatly simplified. This method can be applied to the CCIR System as well as the NTSC System.

Figure 10 shows a block diagram with the sync generator, of the signal path in the camera/VCR.

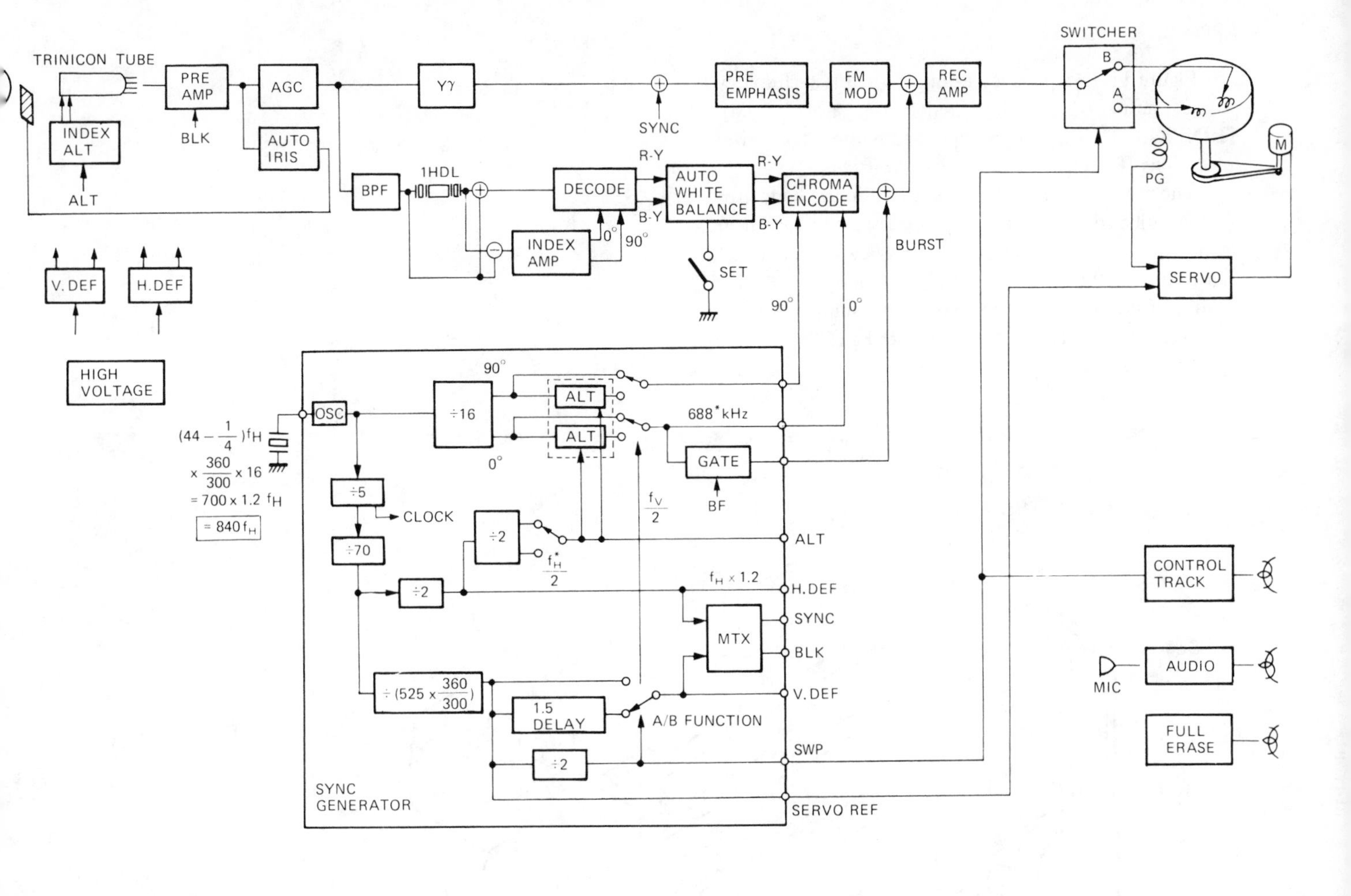

Fig. 10. Block Diagram

VII. CONCLUSION

Using the techniques explained in this paper we were able to make a compact all-in-one recording camera/VCR, using a drum with about 44 mm diameter, that has complete compatibility with all models in the Beta Format. In order to get the very best picture and sound quality, it is limited to the BII mode. Using L-830 Beta cassette it is possible to continuously record 200 minutes in the NTSC System and 215 minutes in the CCIR System, which is plenty of time to record almost all events. This has greatly improved the operability of the video camera, and has vastly increased its usage possibilities. These techniques can be applied to video formats other than the Beta Format. We believe it is bound to create new concepts for new video camera products.

VIII. ACKNOWLEDGEMENTS

We would like to express our deep appreciation to those who offered us valuable advice and support during the development of this product. Those whom we would like to give a special thanks to include: Mr. Norio Ohga, Mr. Fumio Kohno, Mr. Kiyoshi Yamakawa, Mr. Yoshimi Watanabe, and Mr. Yasuo Kuroki for their continuous encouragment and support during the research and development of this project.

APPENDIX

Figure 11 is a product example of this technology put to actual use. This model uses a 1/2″ SMF TRINICON TUBE, and is equipped with a 6× power zoom lens (F 1.2).

The rechargeable battery is encased in the hand grip. With a fully charged battery, it is possible to record up to one continuous hour. An optical view finder, which can fold over the top of the body, when not in use, has been incorporated, to keep power consumption at its lowerest. The body itself weighs 2.48 kg (5 lb 7 oz).

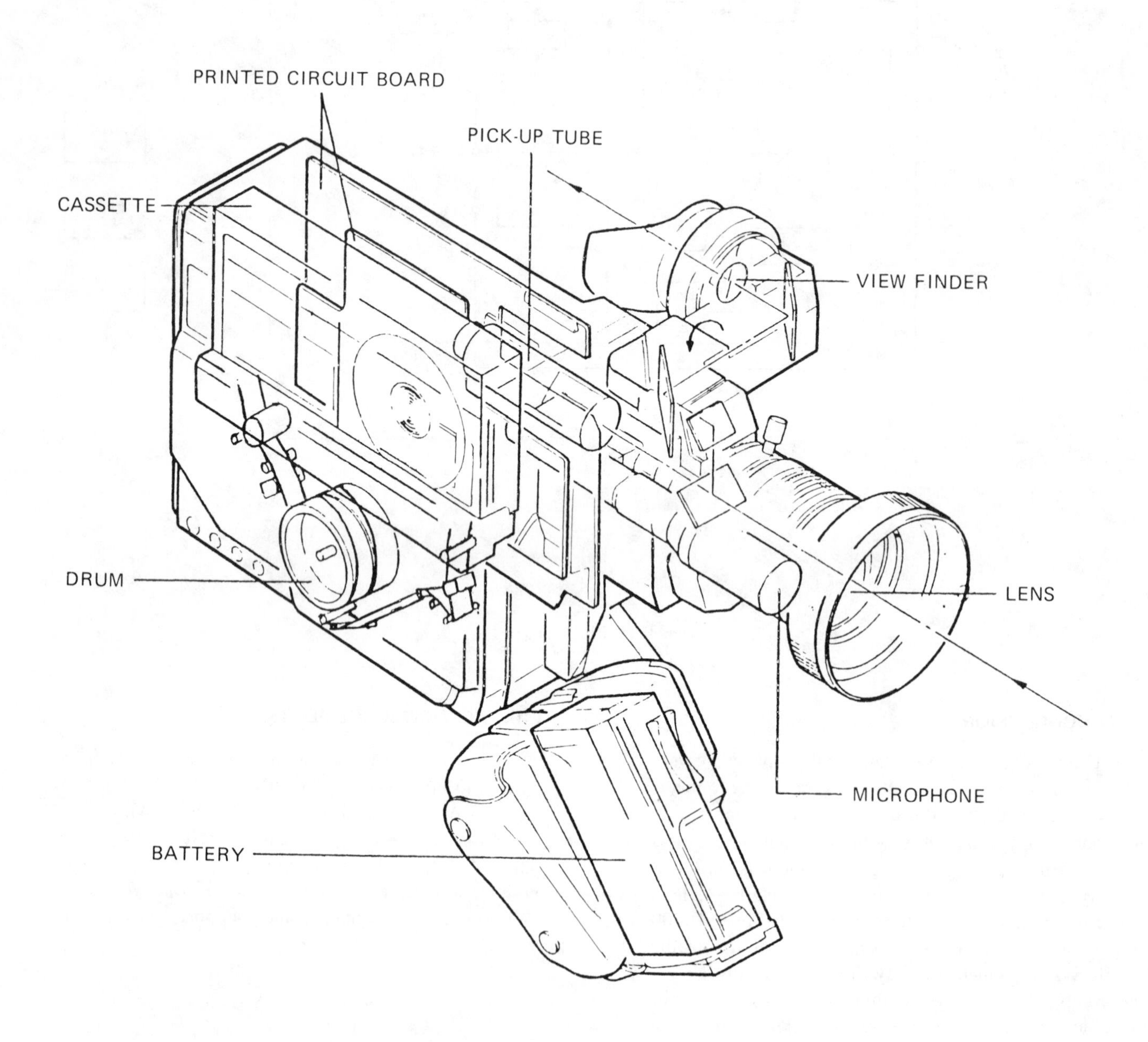

Fig. 11. A Picture of an Actual Product Using This System

Magnetic Tape Duplication by Contact Printing at Short Wavelengths

HIROSHI SUGAYA, MEMBER, IEEE, FUKASHI KOBAYASHI, AND MITSUAKI ONO

Abstract—Magnetic tape duplication by the contact printing method was first developed in 1949, but there were still many difficulties in the way of successful duplication at short wavelength signals of a few micrometers. The *bifilar tape winding system* has been newly developed for successful duplication of signals as short as 2 μm. The optimum strength of the transfer field for maximum output of the duplicated slave tape is exactly the same value even at very short wavelengths. The optimum coercivity of the master tape is about two and half times that of the slave tape. The decrease of the recorded master tape at 2-μm wavelength is within 2 dB for the first duplication with no further decrease after more than hundreds duplications. This system is applied to a video tape duplicator and can transfer automatically a 1-hour program within 2 minutes. The duplicated picture quality is almost undiscernible from the master. This system is also adaptable to the duplication of audio, digital, and any other magnetic tapes or sheet information.

I. Introduction

MASS duplication of magnetic tapes by contact printing was suggested as a possibility from the very early days of magnetic tape recording [1]–[3]. The conventional method was a running contact of both the master and slave tapes during the application of a magnetic transfer field. The reports regarding magnetic tape duplication of sound, however, always showed a very big output decrease at wavelengths shorter than 10 μm [4]–[6]. One of the most important reasons why it is so difficult to duplicate the signal at short wavelengths is the slippage and the trapped air between the two tapes. Conventional magnetic tape duplication by contact printing, therefore, was limited to rather poor quality sound tapes (or sheets) for language education or training. In order to improve the duplication characteristics at short wavelengths, the newly developed *bifilar tape winding system* was applied instead of the conventional running contact method. Our experimental results with this *bifilar tape winding system* and an application for video tape duplication are given.

II. Bifilar Tape Winding System

Conventional magnetic tape duplication by contact printing uses the running contact method as shown in Fig. 1(a). The running tapes must not separate until after the magnetic transfer field is sufficiently weaker than the slave tape coercivity. Otherwise, not only does the duplicated short wavelength signal become out of focus on the slave tape, but also the signal on the master tape should be considerably erased at the first duplication because the self-demagnetization coefficient of master tape becomes larger. Sometimes, the vacuum technique was used to obtain good contact [7]. In any event, perfect running contact of both tapes without any slippage of trapped air is not easily realized. This *bifilar tape winding system*, therefore, is a breakthrough concerning this problem.

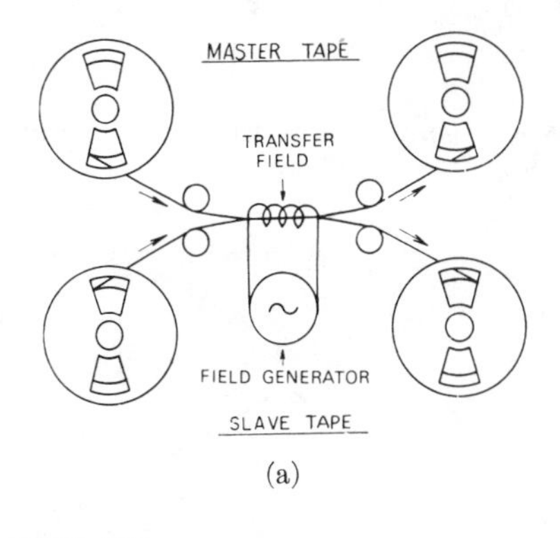

(a)

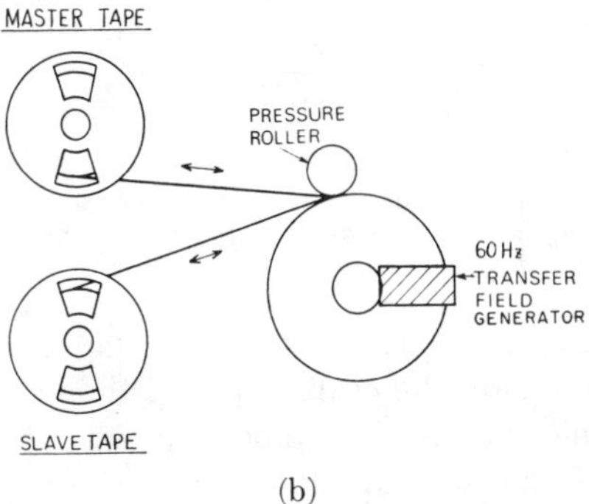

(b)

Fig. 1. Principle of magnetic tape duplicators by contact printing method. (a) Conventional tape running contact method. (b) *Bifilar tape winding system.*

The principle of the *bifilar tape winding system* is very simple. Both the master and slave tapes are wound together with the aid of a pressure roller which presses out the trapped air and makes a very tight contact between both tapes [Fig. 1(b)]. After winding by this system, both tapes have absolutely perfect contact without any slippage. A magnetic transfer field is then applied for only a few seconds on the turning bulk-wound tape. Both tapes are then rewound on their respective reels. A 720-meter (2400-foot) tape, for instance, can be wound within 1 minute with no difficulty. This tape speed is faster than the highest speed of the conventional audio duplicating machine (head-to-head system). The only disadvantage of this system is the "print-through" effect at relatively long wavelengths. When sound tape, which has long wavelength components, is duplicated by this system, for example, the

Manuscript received March 5, 1969; revised May 30, 1969. Paper 17.3, presented at the 1969 INTERMAG Conference, Amsterdam, The Netherlands, April 15–18.

The authors are with the Products Development Laboratory, Matsushita Electric Industrial Company, Ltd., Osaka, Japan.

Reprinted from *IEEE Trans. Magn.*, vol. MAG-5, pp. 437-441, Sept. 1969.

TABLE I
MAGNETIC AND PHYSICAL PROPERTIES OF MASTER AND SLAVE TAPE

Tape	Master Tapes						Slave Tapes	
	A	*B*	*C*	*D*	*E*	*F*	*G*	*H*
Hc (Oe)	1200	915	880	810	750	540	295	285
4πIr (gauss)	1000	1380	1210	990	1090	1240	950	945
Ir/Is	0.62	0.8	0.82	0.83	0.81	0.67	0.70	0.73
Coating thickness (μm)	7.4	5.3	7.1	7.4	7.0	4.3	12.0	4.1

All tapes have 25-μm (1-mil) Mylar base. All master tapes are coated with cobalt–ferrite powder. The magnetic field applied for measurement is 2000 Oe.

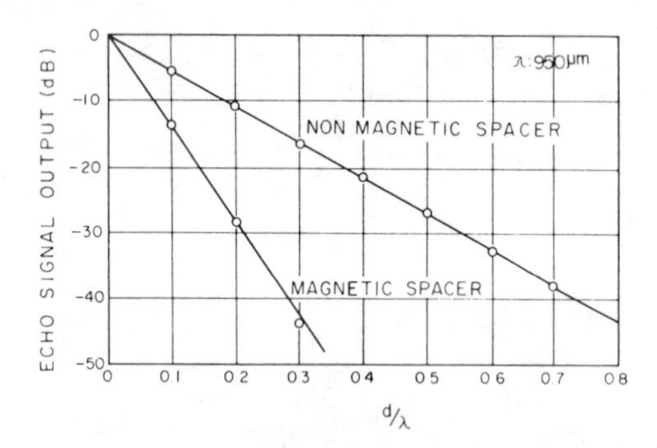

Fig. 2. Decrease of echo signal by a nonmagnetic spacer and by a magnetic coating (10-μm) spacer.

long wavelength signals on the master tape will be transfered not only onto the nearest part of the slave tape, but also onto one or more slave tape layers above and below. Thus, the duplicated tape has many echo signals before and after the actual signal.

One of the simplest ways to reduce this echo effect is insertion of a spacer tape between the two tapes. The resulting decrease of the echo signal by use of a nonmagnetic spacer follows the rule of spacing loss [8] (Fig. 2). If the spacer tape is coated with a magnetic material, the echo effect will be less than half (in decibels) that noted when a nonmagnetic spacer is used. Because very thick spacer tape is not practical, a combination of this system and the conventional running contact method is also conceivable for separate recording of long and short wavelength signals such as for a helical scanning type video tape. (The video tape duplicator will be discussed later.)

The advantages of this *bifilar tape winding system* are summarized as follows:

1) absolutely no slippage and trapped air between the tapes during the application of magnetic transfer field;
2) the tape speed for duplication is limited only by mechanical reasons such as a damage of tape coating (tentative maximum tape winding speed is about 12 m/s (480 in/s));
3) very simple mechanism, with no electronic circuitry, is easy to operate and to make fully automatic;
4) the transfer field can be supplied directly from a commercial line of 60 or 50 Hz.

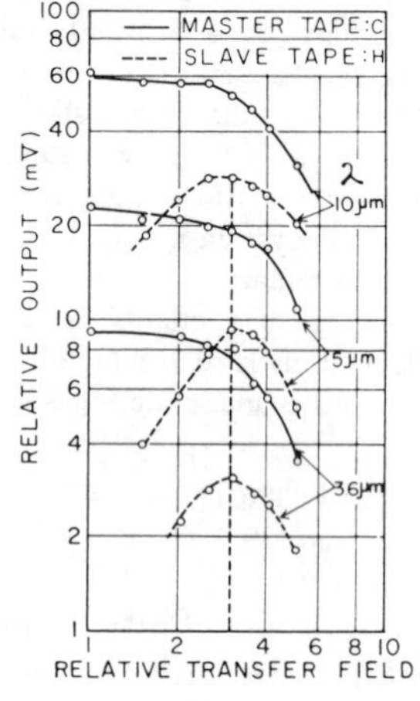

Fig. 3. Recorded master tape outputs and duplicated slave tape outputs, changing transfer field strength at different wavelengths.

III. EXPERIMENTAL RESULTS

The experiments we made were carried out by the *bifilar tape winding system*. All master tapes used were cobalt–ferrite-coated tape which can be controlled for coercivity in wide ranges, such as 540–1200 Oe, with little change of other characteristics. The slave tapes were selected from commercially available γ-Fe_2O_3 audio and video tapes which had different coating thicknesses (Table I). The master tape was recorded with no bias at $3\frac{3}{4}$ or $1\frac{7}{8}$ in/s tape speed by *hot-pressed ferrite* heads [9] of different gap lengths (1, 2.5, or 5 μm) and at first, the recorded currents were chosen to obtain the maximum output at each frequency.

The magnetic transfer field generator in this system was supplied directly from a commercial line of 60 Hz. The alternative current in the transfer field generator was changed from 1–10 amperes, and the duplicated slave tape outputs were observed at different wavelengths (Fig. 3). The optimum transfer field for the maximum output of the duplicated slave tape was exactly the same value even at very short wavelength, and mainly determined by the coercivity of the master tape. This is very important for realization of video tape duplication by the contact printing method. The optimum transfer field is increased according to the coercivity of the master tape naturally, and the duplicated slave tape output will also increase somewhat (Fig. 4). The transfer field mentioned in this paper

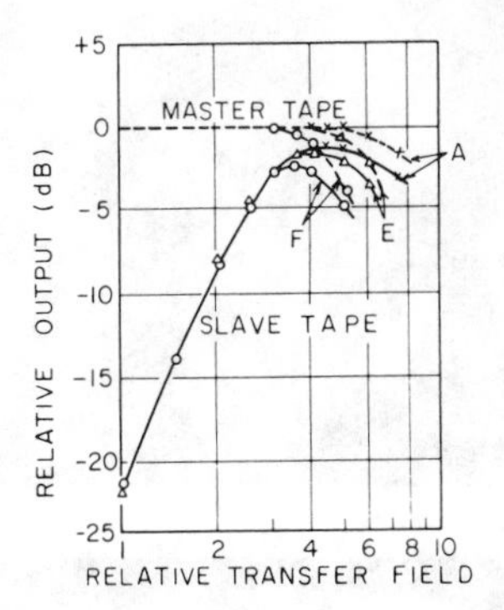

Fig. 4. Different recorded master tape outputs and duplicated slave tape outputs with different transfer field strengths. *A*, *E*, and *F* represent master tapes of Table I.

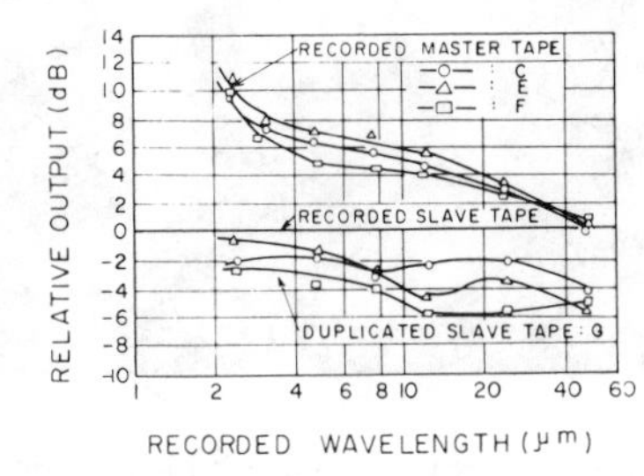

Fig. 5. Different recorded master tape outputs and their duplicated slave tape outputs (recorded slave tape output by the head is normalized at 0 dB).

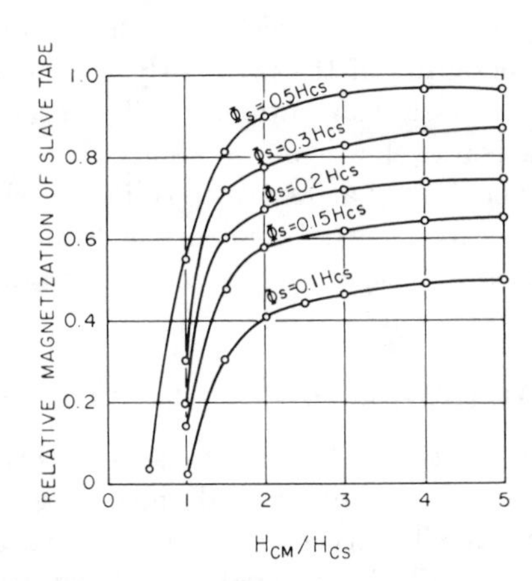

Fig. 6. Magnetization of duplicated slave tape versus coercivity ratio of master tape to slave tape (calculated by a specially designed hysteresis curve simulator). Φ_s represents a strength of a signal field from master tape.

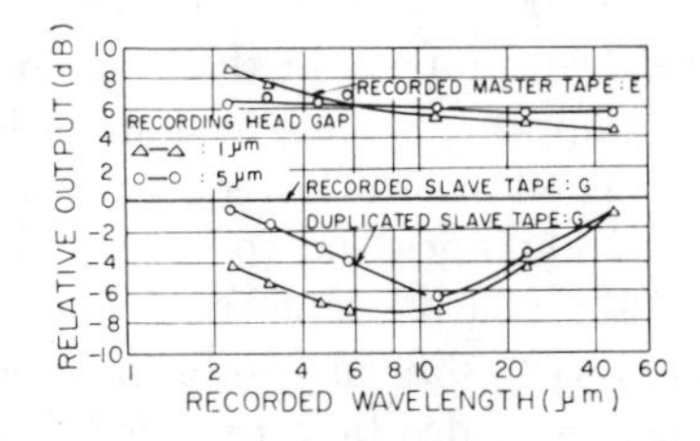

Fig. 7. Recorded master tape outputs by recording heads having different gap lengths and their duplicated slave tape output (recorded slave tape output by the head is normalized at 0 dB).

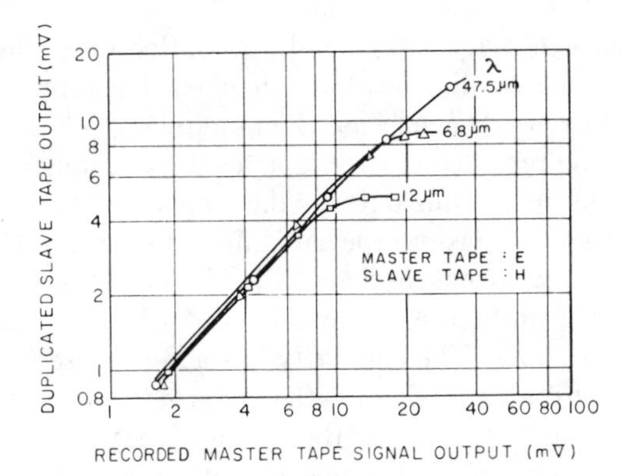

Fig. 8. Recorded master tape outputs versus duplicated slave tape outputs at different wavelengths.

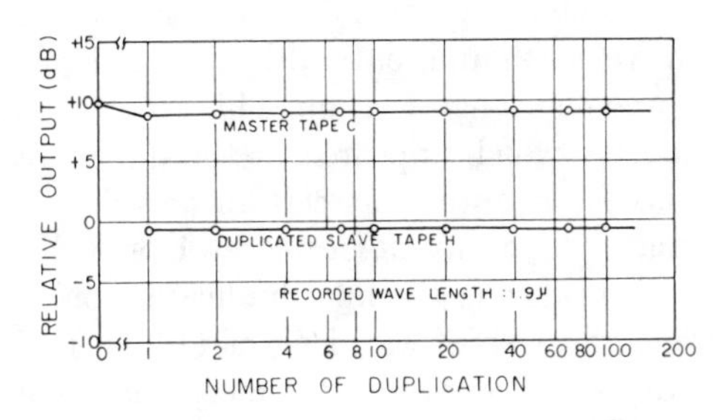

Fig. 9. Master and duplicated slave tape outputs after duplications at a short wavelength (1.9 μm). (Same results are obtained at longer wavelengths.)

is the optimum value for maximum output of the duplicated signal unless otherwise noted.

The duplicated slave tape signals (by contact printing) and the recorded master tape signals (recorded by a 2.5-μm gap length head at $1\frac{7}{8}$ in/s) were compared with the slave tape output (recorded by the same head with no bias, and normalized at 0 dB) as to the maximum output at each wavelength when the coercivity of the master tape was changed (Fig. 5). The higher the coercivity of the master tape, the higher the duplicated slave tape output, as would be expected, but the increase is small. The relationship between the ratio of coercivity of the slave tape to master tape and the duplicated magnetization of the slave tape was calculated with the aid of a hysteresis curve simulator [9], which was made by the combination of an analog computer and a closed magnetic circuitry having a Hall element and a magnetic tape element [10] (Fig. 6). The critical ratio of coercivity is about 2, which is practical, provided the squareness of the master tape is higher than 0.7, although some margin is required. The optimum coercivity of the master tape, therefore, is desirable to be about two and half times that of the slave tape under normal conditions. The results obtained from the hysteresis curve simulator have a close coincidence with the results in Figs. 4 and 5. The reproduced output from the duplicated slave tape in Fig. 7 was only a few decible lower than the directly recorded signal on the slave tape, except very long wavelengths, and had a concavity at around 10-μm wavelength, which was comparable to the coating thickness of the master tape. The wavelength at which the concavity was caused was shifted when the gap length of the recording head was changed. The reproducing head in this experiment was a 2-μm gap length head.

When the recorded master tape level was changed from zero to the maximum output, the duplicated slave tape output from the master tape had different saturation levels at different wavelengths (Fig. 8). The lowest saturation level at 12 μm is the minimum point of the concavity mentioned above. The concavity has a very close relation to the effective magnetization thickness of the master tape. At very long wavelengths, the demagnetization factor of the recorded signal on the master tape is negligible. At a wavelength comparable to the coating thickness of the master tape, however, the effective magnetization thickness of the master tape will become thinner than the recording wavelength and, naturally, the demagnetization factor will become negligible again. This is one explanation of the concavity which appeared in the duplicated slave tape for maximum output at certain wavelengths. The recorded master tape output decreases 1 or 2 dB due to the transfer field during the first duplication only and, of course, even at very short wavelengths (1.9 μm). No more decreases are observed even after more than a hundred duplications (Fig. 9). The life of the master tape, therefore, is determined only by such mechanical problems as damage of the tape coating, increase of drop out and so on.

The decrease of the transferred signal at very long wavelengths is a fatal problem in the contact printing method. The divergence of the remanence magnetization of the master tape is inversely related to the recorded wavelength, and the leakage flux of the master tape at long wavelengths will decrease consequently. Theoretically, the duplicated signal at very long wavelengths, therefore, has 6 dB/oct loss [11]. One solution of this problem is use of a thickly coated tape for the slave. Because the contact printing method is processed under an ideal magnetization, the tape output at long wavelengths is related directly to the coating thickness. The duplicated magnetization at short wavelength becomes thin due to the spacing loss, and so the self-demagnetization factor at short wavelengths will be negligible compared to a signal recorded from a head.

The noise level of the duplicated slave tape is somewhat lower than that of a slave tape, which is directly recorded from a head, and is slightly less than the bulk erased master tape noise.

In theory, there is no modulation noise caused by tape vibration during tape running and slippage between the tapes, with this system, but even a very slight irregularity of the tape surface sometimes causes a kind of modulation noise as a matter of fact. A smooth tape, therefore, will help to improve the short wavelength characteristics and decrease the modulation noise of the duplicated signal, but will result in poorer running performance. It is very important to find a satisfactory compromise concerning this point. Most of the noise of the duplicated slave tape, however, seems to be transferred from the master tape. Therefore the noise level of the master tape is very important for good quality duplication by the contact printing method, because the magnetic transfer field will also contribute to elimination of the previously recorded noise on the slave tape as a conventional bulk eraser, and the duplicated noise level will be determined by the master tape noise level. A very attractive possibility—duplication at very short wavelengths (as short as 2 μm) with this *bifilar tape winding system*—has been obtained incidentally from these experimental results.

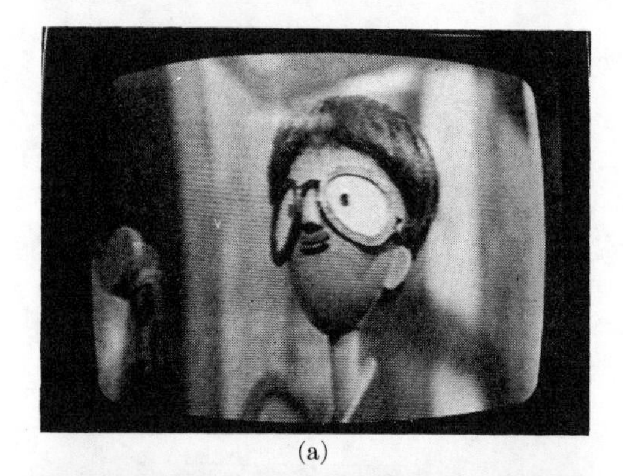

(a)

(b)

Fig. 10. Pictures reproduced by a video tape recorder. (a) Master tape recorded from a video tape recorder. (b) Duplicated by *bifilar tape winding system.*

IV. An Application to Video Tape Duplication

One of the most valuable applications of this system is the mass production of duplicated video tape. Currently, most people have TV receivers in their home, and much information is distributed through the TV set. In this way, video tape recorders will become very important information sources for everyone, though now used only for education and industrial applications. Most video tape recorder are the rotary head type. The relative tape speed between the head and tape is in the range of 10–40 m/s, and the recorded upper frequency is more or less 10 MHz. These tape speeds and recording frequencies are almost the limit of current techniques and so even a two- or threefold increase of the relative tape speed and recorded frequency will be a very difficult problem, especially for a helical-scanning-type video tape recorder. Contact printing, therefore, seems to us the only solution for mass duplication of video tapes on an economical basis. A special video tape recorder has been designed for making mirror-image master tapes. The automatic tape duplicating machine can duplicate a 720-meter (2400-foot) tape within

2 minutes. That is more than 30 times faster than conventional head-to-head-type video tape duplicating machines. A picture reproduced by a conventional video tape recorder (Panasonic NV8100) with a master video tape and a duplicated tape by this *bifilar tape winding system* are compared [Fig. 10(a) and (b)]. Most video tape recorders use a FM system to record the video signal on tape. The reproduced picture S/N, therefore, is not related to the head output directly, and the resolution is generally decided by the carrier frequency of the FM system When the duplicated tape has only a few decibels lower output than the tape output recorded by a head (as shown in Fig. 5), there is no significant difference in the reproduced picture.

V. Conclusion

We have discussed the possibility of duplication of very short wavelength signals, such as 2 μm, by the contact printing method, and have shown an example of application for a video tape duplicator which will be used for mass production of information media in the very near future. Many minor problems, however, still remain:

1) better master tapes—CrO_2, alloy-powder-coated, alloy-plated, and other recording tapes;
2) investigation of better recording methods—recording head, equalization, and recording systems;
3) improvement of the bifilar tape winding machine—winding speed, stability of tape transportation, and mass productivity;
4) more exact theoretical analysis—exact calculation of the demagnetization factor, analysis of the transfer efficiency, analysis of digital transferred signals and so on.

This type of tape duplication will be used very widely in the fields of video, audio (especially cassette tape), digital information, and all kinds of magnetic tape and sheet applications.

Acknowledgment

Matsushita Electric Industrial Company is indebted to Fuji Photo-Film Company for providing the high coercivity master tape used in the experiment. The authors wish to thank T. Nakao and S. Nishimura for giving them the opportunity to develop this duplicator; Dr. T. Nasu for his guidance; and K. Imanishi, Y. Higashida, I. Arimura, H. Goto, and others for their technical support

References

[1] R. Herr, "Duplication of magnetic tape recording by contact printing," *Tele-Tech.*, vol. 8, pp. 28–30, November 1949.
[2] M. Camras and R. Herr, "Duplicating magnetic tape by contact printing," *Electronics*, pp. 78–83, December 1949.
[3] R. Herr and R. Reynolds, "Methods and apparatus for duplicating magnetic tape record members," U.S. Patent 2 738 383, March 13, 1956.
[4] M. Sato and Y. Hoshino, "Magnetic printing characteristics of sound recording sheet applied damped oscillating field" (in Japanese), *J. Electrochem. Soc. Japan*, vol. 28, E80, p. 173, 1960.
[5] M. Sato, "Magnetic materials for magnetic tape duplicating by contact printing" (in Japanese), presented at the 2nd Tech. Group Meeting of Magnetic Recording Materials of IECE of Japan, December 10, 1962.
[6] S. Hokkyo and K. Yamanouchi, "An experiment of sound duplication with an alloy-coated master tape" (in Japanese), *Proc. Tech. Group of Magnetic Recording of IECE of Japan*, MR-64, 1–4, January 26, 1965.
[7] J. R. Morrison and D. E. Speliotis, "The magnetic transfer process," *IEEE Trans. Magnetics*, vol. MAG-4, pp. 290–295, September 1968.
[8] R. L. Wallace, "The reproduction of magnetically recorded signals," *Bell Sys. Tech. J.*, pp. 1145–1173, October 1951.
[9] H. Sugaya, "Newly developed hot-pressed ferrite head," *IEEE Trans. Magnetics*, vol. MAG-4, pp. 295–301, September 1968.
[10] H. Sugaya and F. Kobayashi, "Simulator for magnetic recording process" (in Japanese), *J. Inst. Tv. Engrg. of Japan*, vol. 22, no. 4, pp. 289–295, 1968.
[11] E. D. Daniel and P. E. Axon, "Accidental printing in magnetic recording," *BBC Quarterly*, vol. 5, no. 4, p. 241, 1950.

HIGH SPEED VIDEO REPLICATOR SYSTEM USING CONTACT PRINTING

Nobutoshi Kihara, Yoichi Odagiri, Tsuguo Sato

Sony Corporation

Tokyo, Japan

High speed Industrial systems for mass replications of video tape software have been developed. They are of the contact magnetic printing system, and can produce replicated video tape software copies at a speed of 100 times normal play. This contact magnetic printing technique allows the best method for industrial high speed replications of both audio and high density video signals recorded on the tapes with the 'slant track' format now widely used in home- and professional-VTRs.

In the process of contact printing, it is first necessary to prepare a high quality printing mother tape that carries a high energy recorded magnetization in the precise mirror image pattern of the repliated final tape aimed at.

Mechanism used in this printing system includes a particular transfer drum, along which the mother and the copy tapes are loaded together with both tapes' coated surfaces tightly contacted. An external high frequency transfer bias MMF is interacted upon them while they run at a speed almost 100 times as fast as of the normal Betamax II mode.

The mother must have a coersive force at least three times that of the usual Beta tape, which is around 650 Oe. This is necessary for keeping the mother tape from any noticeable ebbing of its recorded magnetization during repeated process of the printing, and also for the copy tape to gain the appropriate magnetization to produce the normal playback output. About 2000 Oe has been achieved for the newly developed mother tape. In the process of printing the mother tape is pressed against the copy tape so that the magnetic surfaces of the both tapes must be kept in the closest possible contact and without any mutual displacement in order to transfer the magntic patterns exactly and in precision to the copy tape. This has been achieved by applying air nozzle pressure to the tapes at the position where they are subjected to the transfer MMF as they are wrapping around the transfer drum, which is revolving under the force of tape tension. This method, which we call the "Compressed-Air Drum Method" enables the tapes to run smoothly with high speed and without mutual displacement. The bottom edges of the tapes are accurately aligned and the tapes are guided along to the surface of the revolving drum. The skewing which would normally occur at this point due to the difference in radii between the mother and the copy tape has been compensated precisely at the the production of the mother tape.

The whole system of the replicator can be made adaptable for different television standards such as NTSC, PAL, SECAM, etc.

The merits of our system over already proposed "Bifilar-Reel" type contact duplicator are;

1) Audio and video as well as control signals are printed simultaneously in one process, without any hazardous effect of 'over the layer' print-through which often make the replicated program degraded;

2) Uniformity of replicated signal quality from the beginning to the end of the reel (Hard to achieve in Bifilar-Reel system);

3) Large pancake reel multiple-unit mother is easily at hand (Hard);

4) Less liability of mechanical damage to mother tape in the printing process (Damage liable because the mother must be reeled very tight on the bifilar reel);

5) Speedy printing cycle due to continuous flow process (Batch process requires longer cycle time),

Fig. 1 shows the whole process replicating the video software tapes using contact print technique. The mother tape is produced by the special VTR capable of recording metal tape of very high coersivity. This process is shown in the left side of the diagram.

There are two systems for replicating process, pancake type and cassette type, which are shown in the right side.

The mother tape and the blank pancake with 1/2" tape are loaded into the pancake type replicator, which replicates the program at a speed of 100 times normal. The effective printing speed is about 70 times due to mother tape rewinding. Replicated pancake is then loaded into tape winder, which load the replicated tapes into cassettes, thereby automaticaly cutting them by detecting the replicated program end signal contact-printed from the mother tape.

Cassette type replicator can use blank cassette available in the consumer markets as copy tape. Although actual printing speed is the same, effective speed is decreased to 45 times normal due to slower rewind time of the in-cassette copy tape.

Reprinted from *IEEE Int. Conf. Consum. Electron.*, 1983, pp. 72–73.

Fig. 1

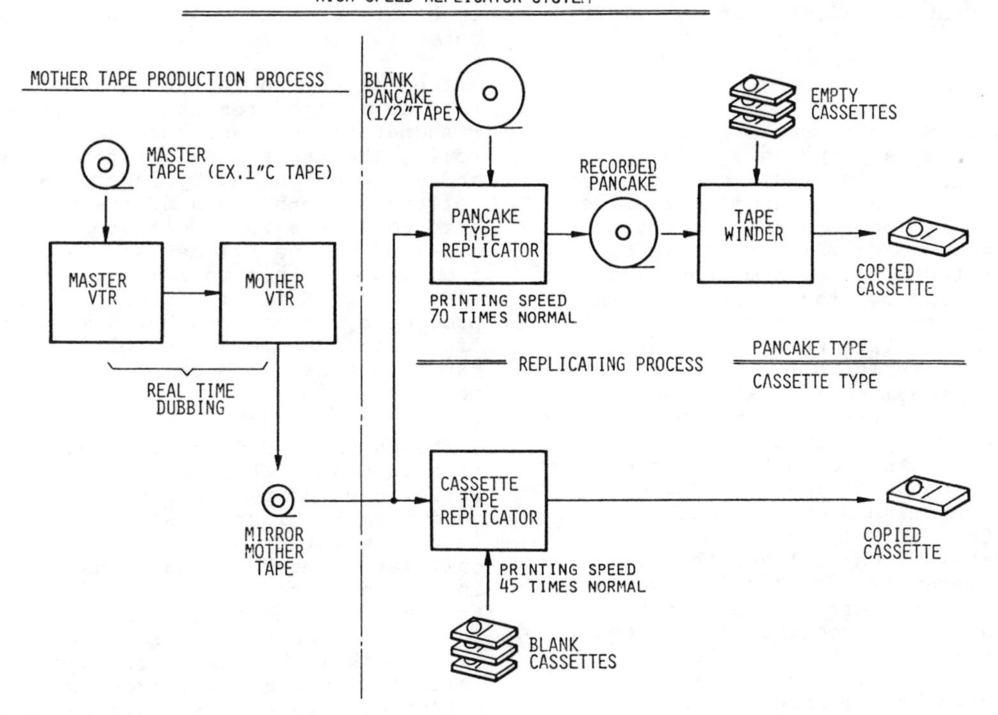

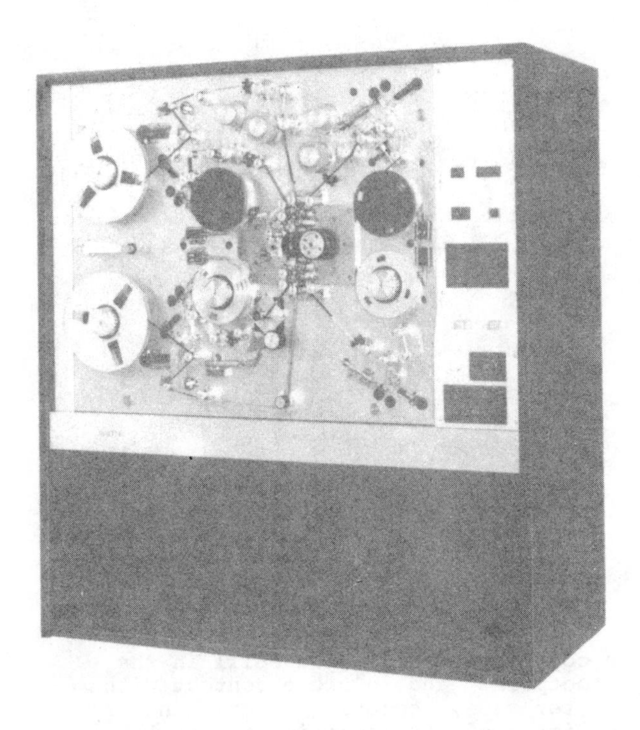

Photograph of the Pancake Type Replicator

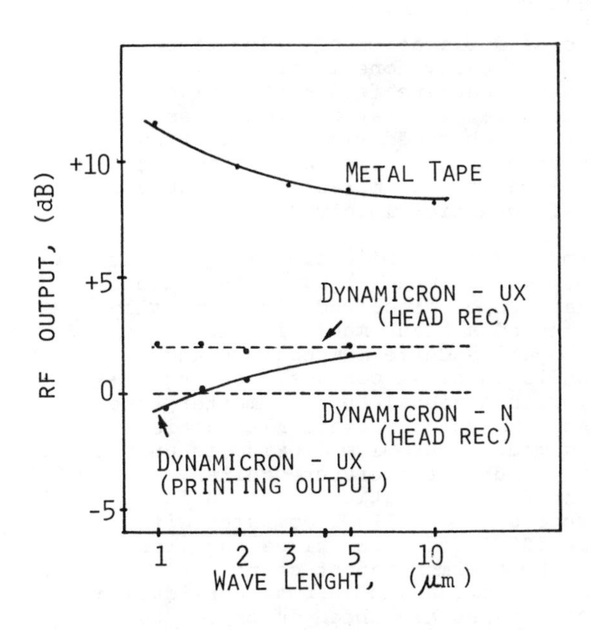

Fig. 2 shows the RF outputs of the mother (metal tape) and the copied tape in relation to the recorded signal wave lengh,thereby Beta Dynamicron-UX tape is used as copy tape.

The outputs of the signals recorded on Dynamicron-N and Dynamicron-UX by identical Betamax also shown as reference.

THERMOMAGNETIC DUPLICATION OF CHROMIUM DIOXIDE VIDEO TAPE

G. R. Cole, L. C. Bancroft, M. P. Chouinard, and J. W. McCloud
E. I. du Pont de Nemours & Co., Inc.
Experimental Station, Wilmington, DE 19898

ABSTRACT

Video program material was duplicated thermomagnetically at copying speeds up to 165 cm/sec, which is 50 times real time for the VHS SP format. Blank chromium dioxide video tape was heated to above its Curie temperature and cooled while in close contact with a master tape bearing a mirror image video signal. The two tapes were held in close contact between a hard transparent polymethyl methacrylate wheel and a slightly deformable elastomeric roll, while heat was supplied by a neodymium YAG laser beam focused through the transparent wheel and the back of the copy tape onto the CrO_2 coating. The 1.06 μm beam was transmitted by the clear polyester base and efficiently absorbed by the opaque black chromium dioxide coating, raising the CrO_2 briefly above its 130°C Curie temperature. On cooling, the CrO_2 layer was permanently magnetized in a mirror image of the master tape magnetic pattern. Heating and cooling took place in the 2.5 mm long nip between the two rolls in about two milliseconds, so the heat transferred to the polyester base was insufficient to deform it. The thermal copies had RF outputs approximately equal to real time copies.

INTRODUCTION

Commercial duplication of video tape programs is usually done in real time with a master machine feeding the video signals to a large array of video cassette recorders. High-speed video duplication by electronic means analogous to commercial audio duplication is impractical because of the high frequencies involved.

High-speed video duplication is possible by thermomagnetic means because the signals do not have to be processed electronically. In this method a blank magnetic tape is heated above its Curie temperature and cooled while in close contact, coating-to-coating, with a signal-bearing master tape. Among the commercially available particulate recording media, chromium dioxide is unique in having a Curie temperature suitable for thermomagnetic duplication. Its Curie temperature of about 130°C compares with more than 500°C for iron oxides and iron alloys. The thermoremanent characteristics of chromium dioxide and thermal duplication of magnetic tapes have been described by Waring[1], Dickens and Jordan[2], Ono, Kobayashi and Sugaya[3], and Berkowitz and Meiklejohn[4]; applications of thermal copying have been disclosed by Greiner, et al[5], Lemke[6], Stancel, et al[7], and Kobayashi, et al[8], among others.

APPARATUS

Some of the earlier attempts at practical thermal duplication of video tapes were unsuccessful because of thermal instability of the polyester base, as evidenced by poor electronic synchronization, and insufficient intimacy of contact between master and copy tapes resulting in limited short wavelength transfer. Both of these problems appear to be solved by the new equipment design concept.

The master and copy tapes are run face-to-face in a high pressure nip between a hard transparent roll and an elastomeric nip roll. The tapes run over tensiometers and are edge guided by tapered alignment rollers on both sides of the nip. The hard roll is a 20 cm diameter transparent wheel of polymethyl methacrylate, and the 5 cm diameter nip roll is made of Adiprene® elastomer. Sufficient force is applied to produce a 2.5 mm long by 25 mm wide "footprint" between the rolls, holding the tapes together with an average pressure of 18 kg/cm^2.

The chromium dioxide coating of the copy tape is heated above its Curie temperature while in the nip by a continuous wave neodymium YAG laser that emits light of 1.06 μm wavelength. The laser beam is collimated, masked, and focused, as shown in Figure 1, to a 13 mm by 0.25 mm rectangle at the point where it strikes the copy tape in the nip footprint (Figure 2). The 1.06 μm wavelength light passes through the anti-reflection coated transparent wheel and the polyester base and is efficiently absorbed by the black CrO_2 magnetic coating. 50 to 60% of the total laser power actually reaches the tape.

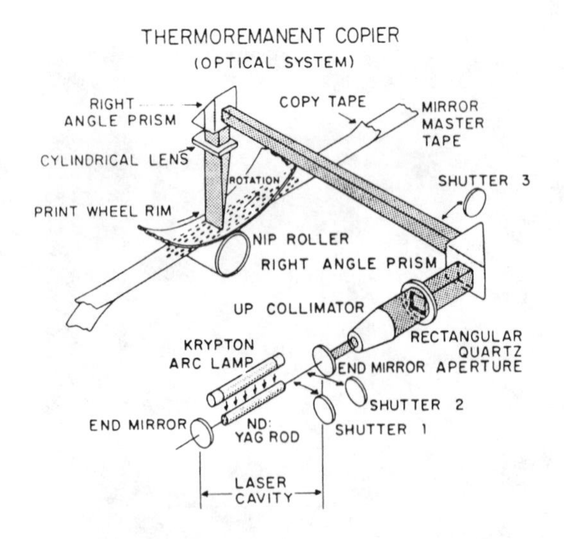

FIGURE 1
Laser Optical Path

The copy tape cools while still in the nip "footprint" and in close contact with the master tape, accepting a permanent mirror image of its magnetic pattern. Good short wavelength transfer is effected by the high nip pressure. The dimensional stability of the polyester is not jeopardized by excessive heating because the laser heat is redistributed rapidly by conduction. Computed temperature profiles across a

Reprinted from *IEEE Trans. Magn.*, vol. MAG-20, pp. 19-23, Jan. 1984.

section through the copy tape, assuming a copying speed of 250 cm/sec, show that the temperatures drop to around 60°C during the millisecond before the tape leaves the nip (Figure 3).

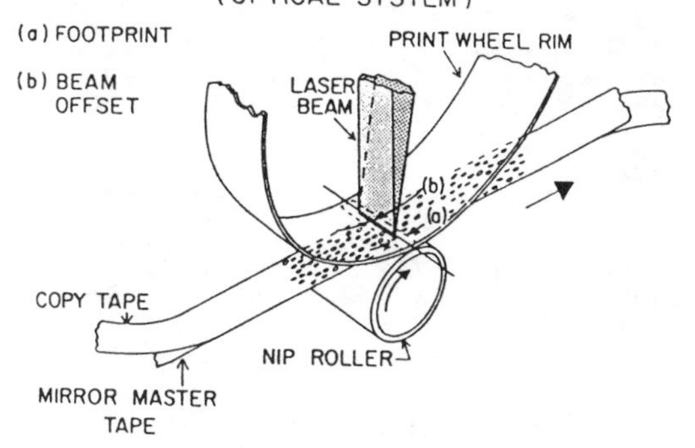

FIGURE 2
Laser beam at tape copying point

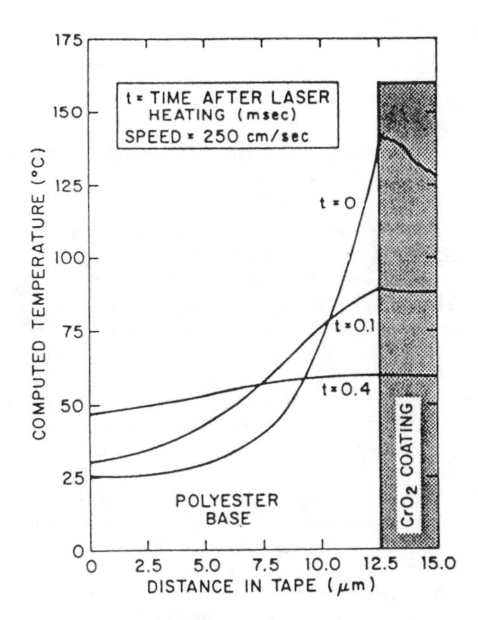

FIGURE 3
Computed temperature profiles for a cross-section of the copy tape, taken perpendicular to the direction of travel, as the tape advances through the nip.

RESULTS

Video

Good quality video copies were made by thermal duplication on chromium dioxide tapes at duplicating speeds up to 165 cm/sec. Copies made from VHS SP (two-hour) format masters were comparable in picture quality and equal in RF output (luminance) to real time copies made by a commercial video duplicator. As is the case with real time copies, picture quality was slightly lower for thermal copies made from a VHS LP (four hour) format master.

Good picture synchronization and skew (typically 2 to 3 microseconds) demonstrated that the polyester base stability was maintained. The good signal transfer and electronic synchronization confirmed that two basic requirements for thermal copying were met: first, that the temperature of the copy tape coating exceeded and then dropped below the Curie temperature of CrO_2 while the tape was in the nip; second, that the polyester base remained below the time-temperature condition for relaxation, where dimensional instability would occur.

The effects of laser power and duplicating speed on the video outputs of the thermomagnetic copies were interrelated. At a fixed copying speed, the video RF outputs increased with laser power until a plateau was reached. As the copying speed was increased, the plateau shifted to successively greater laser powers. This effect is illustrated in Figure 4 for the luminance (RF) output versus laser power at copying speeds ranging from 64 to 165 cm/sec.

The optimum ranges indicated by the RF output curves were too broad. The limits placed by chroma output and signal-to-noise ratio (S/N) turned out to be more restrictive. The luminance output and S/N versus laser power are compared to the chroma otuput and S/N in Figure 5 for a copying speed of 127 cm/sec. Where the luminance data called for a lower limit of 80 watts for laser power, the chroma S/N required a lower limit of 90 watts. Increasing the laser power further had no effect except eventually to cause lateral deformation ("cupping") of the copy tape at excessive power levels.

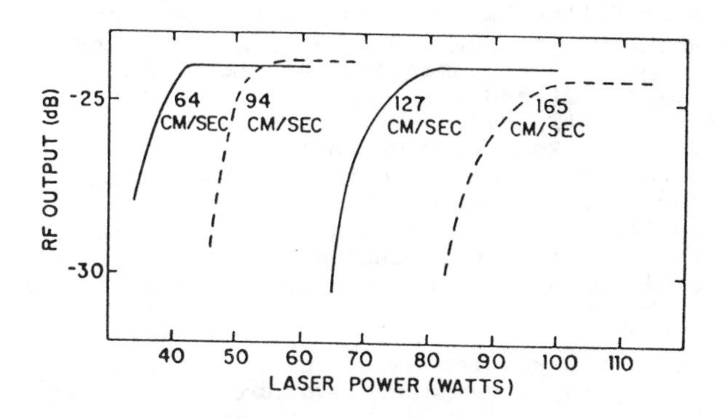

FIGURE 4
The effect of laser power and copying speed on the luminance (RF) output of thermally copied signals

The incidence of video dropouts was not dependent upon any controlled operating variable. The dropouts were the sum of those initially present in the copy tape plus a small fraction added by the duplicating process. Extreme cleanliness of the transparent print wheel rim was essential to minimize process-related dropouts.

Two different types of master tapes were used to make the CrO_2 thermal copies. One was 2000 Oe metal particle tape obtained from Panasonic and Sony; the other was a conventional cobalt-modified iron oxide video tape of about 700 Oe coercivity. The master tapes were recorded on a mirror image video tape recorder. Different write currents were employed for the two types of master tapes.

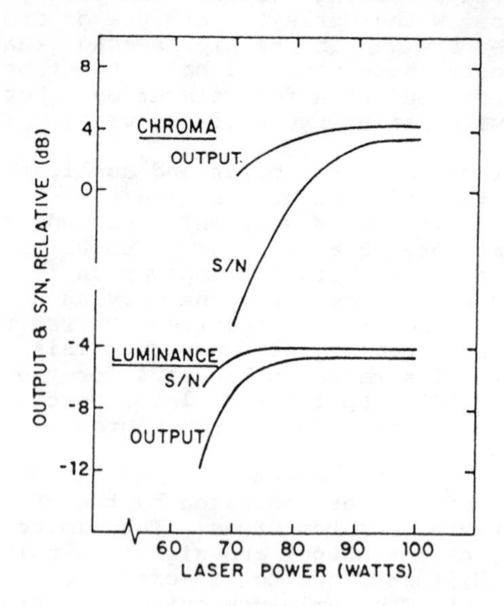

FIGURE 5
Effect of laser power on the luminance and chroma outputs and signal-to-noise ratios of thermally copied signals

Thermal copies made from a metal particle master had luminance (RF) and chroma outputs and S/N approximately equal to copies made by conventional real time duplicating techniques (see Table I). Thermal copies from an "oxide" master (the same tape type as was recorded for the "real time" comparison) had poorer luminance but increased chroma output and S/N. This oxide master was apparently recorded at a write current which was larger than optimum, emphasizing long wavelength chroma output at the expense of the shorter wavelength luminance output. Optimizing the write current was identified as the key step in producing satisfactory thermal copies from an inexpensive, readily available master.

The thermal copies made from both types of master tapes compared favorably with those copied on another type of high speed video duplicator employing anhysteretic magnetic transfer (Table I). In the anhysteretic duplicator, the master and copy tapes were wound in a bifilar arrangement, subjected to a strong ac transfer field, then unwound. The audio signal was applied during the unwind. A very high coercivity (2000 Oe) master tape is essential in the anhysteretic process to prevent erasure of the master by the ac transfer field.

Audio

The audio results from thermal duplication onto chromium dioxide video tape showed an effect of frequency on the magnitude of the signal transferred, showed satisfactory output linearity, and confirmed the operating limits set by the video data already

COPY METHOD AND MASTER TAPE	LUMINANCE		CHROMA	
	OUTPUT	S/N	OUTPUT	S/N
Real time on a CoFe oxide tape	0	49	-3.5	40
CrO_2 thermal copy from high coercivity tape	0	49	-3	40
CrO_2 thermal copy from CoFe oxide tape	-3	45	0	42
Anhysteretic copy on CoFe oxide from high coercivity tape	-4	45	-8	40

TABLE I
Video Output and S/N Comparisons

discussed. The copies were made from a cobalt modified iron oxide (CoFe oxide) video tape with a single audio track, VHS SP format. The duplication speed was 127 cm/sec for the audio data in this section.

Thermal duplication resulted in substantial increases in the signal strength of the copy, compared to that of the master, for the mid-range audio frequencies. The relative audio outputs of the CoFe oxide master tape compared with the thermally duplicated copy tape showed a maximum amplification, due to duplication, of 7.8 dB at 400 Hz. The amplification dropped smoothly at frequencies lower and higher than 400 Hz, with the copy output 3.1 dB greater at 100 Hz, equal to the master at about 8 KHz, and 2.7 dB lower at 10 KHz, as shown in Figure 6. The master

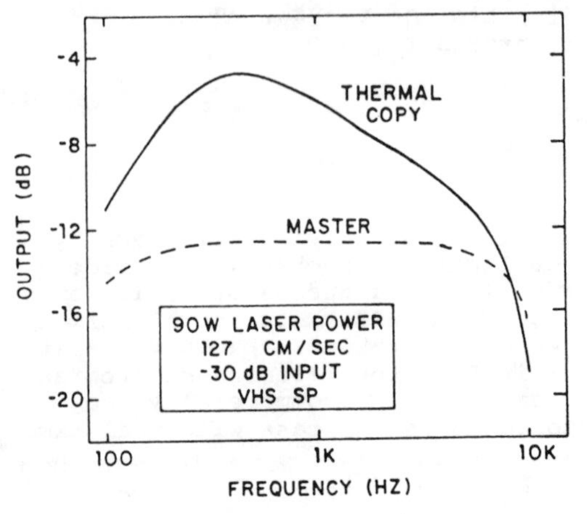

FIGURE 6
Audio sensitivity vs. frequency for thermally copied tape

tape was recorded with a constant amplitude input over the frequency range. This audio response characteristic of thermal copying can be compensated as needed by pre-emphasis in master preparation.

The system linearity was satisfactory over the audio frequency range investigated, 100 Hz to 10 KHz. Both the copy tape output and the master tape output were linear versus the initial input to the master. The results for 2 KHz shown in Figure 7 are representative. The deviation from

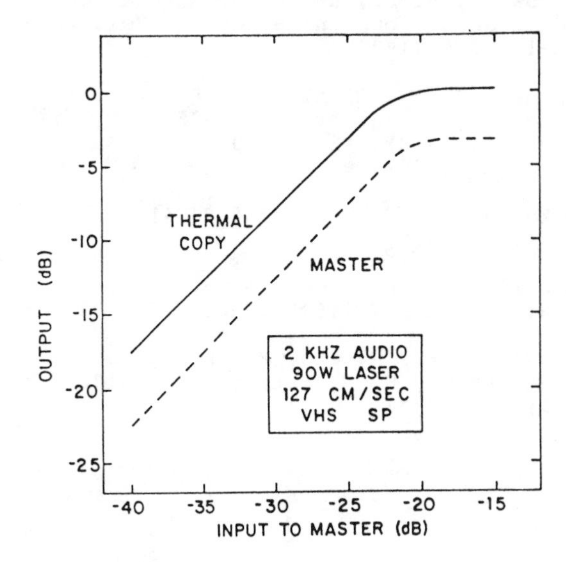

FIGURE 7
Master and copy outputs vs. input to master

linearity at the highest inputs was due to input limiting by the video recorder. The offset from the master tape was consistent with amplification predicted by Figure 6.

The audio output increased as a function of laser power and reached a plateau in much the same way as the video outputs. The audio sensitivity and S/N versus laser power, shown for 1 KHz in Figure 8, are representative for the audio range. As with the video luminance output, the audio sensitivity versus laser power indicated an operating plateau from about 80 to 100 watts.

The total harmonic distortion (THD) at 400 Hz fundamental frequency was more restrictive, establishing 90 watts as the lower limit of the operating plateau. Hence the audio THD joined with chroma S/N as the final determinant of the optimum laser power at the given copying speed. The 90 watt copy had much lower distortion than the 80 watt copy, and was generally comparable to or better than the master tape at equal outputs (Figure 9).

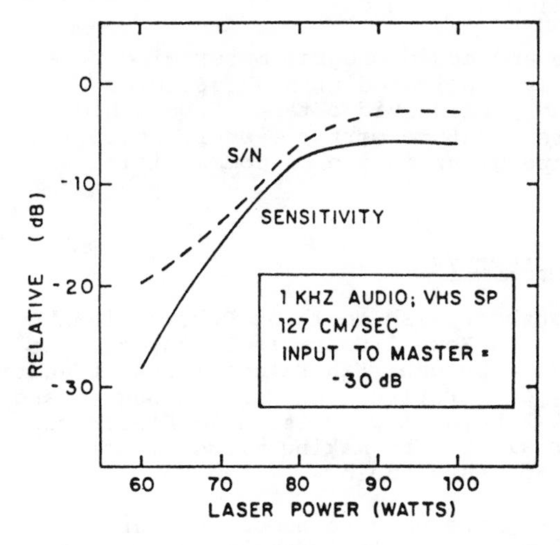

FIGURE 8

Effect of laser power on 1 KHz audio sensitivity and signal-to-noise ratio of thermally copied signals

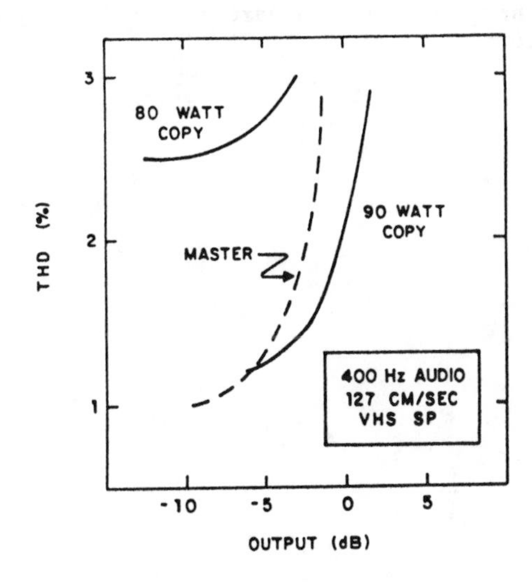

FIGURE 9
Total harmonic distortion (THD) at 400 Hz audio vs. output for a cobalt iron oxide master and thermal copies made at 127 cm/sec and at 80 and 90 watts laser power.

The greater 400 Hz 3% THD output of the copy compared with the master was approximately balanced by the low frequency noise amplification of the thermal copying process. The signal-to-noise ratio, measured as the separation between the output at 3% THD and the A-weighted bias noise level, was 47 dB for the copy compared with 48 dB for the master tape.

CONCLUSION

Video and audio program material can be thermally duplicated at high speed on chromium dioxide video tapes from a high Curie temperature mirror master tape. The copy tape programs are of commercial quality.

ACKNOWLEDGMENTS

The authors wish to thank Robert Pfannkuch of Bell and Howell/Columbia Pictures Video Systems, Nobutoshi Kihara and Yoichi Odagiri of Sony Corporation, and Hiroshi Sugaya and Fukashi Kobayashi of Matsushita Electrical Industrial Co. for making mirror master video tapes available.

We are grateful to a number of our colleagues for their important contributions to the successful development of this thermomagnetic duplicator, in particular to W. J. Lingg for optical design consultations, to O. R. Averitt for electrical design, to W. L. Gardenhour, N. C. Morse and R. J. Bean for mechanical design, and to T. D. Colvard for video tape evaluations.

REFERENCES

1. Waring, R. K. Jr., J. Appl. Physics, 42, 1763 (1971).

2. Dickens, J. E., and L. K. Jordan, Journal of the SMPTE, 80, 177 (1971).

3. Ono, M., F. Kobayashi and H. Sugaya, Intermag. 1972, Paper 19.4; Apr. 10-13, 1972, Kyoto, Japan.

4. Berkowitz, A. E., and W. H. Meiklejohn, IEEE Transactions on Magnetics, Vol. MAG-11, 996 (1975).

5. Greiner, J., W. Eichler and F. Krones, U. S. Patent 3,364,496 (Jan. 16, 1968), RE 28,290 (Dec. 31, 1974).

6. Lemke, J. U., U. S. Patent 3,541,577 (Nov. 17, 1970).

7. Stancel, A. L. Jr., and W. R. Isom, U. S. Patent 3,761,645 (Sept. 25, 1973).

8. Kobayashi, F., M. Ono, M. Yatsugake Y. Fukushima, U. S. Patent 3,824,617 (July 16, 1974).

Author Index

Subject Index

M

N

O

P

Q

R

S

T

U

V

W

Editor's Biography

Theodore S. Rzeszewski (S'63–M'65) was born in Chicago, IL. He received the B.S.E.E. and M.S.E.E. degrees from the University of Illinois, Urbana, in 1964 and 1965, respectively, the Professional degree of Electrical Engineer from the Midwest College of Engineering, Lombard, IL, in 1971, and an honorary doctorate degree in 1973.

From 1965 to 1967 he worked in the Communications Division of Motorola. He joined Hazeltine Research in 1967, and went to the Consumer Products Division (CPD) of Motorola in 1971. After the CPD was purchased by Matsushita Electric in 1974, he served in several capacities there, including Manager of Advanced Development of the Matsushita Industrial Company and Manager of Technical Operations of the Matsushita Applied Research Laboratory. He joined Bell Laboratories in 1982. He is also an Associate Professor at Midwest College of Engineering, where he has taught junior, senior, and graduate level electrical engineering courses. He was Project Engineer on the first microcomputer-controlled frequency synthesizer for television. His other accomplishments include 19 U.S. patents, papers published in the *Bell System Technical Journal*, the IEEE TRANSACTIONS ON COMMUNICATIONS, and the IEEE TRANSACTIONS ON CONSUMER ELECTRONICS, and Best Paper Awards from both the *Bell System Technical Journal* and the IEEE TRANSACTIONS ON CONSUMER ELECTRONICS. He is the editor of the IEEE PRESS book *Color Television.*

Dr. Rzeszewski is a member of the Administrative Committee of the IEEE Consumer Electronics Society. He has also been a member of the Popov Society Delegation, a Chicago Spring Conference Chairman, an invited speaker at the Tokyo Section of the IEEE, a member of the EIA Industry Committees on Teletext and Multichannel Sound, and an alternate member of the ATSC. He is a member of the SMPTE, Phi Eta Sigma, and Eta Kappa Nu.

TK6630. T368 1985 Engin.
TELEVISION TECHNOLOGY TODAY.

3/24/87